ROBERT K. STEUNENBERG AND LESLIE BURRIS

FROM TEST TUBE TO PILOT PLANT

A 50 YEAR HISTORY OF THE CHEMICAL TECHNOLOGY DIVISION AT ARGONNE NATIONAL LABORATORY

Building 205 at Argonne National Laboratory.

Contents

FOREWORD v

1940–1950: THE BEGINNING 1
The Manhattan Project 1
The Atomic Energy Commission and the National Laboratories 7
The Met Lab 8
The Chemical Engineering Division 16

1950–1960: THE NUCLEAR PROMISE 23
The National and International Scene 23
Reactors 26
The Move to Site D 29
New Employees 31
Services 34
Tools of the Trade 37
The New Buildings 39
Division Organization and Management 47
Technical Programs 50

1960–1970: TURBULENT TIMES 89
The Domestic and International Climate 89
Reactors 92
The Division 92
Management 96
Building Addition 98
Tools of the Trade 99
Cenham Glove Boxes 101
Technical Programs 102

1970–1980: TRANSITIONS 139
The National Scene 140
Nuclear Power 140
The Laboratory 145
The Division 146
Technical Programs 160

1980–1990: CMT—NEW NAME, NEW MISSIONS 237
National Affairs 238
The Nuclear Industry 241
The Division 241
Technical Programs 249

1990–1998: INTO THE FUTURE.......... 361
National Affairs.......... 361
Energy and the Environment.......... 363
The Laboratory.......... 365
The Division.......... 366
Technical Programs.......... 371
The Division's Future.......... 471

APPENDIXES
General References.......... 475
Acronyms and Definitions.......... 477
Subject Guide.......... 485
Name Index.......... 495

Foreword

The idea of writing a history of the Chemical Technology Division was proposed in 1994 by Dr. James Battles, who was the Division Director at the time. About the only guidance Jim provided was that he had in mind neither an extended technical report, nor a "puff piece" extolling the scientific and engineering prowess of the organization. The people in the Division do, in fact, have reason to take pride in the many scientific and technological contributions they have made and the excellence of the work. But the intent of this document is to tell the story of how the Division originated and to give the reader a general idea of what research and development work has been done, why and how it was done, and who did it.

There are several reasons why we created this document. A number of the Divisional staff have expressed concern about losing some of the technical knowledge of the old timers who are retiring, particularly the kind of lore or "know-how" that is not normally documented in technical reports. Dr. Battles expressed the thought that newer members of the Division might gain a useful perspective on how the Division came to be what it is now. Others have expressed curiosity as to why various programs were started, redirected, or terminated. Even those people who have worked in the Division for many years often tended to focus on their own research and were not always fully aware of what others were doing.

When we started writing this volume, it was immediately obvious that detailed literature references would be impractical. They would run into tens of thousands. By the same token, descriptions of the technical programs are very abbreviated. Summary reports alone of the Division's work have averaged several hundred pages per year for 50 years; we have attempted to compress the information in these summary reports to considerably less than 5% of the original volume, and probably did so unintentionally in a somewhat uneven manner. We beg forgiveness from any individuals who might feel that their contributions have been slighted. Where there are such cases, it was certainly not out of malice.

This is not intended to be a reference book, nor have we attempted to generate a subject index. We have, however, included a rather detailed table of contents as an appendix. It happens occasionally that somebody has a vague recollection of work done many years earlier which might be germane to a current problem but cannot remember exactly what it was, who did it, or when it was done. There may be times when the detailed contents table could be helpful in tracking down such a source. Also included as an appendix is a list of acronyms with their definitions.

At the start of each chapter, we have included some discussions of national and world affairs that may seem extraneous, but they have had a major influence on the nature of the Division's programs as public perceptions changed and political wind shifts occurred over the years. They also serve as sort of a time line that might help the reader relate the Division's work to the outside world. For example, nuclear power was a popular idea in the 1950s and 1960s, and the national laboratories were mandated by Congress to promote the concept under the Eisenhower "Atoms for Peace" program. Much of the Division's work at that time involved nuclear fuel reprocessing and fast breeder reactors, and our discussions probably reflect a pro-nuclear attitude during that period. During the 1970s, however, it was becoming clear that nuclear power was not going to solve the nation's near-term energy problems, and much of the Division's effort was shifted toward alternative sources (solar and fusion power), improved utilization of existing sources (coal), and conservation (the battery programs). Most of the Division's applied programs at present are directed toward environmental concerns, a major one being the handling and disposal of plutonium and nuclear wastes.

Joe Harmon, head of the Division's Technical Editing Group, put a difficult question to us: "Who would be interested in reading a book of this kind and at what technical level will it be pitched?" The technical level, we believe, is one at which nearly everything would be comprehensible to the average chemist or chemical engineer. Although some of the descriptions of the technical programs include equations, phase diagrams, and jargon that might not be understood by a non-technical reader, much of the material is either non-technical or descriptive. We felt that some technical detail was necessary to illustrate the depth and scope of the work. Our suggestion to the non-technical reader would be to skip through the technical sections lightly and just try to absorb the general drift of what the programs were about. This is not a textbook and there is no exam. We would expect most of the readers to be present and former staff members of the Division, some of the administrative and management personnel, technicians and secretaries, some family members, a few people from other ANL divisions or outside organizations such as universities and contractors who worked on the programs. Some students contemplating a scientific or engineering career might be interested in the type of work that scientists and engineers do in a typical research and development organization.

We have attempted to associate names of investigators with the various programs insofar as possible, but found it to be a difficult task, so there may be omissions or errors. For those, we apologize. A name index is included at the back, where names of non-ANL personnel are italicized.

Finally, we wish to express our deep appreciation to the many individuals who helped us create this volume. Joe Harmon's advice, encouragement, and editing were invaluable. He also contributed a major effort in the production of the final document, as did Maria Contos. Dr. Stephen Lawroski, in particular, provided much oral history, as well as many technical details about the early days of the Division. Dr. Martin Steindler also deserves acknowledgment for his careful review of the entire manuscript and many thoughtful suggestions. Others who deserve special thanks for their help include Jim Battles, George Bernstein, Milt Blander, Ron Breyne, Herb Brown, Loretta Cescato, Sharon Clark, Dennis Dees, Pat Finn, Al Fischer, Steve Gabelnick, Helen Hill, Carl Johnson, Jerry Johnson, Irv Johnson, Tom Kaun, Jim Laidler, Ralph Leonard, Dick Malecha, Vic Maroni, Bill Miller, Leo Morrissey, Jan Muller, Sofia Napora, Paul Nelson, Al Panek, Dean Pierce, Jerry Rathke, Roberta Riel, Laury Ross, Wally Seefeldt, Chuck Seils, Mike Thackeray, Ziggy Tomczuk, and George Vandegrift.

Also contributing to production of the final document were Jane Andrew, Judith Carr, Mary Ann Forys, and Barbara Salbego.

Bob Steunenberg
Les Burris

1940-1950: The Beginning

1940-1950

(top) West Stands at Stagg Field, University of Chicago—site of world's first nuclear reactor and, later, first research on nuclear fuel processing by the ANL Chemical Engineering Division.

(bottom) Walter Zinn (left), first ANL Laboratory Director, and Stephen Lawroski (right), first Director of the Chemical Engineering Division.

1

1940-1950: The Beginning

The Chemical Engineering Division (CEN), now renamed the Chemical Technology Division (CMT), at Argonne National Laboratory (ANL), was formed officially in February 1948. Its roots extend well back into the Manhattan Project, however, where it evolved from a group in the Chemistry Division of the Metallurgical Laboratory at the University of Chicago and became a separate division. A brief review of the Manhattan Project, the Atomic Energy Commission, and the Met Lab should prove helpful in understanding the various events that led to the formation of the Division and shaped its future role as a research and development organization.

THE MANHATTAN PROJECT

It is generally accepted that the atomic age began in Berlin with the discovery by Otto Hahn and Fritz Strassman in 1938 that uranium can undergo nuclear fission. Earlier workers had achieved fission by bombarding uranium with neutrons, but did not recognize it as such because they mistook the fission product, barium, for actinium. Hahn established clearly that the product was barium by separating and identifying its decay product, lanthanum. The thinking at that time was that neutrons might be captured or that they might knock small chips off the nucleus, but gross fissioning of the nucleus into large fragments was not believed to be possible. Hahn realized that such fissioning was the only explanation for the barium, but he was reluctant to publicize such a radical result without some theoretical backup. He informed Lise Meitner, a highly competent theoretical physicist, of his results. Meitner agreed that the barium could be explained only by fission but was troubled as to how it could occur. She and her nephew, Otto Frisch, also an excellent physicist, after much speculation and agonizing, came up with a liquid drop model. According to this model, the uranium nucleus assumes a sort of dumbbell shape in which the binding forces arrange themselves in such a way that fission can take place. Still somewhat nervous about their finding, Frisch approached the eminent Danish physicist, Niels Bohr, who grasped the concept immediately with much enthusiasm.

Bohr sailed for the U.S. shortly thereafter, and upon his arrival announced the discovery on January 16, 1939, at the Princeton Monday Evening Journal Club, a weekly gathering of Princeton physicists. Almost immediately, related work emerged nearly everywhere. At Columbia University, Enrico Fermi, and Leo Szilard measured the absorption of neutrons by uranium, and Bohr and John Wheeler at Princeton performed a classical analysis of uranium fission. Frederick Joliot-Curie in France confirmed the theoretical model experimentally and attempted to produce a chain reaction. Rudolf Peierls in England determined the critical mass for a chain reaction. This burst of activity over a period of only a year or two led to a high confidence level that a fission chain reaction in uranium was possible. This discovery also led to a great deal of speculation about the possibility of a nuclear weapon, which alarmed many of the scientists who were involved in the work. The totalitarian regimes in Europe had created conditions that caused many of their leading nuclear scientists to flee to the U.S., which assured its future preeminence in the field of nuclear research. But the concerns of the scientists over potential nuclear weapons fell largely on deaf ears in the U.S. Government administration until Szilard, Eugene Wigner, and Edward Teller prevailed upon Albert Einstein to write his famous letter to President Franklin Roosevelt on August 2, 1939.

Once the administration realized the significance of potential nuclear weapons, it reacted as governments usually do—it formed a committee to study the problem. At Roosevelt's request, the National Academy of Sciences (NAS) appointed an Advisory Committee on Uranium (ACU), which was chaired by Lyman Briggs, the director of the National Bureau of Standards. Its mission was to coordinate fission research and to evaluate the possibility of developing nuclear weapons. The committee acted slowly and was relatively ineffective. The increasing intensity of the war in Europe in 1940, however, brought about an ever more rapid mobilization of scientific, as well as military resources in the U.S., along with a strong sense of urgency. A new organization, the National Defense Research Council (NDRC), was formed and placed under the leadership of the director of the Carnegie Institute, Vannevar Bush, who was a well known and respected individual in the power circles of Washington at the time. The ACU was reorganized and placed under NDRC, but it still remained indecisive. At this juncture, three Nobel laureates, Harold Urey, Ernest Lawrence, and Arthur Compton, who were members of a NAS review committee, expressed their impatience with the lack of action, causing Bush to superimpose on NDRC a more powerful agency, the Office of Scientific Research and Development (OSRD). This office was given jurisdiction over all war-related research and development. James Conant, president of Harvard and a well-known organic chemist, replaced Bush as the chairman of NDRC and became his deputy at OSRD. Urey, Lawrence, and Compton provided the leadership for a reorganized uranium committee known as the S-1 Section of OSRD.

The U.S. declaration of war in December 1941, plus various bits and pieces of intelligence emanating from Europe that Germany was most likely attempting to develop a nuclear weapon, finally galvanized the American effort into a strong course of action. The S-1 Section placed Compton in charge of the theoretical and experimental studies of fission and nuclear weapons design. Compton wasted no time in recruiting the necessary physicists, chemists, engineers, and other personnel, mostly from universities and industrial research and development laboratories, and organizing them into an entity bearing the code name "Metallurgical Laboratory" or "Met Lab." There was much discussion as to where the new lab should be located. Cases were made for Columbia, Princeton, Berkeley, Cleveland, and Chicago. Nobody wanted to move. Compton made a unilateral decision that it would be Chicago. His arguments were that (1) the University of Chicago was receptive to the idea, (2) Chicago was conveniently located for travel to other sites, and (3) more scientists were available to staff the operation than on the coasts where faculties and graduate students had been drained for other war work. Between March and June 1942, the staff at the Met Lab increased from 25 to 1,250. Much of the experimental program was conducted in space under the West Stands of Stagg Field at the University of Chicago, a rather forbidding fortress-like structure. The University of Chicago Maroons, a football team once coached by the legendary Amos Alonzo Stagg and known as the "Monsters of the Midway," had suffered a series of embarrassing defeats, as had the U. of C. baseball and basketball teams, so they had withdrawn from the Big Ten.

The first mission of the Met Lab was to determine the feasibility of a uranium chain reaction, and this effort was placed under the very able leadership of Enrico Fermi, who was widely accepted as the leading authority in the group. In addition to his excellent theoretical understanding of the subject, he seemed to have an uncanny intuitive feeling for the fission process. Under Fermi's direction, slugs of natural uranium oxide, and later, uranium metal, were placed in blocks of graphite moderator, which were assembled into a stack or "pile" in a squash court under the West Stands. This pile became known as CP-1 (Chicago Pile No. 1). Criticality was achieved on December 2, 1942, at the stage that Fermi had predicted. Compton, at the Met Lab, called Conant at Harvard and told him "The Italian navigator has landed in the New World." Conant asked, "How were the natives?" Compton replied, "Very friendly." This is probably one of the most widely quoted telephone conversations in history, second only to the original one between Alexander Graham Bell and Thomas Watson.

Even before the success of CP-1, it was recognized that a massive national effort would be needed to develop a nuclear bomb on a timely schedule. This effort, which became known as the "Manhattan Project" (officially the Manhattan Engineer District), was initiated in August 1942. Because of the huge size and complexity of the undertaking, it was assigned to the U.S. Army Corps of Engineers, and Brigadier General Leslie Groves became the commanding officer on September 23, 1942. Groves, although not a universally popular individual, to put it mildly, was a highly effective manager who pushed the Manhattan Project forward at a rapid pace.

During the earlier studies of uranium, Arthur Dempster at the University of Chicago had shown that natural uranium, which is predominantly U-238, also contains a small amount of a second isotope, U-235. Alfred Neir, a postdoctoral student at Harvard, then quantified the U-238/U-235 ratio as 139:1. Bohr came up with a theoretical explanation for the fissioning of U-235 by slow, as well as fast neutrons. Leo Szilard and Walter Zinn at Columbia found that two neutrons were produced by the fissioning of a U-235 atom, and somewhat higher values were obtained later by other investigators, showing that a self-sustaining chain reaction was possible.

While these uranium fission studies were still in progress, a new chemical element, plutonium, was discovered in 1940 at the University of California, Berkeley, by Glenn Seaborg and his colleagues, who produced minute amounts of Pu-238 by bombarding uranium with deuterons in their 60-inch cyclotron. Ensuing work showed that Pu-239 was formed readily by slow neutron capture in U-238:

$$ {}_{92}U^{238} + {}_{0}n^{1} \rightarrow {}_{92}U^{239} \rightarrow {}_{1}e^{0} $$
$$ + {}_{93}Np^{239} \rightarrow {}_{1}e^{0} + {}_{94}Pu^{239} $$

The Pu-239 isotope was then found to be even more readily fissionable by neutrons than U-235, so it also became a likely candidate for a nuclear bomb. Furthermore, plutonium, being a different chemical element, could be separated from its parent uranium by a chemical process, which, in principle, is simpler than an isotopic separation. This potential advantage of using Pu-239, however, carried with it the problems associated with its production on a scale sufficient to produce nuclear weapons. A decision was made to pursue both the uranium and the plutonium options for nuclear weapons in the Manhattan Project, and Seaborg joined the Met Lab to head up a major effort on the chemistry and technology of plutonium.

Three new highly secret projects were to be constructed, all in remote areas—Oak Ridge in Tennessee (Site X), Hanford in Washington State (Site W), and Los Alamos in New Mexico (Site Y). The first was Oak Ridge, which consisted of four major installations. The Oak Ridge facilities, operated by Tennessee Eastman, a subsidiary of Eastman

Kodak, had the task of developing and building a plant to recover the fissionable isotope, U-235, from natural uranium, which contains only 0.7% of this isotope, the remainder being non-fissionable U-238. Two approaches were investigated: electromagnetic separation by calutrons, which operate on the principle of a mass spectrometer, and gaseous diffusion, in which gaseous uranium hexafluoride (UF_6) is passed repeatedly through barriers having extremely fine pores. A third method of isotopic separation that was investigated was the use of gas centrifuges, but it was impossible at the time to build units that could operate at the extremely high speeds required to achieve any significant degree of separation.

In the gaseous diffusion process, the U-235 hexafluoride molecules, being just a bit lighter, pass very slightly more readily through the porous structure to provide a U-235/U-238 separation, but a very large number of stages are required to achieve the desired U-235 enrichment. (The composition and structure of the barrier material was one of the most closely guarded secrets in the nuclear program.) A plant for the electromagnetic separation process, Y-12, and one for the gaseous diffusion process, K-25, were constructed. A small thermal diffusion plant, S-50, was also built within K-25 to provide that facility with slightly enriched uranium. Although the plants could provide uranium of any desired enrichment, a level greater than 90% was required for weapons use. Later on, enrichments of approximately 3% were used for power reactors. A major consideration in the siting of the facility at Oak Ridge was a requirement for a prodigious amount of electric power, which could be provided only by a huge utility such as the Tennessee Valley Authority (TVA), to operate the uranium enrichment plant. The fourth installation at Oak Ridge was the Clinton Laboratory (X-10), which was concerned mainly with separation process research. A pilot plant to study plutonium recovery processes was built and operated. The X-10 facility was also used to train personnel for nuclear work at other sites.

Hanford, operated by I. E. du Pont de Nemours and Co., had the responsibility for plutonium production. This was accomplished by three large uranium-fueled reactors located near the Columbia River, which provided the necessary cooling water. During the reactor operation, a small concentration of plutonium is generated in the fuel by neutron capture. Periodically, some of the irradiated uranium (in the form of aluminum-clad slugs) is discharged and processed to separate the plutonium product. Fissioning of uranium also produces some three dozen fission-product elements, each of which has its own unique chemical properties and, most often, several isotopes with different radioactive emissions. Because of the many chemical elements involved, processing this discharged fuel was a formidable task. The uranium and plutonium products had to be recovered separately and free of any significant fission products. The high radiation levels required that the process, down to its final stages, consist of remotely controlled operations behind thick barriers of concrete shielding. The plutonium product was recovered initially by the Bismuth Phosphate process and subsequently by the Redox solvent-extraction process, both of which will be described later. The production of plutonium at Hanford began in 1944.

General Groves placed Robert Oppenheimer in charge of Los Alamos, where construction began in December 1942. The function of Los Alamos was to design, fabricate, and test the nuclear weapons, using the enriched uranium and plutonium from Oak Ridge and Hanford. This effort was more closely connected with the military and required the services of a number of well-known theoretical physicists supported by a large number of technical and service personnel. Oppenheimer later became a somewhat controversial figure during the

Communist "witch hunts" of the early 1950s and his security clearance was withdrawn, but there was never any evidence of disloyalty or wrongdoing on his part and he was later exonerated of all charges.

The Manhattan Project was conducted under the highest possible level of secrecy and with great urgency. It had a military atmosphere that was not always compatible with the personalities of some of the scientists. A major impetus was to end World War II without having to invade the Japanese homeland by amphibious assault, but there was also much concern over the possibility that the Germans might develop a deployable nuclear weapon before the war in Europe was over.

There was a sound basis for such concern, particularly in the early stages of the project. The German scientists had produced nuclear fission in the laboratory. They had also been looking at nuclear fusion and U-235 separations and were approaching criticality in a nuclear pile in a cave at Haigerloch. Their nuclear program was inhibited somewhat by a lack of enthusiasm on the part of Adolph Hitler, who believed the time frame was too long, and even more so by a serious miscalculation in its early stages. In 1941, one of their leading scientists, Walther Böthe, a highly regarded German physicist, greatly underestimated the diffusion path length of slow neutrons in graphite, apparently because graphite of inadequate purity was used in the German studies. Consequently, the German scientists selected heavy water as the moderator, rather than graphite, which was used in the U.S. program. The only significant source of heavy water available to them was at the Vermorsk power plant in southern Norway. Having gotten wind of what was going on, the British mounted a commando attack on that facility, followed by an RAF bombing raid which destroyed it. The British also sank a small ferry that was hauling heavy water to Germany. At the end of the war, an American intelligence force called *Alsos* quickly nabbed all the German nuclear documentation and scientists they could find to keep them out of the hands of the Soviets. (*Alsos* was a thinly disguised code name; in Greek it means "grove.") In addition, the U.S. Army Air Corps bombed the German nuclear production works near Berlin. Thus ended the German nuclear threat. Although General Groves was aware of this fact, he did not pass the information on to the scientists in the Manhattan Project.

The Manhattan Project was spectacularly successful in achieving its immediate objective. Three nuclear weapons were detonated in rapid succession: Almagordo, New Mexico (July 16, 1945), Hiroshima (August 6, 1945), and Nagasaki (August 8, 1945). The Almagordo and Nagasaki bombs were plutonium, and the one at Hiroshima was U-235. The reason for this was that the Los Alamos scientists were certain that the uranium bomb would work because its firing mechanism was straightforward. They were not so sure about the plutonium bomb, which required a more complex configuration and firing mechanism to assure sufficient detonation for an effective weapon, and they felt that a test shot was needed. At that time, the production of U-235 had been so slow that only enough was available for one bomb.

Although the tide had turned in favor of the allies in World War II by 1944, vicious fighting continued well into 1945—the Normandy landing, the Battle of the Bulge, Iwo Jima, Okinawa—and it appeared that many more American lives were yet to be lost. Victory in Europe came on May 7, 1945, but the Pacific war raged on until the nuclear weapons were used and Japan surrendered on August 15, 1945.

Even before Hiroshima and Nagasaki, there was a great deal of controversy, both in the scientific community and in government circles, as to the manner in which the nuclear weapons should be used. Some wondered whether a demonstration shot or the destruction of a military target such as the Japanese naval base on the island of Truk would achieve the objective of ending the war. Others advocated a direct attack on Japan itself to end the war quickly in order to save the lives of American servicemen. There were still others who were concerned about the moral aspects of using the weapons on Japanese civilians.

Harry S. Truman who was thrust into the presidency by the death of Franklin D. Roosevelt on April 12, 1945, had not been told of the existence of the nuclear weapons until that time. Secretary of War Henry Stimson briefed him on the situation. Joseph Stalin, at Potsdam, had committed the U.S.S.R. to enter the war against Japan within 90 days of VE Day. After extensive discussions with Stimson and other government officials, military people including Generals Dwight Eisenhower and George Marshall, and various scientific leaders in the Manhattan Project, Truman made his decision. That may have been the sort of thing he had in mind when he installed the motto "The buck stops here" in the Oval Office. Whether that was the right decision has been a matter of much conjecture and controversy for the last 50 years.

The year 1945 was fraught with many changes. When the war ended, the military forces were demobilized rapidly and the defense budgets were cut even more. Scientists in the Manhattan Project, feeling that their mission was accomplished, and, fed up with the oppressive security, left in large numbers. The question of civilian applications arose and continuing military control of atomic energy was debated. Policy questions about future uses and control of nuclear energy were being raised and the Federation of Atomic Scientists, based primarily at the Met Lab, was formed. Meanwhile, the report, *Atomic Energy for Military Purposes* by Henry D. Smyth, was released to the public. This report contained a surprisingly candid description of the Manhattan Project, but, as pointed out by Seaborg, it made only minor mention of the chemists and engineers at the Met Lab who had done an enormous amount of difficult, painstaking work on the development of plutonium chemistry and technology. (Seaborg rectified this situation in 1994 when he published his book, *The Plutonium Story: the Journals of Professor Glenn T. Seaborg 1939-1946.*) Also in 1945, a new dark cloud loomed on the horizon. Irving Langmuir, a well-known chemist, predicted that Russia would explode a nuclear weapon in five years. He was optimistic; it happened in 1949.

Early in 1946, the U.S. Navy, which had previously played only a minor role in the nuclear weapons program, conducted Operation Crossroads, which consisted of two fission bomb detonations on the Bikini Atoll in the Marshall Islands. The first test, "Able," was an atmospheric shot, and the second, "Baker," was underwater. Just prior to these tests, some of the sailors on U.S. Navy ships returning from World War II were startled to see a bright red battleship among the usual gray vessels in the fleet. Aware of the Hiroshima and Nagasaki shots the year before, they suspected something important was afoot but had no idea what it was. The red vessel turned out to be the U.S.S. *Nevada*, an aging battleship that had been selected to serve as the primary target for Operation Crossroads. The purpose of these tests was to determine how much damage warships would sustain near a nuclear weapon and to evaluate decontamination procedures. The *Nevada* actually survived the test and was finally disposed of by naval gunfire.

During this period, the question arose as to whether the U.S. nuclear programs should remain under control of the military, or if they should be brought under civilian jurisdiction. The military point of view that they should remain in control was introduced into Congress as the May-Johnson Bill, which was defeated. An alternative proposal favoring civilian control (the McMahon Bill) was accepted and passed by Congress in the form of the Atomic Energy Act of 1946.

THE ATOMIC ENERGY COMMISSION AND THE NATIONAL LABORATORIES

The main thrust of the Atomic Energy Act, which became effective January 1, 1947, was to transfer the U.S. nuclear effort from military to civilian control. It created the Atomic Energy Commission (AEC) under the executive branch of the government and the Joint Committee on Atomic Energy (JCAE) under the legislative branch. The original AEC consisted of five prominent public figures: David Lilienthal (Chairman), Lewis Strauss, Sumner Pike, Robert Bacher, and William Waymach. Bacher was the only technical person. The Joint Committee was made up of 18 senators and congressmen, with Senator Bourke Hickenlooper as the chairman. Although the military had relinquished overall management of the nuclear programs, they still maintained a strong influence as one of the four divisions of the AEC.

During the AEC "start-up" period of 1947-1950, a General Advisory Committee (GAC) provided technical advice and guidance for the new organization. Members of this committee were: Robert Oppenheimer (Chairman), Enrico Fermi, Walter Zinn, Isidor Rabi, Glenn Seaborg, Lee DuBridge, James Conant, Eugene Wigner, Frank Spedding, and Norris Bradbury.

This newly formed group of organizations faced a plethora of problems and decisions at the outset. Foremost was the Russian nuclear threat. It was clear that the U.S. would have to maintain a strong nuclear weapons program. The U.S.S.R., using captured German scientists and engineers, had developed a highly efficient technology for the separation and recovery of U-235, based upon ultra-high-speed gas centrifuges, and had produced large quantities of weapons-grade material. The U.S. continued to use the gaseous diffusion plants.

When the AEC was formed, scientists and engineers at the Met Lab and other sites began to propose many new potential uses for nuclear energy, the main one being civilian nuclear power generation. Breeder reactors were already under consideration. The idea of a nuclear-powered locomotive came up, but it was not pursued. Naval propulsion reactors were proposed, and Captain (later Admiral) Hyman Rickover, along with the General Electric Co., designed a nuclear power system for a destroyer escort. Shortly thereafter, work began on nuclear powered submarines and the U.S.S. *Nautilus* was launched in 1954. Work had started on the Nuclear Energy for the Propulsion of Aircraft project (NEPA) and this program lasted until 1961. The use of nuclear explosives for civil engineering projects (later called "Plowshare") was proposed. Biological and medical uses were considered to be highly promising. In addition to these and other applications of nuclear energy, scientists were agitating strongly for a greatly expanded effort on basic nuclear research, particularly in the area of particle physics.

When the AEC was formed, the principal facilities under its management were Argonne National Laboratory (formerly the Met Lab), the Oak Ridge complex, the Hanford facilities, and the Los Alamos Scientific Laboratory (then called LASL). Important work was also being done at various university laboratories, including Berkeley, Iowa State, and Columbia. The Clinton Laboratory was renamed Oak Ridge National Laboratory (ORNL) in 1948.

Several new installations were built during the next few years. Brookhaven National Laboratory (BNL), which grew largely out of the research group at Columbia, was founded on Long Island in 1947. Argonne, Brookhaven, and Oak Ridge, the three original

National Laboratories, are all multipurpose facilities at present. Edward Teller, who has been dubbed the "Father of the H-bomb" and was a staunch advocate of thermonuclear weapons, became dissatisfied with the Los Alamos program and proposed a new laboratory for that type of work. The result was the creation of what is now Lawrence Livermore National Laboratory (LLNL), located near Berkeley. Construction of the National Reactor Testing Station (NRTS) in the Idaho desert was begun in 1949. Fermi, Oppenheimer, and Seaborg felt that three major reactors (a breeder, a materials testing reactor, and a propulsion-type reactor) should be located at Argonne, but Teller pushed through the Idaho site. Several other special-purpose facilities were built in the late 1940s and the 1950s. A list of the principal U.S. nuclear facilities, some of which were built later, and others that are no longer operational, is given in Table 1-1.

In 1946, the Met Lab became Argonne National Laboratory with Walter Zinn as the Director. Zinn had been deeply involved in the Manhattan Project, including the operation of Fermi's original pile, and he served ANL in its early years as a highly competent, dynamic leader. This was undoubtedly the major factor in the AEC's decision in 1947 to center all the nation's nuclear reactor research at ANL. The Laboratory did, in fact, play a leadership, but not exclusive, role in the U.S. reactor research and development programs for several years. In 1947, General Groves approved the purchase of 3,700 acres of land in DuPage County, including the estate of a sausage magnate in the Chicago area by the name of Erwin Freund, as the future site for Argonne. With its new name and this "Site D" property, ANL began to develop into the institution we know now. The first Argonne picnic, which has become an annual event, took place on September 9, 1948.

THE MET LAB

In 1942, Arthur Compton had consolidated nearly all the national atomic research activities at the Met Lab, located at the University of Chicago. Most of the atomic physicists had been working at various East Coast universities, using different types of equipment and experimental approaches, and he felt that the effort should be more closely coordinated in one location. The group he assembled could well have been the largest collection of Nobel laureates and other renowned scientists ever to work together in one laboratory. In spite of the code name "Metallurgical Laboratory," as Laura Fermi once pointed out, there wasn't a single metallurgist in the entire group at the time. The task of this group was to provide the scientific and technical "know how" that was desperately needed in planning and constructing the facilities at Oak Ridge, Hanford, and Los Alamos. As those facilities became a reality, many people were transferred to them from the Met Lab to provide the technical leadership and expertise that was needed to get them into operation.

Life at the Met Lab for the workers and their families was quite different from that at the newer sites where they lived in "secret" company towns in remote locations, basically cut off from civilization. The entire towns, enclosed by fences, were under complete control of the Army, which provided the necessities, but few luxuries, for schools, housing, shopping, and other ordinary needs of a family, and security was at the highest possible level. At Los Alamos, there were no individual mailing addresses, and incoming and outgoing mail was censored. In contrast, most of the Met Lab workers and their families lived in the Hyde Park area of Chicago in rental apartments or houses, and their children attended the Chicago schools. The workers walked or commuted to work by public

Table 1-1. Major U.S. Nuclear Facilities

Multipurpose Laboratories[a]

- Argonne National Laboratory (Univ. of Chicago)
- Brookhaven National Laboratory (Associated Universities, Inc.)
- Oak Ridge National Laboratory (Martin-Marietta)

Uranium Enrichment Plants

- Oak Ridge, TN (Union Carbide): UF_6 diffusion plant
- Portsmouth, OH (Goodyear Aerospace): UF_6 diffusion plant
- Paducah, KY (Union Carbide): UF_6 diffusion plant

Materials Processing Plants

- Ashtabula Feed Materials Plant, OH (Reactive Metals): Fabricate metal parts from depleted and low-enriched uranium for production reactors and bomb parts
- Fernald, OH (National Lead): Same as Ashtabula
- Idaho Chemical Processing Plant (ICPP), Idaho Falls (Allied Chemical): Unburned enriched uranium (mostly from submarines) removed from used fuel rods and sent on for recycling
- Hanford Production Operations, Richland, WA (Rockwell Hanford and United Nuclear): Generate plutonium in reactors, separate and recover it by reprocessing, then send it on for bomb parts
- Savannah River Plant, Aiken, SC (E. I. DuPont): Same as Hanford. Also prepares deuterium as heavy water and makes tritium by irradiation of lithium in the reactors

Weapons Fabrication Plants

- Kansas City Plant, MO (Bendix): Electronic and mechanical weapons parts
- Mound Laboratory, Miamisburg, OH (Monsanto Research): Special small high-explosive components and radioisotope batteries for bombs; uses Pu-238 from Savannah River
- Savannah River Weapons Facility, Aiken, SC (E. I. DuPont): Fabrication of uranium and lithium deuteride parts
- Y-12 Plant, Oak Ridge, TN (Union Carbide): Same as Savannah River Weapons Facility
- Pinellas Plant, St. Petersburg, FL (General Electric): Makes neutron trigger bombs
- Rocky Flats Plant, Golden, CO (Rockwell International): Fabrication of plutonium metal parts
- Pantex Plant, Amarillo, TX (Mason and Hangar-Silas Mason): Fabrication of larger high-explosive parts, assembly of weapons from components, recycle of old warheads

Weapons Research & Development

- Los Alamos National Laboratory, Los Alamos, NM (U. of CA)
- Lawrence Livermore National Laboratory, Livermore, CA (U. of CA)
- Sandia Laboratories, Albuquerque, NM (Western Electric)
- Nevada Test Site, Las Vegas, NV (Reynolds Electrical & Engineering)

Naval Nuclear Propulsion

- Bettis Atomic Power Laboratory, West Miffin, PA (Westinghouse)
- Knolls Atomic Power, Schenectady, NY (General Electric)

---

[a] At the present time, the multiprogram laboratories also include Idaho National Engineering and Environmental Laboratory (INEEL), Lawrence Berkeley National Laboratory (LBNL), Lawrence Livermore National Laboratory (LLNL), Los Alamos National Laboratory (LANL), Pacific Northwest National Laboratory (PNNL), and Sandia National Laboratories (SNL).

transportation and generally lived a rather normal life, except for the long hours and the extreme security. They were, of course, like all other civilians, subject to war-time rationing of gasoline, meat, sugar, butter, shoes, tires, and other items and to the nationwide 35-mph speed limit.

As early as 1943, some of the Met Lab workers began to be concerned about its future prospects. The Met Lab was being used as a training facility for personnel at the new nuclear labs and plants and many of the staff personnel were being siphoned away. It was clear that the weapons work would continue to be taken over by others. Nevertheless, there was still plenty of work that needed to be done, and programs continued at the Met Lab in several areas of nuclear research, including many of the early studies on radiation safety and health physics. The two programs that were most pertinent to the future interests of the Chemical Engineering Division at Argonne were reactor physics and development work on processes for the recovery of plutonium from irradiated uranium reactor fuels.

Most of the reactor research at the Met Lab and, later, Argonne, in the 1940s and early 1950s, involved the "CP" (Chicago Pile) series of reactors listed in Table 1-2.

In 1943, the CP-1 pile was disassembled and removed to a site in the Argonne Woods (now the Red Gate Woods) in Palos Hills about two miles southwest of Willowbrook, where it was enlarged somewhat, renamed "CP-2," and used for further reactor physics experiments. This location was designated "Site A," and is the source of the name Argonne. It is accessible from Archer Avenue through the Red Gate entrance, but the only evidence today of its previous existence is a grassy mound with a small marker.

CP-3 had a higher power level, and employed heavy water instead of graphite as the moderator. It was used primarily for reactor physics research that involved neutron optics studies, cross-section measurements, the effects of oscillation, and other phenomena of interest. After CP-3 had operated for six years, the natural uranium fuel was replaced with an alloy of 2% highly enriched uranium in aluminum; this version was called CP-3'.

People sometimes ask, "Why isn't a CP-4 on the list?" There was, in fact, a reactor design that started out as CP-4 in its early stages, but it eventually became transformed into Experimental Breeder Reactor I (EBR-I), which is another story.

Table 1-2. CP Series of Reactors

Reactor	Location	Power	Fuel	Moderator	Operation
CP-1	West Stands	200 W (max.)	Natural U Metal, Oxide	Graphite	12/2/42
CP-2	Site "A" Palos Park	200 W-2 kW	Natural U Metal, Oxide	Graphite	1943-1954
CP-3	Site "A"	300 kW	Natural U Metal	Heavy Water	1944-1950
CP-3'	Site "A"	300 kW	98% Al-2% Enriched U	Heavy Water	1950-1954
CP-5	Site "D" (DuPage Co.)	1,000 kW	98% Al-2% Enriched U	Heavy Water	1954-1979

The last of the series, CP-5, was similar to CP-3', but it was larger and designed to accommodate a wide variety of users. It was used extensively by ANL scientists from the various divisions, as well as many others from universities and industrial laboratories. Two novel features were of much interest and utility. One was a neutron chopper operating on the same general principle as a time-of-flight spectrometer, which could provide a neutron beam at a specific energy level, thereby permitting cross sections or other nuclear data to be determined as a function of neutron energy. The other feature was a "rabbit" that could be passed through the reactor *via* a pneumatic tube, so a sample could be recovered very quickly for short half-life measurements. Both features were used occasionally by members of the Chemical Engineering Division.

Early in the Manhattan Project, a decision was made to pursue both uranium-235 and plutonium-239 as fissionable materials for nuclear weapons. As mentioned earlier, the uranium-235 could be recovered from natural uranium either by electromagnetic separation or by gaseous diffusion of the hexafluoride; both approaches were pursued immediately at the Oak Ridge installation. The recovery of plutonium was a more challenging problem because it first had to be generated through neutron capture by uranium in a reactor, and then separated chemically from the uranium and fission products. It was this program that eventually spawned the Argonne Chemical Engineering Division.

The first work on this type of separation was performed at the University of Chicago by a small group of chemists, some of whom had been involved in the early studies of plutonium by Seaborg's group at Berkeley in 1940. The initial studies were done in the Kent Laboratory and the George Herbert Jones Laboratory. As the effort expanded, most of the research on plutonium was transferred to a temporary building, called "New Chem," which was erected in 1943 on the northwest corner of the University of Chicago campus. Subsequent work on the engineering aspects of the separations processes was located beneath the West Stands of Stagg Field, with semiworks facilities in the area of the squash court where the original CP-1 pile had been. A second floor housed chemical research and analytical laboratories. (The term "semiworks" refers to small-scale engineering development work on equipment, operating conditions, and general feasibility of process operations as opposed to a pilot plant, which is usually a small prototype of a specific full-size plant.) At that time, and throughout the existence of the Chemical Engineering Division, an analytical laboratory has been essential because of the large number of chemical and radiochemical analyses needed to determine the effectiveness of the separation procedures that were under development.

The first order of business was to develop a process as quickly as possible for the recovery of plutonium that was to be bred in the irradiated uranium fuel of the Hanford reactors. When Seaborg arrived at the Met Lab in 1942, some information was available on the chemistry of plutonium from the work he and his coworkers had done at Berkeley. However, a huge amount of creative, meticulous research was necessary to obtain the information needed to develop a full-scale plutonium recovery process. In addition, the nuclear physicists concerned with weapons design were desperate for data on the physical properties of metallic plutonium, such as density, hardness, and phase transitions. At the outset, only sub-microgram quantities of plutonium, which had been generated by irradiating several hundred pounds of uranium oxide in a 45-in. cyclotron at Washington University in St. Louis, were available. The irradiated uranium was moved from St. Louis to Chicago by personal car or truck in wooden and Masonite® boxes shielded internally with lead bricks and often of questionable integrity.

The only analytical method for plutonium at that time was radiation counting.

The first plutonium to be observed visually was about one microgram of the fluoride that was isolated in pure form on August 20, 1942. Later on, milligram amounts began to be produced, making the work considerably easier. The amount of information on the chemistry and physical properties of plutonium and its compounds that Seaborg and his associates were able to generate in a short time is astounding, especially when one considers the micro scale of the work.

An interesting sidelight of this program was an effort to develop a convenient nomenclature for the various isotopes and materials. A convention that one still runs into occasionally in conversation or the older literature is an isotope naming system in which the actinide isotopes are identified by the last digit of the atomic number and the last digit of the atomic weight; *e.g.*, 28 is U-238, 25 is U-235, 49 is Pu-239, and 39 is Np-239.

The primary task of Seaborg's group at the Met Lab was to develop a procedure for separating weapons-grade plutonium from uranium and fission products. Several avenues were explored, one of which was the Bismuth Phosphate ($BiPO_4$) process. This was a batch precipitation procedure that separated plutonium from the uranium and fission products by a series of $BiPO_4$ precipitations from aqueous solutions. Plutonium was coprecipitated with the $BiPO_4$ in the tetravalent state and left in solution in the hexavalent form. Final purification of the plutonium was achieved by a similar precipitation cycle, using lanthanum fluoride (LaF_3) as the carrier precipitate. Uranium, along with the fission products, was discarded to waste. (A solvent-extraction process was used several years later to recover the uranium.) The process met the immediate objective by recovering plutonium with greater than 95% efficiency and a ten-million-fold removal of fission products, *i.e.*, a decontamination factor of 10^7, and it was put into full-scale operation at Hanford in 1944. The elapsed time between the first visual observation of plutonium (as a fluoride) and full-scale production at Hanford was only two years. This billionfold scale-up from microgram to kilogram quantities in one step was an incredible achievement.

The Bismuth Phosphate process did, however, have serious disadvantages—the multiple batch operations, the inability to recover uranium, the large quantities of process chemicals that were required, and the large volume of process wastes. For these reasons, a search was begun for processes having a potential for higher capacities, improved efficiency, and lower costs, as well as a capability for a three-way separation of uranium, plutonium, and fission products.

Some experience was already available on the use of solvent-extraction processes to extract uranium from leach liquors produced during processing of the ore. In the processing of discharged reactor fuels, the fuel material was first dissolved in an aqueous solvent, which was normally nitric acid (HNO_3). Separation of the actinide elements from each other and from the fission products was then accomplished by repeated extractions between the aqueous solutions and an organic solvent in continuous, multistage equipment such as packed columns. Partitioning of the various elements between the two phases depends on the compositions of the aqueous and organic phases and can be manipulated through the use of oxidants, reductants, salting-out agents, and complex-forming compounds.

Solvent-extraction processes offered the potential advantages of continuous operation in multistage, countercurrent extraction devices in which separation factors are multiplied manyfold to achieve very high fission-product decontamination factors (typically 10^6-10^9), and excellent recovery (>99.5%) of uranium and plutonium. Processes of this type also have the advantage that they avoid the materials handling problems associated with solids.

Preliminary studies of solvent-extraction processes were conducted at the Met Lab. The results were sufficiently promising that an increased effort was justified and extraction columns were set up in the West Stands for this purpose.

Seaborg, realizing that solvent-extraction technology was a specialized field, knew that he had to find an expert to continue the work, and in 1944 he asked Dr. Stephen Lawroski to direct the effort. Lawroski, a recognized authority in solvent extraction, had studied under Professor Merrill Fenske and received a Ph.D. at Pennsylvania State University. He was employed by the Standard Oil Development Company at the time. At the request of the Manhattan District authorities, he was placed on loan to the Met Lab for two years. Later on, he became the original director of the Argonne Chemical Engineering Division.

The initial assignment of Lawroski's group at West Stands was to develop process compositions and equipment to achieve ultra-purification of Hanford plutonium from light element impurities. This requirement stemmed from concern by nuclear weapon designers at Los Alamos that alpha particles from the decay of plutonium would interact with the light elements to produce neutrons prematurely and thereby cause a major reduction of explosive power of nuclear weapons using plutonium. The West Stands group soon demonstrated that solvent extraction could, indeed, achieve ultra-purification of plutonium. That achievement of itself, however, turned out to be useless when it was learned that the Hanford plutonium contained a small, but significant amount of a spontaneously fissioning isotope of plutonium that would still result in the premature presence of neutrons. This meant that the weapons scientists had to develop a plutonium weapon design entirely different from that used for the U-235 weapon. To increase the velocity with which the subcritical masses of plutonium were forced together, explosive charges surrounding the plutonium were shaped into a "focusing lens" configuration that would create a very rapid implosion.

Despite the situation just described, the work of the West Stands group was not terminated. It was instead redirected to take advantage of the potential already demonstrated for solvent extraction by that group and new information on the chemistry of plutonium that had been generated by Seaborg's chemists at New Chem. The redirected effort ultimately culminated in the Redox process, which, after pilot-plant tests at Oak Ridge, was applied in 1951 on a production scale at Hanford to recover decontaminated plutonium and uranium separately. Replacement of the cumbersome Bismuth Phosphate process by the Redox solvent-extraction process at Hanford resulted in an enormous cost saving that paid for the new $50 million plant within only two years.

A simplified schematic flowsheet for the first cycle of a Redox process is shown in Fig. 1-1 to illustrate how the basic separations of uranium, plutonium, and fission products were made. Hexone (methyl isobutyl ketone) was used as the organic solvent, and aluminum nitrate, $Al(NO_3)_3$, served as a salting agent in the aqueous nitric acid phase to increase the distribution of uranium and plutonium into the hexone. In this, as well as in later processes, plutonium was separated from uranium by reducing the plutonium to the trivalent state, in which it strongly favors the aqueous phase. Figure 1-1 shows only the first cycle; in practice, one or two additional uranium and plutonium purification cycles consisting only of extraction and stripping operations are added, since complete partition of the uranium and plutonium occurs in the first cycle. The additional cycles result in much higher fission-product decontamination factors and in product recoveries greater than 99.5%.

Solvent-extraction processes for reactor fuels tend to be generic in nature in that they all involve dissolution of the fuel in acid,

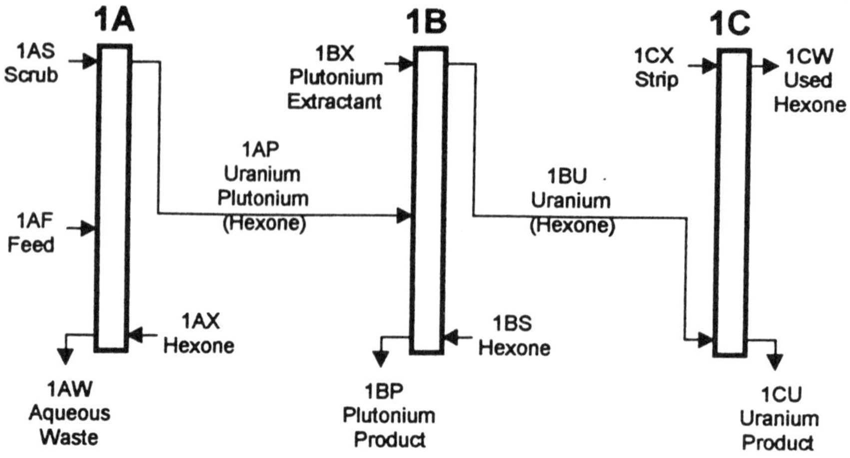

The spent reactor fuel is dissolved in nitric acid and a strong oxidant such as sodium dichromate ($Na_2Cr_2O_7$) is added to convert all the uranium and plutonium to the hexavalent state (stream 1AF). When in the hexavalent state, the uranium and plutonium tend to favor the organic (hexone) (1A) phase, and a scrub stream (1AS) containing aluminum nitrate [($Al(NO_3)_3$)] as a salting agent enhances this effect. The fission products remain in the aqueous phase (1AW), which is discarded as waste. In the 1B column, stream 1BX is an aqueous solution of nitric acid and a reducing agent such as ferrous sulfamate [($Fe(H_2NHSO_3)_2$)], which reduces the plutonium selectively to the trivalent state. Trivalent plutonium favors the aqueous phase and is recovered in the product stream, 1BP. In column 1C, a dilute solution of nitric acid, stream 1CX, is used to extract the uranium from the hexone back into the aqueous phase, which becomes the uranium product stream, 1CU. The used hexone, stream 1CW, is recovered and recycled.

Fig. 1-1. Flowsheet for First Redox Cycle

followed by a series of extractions between the acidic aqueous solution and an organic solvent. The types and efficiencies of the separations that can be made, however, are affected markedly by the compositions of the two solvent phases, and most of the progress that was made in this technology resulted from investigations of a wide variety of organic solvents and complexing and salting agents. The design of the equipment is also important because it determines the throughput rate of the process, the efficiency of the separations and the methods for handling the input and product streams. Basic studies were conducted in which the effectiveness of extractants was correlated with acid-base theory, and the settling rates and interfacial areas of the immiscible liquids were related to the physical properties of the liquids. In the early small-scale engineering studies, the extraction columns consisted of 1-in.-diameter glass pipe packed with glass helix rings about 1/4 in. in diameter. These countercurrent columns, which were up to 20 feet in height, were operated by remote control behind heavy concrete shielding, and were pulsed in some cases. Later on, in the Argonne Chemical Engineering Division, most of the solvent-extraction work was done initially with 1-in.-diameter stainless steel columns packed with 1/4-in. Raschig rings (hollow cylinders). These columns, up to 30 feet in height, were located behind concrete shielding in the high-bay area of Building D-205 and were operated remotely. As the development studies on solvent extraction progressed, other contacting devices such as mixer-settlers and stacked-stage extractors came into the picture because of their high efficiency and flexibility of layout in a full-scale plant.

Following the development of the Redox process, the Canadians developed the Trigly process, which employed triglycol dichloride as the organic phase in the head-end cycle and hexone in the subsequent cycles as in the Redox process. The Butex process, which used dibutyl carbitol as the organic solvent, was developed at Oak Ridge. An advantage of this process is that no additional salting agent is required in the nitric acid solution, so the nitric acid can be recovered by evaporation and the fission-product waste volumes can be reduced. The Butex process is still used at the Y-12 plant in Oak Ridge for the recovery of enriched uranium.

A major advance was made in nuclear fuel reprocessing when workers at Oak Ridge originated the Purex process. The solvent in this process is tri-*n*-butyl phosphate (TBP) diluted with a kerosene-type hydrocarbon such as dodecane. This solvent has a strong affinity for uranium and plutonium and is able to extract their nitrate salts from nitric acid solutions. The Purex process has several advantages over the Redox process—elimination of nonvolatile salting agents, lower solvent volatility and flammability, high chemical and radiation stability of the solvent, and lower operating costs. The Purex process was tested on a pilot-plant scale at Oak Ridge National Laboratory and installed in the plutonium production plants at Hanford and Savannah River. At Hanford, it replaced the Redox process.

The diversity of solvent-extraction schemes that were investigated at the Met Lab, Argonne, and elsewhere after the Redox work was completed is too great for a detailed description of each, but Table 1-3 lists most of the processes, or variations thereof, which have been investigated at one time or another.

Solvent-extraction processes are still under investigation. The current work is concerned mainly with the TRUEX Process, which has the capability to separate the long-lived transuranium elements, such as neptunium, curium, and americium, from fission-product waste streams.

Work on the Redox solvent-extraction processes was continuing at the time the Chemical Engineering Division was formed. According to a report (ANL-4110) issued in January 1948, a month before the new division came into being, the organization of the Process Chemistry Group in the Chemistry Division was as follows:

Group Leader: Herbert Hyman

Assistant Group Leader: John Schraidt

Semi-Works Operations: John Schraidt, Phil Fineman, George Bernstein, Les Coleman, Lee Gaumer, Sherman Greenberg, Dave Jacobson, Jim King, Milt Klein, Harry Litland, John Natale, Laury Ross, Art Shor, and Bill Walters

Laboratory Operations: Harold Evans, Sy Vogler, and Eugene Hausman

Analytical Operations: Al Jonke (head), Olga Fineman, Jodie Hoekstra, and Carolyn Kennedy

This group was under the direction of Dr. Lawroski, who, at that time, was the Associate Director of the Chemistry Division in charge of process development.

Table 1-3. Solvent-Extraction Processes

Process	Solvent	Salting Agent(s)	Irradiation Fuel	Recovered Products
Redox	Hexone[a]	HNO_3, $Al(NO_3)_3$	Natural U	U, Pu
Purex	TBP[b] in Hydrocarbon[c]	HNO_3	Natural U	U, Pu
Butex	Dibutyl Ether or Ethylene Glycol	HNO_3, NH_4NO_3	Natural U	U, Pu
Halex	TBP in CCl_4	HNO_3	Natural U	U, Pu
Hexone 25	Hexone	HNO_3, $Al(NO_3)_3$	Enriched U-Al	Enriched U
Zr Alloy	TBP in Hydrocarbon	HNO_3, $Al(NO_3)_3$	Enriched U-Zr	Enriched U
Thorex	TBP in Hydrocarbon	HNO_3, $Al(NO_3)_3$	Thorium	Th, U-233
Ether	Diisopropyl Ether	$Th(NO_3)_4$, $Al(NO_3)_3$	Th-Al	U-233

[a] Methyl isobutyl ketone.
[b] Tri-*n*-butyl phosphate.
[c] Kerosene-type solvents such as dodecane.

THE CHEMICAL ENGINEERING DIVISION

The story of the Chemical Engineering Division begins with Dr. Stephen Lawroski, who was its original Division Director. Dr. Lawroski received a doctorate in chemical engineering in 1943 from Pennsylvania State University. While doing his graduate work, he was employed as a Research Assistant at the Petroleum Refining Laboratory at State College, Pennsylvania, where he was one of the principal staff members working on high-efficiency packing materials for distillation and solvent-extraction equipment. As mentioned earlier, he spent two years at the Met Lab from 1944 to 1946 on loan from the Standard Oil Development Company (later named the EXXON Research and Engineering Company). There, he directed a group engaged in the development of solvent-extraction processes for the recovery and purification of uranium and plutonium from Hanford plutonium production reactors. This group, under his leadership, was highly productive

Fig. 1-2. Stephen Lawroski

and its work led to the Redox process. In 1946, he returned to the Standard Oil Development Company as Assistant Section Chief of the Manufacturing and Process Section of the Research Division. In September of that year, however, his company recommended him for atomic energy training as an Advanced Professional Trainee at the Clinton Laboratory, where he remained until June 1947. This assignment offered the opportunity to study reactor and separations technology, including the solvent-extraction pilot plant that had been built to test the large-scale Redox process for the Hanford facility.

From Argonne's standpoint, one of the most valuable results of this assignment was that Dr. Lawroski, in part due to his gregarious personality, made many friends among the other engineers, scientists, and trainees. Later on, he persuaded some of these people (Hal Feder, Milt Levenson, Walt Rodger, Les Coleman, and Les Burris) to come to work at ANL. He also became acquainted with a number of other people who later became important contacts in the AEC and the other national nuclear establishments.

From Dr. Lawroski's viewpoint, he would no doubt be the first to agree that by far the most valuable asset he acquired at the Clinton Laboratories was his new bride, Helen, who had been working in their Health Physics Division.

When Dr. Lawroski returned to the Chicago area in July 1947, he accepted employment as head of the Process Development Section and Associate Director of the Chemistry Division (CHM) and the section under Herbert Hyman became one of his responsibilities. Early in 1948, Dr. Zinn approached Dr. Lawroski with the proposal that ANL should establish a Chemical Engineering Division (CEN) with him as director. Thus, the Chemical Engineering Division was born in February 1948. Soon thereafter, Zinn presented him with an interesting choice. As the Laboratory was moving to the present DuPage site, the Chemical Engineering Division was given the option of moving to the new DuPage site within about a year if it were willing to move into military-type Quonset buildings. Or, if it preferred to wait another year, it could have new buildings built specifically to meet its requirements. That was the genesis of Buildings D-205 and D-310. Members of the Division who remember the leaky, drafty Quonset buildings occupied by the Administration, Travel Office, Graphic Arts, Health Services, and other ANL organizations for several decades can appreciate the benefits of the choice that was made. Some of the longer-term employees will remember the annual physical examinations, in which the procedure, after the chest x-ray and blood sample, was for the patient, essentially unclothed, to wait in a small room for 30 to 45 minutes for the doctor to show up. During the winter in that leaky Quonset hut, "cooling your heels" was an understatement.

Fig. 1-3. Herbert Hyman

Building D-310, which was completed first, was designed originally as a structure for experimental processing, storage, and shipping of radioactive wastes. A high-level gamma irradiation facility was added later. This consisted of a so-called "swimming pool" into which irradiated reactor fuel assemblies could be lowered by a crane. The pool was filled with water to provide thermal cooling and radiation shielding. The building was occupied initially by an incinerator and scrubber for processing dry active wastes and various pieces of equipment for treating active liquid wastes. Later on, Building D-310 served as a general-purpose area for a variety of projects that involved large equipment.

Building D-205 was much larger than D-310, and was planned to meet all the other needs of the Division, including engineering and chemistry laboratories, office space, drafting rooms, a library, conference rooms, shops, and a number of other facilities. These buildings will be described in more detail in the next chapter. The exodus of CEN personnel from the West Stands to the DuPage site began in 1950 and was completed in 1951.

Les Burris, who shared an office with Charlie Stevenson, tells of the time when Charlie, deeply involved in the planning of the new buildings, had a habit of laying large blueprints on a table where Les had been working with secret documents. The result was a series of security violations that got Les into some hot water.

Another incident that occurred during this period was the case of a new staff member who had not yet become fully familiarized with the operating procedures. People walking by his office one day noticed smoke emanating from the door. At that time, each office was provided with a red wastepaper basket labeled "BURN" for disposal of classified papers and he had taken it literally. Rumor has it that this happened more than once. Burn baskets, if used today, would most likely come with an operating manual and training sessions.

When the Division was formed, Dr. Lawroski made a policy decision that probably had a more profound effect than any other single factor on the nature of its future work. He believed that process development should be an integrated effort from the test tube to plant design. Thus the major programs often included basic and applied lab-scale research, basic engineering studies, equipment development, engineering design, materials development, pilot plant or semiworks testing, conceptual plant design, and some economic evaluations. With this type of organization, team efforts could include whatever particular talents were needed at any stage of process development, and much of the work could be done in parallel instead of sequentially. It also expedited feedback of problems for further work. The basic chemistry and engineering studies, although directed toward solutions of practical problems, were most often performed with the care and scope necessary to produce quality publications in the basic scientific and engineering journals. At the same time, these resources were available for troubleshooting on problems arising in the process development work.

The combination of engineers and scientists in the Division made for some interesting interactions. A certain amount of good-natured ribbing occurred in which the engineers referred to the chemists as "pharmacists" or something similar, and the chemists referred to the engineers as "pipefitters," but there was a mutual respect between the two groups. In fact, it was not unusual to find an engineer working with test tubes or a chemist assembling pipes and valves. The engineers most often used the English system of measurement, which made sense because the process equipment was almost always sized in those units. The chemists, however, had been brought up with the metric system and their equipment used those units. Both groups had to become bilingual in this respect, and both ANL and the Federal Government to this day are continuing to cope with this problem in the written materials that are issued. This general problem gave the basic scientists a special appreciation for the engineers' penchant for dimensionless numbers.

As a first step in implementing this policy, Dr. Lawroski, in order to complement his own training and experience as a chemical engineer, hired a highly qualified chemist to serve as the Associate Division Director. The man he selected was Dr. Charles Stevenson, who had earned a Ph.D. in organic chemistry at Pennsylvania State University and then worked as a research chemist at the Standard Oil Development Corporation and the Diamond Glass Company. Lawroski and Stevenson had been colleagues and personal friends both at Penn State and at Standard Oil. Charlie was a highly competent, affable individual, and he brought a new dimension to the Division.

Fig. 1-4. Charles Stevenson

The core personnel of the new Chemical Engineering Division were basically those from the Process Development Section of the ANL Chemistry Division plus new hires, including those from the Clinton Laboratory at Oak Ridge. People who were at the Met Lab in the early days and in the Chemical Engineering Division after it was formed include Elton Turk (1942), Milt Ader (1944), George Bernstein (1944), Phil Fineman (1944), John Schraidt (1944), Les Coleman (1946), Harold Evans (1946), and John Natale (1946). Milt, George, John, and Phil were members of SED (Special Engineering Detachment) of the U.S. Army during part of the time. Don Webster, who joined the Division much later and served as an Associate Division Director, had also spent a short time at the Met Lab in 1942-43. Marvin Tetenbaum spent some time at the Met Lab in 1942, returned to New York to obtain a Ph.D., worked at Columbia University for a time, and came to CEN several years later. The people from Oak Ridge (Hal Feder, Walt Rodger, Milt Levenson, and Les Burris) brought with them a great deal of practical experience in radiochemistry and hot pilot-plant operations. In 1949, Richard Vogel, who had received a Ph.D. in physical chemistry at Harvard University and was on the faculty of the Illinois Institute of Technology, was hired as a Senior Chemist, and was destined to succeed Dr. Lawroski as the Division Director several years later. Victor Munnecke, a chemical engineer, became the Assistant Division Director, and was responsible for the administrative and financial affairs of the Division. Ed Peterson had the primary management responsibility for the new buildings.

Once the Division was established, it expanded rapidly, both in manpower and in the scope of the work. Nearly all of the work during 1948 and 1949 continued to be directed toward solvent-extraction processes. A large program under Walt Rodger was concerned with the use of acid-deficient solvent-extraction flowsheets that had been proposed by Oak Ridge and later by Hanford. Some of the studies were done with extraction columns and others with two 20-stage mixer-settler units in which all the stages could be sampled simultaneously to obtain equilibrium data. These were especially useful in constructing equilibrium diagrams for various operating conditions. Individual studies were conducted on the precipitation of plutonium oxalate in columns, the behavior of neptunium, and the possibility of volatilizing ruthenium from solutions by oxidation to RuO_4 with oxygen-ozone mixtures.

Because the breeder reactor concept had become popular both at ANL and in the AEC, interest began to develop in the reprocessing of breeder reactor fuel. Recovery of Experimental Breeder Reactor and Materials Test Reactor fuels had been demonstrated in the Oak Ridge pilot plant. One such ANL program, headed by Les Burris, was the development of a simpler tributyl phosphate (TBP)-methylcyclohexane process for the recovery of highly enriched uranium from experimental EBR cores. This process proved to be capable of achieving the requisite fission product removal (a decontamination factor of 10^5) and uranium recovery (>99.9%) in a single solvent-extraction cycle. Sixteen runs with active feed material from Hanford that were conducted in the shielded columns in the high bay section of Building D-205 showed that the process could meet the requirements. While this work was still in progress, however, the AEC issued an edict that the bulk of the EBR fuel would be processed at the Idaho site, and the TBP process would be used at ANL only for analytical samples and cleanup operations.

That research is covered in a 1950 report, which credits the work to this group of people:

Project Leader: Les Burris, Jr.

Laboratory Group: Richard Vogel, Harold Evans, Morris Beederman, Bob Hildebrandt, Homer Tyler, and Bob Schablaske

Semi-Works Group: Walt Rodger, John Schraidt, John Natale, Lee Gaumer, Ed Hykan, John Loeding, Alex Aikens, Virgil Trice, Ira Dillon, Don Hampson, Norm Levitz, Les Coleman, Les Dorsey, Elmo West, Herb Brown, and Bill Voss

Pulse Column Group: Kegham Varteressian, Milt Levenson, and George Bernstein

Analytical Group: Doug Krause, Betty Reilly, Corky Thompson, Vincent Story, Chuck Seils, Jackie Williams, Larry Marek, Cynthia Hall, John Breeden, and Myron Homa

Some work was performed on the recovery of simulated Mark I naval reactor fuel, which was an enriched uranium-zirconium alloy. A Redox-type process seemed to be the best choice for this type of fuel, but it could not be dissolved in nitric acid because of its high zirconium content. Hydrofluoric acid with aluminum nitrate proved later to be the most promising solvent for this alloy.

One of the early processes initiated in the late 1940s and developed by the Division was of considerable import for recovery of tritium from irradiated lithium-aluminum alloy. Tritium was needed for the development of thermonuclear weapons (H-bombs). Tritium (hydrogen-3) is generated by irradiation of lithium-6 with neutrons, which results in the alpha reaction:

$$_3Li^6 + {_0n^1} \rightarrow {_1H^3} + {_2He^4}$$

Bernie Abraham of the Chemistry Division had proposed the use of lithium-aluminum alloy for this purpose. The tritium and helium recovery process consisted of heating the irradiated alloy to just below its melting point (about 600°C) at which temperature the gases, principally hydrogen-3 (tritium), helium-3, and helium-4, are released. The gases were pumped off, passed over uranium turnings at 800°C to remove any gaseous contaminants such as oxygen or moisture, and then through a palladium barrier to separate the helium isotopes from the tritium. The palladium barrier, a disc in the line maintained at a temperature of several hundred degrees Celsius, was permeable by the tritium, but not by the helium. This process was installed at

Hanford and later in the Savannah River production plant where it has been used for many years.

Development work was also initiated on the fluoride volatility process, in which uranium in the fuel was fluorinated to form uranium hexafluoride (UF_6). The UF_6 is volatile and can be separated from the other fuel constituents by vaporization or distillation. The rationale behind this process was that the decontaminated uranium product is in the form of a fluoride, which is directly suitable for reconversion to the metal, and the fission-product wastes would be a small volume of solid fluorides.

The idea of recovering uranium and plutonium by volatilization of the hexafluorides was not new. As early as 1942, Harold Urey had suggested the possibility of volatilizing uranium as the hexafluoride to separate it from plutonium. That same year Harrison Brown and Orville Hill at the Met Lab fluorinated the tetrafluorides of uranium and plutonium completely to the hexafluorides and suggested the procedure as a method for separating them from fission products. Fluorination studies continued off and on in the Met Lab for several years. Fluorine research was also in progress, particularly on the plutonium fluorides at Los Alamos. In 1944, Seaborg, in a systematic review of the stabilities of the actinide metal halides, concluded by analogy that plutonium hexafluoride (PuF_6) should be marginally stable, which was borne out by later experimental studies. The use of elemental fluorine as a fluorinating agent for metallic fuels did not work out well because of heat-transfer problems and irregular reaction rates. Joe Katz and Herbert Hyman of the Chemistry Division did some preliminary work on the use of halogen fluorides, such as ClF_3, BrF_3 or BrF_5, which are liquids. Bill Mecham and Milt Levenson conducted an experiment in the Chemical Engineering Division in which 10 g of irradiated uranium metal was dissolved in a BrF_3-BrF_5 mixture. The uranium dissolved smoothly and the UF_6 product was distilled off. The gross gamma decontamination factor was 2,000, and over 97% of the plutonium was in the residue. The only detectable fission-product activity in the UF_6 was tellurium. These results were highly encouraging and the fluoride volatility process became a major program in the 1950s.

Work continued on waste processing as the incinerator proceeded to dispose of radioactive combustible wastes from the entire Laboratory. Some development studies also continued on a process for the recovery of waste aluminum nitrate solutions from the Redox process.

By the end of the 1940s, the Chemical Engineering Division had established its identity as a major part of ANL and had become recognized nationally for the originality and excellence of its contributions to nuclear technology. It had expanded both in personnel and in programs to the stage that larger quarters were necessary. The time was ripe to move on to the new buildings at the DuPage site.

An attempt has been made to list in Table 1-4 all the people who worked in the Chemical Engineering Division during the 1940s.

Table 1-4. CEN Personnel in the 1940s

Milt Ader	Cynthia Hall	John Natale
Alex Aikens	Don Hampson	Ed Peterson
George Asanovich	Gerry Harmon	Laurie Peterson
Eunice Banks	Eugene Hausman	Norma Pinches
Horace Baxman	Bob Hildebrand	Roy Post
Helen Bednarick	Jodi Hoekstra	Betty Reilly
Morrie Beederman	Myron Homa	Walt Rodger
Michael Berkman	Ed Hykan	Sy Rosenthal
George Bernstein	Herb Hyman	Laury Ross
Walt Blaedel	Dave Jacobson	Bob Schablaske
Aaron Boyd	Joe Jacobson	Karl Schoeneman
John Breeden	Al Jonke	John Schraidt
Jim Bresee	Bettye Kaplan	Wally Seefeldt
Herb Brown	Lou Kaplan	Chuck Seils
Les Burris	Alec Keday	Irv Shaffner
Artie Butschelder	Carolyn Kennedy	Art Shor
Jack Caster	Jim King	Charlie Stevenson
Norm Chellew	Milt Klein	Gladys Swope
Les Coleman	Corky Kloska	Virgil Trice
Virginia DeGrande	Doug Krause	Elton Turk
Lee Deutsch	Stephen Lawroski	Homer Tyler
Barry Devine	Henry Lee	Kegham Varteressian
Chester Deziehl	Milt Levenson	Richard Vogel
Ira Dillon	Norm Levitz	Sy Vogler
Les Dorsey	Harry Littmen	Bill Voss
Marie Driskell	John Loeding	Roberta Wagner
Harold Evans	Les Mandelstein	Matt Walling
Hal Feder	Larry Marek	Bill Walters
Olga Fineman	Asher Margolis	Elmo West
Phil Fineman	George Mason	Jackie Williams
Lee Gaumer	Lee Mead	Irv Winsch
Jim Gilbreath	Bill Mecham	George Yasui
Sherman Greenberg	Vic Munnecke	Marion Yoshioka

1950-1960: THE NUCLEAR PROMISE

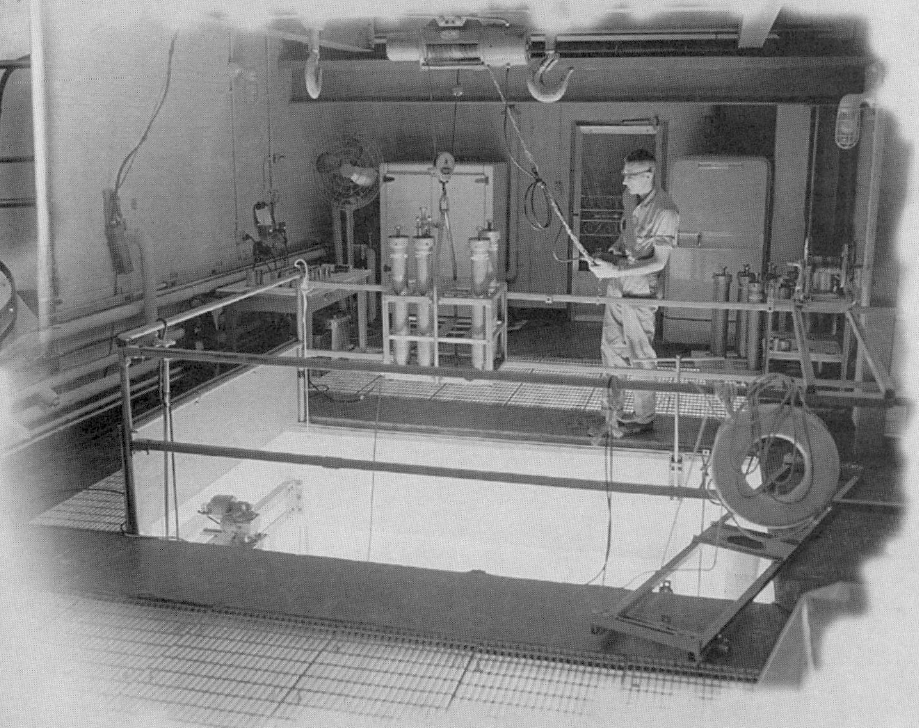

1950-1960

(top) Experimental Breeder Reactor-II (domed structure) with close-coupled fuel cycle facility (right foreground) at the National Reactor Test Station in Idaho.

(bottom, left) Melt refining furnace in which uranium fuel is melted in a ceramic crucible, then poured into a graphite mold. This simple procedure removes fission products to an extent that the fuel can be refabricated and recycled to the reactor.

(bottom, right) The Gamma Irradiation Facility ("Swimming Pool"), located in Bldg. 310. This facility was used to evaluate the effects of gamma radiation on foods and various other materials.

2

1950-1960: The Nuclear Promise

The 1950s was a decade of exploration into new uses of atomic energy, mainly for the generation of electric power by public utilities. Several types of reactors were being considered. Much of the work at ANL was concerned with the potential use of breeder reactors as a long-range means of conserving natural uranium resources. The research and development effort by CEN on nuclear fuel reprocessing was expanded to include pyrometallurgical, fluoride volatility, and aqueous methods in order to accommodate the different types of fuels that might be used in the new types of reactors and to reduce the cost of reprocessing. The Division also broadened the scope of its activities to include other phases of the nuclear fuel cycle by initiating programs on feed materials processing and the treatment of radioactive wastes. Several other programs were started in areas such as the chemical aspects of reactor safety, fluidized bed technology, calorimetry, analytical chemistry research, and determinations of nuclear cross sections. The Geneva Conferences of 1955 and 1958 were particularly significant in that they marked the first large-scale exchange of information on nuclear technology among the nations of the world. A few CEN members also participated in the Symposium on the Reprocessing of Irradiated Fuels, which took place in Brussels, Belgium, in 1957.

THE NATIONAL AND INTERNATIONAL SCENE

Before turning to CEN in the 1950s, we briefly review some key national and world events that relate to the nuclear, reactor technology, and other work that was being done at the national laboratories during this period. Political, economic, and social issues, which are in a constant state of flux, have a major influence on the areas of investigation to be undertaken at the national laboratories and their levels of financial support.

According to some historians, the United States became involved in three new wars in 1950: (1) Vietnam, (2) Korea, and (3) the "Cold War" with the Soviet Union. The U.S. involvement in Vietnam at the time consisted only of sending a 35-man advisory group to assist the French in maintaining their colonial power in the country, but it did not develop into a shooting war until the 1960s. The Korean War, in contrast, began as a United Nations (U.N. "police action" when the North Korean Communist forces invaded South Korea, and it quickly escalated into an undeclared war. The U.N. forces under General Douglas MacArthur had managed to recapture most of the country when the Chinese Communists joined the North Koreans, forcing the U.N. troops to retreat to the 39th parallel. The conflict became a stalemate at the 39th parallel and an armistice was signed in 1953. The Korean War cost more than 54,000 American lives, and 40 years later we still face a belligerent North Korea threatening South Korea, possibly with nuclear weapons. Neither of these two wars had a direct impact on the activities of CEN, but they contributed to a general feeling of dismay in the country over the fact that the sacrifices of World War II had not ended our foreign problems.

The Soviet nuclear threat had become a matter of great concern and the arms race with the U.S.S.R. (the Cold War) was formalized in a National Research Council Report, NSC-58. The Soviets were obviously catching up with the U.S. in nuclear weaponry, and many individuals, including President Truman, felt that this could have happened only through extensive Soviet espionage. Recent information from the Soviet Union indicates a good deal of espionage had, indeed, occurred in the 1940s. Lavrenti Beria, head of the Soviet Secret Police (NKVD), and Igor Kurchatov, the Soviet physicist who supervised their nuclear weapons programs, had access to intelligence from espionage by Klaus Fuchs and others at Los Alamos which indicated that the bomb was possible and included other critical technical information. In 1950, the Americans discovered that Fuchs, a German physicist, who had become a British citizen and worked at Los Alamos, had passed along information from 1942 to 1949. He was sentenced to prison in England in 1950 and released in 1959, when he went to East Germany.

In 1948, Whittaker Chambers, an editor of *Time* magazine, and a former member of the Communist Party and Soviet agent in the 1930s, claimed that Alger Hiss, a former State Department employee, had given him State Department documents to be delivered to the Soviets. Hiss denied ever having known Chambers but he was indicted and served 44 months of a five-year sentence.

Another espionage case that attracted major attention was that of Julius and Ethel Rosenberg, who were accused of relaying vital information about the atomic bomb to Soviet agents. Ethel's brother, David Greenglass, a Los Alamos employee, who had supplied the information to the Rosenbergs, was sentenced to 15 years in prison. The Rosenbergs were sentenced to death and executed on June 19, 1953.

The Hiss and Rosenberg cases provided a springboard for Senator Joseph McCarthy to claim that the State Department was riddled with card-carrying Communists. He accused Presidents Roosevelt and Truman of 20 years of treason and denounced Gen. George C. Marshall. Even after Eisenhower was elected in 1952, McCarthy attacked large numbers of people, many from Hollywood and the news media, generally with unfounded charges. He finally met his match when he accused the U.S. Army of Communist penetration. Joseph Welch, the Army attorney, demolished McCarthy's credibility in widely televised hearings, and McCarthy was later censured by the U.S. Senate.

In 1952, Dwight Eisenhower defeated Adlai Stevenson in a race for the U.S. presidency. The Republicans coined the campaign slogan, "I like Ike," which was so popular that it became part of a song in a Broadway stage production, *Fiorello.* Eisenhower appeared to have an ambivalent attitude toward nuclear energy. When he first took office, he was concerned about nuclear energy contributing to "creeping socialism," but his position shifted and in 1953 he introduced his "Atoms for Peace" program in a speech to the United Nations. During the early 1950s, interest began to develop in commercial nuclear power as an outgrowth of the naval propulsion program, and the AEC became serious about power production. The McMahon Bill was revised in 1954 to: (1) provide for the development of nuclear power by industry, (2) permit international nuclear cooperation, and (3) relax the security requirements somewhat.

Prior to this act, the security classifications had been the same as those in the military: Official Use Only, Restricted, Confidential, Secret, and Top Secret. The McMahon revision provided for the "Restricted Data" classification, and the "L" clearance was instituted as a new category below the "Q" level. When the AEC was formed, AEC and military security clearances were separate and non-interchangeable and that continues to be the policy.

The first U.S. thermonuclear device was exploded in 1952, and the Soviets followed in 1953. Joseph Stalin died on March 6, 1953 and was replaced by Nikita Khrushchev, but this event had little effect on the arms race. The U.S. conducted many nuclear weapons tests, most of them in secret, in the 1950s, particularly in 1957-58. Only very recently (in 1994) was full information on the extent and types of these tests declassified on the authority of Hazel O'Leary, Secretary of the Department of Energy (DOE).

In 1957, the Soviets launched the first ICBM, followed closely by three Sputnik satellites. The U.S. news media became almost hysterical and began making much of the "missile gap." An attempt was made to revamp the educational system with the "New Math" and other innovations, but students and parents alike seemed more confused than edified by the abstract concepts that were being offered, and most of the new approach died out in a few years. Also in 1957, the nuclear-powered aircraft program was given a boost because the U.S. thought the Soviets had one, but that was finally discontinued in 1961, mainly because of shielding, weight, and safety problems. The missile gap fears were alleviated somewhat when the U.S. launched its first satellite, Explorer I, in 1958.

Antinuclear sentiment had begun to develop in the 1950s as a result of tests conducted by the military in the 1940s. United States servicemen had been allowed radiation doses up to 20 R, and some Pacific island natives had received exposures as high as 175 R. In 1957, Ralph Lapp wrote the *Voyage of the Lucky Dragon,* and nuclear doomsday movies such as *On the Beach* began to appear. The symbol,

was adopted by the antinuclear activists in 1958 as the "peace sign." It is based on the semaphore signal code wherein the two diagonal lines in the lower half of the circle represent the letter "N" and the vertical one is "D"—thus, "nuclear disarmament."

The AEC was required to hold hearings on the dangers of fallout in 1957. The first organized interventions in nuclear licensing hearings took place when Detroit Edison along with 20 other firms which had formed the Power Reactor Development Corporation (PRDC) proposed the Fermi I reactor near Detroit. This was to be a 60-MW fast breeder to produce power for Detroit and plutonium for the AEC. Union Leader Walter Reuther and the United Auto Workers were particularly active in these protests. The meltdown of the second core loading in ANL's EBR-I (see next section) was cited repeatedly.

About this time, at the request of the Joint Committee on Atomic Energy (JCAE), Brookhaven generated WASH-740, a study of the potential effects of a nuclear reactor accident. The news media exaggerated the worst and ignored the near-zero probability of such an event. Soon thereafter, the University of Michigan published a report that was even scarier. The insurance concerns raised by these and other studies culminated in the Price-Anderson Act of 1957, which limits the liability of utilities operating nuclear power plants. The JCAE was a strong advocate of civilian nuclear power, and several of the members became quite knowledgeable about the subject. Melvin Price, a congressman from the East St. Louis area in Illinois, was particularly active; he served as chairman of the JCAE for a period of time and was a co-sponsor of the Price-Anderson bill. He was acquainted with Admiral Rickover and was friendly toward Argonne.

In spite of the various problems and protests, an aura of optimism prevailed at the end of the 1950s about the future of civilian nuclear power.

REACTORS

Argonne was in the forefront of reactor development in the early 1950s. Reactor physics experiments were continuing in CP-3'. In 1954, CP-5, which was started up at the DuPage site, became a workhorse for users both within ANL and from outside the Laboratory. Various divisions of the Laboratory were also doing work related to outside reactor development efforts such as the Naval Propulsion Program.

The most innovative program, however, was the ANL work on fast breeder reactors. The first one was EBR-I, which was located in the Idaho desert at the National Reactor Test Station (NRTS). The EBR-I program had two major objectives: to demonstrate the feasibility of power generation and to demonstrate breeding, *i.e.*, a breeding ratio of 1 or higher. Three different cores were used in EBR-I, and the coolant was the sodium-potassium eutectic (known as NaK, pronounced like "knack"). This coolant was used instead of sodium alone because it is a liquid at room temperature (eutectic at -12.7°C, 9.1°F), which makes it easier to handle. An interesting innovation in the EBR-I reactors was the use of electromagnetic pumps for the NaK coolant, which avoided moving parts such as bearings in the liquid metal. They were backed up with mechanical pumps on standby as a safety measure, but proved to be highly satisfactory and were used as the normal operating mode. The three versions of EBR-I are listed in Table 2-1.

The generation of useful electrical power from the atom for the first time on December 22, 1951, was a major milestone in the history of nuclear technology. The fact that this was done in the Mark I version of EBR-I, the first fast breeder reactor, made the feat even more remarkable. The Mark I version was operated for about four years, during which time 4,000 MWh of heat was produced.

The Mark I core had metallic uranium fuel pins and stainless steel cladding, which are incompatible when in direct contact because they form a low-melting eutectic. To avoid this problem, NaK was used in the annulus to separate the two materials but still provide good heat transfer. The Mark I fuel elements were separated by 120° horizontal ribs in the cladding; these were eliminated in the Mark II core. The Mark II core was self-regulating under normal conditions, but instabilities were noted in transient tests, and in a test at high core temperatures and a short reactor period, a partial meltdown occurred on November 29, 1955, probably due to bowing of the fuel pins.

Table 2-1. Versions of EBR-I

Version	Fuel	Blanket	Cladding	Power
Mark I	93.5% Enriched Uranium	Natural Uranium	Stainless Steel	200 kW(e)
Mark II	93.5% Enriched Uranium	Natural Uranium	Stainless Steel	200 kW(e)
Mark III	93.5% Enriched U-2% Zirconium	Natural U-2% Zirconium	Zircaloy	200 kW(e)

Approximately one-third of the core was melted in the interior region; the fuel pins in the outer portions of the core and those in the blanket remained intact. Although this was a relatively minor event in a severe reactor test, the antinuclear activists viewed it with great alarm in the following years when they were protesting the proposed Fermi reactor. The Mark III loading used zirconium as a stabilizing element for the fuel and Zircaloy, an alloy with a higher melting temperature, for the cladding.

The Chemical Engineering Division was given the responsibility of determining whether or not breeding had actually occurred in the EBR-I reactor, and found that it had by a small margin, which proved the principle. These "proof of breeding" experiments are discussed later.

As shown below in Table 2-2, Argonne designed, built, and operated several experimental boiling water reactors (BWRs) during the 1950s, primarily for safety studies.

The ALPR (Argonne Low Power Reactor) was part of an Army reactor-development program. It was designed as a prototype of a packaged power plant that could be used in remote areas. The fuel was an enriched uranium-aluminum alloy; light water served as the moderator and coolant. The reactor could produce 300 kW of electrical power and 400 kW of space heat.

The BORAX reactors were used to investigate some of the safety aspects of boiling water reactors. BORAX-I had a small core of fully enriched uranium-aluminum alloy fuel plates clad with aluminum, and cooling was provided by natural circulation of water. The steam bubbles caused no instabilities and the system was inherently stable under transients. It was finally destroyed intentionally in a simulated "runaway" test in 1953, which caused a small steam explosion. BORAX-II, operated in 1954, was a larger version of BORAX-I. BORAX-III showed that turbine contamination by the steam was not a problem, but it was most remembered by the fact that it was used to light up the town of Arco, Idaho, on July 17, 1958. In 1956, BORAX-IV was operated with a thorium-uranium oxide fuel mixture. BORAX-V, in 1962, demonstrated the capability for supplying a conventional turbine with superheated steam. Argonne continued on with several other BORAX-type experiments to study various aspects of boiling water reactors.

Table 2-2. Argonne Boiling Water Reactors in the 1950s

Reactor	Power	Type	Location	Mission
ALPR	3 MW(t)	BWR	NRTS	Army Program
BORAX-I	1,200 kW(t)	BWR	NRTS	Safety Studies
BORAX-II	6,400 kW(t)	BWR	NRTS	Safety Studies
BORAX-III	12 MW(t)	BWR	NRTS	Safety Studies
BORAX-IV	20.5 MW(t)	BWR	NRTS	Safety Studies
BORAX-V		BWR	NRTS	Safety Studies
EBWR	20 MW(t) 5 MW(e)	BWR	DuPage	Small-Scale BWR Prototype

The BORAX studies culminated in construction of the Experimental Boiling Water Reactor (EBWR), which was a small prototype of a commercial boiling water reactor for commercial power generation. Located at the DuPage site, EBWR became operational in December 1956, and ran on the Commonwealth Edison network. After the demonstration, it was used in a joint ANL-Hanford Plutonium Recycle Program to obtain information on the use of plutonium as a fuel in light water reactors. In September 1965, EBWR began running at 70 MW, and then at 100 MW for a brief time, with plutonium as the principal fuel. It was shut down in 1967 when the mission was completed.

Another major ANL event in the 1950s was the beginning of construction work on the 12.5-BeV Zero Gradient Proton Synchroton (ZGS) on June 22, 1959.

The first nuclear-powered submarine, the U.S.S. *Nautilus*, was launched on January 21, 1954. The pressurized water reactor used in this vessel was based on ANL concepts and designs and was built under extremely stringent engineering specifications imposed by Admiral Rickover. Its performance was outstanding. It logged about 105,000 statute miles, mostly submerged, before the first refueling. About three years later, the Navy demonstrated the use of sodium-cooled reactor technology with the Sodium Intermediate Reactor (SIR), which was a prototype submarine propulsion reactor. This reactor, developed by the Knolls Atomic Power Laboratory (KAPL), was installed in the U.S.S. *Seawolf*, which operated from 1957 to 1959. The Navy abandoned this approach, however, after problems developed with sodium leakage into the *Seawolf's* steam reheaters. Construction of a nuclear-powered merchant ship, the N.S. *Savannah*, began in 1956. The ship was built and operated as a demonstration, but it generated little interest in the shipping industry or elsewhere. Now moored at Charleston, South Carolina, it is used as a museum. In 1959, the U.S.S.R. launched a nuclear powered icebreaker, the *Lenin*.

Rickover also had a major role in the construction of the Shippingport, Pennsylvania, pressurized water reactor, which was a joint project of the Duquesne Power & Light Co., Babcock & Wilcox Co., and Stone & Webster, Inc. This was the first civilian power reactor, and it operated from 1957 to 1982. A replica of its core was displayed at the Geneva Conference in 1958. Consolidated Edison, Inc., followed with the Indian Point reactor and the Commonwealth Edison Co. with Dresden-1. By the end of the 1950s, industry had developed a strong interest in nuclear power, and the Westinghouse Electric Co. offered the first guaranteed-price, turnkey power reactor.

In 1954, the first nuclear-generated electricity in the Soviet Union was produced by the 5-MW(e) Obninsk light-water, graphite-moderated reactor.

The 1950s were marred by two reactor accidents that were more significant than the EBR-I incident. In 1952, a meltdown and hydrogen explosion occurred in the NRX reactor at Chalk River, which is located in an isolated area in the Province of Ontario, Canada. The Chalk River reactor used natural uranium fuel with heavy water as the moderator. This proved to be more of a mess than a disaster; it was cleaned up and the reactor was back in operation in 14 months.

A more serious incident took place on October 8, 1957, when the British Windscale reactor, a graphite air-cooled thermal production reactor with aluminum-clad uranium fuel, caught fire, producing fallout in England and low, but detectable levels of radiation in France, Germany, and the Low Countries. Most of the public concern in England about this event was over iodine-131 contamination of cows' milk. The Wigner effect, a buildup of stored energy in graphite when it is irradiated by neutrons, was identified as the initiating factor in the temperature excursion that caused the fire.

In 1957, a very serious accident occurred at a nuclear weapons factory about 12 miles from the city of Kyshtym in the Ural Mountains. Over 10,000 people were forced to evacuate the contaminated area. For many years, the Soviets attempted to keep the event under wraps, but the rest of the world knew that something catastrophic had happened. It appears that a large quantity of nuclear waste material underwent a violent explosion, but there still seems to be some uncertainty as to whether it was caused by nuclear criticality, a chemical reaction, or both.

THE MOVE TO SITE D

When Site D was acquired for relocation of ANL, it consisted of about six square miles of land bounded roughly by Highway U.S. 66 (now I-55) on the north, Cass Avenue on the east, 91st Street and Bluff Road on the south and Lemont road on the west. The purpose of the large area was to create a buffer zone around the laboratory both for safety and security. In the early 1970s, several hundred acres of the land was made available as federal surplus property, and in 1973, under the Great Legacy of Parks Program, an additional 2,433 acres was transferred to the DuPage County Forest Preserve District, which added it to the Waterfall Glen Forest Preserve. The Argonne site now consists of about 1,700 acres.

Site D was created by purchasing properties from local farmers, along with the Freund estate, and consolidating them into a single entity. When the U.S. government acquired the land for Site D, it was removed from the DuPage County tax rolls because it was no longer private property. To compensate the DuPage County for this loss of income, the federal government has been making annual payments in lieu of taxes.

The land is generally flat, with some gently rolling areas, and is traversed by Sawmill Creek. A few magnificent old oak trees are in evidence. To enhance and preserve the property, a project was undertaken in 1953 to plant a million red, white, and jack pine trees, a formidable task that was completed in 1955. At present, much of the land is forested, with hardwoods and other deciduous trees thriving among the pines.

Everybody at Argonne is aware of the white deer. They came with the Freund estate, but their origin is uncertain. At one time there was concern that they might not survive because of disease, but they seem to have recovered and are thriving. The Argonne Guest Facility was opened in February 1958, and first-time visitors stepping out of the door in the morning were sometimes astonished to find themselves in the company of one or more all-white deer.

In the 1950s and for several following years, most of the buildings in the East Area were Quonset huts. This, along with the guard posts, exposed steam lines, and road layout, made the area look much like a WWII Army or Navy base. The 200 Area, with the new brick buildings around the inner circle, had more of a civilian campus character.

In 1950, the Chemical Engineering Division began to move to the DuPage site, starting with Bldg. D-310, and continuing with the major part of the move to the main building, D-205. Moving the site from Chicago to DuPage County made it necessary for nearly everybody to commute to work, as there was no public transportation to the Laboratory. This was a special problem for a few families, who, having lived in a large city all their lives, did not own a car and had not learned to drive. Many of the employees remained in Chicago, while others, especially the new people who were being brought on board, sought housing in the suburbs. For the first year or two, ANL operated a bus system with routes from Chicago and some of the suburbs to the Laboratory. Alice Graczyk sold the 35-cent tokens in her L-Wing office. One driver, in particular, on the Chicago-ANL bus, who was a ventriloquist, sometimes startled the passengers by opening and closing the door and making it sound as if somebody outside were yelling to get aboard. Another route, which went to Downers Grove, Lisle, and Naperville, had a regular passenger who sat in the back quietly strumming a guitar and singing Western ballads. In 1953, the ANL bus system became a casualty of budget cuts and was discontinued.

Most of the employees were relatively young (in their twenties and thirties) at that time and could not afford two cars, so car pools became popular. The optimum size for a car pool is a complex problem, and was the subject of much discussion. Viewed simply, a two-person pool should decrease an individual's driving by 50%, a three-person pool by 67%, *etc.*, and little additional benefit would result from going to four or five members, especially considering the additional time required to pick up and discharge everybody on both ends. Some pools had five members so a person would drive the same day each week. But then, one had to factor in the probability that somebody would be sick, on vacation or travel, or would oversleep or have a late meeting. Sometimes there were personality clashes. In spite of these problems, many established car pools have been operating for decades; they offer the opportunity for humor, gossip, relaxation, and technical discussions and have created many close friendships. They were a godsend during the fuel shortages of the 1970s.

When the Division moved to DuPage, Bill Mecham, a chemical engineer, owned a 1931 Rolls-Royce, which he drove to work. The windowsills of that large, black car were on about the same level as the roofs of the other cars in the north parking lot. Occasionally one could see Bill riding "high in the saddle" above the other cars as he cruised through the lot looking for an oversized parking place. He once mentioned a couple of "fender benders" he had had in the Hyde Park area with that car; according to Bill, it was the other cars' fenders that did all the bending.

The area surrounding Argonne in the 1950s was much less populated than it is now, and the only four-lane road was the legendary U.S. Route 66 at the north edge of the site. During heavy snows, the plowing was less efficient than it is now, and getting to work or back home could be chancy. On at least one occasion, Dr. Lawroski's car pool had to take shelter in a nearby farm-house on Lemont Road until the situation improved, and there were several times that many people didn't get back home until 9 or 10 p.m. because of heavy snow or freezing rain.

There are at least two CEN car pools that deserve longevity awards. One is the famous Park Forest pool, which, in 1953, consisted of Milt Ader, Hal Feder, Bob Larsen, Charlie Stevenson, and Martin Steindler. That car pool has functioned more than 40 years with various participants. According to some of its members at the time, riding with Feder, who had just learned to drive, was an unnerving experience. The other car pool of note was the one from Wheaton, consisting of Paul Nelson, Martin Kyle, Terry Johnson, and Les Coleman.

Most of the CEN employees began to seek housing in communities within a reasonable commuting distance from the Laboratory. These communities were a diverse lot. Many of them were located along the various commuter rail lines such as the Northwestern, Burlington, Illinois Central, and Santa Fe, which fan out from Chicago. Some of the communities were well established, others were smaller rural towns that were beginning to grow, and many were basically "bedroom communities" that had sprung up after WWII. Some of these communities were apprehensive about the large influx of Argonne employees, so the Laboratory sent out advance people to explain that these scientists and engineers generally behaved themselves, had been screened for criminal records, often went to church, and, most importantly, paid their bills on time. Some racial problems arose, but they eventually became defused.

Housing, especially rental apartments, was not easy to find, and the Laboratory assisted people who needed help. For those who were interested in buying property, the Laboratory provided the services of Byron Kilbourne, who was well-versed on the market and real estate values in the area, as well as the details and pitfalls of home construction. He would not normally seek out properties, but if an employee found one he thought he might want to buy, Kilbourne was most accommodating in going out and looking over the property and making a solid recommendation. He often did this several times for a particular individual. He was a tough evaluator, and frequently shattered a family's dreams about some house they had found, but one could be sure that any property and price he approved was a good deal. On one occasion, a potential buyer had his eye on a beautiful wooded lot in the Green Acres area of Naperville. Kilbourne looked it over and said, "Dig a hole a foot square and two feet deep, fill it with water and see how long it takes for it to drain away." The man dug the hole, came back the next morning to fill it with water, and found that it was already half full. The deal was off.

As the ANL employees moved into their new homes, their social lives tended to become more oriented toward their own communities. It wasn't long before many of them became involved with local community affairs. Many joined churches, service clubs such as the Jaycees, Kiwanis, or Rotary, or special interest groups, and some were active in school activities. Argonne people then started being appointed to various advisory groups or elected to public offices such as city councils, school boards, park district boards, and others. The suburbs all have a great need for coaches and officials to handle all the organized athletic programs for children, and many Argonne fathers became involved in these activities. Two future CEN division directors, Les Burris and Martin Steindler, served as school board presidents. Dr. Lawroski never ran for public office, but he achieved what is considered an even more prestigious position in Naperville. He was accepted into a small, elite group of individuals that included the current mayor and some other highly influential people in town. They had breakfast every morning at a downtown drugstore counter, each with a reserved stool. Within this group, he was known as "The Professor."

NEW EMPLOYEES

A new staff member coming into the Division in the 1950s was subjected to a more or less standard routine. For the chemists, the first assignment was to spend a few weeks in the Analytical Chemistry Laboratory. Doug Krause was in charge of that group, and Betty Reilly was the instructor. Betty was a good-natured, patient teacher, and this training experience was both pleasant and highly instructive. Many of the people had not dealt with radioactivity before, and this was an excellent introduction to handling hot materials and counting techniques. It also gave one a chance to become acquainted with the people who would be doing their analytical work later and to gain an appreciation for their problems. New staff employees were expected to participate in training courses on reactor technology or similar subjects that were being offered by the Laboratory from time to time. Newly hired laboratory technicians and operators were normally well-skilled, but required some on-the-job training to become familiar with the unique problems of dealing with radiation, security, and other specialized aspects of a nuclear research facility.

One of the most rewarding aspect of working in CEN was, and still is, an opportunity to interact with a wide variety of technical people both within and outside the specific group to which one was assigned. Sharing of ideas, techniques, and equipment among the various groups of the Division was extensive. There was also a lot of cooperation among the divisions of the Laboratory, especially during the development and construction of EBR-II and the Fuel Cycle Facility. Nearly the whole laboratory was involved in that project in one way or another. Over the years, CEN has had particularly close ties with the Reactor Engineering, Reactor Analysis and Safety, Chemistry, and Metallurgy Divisions. (Due to expansion in the scope of the work, the Metallurgy Division has a number of successors, including Solid State Science, Materials Science, Materials Components, and Energy Technology.)

There were also many interactions with the academic world. Several of the ANL staff members had come from teaching positions at universities. The Division had connections with departments at a number of universities, and many individuals, including some CEN members, completed the experimental part of their thesis work at ANL in a cooperative arrangement with a university. Opportunities were provided for students and faculty members to work in the various CEN programs on temporary assignments. The Division used consultants from university faculties when some particular expertise was needed. Staff personnel from CEN were occasionally invited to present seminars to university departments, sometimes in connection with recruiting trips. Staff members from CEN were often involved with various industrial organizations, some of which were potential customers for the technology that was being developed. An example of this was the UF_6 production plant that was built by the Allied Chemical Co. at Metropolis, Illinois. In some cases, the Division contracted with commercial firms to provide specialized services or equipment.

The CEN staff people frequently presented papers at national meetings of professional societies such as the American Institute of Chemical Engineers (AIChE), the American Chemical Society (ACS), the American Nuclear Society (ANS), and the Geneva Conferences. (Some of the first visitors to Europe came back home using the British pronunciation of "processes" and the French pronunciation of "centimeters"; a few were wearing berets.) Intersite visits to the other laboratories having similar interests were common, and topical meetings were arranged by the AEC in some cases. The people involved in particular areas of work, after a few years, seemed to develop a kind of camaraderie with their counterparts at other institutions both in the U.S. and overseas. These interactions extended to the technical staff at AEC Headquarters, with the result that they became personally acquainted with some of the CEN staff. An interesting aspect of these various interactions was that if one transferred to a different CEN program, the whole process was repeated. One had to not only become technically proficient in the new area, but also get acquainted with others doing related work both at ANL and in the outside world.

Everybody had to become familiar with the radiation safety rules and regulations and to learn how to use film badges, dosimeters, and monitoring instruments. Special safety shoes that were colored bright yellow were required when one was in the radiation areas of the building. It didn't happen at all frequently, but there were a few occasions when someone would be walking down a street in his town and suddenly realize to his horror that he had forgotten to change his shoes; it was hard to be inconspicuous. Although this was a technical violation of the safety rules and highly embarrassing, it didn't create a real hazard because all personnel had to check their feet to leave the building.

Because many of the CEN programs entailed work with radiation, urine samples were requested rather frequently for bioassay purposes. Empty sample bottles were provided for the individual to take home overnight and bring back full in the morning, when they would be picked up. The containers for these sample bottles were identical to the black metal lunch boxes that were used by quite a few people who brought their lunch to work. There are undocumented rumors that such boxes have been accidentally switched on occasion, with the result that the bioassay lab received a ham sandwich and the worker found a bottle of urine for his lunch. There was one instance in which a new employee received his first request for a urine sample during a certain week. His understanding of the instruction was that they meant every day of that week; the Division had a call from Bioassay, asking that the nature of their request be explained to him more clearly.

Safety has always been a paramount concern in CEN. John Schraidt was the first Division Safety Officer, and he took the job seriously. Anybody who was caught twice without safety glasses in the laboratory was threatened with dismissal. New employees were trained to use the different types of fire extinguishers, fire blankets, safety showers, self-contained breathing apparatus, eyewash fountains, and other safety equipment. Instructions were given on "Dial 13" (now "911") emergency line. A fire brigade was set up to handle emergency situations in the building, and a safety committee with a rotating membership was organized to conduct routine safety inspections throughout the building. A special committee was appointed to review any new experimental setups and procedures for potential hazards and to develop preventive measures if necessary. Safety information and directives were provided to staff members continuously by the AEC, the Laboratory, and the Division. The scope of the safety activities is far too large to be described here, but the above-mentioned practices are still in effect, and they have proved their value over the years. The Division has received numerous commendations for large numbers of man-hours worked without a disabling injury. During one period of time, all employees were presented with rather nice gifts, such as card tables, hand lanterns, and home fire extinguishers when the Division had completed a certain number of employee-years without a lost-time accident.

Fig. 2-1. John Schraidt

Security was another important element in the training of new employees. In the 1950s, practically everything a staff member did was classified as secret. All the experimental procedures and results were to be entered into secret notebooks, with each page signed by the investigator and witnessed by two other individuals who had either observed the experiment or would state that they had read and understood the results. The regular CEN progress reports, as well as the internal weekly or monthly reports, were secret. Reports on most of the individual investigations, however, could be sent to Hoylande Young, the Director of Technical Information, where they were cleared for publication in the open literature or presentation at a meeting. Dr. Young, who had a Ph.D. in chemistry and had worked as a Senior Chemist on the plutonium project in the Met Lab, was always most cooperative in expediting the declassification process and suggesting changes if they were necessary.

The fact that the offices were usually occupied by three staff members who shared a file cabinet with a combination lock created some problems. The file was to be locked when nobody was in the office, but when people left for the day there was sometimes a slip-up that resulted in a security violation. The night security officer, not knowing who was the culprit, would arbitrarily put down any one of the occupants' names. On a few occasions, some new employee, usually a chemist, would convert the three numbers of the file combination to symbols for the elements of those atomic numbers and write them down in some inconspicuous place. The guards had that one down pat; they could read the periodic table as well as anybody else, and that was a sure way to get a violation. The punishments for a security violation varied. One might be summoned for an interview with the division director to explain why the file was not locked. Forgetfulness was not a good answer, nor was there much of anything else one could say that would be very convincing. In fact, there was usually a good chance that you were not the guilty party, but you couldn't be sure. It was a bit like explaining to your wife why you locked her keys in her car—there was no good answer. Another punishment that was invoked on occasion was a week's assignment to go through all the offices in the building at the end of the day and make sure that everybody's file was locked. If an employee developed a pattern of repeated violations, it became a serious problem.

On February 5, 1951, the nationally known radio commentator and newspaper columnist, Paul Harvey, decided to get a scoop on Argonne's lax security measures by climbing over the fence along the outer perimeter of the laboratory site. Unfortunately for him, the security force was there to greet him, and the Chicago media had a ball with the story. When Bldg. 205 was first occupied, there was an 8-foot perimeter fence around the building, and the only access was through a single guard post. Evidence of that guard post still remains in the form of the concrete steps with iron railings at the south end of the north parking lot. Later on, for a period of time, a guard was posted in the Bldg. 205 lobby. The Physics Building (D-203) and the CP-5 reactor were declassified in 1953, followed by the Chemistry Building (D-200) in 1955.

One of the responsibilities of the security guards was to patrol the buildings at night to assure the physical security of the building, and also to watch for any problems in the laboratory areas. The most common problem by far was with cooling water lines that had lost their integrity for one reason or another, causing flooding of the area. Names and phone numbers of workers responsible for each lab were posted on the door, and almost every scientist or engineer who was doing laboratory work in those days has probably had at least one of those unnerving, middle-of-the-night phone calls from a frantic security man asking what he should do. If it was your own equipment, you could usually tell him over the phone how to shut it down safely (such as turning off a furnace before the cooling water supply). Often it was somebody else's equipment and you would know who should be called, but on some occasions it was necessary to jump into your car, race out to ANL and try to cope with the situation.

SERVICES

A wide variety of supporting services was, and still is, available to Laboratory employees. Radiation Safety was one of the most important ones. Everybody was required to wear dosimeters and film badges, which were read by the Radiation Safety personnel. They had many other duties, surveying laboratories and equipment, checking items going out of the building, maintaining the hand-and-foot counters, providing advice and assistance in the design and operation of equipment, surveying wastes—to name a few. The radiation safety personnel were not members of CEN, but they tended to have long assignments to particular buildings and became regarded as co-workers. Some of the "Health Physics" personnel most closely connected with the Division in the early days included Ken Woods, Walt Smith, Ted Allen, and Frank Marchetti.

Another service was the Travel Office, which was a little different in the early 1950s than it is now. On the travel request, one had the choice of rail or air transportation. Train travel was still used widely, although flying was rapidly becoming the mode of choice. The trains had not been improved a great deal after WWII, and some of them were still pulled by steam locomotives. One particularly quaint train was the Long Island Railroad route from New York City to Patchogue, New York, which was the normal destination for a Brookhaven visit. Air conditioning was not universal. Many CEN people made long train trips to destinations such as the Idaho site, Hanford, Los Alamos, Oak Ridge, and Washington, D.C. The Laboratory would provide normal Pullman accommodations on a train, but those in the know usually contributed a small amount of their own money to get a double bedroom. One of those trips, together with a meeting, could easily occupy a week or longer, but it did provide extra time to work on a presentation. The planes were propeller models (DC-3, DC-6), slow and noisy by present-day standards, and they served each passenger a small package of cigarettes with the meals. Some employees, including Walt Rodger, who felt that flying was an unnatural act for a human being, refused to have anything to do with it. (Surprisingly, some years later Walt became a private flying enthusiast.) Rental cars had not yet become commonplace, so people depended on taxis and local public transport systems.

All the air travel was out of Midway Airport until O'Hare was completed in the mid-1950s. Transportation between one's home or the Laboratory and the airport was by government car with an Argonne driver. Those drivers seemed to know everybody at ANL. Their relationship with the Chicago police at O'Hare was interesting. The police officers were in the habit of whacking the fenders of cars with their nightsticks to keep the traffic moving, but they apparently suspected that this might not be advisable for a U.S. Government car. The cars also met the arriving ANL passengers on the upper (departure) deck at O'Hare, a practice that could earn the average driver a traffic ticket.

Originally, the only cafeteria was in Building 2 in the East Area. For many people, bringing their own lunch was more convenient because of the distance to the East Area, which required riding a shuttle bus. There were few restaurants in the surrounding area at that time. In addition, engineering projects at the time were frequently operated around the clock. For these reasons, a cool room for the storage of lunches was provided near the lobby of Bldg. 205. An additional cafeteria, which was installed in Bldg. 203 where the Central Library is now, was better situated for Bldg. 205 occupants. Finally, several years later, a new cafeteria (Bldg. D-213) was built to serve the entire laboratory. The cafeterias were operated by ANL, rather than by an outside contractor as is the case now. A lunch consisting of an entree, two side dishes, and a roll was 65 cents in the 1950s.

For those who did "brown bag" it, Building 205 had a large, attractive lunchroom on the second floor of L-Wing. Bridge, pinochle, and chess games were popular at lunchtime. One particularly cutthroat bridge group consisted of Al Glassner, Bob Larsen, and Hal Feder plus anybody else they could nail for a fourth. Some individuals, notably Walt Rodger, John Loeding, and John Schraidt, were famous for their gargantuan lunches, which consisted of three or four full-size sandwiches along with the usual fruit and dessert. None of them seemed to gain any excessive weight. The 205 lunchroom could be opened up to a large conference room, and this arrangement was used a few years for the traditional CEN Christmas parties, which were lively affairs. In 1953, the Laboratory Director imposed a definite ban on alcohol at such parties. The 1954 party was a more somber affair, the main entertainment being a cake-decorating demonstration by Bill Sovereign, whose family operated a bakery in Naperville. The Christmas parties were later moved off site, and then, some years later, they were returned to the building in the form of a potluck luncheon and some entertainment. Annual CEN picnics were instituted and proved to be highly successful because families, as well as the employees, were invited, and Argonne Park had the facilities for games and other activities. Eventually, the lunchroom was relocated to the service floor where it is now.

Nearly all the experimental work at CEN required the services of designers, draftsmen, machinists, and, to a lesser extent, glassblowers. The technicians and staff personnel could handle the more mundane work of this type, but real expertise was needed for many of the projects. In the early 1950s, Bill Voss and Tom Denst were full-time machinists who worked for CEN although they were officially a part of Central Shops. These artisans tended to be a bit stand-offish and gruff at times with a new staff member until they had a chance to size him or her up. If the new person was reasonable and appreciative of their work, however, things would work out well, and after a few months they would knock themselves out to be helpful. If asked, the machinists would often contribute ideas during the design stage of a piece of equipment that would make it easier to fabricate and use. In the early days, most of the glass blowing was done in the Chemistry Division, and any specialized work was sent to Central Shops. Later on, Bill Schulze, a professional glass blower, took care of the Division's needs, and performed several other important functions, including the coffee facility. John Schraidt was in charge of the design work and was extremely helpful in working with the staff on their equipment designs. Dick Malecha and Johan Graae, on assignment from Central Shops, were also responsible for much of the design work and made major contributions to the equipment and facilities available to CEN. Dick later became a member of CEN. Harry Smith, a pleasant, cooperative individual, was the head draftsman; he also was most helpful in arranging for drawings that were required for the shop work and other purposes.

Special Materials was another organization that interacted closely with CEN. Then, as now, all fissile and fertile materials, including uranium, plutonium, and thorium, were logged in when received, and a careful accounting similar to financial bookkeeping was required throughout their use until the time they were returned to the Special Materials Division. Other nuclear-related and expensive materials such as beryllium, zirconium, and platinum were subject to similar accounting and auditing. In many cases, the group leader was held responsible for all the special materials used by the group in order to simplify the bookkeeping.

The Division had a limited library when it occupied the new building. A technical records room with a vault for classified materials was situated on the second floor of L-Wing, but one had to use the Chemistry Division library in Bldg. 200 or the Central Library for access to most journals and reference books. It was rumored for a long time that one of the items in the vault was a pre-war issue of the *Saturday Evening Post*, which had an article about the possibility of a nuclear weapon with enough technical veracity that it was recalled by the government and classified secret. Since that time the CEN Library has grown much larger, and has had to move several times to accommodate that growth. It finally evolved into a permanent, attractive, and well-equipped facility at the east end of L-Wing. Augustella Thompson was the librarian for many years. She also managed the classified document room. In recent years, this function has been ably performed by Sharon Clark, Paulette Windsor, and others.

The Division has always enjoyed an excellent reputation with respect to the quality of its progress reports and other publications. That stems primarily from the quality of the work done by the staff and leadership of the management, but much credit is due to the technical editors. Joe Royal was hired as the first full-time technical editor and he set high standards. He was not a "blue pencil" editor. Instead, he flagged items with numbers keyed to a list of questions and comments that sometimes exceeded the length of the manuscript, but they were always germane and often educational. Joe, who at one time was associated with the Met Lab, came to CEN from the American Medical Association. He had a Ph.D. in chemistry from Berkeley and was able to comment cogently on the technical content, as well as the writing. Prior to Joe's arrival, the associate division directors were doing most of the final technical editing. The technical editors not only ensured that the Division produced high-quality writing, but also improved the writing skills of the staff through their extensive comments and conversations. Several other people have served in the Technical Editing Group, both under Joe and later on, including Tom Cramer, John Simmons, Gwen Kesser, and Susan Barr. In addition to the Division technical editors, some of the larger groups had their own editors to help with the workload. Among these individuals were Sy Vogler, Milt Ader, Jack Arntzen, and Ellen Hathaway. In more recent times, this tradition of excellence has continued under the direction of the current Division editor, Joe Harmon. Maria Contos has made a major contribution in organizing, compiling, and documenting lists of almost all the Division's publications since its inception.

For a short period of time after the Division moved into Bldg. 205, a full-time nurse was available on the premises. It turned out that she had little to do other than dispensing an occasional Band-Aid® or aspirin, so that service was dropped. Everybody was required to take the annual physicals, and all injuries, no matter how minor, were to be handled by Health Services. If an individual working in a hot lab had even a minor cut, it was checked for radioactive contamination. On one occasion, Bob Larsen nicked his finger slightly while working in a plutonium glove box, and the Health Division found no contamination but suggested that he should wear a finger cot for a few days just as a precaution. He told them that he couldn't do that because he was a Catholic.

The other services available to the employees, such as Graphic Arts, the Credit Union, Procurement, and others were much the same as they are now. Lee Mead, a big genial man who had previously been a guard at the Met Lab, was in charge of procurement, assisted by Marie Driskell, who always seemed to be on the paging system.

TOOLS OF THE TRADE

Considering that William Shockley and his co-workers invented the transistor in 1948, it is not surprising that solid-state electronic devices were unheard of in the early 1950s. Calculations were usually done with slide rules, nomographs, electromechanical desk calculators (Frieden, Marchant and Monroe were popular brands), or published tables of logarithms and other functions. (Many of those tables had been generated or updated by unemployed mathematicians under the WPA program during the Great Depression.) Ward Hubbard, who was raised in China, occasionally used an abacus and was proficient with it, but that didn't seem to catch on with the other staff people. One salutary effect of the slide rule was that it usually couldn't generate more significant figures than the data warranted. Anybody who has not shared an office with someone doing a least squares fit with one of those noisy electromechanical calculators cannot fully appreciate the term "grinding out data." The Division management was reluctant to purchase additional calculators, the attitude being that the "real" work was done in the laboratory and computations were a minor aspect of the research. Another factor leading to this reluctance was a concern that a large investment might be wasted on a system that could become obsolete almost overnight. Irv Johnson recalls, "Ward Hubbard and I had to scheme to get a Frieden calculator for our joint use. Ward was able to salvage a worn-out calculator that an accounting office had thrown away. After having it serviced at least once a week, Dr. Vogel finally gave in and allowed us to purchase a new Frieden."

Argonne was one of the national leaders in computer research and development. On January 28, 1953, members of the Physics Division completed their first electronic digital computer, the AVIDAC (Argonne's Version of the Institute's Digital Automatic Computer). Patterned after a machine at Princeton Institute, it cost $250,000, and used 2,500 vacuum tubes, 8,000 resistors, and 3.5 miles of wire. The memory consisted of electric charges on the inside face of a cathode-ray tube and required continuous renewal. The following September, this group completed the ORACLE, a similar, but larger machine to be used at Oak Ridge. It cost $350,000, and had 3,500 tubes, 20,000 resistors, 7 miles of wire, and a cathode-ray-tube memory. One can't help but suspect that the name of the machine, Oak Ridge Automatic Computer Logical Engine, was conjured up to fit the acronym. These early machines, although very useful and remarkable for their time, didn't begin to approach the capabilities of today's ordinary desktop computer. Computer technology advanced rapidly in the 1950s, however, and the reactor engineers began using the Univac machine routinely. By about 1960, ANL had obtained large mainframe digital computers from IBM. As an interesting sidelight, a number of analog computers were built at ANL; these were excellent simulators for reactor control systems and were adapted for reactor operator training. Lou Baker and his group used the PACE analog computer at the Applied Mathematics Division (AMD) for some early theoretical calculations in the metal-water reaction program.

It was only near the end of the 1950s that CEN personnel began to use the central computer facilities in AMD. The procedure for having a job done was first to explain the required computation exactly to an AMD programmer, who would then write the program and develop a set of data forms, which were sometimes rather arcane. The completed forms would be submitted to AMD (with the cost code) and the results would eventually be printed out. This entailed a lot of running back and forth between Buildings 205 and 221; Dean Pierce brought an old "beater" bicycle to ANL expressly for this purpose. At about this time, several CEN people began to take courses in programming languages (mostly Fortran) so they could write their own programs. It was not until the 1960s, though, that the use of computers by CEN personnel began to flourish.

The use of computers in conjunction with experimental equipment was much the same. Carl Crouthamel's group, who employed 256-channel analyzers to sort out gamma-ray spectra, was one of the earliest to use computers in conjunction with experimental equipment. This whole room was full of electronics, mainly vacuum tubes.

The situation was much the same with laboratory equipment. There were no integrated circuit devices or electronic digital readouts. Chemists used the classical two-pan balances that required averaging of several swings of the pointer around zero. Temperatures were usually determined with thermometers, by measuring a thermocouple output with a portable (usually Leeds & Northrup Type K®) potentiometer or with a strip-chart recorder. Platinum resistance thermometers were used for high-precision temperature measurements. Oscilloscopes were available, but were primitive by today's standards. Bill Olsen had an instrument shop and storage area on the service floor. Bill was a cooperative individual, and through a combination of doing repairs and shuffling equipment around as needed, he helped many programs through crises.

There is probably still enough vintage equipment around the building to create a small museum; one cannot help but admire the beauty and craftsmanship of some of that old laboratory equipment. Its polished brass and wood had much more esthetic appeal than the gray or beige plastic and metal housings that are now in vogue.

There is probably no one group of people who have benefited more from the technical advancements of the last 40 or 50 years than the secretaries and administrative personnel. Fortunately, changing an old-fashioned typewriter ribbon is now a lost art, although it did offer the young men an occasional chance to perform a gallant act for one of the secretaries, who could most likely have done it faster and better herself. Duplication was by mimeograph or carbon copies (we still see "cc" occasionally on distribution lists produced by a laser printer). Ditto® then came into general use, and, although more convenient than mimeograph, it was still messy and corrections were a pain in the neck. Purple fingers were the norm with the secretarial staff, and if they weren't careful, it wasn't necessarily confined to the fingers. To check typos, one person read the material aloud while another one checked the text. The only spelling checker was the dictionary. The secretaries had to learn a lot of technical terms and jargon. Dictaphones were available, but not generally used by the technical staff. Almost everything was transcribed from handwritten material, much of which was nearly illegible. It was not uncommon for someone to take a hand-written note to the writer's secretary to have it translated. Manual typewriters began to be replaced by electric models, but it was not until the 1960s that the IBM Selectric® typewriter became available. It was popular because of the ease of use and the capability for different fonts by changing the type balls, which some found, to their dismay, were quite fragile. Finally, to add to the problems, nearly all the material a secretary typed in those days was classified and had to be handled as such. The ubiquitous ballpoint pen, which now seems to have been around forever, was introduced to the general public with considerable fanfare in the 1950s, a major claim being that it could write under water.

In the administrative areas, there were a variety of electromechanical "business machines" such as the Addressograph®, which fulfilled the needs of the payroll, accounting, and other such groups. These, too, were ripe for change to the electronic age.

THE NEW BUILDINGS

Building D-205

Building D-205 was constructed during the period 1949-1950 (Fig. 2-2). The architect-engineer firm was Voorhees, Walker, Foley, and Smith of New York City, and the layout and design were developed primarily by Steve Lawroski, Charlie Stevenson, John Schraidt,

Fig. 2-2. Building 205

and Ed Peterson. A plan of the building as it now exists is shown in Fig. 2-3. The original structure did not include the following, which were added in the years indicated:

1. The Senior Cave (K-Wing) in 1956.
2. X and W Wings plus the extensions of A and B Wings in 1961-62.
3. Y-Wing in 1975-76.
4. Environmental Testing Annex in 1982.

Otherwise, the building was much as it is now.

When Building 205 was first occupied, the administrative functions were mostly in L-Wing, which was laid out somewhat differently than it is at present. Figure 2-4 shows the original plan of L-Wing as one entered the building from the north through the front doors. On the ground floor, the Division Director's office, occupied by Dr. Lawroski, was at the northeast corner. Charlie Stevenson, the Associate Director, and secretaries Virginia DeGrande, Evelyn Rafacz, and Florence O'Neil were in the adjoining offices. A conference room and an area containing the mailroom and space for stationery supplies and duplicating equipment were across the hall. The offices in the west part of L-Wing were occupied by other CEN administrative and management personnel, secretary Alice Graczyk, a Special Materials office, the Site Administrator (Ed Peterson), and a design group consisting of John Schraidt, Johan Graae, and Dick Malecha. Across the hall was a drafting room, with Harry Smith in charge, and the washrooms. The upper level of L-Wing had a large conference room with a folding partition that could be opened up to a spacious lunchroom with limited service facilities. West of the conference room was a technical records room, which was a forerunner of the CEN library, and a vault for classified materials.

For many years, most of the CEN administrative offices were relocated to A- and C- Wings, and L-Wing was rearranged and occupied by the ANL administration and various other groups, depending on the relative space requirements of the organizations. As the Division expanded with larger and more diverse programs, so did the need for more extensive design and drafting space. As a consequence, the upper level L-Wing lunchroom with its large area and excellent natural lighting was converted into a drafting room.

At one time, Robert Laney, the Associate Laboratory Director, and his staff occupied L-Wing and Room L-252 became known as

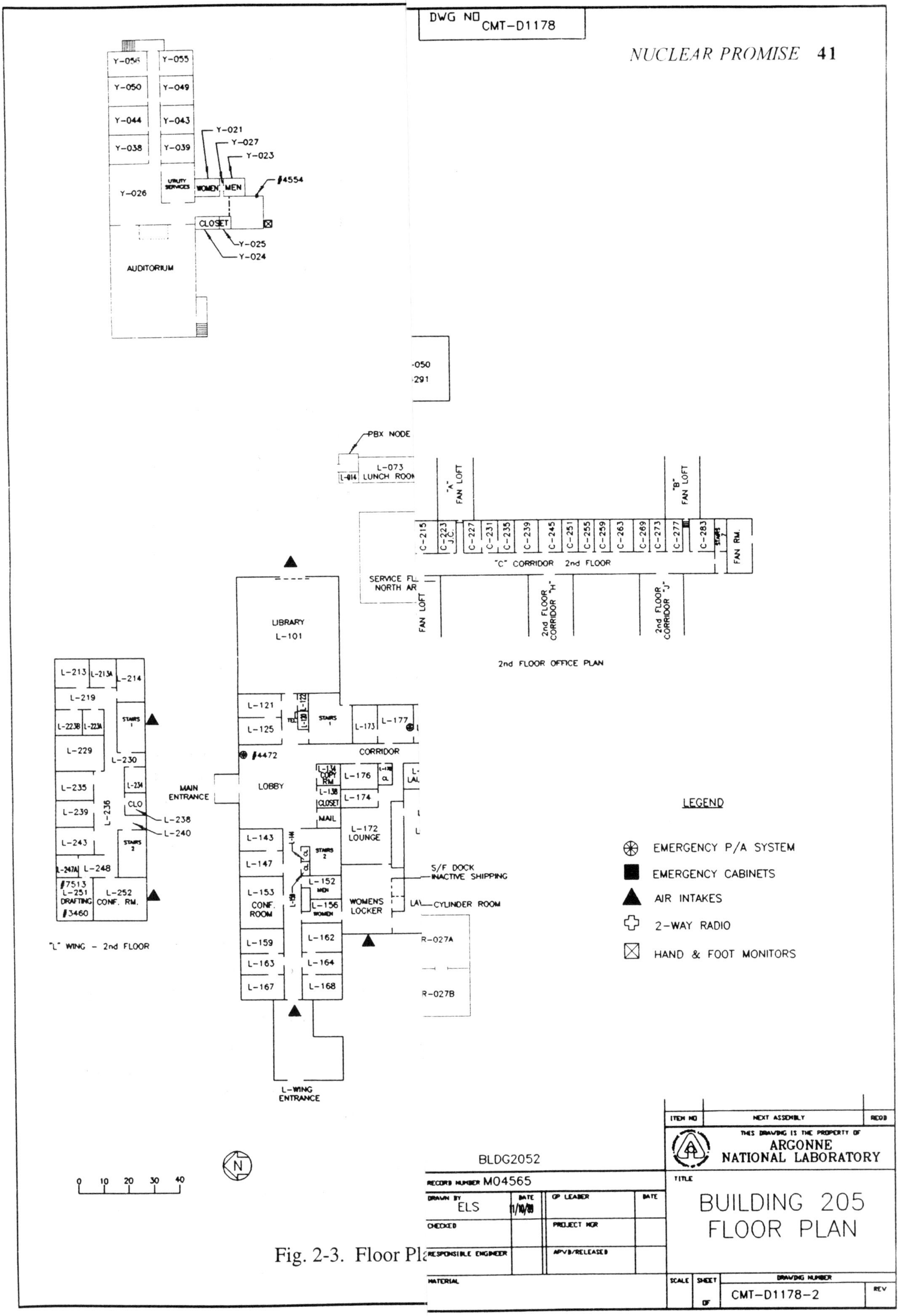

Fig. 2-3. Floor Pl

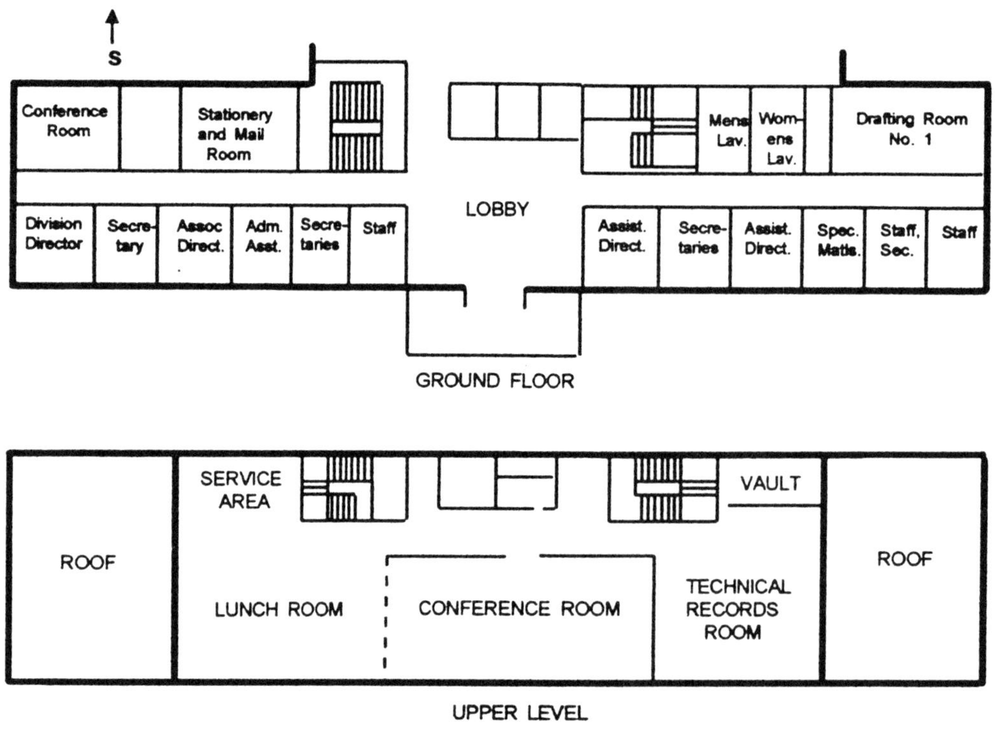

Fig. 2-4. Building 205 L-Wing in the 1950s

"Laney's Conference Room"; some old-timers still call it that. There was also a period when Robert Duffield claimed L-153 as a Laboratory Director's conference room, complete with carpeting and a teak wastepaper basket that attracted some comment. At present, L-Wing is occupied largely by CEN personnel and the library. The small offices where the mailroom and the copying machine are now located were originally intended to serve as two interview rooms and space for a future elevator, which was never installed.

The original building plan was based on the assumption that the Division would be involved in extensive work with radioactive materials, and the space now occupied by Jan Muller, Ron Tollner, and the secretaries was all allocated to health physics. The space where Ray Wolson's group is located was designated as a "control room."

C-Wing is now used largely according to the original plan, with the exception that the present radiation safety office once housed a first-aid facility and ventilation equipment. Some of the ground floor offices near A-wing were expected to serve as small machine and glass-blowing shops for the staff personnel, but they were never used for that purpose.

A- and B-Wings were planned as conventional chemistry laboratories and offices. Hauserman partitions® made of metal were used because of their modular construction, which made it relatively easy to rearrange the layout when the need arose. (The occupants soon discovered another handy feature—magnets could be used to attach various items to the walls.) When the building was occupied, there were normally three staff people per office. The office furniture was (and still is in many cases) standard government issue gray-colored metal desks, chairs, tables, bookcases, *etc.,* that are identical to those used on most Navy ships, except they are not welded to the floor. Most of the offices at ANL tend to be rather Spartan, with concrete block or metal walls, tile floors and concrete ceilings, but they serve their purpose well.

The A-Wing laboratories were used for research, primarily on solvent-extraction processes. Martin Steindler recalls much of the work going on in A-Wing at the time. Room A-101 was set up to accommodate Alberta Hoover's glassware washing for the Analytical Laboratory. Hal Feder, Norm Chellew, Ken Rhode, Don Hampson, and Milt Ader did much of the early pyrometallurgical research in A-109. Sy Vogler was dissolving enriched uranium from plastic planchets, and Bob Larsen with Roberta Shor was working on the problem of explosions when uranium-zirconium alloys are dissolved in nitric acid solutions. Chuck Seils and Bill Sovereign operated a plutonium analytical facility in A-133. (Bill later moved to the Idaho site.) Room A-141 was a plutonium lab where Martin Steindler, Fred Linzer, and Karl Schoeneman were doing fluorinations of fused salt materials and some thorium fluoride phase work. Max Adams and Dave Steidl came later into A-141 and worked in a Blickman hood in A-133 with bromine fluorides and uranium-plutonium alloys. Steindler also remembers a special project Feder had Don Fredrickson doing that used cyanide as a reagent, which generated an enormous flap with the medical people that went to upper management. Steindler still has the acetonitrile they gave him as an antidote for cyanide exposure.

B-Wing was devoted almost entirely to analytical chemistry. When the building was being designed, consideration was given to a third wing extending east from C-Wing the same as A- and B-Wings, but Dr. Lawroski agreed with Dr. Zinn that it could be eliminated in view of the budget limitations.

The high-bay areas, G-, H-, and J-Wings, were designed for engineering research with highly radioactive materials. Unlike most of the rest of the building, where there is a service floor beneath the working areas, these laboratories were built directly on undisturbed soil to support the weight of heavy shielding and equipment. Part of the rationale for the high ceilings (about 25 feet in G- and H-Wings, and 50 feet in J-Wing) was the expectation that much of the work would involve tall solvent-extraction columns. Heavy shielding in the form of high-density concrete was erected in several of the laboratories to accommodate such columns. Anybody who has had the task of drilling a hole through that concrete shielding to provide access to one of the cells has a special appreciation of its hardness and density. Traveling bridge cranes are used to handle heavy equipment in the high-bay areas. As it turned out, the process development work gradually shifted away from solvent extraction to other types of processes that required glove boxes and large walk-in hoods. Nevertheless, the shielded cells continued to be useful for a wide variety of research projects over the years.

The part of D-Wing that extends south from C-Wing contains the machine shop and a drafting room. The other part of D-Wing, situated at the southwest corner of the building, is occupied by various service facilities, including shipping and receiving docks, solvent storage, an electric transporter, and battery-charging station.

The building had Special Materials vaults in active use, with a criticality alarm that was tested periodically. The vaults are still there, but are no longer used to store special materials.

E-Wing, which is situated on the other side of the corridor, consists of the stockroom, a materials-storage area, and open space for shop operations such as welding and sheet metal work. In the early days of the Division, the stockroom was larger and had a much wider variety of supplies. Esmer Zeno, who was everybody's friend and liked to talk about his most recent "miseries," operated it. The early programs such as the development work on solvent extraction, pyrometallurgical processes for EBR-II fuel, and the fluoride volatility processes all involved the construction of large, complex equipment, often using rather exotic metals and other materials. The materials storage area in E-Wing was well supplied with hardware such as nuts and bolts, tubing connectors and fittings, Unistrut®, pipe and tubing, welding rods, and various sheet metals, as well as other commonly used supplies. As time passed, the need to reduce inventory costs and space resulted in a marked reduction of the supplies in the stockroom and materials-storage area.

F-Wing and parts of E-Wing include several laboratories that have been used mostly for special-purpose operations such as the microprobes and metallographs. R-Wing is made up of offices. These offices are generally occupied by staff people working in E- and F-Wings, and were used for many years by the CEN Editorial Group.

In 1956, the Senior Cave (Fig. 2-5) was added to the west side of Bldg. D-205 opposite G- and H-Wings, and it, along with its various service areas, was designated "K-Wing." The cave consists of three heavily shielded cells, each one equipped with viewing windows and master-slave manipulators. The original Mod 3 electronic manipulators, which were designed and built at ANL, attracted a lot of interest. They were often demonstrated for visitors and were always a favorite both with children and adults at the ANL open houses. One shortcoming of the original model was that the operator had no sense of touch when grasping an object. This was corrected by adding a feedback system so one could feel a resistance to the force being applied. Another interesting nicety that was added was small metal "fingernails" on the square tips of the rubber manipulator fingers to assist in picking up small objects. Later on, commercial manipulators became available and are now being used.

The requirement for safe handling of radioactive materials imposed many design considerations in the planning of the building.

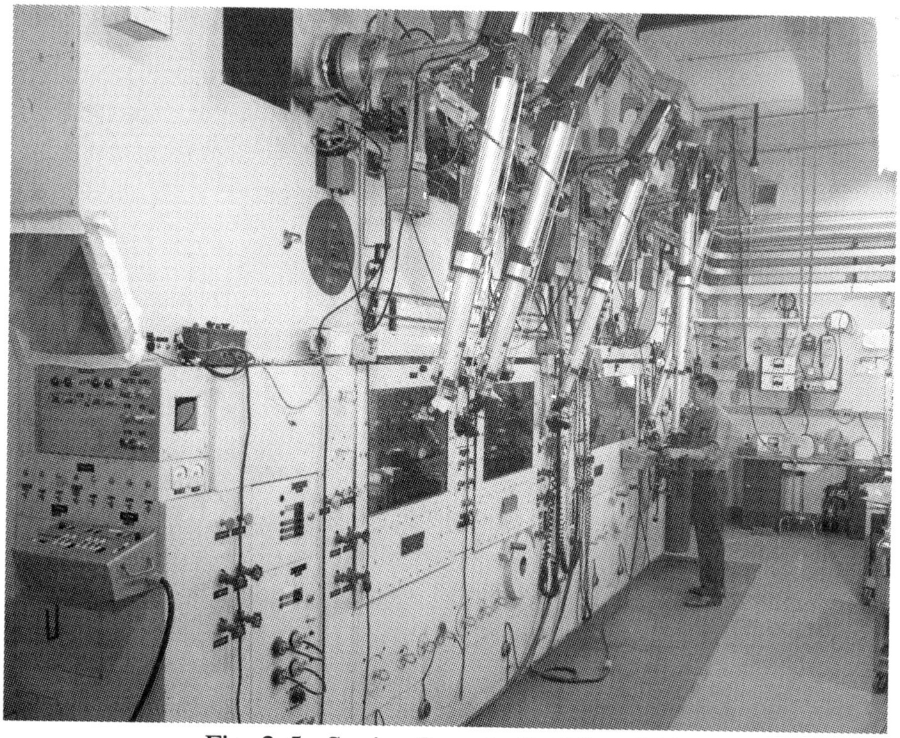

Fig. 2-5. Senior Cave in Building 205

One of the major ones was the ventilation system. Unlike commercial buildings where the air is recirculated, Building 205 uses a once-through system. Part of the outside air is drawn into the offices and other non-active areas, and exited through the laboratories from where it passes through the fan loft before it is exhausted to the outside. In the fan loft, the air passes through ultra-high-efficiency filters, which are sometimes called "absolute" or High Efficiency Particulate Air (HEPA) filters. These filters remove practically all the particulate material in the air, but not radioactive gases that might be released. With this arrangement, heating and cooling become complicated and expensive. For several years after the building was constructed, there was no air conditioning, and one could not open the windows, so it was often uncomfortably hot during the summers. On hot, humid days, condensed moisture on the overhead cool water lines in the laboratory areas dripped on the workers and their equipment. Some researchers used plastic tents to protect critical equipment. A few lucky people had equipment that required air conditioning, which was provided, and their labs tended to be visited frequently by the less fortunate on hot days.

Safety considerations dictated that the ventilation system must continue to function during power outages. To supply this need, as well as those of other critical systems in the building, emergency backup power was supplied by a 900-horsepower, 12-cylinder diesel engine on the service floor. As was usually the case, Milt Levenson had a story about that engine. He claimed that it was salvaged from a decommissioned LST (Landing Ship Tank) from WWII and dropped in San Francisco Bay in the process before it was completely overhauled and sent to Argonne. It was replaced later by a new unit.

Building 205 originally had three water systems: (1) domestic water, which was piped to all drinking fountains, washrooms, locker-room showers, eyewash fountains, and most safety showers, (2) the laboratory water system, which provided water to all the engineering areas and the laboratories for general-purpose use and cooling of equipment and apparatus, and (3) a system that consisted of two distilled water supplies, one for the high-bay engineering areas and the other for the A-, B-, and G-Wing laboratories. The distilled water system has now been supplanted by a relatively new deionization system for all areas. Three other new water systems have been installed in the building. A canal water system, in which water from the Chicago Sanitary Canal is filtered and supplied to ANL, is now used for cooling building operating equipment. A central cooling water recirculating system is used for the building air-conditioning equipment and for cooling some of the large laboratory-support equipment. The third is an in-house cooling water recirculation system that is piped to all engineering and laboratory areas to augment the existing laboratory water and to help reduce the water consumption in Building 205. The drainwater from the laboratories goes to retention tanks in a sub-basement, where it can be held and monitored for radioactivity or other contaminants before it is released into the Laboratory sewage system.

The laboratories are equipped with all the usual services such as hot and cold water, deionized water, natural gas, compressed air, vacuum, and nitrogen. Electrical power is available routinely at 120 and 208 V, and higher voltages can be provided where needed. One problem that developed with the Hauserman partitions in A- and B-Wings was that they came prewired with installed fuses for 208-V, single-phase power, to the chagrin of some who had to install industrial equipment that required 220-V, three-phase service. There was another problem with compatibility of connectors in the electrical boxes.

Building D-310

Building D-310 was completed and occupied before Bldg. D-205, but it was always an adjunct in that the Division headquarters was never located there. It is situated on the southeast corner of the intersection of Meridian and Rock Roads. As was the case for Bldg. D-205, Vorhees, Walker, Foley, and Smith of New York City were the architect-engineers. The construction plans refer to the structure as an "experimental waste processing, storage, and shipping facility." Figure 2-6 is a simplified version of the layout of Bldg. D-310, which was designed mainly to accommodate semi-works and radioactive waste-disposal studies. The building contained a machine shop area and five laboratories for supporting work. The laboratories were essentially the same as those in Bldg. D-205.

The extensive open area in this building, with a high ceiling and balconies, permitted work with large pieces of equipment. A three-ton bridge crane was provided at the loading platform to handle heavy shielding and equipment. The service floor, in addition to providing services for the main floor operations, included some of the operating equipment. For example, there is now a steel plate on the main floor that covers an opening where an incinerator extended from the service floor up into the main floor area.

Also underground was the "swimming pool" where highly radioactive fuel assemblies were located under water to provide shielding and still permit visual observations. This facility, designed under the direction of Phil Fineman, was used for high-level gamma irradiation experiments with food and other materials. This facility was located underground just south of the main building. Various types of gamma irradiations, including food-preservation studies that were conducted in this facility are described elsewhere.

Although Bldg. 310 was used intensively by CEN for several years, the Division's work shifted gradually to other projects that could be handled more efficiently in Bldg. 205. The waste-disposal work continued for many years, however, primarily as a facility for incinerating combustible dry active wastes and treating radioactive liquid wastes produced at ANL. The building was also used extensively to develop and test equipment for use in the EBR-II Fuel Cycle Facility (FCF), which was being built adjacent to the reactor at ANL-W. The Chemical Engineering Division played a large role in this development; it was one of

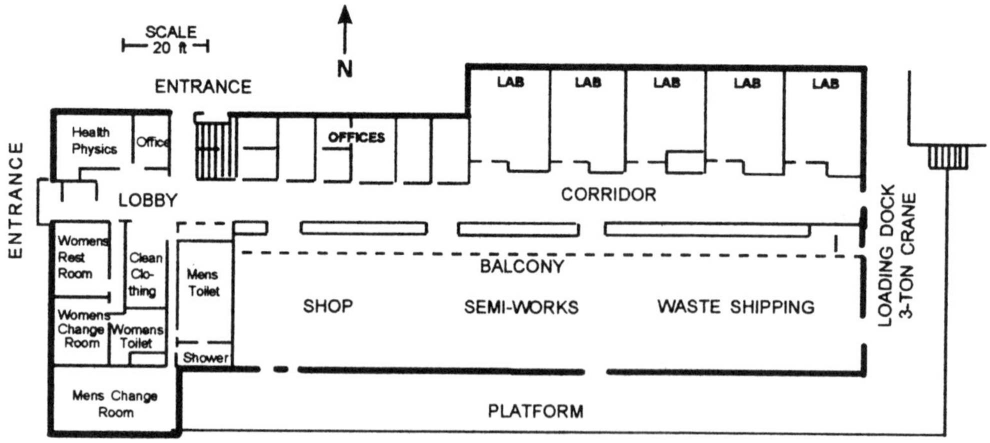

Fig. 2-6. Floor Plan of Building 310

CEN's major programs up until the FCF began its operation in the 1960s. Other CEN work included a large glove-box installation to purify molten chloride salts for use in pyrometallurgical fuel reprocessing research and development (Fig. 2-7).

Herb Brown recalls that in the early years of Bldg. D-310 operation "the fan loft contained a large, high-horsepower, very noisy lobe blower that could be heard at a great distance from the building. Dr. Lawroski let it be known that as he arrived at Bldg. D-205 he could hear the blower running at Bldg. D-310. It was mutually agreed by all CEN people in 310 that the first one to arrive in the morning had it as his sworn duty to start the blower."

Over the years, the nature of the CEN programs changed in such a way that its occupancy of Bldg. D-310 diminished to the point where it became impractical for the Division to retain the primary responsibility for the building. It is now under the jurisdiction of the Reactor Engineering Division (RE). However, some CMT work is being done there, mainly on equipment development and testing in support of the ANL-W program on the electrochemical processing of EBR-II fuel.

DIVISION ORGANIZATION AND MANAGEMENT

The management style in CEN could probably be best described as basically corporate in nature with something of an academic flavor. Dr. Lawroski was a strong leader and was demanding of the Division personnel. At the same time, he had a good sense of humor and took good care of his employees. He insisted that the staff personnel continue to further their technical knowledge. As an example, when a division seminar was in progress, Vic Munnecke, the Assistant Director, would roam through the building and closely quiz anyone in his office or lab as to why he wasn't at the seminar. Most of the seminars were

Fig. 2-7. Pyrochemical Facility in Building 310

given by staff members on their own work, and they were often challenged by one or two people in the audience who had made it a point to read up on the subject in advance and came in "loaded for bear." Hal Feder, in particular, often claimed that the speaker had overlooked some pertinent information that was in the literature, but he could never remember just where. After the initial irritation had worn off, this probably had a beneficial effect in causing the speaker to comb the literature on his subject more thoroughly.

The secretarial staff and the technicians were paid on an hourly basis and had to punch a time clock. (At one time, some people were playing "paycheck poker" games, which were based on the serial numbers of their checks; this was not encouraged by the Laboratory.) The technical staff were paid monthly and were not clocked, so a few began to trickle in a bit late in the morning. Dr. Lawroski solved this problem effectively by standing in the lobby in the morning and looking alternately at each late arrival and his wrist watch. He didn't have to say anything; they got the message. He also had an interesting habit of addressing individuals directly as "Mac," "Pal," or "Chum" when he was not pleased about something. There was some uncertainty as to just what these terms meant, which added to their effectiveness, but there seemed to be general agreement that "Chum" was the most ominous. When Dr. Lawroski was visiting a laboratory or was at a meeting and asked a question, the "askee" would have been well-advised to give the best straight answer he could. Any bluffing or double-talk was sure to result in further questions until he was in a very deep hole.

Dr. Lawroski made it a point to acquaint his more senior staff people with visitors and important people at meetings. This was beneficial to CEN's relationships with the AEC and other organizations. The AEC at that time seemed to trust the Laboratory management to handle the technical programs, and there was little or no micromanagement. The "189s" (budget justifications to the AEC) were simple one- or two-page documents which were all written by Vic Munnecke, the Assistant Division Director. Monitoring of the individual programs by the AEC generally consisted of a man from AEC Headquarters dropping in once a year or so for an informal chat and asking how things were going. One had more of a sense of support than any kind of a threat.

Early in the 1950s, the University of Chicago began holding annual reviews of the Division's performance during the preceding year. The people selected for these committees were well-known, highly regarded individuals from industry, academia, and other national laboratories, and included Henry Taube, who became a Nobel Laureate. Over the years these reviews have been generally commendatory and supportive of the Division's management and staff. There have been instances, however, where the recommendations of the Review Committees and the mandates of the AEC (or DOE) have been at odds, usually over the amount of basic research the Division should do in support of the applied programs. The annual Review Committee visits continue to require a lot of work in preparing the handouts of the slides and conducting dry runs of the talks, all of which leads to a last-minute mild state of panic both for the technical staff and the secretaries. A social evening during the review process has proved especially beneficial in allowing the Division staff members to talk with the Review Committee members on an individual basis. Poster sessions have also been used for the same purpose.

Group meetings and reports were weekly and Division meetings were held monthly on Friday afternoons in the early 1950s. With the lack of air conditioning, it was not unusual on a warm summer day to see somebody drop off to sleep during a Division seminar or meeting and suddenly "break his neck" waking up. Walt Rodger always wore dark glasses. Over the years, the group meetings became semimonthly, and then finally monthly, as they are now.

Individual secretaries were provided for the upper management and administrative personnel. Most of the secretaries were assigned to a section head and functioned as a "den mother," doing the typing and other work for all the people in that section. New arrivals in the Division soon learned who the upper management people were because they were always paged with the title "Dr." or "Mr." while the first and last names were used for everyone else.

When the Division moved to the DuPage site, Dr. Lawroski was the Division Director, Charlie Stevenson was the Associate Director, and Vic Munnecke was the Assistant Director. Walt Rodger was in charge of the engineering programs and Richard Vogel had the chemistry work. In 1954, Charlie departed to become the Technical Director of the Idaho Chemical Processing Plant (ICPP), at which time Rodger and Vogel became Associate Division Directors. In July 1958, Octave J. DuTemple, a chemical engineer in CEN, left the Division to become the Executive Director of the American Nuclear Society (ANS), which had about 3,000 members at the time. The society, which was formed in 1955 with Walter Zinn as President, had just moved from Oak Ridge to the John Crerar library in Chicago, and later established its current headquarters in LaGrange Park, Illinois.

Fig. 2-8. Walton Rodger

Dr. Rollin Taecker, a professor of Chemical Engineering from Kansas State University, took a sabbatical leave in 1953 working in CEN, and then returned to Argonne in 1955 to become the director of the International School of Nuclear Engineering (ISNE). The school, which opened on March 14, 1955, was an ANL organization fostered by the AEC to further the Eisenhower Atoms for Peace program, and it was headed up initially by Dr. Norman Hilberry. Its mission was to provide, in cooperation with universities, technical training in nuclear engineering for foreign and some American engineers and scientists. Several CEN staff personnel participated in the activities of the school by presenting lectures or seminars.

Among the temporary employees from academia, Dr. Joel Hildebrand, a Professor Emeritus from the University of California at Berkeley, who spent two months with CEN in 1953, was probably the most prestigious and also the most interesting. Although not a Nobel Prize winner, Professor Hildebrand received almost every other prize a chemist could. He liked to refer to Bob Steunenberg and others who had done their thesis work under his students as his "academic grandsons." One of the buildings at the University of California campus in Berkeley is named in honor of Hildebrand.

On February 20, 1957, Dr. Hilberry became the Laboratory Director, succeeding Dr. Zinn, who then became a vice president at Combustion Engineering, Inc. Hilberry was the person at the original test of CP-1 who had the assignment of standing on top of the pile with an axe to cut a rope suspending a "scram" safety rod that would drop into the assembly in case something went wrong. He was a highly competent scientist with a "down-home," easygoing manner and an impish sense of humor. One of his comments was that perhaps ANL should buy the Chicago Bears so as to gain full academic status. If you were at lunch at the cafeteria, he was likely to sit down and join your group for some conversation whether he knew you or not.

Dr. Lawroski's accomplishments and leadership received special recognition in 1959, when he was appointed Coordinator of Engineering Research and Development Programs at ANL. One important part of this position was to serve as Chairman of Argonne's Reactor Engineering Division Steering Committee. During the 1950s, he held many other positions of responsibility, both at ANL and nationally. He served on the ANL Scientific Personnel Committee, occupying the position of chairman for several years. He was a member of the Visiting Committee for the Nuclear Engineering Department and Reactor Division of Brookhaven National Laboratory. He was also Chairman of the American Standards Association Committee N5 and represented the Association Committee at the 1959 meeting of the International Organization for Standardization in Warsaw, Poland. He was a member of U.S. Fast Reactor Teams during which time he visited nuclear sites in the United Kingdom and in Europe. In 1956-57, he was Chairman of the Nuclear Technology Subdivision of the Industrial and Engineering Chemistry Division of the American Chemical Society. In the American Nuclear Society he served on the Board of Directors for a three-year term, was a member of the Executive Committee in 1958-59, and was Chairman of the Planning and Coordinating Committee during 1960-61. He also served on the Admissions Committee of the ANS. He was a member of the General Advisory Committee (GAC) for the AEC, and served later on the NRC Advisory Committee for Reactor Safeguards. After leaving the Division, Dr. Lawroski continued to garner many additional awards, and he was elected to the National Academy of Engineering in 1969.

Dr. Lawroski's brother, Harry, also a prominent figure in the nuclear business, was President of the American Nuclear Society in 1980-1981.

TECHNICAL PROGRAMS

Aqueous Processes

DETERMINATION OF BREEDING GAIN IN EBR-I

EBR-I, the first fast breeder reactor in the world, was built to demonstrate the feasibility of fast breeder reactors and the potential of the breeding concept. The CEN Division was concerned with two parts of the EBR-I program: first, the development of a chemical process for recovering enriched uranium from the reactor core; second, a determination of the conversion ratio (plutonium generated/U-235 consumed) of the reactor.

In giving ANL permission to design and construct EBR-I, the AEC stipulated that the enriched uranium in the core, which had been borrowed from the military stock, would have to be returned completely free of fission products and at a specific time. Development of a process to recover and decontaminate the enriched uranium in the EBR-I core began in 1949 in the West Stands at the University of Chicago and required about 60 man-years of effort. The result was a solvent-extraction process utilizing tributyl phosphate (TBP) diluted with carbon tetrachloride as the organic solvent. The process was then installed inside one of the 8- × 10-foot cells shielded by 18 inches of high-density concrete in Bldg. 205. A 40-foot-high, 1-in.-dia stainless steel column was used for the extraction-scrubbing step and a similar 35-foot-high column for uranium stripping. This plant was to have been ready for operation by the end of summer in 1951, but a decision was made by the AEC that the EBR-I core, other than samples removed for the work described below, would be reprocessed at the Idaho Chemical Processing Plant (ICPP).

The objective was to determine the conversion ratio (fissile material produced/fissile material consumed) in the reactor. If the ratio is greater than unity, breeding has occurred and the difference is the "breeding gain." Samples of the core fuel rods and the blanket rods were selected in a manner that would allow calculations of the average core burnup and the plutonium content of the blanket. The samples were dissolved in nitric acid for analysis. Eighty-five core samples and 35 blanket samples were analyzed.

The U-235 burnup in the core samples was based on analyses of the dissolver solutions for cesium, whose fission yield had been well established. The dissolved blanket samples were analyzed for plutonium. Because some U-238 is fissioned at high neutron energies, a method was developed in cooperation with the Chemistry Division to distinguish between U-238 and U-235 fission by measuring the Ru-106/Cs-137 ratio.

The conversion ratio for this first core loading of EBR-I was found to be 1.00 ± 0.04 atoms of plutonium for each atom of U-235 consumed, which is equivalent to a breeding gain of 0 ± 0.04. This result showed clearly that plutonium breeding had reached the break-even point in EBR-I and that a significant breeding gain should be achievable with more advanced reactor designs and with plutonium as the fissile material.

Les Burris was the project leader for the process development work, and others who were involved in the engineering development included John Schraidt, John Natale, John Loeding, Virgil Trice, Ira Dillon, Norm Levitz, Les Coleman, Herb Brown, Don Hampson, Sy Vogler, and Bill Voss (from Central Shops). Milt Levenson was in charge of the design and start-up of the facility and made the calculations required to determine the conversion ratio. Wally Seefeldt directed the operations, which were carried out by Les Dorsey, Artie Freeman, and Bill Spicer. An enormous analytical effort on this project was carried out under the direction of Dick Vogel; some of the individuals contributing to this effort were Carl Crouthamel, Bob Schablaske, Chuck Seils, Myron Homa, and Jackie Williams. Tony Engelkemeier and Arthur Jaffey from the Chemistry Division, Ted Novey from High Energy Physics, and Dave Hess from Physics also participated in the program.

THE HALEX PROCESS

Because of the Cold War with the Soviet Union and the resultant arms race, the U.S. moved expeditiously in the late 1940s and early 1950s to expand its capabilities for the production of weapons-grade plutonium for nuclear weapons and tritium for thermonuclear weapons (the hydrogen bomb). To augment existing production capabilities at Hanford, the U.S. authorized construction of the Savannah River Plant (SRP) in South Carolina. Located within the large plant area were the production reactors, two large solvent-extraction plants for recovery and purification of uranium and plutonium, facilities for the recovery of deuterium and tritium, and research and development laboratories. E. I. du Pont de Nemours and Company operated the plant.

During construction of the plant, Du Pont operating personnel were assigned to various national laboratories for training. About 50 Du Pont trainees spent up to three years in the various divisions at ANL—in Chemical Engineering for solvent-extraction technology, in Reactor Engineering for reactor design and operation, in Chemistry for studies of organic solvents, and in Metallurgy for metallurgical studies of fuel fabrication and behavior in a reactor.

The Purex process, which uses tributyl phosphate (TBP) diluted with dodecane, was chosen for the uranium/plutonium recovery process. This process had been discovered in 1949 at Oak Ridge National Laboratory (ORNL). Early work on it was also conducted at the Knolls Atomic Power Laboratory (KAPL) in Schenectady, New York. It was subsequently installed in the processing plants at Hanford and Savannah River.

In the early stages of Purex process development, Du Pont became concerned about potential flammability of the dodecane solvent and sought a nonflammable alternative. Carbon tetrachloride was eventually selected, and the ANL Chemical Engineering Division was assigned the responsibility for demonstrating the Purex process, using carbon tetrachloride as the diluent. To distinguish this process from Purex, it was named "Halex."

Although carbon tetrachloride is not flammable, it raised other concerns. One was the extent of its radiolytic decomposition in high radiation fields to yield chloride ions. At concentrations above about 10 ppm, chloride embrittlement of stainless steel can occur. Another concern was the chemical toxicity of carbon tetrachloride, which requires precautions to prevent inhalation of vapors by workers. In the ANL studies, a thin (approximately 1-inch-thick) layer of water on top of the solvent in the feed tanks was used to limit vaporization of the carbon tetrachloride. The feed tanks were also vented to the Bldg. D-205 vent system. Finally, although not a concern, the high specific gravity of carbon tetrachloride (about 1.4) made the solvent the "heavy" phase, which is just the reverse of the situation in the Purex process where the organic solvent is the "light" phase. These concerns and the changes in the physical properties of the solvent warranted a pilot-plant demonstration under representative plant conditions.

To demonstrate the Halex process, a pilot-plant facility was installed behind 18-inch-thick concrete shielding walls in two adjacent 12-foot-high bays in the Bldg. D-205 G-Corridor (G-102 and G-118). The first cycle (IA) contactor banks were installed in G-118. A single set of extraction and stripping contactors was used alternately in G-102 for the second uranium and plutonium purification cycles. Fully irradiated Hanford fuel slugs were dissolved in one of the three dissolvers in J-117 to give a typical uranium feed solution.

Banks of mixer-settlers obtained from the Standard Oil Development Company were used as the contactors. These simulated the pump-mix mixer-settlers that had been selected by Du Pont for use at Savannah River, but they lacked the capability to pump the immiscible aqueous and organic solvent phases to the next stage. Therefore, each bank was tilted about 15 degrees to provide gravity flow of the immiscible fluids through the bank. The Standard Oil Development contactors were either 15- or 20-stage units, the former being used for the stripping operation and the latter for extraction-scrubbing and uranium-plutonium separation.

Many runs were made in the G-102/G-118 complex. The Halex process performed very well. Fission-product decontamination factors were equal to, if not higher, than those achieved with the Purex process. Complete separation of the uranium and plutonium was obtained in the IB unit, and high recoveries (>99%) of uranium and plutonium were achieved. The carbon tetrachloride proved to be highly resistant to radiolytic decomposition. Chloride ion concentrations in the high-level waste were 10 ppm or less. Despite the successful demonstration of the Halex process, Du Pont decided to go with the conventional Purex process, having become convinced that dodecane posed no significant fire hazard after all. Interestingly, carbon tetrachloride was used for many years at Hanford in the Recuplex process for recovering plutonium from scrap and recycle material.

Les Burris was the Project Leader for the Halex process demonstration. Major responsibilities for the design of the facility and its subsequent operation were borne by

John Loeding and Virgil Trice. The other participants in the program were mostly the same ones who had worked on the breeding gain effort. Again, much of the credit for the success of the Halex demonstration is due to the Analytical Chemistry Group, who, during the course of the program, analyzed the hundreds of samples required to evaluate the process performance.

AQUEOUS PROCESSING OF ALLOY FUELS

Several uranium alloy fuels for specialized reactor applications began to appear in the early 1950s. These were generally combinations of enriched uranium alloyed and/or clad with metals such as zirconium (or Zircaloy), aluminum or stainless steel, and they could not be dissolved readily by nitric acid, which was the standard procedure for metallic uranium fuel slugs. These fuels usually consisted of a small amount of highly enriched uranium in the other metal. Plutonium was not involved because the limited amount available at the time was needed for weapons.

Research and development work on the uranium-aluminum fuel resulted in a process in which mercury was used as a catalyst to dissolve the fuel in nitric acid. The zirconium-alloy fuels were more of a challenge, and this was addressed by CEN in a research effort that extended from about 1949 to 1954. This program was aimed at the fuels used in the Submarine Thermal Reactor (STR), which were of two types. One was 99.39% zirconium and 0.61% uranium; the other, in which Zircaloy was alloyed with the uranium, was 97.9% zirconium, 1.46% tin, and 0.64% uranium. Vic Munnecke and Elton Turk did the early work in this area in CEN. A general problem with nitric-acid based dissolutions was that explosions occurred under some conditions. An expanded effort was undertaken in which it was shown that the explosions could be avoided by the use of fluorides to complex the zirconium ions. Bob Larsen, Al Jonke, Harold Evans, Roberta Shor, Sy Vogler, and Elton Turk conducted these studies. Turk later transferred to the ICPP. The CEN work on processing methods for zirconium-alloy fuels was a major contribution to the technology, and it was used as the basis for scale-up studies, followed by plant-scale operation at ICPP, where it was used many years for the recovery of enriched uranium from naval fuels.

CONTACTOR DEVELOPMENT

In the late 1940s and early 1950s, the search for more selective solvents for recovery and purification of uranium was accompanied by increased efforts on the development of more efficient contactors. A major driving force was a need to reduce the height of the processing canyons, which had thick walls of concrete shielding, and their cost was essentially proportional to the height. More efficient low-height contactors became even more important when it was found that the height of an equivalent theoretical stage (HETS) was greater in the Purex process than in the Redox process, which required 40-ft-high towers packed with Raschig rings. The prospect of using 50-ft-high towers (or columns) for the Purex process was not appealing.

Hanford's solution was to use pulsed sieve-plate columns. On each pulse, the solvent (discontinuous) phase was driven through the next of an array of sieve plates, breaking the solvent into fine bubbles, and thereby increasing the surface area for mass transfer of uranium and plutonium. The pulsed columns worked very well and were used in the Hanford plant until it was shut down.

The people at Savannah River chose to use banks of mixer-settlers. Each stage in a bank consisted of a mixing region in which the two phases were vigorously agitated together and a settling region where they disengaged into two layers. The mixers provided sufficient pumping action to move the two immiscible phases into the next stage. As a result, a mixer-settler bank could be operated in a horizontal position, thereby reducing the required canyon height. Mixer-settlers have been in successful operation at Savannah River for over 30 years.

The small mixer-settler units used by CEN to demonstrate the Halex process were a close replica of the Savannah River units, but they were tilted slightly to induce gravity flow of the immiscible liquids through each bank. Argonne also built and demonstrated the Stacked Plate Contactor, a high-throughput contactor that had been designed by Merrill Fenske of Pennsylvania State University. Kegham Varteressian ("Varty") and George Bernstein conducted the ANL demonstration. A contactor had already been selected for the Savannah River plant, so further development of the Stacked Plate Contactor was not pursued.

One drawback of mixer-settlers is the large holdup of solvents in each stage. The long exposure of the solvent to the highly radioactive aqueous phase maximizes the opportunity for radiation damage to the solvent. This concern eventually led to the development of a centrifugal contactor capable of high throughput and low solvent hold-up, which was pioneered at Savannah River. The Savannah River work provided a basis for later intensive development work on centrifugal contactors at Argonne for application to high-burnup fast breeder reactor fuels.

Pyrometallurgical Processes

PYROMETALLURGICAL RESEARCH

By 1950, a strong interest had developed in the possible use of pyrometallurgical methods for reprocessing reactor fuels. Their potential advantages of compactness, simplicity, fewer criticality problems, and potentially low cost are discussed in the following section on process development. Exploratory and basic studies of pyrometallurgical separations were performed primarily in Hal Feder's group, which also had the responsibility for some other projects. The pyrometallurgical investigations at ANL were targeted almost entirely toward EBR-II and were therefore concerned mainly with recovery of enriched uranium metal from the core, where plutonium was of secondary interest, and recovery of plutonium from the metallic uranium blanket, which was a longer-range concern.

Feder, because of his excellent background in the nuclear area, knowledge of the pertinent literature, and understanding of thermodynamics and other theoretical principles, was a highly effective leader in this effort. (An area of special experience that he admits to, but never advertised widely, resulted from a stint in the Army where he was doing research on the use of banana peels as a lubricant for the ways used in launching ships.) A review article he wrote in the *Reactor Handbook* classifies pyrometallurgical separations into the following categories:

1. Fractional crystallization, with and without a liquid metal solvent
2. Fractional distillation
3. Liquid metal partition
4. Selective oxidation
5. Cyclic oxidation-reduction

Argonne was one of the leading laboratories in pyrometallurgical research at the time, but similar work was being done at several other sites, including Ames Laboratory, Atomics International, Brookhaven, the Canadian Atomic Energy Authority, Harwell in England, Los Alamos, and Oak Ridge. Los Alamos probably had the most comprehensive effort for two reasons, one being that their weapons production program used pyrometallurgical methods. The other reason was that they had been developing the Los Alamos Molten Plutonium Reactor Experiment (LAMPRE), a reactor concept that utilized a molten Pu-9.5 at.% Fe alloy as the fuel. Among the methods that Los Alamos investigated for plutonium purification were liquation and filtration, oxide drossing, carbide slagging, halide slagging, halide conversion cycles, electrorefining, recrystallization from mercury, and liquid-liquid metal extraction. Brookhaven had proposed a liquid metal fuel reactor (LMFR) in which the fuel was molten bismuth containing 0.1-0.2% uranium, 250 ppm zirconium, and 350 ppm magnesium. The ternary eutectic salt 50 $MgCl_2$-30 NaCl-20 KCl (mol %) with a melting temperature of 376°C was used to extract fission products from the molten metal fuel. This salt composition was used in much of the work at ANL where it was referred to informally as "Brookhaven salt," much to the annoyance of Dr. Vogel. Oak Ridge conceived the Hermex process, in which metallic uranium was decontaminated by selective precipitation from mercury. Some of the other laboratories investigated potential fuel-reprocessing methods for thorium, as well as uranium and plutonium. Feder kept well abreast of these other programs and stayed in close touch with those doing the work. Because of the reputation he had developed in this field through these connections and his own program at ANL, he was often sought out as an authority to write review articles on the subject.

Members of Feder's group who were involved with the pyrometallurgical work in the early 1950s included Milt Ader, Karl Anderson, Paul Bergland, Norm Chellew, Guy Elliott, Don Fredrickson, Al Glassner, Irv Johnson, Al Martin, Marcel Nathans, Ralph Nuttall, Ken Rhode, Charlie Rosen, Ronald Uhle, Sy Vogler, Charlie Wach, and Bob Yonco. Jim Knighton, who was on loan from the American Smelting and Refining Company (ASARCO), later became an ANL staff member. Don Hampson and George Bennett contributed to the engineering aspects of this effort.

The earliest ANL studies were conducted on melt refining. Previous work had shown that many of the fission products can be removed from irradiated metallic uranium fuel simply by melting the fuel in an oxide ceramic crucible and holding it in the molten state for a few hours. The noble gases (Xe and Kr) and other volatile fission products were released or vaporized and the electropositive fission-product elements such as the rare earths and alkaline earths reacted with the crucible material to form non-volatile oxides. The noble metal fission products (Zr, Nb, Mo, Tc, Ru, Rh, Pd) remained in the molten uranium. Plutonium recovery was not a major concern in the recovery of EBR-II driver fuel, but some preliminary experiments showed that the plutonium/uranium ratio in the oxide product was about twice that in the metal phase. The following crucible materials were studied: alumina (Al_2O_3), magnesia (MgO), beryllia (BeO), thoria (ThO_2), and zirconia (ZrO_2). Both alumina and magnesia reacted stoichiometrically with the reactive fission-product metals (for example, $2\ Ce + Al_2O_3 \rightarrow Ce_2O_3 + 2\ Al$), and the reaction proceeded at a constant rate. The aluminum metal product from the reaction dissolved in the liquid uranium. With a magnesia crucible, metallic magnesium was vaporized. Beryllia, thoria, and zirconia, however, reacted to form suboxides in the crucible wall (identified by x-ray diffraction), and the rate increased with time. Little or no fission-product iodine was released. The molten uranium wet none of the crucible materials except Al_2O_3, which was wet only if the uranium contained dissolved molybdenum and ruthenium. The molten uranium in the various crucibles had a contact angle of 135° and its surface tension was 800 dyn/cm. Contamination of the uranium product by the crucible materials was in the order Al>Be>Th>Zr. Zirconia was chosen as the best crucible material for melt refining, and thermal shock problems were eliminated by stabilizing it with 5 wt% calcia (CaO).

Attention was given to possible methods for recovering bred plutonium from the EBR-II blanket material, using liquid magnesium as an extractant. Ralph Nuttall found that a surprisingly high plutonium recovery could be obtained by agitating hydrided-dehydrided U-1 wt% Pu in liquid magnesium. Some initial studies were conducted on the extraction of plutonium from molten uranium into a molten chloride salt (halide slagging).

The most extensive, systematic studies by this group in the 1950s were on the chemistry of liquid metal solvents, which had potential application both for EBR-II blanket processing and for the recovery of fuel values from melt refining residues. Solubilities of U, Pu, Th, and most of the important fission-product elements in liquid zinc were determined. Similar studies were done with liquid cadmium. Through a combination of temperature coefficients of solubilities, identifications of equilibrium solid phases, electrochemical and effusion measurements, coprecipitation experiments, and phase-diagram definitions, a large body of high-quality thermochemical data was generated. This work, which continued into the 1960s, was essential for much of CEN's ongoing pyrochemical process development work. In addition, it was a major contribution to the literature on the chemistry of liquid metals.

PYROMETALLURGICAL PROCESS DEVELOPMENT

Engineering development of pyrometallurgical processes for recovery of spent fuel discharged from fast breeder reactors was undertaken in the mid-1950s to meet the special needs of fast breeder reactor fuel cycles. These reactors, which use plutonium as a fuel, generate more plutonium than is consumed. As mentioned before, the amount of excess plutonium generated per cycle is called the "breeding gain." The doubling time is the time required to double the amount of fuel in the fuel cycle (that in the reactor plus that out of the reactor in storage, processing, and refabrication of fuel for return to the reactor). Minimizing the fuel in the out-of-reactor fuel-cycle operations reduces the doubling time needed to start a new reactor.

Metal fuels (uranium-plutonium alloys) provide the greatest breeding gain compared with other potential fast reactor fuels such as mixed uranium-plutonium carbides, nitrides, and oxides. Because of the successful experience with metal fuels in plutonium-production reactors and the developed technology for producing and fabricating uranium metal, metal fuels were the natural choice by ANL in its early development of fast breeder reactors.

For discharged fast breeder reactor fuels, solvent-extraction processes such as Redox and Purex were not attractive. These fuels required long storage (cooling) times of about two years before processing by solvent extraction to avoid excessive solvent degradation (radiation damage), which results in less effective decontamination of the uranium and plutonium products. Moreover, the complexity of the tail-end operations required to convert the uranium and plutonium in the dilute nitric acid product of the solvent-extraction process back to the metals increased the out-of-reactor plutonium inventory and the cost of the fuel cycle.

Another important factor in the design of processes for fast breeder reactor fuels is that high decontamination of the processed fuel is not required. The performance of fast reactors is affected very little by the presence of impurities, *e.g.*, residual fission products or alloying elements, in the fuel. In contrast, the performance of a slow (thermal) neutron reactor, *i.e.*, a light-water-cooled reactor, is seriously impaired by the presence of certain impurity elements in the fuel. Therefore, fuel-recovery processes for these reactors must provide very high fission-product decontamination.

While low decontamination is acceptable for the processing of discharged fast breeder reactor fuels, some removal of every fission product must be achieved to avoid unrestrained buildup of any individual fission-product elements as the fuel is recycled repeatedly. Low decontamination of processed fast reactor fuel is actually advantageous for another reason. The intrinsic high radiation levels in the processed fuel strongly discourage its clandestine diversion to weapons production or terroristic purposes.

Pyrometallurgical processes appeared to be well suited to fast reactor fuels. Because organic materials are absent, they could accommodate the high radiation levels of short-cooled fuels. In addition, they offered the opportunity to avoid cumbersome product-conversion steps by keeping the fuel in the metallic state throughout the process. It should be noted here that, although plutonium was the fissionable material of interest, U-235 was employed in EBR-I core to provide sufficient reactivity to operate the reactor. Because significant amounts of U-235 were involved, the early process development was concentrated primarily on the recovery of U-235.

The first pyrometallurgical process that was investigated was zone melting, also called zone refining. In this process, a narrow molten zone, created by a movable heater, slowly traverses a long bar of metal. Impurities that favor the solidus in a phase diagram move counter to the direction of travel of the molten zone. Those that favor the liquidus move in the same direction as the molten zone. It is usually necessary to make a very large number of passes of the molten zone to achieve significant movement of the impurities to the ends of the bar where they can be removed by cropping off the end sections. Zone melting had been invented by William G. Pfann of Du Pont, who used it successfully to purify metals such as silicon and germanium for use in the electronics industry. The method was unsuccessful, however, for purifying uranium because the solidus/liquidus ratios of many fission products (especially the noble metals) at equilibrium were too close to unity to produce significant separation toward the ends of the bar at practical rates. The failure to separate fission products, the very long processing time under a high-purity inert atmosphere, and the requirement for a supplemental process to recover plutonium from the cropped ends forced abandonment of the process. This program did, however, result in a useful basic contribution to the mathematical modeling of zone melting. Les Burris, Ira Dillon, and Charles Stockman did this work.

Attention was then turned to a simple slagging-type process in which the fuel was melted in a calcia-stabilized zirconium oxide (ZrO_2 + CaO) crucible, held at a temperature of 1400°C for four hours and poured into a graphite product-receiver mold. The behavior of fission products during melt refining is shown in Fig. 2-9, and the melt-refining furnace in Fig. 2-10.

An appreciable fraction of most fission products is removed by the melt refining process, but noble metals (*e.g.*, ruthenium, rhodium, and palladium) and molybdenum are not removed. To prevent the buildup of these elements as the fuel is recycled through the reactor, a small fraction of the fuel, known as "dragout," is removed for separate reprocessing. The skull material remaining in the melt refining crucible, which constitutes 5 to 10% of the processed fuel, serves adequately as the dragout material. Details of the dragout process are presented later.

The melt-refining process was demonstrated at full scale (10 kg of U per batch) in prototypical plant equipment in Bldg. 205. Subsequently, it was installed in the EBR-II Fuel Cycle Facility at ANL-W, where it was used successfully for five years to recover and recycle fuel discharged from EBR-II.

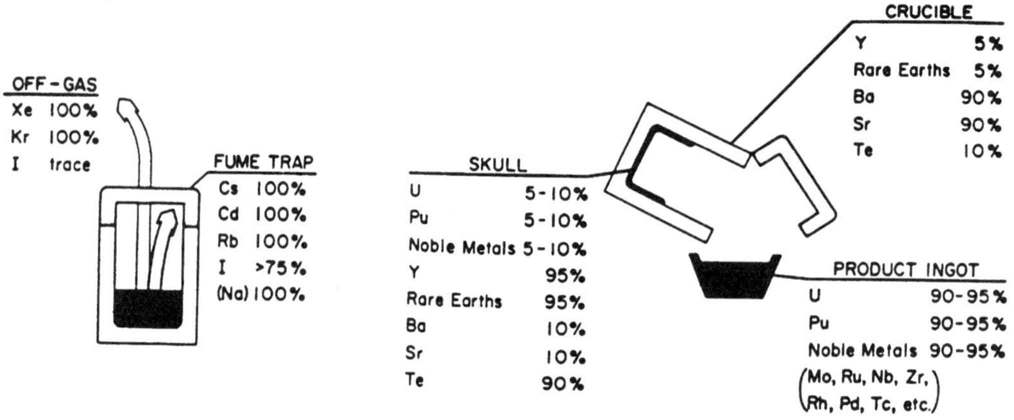

At the high operating temperature of 1400ºC, strontium, barium, and rare earths, which form very stable oxides, are removed in a reaction layer on the wetted surface of the zirconia crucible. The noble gases (krypton and xenon) and some iodine are released when the fuel is melted. The volatile fission products (cesium and rubidium), and sodium, inserted into the fuel elements as a heat-transfer ligament, are volatilized. The noble metal fission products such as ruthenium, rhodium, palladium, and molybdenum are not removed. Their continuing buildup in the recycled fuel is prevented by removing a small fraction of the fuel, called "dragout" in each cycle for separate processing. The crucible skulls serve as the dragout stream.

Fig. 2-9. Fission-Product Removal by Melt Refining

Fig. 2-10. Melt Refining Furnace

Two interesting synergistic developments occurred in the course of the research on melt refining. One was the injection casting method for producing fuel pins. It was an outgrowth of the method used for sampling molten uranium. Samples of the uranium were taken by drawing a small amount of the liquid into a 1/4-inch-diameter Vycor® (quartz) tube. Noticing the perfectly formed cylindrical segments of solidified uranium, Milt Levenson conceived the idea of using precision-bore Vycor tubes coated with a thoria (ThO_2) wash to cast fuel pins for EBR-II. So was born injection casting for EBR-II fuel. In the casting procedure, an array of open-ended Vycor tubes (closed at the top) suspended above a crucible of molten fuel alloy within an evacuated bell-jar furnace was plunged into the melt as the furnace was simultaneously pressurized. Molten fuel was forced up into the tubes to a height of 16 to 18 inches. After cooling, the perfectly formed fuel rods were recovered by crushing the Vycor. The rods were cropped to a 14-inch length, giving fuel pins, which were then clad with stainless steel and incorporated into fuel subassemblies for return to the

Fig. 2-11. Milton Levenson

reactor. The entire sequence of operations was performed in the EBR-II Fuel Cycle Facility.

The second synergistic development was a reactor fuel alloy called "fissium." Fissium is the steady-state composition of fuel resulting from dragout (removal) of a small fraction of the fuel for separate processing to extract noble metal fission products and reclaim the uranium. The steady-state composition depends on the fraction of the fuel removed per cycle and the fuel burnup (assuming all the noble metals are completely removed from the dragout fraction). For a fuel burnup of 3 wt% of the uranium and a 7% dragout, the steady-state total concentration of noble metals is about 5 wt%. Fuel of this composition proved to have a remarkable resistance to radiation damage. Unlike uranium metal, which grows uniaxially (lengthwise) under irradiation, fissium undergoes a growth that is much smaller and equiaxial. The fissium fuel was found later to be capable of sustaining burnups in excess of 10 wt%.

After melt refining, between 5 and 10% of the charged fuel remained in the crucible as unpoured metal. Because the wetted surface had a cup-like shape, the unpoured material was called the "skull." It was decided that this material would constitute the dragout fraction. It was removed from the crucible by oxidizing the uranium to UO_2, which was dumped from the crucible as a powder.

Owing to their batch nature, pyrometallurgical processes were considered best suited to small, on-site processing plants serving one, or at most, a few reactors. The EBR-II Fuel Cycle Facility built adjacent to the EBR-II epitomized this concept. An early question was how to reclaim uranium from the skull material and maintain the concept of small, on-site plants. The large facilities required for solvent-extraction processes subverted the concept of compact, self-contained fuel cycles. Therefore, a pyrometallurgical process was sought for recovery of the skull material, and the idea of using liquid metal solvents and molten salts as processing media was advanced. The use of these media opened up a new frontier of process development and spawned research and development programs that have, with some interruptions, extended into the 1990s.

Solvent metals had to be able to dissolve uranium and to have sufficiently high vapor pressures at elevated temperatures to allow their ultimate removal by vaporization. The molten salts had to possess high chemical stabilities, reasonable liquid ranges at temperatures of process interest, and also sufficiently high vapor pressures at elevated temperatures for removal by vaporization. The most promising candidate metal solvents were zinc, cadmium, and magnesium. The molten salt systems of greatest interest were combinations of alkali and alkaline earth chlorides.

Several processes involving the use of liquid zinc as a solvent had been suggested and were generally referred to collectively as "pyrozinc processes." A typical pyrozinc process is illustrated in Fig. 2-12. These

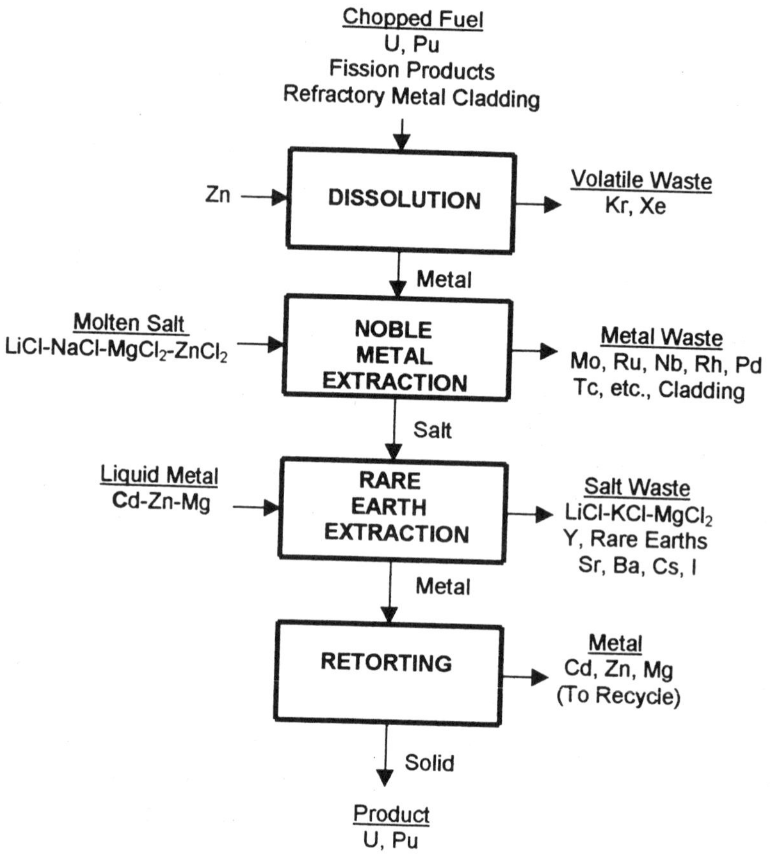

The Pyrozinc process is appropriate for metallic reactor fuels. The fuel elements and their refractory metal cladding are chopped into short lengths and dissolved in molten zinc. Fission-product gases, Kr and Xe, are released in this step. The zinc solution, which contains U, Pu, and fission products, is then contacted with a molten salt such as $LiCl$-KCl-$MgCl_2$ containing $ZnCl_2$. The $ZnCl_2$ oxidizes the U, Pu, alkali metal, alkaline earth metal, and rare earth fission products to their chlorides, which are extracted into the salt phase. The liquid metal phase then contains the noble metal fission products (Mo, Ru, Nb, Rh, Pd, Tc) and the refractory cladding metals (typically V, Ti, Zr, W). This metal phase is discarded as a waste. The salt phase, which contains the chlorides of U, Pu, alkali metals, alkaline earth metals, and rare earths, is contacted with a liquid Cd-Zn-Mg alloy which reduces the U and Pu selectively to the metals, leaving the remaining fission-product elements in the salt phase, which is a waste stream. The purified U-Pu metal product is recovered by retorting off the Cd, Zn, and Mg solvent metals, which are recycled.

Fig. 2-12. Typical Pyrozinc Process

processes are predicated on the following observations:

1. The solubilities of nearly all common metals in zinc at 700-800°C are sufficiently high to dissolve reactor fuel elements in a single step.
2. Uranium oxides can be dissolved in zinc by the addition of magnesium, which reduces the oxides.
3. Uranium metal can be recovered from the zinc either by retorting or by the addition of magnesium to form a metallic uranium precipitate.
4. In a molten fluoride salt containing uranium chloride, uranium can be transferred electrolytically from a metal anode to a liquid zinc cathode.

One example of a pyrozinc process, the EBR-II "Skull Reclamation Process," utilized liquid zinc to leach the noble metal fission products from the skull oxides, and then a Zn-5 wt% Mg solvent with a $MgCl_2$-rich molten salt phase to reduce the UO_2 and dissolve the metallic uranium product. Selective precipitation of the uranium product was also used as a purification step. Further development work on the skull reclamation process, which continued into the 1960s, is discussed in the next chapter.

A simple, ingenious process was also developed for concentrating plutonium bred in the U-238 blanket of a fast breeder from less than 3 wt% to greater than 30 wt% in uranium. In this process, the blanket material was dissolved in a Zn-12 wt% Mg solvent at 800°C (the uranium solubility was about 20 wt%). Uranium, which is insoluble in liquid magnesium, was then selectively precipitated by the addition of magnesium, leaving the plutonium in solution. The solid and liquid fractions were then separated and processed to yield metallic uranium and plutonium products. While intended for installation at EBR-II, the blanket process was a casualty of the AEC decision to abandon metal fuels for fast reactors.

For the skull reclamation process, the requirement to reduce UO_2 to the metal resulted in a major research effort on reductions of UO_2, PuO_2, and even ThO_2 (of interest mainly from a basic research standpoint). Magnesium was selected as the reductant with zinc or cadmium as the solvent for the reduced metals. The molten salt phase was usually a LiCl-KCl-$MgCl_2$ mixture. (Multicomponent salts are required to provide a liquidus temperature low enough for process use.) The reduction is driven, in part, by the low chemical activities of uranium, plutonium, and thorium in the solvent metals. Plutonium dioxide proved to be the easiest oxide to reduce because PuO_2 forms an oxychloride that is soluble in the molten salt, thereby allowing its ready access to the magnesium reductant. Uranium dioxide was more difficult to reduce, but complete reductions were achieved under suitable conditions. Even ThO_2, which is the most stable of the three oxides, was reduced with particular salt compositions. The best and most used reductants were obtained with zinc as the liquid metal solvent. The ability to reduce UO_2 and PuO_2 turned out to be important in the development of processes for oxide fuels in the 1960s and, later, in the 1990s.

Milt Levenson and John Schraidt followed the engineering development program closely, but Les Burris was in charge of most of the work. Among the people working on the engineering research and development effort were Ray Beck, George Bennett, Tom Cannon, John DeKany, Les Dorsey, Jim Hesson, Terry Johnson, Ernie Johnston, Henry Lavendel, Paul Nelson, John Pavlik, Wilfred Pehl, Dean Pierce, Alfred Schneider, Irv Winsch, and Jerry Wolkoff.

THE EBR-II FUEL CYCLE FACILITY (FCF)

The successful operation of several experimental fast breeder reactors together with the promising results from the pyrometallurgical fuel reprocessing studies paved the way for a demonstration of an on-site, closed-cycle pyrometallurgical reprocessing facility coupled to the 62.5-MW(t) Experimental Breeder Reactor that was to be built by ANL at the National Reactor Testing Station (NRTS), now called INEEL (Idaho National Engineering and Environmental Laboratory). A closed-cycle process of this type offers a number of potential advantages over other methods for reprocessing the metallic fuel used in this type of fast breeder. Consolidation of all the reprocessing steps, including refabrication of the recycled fuel, into a closed, on-site operation should be cost-effective due to the low out-of-reactor fuel inventory, the simplicity of the reprocessing operations, and the low waste volume. Because fast breeders are less affected by fuel impurities than thermal reactors, only a modest fission-product decontamination factor of three or so is required.

The EBR-II Fuel Cycle Facility was a major project for CEN (Fig. 2-13). The magnitude of the project was too great for an adequate description here, but it has been covered in detail in a comprehensive book by Charlie Stevenson, entitled *The EBR-II Fuel Cycle Story*, which was published by the American Nuclear Society in 1987. The melt-refining process was chosen for this pilot-plant demonstration. As mentioned earlier, research studies were conducted on a number of potential process concepts in which the bulk of the spent fuel would remain in the metallic state throughout the process, thereby avoiding

Fig. 2-13. Experimental Breeder Reactor-II with Adjoining Fuel Cycle Facility

chemical conversion steps, which tend to be complex and costly. The melt refining process appeared to be the most promising approach, based on the earlier research and development work on various pyrometallurgical processes.

Detailed designs of EBR-II and the Fuel Cycle Facility were initiated in 1957, construction began that same year, and the facility was completed in 1962. The first fuel was processed in 1964. Although the design of the Fuel Cycle Facility was spearheaded by the ANL Chemical Engineering Division, it was a major interdivisional cooperative effort involving the Reactor Engineering, Metallurgy, and other divisions. Within CEN, the principal engineers handling the project were Milt Levenson, John Schraidt, George Bernstein, Johan Graae, Les Coleman, and Don Hampson. Many others, including Tom Eckels, Dick Malecha, Phil Fineman, Al Chilenskas, and Jim Hesson were also involved. In fact, it seemed as if almost everybody in the Division sooner or later played some role in the project. There was a great deal of traveling between Chicago and Idaho during this period, and several CEN personnel spent months or even a year or more in residence at the site.

As shown in Fig. 2-14, the plant consisted of two large shielded cells and associated equipment. The overall building dimensions were 135 by 170 ft. The rectangular cell on the left was a rather conventional shielded "cave" with an air atmosphere, and it was equipped with the usual windows, manipulators, cranes, *etc.* This cell was used for partial disassembly of the fuel subassemblies and assembly of new ones, as well as the canning of scrap and equipment maintenance.

The second cell was circular and had a high-purity argon atmosphere to permit the handling of pyrophoric materials such as uranium, plutonium and sodium without other special precautions. The cell was an annular structure 72 ft in diameter. The process equipment was contained within the shielded annulus, the operations being conducted by personnel in the area outside the annulus through the use of shielding windows, manipulators and cranes. A shielded control

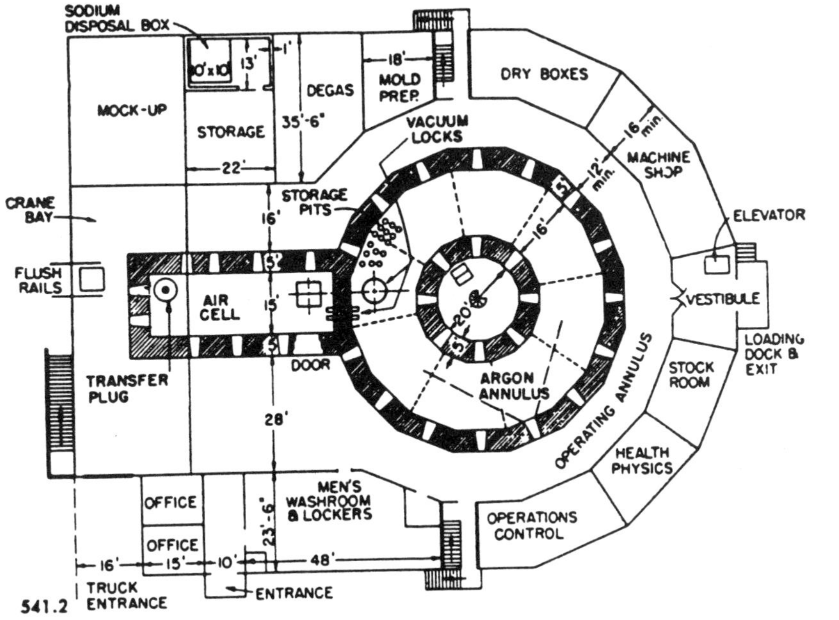

Fig. 2-14. EBR-II Fuel Cycle Facility

room at the center of the annulus permitted observation of the cell from the inside. A sub-cell was used for various service activities.

The rather unorthodox circular cell design (attributed to Johan Graae) was based on several considerations, the principal one being that by mounting the manipulator and crane bridges to swing around a central pivot, all locations in the cell could be reached by any of these units without their interfering seriously with one another. The design also worked well for the process train where the discharged fuel came in through the transfer lock from the air cell, progressed through the various processing steps as it went around the circle, and left the cell through the same transfer lock ready for reassembly and insertion into the reactor.

A large development effort was required for this facility because of its unique features, some of which are illustrated by the sectional view in Fig. 2-15. Such simple equipment components as graphite bearings in motors had to be modified because the graphite behaved more like an abrasive than a lubricant in the ultra-dry atmosphere. Mineral insulation was used in electrical cables to avoid radiation damage. The shielding windows consisted of six thick layers of radiation-resistant (non-browning) glass, which were optically coupled by oil laminations between the layers. Since the inner layers were still expected to darken under the anticipated radiation levels of 10^6 R/h, they were designed for periodic replacement, and heavy steel shutters on the inside were used to protect the glass when the windows were not in use. High-intensity lighting was required due to attenuation of light by the windows. This was accomplished with 1-kW mercury-vapor lamps (72 in the argon cell and 24 in the air cell). The inside of the annulus was lined with galvanized steel, which was shot-peened to minimize diffusion of gases through the cell walls. A special purification system was needed to maintain the argon atmosphere at a water concentration below 5 ppm, with the oxygen below 100 ppm and the nitrogen at 5 vol% or less. (These stringently low levels were relaxed somewhat later when it was found that they caused metallic sodium to stick to metal surfaces.) These were just a few of the typical problems that required extensive design and development work.

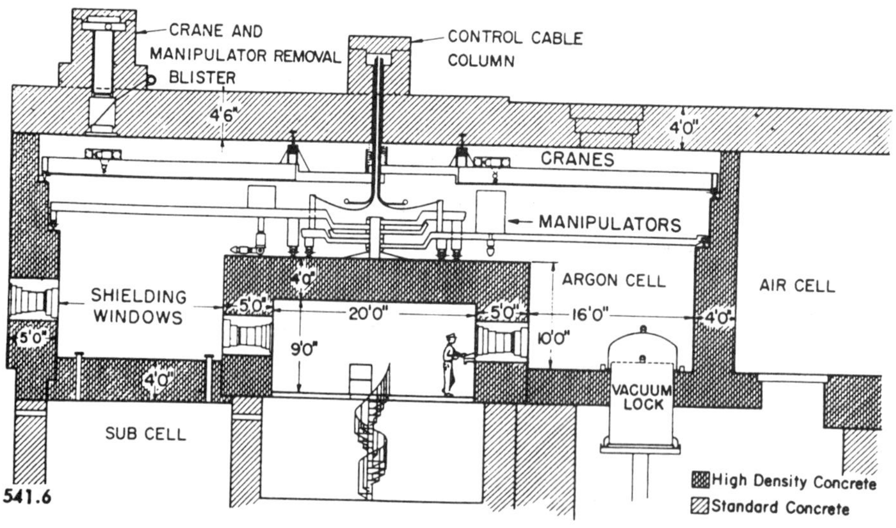

Fig. 2-15. EBR-II Fuel Cycle Facility (Sectional View)

Because fuel processing in the Fuel Cycle Facility did not begin until 1964, this subject is deferred to the next chapter.

Fluoride Volatility Processes

Brief reference was made in the previous chapter to some preliminary studies that were conducted on fluoride volatility processes for the recovery of uranium from spent reactor fuels. This work was expanded into a major CEN program in the 1950s, and several new staff personnel who had previous experience in fluorine chemistry (Glenn Schnizlein, Bob Steunenberg, Larry Stein, and Roger Jarry) were added to the staff. Also, Joe Katz, Herb Hyman, and Irv Sheft of the Chemistry Division (CHM) continued to participate in the program, mainly in an advisory capacity, for several years.

When Schnizlein arrived on the scene, one couldn't help being impressed by his height—6 ft 8 in. One day in the laboratory, he demonstrated that with just a little stretching he could achieve a seven-foot reach between his fingertips. He also had a craggy Lincolnesque visage, and, later in life, he appeared in many parades and other public events around DuPage County in a top hat and tails as Abraham Lincoln, along with his wife, Lois, as Mary Todd.

Jack Fischer, who joined the volatility group soon after arriving at ANL, was another one of the more memorable CEN staff personalities. He was a highly competent physical chemist and turned out a lot of excellent work, but frequently had an abrasive manner. Although the technicians or assistants who worked for him didn't particularly appreciate that trait, they saw a certain humor in the situation. He had a short fuse, and one day when a driver sat in a car blocking the crosswalk in front of the building, Jack, not saying a word, opened the back door, slid across the seat, exited from the other back door, and proceeded on his way, leaving both doors open, much to the amusement of bystanders. On another occasion, during an American Chemical Society meeting in Minneapolis, he curbed a city bus that had cut him off in the traffic, leaving the bus driver dumfounded. In spite of his idiosyncrasies, Jack had many good friends in the Division, and, unlike many people with his personality traits, he could often laugh at himself.

The major emphasis of the fluoride volatility program at that time was on the recovery of enriched uranium from metallic fuels, using a strong fluorinating agent to convert the uranium to UF_6, which is a volatile compound. The need to recover plutonium, as well as uranium, was recognized, but it was not a top priority. A major advantage claimed for the process was that the UF_6 product could be returned directly to a diffusion plant for reenrichment, thereby avoiding a series of chemical conversion steps that are required by solvent extraction or other aqueous processes. By the same token, UF_6 is easily reduced to UF_4, which is the usual precursor for the metal-production process where uranium metal is the desired product.

The choice of a fluorinating agent involved several considerations. Elemental fluorine gas converts metallic uranium to the hexafluoride, but the highly exothermic reaction and the absence of a condensed phase to remove the heat make temperature control difficult. The following halogen fluorides in Table 2-3 were considered for the fluorination step.

Table 2-3. Fluorinating Agents

Fluorinating Agent	Boiling Point, °C
Chlorine monofluoride (ClF)	-101
Chlorine trifluoride (ClF_3)	12
Bromine trifluoride (BrF_3)	126
Bromine pentafluoride (BrF_5)	41

The bromine fluorides were selected for the work at CEN because they are liquids at convenient working temperatures. Soon after work on the fluoride volatility process was started at ANL, similar programs were initiated at Brookhaven (BNL) and Oak Ridge (ORNL). The Brookhaven people, like those at ANL, selected the bromine fluorides as the fluorinating agents, while the Oak Ridge workers preferred the chlorine fluorides.

Although the reactions of liquid halogen fluorides with uranium are somewhat less energetic thermochemically than those of fluorine, they, being condensed phases, pack a very large amount of fluorinating power into a small volume. Like fluorine, they must be handled in materials such as nickel, Monel, and fluorinated plastics such as Teflon® (polytetrafluoroethylene) or Fluorothene®, also known as Kel-F (primarily polychlorotrifluoroethylene). Fluorothene and Kel-F were especially useful in laboratory experiments because they are somewhat transparent and were frequently used as cold traps and sight glasses. In corrosion studies, nickel, Monel, and K-Monel showed excellent resistance to BrF_3 attack. Inconel and Duranickel were also satisfactory. Stainless steel, low-carbon steel, and aluminum were used in some less critical applications. All metals that withstand fluorine attack do so by forming a protective layer of fluoride on their surfaces. The unavailability of suitable valves for fluorine and halogen fluorides was a problem in the early stages of the program; this difficulty was overcome by contracting with Hoke Incorporated to manufacture valves made of nickel with Teflon® gaskets, which were purchased in large quantities. Some of those special valves may still be in use in Bldg. D-205. Several other special types of apparatus, including pressure gauges with Monel Bourdon tubes, differential pressure transmitters, and thermal flowmeters were procured or developed for the fluoride volatility work.

Elemental fluorine is normally received in full-size cylinders containing the compressed gas at 400 psi. With the proper equipment and know-how, high-pressure fluorine can be handled safely, but it demands a great deal of respect. Opening the valve on those cylinders was always a tricky operation, because the valve tended to stick and could not be throttled easily. The result was a sudden burst of 400-psi fluorine, which can burn its way through many metals and most other materials. Old-timers from university and industrial laboratories had a practice of keeping their bare hand on the valve bonnet, so if anything went awry they could detect the heat immediately and shut it off. That was not an acceptable practice at ANL, where the cylinder was enclosed in a heavy steel box, and the valve was operated remotely by a steel rod that extended through the top of the box and terminated with a T-shaped handle. On occasion, when a cylinder was opened a lead gasket between the cylinder outlet and a fluorine supply line would fail, releasing the entire contents of the cylinder. The nickel tubing used to pipe high-pressure fluorine required careful handling. A spot of grease or oil could react with the fluorine, creating a small hot spot that caused the tubing itself to burn in the fluorine, producing flame and molten metal droplets. The people building the fluoride volatility pilot plant discovered that even sharp bends in the tubing can cause a fire. Hal Feder was always greatly concerned about any release of fluorine, and he could detect its pungent, chlorine-like odor long before anybody else could. Bob Steunenberg once unintentionally discovered that a release of just a few milliliters of fluorine gas in a hood in G-134 would bring Hal roaring out of his office "I smell fluorine!" at the end of A-Wing within five minutes or so.

The halogen fluorides are also hazardous materials—they react explosively with organics and many other materials. On a couple of occasions in G-134 a very small amount of BrF_3 was pumped accidentally into a rubber vacuum-pump line, and when it contacted the rubber hose it detonated with a sound like a shotgun blast. When the original BrF_3 fluoride volatility pilot plant was being built in G-134, a tradesman would occasionally drop a pipe or other large object on the floor behind someone working with halogen fluorides in a Blickman hood, shattering his nerves. Metal equipment used to contain fluorine and/or the halogen fluorides was degreased meticulously, dried, and prefluorinated carefully both for safety reasons and for prevention of volatile fluoride loss by reduction on the container walls.

Safety was a major concern because exposure to the bromine fluorides, as well as fluorine and bromine themselves, can result in very serious burns. Early in the program, members of the engineering and research groups were sent into Chicago and fitted with leather jackets and pants as protective clothing. These were used together with gauntlet leather gloves and face shields during much of the work. At the time, the medical profession had only limited experience in handling burns from fluorine, hydrogen fluoride, and the halogen fluorides because they had not yet been used widely in industry. After much discussion, a decision was made that a saturated magnesium sulfate ($MgSO_4$) solution was the best first aid measure, and large bottles of this solution were located in the areas where these materials were in use.

When Martin Steindler first joined the Division and was being shown around the laboratories, a few people in their protective garb were on their hands and knees on the floor in Laboratory G-134 pounding on small samples of frozen BrF_3 to see if they would detonate. That science-fiction-like scene may be the reason he went to work in Hal Feder's group in another program initially, but he eventually became the head honcho and leading expert in the CEN fluoride volatility research. Notwithstanding the large number of people in this program at ANL and the wide variety of operations, nobody was ever injured by these materials.

The basic reaction for the fluorination of uranium by BrF_3 is

$$2\ BrF_3 + U \rightarrow UF_6 + Br_2$$

Nearly all the fission products are also converted to the fluorides. Most of them form solid, non-volatile fluorides, some form very volatile fluorides, and two or three form fluorides having vapor pressures not far different from that of UF_6, which can then be separated from them by fractional distillation. (Because UF_6 has a triple point of 64°C at 1137 torr, a small pressure is required to maintain it in the liquid state.) The principal fission-product species that may be present after the fluorination are shown in Table 2-4.

The only fission-product fluorides having vapor pressures close to that of UF_6 are those of As, I, Mo, Tc, and Te. The vapor pressure of PuF_6 (b.p., 62°C at 1 atm) is near that of UF_6, which has a sublimation point of 56.5°C at 1 atm. The free energy relationships, however, are such that BrF_3 converts the uranium to UF_6 and the plutonium to PuF_4, which remains with the non-volatile fluorides.

The primary fluorination reaction of uranium with BrF_3 is complicated by the fact that bromine product reacts with the BrF_3 to form BrF, which exists as a gas in a mobile equilibrium with the BrF_3-Br_2 mixture:

$$BrF_3 + Br_2 \leftrightarrow 3\ BrF$$

At the time this work was being done, little was known about BrF, the only evidence for its existence being some rather obscure spectroscopic data in the literature. Pressure measurements and spectrophotometric studies of BrF_3-Br_2 mixtures by Bob Steunenberg and George Redding in CEN showed that the above reaction occurs, and that the equilibrium reaction produces sufficient BrF to be of

Table 2-4. Volatilities of Fission-Product Fluorides

Non-volatile Fluorides			Volatile Fluorides (b.p., °C)			
AgF	CsF	SnF_2	AsF_3	63	RuF_5	313
BaF_2	LaF_3	SnF_4	AsF_5	-53	SbF_3	319
CdF_2	$(RE)F_3$[a]	SrF_2	IF_5	100	SbF_5	150
CeF_3	RhF_3	YF_3	IF_7	4[b]	TeF_4	284
CeF_4		ZrF_4	MoF_6	35	Te_2F_{10}	59
			NbF_5	229	TeF_6	-39[b]
			$Tc\ F_6$	55		

[a] Rare earth fluoride.
[b] Sublimation point.

concern in a process. Larry Stein in the Chemistry Division (CHM) later refined these results with further experimental studies. Although the generation of BrF was of concern, this reaction can be suppressed by the addition of fluorine or BrF_5, and a step of this type was incorporated into some of the fluoride volatility process flowsheets.

The first version of the fluoride volatility process to be investigated at ANL was designed for metallic fuels and utilized BrF_3 as the fluorinating agent. The basic process was fairly simple, as illustrated in Fig. 2-16. The metallic fuel slugs were dissolved in BrF_3, the volatile UF_6 product and fission-product fluorides were collected by condensation, and then separated by fractional distillation. The non-volatile fission products and the PuF_4 remained in the dissolver vessel as solids. Two options were invoked for recovery of the plutonium. One was to fluorinate the dissolver residue with elemental fluorine and collect the volatilized PuF_6. The other option was to dissolve the residue in a small amount of aqueous solution for recovery in an existing solvent-extraction plant.

This version of the volatility process was investigated both in the laboratory and in pilot-plant studies. The actual process was considerably more complex than shown in Fig. 2-16 because of numerous recycle and refluorination steps. Pilot-plant runs with irradiated fuel slugs gave an excellent fission-product decontamination factor greater than 10^8. No effort was made to recover the plutonium, since the main objective was to recover the enriched uranium. These results showed that the fluoride volatility process had the potential for excellent recovery and decontamination of enriched uranium from metallic fuels. At ORNL, E. L. Nicholson published a study indicating that fluoride volatility processes might have a significant economic advantage over aqueous processes, but that they were not accepted as a main-line process for two major reasons: (1) Large-scale Purex plants already in operation could do the same job, and also had the built-in facilities for plutonium recovery. (2) It was beginning to become apparent that civilian power reactors would use oxide, rather than metal fuels.

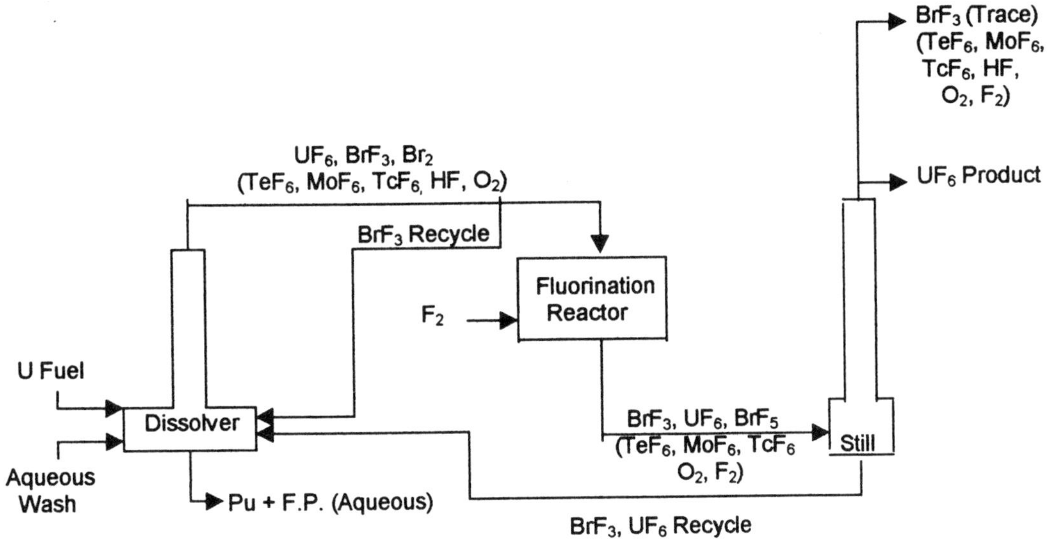

This simplified flow diagram shows a fluoride volatility process for recovering decontaminated uranium as the hexafluoride from irradiated uranium metal fuel. The fuel slugs are declad mechanically or chemically and dissolved in BrF_3, which comes from a recycle stream. This stream also contains UF_6, which increases the dissolution rate of the uranium. The main products from the dissolver are UF_6, BrF_3, Br_2, and BrF. The TeF_6 is a fission-product fluoride; HF and O_2 may be present as trace impurities. A fractionating column on the dissolver permits recovery of some of the BrF_3 for direct recycle. The other fraction from this column is treated with F_2 to reconvert the Br_2 and BrF to BrF_3 and a small amount of BrF_5. The resulting mixture enters a fractional distillation column, which separates the UF_6 product from the more volatile impurities and the BrF_3, which is recycled to the dissolver with some of the UF_6. During the dissolution, plutonium is converted to solid PuF_4, which remains in the dissolver with the non-volatile fission-product-fluoride waste. The plutonium is recovered by an aqueous wash for recovery and purification in a solvent-extraction plant. Later versions of this process included treatment of this residue with F_2 to recover the plutonium as volatile PuF_6.

Fig. 2-16. Bromine Trifluoride Process

Enthusiasm for processes involving the dissolution of metallic uranium in BrF_3 was also dampened somewhat by an ingenious set of experiments conducted by Larry Stein, prior to his transfer to the Chemistry Division. The highly exothermic nature of the reaction was recognized, and many studies had shown that the reaction was smooth and controllable as long as the metal was covered by liquid BrF_3 to conduct heat away from the reaction site. Larry set up an apparatus in which color motion pictures were taken of a metallic uranium sample exposed only to the interhalogen vapor. The metallic uranium sample became incandescent, and, except for the red bromine fumes, looked very much like a pat of butter melting in a microwave oven. Larry repeated these experiments with other halogen fluorides and obtained similar results.

Brookhaven National Laboratory (BNL) had a program, smaller but similar to the one at ANL, in which metallic uranium slugs were dissolved in BrF_3 and the UF_6 product was purified by distillation. An explosion occurred in their pilot-plant unit, blowing out a wall and throwing a panel board against one of the workers, who was injured. Several people from the volatility program at ANL aided in an investigation, and the general conclusion was that a uranium slug had hung up on a slug "feeler." The phenomenon that Stein had observed then occurred, dropping molten uranium into the liquid BrF_3, causing the explosion and fire.

On one occasion, a decision was made at ANL to conduct a mock incident drill in the fluoride volatility pilot plant in Laboratory H-126. To make it as realistic as possible, the ANL emergency units, including Health Services, were given very limited, if any, advance notice of the drill, and only a half dozen or so people in CEN were informed of it. Bob Kessie was to play the role of a victim who needed to be rescued from one of the upper levels of H-126, and a couple of smoke bombs were activated to add realism. The high-bay area was filled with smoke, which nearly eliminated the light and visibility. The "Dial 13" system was activated.

Fire trucks arrived and the building was evacuated. The ANL physician from Health Services arrived *via* a high-speed ambulance ride and was said to have been in a state of near shock. Meanwhile, Kessie, who is very near-sighted, was wearing a gas mask with non-prescription lenses, and, not being able to see a thing, came close to falling over the railing. The CEN Fire Brigade was having similar problems. After a while, the building ventilation system cleared the smoke and things returned to normal except for some frayed tempers. The drill actually proved to have been quite useful in that it revealed a number of improvements that were needed both in the CEN emergency planning and in the operation of the ANL site-wide emergency communications procedures.

In spite of the lack of an immediate application, the work on fluoride volatility processes for metallic fuels generated a large body of useful information on the basic chemistry and the technology of dry processes using fluorine. Some of the individuals who worked on the chemistry of the process included Jim Bingle, Ron Breyne, Octave DuTemple, Jack Fischer, Lee Gaumer, Bob Hildebrandt, Herb Hyman, Milt Levenson, Bob Liimatainen, Ray Long (Ph.D. student), Walt Ludewig, Bill Mecham, Glenn Schnizlein, Paul Seufzer, Irv Sheft, Bill Shinn, Dave Steidl, Larry Stein, Verne Trevorrow, Homer Tyler, and Warren Wade. The Group Leader was first Joe Katz, followed by Dr. Vogel, and then Bob Steunenberg. The pilot-plant personnel included Lee Gaumer, Jeff Goring (Union Carbide Corp.), Bob Kessie, Bob Liimatainen, Walt Rodger, Wally Seefeldt, and Jonathan Thigpen. Milt Levenson was the first Group Leader, followed by Bill Mecham.

In the mid-1950s, work was started on the recovery of plutonium by fluorination of the dissolver residues and other materials to produce PuF_6. Although this compound was known to exist and some information was available on its properties, it was clear that a significant research effort would be required to obtain the definitive data needed to consider its use in a practical process. Research on PuF_6 was a challenging task because of its marginal stability with respect to the reaction

$$PuF_6 \leftrightarrow PuF_4 + F_2$$

Plutonium hexafluoride (PuF_6) is very readily reduced and can be handled only in scrupulously clean, prefluorinated metals such as nickel. When PuF_6 is stored in a vessel, this disproportionation reaction is enhanced by the alpha activity of the plutonium. Previous research on the properties of PuF_6 had been conducted by Art Tevebaugh and others at Ames Laboratory and by John Malm and Bernie Weinstock of the ANL Chemistry Division. Al Florin at Los Alamos was the first investigator to carry out a systematic research study on the subject. The CEN researchers concentrated more on the process-oriented problems. Martin Steindler and Dave Steidl performed a careful investigation of the fluorination rates of PuF_4 by fluorine and continued on with a variety of studies. Max Adams looked at the thermal stability of the material and methods for performing quantitative transfers. Sy Vogler also contributed to this work. These efforts expanded into several other studies, including the fluorination of oxides, which laid the groundwork for fluoride volatility processes aimed at oxide reactor fuels.

Also, in the mid-1950s, an interest developed in the use of molten fluoride salts as media for processing various kinds of enriched uranium alloy and oxide fuels in which plutonium was not produced in sufficient amounts to warrant its recovery. These fuels generally consisted of U-Al and U-Zr alloys, UO_2 clad with stainless steel, and molten fluoride salts containing the lower-valent uranium fluorides. In one such process, U-Zr alloy fuel elements were immersed in a NaF-ZrF_4 melt at 600°C. Anhydrous HF was then bubbled through the melt, converting both the uranium and the zirconium to the soluble tetrafluorides. Fluorine was passed through the melt to convert the uranium to UF_6, which was vaporized and collected. This product could then be decontaminated by distillation or by selective sorption-desorption cycles using solid NaF beds. Argonne and ORNL both worked on this process concept, with ANL using the distillation step while ORNL preferred the sorption-desorption approach. Both fluorine and the halogen fluorides were investigated as fluorinating agents. Various melt compositions were also studied.

As might be expected, severe corrosion problems arose in the molten salt processes. Graphite and a number of fluorine-resistant metals were used as the container material, and some work was done at ANL on a "frozen wall" approach. In pilot-plant experiments at ANL, two different dissolvers were used. The first, Mark I, was made of A-nickel, and the second, Mark II, was low-carbon nickel. These runs included only the dissolution and fluorination steps, since the decontamination of UF_6 by distillation had been demonstrated earlier. The pilot plant was operated with 200 lb of NaF-ZrF_4 molten salt at 600°C; the fuel charged to the system was 14 lb of Zr-1 wt% U alloy. In six hours, 90% of the metal had dissolved. After three more hours of HF sparging, the dissolution was complete, and 99.5% of the uranium was recovered after 1.3 h of fluorination with BrF_5. The personnel most involved with the molten salt laboratory and pilot-plant work were Jack Fischer, Jeff Goring, Bob Kessie, Bob Liimatainen, Bill Mecham, Walt Rodger, and Sy Vogler. Wally Seefeldt did an outstanding job of coping with the materials problems posed by this process. Bob Steunenberg was the Group Leader, and, as with the other development programs, Dr. Lawroski and Dr. Vogel kept in close touch with this project and made many useful suggestions.

Work on fluoride volatility processes extended on into the 1960s; this continuing effort is covered in the next chapter.

Feed Materials Processing

CONVERSION OF URANYL NITRATE SOLUTION TO URANIUM OXIDE

One of the steps in the nuclear fuel cycle entails the conversion of uranium nitrate solutions from solvent-extraction purification processes to solid uranium trioxide (UO_3). In the past, this conversion had involved a labor-intensive batch process in which molten uranyl nitrate hexahydrate [$UO_2(NO_3)_2{\cdot}6H_2O$)] was heated in agitated pots. A continuous process employing a fluidized bed was developed and demonstrated on a pilot-plant scale by a group at CEN. The uranium nitrate solution was sprayed into a bed of UO_3 that was agitated by air flow and maintained at a temperature of 300-350°C. Under these conditions, the liquid feed dispersed itself over the particle surfaces and decomposed to the UO_3 product, which was withdrawn from the bottom of the bed to maintain a constant bed level. The off-gases were passed through filters or cyclone separators to recover entrained uranium oxide particles. The equipment used for this work was similar to that described later in a discussion of waste calcination in fluidized beds.

The feasibility of the process was demonstrated by a large number of development runs in which several thousand pounds of UO_3 was produced in a 6-in.-diameter stainless steel calciner. The capacity of the equipment was about 100 lb/h. During the runs, many useful data were collected on the effects of operating variables. This work was followed by scale-up studies at the Mallinkrodt Chemical Works. The principal investigators in the program at CEN were Dr. Lawroski, Al Jonke, John Loeding, Ed Petkus, and Rollin Taecker.

PRODUCTION OF REFINED URANIUM HEXAFLUORIDE

At the suggestion of Dr. Lawroski, a group of CEN staff members began to look into the possibility of eliminating some of the purification steps required to convert uranium ore concentrates into refined UF_6 suitable for direct introduction into the diffusion plants for uranium isotope separation. The compositions and physical properties of the ore concentrates varied from plant to plant, depending on the type of uranium ore and the technology used to recover the uranium values. Two general types of concentration processes were in use at the time: acid leach and carbonate leach. In some cases a prior salt roast was used to convert the uranium to a soluble form. The ore concentrates were produced by precipitating diuranate from the leach liquors with ammonium or sodium hydroxide. The precipitates were then filtered, dried, and calcined. The concentrates consisted primarily of uranium oxides or diuranates (*e.g.*, $Na_2U_2O_7$), assaying at least 70% as U_3O_8 in most cases. The major impurities were oxides of the gangue elements, *e.g.*, V, P, Mo, S, Fe, Na, Cu, Ni, Pb, Bi, Sb, As, Sb, Ca, Si, Al, Cr, Mg, and Mn.

In the existing feed materials plants, various purification methods such as solvent extraction, ion exchange, and selective precipitation were used to produce refined uranium oxides. These oxides were then reduced to UO_2 with hydrogen (or cracked ammonia), converted to UF_4 with anhydrous HF, and then treated with elemental fluorine to produce UF_6. The overall fluorination sequence is:

$$UO_3 + H_2 \rightarrow UO_2 + H_2O \text{ or } U_3O_8 + 2\,H_2 \rightarrow UO_2 + 2\,H_2O$$

$$UO_2 + 4\,HF \rightarrow UF_4 + 2\,H_2O$$

$$UF_4 + F_2 \rightarrow UF_6$$

Some of the uranium oxides in the first two reactions may be in the form of uranates. Several types of solid-gas contacting equipment, such as vibrating-tray reactors, horizontal screw-feed reactors, and fluidized beds were being used by the processing plants for these operations.

The process proposed by CEN, illustrated in Fig. 2-17, eliminated the chemical purification steps and introduced the ore concentrate directly into the hydrogen reduction, hydrofluorination, and fluorination sequence, using continuous fluidized-bed contactors for all the steps. These steps removed most of the gangue elements, and any remaining impurities were eliminated by fractional distillation of the UF_6 product. This process did, however, require a preliminary size preparation step to produce a material

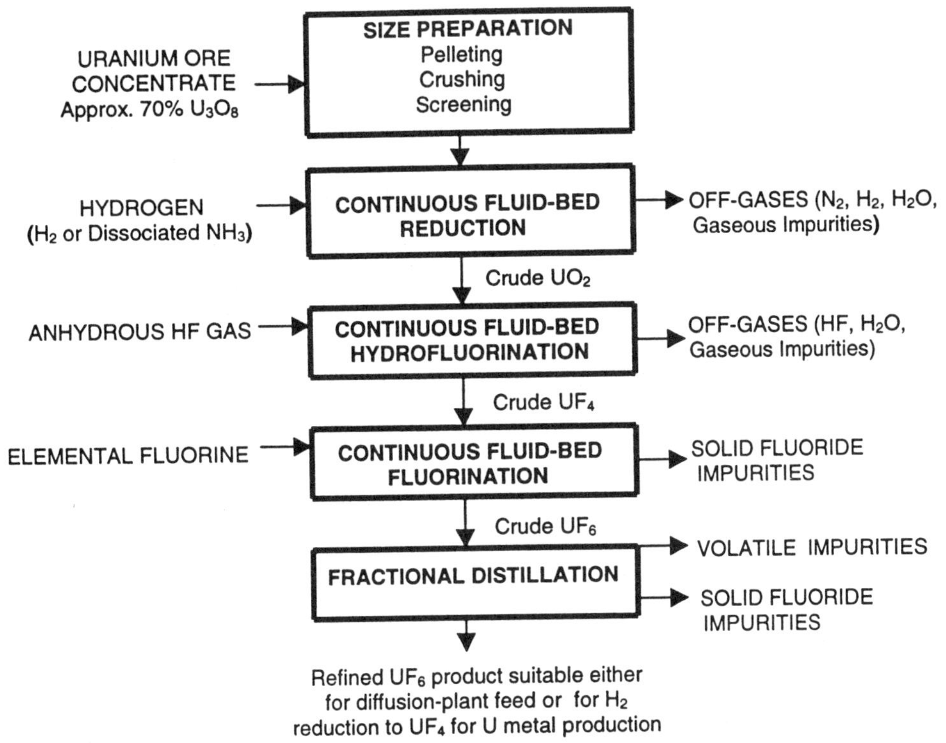

The physical and chemical characteristics of the uranium ore concentrates can vary considerably, depending on the source of the ore and the concentration process. The uranium may be present in the form of uranic oxide (UO_3), urano-uranic oxide (U_3O_8), or diuranates such as $Na_2U_2O_7$, and normally assays at 70 wt% or more as U_3O_8. To convert the material to a particle-size range suitable for use in fluidized beds, coarse material is crushed or ground, and fine material is pelletized or briquetted. Hydrogen or cracked ammonia is used to convert the uranium oxides to UO_2 in a fluidized bed, and then anhydrous hydrogen fluoride (HF) is passed through the bed to convert the UO_2 to UF_4. These two steps remove some troublesome impurities such as silicon, boron, and sulfates, which appear in the off-gases. The crude UF_4 is treated with elemental fluorine to convert it to volatile UF_6, which is collected from the off-gas. Most of the gangue-element fluorides are non-volatile and remain in the bed. Any volatile fluoride impurities are then removed from the uranium by fractional distillation. The reduction step was operated at about 575°C and the fluorination step around 450°C.

Fig. 2-17. Production of Refined Uranium Hexafluoride from Ore Concentrates suitable for fluidization. The sizing procedures depended on the nature of the particular ore concentrate, but they included combinations of crushing, screening, briquetting, and pelleting.

Initial studies of the conversion of uranium oxides to UF_4 were conducted by Norm Levitz, Al Jonke, Al Litty (General Chemical Co.), and Ed Petkus, using single-stage fluid beds 3 and 6 in. in diameter, and additional demonstrations of the process were made in a pilot plant consisting of a 5-in.-diameter, four-stage stainless steel reduction reactor connected in series with a 6-in.-diameter, five-stage Monel hydrofluorination reactor. The two reactors operated at 575°C and 450°C, respectively. These and other similar tests were, in general, very satisfactory, and showed that the rapid reactions in the fluid beds would permit the use of single-stage beds.

John Vogel, Bob Steunenberg, and Oscar Sandus worked on the fluorination of UF_4 to UF_6. Although plant-scale tower fluorinators were in general use for this conversion, it was doubtful whether they would be suitable for the impure UF_4 because the tower reactors use a high-temperature fluorine flame reaction that would most likely result in product sintering and caking.

Initial studies used a 1-in.-diameter Monel reactor with a bed temperature of about 450°C. Different concentrations of nitrogen were used to dilute the fluorine, and CaF_2 was added to maintain the bed of solids as the UF_4 was consumed. A pilot-plant fluorinator consisting of a Monel 2.5-in. reactor topped by a 6-in. disengaging section was operated in conjunction with a condenser that was 10 ft tall and 3 in. in diameter. The fluorinator was operated at temperatures of 350-500°C and gas velocities of about 0.5 ft/sec. The results of the tests were considered to be satisfactory, and the only significant impurities in the UF_6 were vanadium and molybdenum. These two impurities were not unexpected because they form the volatile compounds VOF_3, VF_5, and MoF_6, which have vapor pressures that would cause them to condense with the UF_6. They can, fortunately, be separated from the UF_6 by fractional distillation.

The UF_6 distillation studies were performed by Bill Mecham, Bob Liimatainen, Bob Kessie, and Verne Trevorrow. Very little information on the vapor pressures of VOF_3 and VF_5 and no vapor-liquid equilibrium data on the systems VOF_5-UF_6, VF_5-UF_6, and MoF_6-UF_6 were available in the literature. The vapor-pressure data were generated by CEN. Workers at the Allied Chemical and Dye Corporation, who were interested in using the process, obtained vapor-liquid equilibrium data. The solubility of VOF_3 in UF_6 was found to be low, indicating that periodic removal of the VOF_3 from the UF_6 condenser might be necessary. Preliminary pilot-plant experiments were conducted with a 2-in.-diameter, 16-ft-long distillation column. The nickel column consisted of a still pot, tower, condenser, overhead receiver, and charging and sampling facilities. The runs involved approximately 100-pound charges of UF_6 spiked with the molybdenum and vanadium impurities. Overall results from the UF_6 distillation studies indicated that it would be a practical process operation.

The ANL results were augmented by work performed under the direction of Sy Smiley at the Oak Ridge Diffusion Plant (Y-12) on the fluorination of crude UF_4 and on UF_6 distillation.

A number of additional CEN people were involved in this project, including Dr. Lawroski, Dr. Vogel, and Dr. Rodger, who provided technical guidance, Wally Seefeldt, who investigated the corrosion aspects of the process, and the Analytical Chemistry Laboratory. This turned out to be a highly successful program. The process was placed in a full-scale plant operation at Metropolis, Illinois, by Allied Chemical, which has been a major UF_6 production facility for about 40 years.

Although this particular program was aimed at the recovery of UF_6 from ore concentrates, the technology involved in any one or more of the individual process steps is directly applicable to various conversions of high-purity materials that might be needed in other types of fuel-cycle operations that require chemical conversions.

Waste Treatment

INCINERATION OF RADIOACTIVE SOLID WASTES

One of the first major projects to be undertaken by CEN after the move to Site D was a pilot-plant demonstration of an incinerator for the disposal of low-level combustible radioactive waste materials generated at the Laboratory. These wastes consisted of materials such as paper, clothing, wood, filters, rubber gloves, *etc.* that had been discarded as DAW (dry active waste). The activity level in this material was typically in the range of 30-50 mR/hr.

This operation, which was conducted in Bldg. D-310 during the period of 1950-1953, had three objectives:

1. To design and construct an incinerator that would handle the combustible waste generated at ANL
2. To obtain operating data that would point to design modifications for increased efficiency
3. To provide a prototype for any future incinerator projects

The entire installation was heavily instrumented because it had the dual purpose of generating pilot-plant data and serving as a routine production facility.

A. D. Little of Cambridge, MA, was contracted to provide the basic design of this pilot plant. The primary unit was the incinerator furnace, which was made of Type 330 stainless steel, and consisted of a cylindrical section 5 ft in diameter and about 10 ft high atop a 4-ft-long conical bottom that tapered to a diameter of 8 in. at the bottom. The furnace contained a cast iron grate system, and its top was equipped with a waste-charging system of interlocking doors. The combustion system consisted of three gas burners at the periphery of the furnace, and primary air was blown through four 4-in.-diameter ducts located 4 in. below the grate. Secondary air was introduced through four 1-in. tangential pipes 16 in. above the grate to generate turbulence for good mixing. The cone at the bottom contained water to collect the fly ash and terminated in a valve to release the wet material into collection bags.

The upper portion and top of the furnace were jacketed with a carbon steel shell to provide an annulus that served as a boiler for heat removal; the resulting steam was condensed in three cooling units in the fan loft. The steam-condensing capacity of these cooling units was the limiting factor on the throughput of the furnace. The operating temperatures of the furnace ranged between 850 and 1450°F, with an average of about 1150°F.

An extensive off-gas treatment system was used to prevent the escape of any radioactive particulates to the environment. This system included a Schreiber-Bartolucci vane plate washer to remove large particles, and a Pease-Anthony venturi followed by a Peabody scrubber. An AEC filter was used for final cleanup.

The installation operated for 20 months, incinerating 16,000 ft^3 of waste with a volume reduction of 95%. The normal throughput was 17 ft^3/hr. In one test, the operation was run continuously for 120 hr. The decontamination factor for the exhaust gas was 3×10^7, which resulted in an activity level lower than that of the outside ambient air. The reduction in volume decreased the cost of storing the solid waste by about 50%.

During the development and operation, a few problems such as plugging and caking were encountered, but these were not major and were corrected. As might be expected, corrosion was identified as a long-range problem that would require attention; some of this was attributed to the sulfate in the large amount of paper that was burned.

Don Hampson, Ed Hyken, and Walt Rodger had the major responsibility for this work. Herb Brown, O. Levison, Wilfred Pehl, and Don Hulet were the principal operators of the equipment; C. Bullinger provided design assistance and John Schilb was responsible for the analytical work.

This program was successful in meeting its three objectives; a report (ANL-5067) was issued in 1953; and the installation was dismantled in the mid-1950s to make room for other projects.

FLUIDIZED BED CALCINATION OF AQUEOUS WASTES

In a report issued by the AEC in 1957 (WASH-742), the inventory of high-level liquid wastes in tankage in the U.S. was stated to be greater than 60 million gallons, and another 10 million gallons was expected within the next two years. Most of this waste was from the Hanford plutonium production plants, and this method of storage was to have been a temporary stopgap measure. As we all know, these liquid wastes are still a problem and a subject of national embarrassment.

In the mid-1950s, the staff at CEN came up with the idea of using the Division's expertise in fluidized bed technology to investigate calcination as a method to convert the liquid wastes into a more innocuous solid material. This was a joint undertaking between CEN at ANL-E and the Phillips Petroleum Company at the Idaho site. The personnel involved in this program were Walt Rodger, Al Jonke, John Loeding, Bob Larsen, and Dr. Lawroski at ANL, and E. S. Grimmett, J. I. Stevens, and Charlie Stevenson at Phillips.

At the Idaho Chemical Processing Plant (ICPP), highly enriched uranium fuels were reprocessed to recover the fissionable material, with the principal high activity waste coming from the processing of MTR- (Materials Test Reactor-) type U-Al alloy fuels. The resulting acidic $Al(NO_3)_3$ solution was being stored underground in stainless steel tanks. Parallel efforts were conducted at the two sites, using fluidized bed calciners. The fluidized bed material was Al_2O_3 particles, although almost any inexpensive, inert solid could be used, as, for example, for a Purex waste, where little solid material would be generated during the calcination.

Four pilot-plant units were built and operated, two at ANL and two at Phillips. All were of a similar design, in which the fluidized bed was supported on a porous plate that also served as a gas distributor. Preheated air was directed upward through the bed, causing it to behave much like a vigorously boiling liquid. This mixing action provides excellent gas-solid contact and heat transfer to the vessel wall. The bed was maintained at temperatures in the general range of 400 to 500°C either by heating the vessel wall or by heat-transfer tubes within the bed. The liquid waste solution was injected into the bed by several spray nozzles around the periphery of the calciner in a horizontal plane near the bottom of the unit. As the liquid droplets contacted the bed particles, they were flash dried and deposited as the oxides. The strongly agitated bed provided sufficient attrition of the solid particles to produce new nuclei so as to prevent a continuing increase in the overall particle size. The product was withdrawn continuously from the bottom or from an overflow pipe. The calciner off-gases, which consisted mostly of water vapor and nitrogen oxides, were passed through high-efficiency filters or liquid scrubbers to remove entrained radioactive oxide dust. Special provisions were required to remove fission-product ruthenium, which forms volatile compounds under these conditions.

After some preliminary experiments with a 3-in.-diameter stainless steel unit, two 6-in. calciners were operated at ANL, one for inactive studies and the other for hot runs behind shielding. External electrical heaters maintained the bed temperature in both units, and two filters that could be blown back alternately were used for the off-gases. At Phillips, a 6-in. calciner, of a generally similar design as the ANL units, was used to process up to seven liters of $Al(NO_3)_3$ solution per hour, and a larger unit with a cross-sectional area of 4 ft^2 was designed to process up to 100 liters per hour.

The hot runs at ANL showed that all the fission products except ruthenium remained in the solid bed. Various measures for suppressing or handling the ruthenium activity, including the addition of CO to the fluidizing air, were investigated.

Operation of the larger pilot-plant unit at ICPP was so successful that the unit began to be used routinely as a plant for liquid waste disposal in 1963, and this operation continued until 1981 when a newer calciner came on line. During that 18-year period, this facility processed more than 4 million gallons of liquid waste into about 77,000 ft^3 of granular solids, thereby decreasing the waste volume by a factor of 7 to 10. This was the first facility in the world to convert radioactive liquid wastes into solids on a plant scale, and the achievement was recognized at a ceremony on March 17, 1994. The facility, currently operated by the Lockheed Idaho Technologies Co., was named by the American Nuclear Society as a National Historical Landmark. The early development of this process was one of the major achievements of CEN.

ION-EXCHANGE STUDIES

Operation of a research and development laboratory which handles a wide variety of radioactive materials inevitably produces large volumes of low-level aqueous waste. In Bldg. D-205, the drains from all the laboratory sinks empty into 1,500-gal retention tanks where the wastewater can be monitored to make certain that unacceptable levels of radioactivity are not sent to the ANL wastewater treatment plant. If the levels exceed the standard for normal disposal, the water is sent to a special treatment facility for cleanup. Safety rules forbid the dumping of radioactive material into laboratory sinks, so this is a backup precaution.

The removal of low levels of radioactive species from wastewater is, however, a more general problem, which was addressed by Gladys Swope and her co-workers in a study of cation-exchange methods. This study was aimed primarily at mixed fission-product activities in tap water. The exchange media were of the styrene-base sulfonic acid type, which are commonly used for water softening.

The results showed, in general, that 75-80% of the beta-gamma activity was removed up to the point of hardness (calcium and magnesium) breakthrough, which was about 260,000 gal per cubic foot of resin. Flow rates up to 10 gal/min per cubic foot of resin were achieved without loss of performance. As might be expected, strontium-90 was removed up to the point of calcium/magnesium breakthrough. Prior to breakthrough, the principal limitation on performance was due to cesium and ruthenium leakage; these activities were the limiting factor on the efficiency of the ion-exchange approach. Total rare earths were removed both before and after hardness breakthrough. The overall conclusion was that water containing low levels of fission-product activity requiring removal factors of 80% or less could be processed economically.

The disposal of radioactive liquid wastes became a routine operation in the 1950s and extended well into the 1960s. Some of the people involved in this work, which was also under Swope's direction, were Joe Harast, Bruce Kullen, Ken Brewer, Ruth Juvinall, and Carl Ryberg. Several different types of operations were performed, depending on the nature of the particular waste: evaporation and concentration, filtration, cation exchange, absorption in Vermiculite®, and neutralization. The quantities of liquid waste were typically in the range of about 30,000-40,000 gallons per month.

Extensive support from the Analytical Laboratory was required both for the research studies and for the routine waste-processing operation. Doug Krause was in charge of the analytical work at that time, and the analyses were performed by Ruth Juvinall and Arden Schilb. Elton Turk provided technical advice.

When Gladys Swope left ANL later on, she established a consulting business in the Chicago area on water treatment.

GAMMA IRRADIATION FACILITY

Scientists and engineers in the early 1950s felt that there must be some practical uses for the highly radioactive fission products that were being produced in reactors and came up with a variety of suggestions. One of these was irradiation with gamma rays to preserve food, since gammas kill bacteria, but induce no activity into the irradiated material. To pursue this and other studies of the effect of gamma radiation on materials in general, the Argonne High Level Gamma Irradiation Facility was constructed as an underground annex on the south side of Building D-310 (Fig. 2-18). The facility was a water-filled canal 28-ft long, 14-ft wide, and 24-ft deep, in which spent fuel rods from the Materials Test Reactor (MTR) and CP-5 were contained in racks at the bottom. The level of the water, which served both as shielding and as a coolant, varied from 16 to 20 ft. The water was demineralized and circulated constantly through an ion exchanger at 800 gallons per hour at a temperature of 76°F. The facility was designed under the direction of Phil Fineman, and it went into operation on August 5, 1955.

There were three fuel-rod racks. One consisted of a honeycomb of 12 fuel rods with six 4.25-in.-diameter sample ports where flux intensities as high as 3.5×10^6 rad were achieved. The second one could accommodate a sample as large as 20 in. in diameter and 30 in. high. It was surrounded by eight fuel rods at least 150 days old, so the flux was much lower—about 15,000 rad/h. The third rack had the same size sample ports as the first one (4.25×28 in.) with four fuel rods that had been cooled even longer. The gamma dose intensity in this unit was about 200,000 rad/h. Provisions were made to rotate the samples in the first two racks. The oxidation rate of ferrous sulfate ($FeSO_4$) solution in dilute sulfuric acid was used to determine the radiation doses at various locations.

This facility was popular with visitors. When the room lights were turned off, the blue glow from the Cherenkov radiation was impressive. On display were also several food samples that had been irradiated and kept for several months, including some bananas that looked almost as good as new. These food-irradiation studies were supported by the Army Corps of Engineers and the Quartermaster Food and Container Institute for the Armed Forces. Some CEN personnel used this facility for research. Alfred Schneider used it for his Ph.D. thesis, and Martin Steindler and Dave Steidl carried on gamma decomposition studies on PuF_6. A few glass items that had been colored by the gamma radiation were given out

Fig. 2-18. Gamma Irradiation Facility in Building 310

as souvenirs after a Geiger counter was used to show the visitors that the glass was not radioactive.

Gladys Swope was also the principal operator of this facility under the direction of Walt Rodger. Gladys was a rather assertive individual, and she seemed to believe that she was immune to radiation. At one time, she objected to the rule that she must wear a dosimeter and film badge, but apparently decided that it wouldn't do any harm and would keep her supervisors happy. In that connection, one time she called her boss, Walt Rodger, on the phone and put him on hold when he answered; he wasn't a bit happy. Even today, that's not considered to be good protocol and Walt was livid.

THE "HOT ROCK"

Another approach to the use of fission-product radiation was investigated by mixing the radioactive fission products into concrete, which was cast in the form of a small, hollow cylinder. The concrete cylinder was surrounded by lead and concrete shielding, and the hollow core was designed to accept samples for irradiation testing. This device, which was fabricated by John Loeding, Ira Dillon, Ed Petkus, George Yasui, and Walt Rodger under the direction of Dr. Lawroski, was sent to the Massachusetts Institute of Technology for studies of food irradiation.

The possibility of using mixed fission-product activity for large-scale applications such as sewage treatment and heat sources has, in general, been unsuccessful for several reasons. Two major problems are the shielding requirements and the rapid decay rate of fission products at cooling times sufficiently short to provide the high energy output that is needed. Small amounts of a few specific isotopes, however, can be recovered and used for various purposes.

Fluidized Bed Technology

Fluidized beds were used extensively in the research and development programs in CEN. This technology was not new. It had been used widely in the petroleum industry and elsewhere, but CEN was a pioneer in adapting it to a wide variety of new applications in the nuclear fuel cycle. The diversity of these applications is illustrated by the following list of processes that were investigated in the 1950s:

1. Calcination of radioactive waste solutions to produce a much smaller volume of less innocuous solid waste for storage or further treatment
2. Denitration of uranium solutions to form solid uranium oxide
3. Conversion of U_3O_8 to UO_2
4. Conversion of UO_2 to UF_4
5. Conversion of UF_4 to UF_6
6. Combinations of the above steps to convert uranium ore concentrates to refined UF_6

Later on, in the 1960s, fluidization methods were used for even more applications, including direct fluorination of oxide reactor fuels, preparation of uranium carbide and nitride fuels, processing of uranium-zirconium alloy fuels, and control of sulfur emissions in the burning of coal by limestone additions.

A fluidized bed is a bed of granular solids through which an evenly distributed stream of gas is passed at a velocity sufficient to cause partial suspension of the particles. Such a bed takes on the general appearance of a vigorously boiling liquid. The free movement of the particles results in their continuous agitation and mixing. This movement, along with the large surface area of the particles, creates a highly favorable condition for gas-solid reactions.

Fluidized beds have several favorable features that are particularly useful for processing radioactive materials, where remote operations are required. Plant operations are simplified by the fact that the fluidized solids can be handled much like a liquid. The bed requires no internal mechanical moving parts, temperatures tend to be uniform, and heat transfer to the vessel walls is excellent due to the high thermal conductivity of the bed. Reaction rates are high because of the continuous mixing and the large gas-solid surface areas.

The disadvantages relate mostly to the particular application. The solid particles cannot be too fine or too coarse. Sizing operations on the starting material such as pelleting, briquetting, or crushing may be required. Caking can occur as a result of sintering or other types of agglomeration. Attrition or chemical conversion may produce fines that plug the off-gas filters. (Alternating blowback between two or more filters and the bed has been used to handle plugging in some instances.) Finally, as in any process equipment, corrosion and erosion are sources of potential problems.

Although most of the work with fluidized beds in CEN was related to one specific process or another, systematic engineering research studies were also conducted to investigate the basic characteristics of these systems. Bed size and geometry were important not only to the operating characteristics, but also to nuclear criticality considerations in some systems. Other factors such as gas dispersion techniques and flow rates, particle-size distribution, densities, methods of introducing liquids, and heating and cooling methods are important in establishing optimum equipment designs and operating conditions. When one considers the complexity of the factors that can arise in a fluidized bed operation, it is a tribute to the workers that CEN was so successful in applying this technology to such a wide variety of applications. Basic studies of fluidized beds continued into the 1960s.

Reactor Safety

Two programs designed to provide basic information on the safety aspects of various operations employed in nuclear technology were initiated by CEN in the mid-1950s. The first was aimed at fire and explosion hazards resulting from ignition of metals such as uranium, zirconium, plutonium, and thorium in air or oxygen atmospheres. The other program was directed toward reactions of metallic nuclear fuels with water that could occur in reactors as a result of nuclear transients or loss of cooling.

METAL-AIR REACTIONS

The initial studies were made with uranium and zirconium. Both the ignition behavior and the subsequent burning characteristics were investigated. Metal powders, turnings, foils, wires, and small solid bodies were ignited in air and oxygen atmospheres. Turnings were of special interest because fires had occurred frequently in machining operations. Although high specific area was established as a major factor in lowering the ignition temperature and increasing the burning propagation rate of uranium and zirconium, other variables that were studied included purity of the metal, alloying additives, surface and gas contaminants, and surface etching. During these studies, a complementary investigation was conducted on the fundamental kinetics of oxidation.

In 1958, a report (ANL-5974) was issued on the ignition behavior and kinetics of oxidation of uranium, zirconium, plutonium, thorium, and the binary alloys of each. Subsequent work was geared more toward the burning process after ignition had occurred. Electron-diffraction studies were performed on partially burned specimens to reveal the nature of the oxide layers on the metal surfaces.

Ignition and burning-propagation-rate experiments were performed with foils of binary plutonium alloys containing 2 at.% Al, Fe, C, Ni, Mn, and Cr. Manganese lowered the ignition temperature and aluminum raised it.

Dr. Vogel and Glenn Schnizlein were the principal investigators in this program; other workers included Jim Bingle, Don Fischer, Len Leibowitz, Larry Mishler, Phil Pizzolato, and Marv Tetenbaum.

METAL-WATER REACTIONS

A condenser-discharge technique was used to obtain fundamental reaction-rate data under conditions that could be expected to prevail during a serious accident in a nuclear reactor. Either a nuclear runaway or a loss of coolant in a water-cooled nuclear reactor could result in contact of very hot fuel and cladding metals with water or steam. In the condenser-discharge experiments, metal wires were quickly melted and dispersed in a water-filled cell by a surge current from a bank of capacitors. A surprisingly large amount of information can be obtained by this relatively simple technique. The energy input to the wire was used to calculate the initial metal temperature; the transient pressure indicated the reaction rate; the quantity of evolved hydrogen gave the extent of the reaction; and the particle size of the residue revealed the surface area exposed to the reaction. An analysis of the data based on known laws of metal oxidation and a simple heat-transfer model produced rate constants that could then be used for a detailed analysis of any other particular system. It is interesting to note that these calculations were made in cooperation with the Applied Mathematics Division (AMD) using an analog computer program. Capacitor-discharge experiments were conducted with uranium and zirconium, and the reaction rates in both cases showed parabolic behavior. At higher temperatures (2600-2700°C), where zirconium oxide reaches its melting point, the reaction became explosive.

A pressure-pulse method involving momentary contact of water vapor with molten metal in a crucible was also being developed to investigate metal-water reactions. Reliable techniques for using this method were still being developed at the end of the 1950s.

In-pile experiments were conducted to investigate metal-water reactions initiated either by a nuclear reactor transient or by a loss-of-coolant incident in the presence of an intense neutron and gamma radiation field. The initial work was done by inserting an autoclave containing the samples into the CP-5 reactor; subsequent tests were conducted in the TREAT (Transient Reactor Test) facility. Data were obtained on unclad uranium metal pins, Zircaloy-2-clad metal core pins, and oxide and cermet core pins jacketed with aluminum, stainless steel, and Zircaloy-2. The amount of metal reacted was obtained from hydrogen analyses, using a mass spectrometer, and the effects of the transients were evaluated by direct physical examination supplemented by metallographic methods.

The principal investigators in the metal-water program were Walt Rodger and Lou Baker. Bob Liimatainen, Don Mason, Peter Martin, and Ray Warchal did the condenser-discharge work, and Professor Martin Kilpatrick of the Illinois Institute of Technology served as a consultant. Dick Ivins, Marshall Deerwester, Bob Liimatainen, and Frank Testa were involved in the in-pile work.

Calorimetry

The calorimetry program in CEN began in about 1957 as a part of the pyrometallurgical research group under Hal Feder. At that time, the U.S. Bureau of Standards in Washington, the U.S. Bureau of Mines at Bartlesville, Oklahoma, and various universities were doing most of the calorimetric work in the U.S. It seemed a bit odd that an organization such as CEN would enter that field, but there was a good reason for it. Combustion calorimetry up to that time had normally been done with oxygen, and the realization struck that the use of fluorine, a much more powerful oxidizing agent, could extend the combustion method to include many materials that could not be burned satisfactorily with oxygen. Fluorine will normally convert all the elements in a substance to fluorides of their highest valence states in a clean reaction. Many of the materials that were investigated in this program, such as zirconium, were important in nuclear technology. Fluorine-combustion calorimetry was an attractive prospect, but it offered some difficult challenges in the handling of fluorine for this particular application. Nearly all the necessary technology and "know how" to meet these challenges were already available from the CEN fluoride volatility program.

The Division was fortunate in being able to hire Ward Hubbard to lead the effort. Ward, who was employed by the Bureau of Mines at the Bartlesville Petroleum Research Center at the time, was already a nationally recognized calorimetry expert when he came to Argonne. This expertise, coupled with the CEN fluorine technology, was ideal for this program. Ward was also one of the certifiable characters in the Division; he was always up to something. One of his first outside interests was learning to fly. Once he got his license, he liked to take the secretaries for rides, some of whom thought it had been a death-defying experience. Later on, he graduated to gliders and became quite successful in national competitions of high-performance sailplanes. Another personal quirk of Ward's had to do with food. Several people in the Division had gardens with the usual over-production of zucchini squash. Ward was always willing to take it off their hands, and nobody could figure out what anyone could do with that amount of squash. He also seemed to be the principal instigator of an ongoing weight-losing competition that went on for several years and involved various people, including John Ackerman and Paul Cunningham. The weighings were conducted on the platform scale at the end of H-Wing. It was rumored that the grand prize was to be a bottle of high-quality scotch, but nobody ever qualified for it. Jerry Johnson claims that Ward kept a personal weight-loss chart in which the first derivative of the curve was always zero.

The first calorimetry laboratory was set up in J-137. Provisions were made for both fluorine and oxygen combustions. The oxygen calorimeter was a platinum-lined, rotating unit so the products could be collected in a solution. The fluorine bomb was made of nickel. Members of the group at that time included Don Fredrickson, Elliot Greenberg, Ralph Nuttall (probably the quietest person ever to have inhabited Bldg. 205), Ed Rudzitis, Jack Settle, and Steve Wise. Larry Stein, who had transferred from CEN to the Chemistry Division in the mid-1950s, served as an advisor for the fluorine work.

Some of the early work involved calorimetric combustions of molybdenum in fluorine to form MoF_6. Sixteen combustions of oxygen with MoS_2 and equivalent mixtures of molybdenum and sulfur were used to obtain the enthalpy (heat) of formation of MoS_2. The heats of combustion of TiS_2 in oxygen and the heats of formation of ZrF_4 and BF_3 were measured by fluorination of the elements. In the 1960s, the scope of the calorimetry program expanded to include a wide variety of materials.

The Analytical Chemistry Laboratory

Throughout the history of the Chemical Engineering/Technology Division, one of the most vital parts of the organization has been the Analytical Chemistry Laboratory. Nearly every programmatic effort over the years has relied on their services in one way or another, and their contributions have not always been fully recognized in publications by the other groups, probably because analytical work often tended to be regarded only as a service. The group has sometimes been referred to as the "routine analytical laboratory," but many of the samples they have analyzed have been anything but routine due to interferences, matrix components that were difficult to dissolve, and other complications. Procedures often had to be modified or new ones developed to meet the particular need.

Doug Krause was the Group Leader when the Division moved to Bldg. D-205, and the people who were in the group during the 1950s are listed in Table 2-5. This unofficial list is based on the collective memories of a number of people, some technical reports, and various items in the *Argonne News* at the time, so it may not be complete. In addition, some individuals were occasionally assigned to other groups.

Alberta Hoover was the "dishwasher," situated in Laboratory A-101. Some of the individuals developed specialties; for example, Bob Schablaske became an X-ray specialist, and Myron Homa began to concentrate on gas chromatographic analyses.

The Analytical Chemistry Laboratory occupied almost all the laboratory space and all the office space in B-Wing, except the two rooms at the end, which were used for research by Carl Crouthamel and his group. Some of the offices were converted to laboratories for instruments such as the X-ray equipment. Because some of the samples to be analyzed were too "hot" to be handled in open laboratories, a "Junior Cave" was installed in G-102 where work could be done behind shielding. Bill Sovereign and John McCown and, later, Ray Popek were the principal operators of the Junior Cave. (The term "Junior Cave" could lead to confusion in that it was applied both to the facility in Bldg. D-205 to distinguish it from the Senior Cave in K-Wing, and to a remotely operated analytical setup used much for the same purpose at the Idaho site in connection with EBR-II).

Table 2-5. Analytical Chemistry Group in the 1950s

David Anthers	Beatrice Hjelti	Betty Reilly (Peterson)
Steve Banacek	Marilyn Hlavnicka	Mary Robinson
Jim Bingle	Jodie Hoekstra	Laury Ross
Lee Deutsch	Myron Homa	George Sato
Antoinette Engelkemeier	Miriam Jurlow	Bob Schablaske
Alice Essling	Ruth Juvinall	Arden Schilb
Douglas Fairgrieve	Gwen Kesser	John Schilb
Florence Ferry	Doug Krause	Chuck Seils
Stan Flikkema	Eugene Kucera	Bill Sovereign
Irene Fox	Gene McCloud (Kucera)	Bob Sweezer
Carol Garsky	John McCown	Jackie Williams
Ruth Hanna	Ray Popek	

The Analytical Group in the 1950s had a wide range of capabilities, particularly considering the state of the technology at the time. Much of the work involved radiation counting, using Geiger counters or a single-channel gamma analyzer for specific isotopes. X-ray diffraction and fluorescence were valuable tools for compound identification and semi-quantitative determinations of elements. Standard gravimetric and volumetric methods (often EDTA titrations) were used extensively. A Beckman Model B colorimeter and a Beckman Model DU spectrophotometer were available for quantitative determinations of certain elements. The gas chromatograph was used for a variety of analyses, including glove-box atmospheres.

Because most of the research and development programs in the Division were concerned with the processing of spent reactor fuels, a large fraction of the analytical work involved the actinide elements, particularly uranium and plutonium, and a wide spectrum of fission-product isotopes. Due to the wide scope of the work, a great deal of effort was required to prepare and maintain standard solutions and to calibrate the equipment.

The standard procedure was for the person submitting the sample to fill out a card to indicate what analyses he or she wanted, the expected concentration range, and any other elements or matrix materials that might be present. A big problem for the researchers in many cases was the difficulty in obtaining representative samples, which was their own responsibility, and they occasionally blamed the analytical lab for unexpected results caused by their inadequate sampling procedures. On one occasion, Jim Knighton received a result that was nearly exactly one-half the value he expected. His group leader suggested that he talk with the analyst about it, and they did, indeed, find that a factor of two error had been made in taking an aliquot—a very rare occurrence. From that time on, however, Jim repeatedly tried to figure out how an incorrect aliquot could account for every obviously bad data point, and it was hard to convince him that they just don't take weird aliquots such as, for example, 5/8ths.

In 1954, Bob Larsen became the Group Leader of the Analytical Laboratory. He had joined Argonne in 1951 and had been working on dissolution procedures for alloy fuels. He made an important contribution to aqueous fuel reprocessing by determining the cause of explosions mentioned earlier when U-Zr alloy fuel was dissolved in nitric acid and showing that they can be avoided by adding a suitable amount of hydrofluoric acid to the system (ANL-5135). Bob was an excellent chemist and one of the memorable characters in the Division. He had lost the use of his legs through a bout with polio, and navigated about the building with a pair of canes and a stool on wheels. This disability seemed to have no effect either on his ability to work or his robust sense of humor. One thing in particular he is remembered for is his inexhaustible supply of limericks, none of which are printable. His loud voice and boisterous laugh once led Carl Crouthamel to comment that "Larsen reminds me of the Caracas bull—noted for its bellowing."

In some cases, new or modified analytical methods had to be developed. Laury Ross played a major role in much of this work. An example is an analysis for fission-product ruthenium, which was reported by Bob Larsen, Laury Ross, and Gwen Kesser.

In addition to his responsibility for supervising the Analytical Laboratory, Larsen maintained an ongoing interest in neutron cross sections and in burnup analysis of nuclear fuels. He became a nationally recognized authority in these areas and was one of the organizers of "round-robin" arrangements in which results from several laboratories on identical samples were compared.

In 1957 Bob Meyer, who had been involved primarily in spectrophotometry and X-ray analyses, joined the Analytical Group, where his specialties were spectrophotometry, polarography, EDTA titrations, radiochemistry, activation analysis, burnup determinations, and computer programming. He served as an assistant to Larsen, and several years later became a Group Leader in the Sodium Technology Program.

Before the 1950s, new analytical instrumentation had been slow to develop, but that situation began to change rapidly. Multichannel analyzers, coincidence counters, and lithium-drifted germanium (pronounced "jelly") detectors became available, greatly expanding the capabilities of radiation counting. Another example was a Cary Model 10 recording spectrophotometer that was procured by the Division. Alternatives to the old two-pan analytical balances, *e.g.*, the Mettler and Sartorius models, were introduced, and these were followed later by the fully automatic electronic balances. Solid-state digital readout instruments were not yet available in the 1950s, but were just around the corner.

Almost all of the research and development programs in the Division depended heavily on the Analytical Chemistry Laboratory for their experimental results. One interesting facet of this situation was that a large number of samples that contained uranium, plutonium, or other special materials were generated in the various experimental programs, and these ended up in waste solutions from the analytical lab. Nobody knew the exact amounts of these materials in the samples until they were analyzed, which created a complicated bookkeeping problem. A combination of good record keeping and Larsen's cooperation with the other group leaders fortunately produced material balances that were acceptable to the Special Materials people.

One unusual assignment given to the Analytical Chemistry Laboratory in the early 1950s was to train a group of young Du Pont people who were to become analytical laboratory technicians at the new Savannah River facility near Aiken, South Carolina.

Analytical Research and Nuclear Constant Measurements

Also located in B-Wing were a number of research programs that were somewhat related to the Analytical Chemistry Laboratory, but were not a part of it. These were originated for the most part by Carl Crouthamel, who had come to Argonne as a Senior Scientist from

Ames Laboratory in 1950. He was an expert in gamma-ray spectrometry and was also interested in the development of analytical methods for the various constituents of nuclear fuels and in capture-to-fission cross-section ratios for isotopes in fast breeder reactors. He played a major role in the acquisition of new analytical instrumentation as it came on the market.

Working with Crouthamel in the early 1950s was Carl Johnson, who left the Laboratory for a couple of years to obtain a Ph.D. degree at Michigan State University. Crouthamel and Johnson worked on scintillation spectrometry for fission-product analysis and developed spectrophotometric methods for the determination of uranium, molybdenum, technetium, tungsten, titanium, and niobium, using thiocyanates in an acetone medium. Some work was also done on paper chromatographic analyses of irradiated uranium. Various other individuals were involved in analytical development; for example, Sy Vogler and Roberta Shor worked out a procedure for zirconium analysis under Dr. Vogel's direction.

Stan Flikkema, who also appeared on the scene in the 1950s, was best known by his colleagues as a perfectionist, perhaps to a fault. He was engaged in several different areas of work, including studies of zirconium dissolution explosions, X-ray spectrophotometric methods for uranium and plutonium in solution, optical microscopy, and precision weighing on the microbalance. A revealing story about Stan's personality concerns a trip he made to the stockroom to get a bottle of ink. When Esmer Zeno produced the bottle, Stan became quite exercised and refused to accept it because it was slightly dusty. Esmer, with his usual tact, disappeared behind the shelves, polished up the bottle, brought it back and told Stan "Here's a clean one."

Crouthamel, Larsen, and others were interested in fast neutron cross sections because most of the existing information was on thermal systems, and data were needed for the EBR-II project. The Division was in an especially good position to do this research because it had access to irradiated fuel from EBR-I and a capability for identifying specific isotopes by radiochemical and counting techniques. Peter Kafalas, who was a specialist in this type of work, joined the group in the mid-1950s. Much of his effort was directed to neutron capture/fission ratios for uranium and plutonium and fission yields of cesium-137, which were important in determining breeding gain.

Much of the cross-section work was done in CP-5. On one occasion, Crouthamel had an irradiation going when something in the convertor system broke, contaminating a number of graphite stringers. The only practical way to remove the stringers was by manual manipulation and, as was frequently done in those days, a large number of people were used so no one person would receive more than an allowable radiation dose of a few seconds. Martin Steindler reminisces, "I lost a good pipe because I had stuck it into my back pocket and then added the gloves I had used to move the graphite, which turned out to be hot, but I didn't find this out until I got to the monitors in Bldg. 205. The pipe and the gloves, alas, had both become DAW (dry active waste)."

Especially Significant Publications

Because the 1950s were highly productive years for CEN, the journal publications, ANL topical and progress reports, book chapters, *etc.*, are far too numerous to cite in detail. A few of these, however, deserve special mention as landmark sources of information that were cited widely and used throughout the world's nuclear energy programs.

Much of the work done by members of CEN in the 1950s is reported in considerable detail in the proceedings of the 1955 and 1958

Geneva Conferences on the Peaceful Uses of Atomic Energy and the proceedings of the Brussels Conference in 1957. Reviews of CEN work may also be found in the *Reactor Handbook*, 2nd. Ed., Vol. II (Interscience), and in various volumes of the *Progress in Nuclear Energy Series* (Pergamon Press).

In the 1950s, thermochemical information on the compounds involved in fuel reprocessing, particularly for pyrometallurgical and fluoride volatility processes, was, in many cases, very sketchy or missing altogether. Alvin Glassner assembled essentially all the thermochemical data available at the time in a topical report, "A Survey of the Free Energies of Formation of the Fluorides, Chlorides and Oxides of the Elements to 2500 K," which was issued as ANL-5107 (August 1953). An updated and corrected version of this report appeared as ANL-5750 in 1957. These compilations, of necessity, contained many thermochemical values that were estimated, derived from theoretical principles, or based on analogy and were of questionable accuracy. Because of this, they were maligned later on by the calorimetrists and other "pure" scientists, but at the time they were extremely useful to those who had to predict how the individual actinide elements and fission products would behave in a process, and almost any information was better than none. On the whole, the actual performance of the processes turned out to be fairly close to what Glassner's compilation had predicted.

Because the fission-product spectra in fast reactors differ significantly from those for thermal reactors, information was needed to assess the impact of this fact on the reprocessing of EBR-II and other fast reactor fuels. Les Burris and Ira Dillon met this need by extensive fission-yield calculations that culminated in two ANL reports: "Estimation of Fission Product Spectra in Fuel Elements Discharged from the Power Breeder Reactor and EBR-II," ANL-5334 (October 1954) and "The Estimation of Fission Product Spectra in Discharged Fuel from Fast Reactors," ANL-5742 (July 1957). The distribution of activities was given for various cooling times of interest. Calculations of this kind are routine today, given the computer programs and data banks on cross-sections as a function of neutron energy, *etc.*, that are available, but they were long and tedious at that time. The results were essential in designing the shielding and heat-removal equipment for processing EBR-II fuel.

A book that was assembled and edited by Carl Crouthamel in the 1950s and published in 1960 is *Applied Gamma Ray Spectroscopy*, C. E. Crouthamel, ed., Pergamon Press (1960). One of the several valuable uses of this book was the identification of fission-product elements by the energies of their gamma ray emissions.

Perhaps the most widely recognized publication by CEN was a quarterly journal, *Reactor Fuel Processing*, which began in February 1958. In the late 1950s, several different types of reactor fuel were being touted, *e.g.*, metal, oxide, carbide, and even more schemes for processing them were proposed. The purpose of this journal was to "assist those interested in keeping abreast of important developments" in the reactor fuel reprocessing field. Dr. Lawroski had proposed the idea, and produced the publication for the AEC with the help of several CEN staff members. The information was timely, and one particularly interesting thing about it was that much of it appeared in unclassified form for the first time in that journal. In 1967, *Reactor Fuel Processing* was merged with *Power Reactor Technology*, which was also written by ANL staff, to a new quarterly called *Reactor and Fuel Processing Technology*. At this juncture, the nature of the publication changed in that it consisted of review articles rather than reports of current research and development. The Atomic Energy Commission (AEC) took over the publication in 1968 and it became *Reactor Technology* in 1970.

During the late 1950s and early 1960s, one of the largest and most important programs conducted by CEN was its role in the design and construction of the EBR-II Fuel Cycle Facility in Idaho. The excellent book that covers this work in detail, *The EBR-II Fuel Cycle Story*, written and compiled by Dr. Charles E. Stevenson, and published by the American Nuclear Society, La Grange Park, Illinois in 1987, was mentioned earlier.

1960-1970: Turbulent Times

1960-1970

(top) Cutaway view of Fuel Cycle Facility. In the circular building (argon cell), fuel retrieved by dismantling EBR-II subassemblies in an adjacent air cell was processed and fabricated into new fuel elements for recycle to the reactor.

(bottom) Inert atmosphere glove-box facility in Bldg. 310, where large-scale studies were conducted on pyrometallurgical processes for EBR-II fuel.

3

1960-1970: Turbulent Times

The sixties was a decade of considerable turbulence in the United States and abroad. Dissention was in the air with four assassinations, bitter civil rights confrontations, war, political upheavals, international threats, riots, student uprisings, and challenges to the national mores of the time. The ANL Chemical Engineering Division was fairly well insulated from most of these problems, but it also underwent some major changes. In 1963, Dr. Richard C. Vogel succeeded Dr. Lawroski as the Division Director. The programs on aqueous fuel reprocessing had all but disappeared, but the fluoride volatility work continued at a strong pace with its main emphasis shifted toward oxide fuels, which were becoming the standard for commercial power reactors. A greatly expanded effort was placed on pyrometallurgical processes for the EBR-II Fuel Cycle Facility. This effort included not only the main-line melt refining process for the metallic EBR-II fuel, but also auxiliary processes for recovering melt refining skulls, extraction of plutonium from the metallic uranium blanket, and liquid metal-molten salt processes for other types of fuel. The chemical and engineering aspects of the liquid sodium coolant used in EBR-II became a major research program. The metal-air and metal-water reaction studies, calorimetry program, fuel-preparation studies, and high-temperature materials investigations continued to flourish. Basic chemical and engineering research studies were continued. On January 26, 1968, the AEC informed ANL that the AEC laboratories would be used for public health and environmental research, and on December 1, 1969, the Center for Environmental Studies was established at ANL. This was a forerunner of the Laboratory's environmental divisions. Shifts in the AEC priorities and accompanying funding cuts eliminated nearly all of the fuel-reprocessing work in the Division, necessitating some layoffs and transfers of personnel to other ANL organizations, including the group at Idaho. Some slack was taken up by new programs on batteries, coal combustion, and other work. A third Geneva Conference on the Peaceful Uses of Atomic Energy was held in 1964; CEN was well represented by papers on a variety of subjects.

THE DOMESTIC AND INTERNATIONAL CLIMATE

The American public, already concerned about the escalating possibility of a nuclear war with the U.S.S.R., was shocked by the news that a U-2 spy plane piloted by Gary Powers had been shot down by the Soviets on May 1, 1960. This incident killed plans for a Paris summit conference with Nikita Khrushchev, thereby increasing the tensions. The U.S.S.R. was known to be producing nuclear weapons and thought to be well ahead of the U.S. in the development and production of intercontinental ballistic missiles (ICBMs). Civil defense became an important subject and a few U.S. citizens, including Ed Rudzitis in CEN, constructed fallout shelters.

John F. Kennedy defeated Richard M. Nixon in the 1960 presidential election. Kennedy, keenly aware of the nuclear arms race and the so-called "missile gap," had a positive attitude toward U.S. nuclear and space programs. Early in his administration, Cuba became a problem with its confiscation of U.S. property, followed by the Bay of Pigs and the "Cuban Missile Crisis." This confrontation with the U.S.S.R. finally resulted in a partial nuclear test ban treaty and the "hot line" between Washington and Moscow.

On November 22, 1963, the nation was stunned by the assassination of President Kennedy at a parade in Dallas. At Argonne, the employees were given the rest of the afternoon off. The suspected assassin, Lee Harvey Oswald, was shot and killed by Jack Ruby, a Dallas nightclub owner.

Lyndon B. Johnson, the Vice President, assumed the presidency upon Kennedy's death and won the 1964 presidential election in a race against Barry Goldwater. On the domestic front, his concerns were mainly with civil rights and a collection of programs that was dubbed "The Great Society." Johnson's administration, however, was plagued by the highly unpopular Vietnamese war, race riots, and student uprisings as well as many other problems, and he opted not to run for reelection.

Richard M. Nixon was victorious in his presidential campaign against Hubert Humphrey in 1968. Nixon's forte was foreign affairs; he was successful during his first term in winding down the Vietnam war somewhat and in establishing relations with China. Neither Johnson nor Nixon seemed to have a strong pro- or anti-nuclear bias, but both recognized the necessity to maintain a strong nuclear defense and accepted the idea of nuclear power generation. During the Democratic primary campaign in 1968, Senator Robert F. Kennedy (John's brother and former U.S. Attorney General) was assassinated by Sirhan B. Sirhan in Los Angeles. Sirhan was convicted of murder.

Civil rights came to the forefront as a major movement in the 1960s. The decade began with peaceful demonstrations and massive rallies, although many of the participants were arrested. In 1965, Malcolm X, a black nationalist leader, was shot to death at a rally in Harlem, and the race issue exploded with extensive riots, first in the Watts section of Los Angeles, and later in several other cities. On April 4, 1968, Dr. Martin Luther King, Jr., was assassinated by James Earl Ray in Memphis, and this event ignited further rioting.

The Vietnam War began to heat up in 1964 when U.S. destroyers were supposedly attacked by North Vietnamese torpedo boats in the Gulf of Tonkin. The U.S. responded by bombing North Vietnam and sending the first combat troops to Vietnam in 1965. The U.S. troop strength escalated to 525,000 in 1967, and then to 540,000 in response to the North Vietnamese Tet offensive. That same year the North Koreans seized the crew of the U.S.S. *Pueblo*. Much dissatisfaction was expressed by U.S. citizens that the South Vietnamese were not doing their part in the war, so a "Vietnamization" was initiated in which the U.S. troop strength was reduced to 340,000 by 1970.

As if all the above troubles were not enough, the nation had to cope with the "Hippie" movement, Vietnam War protesters, and various student dissidents. Many young people fled to Canada to avoid the U.S. draft while others demonstrated nationwide against the Vietnam War, occupying campus buildings and destroying properties they regarded as targets. A graduate student was killed by a bomb at the University of Wisconsin.

The U.S. space program continued with the launching of Echo I, the first communications satellite, and Tiros I, the first weather satellite. The Soviets fielded two manned space missions in 1961, and the U.S. followed with its first suborbital flight by Alan Shepard. A series of space flights both by the U.S. and the Russians was conducted during the period. A tragic accident occurred in 1967 when three U.S. astronauts were killed in a test launch fire, and that same year a Soviet cosmonaut died during a reentry. The "space gap" was being closed rapidly by the Americans, especially with the introduction of the Apollo system in 1968, and on July 20, 1969, Neil Armstrong became the first man to walk on the moon.

Large-scale nuclear power finally came into its own in the 1960s. To the public, nuclear power was a new and arcane technology, and Alvin Weinberg, the director of Oak Ridge, referred to leaders in the field as a "nuclear priesthood." In 1963, Jersey Central Power and Light ordered a 620-MW plant and Fermi-I went critical. The General Electric Co. projected a cost of 4.3 mill/kWh for power produced in their Oyster Creek turnkey operation. The first big year for power reactor orders was 1965. The U.S. utilities ordered nine units [>6,000 MW(e)] and 16 other units were ordered by other nations. The trend continued with 20 reactors in 1966 and 31 in 1967, but began to wane with 16 in 1968 due to the increasingly long construction schedules and escalating costs. Argonne's EBR-II, a 62.5-MW(t) experimental fast breeder reactor, began operation in 1964. The "(t)" in MW(t) indicates thermal power, or heat; and "(e)" is used similarly in MW(e) to indicate electrical power.

Nuclear power was also becoming popular in foreign countries. In 1961, the first German power reactor, the Kahl plant supplied by the U.S., went critical, and Canada began a nuclear power demonstration with their heavy-water CANDU reactor in Ontario in 1962. Japan's JPDR 12.5-MW(e) boiling water reactor (BWR) demonstrated their first generation of nuclear power. The first Soviet RBMK reactor was commissioned in 1964. In 1967, France switched from gas-cooled to pressurized water reactors (PWRs). Belgian utilities, in 1968, ordered their first two power reactors, Doel-1 and Tihange-1, and nuclear electricity was produced for the first time in Spain and the Netherlands. Argentina contracted with Siemens in West Germany for their Atucha-1 reactor.

Some other nuclear highlights of the 1960s: The 1961 "Project Gnome" test was the first peaceful nuclear explosion under the Plowshare program. The N.S. *Savannah* went critical in 1962 and began commercial operations in 1965, but it generated little interest in the shipping industry or elsewhere, and was decommissioned in 1967. (The prefix "N.S." refers to a nuclear-powered civilian vessel in the same way that "S.S." denotes a steamship and "M.S." a motor ship.) Now moored at Charleston, South Carolina, the *Savannah* serves as a museum. In 1965, SNAP-10A became the first space reactor and in 1969, SNAP-27, a nuclear thermoelectric generator, was placed on the moon by Apollo-12 astronauts. The West Valley reprocessing plant, in which Walt Rodger played a major role, was opened in 1966. That same year, Weston, Illinois, was selected for the new accelerator facility that was to become known as "Fermilab." Preservation of certain foods by irradiation began to be accepted, starting with irradiated potatoes in Canada and the U.S.S.R. Other food irradiation studies continued in the U.S. under the auspices of the AEC and the Army.

Two significant reactor accidents occurred in the U.S. in the 1960s. In 1961, the first U.S. reactor fatalities resulted from a steam explosion at the SL-1 reactor at the Idaho Test Site, which killed three servicemen. The SL-1 was a prototype 3-MW(e) military reactor used for training, and the accident was attributed to a human operational error, but the details are not fully known (or at least made public). Recovery and handling of the bodies involved problems not faced before by emergency personnel because of the high radiation levels. The other accident, in 1966, was a meltdown of the metallic Fermi-1 core, which overheated when a vane in the liquid metal coolant system became dislodged. Two nuclear submarines were lost at sea due to accidents: the U.S.S. *Thresher* in 1963, and the U.S.S. *Scorpion* in 1968. It is not known whether these accidents were nuclear in nature. In 1969, a serious fire at the Rocky Flats Plant produced widespread plutonium contamination within the plant and to a lesser degree in nearby outside areas.

The antinuclear activists continued with protests, denouncing nuclear power and stalling reactor licensing wherever they could. The WASH-740 report was updated in 1964 with conclusions that were even more frightening than the original ones. The Atomic Industrial Forum (AIF) recommended that the conclusions not be published, and they were kept under wraps until David Comey, a nuclear critic, pried them loose under the Freedom of Information Act in 1973. Ernest Sternglass, in 1969, claimed that nuclear power would cause 400,000 infant deaths due to fallout. Arthur Tamplin from the AEC staff said that number was at least 100 times too high and he was backed up by John Gofman at Lawrence Livermore. The Sternglass-Tamplin-Gofman numbers were debated at length, but could neither be proved nor disproved conclusively because they were based on uncertain probabilities of accidents and gross extrapolations of data that were of questionable validity. The problem of nuclear wastes also began to rear its head seriously in the late 1960s.

REACTORS

The experimental reactor EBR-II first achieved dry criticality on September 30, 1961, and began operation with the sodium coolant on November 11, 1963. It produced its first electricity on August 18, 1964, and was dedicated, along with the adjoining Fuel Cycle Facility, on September 13, 1965. Most of the other new ANL reactors were special-purpose facilities that were not directly related to the programs in CEN other than the reactor-safety studies. BORAX-V, mentioned earlier, which was designed to feed superheated steam into a turbine, went into operation in 1962. JUGGERNAUT, a low-power [250 kW(t)] reactor that was started up on January 11, 1962, and designed mainly for nuclear research, was used to take some of the heavy research load being supported by CP-5. JANUS, another 250-kW(t) reactor, had two faces (hence its name), which provided two different levels of radiation. It reached criticality on August 3, 1964 and was the first reactor dedicated to biological research. A series of Zero Power Reactor (ZPR) critical assemblies was built at NRTS in Idaho to simulate reactor configurations for engineering studies. The Zero Power Plutonium Reactor (ZPPR), which went critical on April 16, 1969, was used for physics studies of liquid metal fast breeder reactors (LMFBRs) and could simulate such reactors having power levels as high as 1,000 MW(e).

THE DIVISION

The Chemical Engineering Division entered the 1960s with Dr. Lawroski as the Division Director, Dr. Richard Vogel and Dr. Walton Rodger as Associate Division Directors, and Victor Munnecke as the Assistant Division Director. In 1963, Dr. Lawroski became an Associate Laboratory Director. Although he was no longer a member of CEN, it was still under his jurisdiction, and he continued to maintain close contact with the CEN management and staff personnel. In 1969, Dr. Lawroski became a member of the National Academy of Engineering. Dr. Vogel took the helm from Dr. Lawroski as CEN Division Director in 1963. Dr. Rodger left ANL in 1960 to become a partner in a consulting firm, McLain-Rodger Associates, joined Nuclear Fuel Services as Vice President for Research and Development in

1962, was the General Manager of the West Valley Fuel Reprocessing Plant (W. R. Grace Co.) from 1962 to 1964, and then returned to consulting. Vic Munnecke departed from ANL in 1964, and was replaced by Frank Maston until Everett Proud arrived in 1966. Before joining CEN as a Section Head in 1963, Dr. Arthur Tevebaugh had been a chemist at the Knolls Atomic Power Laboratory (KAPL), operated by the General Electric Company, and was involved in fuel-cell development at the General Electric Research Laboratory. He was appointed Associate Division Director in 1969. Donald Webster, a chemical engineer from the Savannah River Laboratory (SRL), operated by the E. I. du Pont de Nemours and Co., was also appointed Associate Division Director in 1969. He had acquired extensive nuclear experience at the Met Lab, Oak Ridge, Hanford, and Idaho Falls before joining SRL.

In these new assignments, Tevebaugh was responsible for the electrochemical programs, high-temperature properties studies, calorimetry, and basic physical and chemical research. Webster directed the fluidized bed and coal-combustion work, engineering equipment research and development, and nuclear fuel reprocessing studies. Les Burris was a Section Head for pyrochemical processes and, in 1966, was appointed head of the Fuel Recycle Section of the Liquid Metal Fast Breeder Reactor Program Office at ANL. Bob Steunenberg replaced Les when he was transferred. Upon his return to CEN in 1969, Les became an Associate Division Director and Program Manager of Sodium Technology with Paul Nelson and Fred Cafasso, respectively, as Section Heads for Engineering and Chemistry. The following individuals served as Section Heads at one time or another in the 1960s:

Les Burris	Al Jonke
Fred Cafasso	Paul Nelson
Elton Cairns	Bob Steunenberg
Carl Crouthamel	Art Tevebaugh

Management changes were not limited to CEN in the 1960s; two new ANL Laboratory Directors came on the scene. Norman Hilberry announced his retirement, and Dr. Albert V. Crewe became the third Laboratory Director on November 1, 1961. Dr. Hilberry was elected president of the American Nuclear Society for the year 1965-1966. Dr. Crewe, born in England, had been a professor of physics at the University of Chicago with special interests in particle accelerators and high-resolution electron microscopy. He had served as Director of the Particle Accelerator Division (PAD) at ANL from 1958-1961. During these periods, he continued to conduct research on high-magnification electron microscopy at the University, and his work received wide recognition in the scientific community. At this time, interest in the EBR-II project was at its peak. Dr. Crewe, although not as outgoing as Hilberry, was approachable and seemed to relate reasonably well to the ANL staff. Crewe stepped down and later became the Dean of Physical Sciences at the University of Chicago. He was replaced on November 1, 1967, by Dr. Robert Duffield, ANL's fourth Laboratory Director. Dr. Duffield had been at the Los Alamos Scientific Laboratory (LASL), operated by the University of California, during the Manhattan Project. Duffield came to ANL during a difficult time when Laboratory funding was being reduced drastically, particularly for EBR-II and other fast breeder reactor programs, causing numerous layoffs. He reminded the ANL staff that research and development jobs funded by the government offer no guarantee of continuing employment; the staff people, of course, knew that, but didn't appreciate being reminded of it.

By 1960, the many people who had joined the CEN staff some ten years earlier had mostly settled down in a suburban life with a family, a mortgage, and often a dog, and seemed to be largely immune to the influence of the Hippies and Flower Children that was sweeping the country.

An ANL milestone of sorts was reached in 1961 when Branko Dokmanovic was issued Badge No. 10,000. With a few exceptions prior to 1952, ANL payroll numbers have been issued serially, and are never reassigned, so one can get a general idea of a person's length of service from his or her badge number. As one might guess, Badge No. 00001 was issued to the first Laboratory Director, Walter H. Zinn.

During the 1960s, a group of Senior Scientists formed the Argonne Senate. The stated objectives were to exchange technical information and improve the quality of ANL research programs, but some felt that the motivation was basically to inject more of an academic atmosphere into ANL, possibly with perks such as tenure and sabbaticals for senior staff personnel. This movement was viewed somewhat coolly both by the ANL administration and by those staff members who were not at the senior level. Making almost all the staff eligible for the Senate solved the latter problem, and the movement quieted down after a few years.

Another interesting event occurred when the American forces fighting in Vietnam were having an especially difficult time coping with the unorthodox tactics of the North Vietnamese army, such as their tunnel systems. Dr. Crewe, apparently in response to a Government request, called a meeting of the ANL staff and requested them to generate some technical solutions that might aid the American soldiers. Some people tried to comply, but it became quite obvious that ANL staff were neither well-versed in jungle combat nor in coming up with instant solutions. (The time available for implementation of any new ideas was two or three weeks.)

By 1960, some changes had begun to occur in the major programs of the Division. In the area of aqueous fuel reprocessing, existing processing methods were already in place for large-scale applications, and the Idaho Chemical Processing Plant (ICPP) was able to handle the lower-volume specialized fuels. Consequently, ANL could no longer justify a large research effort in this field.

In the pyrometallurgical program, development work on melt refining was nearly completed and process equipment was being installed in the FCF at the EBR-II site. The term "pyrometallurgical" was applied to these processes because the reactor fuel remained in the metallic state throughout the process. When the research effort was redirected nearly entirely to processes for recovering uranium from the melt refining skulls and extracting plutonium from the blanket, the term "pyrochemical" came into use to reflect the fact that many of the operations such as liquid metal-molten salt extractions, precipitation of products from liquid metal solutions, transport of fuel constituents as chlorides in molten salts, and distillation were more typical of chemical than metallurgical processes. Jim Battles, however, argues with some justification that the term "pyrometallurgical" is still valid because operations of this kind are used in the metallurgical industries. His opinion may be colored a bit by the fact that he is a metallurgical engineer. More recently, the term "pyroprocess" has come into vogue to cover all the bases, but it might be misleading, too, in that it implies any high-temperature process.

The emphasis in the fluoride volatility program was changing toward oxide fuels, the use of elemental fluorine, rather than an interhalogen as the primary fluorinating agent, and improved schemes for plutonium recovery. Much discussion was going on in the 1960s as to relative merits of metal, oxide and carbide reactor fuels and the relative suitability of aqueous, pyrochemical and fluoride volatility processes for the various fuels. This question was addressed in a report (ANL-7137) published by Milt Levenson, Virgil Trice and Bill Mecham in 1966. The report did not end the discussions, but did shed some light on the question. One of the uncertainties was in the types of reactors that would prevail in the future, *i.e.*, fast breeders (metal fuel), boiling and pressurized water reactors (oxide fuel), or high-temperature, gas-cooled reactor (carbide fuel). Oxide-fueled fast reactors were also under development in foreign countries.

The reactor safety work on metal reactions with air, water and other substances continued until the mid-1960s under Lou Baker and Dick Ivins, when a decision was made to transfer the program to the Reactor Analysis and Safety Division (RAS). At about the same time, the ANL sodium chemistry and technology programs were consolidated into a single entity in CEN.

Carl Crouthamel instigated a new area of work for CEN, which grew into a major program, when he became interested in regenerative galvanic cells. This work was supported strongly by Art Tevebaugh, who had been working on fuel cells at General Electric. This program was greatly enhanced by the arrival of Dr. Elton Cairns from the General Motors Research Laboratory. Elton was an extremely competent electrochemical engineer who had done his Ph.D. thesis work under Prof. Charles Tobias (a member of the CEN Review Committee for a period of time). He also seemed to know everybody who was doing serious electrochemical work in the U.S. and abroad and was well connected with key people in the government agencies. These connections were most helpful in seeking funding for the program.

The calorimetry work continued to thrive under Ward Hubbard and gained a valuable new staff member when Pat O'Hare, born and educated in Ireland, appeared on the scene. The Analytical Chemistry Laboratory, as well as the analytical research and nuclear constant studies, continued. Ben Tani became responsible for the X-ray laboratory. Jim Battles joined ANL in 1964 and began working with Russ Edwards on the chemical vapor deposition of tungsten on complex W-UO_2 cermets (part of the ROVER project for nuclear powered space vehicles) and other studies including high-temperature mass spectrometry of potential nuclear fuels. Hal Feder had the responsibility for several research programs, including Irv Johnson's extensive studies of liquid metal solubilities and thermodynamic properties of the solutions. As a Senior Scientist with wide-ranging experience, Hal served as sort of a mentor (and critic) for many of the staff members.

Feder was a stickler for accuracy and challenged almost every detail when one was writing a journal article; the authors began referring to this painful editing process as "Federization." Martin Chasanov, Marv Tetenbaum and Len Leibowitz investigated the thermodynamic properties of refractory metal compounds at high temperatures. John Gabor and others conducted a systematic study of the properties of fluidized beds, which became a major area of expertise for CEN. Vic Maroni, who also arrived in the 1960s, began his work on lithium-tellurium electrochemical cells and conducted spectroscopic studies of molten salt solutions. A significant effort was devoted to methods for the preparation of carbide reactor fuels.

The Division was requested to perform two special jobs supported by outside funding. One was the preparation of a small amount (30 grams) of isotopically pure ^{24}Mg metal from the oxide. This was accomplished successfully by Norm Chellew in his usual methodical, very careful (and slow) way with no significant loss of the valuable isotope. The other was to prepare high-purity ^{238}Pu metal suitable for space applications. That was a difficult job because the alpha activity of ^{238}Pu is about 300 times that of ^{239}Pu, and the neutron emissions from (α-n) reactions required special shielding. Jack Fischer and Paul Nelson took the lead in this project. At one point, Bob Steunenberg, who was then the Section Head, commented that this was a lot more demanding and expensive than working with ordinary plutonium. Les Coleman, the Division Safety Officer, replied sourly, "There is no such thing as 'ordinary' plutonium."

In the late 1960s, a young technician who decided to take some courses at the Illinois Institute of Technology in Chicago caused some excitement in the Division. Being concerned about his personal safety in that neighborhood at night, he started carrying a loaded derringer pistol in his pocket, which was understandable, but strictly forbidden by Laboratory rules. One day, in the men's locker room of Bldg. 205, the gun accidentally discharged, wounding the man in the forearm and scaring everybody in the area. Les Coleman, the Division Safety Representative, was highly upset, in part due to the fact that he had to deal with the DuPage County Sheriff. The employee was fired, due to the serious infraction of the regulations. Dr. Vogel's comment on the matter was that he would tolerate "no further gunplay in the locker room."

MANAGEMENT

One of the first things Dr. Vogel did when he became Division Director in 1963 was to move the Division office from L-Wing down to A-Wing. Although perhaps a symbolic move, it did in fact seem to bring the management and staff people closer together. Dr. Vogel, like Dr. Lawroski, was a demanding manager, and he insisted on quality work and reporting. Group reports were relaxed to a monthly schedule but were expected to be more comprehensive and of higher quality. Group Leaders almost routinely received copies of their preceding monthly reports with Dr. Vogel's comments written in the margins. It was obvious that he was staying on top of the work in the Division. He was a strong supporter of Joe Royal's Technical Editing Group. Just a few days after Dr. Vogel had moved into his A-Wing office, the Radiation Safety people detected a small spot of activity on one of the chairs in a routine survey. He became quite

Fig. 3-1. Richard Vogel

upset over the matter, and everybody who had been in the area was called in to have his pants checked, but no culprit was found. He sometimes kidded about putting up a print of El Greco's menacing painting, *View of Toledo*, behind his desk to intimidate visitors to his office.

The Assistant Division Director had a strong influence on the general tone of the Division because he handled the mundane administrative affairs and interacted with everybody. In the early 1960s, everyone was used to Vic Munnecke, who had been there since the Division's inception, and was a fairly "laid back" person. Frank Masten, who replaced Vic, was a businesslike individual who did his job rather quietly. Everett Proud, who replaced Frank, however, shook things up a bit. He had been at KAPL at one time (along with Tevebaugh) and had held several responsible management positions before he came to ANL from the Sperry Rand Research Center. He had a gruff manner, and his appearance and demeanor reminded one of a tough military officer (which he had been in WWII, and then in the Army Reserve). When one approached him for something such as a blackboard or bookcase, for example, Everett's immediate response was usually to growl "no way"—and then it would suddenly appear a few days later. Actually, he was effective in solving administrative problems and went to bat for CEN people on many occasions. At one time, office occupants were allowed to choose the color when their rooms were repainted, and a secretary at the east end of A-Wing selected a brilliant flamingo pink. The next morning, when Dr. Vogel arrived at work, the sun was creating a hot pink glow all down the length of A-corridor, and he told Everett to have that office repainted by the end of the day. It was. The same policy applied to the laboratories, and when Bill Walsh's lab needed repainting, he had it done in a subdued version of the blue and gold colors of the University of Notre Dame. Some wondered if Jim Battles in the lab next door would opt for Alabama crimson, but he didn't.

Argonne, like business firms, was attempting to stay on the leading edge of new management trends, as they became popular. Many CEN supervisory personnel attended training sessions, seminars, and short courses both on and off-site. Carter Johnson of the Personnel Department, an enthusiastic, friendly young man, handled most of these activities and became acquainted with many CEN people. When the Program Evaluation and Review Technique (PERT) along with the Critical Path Method (CPM) was just coming into its own in the early 1960s, John Vogel introduced these techniques to the CEN staff in one of the most entertaining seminars in the history of the division. He went into considerable detail in illustrating how these planning and management tools could be used to set up a moonshining operation in the Tennessee hills. The next major fad in management circles was MBO (Management by Objectives), and the group leaders and section heads were duly instructed in these methods.

Up to about 1960, the regular staff personnel had little direct interaction with the AEC managers in Washington, but that began to change gradually. Those managers were beginning to require a more detailed accountability in the form of program reviews, special progress reports, *etc.*, but the situation had not yet become what one would call onerous.

A major shock wave struck the entire nuclear community in 1965 when Milton Shaw was appointed Director of Reactor Development and Technology (RDT) at AEC Headquarters. He was an alumnus of Admiral Rickover's operation, and told the Joint Committee on Atomic Energy (JCAE) that the AEC laboratories under his jurisdiction "would be converted from a research orientation to one stressing disciplined engineering application." He felt that, once the light reactors were commercialized, the federal research effort should be placed on advanced concepts such as the fast breeder. Research was downplayed in favor of engineering. The Loss of Fluid Test (LOFT), which was to have been a sacrificial reactor to study safety problems in light reactors, was rescheduled to test standards of reactor design. Shaw believed there could be no safety problems if the engineering was done right. The strong deemphasis of research and safety upset many people, particularly at NRTS and ORNL, and the safety issue escalated to a national level in the JCAE, Congress, the nuclear industry, and the general public. When Shaw was appointed, ANL felt the effects almost immediately. The Chemical Engineering Division began receiving several directives per day. Early in the Shaw regime, Milt Levenson served as sort of a lightning rod for the Division because he had intimate knowledge of the EBR-II project; he was also articulate and quick on his feet. The net effect, however, was that the fast breeder fuel processing work in CEN and other laboratories began to wane and was essentially phased out by the end of 1969. Representatives from various laboratories who attended a conference titled "Symposium on the Reprocessing of Nuclear Fuels" at Ames, Iowa, in August 1969 felt as if they were participating in a technical wake.

Shaw was not the only harbinger of heavy-handed micromanagement that was going to be emanating from Washington. The cognizant AEC person for the sodium programs in the late 1960s had an authoritative, hypercritical attitude toward the ANL work.

As mentioned earlier, project management and control methods began receiving greater emphasis in the 1960s. Detailed milestone schedules were (and still are) required in proposals for future work and in the monitoring of current projects. Although a legitimate management tool when applied responsibly, they give the researchers a problem of "inventing on schedule," particularly where highly creative work is needed. The predictable tendency of the funding agencies was to accept a proposed set of milestones, but then insist that they be met in a shorter time and on a lower budget. There was sometimes a temptation to accept a demanding milestone schedule with inadequate funding in order to keep a program alive.

BUILDING ADDITION

By the 1960s, the Division had outgrown its available space. Offices were filled to capacity, the B-Wing labs were occupied by the analytical work, and the increasing volume of bench-scale work had filled all the space in the A-Wing labs. The high-bay areas, originally designed for solvent-extraction studies, had been adapted to other work. A new addition, located at the ends of A- and B-Wings, which was built in 1961-1962, covers an area of about 18,000 square feet. The core of the new addition consisted of eight large general-purpose laboratories (one of which was divided in half), which were designated as "X-Wing." X-Wing was surrounded by extensions of the A- and B-Wing corridors and offices plus a corridor with a string of about 30 offices and a conference room across the east end, which was named "W-Wing." A service floor was built under the entire addition. Although a number of purification systems for glove-box atmospheres, various other equipment installations, and storage areas are located in this service area, it still has sufficient open space to serve as a tornado shelter for Bldg. 205.

The laboratories were provided with all the usual services and were sufficiently capacious to be used either for bench-scale work or for larger scale operations requiring large glove boxes and vacuum-frame hoods. Most of this laboratory space has been devoted to the basic research, electrochemical, and calorimetry programs.

The new wing closed off the open area between A- and B-Wings, creating the present courtyard. This introduced some new maintenance problems in that lawnmowers and other equipment had to be brought in through the building's corridors. After a while, weeds began to take over the area, which became rather unsightly. Everett Proud decided at one time to solve the problem by applying a heavy mulch of cocoa nut hulls, which were highly odoriferous and were located under the building air intakes. For a few days, the building occupants felt as if they were being gassed, but the sun, air and rain disarmed the hulls after a few days and the problem disappeared. Some years later, Herb Brown, aided by several volunteers and with the cooperation of ANL grounds people, installed the present arrangement of walkways, planters, landscaping and picnic tables, which finally transformed the courtyard into an attractive asset, although it has deteriorated somewhat in recent years.

TOOLS OF THE TRADE

Everybody in the Division was affected by the many technological innovations that were introduced to the marketplace in the 1960s—Xerox® copiers, IBM Selectric® typewriters, electronic analytical instrumentation, digital readouts, and computers that, although not yet quite "user friendly," were no longer downright hostile. Color television was maturing, and several CEN people assembled large-screen sets at home from Heathkits®, which were much less expensive than the retail models if one didn't count the assembly time. The circuits still used vacuum tubes, and nearly every drugstore had a tube tester along with a convenient supply of overpriced new tubes for sale.

A major breakthrough in computer miniaturization occurred in 1958, when an American engineer, Jack Kirby, devised the first truly integrated circuit. His prototypes consisted of a germanium wafer with integrated transistors, resistors, and capacitors. By the 1960s, the technology had developed rapidly, using less expensive silicon chips with many more components on each chip. The first large-scale integration (LSI) used hundreds of components on one chip. The next step was very large scale chips with 100,000 components, and today it appears that ultra large scale integrated chips will contain as many as 10 million components on a 1-cm-square chip. Although integrated circuits began to find wide application in the 1960s, they did not develop into computer microprocessors until the 1970s.

By about 1960, CEN staff people were beginning to use the large mainframe computers in the Applied Mathematics Division (AMD). Their first big commercial machine was an IBM 704, and before long they converted to a CDC 1600. Dean Pierce describes the situation at the time: "I had repetitive calculations to reduce the data from many dissolution-rate experiments. The programmer from AMD worked with the equations that Shelby Miller and I presented and wrote a Fortran program to do the calculations. She debugged the program and gave us a deck of cards and a format for our data. We had their keypunch operators punch our data cards. They always double-checked their work and I never had an error."

Several people began to take computer courses, mostly at AMD, but also at schools such as the College of DuPage, to learn programming. The main languages were Fortran and Basic and AMD had an extensive library of scientific subroutines. Packaged software as we know it was not available, and programming was nearly from scratch. Some individuals found that programming could be fun to the point of becoming addictive. A few became hobbyists who put together improvised computers at home and had to learn about machine and assembly languages to make them work. Irv Johnson was one of those people.

The first real computer capability within CEN came about when four Wang desk calculators were purchased and put into operation. These could handle the normal math functions such as exponentials and logarithms, polynomial equations, and trigonometric functions, and the programs were saved on punched cards. The Wangs were useful for stage calculations and other purposes at the time, but they could not serve very many people and seemed to be subject to quite a bit of down time. As CEN people requested more computer availability, Mel Foster was appointed to seek solutions to the problem. One such solution was a real-time phone-line connection between a commercial General Electric computer and Teletype® machines in CEN. A system called "Rescue" allowed one to use either Fortran or Basic. Access to the mainframe computer at AMD was becoming more convenient, and Stanley Cohen of ANL developed a more-or-less plain language program called "Speakeasy." About that same time, cardpunches became available in Bldg. 205. Martin Steindler tells an amusing story about that. He had purchased a tire at a Standard Oil station on a credit card and sent in the payment along with the enclosed remittance card when the bill arrived. The next few months, he kept receiving the same bill and no phone calls or letters could stop the process. Finally, in desperation, he brought the last remittance card to work, added a few random keypunches and sent it in. Shortly thereafter, he got a frantic phone call with the message that his bill was settled and an admonition to never, ever do that again. When AMD got their CDC machine, Mel set up a satellite station in CEN where input could be sent to AMD by punch cards or paper tape and the output could be picked up in Bldg. 205, which saved a lot of running back and forth.

The introduction of xerographic copying was a major innovation of the 1960s. The concept of xerography (from the Greek, meaning "dry writing") was invented in 1938 by a physicist, Chester F. Carlson, who sold the commercial rights to the Haloid Company in 1947, and 13 years later the company, later named the Xerox Corporation, introduced its first office copier to the market. Prior to the introduction of the Xerox copier, a variety of thermographic and chemical copying systems had made a brief appearance, but these were expensive, slow, and required special papers that had several shortcomings (gray backgrounds, curling, thick paper, slick surfaces that were difficult to write on, and unpleasant odors). It wasn't long before nearly all the office copying in CEN was being done with Xerox machines or clones produced by other manufacturers. For large layouts such as shop drawings, "whiteprint" processes such as Ozalid® had largely supplanted blueprints.

By the mid-1960s, solid-state technology was coming on strong and all kinds of new electronic instruments and devices were becoming available. This made it possible for researchers to rig up systems for automated data collection. An early example of this was an oscillating-cup viscometer that Dean Pierce was using to measure the viscosities of molten metals in which the viscosity could be calculated from the decrements of the oscillations. Lad Prucha (Doreen's husband) wired a cardpunch to the system to produce a card deck for computer input. Lou Baker and his people in their reactor-safety experiments, as well as researchers in the coal-combustion program were using automated data-collection systems. Real-time output of processed data, however, was still awaiting the availability of desktop computers.

Because of the increasing importance of electronic devices in the experimental programs, several individuals availed themselves of a course by Heathkit aimed specifically toward scientists and engineers. This course consisted mostly of hands-on experience but also included some textbook material.

For the secretaries, the big event of the 1960s was the introduction of the IBM Selectric typewriter, in which the printing element was a ball that moved back and forth and rotated to print the desired character while the carriage remained stationary. The balls were easily replaceable (also breakable), so font changes were easy. At this point, we were about half way through the typewriter sequence:

Mechanical → Electric → Selectric → Word Processor → Computer

CENHAM GLOVE BOXES

In the late 1950s, a simple, inexpensive, versatile glove box design for work with plutonium and other hazardous materials was developed by Dick Malecha, Harry Smith, John Schraidt, John Natale, Norman Ross, and Herb Brown. The acronym, "CENHAM," is derived from its full name, <u>C</u>hemical <u>En</u>gineering <u>H</u>ood, <u>A</u>lpha <u>M</u>odular. A major design goal was modularity. Modular structural elements allowed construction of glove-box arrays of various lengths and heights to accommodate the needs of a wide variety of experimental programs.

A drawing of a two-module glove box of this type is shown in Fig. 3-2. The basic module is 42 inches in depth, width, and height. It rests on a 36-in.-high steel angle frame. The safety glass windows are 3/8-in. thick. The internal ventilation system allows either for a once-through or a circulating gas atmosphere with an external purification system. The rubber gloves (not shown) are attached to the glass windows by standard ANL 8-in.-dia plastic glove rings sealed to the glass. When the gloves are not in use, the ports are covered by aluminum covers similar to a flanged, deep-dish "cake pan."

A salient feature of the CENHAM glove box is the use of automobile-type weather-stripping to seal the safety glass windows to the metal frames of the box. The glove boxes are usually operated at a slightly negative pressure (about –1 in. of water pressure) so that any air leakage is into the box. In-leakage rates of air have been very low with automobile-type window seals. Moreover, the windows can be removed and replaced relatively quickly. Depending on the requirements of the experimental work, a variety of inert atmospheres, *e.g.*, argon, helium, nitrogen, or dry air can be maintained in the glove box.

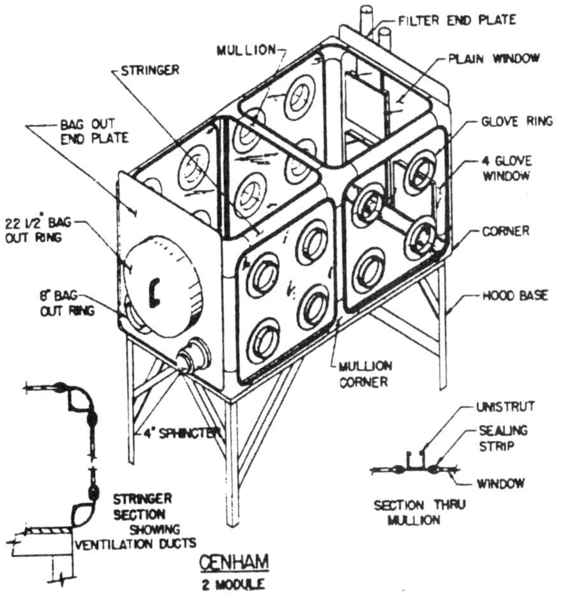

Fig. 3-2. Two-Module CENHAM Glove Box

Other features of the glove-box assembly shown in Fig. 3-2 are the following:

1. Three ports for transferring materials in and out of the box—a large 22.5-in. lock, an 8-in. port, and a small 4-in. sphincter port. In using this latter port, a 4-in.-dia canister is pushed through wipe seal rings until the opening of the canister is just inside the box. When the canister is full, it is pushed into the box by a replacement canister and later bagged out through one of the large ports.
2. A filter housing in which high efficiency filters were installed to filter exhaust gas.
3. Electrical feedthroughs and pipe couplings for connection to external fluid sources (usually water for cooling). Inexpensive construction materials such as carbon steel sheet or plate were used for the box structure.

Many different accessories have been attached to the boxes, depending on the requirements of the work. In some cases, small electric hoists are installed on rails in the boxes to raise, lower, or move heavy items. Furnaces that open into glove boxes are often attached to the floor of the box by water-cooled flanges. Large air locks are often attached to the ends of boxes containing inert atmospheres.

Figure 3-2 shows a simple two-module assembly. Much larger glove box units, some multitiered, have been constructed. The CENHAM glove boxes are in wide use at ANL, and can be seen in many of the laboratories throughout Bldg. 205.

TECHNICAL PROGRAMS

EBR-II Fuel Cycle Facility

There is little question that the most exciting program for the Chemical Engineering Division in the 1960s was the construction and operation of the EBR-II Fuel Cycle Facility (FCF) at the Idaho site. By 1960, CEN and other divisions at ANL-E had completed most of the research and development work that led to this facility. (The practice had begun of referring to the ANL site in Illinois as Argonne East, ANL-E, and the site in Idaho as Argonne West, ANL-W.) Detailed design work by the H. K. Ferguson Company had been initiated in 1957, construction work was started in 1959, and the facility was completed in 1962. The EBR-II reactor went critical in 1963 and was placed in full operation at reduced power in 1964. The power was increased stepwise until it reached the design level of 62.5 MW(t) [20 MW(e)] in 1969.

The Chemical Engineering Division had primary responsibility for the FCF during its construction phase. Charlie Stevenson, who had been in charge of the Idaho Chemical Processing Plant (ICPP), operated by the Phillips Petroleum Company, was appointed Manager of the EBR-II Fuel Cycle Facility in 1960, and held that position until he returned to CEN as a Senior Chemical Engineer in 1969. Milt Levenson was the Resident Lead Project Manager in 1961-1962, followed by John Schraidt in 1962-1963 and Neill Carlson in 1963-1964. In 1964, operating responsibility for FCF was transferred to the ANL-W personnel.

Several CEN personnel, including Don Hampson and Phil Fineman, were transferred to the Idaho site for permanent positions. Many others had shorter-term assignments at the site, and there was a great deal of commuting both by air and rail between Chicago and Idaho Falls, which is somewhat off the beaten track. Those transferees who had always lived in an urban environment generally experienced some initial culture shock with the remoteness of the area, the pervasive Mormon influence, and lack of amenities a big city has to offer. Most of them, however, began to enjoy the excellent outdoor recreational opportunities, such as camping, fishing, hunting, boating, and skiing that the area has to offer. Almost everybody used the buses to commute the 30 miles or so between Idaho Falls and the EBR-II site; bridge games on those buses are legendary, with the players occasionally overshooting their destinations to play out a hand.

Approximately 160 ft from the EBR-II-FCF complex, there was an additional building that housed the analytical laboratories and other auxiliary operations. The analytical facilities included six 20-Ci, 1-MeV gamma radiation caves with American Machine and Foundry (AMF) Mod 8 manipulators and two conventional wet analytical laboratories with fume hoods for hot work. Bill Sovereign, John McCown, and Earl Ebersole were the principal operators of this facility.

The layout and a brief description of the Fuel Cycle Facility were given in the previous chapter. The following is a more detailed description of the fuel-reprocessing operations. The discharged fuel assemblies, after cooling in the reactor, were passed from a shielded air cell to the circular argon cell and continued in a counter-clockwise direction (as viewed from above) through a sequence of workstations. The new subassemblies were then transferred back out through the air cell for reinsertion into the reactor. The fuel-processing operations, all of which were conducted remotely behind 5-foot-thick walls of concrete shielding, were as follows:

1. After a minimum of 15 days of cooling, the fuel subassemblies were moved, one at a time, to the air cell in the Inter-building Transfer Coffin. In transit, each subassembly was washed thoroughly with water to remove adhering sodium coolant, and was then dried.
2. The subassemblies were disassembled, and the fuel elements (stainless steel clad fuel pins) were separated from the other components, which were discarded. The fuel elements were transferred to an argon cell for further processing.
3. The stainless steel cladding was removed from the fuel pins by a spiral cutter, and the pins were chopped into 3/4-in. segments.
4. A 10-kg batch of pin segments was charged into an induction-heated CaO (lime)-stabilized zirconia (ZrO_2) crucible where it was melted and held at 1400°C for three hours. The liquid metal was then poured into a graphite mold with two small holes in the bottom to form an ingot with protrusions, which were broken off and used as analytical samples.
5. The melt refining "skull," a mixture of unpoured metal and oxides, was converted to a powder by oxidation at 700°C through a controlled addition of oxygen, and the resulting powder was poured out of the crucible.
6. The U-235 content of the product ingot from melt refining was brought back to the original concentration by adding a small amount of highly enriched uranium. These materials were melted together in a graphite crucible coated with yttria. New fuel pins were then formed by evacuating the system;

lowering precision-bore Vycor (quartz) tubes, closed at the top and containing a thin thoria (ThO_2) wash on the inside surface, into the liquid metal; and repressurizing to force the metal up into the tubes. This procedure was called "injection casting." After cooling, the Vycor was removed from the pins by crushing, and the pins were cropped to the desired length.

7. New fuel elements were fabricated by inserting the uranium pins into stainless steel jackets, adding the sodium thermal bond in the annulus, and sealing the jackets by welding. The finished fuel elements were subjected to a series of rigorous quality-control tests.
8. The fuel elements were inserted into new subassemblies for return to the reactor.

The melt-refining process resulted in a fission-product decontamination factor of about three, the low value being due to fact that the noble metals were not removed. Composite results from a large number of runs in the FCF, shown in Table 3-1, give the distribution of the various fuel constituents after melt refining. The recoveries of uranium and plutonium ranged between 90 and 95%. Noble metals were not separated from the uranium by melt refining but were maintained at a satisfactory equilibrium concentration by the 5-10% of the metal removed with the skull, which served as a "dragout" stream. The fission-product gases, Xe and Kr, were collected for controlled disposal. The melt-refining crucible had a Fiberfrax® (formulated asbestos) lid, sometimes called a "top hat," that collected vaporized species by a combination of condensation and chemical reaction. A "skull reclamation" process had been developed and tested on a pilot-plant scale in CEN. Some skull-reclamation equipment was installed in the FCF, but it was not used due to an AEC policy decision to recover the skull material by aqueous processing in ICPP. Work on a process for recovering plutonium from the blanket fuel was also well on the way toward application, but it, too, was terminated.

Overall, the FCF demonstration was an outstanding technical success and a tribute to the ANL engineers and scientists who developed it. The EBR-II operated for about five years, using the recycled fuel, which

Table 3-1. Melt Refining Results from the EBR-II Fuel Cycle Facility

Off-gas	Fume Trap	Skull	Crucible	Product
Xe 100%	Cs 100%	U 5-10%	Y 5%	U 90-95%
Kr 100%	Cd 100%	Pu 5-10%	RE 5%	Pu 90-95%
	Rb 100%	NM 5-10%	Ba 90%	NM 90-95%
	I >75%	Y 95%	Sr 90%	
	(Na) 100%	RE 95%	Te 10%	
		Ba 10%		
		Sr 10%		
		Te 90%		

(Na) = Sodium from thermal bond.
NM = Noble metal fission products.
RE = Rare earth fission products.

amounted to about 35,000 fuel elements. This corresponds to about five reactor-core replacements with new fuel. The FCF demonstration was discontinued in 1969, and outside sources were used to supply the new fuel; the spent fuel was reprocessed at ICPP.

The CEN Division continued to function as a support organization for designing special equipment and procedures, conducting tests, and troubleshooting, with special responsibility for melt refining and cell operations. This work included measurement of the air in-leakage to the argon cell, development of special lighting equipment, design of shutters for the shielding windows to protect them from radiation damage, testing of materials (electrical insulators, for example), development of continuous purification systems for support glove boxes located outside the cell, and development of auxiliary process equipment.

The major pieces of equipment that were developed by CEN were the Interbuilding Transfer Coffin and skull oxidation equipment. The transfer coffin was needed to transport discharged fuel assemblies from the reactor to the Fuel Cycle Facility—a distance of about 100 feet. The coffin, as it finally evolved, was a heavy (20-ton) lead-shielded cask with provisions for continuous removal of fission-product decay heat during storage and transit. It also served as a chemical reaction vessel in which provisions were made to convert sodium adhering to the fuel elements and subassembly surfaces to sodium hydroxide, which was subsequently removed by flushing with water. The sodium was converted to its hydroxide by adding humidified air to the circulating argon coolant until the reaction was completed. This procedure was necessary because direct addition of water to the metallic sodium could have resulted in small fires and explosions.

The skull oxidation technique was used to recover the heel and uranium that coated the wetted surfaces of the zirconia crucible (skull) after the purified metal had been poured into the mold. Conditions were developed for controlled oxidation of the skull at a temperature of 700-750°C by slow addition of oxygen to the argon atmosphere of the melt refining furnace. The oxide powder formed by this operation was then dumped from the crucible and sent to the ICPP for processing.

This ongoing support effort for the Fuel Cycle Facility was directed by Milt Levenson, Les Coleman, and John Schraidt. Many people were involved, including George Bernstein, Al Chilenskas, Tom Eckels, Johan Graae, John Ludlow, Bill Miller, Mike Slawecki, and Howard Stethers.

In 1969, EBR-II's role was changed from that of a fast breeder reactor demonstration to one of an irradiation facility for the national fast breeder program. Argonne had proposed a reactor named "FARET" for that purpose, but a potentially more advanced facility, the Fast Flux Test Facility (FFTF), was approved for construction at the Hanford Engineering Development Laboratory (HEDL) near Richland, Washington. A decision was made that EBR-II would serve this function until FFTF was ready. The emphasis had begun to shift to oxide fuels in the fast breeder program on the basis of safety and achievable burnup. More recent developments of metal fuels, however, have eliminated the burnup advantage of oxide fuels, and the same is true for safety because of the inherent nuclear stability of sodium-cooled, metal-fueled reactors under abnormal conditions.

In 1970, the FCF was redesignated FEF (Fuel Examination Facility), and almost all the fuel processing equipment was removed. Since then, a new, larger fuel examination facility, called HFEF (Hot Fuel Examination Facility) was built. It is now known as HFEF/N(north) and the original FCF is HFEF/S (south).

Pyrochemical Process Development

Development of the skull reclamation and blanket processes for metallic fast breeder reactor fuels continued at CEN for several years into the 1960s. A large glove-box facility was constructed in Bldg. 205 for demonstration of an integrated Skull Reclamation Process. This was followed by construction of a Skull Reclamation Process pilot plant in a large inert atmosphere enclosure in Bldg. 310. A similar sequence had been planned for the Blanket Process, but work proceeded only through construction and operation of a small engineering-scale unit in Bldg. 205. All work on these processes ceased at the end of September 1965, when the AEC stopped development of metal fuels and concentrated the funding on mixed UO_2-PuO_2 fuels. The shift to mixed oxide fuels was made to take advantage of the technology and commercial base that was in place for the fabrication of oxide fuels for light water reactors.

The Division then turned its attention to development of processes for mixed oxide (UO_2-PuO_2) fuels, using liquid metal and salt solvents and taking advantage of the separation concepts that had evolved in development of the EBR-II Skull Reclamation and Blanket Processes. These concepts were based on differences in the solubilities of the fuel constituents in liquid metal solvents and in their partition behavior between liquid metal and salt solvents.

Under the direction of Bob Steunenberg and Les Burris, several process flowsheets applicable to mixed oxide and also to mixed carbide (UC-PuC) fuels were developed. A central feature of these processes was a separation of plutonium from uranium and noble metal fission products by salt transport of the plutonium. A molten salt is cycled continuously between a donor metal alloy, in this case copper-33 wt% magnesium, and a receptor alloy, zinc-5 wt% magnesium. When the system reaches equilibrium, the activity of plutonium is the same in the two liquid metals.

The plutonium activity, a_{Pu}, is related to the concentration and activity coefficient by the relationship:

$$a_{Pu} = x_{Pu} \gamma_{Pu}$$

where x_{Pu} and γ_{Pu} are, respectively, the concentration and activity coefficient of the plutonium. Because the activity coefficient of plutonium in the zinc-magnesium solution is about 1/10,000th of that in the copper-magnesium solution, its concentration in the receptor alloy is about 10,000 times that in the donor alloy at equilibrium. Therefore, essentially all the plutonium transfers to the receptor alloy.

As shown schematically in Fig. 3-3, a salt-transport separation involves two vessels containing liquid metal solvents which are connected by means of a molten salt that circulates between them. Both liquid metals contain magnesium, and $MgCl_2$ is a constituent of the salt phase. Uranium and the noble metal fission products are essentially insoluble in the Zn-Mg alloy and remain in

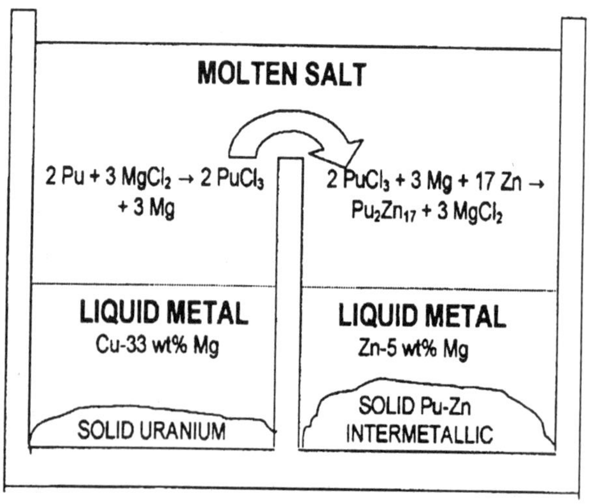

Fig. 3-3. Salt Transport Separation

that vessel as precipitated solids. The plutonium, because of its much higher activity coefficient in the Cu-Mg alloy than in the Zn-Mg alloy, reacts with the salt and is transported as $PuCl_3$ to the Zn-Mg alloy, where it is reduced back to the metal by the magnesium. The alkaline earth and rare earth fission products mostly remain in the salt. Uranium (with the noble metals) is recovered by decanting the supernatant Cu-Mg alloy, washing with magnesium to remove residual copper, and retorted to remove the adhering magnesium. Plutonium is recovered by vaporizing the zinc and magnesium.

A conceptual process based on salt-transport separations is shown in Fig. 3-4 to illustrate a potential application. Numerous variations could be made in this flowsheet, depending on the requirements for a particular fuel. For example, the liquid zinc decladding step might be replaced by mechanical or oxidative decladding methods.

At this stage of the program, Irv Johnson and his colleagues had generated a very extensive body of phase relationship and thermodynamic data on liquid metal solutions. This information, coupled with a large collection of data on the partitioning of

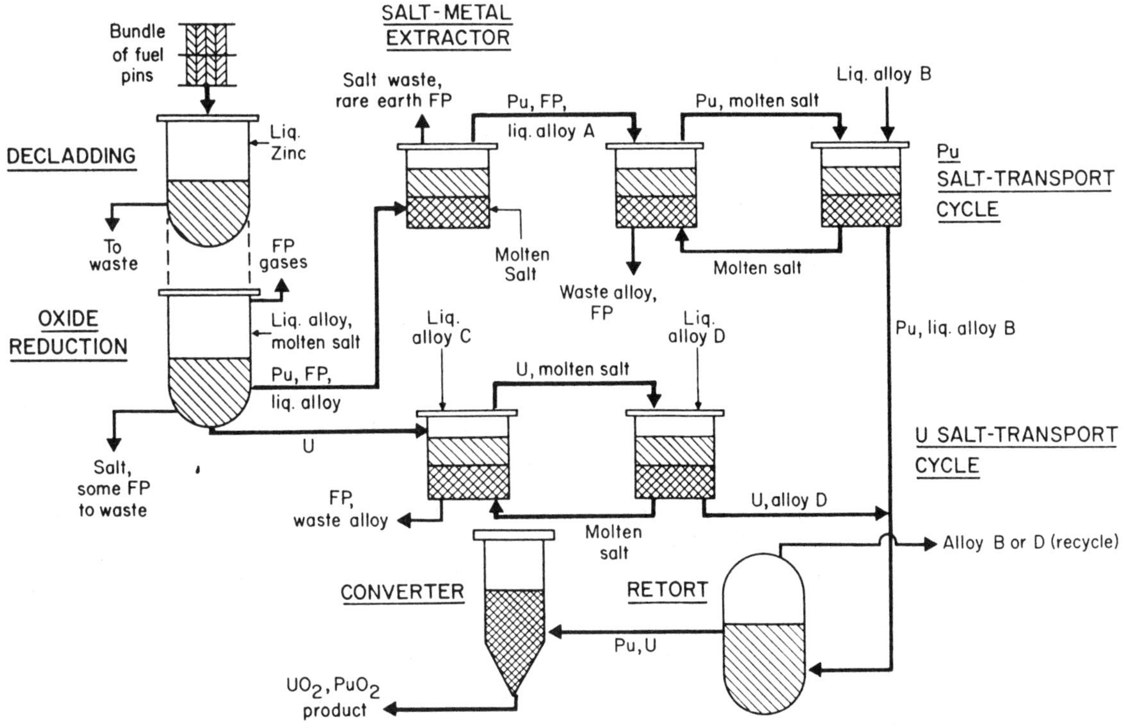

The fuel is first declad by dissolving the cladding in liquid zinc as shown here, or by mechanical or oxidative decladding methods. The mixed oxide fuel is then reduced by a liquid Cu-Mg-Ca alloy with a molten salt flux. Most of the alkaline earth and rare earth fission products stay in the salt flux. The liquid metal solution containing the plutonium becomes the donor alloy for a salt transport separation in which the plutonium is recovered in a Zn-Mg solution and the uranium, along with the noble metal fission products, is recovered from the Cu-Mg alloy. The product solutions are then retorted to remove the solvent metals, leaving metallic uranium and plutonium as the products. The uranium and plutonium are combined in the desired proportions and oxidized to form a mixed oxide powder for fabrication of new fuel.

Fig. 3-4. Pyrochemical Process for Fast Breeder Reactor Oxide Fuels various elements between liquid metal and molten salt solutions, generated mostly by Jim Knighton and co-workers, provided a wide variety of possibilities for the development of new processes. Several variations of the Salt Transport Process were developed. For example, a countercurrent liquid metal-molten salt extraction battery was developed to make the process more efficient. An alternative decladding step was oxidative decladding (also under consideration for the fluoride volatility processes). In this step, the stainless steel cladding was punctured or slit, after which the UO_2 was oxidized to U_3O_8, which expanded the cladding and pulverized the oxides, allowing their separation from the cladding. The U_3O_8 was then reduced to UO_2 by hydrogen to provide the feed material for the subsequent reduction to the metal. Mixed carbide fuels could be accommodated by first converting them to the oxides. The process would then be the same as that for the oxide fuels.

Construction of a glove-box facility to demonstrate the Salt Transport Process was terminated by the AEC (as were many other programs in other AEC laboratories) so the AEC could concentrate its funding resources on development, design, and construction of the Fast Flux Test Facility (FFTF) at Hanford.

The Group Leader for engineering development of the Salt Transport Process was Dean Pierce. Other participants in the program were Jack Arntzen, George Bennett, Tom Cannon, Terry Johnson, Jim Knighton, Paul Mack, Martin Kyle, Kazuo Nishio, Tony Rashinskas, John Schilb, Gale Teats, Ken Tobias, Bill Walsh, Don Wenz, Irv Winsch, and Ray Wolson.

Chemistry of Liquid Metals

The Division emerged from the 1950s with a strong program in place on the chemistry of liquid metals under the leadership of Hal Feder and Irv Johnson, which was initially directed primarily toward liquid metal solvents for nuclear fuel reprocessing. Starting in the early 1960s, the research expanded into two other major areas, the use of liquid sodium as a reactor coolant and liquid lithium as an electrode material in regenerative galvanic cells and secondary (rechargeable) batteries. Both of these blossomed into major programs in the 1970s.

The continuing work on liquid metal solvents for pyrometallurgical fuel reprocessing was concerned mostly with liquid zinc, cadmium, magnesium, and copper as solvent metals for the skull reclamation and blanket processes being developed for EBR-II. Extensive data were obtained on the solubilities of the actinide and fission-product elements in zinc, cadmium, and to a lesser extent, binary solvent systems such as zinc-magnesium. Most of the data were obtained by taking filtered samples. In some cases partial or complete phase diagrams were obtained by sampling, thermal analysis, and emf measurements. Equilibrium solid phases were characterized by X-ray diffraction and metallography. Solubility data, together with emf and effusion measurements, provided information on the thermodynamic characteristics of the systems. In many cases, the equilibrium solid phases were intermetallic compounds.

An important example of this from a process standpoint is illustrated in Fig. 3-5, where the solubility of uranium in liquid cadmium is shown as a function of temperature. At temperatures up to 471.5°C, the uranium solubility increases with temperature and the equilibrium solid phase is UCd_{11}; above that temperature (the peritectic point), the solid phase is uranium metal with a retrograde solubility.

Other interesting observations were made with these liquid metal solvents. One is that tracer levels of solutes exhibited much the same coprecipitation behavior that is observed

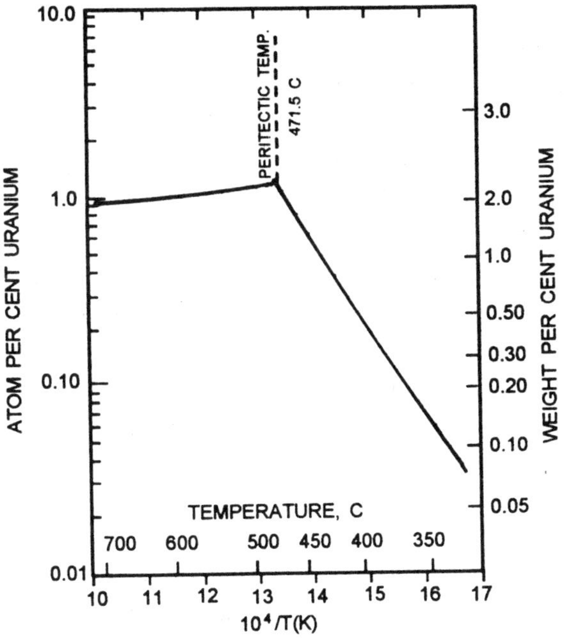

Fig. 3-5. Solubility of Uranium in Liquid Cadmium

in aqueous solutions, *i.e.*, they obeyed the Doerner-Hoskins Law:

log [(tracer in sol'n.)/(total tracer)]
= λ log [(carrier in sol'n.)/(total carrier)]

In general, the values of λ were near unity when the solid phases of the carrier and the coprecipitated material had similar crystal structures. Another interesting observation was that the solubilities of metallic elements in liquid zinc and in liquid cadmium showed a periodicity similar to that of the periodic table when they were plotted against their atomic numbers. The theoretical reason for this is not fully understood, but plots of this type were sometimes useful in gaining some idea of how a fission-product element might behave in a process when no other information was available.

Studies were conducted on the partitioning of uranium and other elements of interest between immiscible liquid metal phases such as aluminum-cadmium, lead-zinc, and bismuth-zinc. During the latter 1960s, the pyrochemical research work shifted largely to separations involving differences in the distribution behavior of fuel constituents between liquid metal and molten chloride solvents. The information from the above studies, together with distribution-coefficient measurements, provided the engineers with an excellent data base for process design work; it was also one of the most systematic, comprehensive programs ever to have been done on the basic chemistry and thermodynamics of liquid metal solutions. A major effort was also devoted to determinations of the distribution behavior of many elements between liquid metal solutions and molten salt phases. The distribution coefficient depends upon a chemical equilibrium when two elements are present in an oxidized form (such as a chloride) in the salt solution and a reduced form (the metal) in the liquid metal solution. For example, the distribution of uranium between liquid Mg-Zn and a salt containing $MgCl_2$ depends upon the equilibrium reaction:

$$U(Zn\text{-}Mg) + 3/2\ MgCl_2(salt) \leftrightarrow UCl_3(salt) + 3/2\ Mg(Zn\text{-}Mg)$$

The thermodynamics of reactions of this type have been reported extensively by Irv Johnson. Jim Knighton and his co-workers obtained experimental data for a large number of systems in which the liquid metal phases were zinc, cadmium, magnesium, and aluminum, alone or in mixtures, and the molten salt phases were mixtures of alkali metal and alkaline earth chlorides. Figure 3-6 shows the large differences in distribution coefficients of rare earth elements and some actinide elements (uranium, plutonium and curium) between molten $MgCl_2$ and liquid Mg-Zn mixtures. The solute-element distributions are changed markedly by altering the composition of the liquid metal phase, this figure being an example of one of the many systems that were studied. Perhaps the most

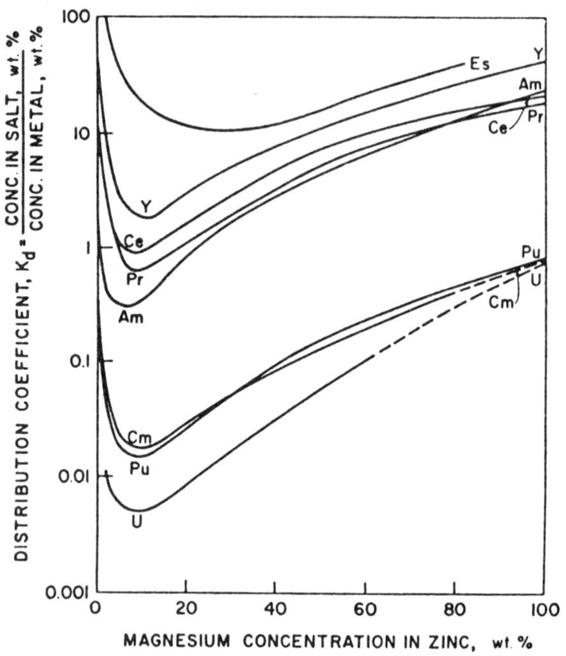

Fig. 3-6. Separation of Rare Earths from Actinides

"far-out" example of this type of separation is represented by a patent Jim Knighton and Bob Steunenberg obtained for the separation of californium from einsteinium. Although the separation worked like a charm, it is an excellent candidate for the patent least likely to ever succeed for commercial application. This work did, however, attract the interest of Glenn Seaborg because of its thermochemical implications.

Many individuals were involved in the foregoing work—Irv Johnson, Karl Anderson, Martin Chasanov, Pete Hunt, Jim Knighton, Al Martin, Ken Tobias, and Bob Yonco. Ewald Veleckis and Erv Van Deventer conducted the effusion studies. Bob Schablaske and Ben Tani played a crucial role in the phase identifications.

One person who livened things up quite a bit in the liquid metal research program was Gideon Hirsch, who came from Israel to spend a year or so at ANL. He was a very bright individual with an overwhelming personality that was basically incompatible with a structured organization. Upon his arrival, he displayed a rubber stamp with his name, identifying him as a physical chemist, educator, philosopher, political scientist, and several other qualifications. He claimed that he essentially had a Ph.D. with only some thesis work to complete, but the exact status of it seemed to be a moot point. During most of his assignment, he worked in Irv Johnson's group on the effects of attractive forces within liquid metal solvents. During his stay at ANL, he was given to doing things such as going on unauthorized travel and submitting the bill to ANL after the fact. He seemed to range all over the Chicago area, became acquainted with Studs Terkel, and visited Eric Hoffer on the West Coast. He requested an audience with Linus Pauling, but no one knows whether it took place. He kept trying to make appointments with the Lab Director Dr. Crewe, who, being an English gentleman, put up with two or three of them. Gideon finally left ANL, much to the relief of Everett Proud, went to work for a couple of companies, and eventually surfaced as a Wall Street broker. This memo, which he wrote just before leaving ANL, was typical:

March 16, 1967

TO: A. Crewe — Laboratory Director
FROM: G. Hirsch — Chemical Engineering

I would like the idea of giving a farewell seminar to the Laboratory on "Applied Philosophy."

If the subject is mine, I will choose, "Physical Chemistry, Society, Entropy, God, Computers, and Human Beings."

G. Hirsch

It might have been an interesting seminar, but apparently Dr. Crewe was not receptive to the idea.

The second area of liquid metal chemistry research, which began about 1963, led by Fred Cafasso in Irv Johnson's group, was concerned with metallic sodium. The main interest in sodium was its role as a coolant for EBR-II and other liquid metal cooled reactors, where impurities in the sodium such as oxygen, carbon, hydrogen, and nitrogen might result in corrosion or buildup of solid phases in the primary and secondary coolant systems. The initial studies were on carburization, purification, and analysis of liquid sodium. Thermodynamic studies were made on potential sodium reactions and on binary alkali metal systems. Analytical methods were worked out for oxygen and carbon in sodium, and potential purification schemes were investigated. During this period, some basic research was performed on the thermodynamics of liquid binary alkali metal systems, and ultrasonic methods were used to investigate possible complex species in liquid binary alkali metal systems. Another piece of work that deserves mention in this connection is a study in which Ira Dillon determined the critical properties of several alkali metals.

In 1968, an applied sodium technology program was initiated to develop methods and processes to handle, analyze, treat, and maintain a purity of sodium suitable for a reactor coolant. This expanded effort included the following chemistry studies:

1. Development of methods for handling and studying liquid sodium from its melting point to approximately 800°C under conditions that preserve the integrity of the liquid metal.
2. Development and refinement of analytical methods for the determination of carbon, oxygen, nitrogen, hydrogen, and tritium in liquid sodium.
3. Development for in-line monitoring of these impurities in liquid sodium.
4. Investigations of carbon behavior—reactions with other elements, characterization of compounds formed, and transport of carbon through liquid sodium systems (decarburization of ferritic steels, and carburization of austenitic steels) that would change the physical properties of the steels.

Some of the chemical studies involved the decarburization of steels through the formation of disodium acetylide (Na_2C_2). Exposure of Fe-0.11 wt% C alloy to sodium at 400 and 650°C resulted in formation of the acetylide. The solubility of helium in liquid sodium was measured, and the role of surface diffusivity in its mass transfer in sodium was studied by immersing samples of α-iron in liquid sodium at 800°C and examining the surfaces of the specimens.

The individuals participating in the sodium chemistry program were Hal Feder, Fred Cafasso, Jim Althoetmar, Karl Anderson, Sanat Dhar, Jim Eberhart, Pete Hunt, Loretta Jones, V. Khanna, Wally Kremsner, Charles Luner, Dick Murray, George Redding, Ewald Veleckis, Don Vissers, and Bob Yonco.

In the energy-conversion program, liquid sodium-lead, sodium-bismuth, lithium-bismuth, and lithium-tin alloys were used as positive electrode materials in thermally regenerative galvanic cells, which are described in the section on energy conversion. Liquid lithium was also being considered as a blanket/coolant material for fusion reactors, which became a subject of interest to CEN in the 1970s. From a theoretical standpoint, liquid lithium metal, because of its highly electropositive character and lightweight, would be an excellent positive electrode material for a storage battery. Attempts to use it, however, were largely thwarted because of its aggressiveness toward insulating materials and its solubility in LiCl-based molten salt electrolytes, which led to high self-discharge rates. These problems were solved by alloying the lithium with aluminum to reduce its activity, but at the expense of somewhat lower performance.

The basic research program on lithium included the following:

1. Extensive studies were made of the thermodynamic properties, including phase relationships, solubilities, and free energy functions of the systems Li-Se, Li-Te, Li-Sn, Li-H, Li-T (tritium), Li-C, Li-O, and Li-Al.
2. Experimental studies were conducted on the handling of liquid lithium from its melting point (179°C) to approximately 1000°C under conditions that preserved the integrity of the metal.
3. Analytical methods were developed or refined for the determination of H, O, N, and C in liquid lithium. Most of the facilities for these methods became an established part of the Analytical Chemistry Laboratory.
4. The compatibility of liquid lithium with molten halide salt systems of various compositions was investigated to select the salts that could be used as electrolytes with lithium electrodes.

The liquid lithium work, headed up by Art Tevebaugh and Carl Crouthamel, was performed by Elton Cairns, Kathleen Davis, Al Fischer, Mel Foster, Ellen Hathaway, Bob Heinrich, Stan Johnson, Gene Kucera, Vic Maroni, Jim Peck, and Bob Shablaske.

Fluoride Volatility Processes

In 1960, the development of fluoride volatility processes consisted of three activities:

1. The Direct Fluorination Process for the recovery of uranium and plutonium from Zircaloy-clad uranium oxide fuel elements of the type used in the Dresden power plant.
2. The Aqueous-Dissolution-Fluorination (ADF) Process for enriched uranium-zirconium alloy fuels, which involved aqueous dissolution of the fuel, fluidized-bed drying of the solution to a dry powder, and fluorination to recover the uranium as UF_6.
3. The Fused Fluoride Volatility Process in which zirconium-uranium fuel elements were dissolved in molten NaF-ZrF_4 by passing a mixture of HF and fluorine through the system to recover the uranium as UF_6.

Although the technical feasibility of the latter two processes had been fairly well demonstrated, further work indicated that the most promising approach was to use gaseous reagents on the solid fuel materials in fluidized beds. Such beds offer many advantages, including control of the reagent concentrations and addition rates, excellent heat-transfer characteristics, and ease of remote materials handling. Therefore, work on the direct fluorination process for oxide fuels was expanded and development studies were started on fluidized-bed processes for the alloy fuels as well. During the 1960s Al Jonke was responsible for the engineering development; Jack Fischer was the Group Leader of the laboratory program until about 1964, when Martin Steindler was assigned to that position.

LABORATORY INVESTIGATIONS

The two largest efforts in the laboratory studies were directed toward the fluorination of oxide fuels and the chemistry of PuF_6. Some early experiments were conducted on the possibility of using sulfur tetrafluoride (SF_4) as a fluorinating agent for UF_4, PuO_2, PuF_4, UO_3, U_3O_8, UO_2, and UO_2F_2. The conclusion was that SF_4 acted as a fluorinating agent, but not as an oxidizing agent with these materials. Uranium hexafluoride was produced only when the uranium was present initially in the hexavalent state, *e.g.*, UO_3 or UO_2F_2, and the by-product of the reaction was thionyl fluoride (SOF_2). Additional studies showed that SF_4 reduced PuF_6, but not UF_6, to the tetrafluoride:

$$SF_4 + PuF_6 \rightarrow SF_6 + PuF_4$$

This reaction offered the possibility of using SF_4 as a chemical trap for removing PuF_6 from a gaseous mixture of UF_6 and PuF_6 by converting the PuF_6 to solid PuF_4.

The principal laboratory investigations on the direct fluorination process were directed toward the initial fluorination step. Direct fluorination of UO_2 to UF_6 had been demonstrated, and it was found that PuO_2 undergoes the same reaction with fluorine at 400°C. Various materials, including MgF_2, ZrF_4, CaF_2, and alundum (a refractory form of Al_2O_3), were tested as fluidized bed media for the fluorination reactions. The best results were obtained with an alundum bed in which the uranium was first removed to a satisfactory level, followed by a 10-hour fluorination at 350-550°C to remove 95-99% of the plutonium. A two-zone fluorinator was devised in which the lower bed contained the UO_2 fuel pellets and alumina and the upper bed contained only alumina. Oxidation with an oxygen-nitrogen mixture converted the UO_2 to U_3O_8 fines, which were transported into the upper bed where they were fluorinated. Systematic studies were made of the effects of bed height and temperature from 450 to 500°C, and 99.5-99.9% fluorination of the uranium was achieved. These studies were done with a 1.5-in.-dia fluidized bed reactor. The experiments used mixtures of fluorine, nitrogen, and oxygen as the fluidizing gas. The nitrogen served as an inert diluent to control the rate of the reaction, and the oxygen enhanced the fluorination by decreasing the particle size of the uranium oxides, thereby increasing their surface area. When the fluorinations were performed, the uranium tended to be fluorinated first, followed by the plutonium. Some degree of separation of the two was accomplished by this effect and could be enhanced by manipulating the composition of the gas stream.

The marginal stability of PuF_6 presented both difficulties and opportunities in the development of a process flowsheet. The vapor pressures of UF_6 and PuF_6 are roughly the same, so they can be volatilized together to separate them from most fission products, and then be separated by selective decomposition of the PuF_6 to form non-volatile PuF_4. The objective was to remove all the plutonium from the original fuel, separate it from the uranium and fission products, and then recover it as a purified product stream for reuse. The studies that were conducted on the chemistry of PuF_6 were far too extensive to describe here in detail, but some of the highlights are as follows:

1. Equilibrium constants for the reaction $PuF_4 + F_2 \leftrightarrow PuF_6$ were determined.
2. The selective reduction of PuF_6 by SF_4 and by bromine was evaluated.
3. Selective thermal decomposition by the reaction $PuF_6 \rightarrow PuF_4 + F_2$ was also investigated as a means of recovering plutonium from UF_6-PuF_6 mixtures. A detailed study was made of the kinetics and mechanism of the thermal decomposition of PuF_6 over a temperature range of 140 to 173°C. The experiments were conducted in packed and unpacked vessels to determine the effect of surface area.
4. Methods were developed for quantitative transfers of PuF_6. In experiments with 10-100 g of PuF_6, 98% transfers

from trap to trap were achieved. Fluorine had little effect in suppressing losses.

5. Extensive data were obtained on the self-decomposition rate of PuF_6 caused by the alpha activity of the plutonium.
6. Studies were conducted on the decomposition of PuF_6 by gamma radiation both in the absence and presence of other gases. The G values (molecules decomposed per 100 MeV absorbed) for PuF_6 was 7.5. This value was not affected significantly by the presence of helium, but krypton, oxygen, or nitrogen increased it.
7. X-ray diffraction results showed that the lattice constants for UF_6, NpF_6, and PuF_6 decreased in that order.
8. The absorption spectrum of PuF_6 was determined.
9. Solid-liquid and liquid-vapor equilibria for the system UF_6-PuF_6 were measured.
10. Various solid fluoride compounds were evaluated for use in chemical traps to separate UF_6, PuF_6, NpF_6, and volatile fission-product fluorides from one another. Beds of solid NaF, LiF, and MgF_2 were found to be the most useful for this purpose.

Some laboratory work was done in support of the process development effort on zirconium-enriched uranium alloy fuels containing 5 and 30 wt% uranium. The optimum recovery of uranium was obtained by treating the alloy with HCl gas in a fluidized bed of alundum to convert the zirconium to $ZrCl_4$ (sublimation point 331°C at one atmosphere) and the uranium to nonvolatile lower chlorides. Uranium recoveries as UF_6 of 99.5% were achieved when using either fluorine at 350°C or chlorine monofluoride (ClF) at 200°C as a fluorinating agent. It was also shown that HF gas would convert the $ZrCl_4$ to ZrF_4 to decrease the amount of fluorine required to convert the uranium to UF_6.

A number of other smaller research areas such as the kinetics of the reaction of fluorine with nickel and the confinement and treatment of PuF_6 in process off-gases were also explored in the laboratory program. The individuals involved in the fluoride volatility laboratory work included Peter Alfredson, Tom Baker, Tom Gerding, Bill Gunther, A. Hariharan, Roger Jarry, Carl Johnson, Bob Kessie, George Manevy, Howard Porte, George Redding, Jim Riha, Jim Savage, Bob Shablaske, Bill Shinn, Roberta Shor, Dave Steidl, Martin Steindler, John Stockbar, and Verne Trevorrow. Alfredson was on assignment from Australia, and Manevy was from the French Atomic Commission (CEA).

As an interesting sidelight, John Stockbar was moonlighting as a local police magistrate in Lockport at about this time.

PROCESS DEVELOPMENT

Aqueous Dissolution Fluorination and Fused Salt Processes. Both of these approaches were under consideration for the processing of fuels consisting of a small amount of highly enriched uranium alloyed and clad with metals such as zirconium and aluminum. Because of the high enrichment and small amount of the uranium, the plutonium levels were too low to warrant its recovery.

The Aqueous Dissolution Fluorination (ADF) process involved (1) aqueous dissolution of the fuel, (2) fluidized-bed drying of the solution to a dry powder, and (3) fluorination to recover uranium as the hexafluoride. These basic steps were demonstrated with about 95% recovery of the uranium. This scheme appeared to be technically feasible, but other process concepts without the aqueous dissolution appeared more attractive, and this program was discontinued.

The fused salt process, also for enriched uranium alloy fuels, consisted of immersing the fuel elements in a molten NaF-ZrF_4 molten salt and sparging the system with hydrogen fluoride gas at temperatures of 600-700°C, which converted the fuel constituents to the fluorides. The resulting UF_4 was then converted to volatile UF_6, which could be recovered in a cold trap. Due to the highly corrosive nature of the system, a graphite dissolver was required, and a frozen wall technique was developed. This technique consisted of heating the molten salt internally while the vessel wall was kept at a temperature such that a layer of solid salt was maintained on its inner surface. This process concept also seemed to be feasible in principle, but there were concerns about its operation and maintenance in a fully remote facility, and the effort was discontinued.

At this stage of development of fluoride volatility processes, it appeared that the greatest promise lay in all-volatility schemes using fluidized beds and gaseous reactants.

Direct Fluorination Process for Alloy Fuels. Both Brookhaven (BNL) and ANL were involved in this development, and it was aimed primarily at alloy fuels that contained low levels of highly enriched uranium in zirconium, although some additional work was done with aluminum and stainless steel systems. The basic steps of this process were as follows:

1. Decladding and conversion of the zirconium and uranium to their tetrachlorides by reaction with gaseous HCl in a fluidized bed of alundum at 350-450°C.
2. Conversion of the tetrachlorides to the tetrafluorides by passing gaseous HF through the bed at 270-350°C.
3. Volatilizing the uranium with fluorine gas at 200-500°C to produce UF_6.

In the early work on this process, many scoping studies were made on decladding reagents and conditions, development of suitable fluidization equipment, and optimization of operating conditions.

A pilot plant with a 6-in.-dia fluidized bed unit and a 30-kg capacity was constructed, in which the direct fluorinations were conducted. Eighteen runs were made, during which 24 fuel elements containing either natural uranium-zirconium or uranium-aluminum alloy were tested. Performance of the equipment was generally trouble-free. The throughput rate was 3 kg/hr for both fuel alloys, and uranium recoveries were greater than 99%.

This work was followed by high-activity-level experiments with irradiated fuels conducted by Al Chilenskas and co-workers in the Senior Cave facility. These fuels included high-enrichment uranium-Zircaloy and uranium-aluminum alloys, low-enrichment uranium oxide, and mixed uranium-plutonium oxide. The objective was to obtain information on (1) the distribution of fission products, (2) decontamination factors that might be realized, and (3) the effects, if any, of irradiation and fission products on uranium recovery.

For the alloy fuels, a simple two-step procedure was used in which the fuel was treated with HCl in a 1.5-in.-dia fluidized bed to form volatile chlorides of zirconium or aluminum, which were vaporized. The non-volatile uranium chlorides (UCl_3, UCl_4) were collected in a packed bed filter and then treated with a nitrogen-fluorine mixture to form UF_6, which was collected on sodium fluoride (NaF) pellets. Further purification of the UF_6 could be achieved by subsequent desorption and collection in cold traps. The zirconium or aluminum chloride in the HCl off-gas was collected as a solid in a condenser, and the HCl was removed by a filter-scrubber arrangement. Fluorine was removed from the nitrogen-fluorine off-gas by a bed of activated alumina (Al_2O_3).

The removal of uranium from the reactor and filter beds was similar to that found in the pilot-plant operation, which suggested that the pilot-plant data could be used for irradiated fuels. Fission-product iodine and krypton were both volatilized completely during the hydrochlorination step. Fission-product cerium, cesium, and strontium were essentially non-volatile and remained in the reactor and filter beds. The ruthenium was basically non-volatile—less than 1% volatilized and that was collected in a magnesium fluoride (MgF_2) fission-product trap. The fission products that were predominantly volatile were molybdenum, technetium, and tellurium. About 21% of the neptunium was found with the uranium. The NaF trap removed most of the remaining activity, giving high overall decontamination factors in the range of 10^6 to 10^7. The results of this work, together with the pilot-plant experience, indicated that the direct fluorination process for alloy fuels was a viable candidate for plant-scale application.

Direct Fluorination Process for Oxide Fuels. Work had started in the late 1950s on direct fluorination schemes for processing uranium oxide fuel of the type used in commercial power reactors. This is low-enrichment (3% or so) uranium dioxide, which is in the form of pellets typically one-half inch in diameter and length. The pellets are clad with Zircaloy or stainless steel tubing to form fuel elements. Plutonium is bred in the fuel during irradiation and is present in low concentrations along with fission products.

The early engineering experiments were conducted with UO_2 pellets in beds of fluidized MgF_2 that were 3, 6, and 9 in. in diameter. The fluorination reaction is highly exothermic, so provisions had to be made both for control of the reaction rate and for heat removal from the bed. Fortunately, the fluidized bed was very effective in conducting heat to the vessel wall, which was air-cooled. (The excellent thermal conductivity of fluidized beds was well known, but little information was available as to how effective they would be in a bed packed with fuel pellets, so a large effort was put into basic studies of this type of situation.) The investigators also found that the reaction rate could be regulated easily by controlling the rate of fluorine addition, and development was initiated on an automatic temperature-control system, using this concept. Although the reaction rate was proportional to fluorine concentration, the efficiency of fluorine utilization was essentially independent of it. Several materials, including MgF_2, ZrF_4, CaF_2 and Al_2O_3 (alundum), were tested as the granular materials in the fluidized bed experiments. Alundum showed the best overall performance, and it was used throughout the remainder of the program with good results.

The fluorination of UO_2 in a fluidized bed, because of its heterogeneous nature, introduced a large number of variables that had to be addressed in the engineering work. For this reason, many experimental studies had to be conducted on a meaningful scale during process optimization. One such variable was the composition of the fluorine-bearing gas used in the process. The early runs were made with nitrogen as the diluent. In this case, the fluorination reactions were:

$$UO_2 + F_2 \rightarrow UO_2F_2$$
$$UO_2F_2 + 2\,F_2 \rightarrow UF_6 + O_2$$

The two reactions occurred simultaneously, and the UO_2F_2 spalled off the pellet as it was formed. The use of oxygen as a diluent converted UO_2 to U_3O_8, which also spalled off the pellet, and some of this effect may have occurred due to the oxygen produced by the fluorination reaction. To take advantage of these reactions, a two-zone bed design was adopted, in which the UO_2 pellets underwent the primary fluorination as a packed bed in the lower zone, and the finely divided intermediate products rose to the upper zone. Both zones contained fluidized alundum. This arrangement eliminated caking and channeling problems that had been encountered in the single-stage beds. The beds were equipped with blowback filters to return fines to the fluidized region.

Several methods of fuel decladding were evaluated. In some of the early work, Zircaloy cladding was removed chemically by a gas mixture of 10% HCl-90% HF at 500°C, which converted the zirconium to the tetrafluoride. An HF-O_2 mixture was used to remove the stainless steel cladding. Another approach was oxidative decladding, in which the cladding was punctured or slit to admit oxygen, which converted the UO_2 to U_3O_8 (a method also tried in the pyrometallurgical processes). The resulting powder in the ruptured fuel segments was then fluorinated to recover the uranium as UF_6. To achieve complete recovery, measures such as gas pulsing, reduction, and reoxidation of the material, and mechanical vibration were investigated. It appeared that several decladding options were feasible, depending on the nature of the fuel.

Some work was done on the use of BrF_5 as a fluorinating agent. This compound converted U_3O_8 to UF_6, and the PuO_2 to PuF_4. This offered the possibility of separating the uranium and plutonium by a two-stage fluorination in which BrF_5 was added first to fluorinate the uranium, followed by fluorine addition to volatilize the plutonium. The BrF_3 by-product in this case could easily be refluorinated to BrF_5 in a separate step.

Engineering-Scale Alpha Facility. A particularly impressive achievement of the fluoride volatility program was the design, construction, and operation of an alpha facility where work with plutonium could be done on a pilot-plant scale. John Vogel had the primary responsibility for this job, and he used the Critical Path Method (CPM), a project management procedure that was relatively new at the time, in planning the design and construction of the facility. A simplified diagram of the facility is shown in Fig. 3-7. The major components were two alpha-containment boxes to house the process equipment. The larger box was 17.5 ft high, 27 ft long, and 3.5 ft wide; the other was 10 ft high, 13 ft long, and 3.5 ft wide. Although the process was to be operated remotely, the boxes, which were based on the standard CENHAM glove-box design, were equipped with windows and gloves, and platform elevators were provided for the workers. The boxes were equipped with bagout ports, electrical service, air line inlets, filters, and sphincter openings. The larger box, which was to house the main process equipment, had two large bagout ports under an overhead crane, and it was located inside a negative pressure cell as a further safety precaution. In addition to the two boxes, the facility also included a large control panel.

The major equipment items in the larger box were a fluidized-bed fluorinator (3-in. dia and 6-ft long); two series-connected 4-in.-dia,

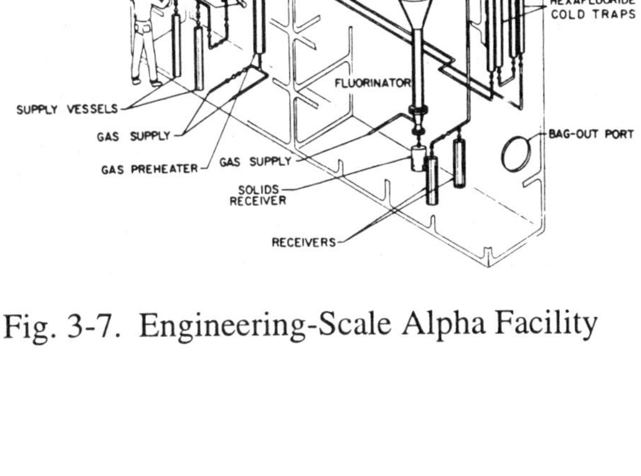

Fig. 3-7. Engineering-Scale Alpha Facility

10-ft-long condensers chilled by trichloroethylene; and 4-in.-dia Al_2O_3 traps. The smaller box contained metering equipment for the fluorine, oxygen, and nitrogen streams. A particular concern with the fluoride volatility processes was the hazard associated with PuF_6, which is chemically and radiologically toxic, is extremely reactive chemically, and exists in a gaseous form. Any possibility of its release to the outside had to be eliminated. Bob Kessie conducted extensive tests and studies on this problem and found that a combination of hydrolysis by humidified air and filtration was effective in removing PuF_6 from exit gas streams. This work involved an in-depth study of the mechanisms involved in the hydrolysis reaction and filtration.

Work on the alpha facility began in 1962, and it was ready in 1965 for shakedown runs, which were conducted with 9-kg batches of UO_2. One impressive achievement was the production of PuF_6 from PuF_4 on a kilogram scale. A 2.3-kg batch was fluorinated in the fluidized bed with product recoveries of 79, 101, and 99% and rates of 2-6 lb PuF_6/hr per square foot. In 1965, the AEC had decreed that development of the fluidized-bed fluoride volatility process was to become a cooperative effort by four AEC labs: Argonne, Brookhaven, Oak Ridge National Laboratory, and the Oak Ridge Diffusion Plant. There was also a great deal of industrial interest in the process at the time. In 1968, ANL was directed to change the emphasis from light water reactor fuels, which contain about 0.5-1.0% plutonium, to fast breeder fuels, where the plutonium content is about 20% in the core and 4-5% in the blanket. Conceptual flowsheets were devised for this application (Fig. 3-8). In 1969, the fluoride volatility processes were included in the national policy decision to discontinue essentially all of the nuclear fuel reprocessing work in the U.S.

Conversion of Uranium Hexafluoride to Uranium Dioxide. The final product of fluoride volatility processes is uranium hexafluoride (UF_6), which may also contain plutonium hexafluoride (PuF_6), depending on the type of fuel being processed. In the 1960s, the fuel of choice for most reactors was uranium dioxide, so there was a potential need for a process to convert the uranium hexafluoride to the dioxide in a ceramic-grade form that was suitable for the refabrication of new fuel. The method that was selected for development work was a combination of steam pyrohydrolysis and hydrogen reduction in fluidized-bed reactors. The basic reactions are

$$UF_6 + 2\,H_2O \rightarrow UO_2F_2 + 4\,HF$$
$$UO_2F_2 + H_2 \rightarrow UO_2 + 2\,HF$$

It was discovered early in the program that these reactions could be conducted simultaneously with a mixture of steam and hydrogen, or sequentially as shown. Fluidized beds were particularly well suited for this application because of their capabilities for good heat removal, rapid reactions resulting from the large solid-gas surface area, and their flexibility in controlling the conditions to achieve the desired particle-size distribution of the product.

Both the simultaneous and sequential approaches were investigated using 3-in.-dia fluidized bed reactors and static bed depths up to about 21 in. Most of the work was done first on the two-step sequential approach because of its greater flexibility, the main point being that it could be run at two different temperatures. A great deal of effort was required to solve problems such as bed caking and to determine optimum temperatures, bed depths, and the velocities and compositions of the fluidizing gas. The best results were obtained when the first step was operated at 230°C with a superficial gas velocity of 0.75 ft/sec, a static bed depth of

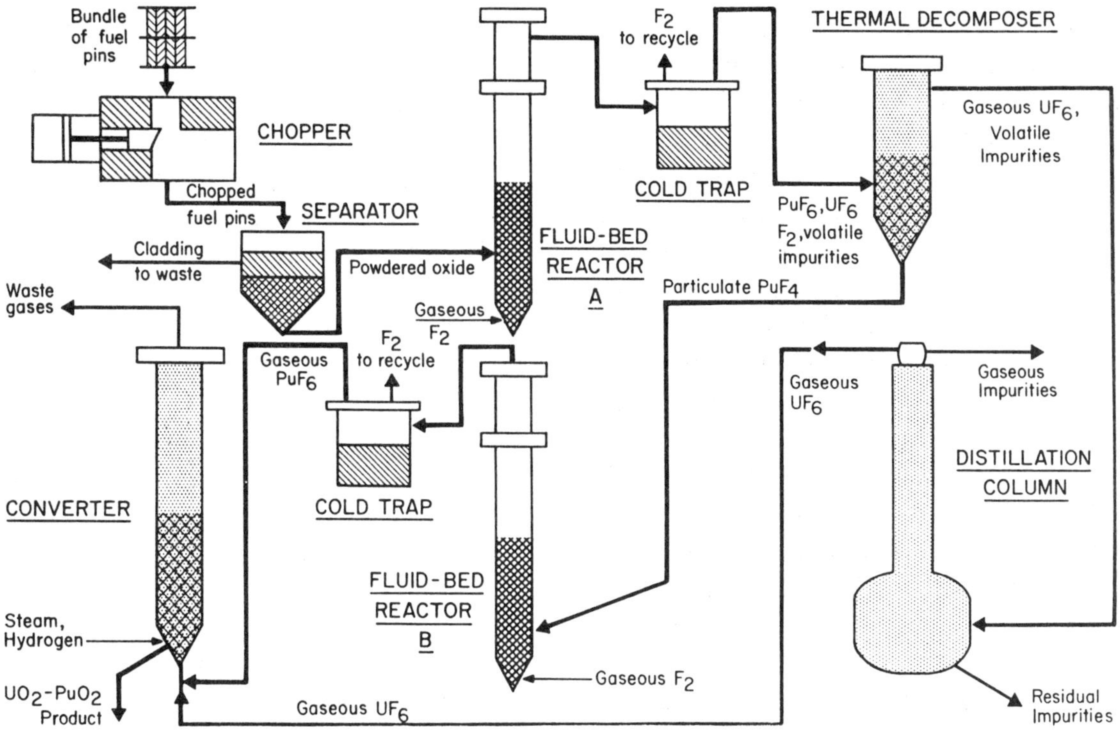

The UO_2-PuO_2 fast breeder reactor fuel pins are first chopped into short lengths, and the oxide fuel in the form of powder is separated from the metallic cladding material. The oxide fuel is charged into a fluidized bed of alumina (Al_2O_3) powder where the uranium and plutonium oxides are converted to their volatile hexafluorides by a gas stream containing elemental fluorine. The uranium is volatilized first, followed by the plutonium hexafluoride. Both are collected in condensers where the gaseous fission products, krypton and xenon, and the volatile fission-product fluorides (those of niobium and ruthenium), are removed. The uranium and plutonium hexafluorides enter a thermal decomposition step, where the plutonium hexafluoride is reduced to the non-volatile tetrafluoride. The uranium from this step is purified by a distillation-sorption procedure, and the plutonium is recovered by fluorinating the tetrafluoride back to the hexafluoride. The purified uranium and plutonium hexafluorides are then subjected to a hydrolysis reaction with steam and hydrogen to form a $(U,Pu)O_2$ powder suitable for refabrication of new fuel pellets.

Fig. 3-8. Fluoride Volatility Process for Fast Breeder Reactor Oxide Fuels

15 in., 245% excess steam over the stoichiometric requirement, 15% recycle of solids, and an average bed particle size of about 350 µm. Favorable conditions for the second step were a temperature of 650°C, a 50-50 mixture of steam and hydrogen, and a static bed depth of 21 in. The fluoride content of the product was less than 200 ppm, which is well below the acceptable maximum of 300 ppm.

The group then turned its attention to the one-step process. The effects of various operating variables were determined, with special attention being given to the particle-size distributions and densities of the products. Sintering in hydrogen was studied as a means of increasing particle density. The results indicated that either the one-step or the two-step process could be used to convert UF_6 to UO_2.

This work, which was performed by M. Jones, Irv Knudsen, Norm Levitz, Don Raue, and Takeo Tamura, is summarized in ANL-6902.

Calcination Studies. The calcination work, although it did not involve a fluoride volatility process, is included in this section because the fluoride volatility programs and other programs using fluidized-bed technology were closely connected and coordinated under Al Jonke's leadership. The following calcination studies, performed by John Loeding, Norm Levitz, Charlie Schoffstall and John Kincinas, were unusual in that a major objective in most process development programs is to scale up the operations for application in a large plant. In this case, the objective was to calcine plutonium or highly enriched uranium solutions in small columns to avoid criticality problems. The small units also reduced the quantity of off-gases to be handled.

The fluidized bed calciner was a 2.25-in.-dia column with a conical bottom in which nozzles were mounted vertically in the apex of the cone to spray the atomized solution upward. The spray nozzles and decomposition of the feed material provided most of the fluidizing gas. A dual blowback bayonet filter system was used at the top of the column. After some initial optimization work, 20 consecutive startups were made successfully with runs longer than seven hours. After some minor equipment modifications, runs of 5-14 hr were made at temperatures of 425-450°C, with frequent blowback every 3 or 4 min. The bed depths were 10-15 in. with residence times of 5.2 and 7.8 hr, respectively. This work had to be suspended due to manpower needs elsewhere, but the results showed that fluidized bed calcination was possible in small, criticality-safe equipment.

Basic Fluidization Studies. A large body of information on the care and feeding of fluidized beds had been and was continuing to be generated in the applied programs, and CEN was developing a reputation for its expertise in this technology. It was appropriate to devote some systematic investigations of more general applicability and greater theoretical depth. In 1960, Ken Williamson, a Ph.D. student from Pennsylvania State University, began a study of the downward transport of solids in a multistage fluidized bed without downcomers, using external vibration. The bed was contained in a three-stage, 6-in.-dia Pyrex and metal column. Individual external vibrators were arranged so that the vibrational energy was transmitted independently to each of the three bed support plates in the column by a metal rod. Glass beads served as the bed material. The throughput rate was controlled by the percentage of "on time" of the vibrators. A correlation was obtained for throughput rate as a function of feed particle density, shape, and size; column cross-sectional area; density and superficial velocity of the fluidizing gas; and depth and size of beads used on the bed support plates.

The second phase of Williamson's work was a study of the effects of operating variables on the adsorption of water by silica gel. Murphree stage efficiencies (ratio of actual change to that if the vapor composition reached equilibrium in the stage) were near 100%. These data permitted calculations of mass-transfer coefficients.

In 1963, John Gabor and John Savage began basic studies on the mixing of fluidized particles in beds packed with larger solid bodies. An example of this situation is the fluorination of uranium dioxide fuel pellets mentioned earlier, where the fuel pellets were too heavy to be fluidized and, therefore, constituted a packed bed in the column, while the fluidized material in the void parts of the bed was granular alumina. A major function of the alumina was to remove the heat of reaction from the bed by transferring it to the walls of the column. In the basic studies, copper and nickel shot were used as the particles. (This permitted the use of a magnetic technique for separating the two metals as an analytical method.) The rate of mixing was proportional to the size of the packing, which is consistent with a random-walk model for eddy diffusion in flowthrough packed beds. Mixing rates were higher in the absence of the fixed packing, and they increased with increased height of the fluidized bed. With fixed packing, baffling resulted in more uniform fluidization, and mixing was independent of bed height. An empirical correlation was developed for beds containing spherical packing, which relates the diffusivity of the fluidized solids to the packing size, the fluidized particle size, and the gas-flow rate.

An investigation was then started on heat conduction in the lateral direction in fluidized-packed beds and on longitudinal transport of particles. The thermal conductivities were correlated by an equation involving the density and heat capacity of the fluidized particle, the diameter of the fixed packing, and the fluidizing gas velocity and minimum fluidizing gas velocity, both corrected for packing void fraction.

A basic study was undertaken by Manfred Baerns and Devabhaktuni Ramaswami on the fluidization of fine particles of a size less than 50 μm. In this size range, frictional and interparticle attractive forces become significant and tend to cause channeling in the bed.

John Gabor and Ramaswami conducted a theoretical study of fluidized particle motion caused by a rising bubble and developed a computer program to simulate the eruption of a bubble at the bed's surface. John Holmes and Lowell Koppel, a faculty member from Purdue University, developed models to predict heat-transfer coefficients at the wall of a fluidized bed and conducted experiments on particle mixing and the relationship between particle residence time and heat transfer.

Ramaswami, Masaru Kobayashi, Bill Brazelton, and Charlie Schoffstall undertook a study to compare the behaviors of pulsed-flow and straight-flow fluidized beds and to determine the best design for the feed-gas distributor. Better mixing was obtained with the pulsed-flow beds, and a simple cone gas distributor proved superior to a sintered metal distributor for the pulsed beds.

Physical and Thermodynamic Properties of High Temperature Materials

In 1963 a systematic effort was begun to determine the thermodynamic properties of high temperature nuclear reactor fuels and materials under the direction of Russ Edwards in Hal Feder's section. Russ, a wiry, muscular individual, enjoyed outdoor athletic activities and, as a result, chose to live in the Indiana Dunes area, which meant about a 60-mile daily commute each way in his Volkswagen "beetle." On one occasion, he was seriously injured in an auto accident that few people would have survived, but apparently due to his excellent physical condition he recovered completely. A somewhat professorial type, he was exacting about his experimental work and liked to interpret his results in terms of theoretical principles.

For the first few years, this program was concerned almost exclusively with the phase relationships and thermodynamics of uranium dioxide, which was the fuel of choice for most reactors under consideration at that time. These studies were aimed primarily at the U-UO_2 region of the uranium-oxygen system.

The initial work consisted of five activities:

1. Effusion vapor pressure measurements by Malleseety Chandrasekharaiah and Phil Danielson, using a Knudsen effusion apparatus.
2. Mass spectrometric effusion studies by John Reishus and Phil Danielson, using a Bendix time-of-flight mass spectrometer to measure species in the gas phase.

3. Transpiration studies by Marv Tetenbaum and Carol Clifton, in which an inert gas was passed over a heated sample to determine the identity and composition of species in the gas phase.
4. Phase-diagram investigations by Al Martin, Leo Yannopoulos, and Frank Mrazek, using conventional thermal analysis and metallographic techniques.
5. Interdiffusional studies by those same individuals to evaluate the compatibility of refractory metals such as tungsten, rhenium, and molybdenum as containment and support materials for the fuel materials under investigation.

Some of this work was done as a collaborative effort with Ray Ackerman and Bob Thorn of the Chemistry Division, who made certain equipment available for the purpose. This integrated approach to the uranium dioxide studies gave the work a wide scope and permitted cross-checking of results obtained by different methods.

A U-UO_2 phase diagram that evolved from these studies is shown in Fig. 3-9.

An outstanding characteristic of this system is a wide liquid miscibility gap above the monotectic temperature of 2470°C. At the monotectic temperature, the O/U ratio units for the three condensed phases in mutual

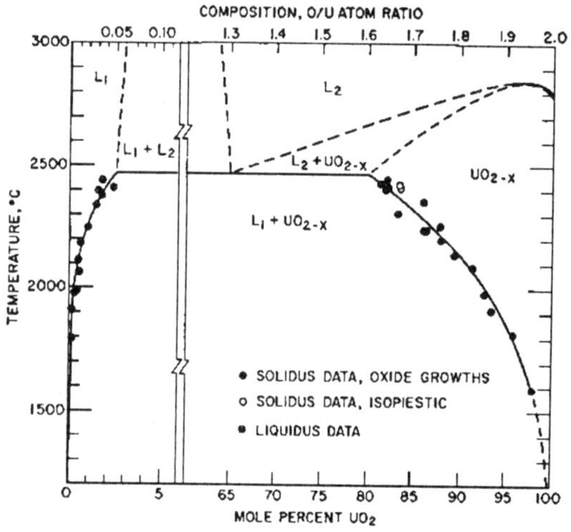

Fig. 3-9. Uranium-UO_2 Phase Diagram

equilibrium are liquid uranium, 0.05; monotectic liquid, 1.30; and hypostoichiometric UO_2, 1.61. Through a combination of the effusion studies, transpiration experiments, and mass-spectrometric measurements, the vapor pressures and compositions for the system shown in Fig. 3-9 were characterized in detail. This information was used to develop a thorough thermodynamic description of the system, which was valuable both as a basic scientific contribution and as a guide in interpreting phenomena observed during examinations of irradiated reactor fuels.

After his arrival at ANL in 1964, Jim Battles joined Russ Edwards in a study of the vapor deposition of tungsten claddings on complex W-UO_2 cermets. This was a part of the ROVER project for nuclear-powered space vehicles, which was a highly classified program at that time. Other studies were undertaken on systems involving refractory metal oxides. A study of the system U-W-UO_2 showed that the miscibility gap on the U-UO_2 side extended into the interior of the ternary phase diagram. Paul Blackburn came to CEN in 1967, and replaced Russ Edwards as the Group Leader in 1968. The program was expanded to include plutonium to provide information on potential oxide fuels for LMFBR applications. This effort included phase-diagram studies of U-Pu-O by Al Martin and Frank Mrazek; mass-spectrometric vaporization investigations by Jim Battles, Gordon Gunderson, John Reishus, and Bill Shinn; effusion measurements by Phil Danielson, and transpiration studies by Paul Blackburn and Steve Banaszek.

For the mass-spectrometry work with plutonium, a glove box was designed and constructed to house the source end of a time-of-flight mass spectrometer, which included a high-temperature Knudsen cell as the inlet system. The initial phase of this study, by Jim Battles, John Reishus, and Gordon

Gunderson, was to find a material for the Knudsen effusion cells that was completely compatible with the uranium and plutonium oxides at high temperatures (>2000°C). The compatibility studies showed that tungsten and molybdenum react with $UO_{2.08}$ at relatively low temperatures, as evidenced by vapor species such as W_3O_9 and Mo_3O_9. A study of the Re-O system showed that an unusual vapor species, Re_2O_7, was the predominant species at 800°C. Vapor pressures of Re_2O_7 over the two systems Re-ReO_2 and ReO_2-ReO_3 were determined. Further compatibility studies identified iridium as being completely compatible with the uranium and plutonium oxides, and it was used in all the subsequent mass-spectrometry work. Based on this work, iridium was also selected as the cladding material for $^{238}PuO_2$ power sources in space applications. Reishus left ANL in 1967 to obtain a doctor of divinity degree, and Gunderson left for other opportunities. About that same time, Bill Shinn returned to the group.

During the period 1967-70, the mass spectrometric-Knudsen diffusion studies were completed on the systems Pu-O, U-Pu-O, and Na-U-O. Jim Battles and Bill Shinn determined the partial pressures of plutonium oxide vapor species, and enthalpy data were derived from the results. These were all first-of-a-kind studies. Complementary investigations of the U-Pu-O system included phase-diagram studies by Al Martin and Frank Mrazek, who established the metal-rich boundary of the U-Pu-O fluorite phase, using equilibration and metallographic methods, effusion measurements by Phil Danielson, and transpiration experiments by Paul Blackburn and Steve Banaszek. Paul Blackburn and Hong Yih Chen investigated possible reactions between cesium oxide and iodine that might occur in irradiated fuel. Their initial results suggested a possible Cs-U-O ternary compound.

Although the primary emphasis at the time was on oxide fuels, both for light water and fast breeder reactors, there was also some interest in the carbides, sulfides, and phosphides of uranium and plutonium, all of which are highly stable ceramic materials. Reishus, before leaving ANL, had conducted some studies on UN and UP. Mass-spectrometric Knudsen effusion studies were planned for the systems Pu-C, U-Pu-C, PuP, and PuS. Jim Battles, Russ Edwards, and Bill Shinn measured the pressures of plutonium gas over the two-phase regions of Pu_2C_3-C, Pu_2C_3-PuC_2, and PuC_2-C and determined the enthalpies of the formation reactions. Marv Tetenbaum and Pete Hunt used transpiration methods to measure the carbon activity and the total pressures of binary carbon species in the vapor phase as a function of temperature and composition. These data provided information on the locations of phase boundaries in the U-C system at high temperatures and estimates of the free energies and heats of formation of UC. Further work on these systems was terminated in 1970 as a consequence of priorities established for the LMFBR programs.

Reactor Safety

Through the mid-1960s, the Chemical Engineering Division continued a large reactor safety program to determine the nature and extent of chemical reactions that could occur during a reactor accident. The purpose of these studies was to provide information that would aid in minimizing the hazards of such accidents. The main areas of research were (1) metal oxidation and ignition kinetics, (2) metal-water (steam) reactions, and (3) studies related to fast reactor safety.

METAL OXIDATION AND IGNITION KINETICS

Work in this area was focused on the kinetics of air oxidation of metallic uranium and plutonium up to temperatures approaching their melting points. Oxidation rates for both metals were determined as a function of temperature and characterized (parabolic or linear), and the protective effects of oxide films were assessed. An interesting phenomenon in the case of plutonium, which undergoes six phase changes between room temperature and 600°C, was a change in the plutonium oxidation rate when a change of phase occurred.

METAL-WATER REACTIONS

In water-cooled reactors, coolant failure or a severe nuclear excursion could cause the reactor core metals to melt and disperse rapidly in water. Chemical reactions between the dispersed metal particles and the water could release an amount of energy approaching or exceeding that released by the fission process during a nuclear excursion. Realistic estimates were needed of the rate and extent of chemical reactions during hypothetical reactor accidents.

Among the metal-water reactions that were studied were stainless steel-steam at 1400°C, steam at 1500°C with UO_2 fuel elements clad with stainless steel or Zircaloy-2, and UO_2-steam at 1500°C. Small autoclave experiments were conducted in the Transient Reactor Test Facility (TREAT) facility, a pulsed thermal reactor used for safety studies. This reactor, located at the National Reactor Test Station (NRTS), went critical in 1958. The purpose of these experiments was to determine the reaction between fuel and water during reactor excursions. Several fuels were studied, the two principal ones being UO_2 fuel-pin clusters having Type 304 stainless steel and Zircaloy-2 cladding. The resulting data were used to make predictions of the extent of metal-water reactions that would occur during a loss-of-coolant accident in a water-cooled reactor.

FAST REACTOR SAFETY STUDIES

The purpose of these studies was to determine the consequences of core melting and rapid dispersion of reactor materials in the liquid sodium coolant due to coolant failure or a rapid reactor excursion. In one set of experiments, small samples of metals heated to temperatures between 1500 and 2400°C were dropped into liquid sodium, which was at 250°C. The list of metals comprised zirconium, vanadium, nickel, Type 304 stainless steel, molybdenum, tantalum, and uranium. None of them reacted with the sodium, but the sudden quenching in sodium produced severe fragmentation, especially with the stainless steel. Similar behavior was observed with UO_2. Autoclave experiments were also conducted in TREAT to measure the expulsion of sodium during sharp reactor transients and to determine the extent of fuel fragmentation. Such information was used to estimate the migration of fuel particles in a fast reactor during and following an excursion. Studies were also made of sodium-air and sodium-water reactions using sodium spray-injected into a reaction vessel to simulate the violent expulsion of sodium from a reactor core.

At the end of Fiscal Year 1968 (September 30, 1968), the Reactor Safety program was transferred to the newly formed Reactor Analysis and Safety Division (RAS).

Lou Baker headed up the Reactor Safety program. Art Tevebaugh and Dick Ivins were also involved in management of the program. Other participants were Don Armstrong, John Barghusen, Charles Barnes, Jim Bingle, Jim Boland, John Cassulo, Martin Chasanov, Dae Cho, Marshall Deerwester, Don Fischer, Bill Gunther, Larry Harrison, Jim Hesson, Len Leibowitz, Bob Liimatainen, Larry Mishler, Kazou Nishio, John Pavlik, Jim Peck, Chester Rogers, Martin Silverman, Carl Stretch, Dave Swift, Frank Testa, Kegham Varteressian, and Bob Wilson.

Chemistry of Irradiated Fuels

A program was initiated, primarily by Carl Crouthamel, with the objective of obtaining fundamental chemical data on irradiated fast reactor fuels to gain an understanding of the chemical processes that occur within the fuel during the irradiation. The initial studies were directed to carbide and oxide fuels. This work required a number of specialized facilities and equipment, and an inert atmosphere enclosure was installed in one cell of the Chemistry Division cave facility in Bldg. 200.

An impressive array of new instruments was assembled, including a Leitz remote metallograph, a spark-source mass spectrometer, a 4096-channel analyzer with two solid-state detectors for γ-ray spectroscopy, a surface ion-emission spectrometer, and a shielded electron microprobe. The first studies, made by Norm Chellew, Max Adams, and Chuck Honesty, were on pressed-and-sintered and arc-melted uranium monocarbide and on irradiated uranium dioxide. Microdrilling and laser sampling were used to obtain concentration profiles of the fuel constituents. The electron microprobe proved to be an excellent tool for the fuel examinations. Results from a stainless-steel-clad UO_2-20% PuO_2 fuel pin irradiated to 3.7 at.% burnup in EBR-II showed a fuel-cladding interaction, fission-product iodine at the fuel-cladding interface, and iron from the cladding in the fuel matrix. Somewhat similar effects were observed with a UC-20% PuC fuel pin.

The microprobe operators developed a procedure by which "maps" of the sample were made in which each element was represented by a particular color. This feature was useful in visualizing the migration of elements that had occurred during irradiation of the sample. These rather spectacular colored pictures of the sample surface were also very useful for public-relations purposes.

Some work was also done with tagging methods to identify positions of fuel failures in a reactor. One method involved spiking each fuel element in a subassembly with a unique ratio of two sodium-soluble radioactive isotopes that could be detected in the coolant sodium in the event of a leak. The other method involved the use of two xenon isotopes, again in unique ratios, which could be detected in the cover gas over the sodium coolant.

In 1969, the group, still under the direction of Carl Crouthamel, had expanded to include Harry Edwards, Ping-Kay Hon, Laury Ross, and Gus Staahl. Carl Johnson, Norb Stalica, Chuck Seils, and Karl Anderson did the electron microprobe work; Bob Meyer and Carl Johnson handled the sodium-soluble tagging studies; the xenon tagging was a joint effort between George Bernstein and Les Coleman at CEN and the EBR-II staff at the Idaho site.

Preparation of Fast Reactor Fuels

The preparation of metallic and oxide fuels for fast breeders was mentioned in the previous discussions of pyrometallurgical and fluoride volatility processes, respectively. Work was also done on several other more exotic materials that were under consideration as potential fuels. In 1960, Carl Crouthamel and Bill Knapp terminated a study they had been doing on a "Slurrex" process for making a ThO_2-10% UO_2 ceramic fuel, which involved a gel precipitation step and firing the resulting material to 1000°C.

Carbide fuels were being considered both for high temperature gas-cooled reactors (HTGRs) and certain other types of reactors. As an outgrowth of the pyrometallurgical program, work was initiated on schemes for producing mixed uranium-plutonium carbide fuel material by precipitation from liquid metal solutions. The first experiments used hydrocarbon gases (methane, ethane, propane, acetylene) as the carbon source; these gases were bubbled through a solution of uranium in molten zinc-magnesium. In later work, activated charcoal and graphite were added directly to the solution. The procedure consisted of adding the carbon to a Zn-12 wt% Mg-15 wt% U solution and agitating the system for several hours at 800°C. The product was recovered by pouring off most of the supernatant liquid metal and removing the remainder by vacuum retorting. These studies were done in cooperation with George White of MET (the Metallurgy Division, now Materials Science), who evaluated the fabricability of the products. The material sintered readily and made excellent fired pellets with densities greater than 90% of theoretical value. The procedure was scaled up to produce 0.5-1.0-kg batches of material for the fabrication tests by MET.

A fluidized-bed process for the preparation of uranium monocarbide (UC) was also investigated. In this process, 300- to 600-g batches of (U-15 wt% Pu) fuel pins were hydrided at 250°C by a hydrogen-2 vol% methane mixture to form a fluidizable powder. The material was then dehydrided by continuing the operation at a temperature of 500°C with the same gas mixture. The product from this step was fluidized with a hydrogen-6 vol% methane mixture at 800°C and 2 atm pressure. The hydrogen/methane ratio was critical in avoiding the formation of plutonium sesquicarbide (Pu_2C_3), but, if necessary, this compound could be reduced back to the monocarbide by treating the product with hydrogen. Several batches of (U-20% Pu)C were prepared and fabricated into pellets by pressing and sintering.

A second method of preparing carbide fuels from the oxides entailed the reaction of UO_2 with graphite at 1500-1800°C in a bed fluidized by an inert gas. There were some problems with sintering of the UO_2, but a projection was made that a 30-kg/day plant-scale operation could be achieved with a 12-in.-dia fluidized bed. Preliminary work was also done on the use of a plasma torch to conduct this reaction, but the program was terminated due to a fading interest in carbide fast reactor fuels.

A small program was directed to the conversion of uranium metal to the dioxide to prepare oxide fuel for a breeder reactor from the metallic uranium product of a pyrometallurgical process. The uranium was first hydrided-dehydrided in a fluidized bed, and then treated with a CO_2-inert gas mixture for 4-9 hr with the temperature increasing from 400 to 825°C. When the carbon content of the product was found to be too high, a partial nitriding reaction prior to the CO_2 addition resulted in a stoichiometric product with an acceptable carbon content (<100 ppm).

Some work was done on the preparation of uranium monosulfide, another refractory material that was in contention as a fast reactor fuel. This preparation involved: (1) hydriding-dehydriding of uranium to form a fine powder, (2) heating the powder with the stoichiometric amount of hydrogen sulfide (H_2S) at 400-500°C, and (3) heating the resulting mixture to 1900°C to convert any unreacted uranium and higher uranium sulfides to the monosulfide. About 2 kg of this product was prepared in 100- to 300-g batches to provide the MET Division with material for fabrication and irradiation tests.

Finally, research was conducted on a paste of uranium nitride (UN) in liquid sodium, which had been proposed as a semi-fluid blanket for fast breeder reactors. The idea was that this material could be circulated through tubes, which would serve as heat exchangers and would have some advantages over solid blankets: (1) ease of fission-product gas removal, (2) the possibility of frequent or continuous processing, and (3) reduction of thermal gradients in the blanket. The UN was prepared by exposing uranium shot to nitrogen first at 900-1050°C, followed by several hours at 1350°C. When mixed with sodium, this material formed a paste that had the degree of fluidity needed for the purpose.

Terry Johnson took the lead on the initial carbide studies, assisted by John Trischan. Dean Pierce, John Lenc, John Pavlik, and Marion Bowden worked on the precipitation studies, and Irv Winsch, Tom Cannon, and John Schilb handled the plutonium experiments. Don Armstrong, G. Gorth, Dave Grosvenor, John Holmes, Paul Krause, Paul Nelson, John Pavlik, Charles Payne, Ed Petkus, Nick Quattropani, D. Ramaswami, and Sy Vogler did the fluidized-bed and plasma-torch studies. Sy Vogler and John Trischan carried out the sulfide studies, and Paul Nelson, Martin Chasanov, and Clarence Lehmann did the work on paste blankets.

Preparation of High-Purity Plutonium-238 Metal

The Division was requested to investigate the potential use of pyrochemical methods for the recovery and purification of plutonium-238 in various materials. This isotope, which was used as a power source in the space program, has about 300 times the alpha activity of the plutonium-239 associated with fuel reprocessing and therefore required special handling procedures. The major hazard was neutrons produced by (α,n) reactions with impurities such as fluorine, oxygen, and other common light elements. Paul Nelson, Jack Fischer, Jim Haley, John Lenc, and John Schilb conducted these studies.

A special glove-box facility was adapted for the work by using water as shielding for protection against the neutron emissions. Jack Fischer solved a problem that had plagued many people doing glove box work. Every time a researcher weighed something, read a pressure, or observed something significant, he or she had to disengage from the gloves, check their hands, make an entry in the notebook, and then reinsert their hands back into the gloves. Jack installed a voice-activated tape recorder to record the data orally.

Two laboratory-scale runs were made to demonstrate the recovery of metallic plutonium-238 by the reduction of $^{238}PuO_2$ in liquid zinc-magnesium or zinc-calcium, followed by vacuum distillation of the solvent metals. Most of the undesirable elements that cause (α,n) reactions were removed by this procedure. Two other experiments were conducted with $^{239}PuO_2$ microspheres (a stand-in for $^{238}PuO_2$) by the above procedure, except that the plutonium was subjected to a salt-transport separation and recovered by retorting. A satisfactory recovery of plutonium and a good separation from the light elements was achieved in both experiments.

In two further runs, an attempt was made to recover high-purity plutonium metal from a $^{239}PuO_2$-molybdenum cermet fuel. The cermet was disintegrated successfully, but the plutonium product contained excessive copper from the $CuCl_2$ used in the salt-transport separation. Work on this project was discontinued June 30, 1969.

Determination of Nuclear Constants

With the prospect of EBR-II coming on line soon, there was a need to have detailed, accurate information on pertinent cross sections for reactor physics calculations involving fast neutrons. Much of this information was available, but there were gaps that needed to be filled and questionable values that needed be checked. To perform these calculations, one needs to know (1) the neutron energy spectrum of the reactor as a function of age and location in the core, (2) capture and fission cross sections as a function of neutron energy, and (3) the production rate of nuclides from other nuclides. Cross-section data are required both for the fuel constituents and the other components such as control and structural materials.

Most of the early studies in the 1960s made use of Van de Graaff generators in the Reactor Engineering (RE) and Physics (PHY) Divisions, which could produce monoenergetic protons at a prescribed energy level. The proton beam was then directed to a lithium target to produce monoenergetic neutrons through the reaction $Li^7(p,n)Be^7$. The range of neutron energy levels was 0.4 to 1.7 MeV. The capture cross sections were derived by counting the decay rate of the activation product. This count was done with an alpha pulse-height analyzer in conjunction with a 256-channel analyzer. The detector was a thallium-activated NaI crystal.

The Mark III version of EBR-I was used to determine neutron capture/fission cross sections of U-233, U-235, U-236, Pu-239, and Pu-240. The samples were irradiated for 800,000 kWh. The fission cross sections were obtained by analysis of fission-product Cs-137, which is produced at a rate essentially independent of neutron energy. Technetium-99 was shown to share this characteristic and to be an alternative isotope for monitoring fission. This observation was of interest because cesium was removed during the melt refining process for EBR-II fuel, whereas technetium remained with the uranium and plutonium. The capture cross sections were established from mass spectrometric analyses of the products. Capture/fission cross-section ratios were measured for specimens from various locations in the EBR-I reactor, and they varied significantly. For example, the values for U-233 ranged from 0.06 in the core to 0.13 in the outer blanket. A computer program was developed to reduce the EBR-I data to monoenergetic neutron cross sections. Following the EBR-I irradiations, arrangements were made to insert samples of U-233, U-235, U-238, Pu-239, and Pu-240 into EBR-II for similar measurements.

The rare earth elements were of interest as possible control materials. Neutron capture cross sections were determined for Eu-170, Gd-158, Pr-141, Lu-175, La-139, Y-84, Y-89, Yb-176, Ti-50, Rb-85, Rb-87, Sr-84, Sr-86, and Sr-87, using the Van de Graaff machines. Radiative capture and scattering of neutrons by structural materials such as molybdenum and niobium were also determined.

Integral capture cross sections were determined by irradiating specimens of stainless steel and other structural metals in EBR-II and examining the reactor neutron energy spectrum. Further capture-to-fission ratios were determined by irradiations of various uranium and plutonium isotopes, and experiments were conducted to measure the fast fission yields of tritium, hydrogen, and helium as a function of neutron energy.

During the course of this work, an ongoing effort was maintained on the development of new counting techniques. A liquid scintillation system employing a photoelectric tube and amplifiers was developed for 4π beta counting, and in 1967 a lithium-drifted germanium detector was installed.

Carl Crouthamel was the principal investigator during most of this work, but Don Stupegia and Ed Dewell handled this responsibility during the two years that Carl was absent from ANL. Norm Dudey, Bob Heinrich, Al Madson, Chris Gatrousis, Jackie Williams, and Gene Kucera also made major contributions to this program.

Calorimetry

The calorimetry program continued with the objective of determining the thermodynamic properties of substances of interest in high temperature chemistry and nuclear technology. This was a major part of the Division's overall effort on the physical chemistry and thermodynamics of high-temperature materials, which also included phase studies, property measurements, and other activities. The specific mission of the calorimetry group was to determine the room-temperature and high-temperature enthalpies (heats) of formation.

In using combustion calorimetry to determine the enthalpy of formation of compounds, accurate data must be available on the enthalpies of formation of the reaction products. For example, a determination of the enthalpy of formation of US (uranium monosulfide), using fluorine combustion calorimetry, involves the reaction

$$US + 6\,F_2 \rightarrow UF_6 + SF_6$$

The value for the enthalpy of formation of US can be no more accurate than the values available for UF_6 and SF_6. For this reason, much of the effort in the calorimetry program was aimed toward the enthalpies of formation of fluorides of the individual elements. Although literature data were available on most of these fluorides, many of the values had been obtained by indirect methods or were questionable for some other reason and were therefore redetermined experimentally by the fluorine combustion method. These elements included Al, B, C, Cd, Gd, Ge, H, Hf, Ho, I, La, Mg, Mo, Nb, Ni, P, Ru, S, Se, Si, Ta, Th, Ti, U, W, Y, Zn, and Zr.

Calorimetric measurements are inherently difficult not only because they require very careful, precise techniques and equipment, but for many other reasons as well. The starting material must be of high, or at least exactly known purity. Some substances are difficult to ignite. In most cases, a thin wire was used to start the reaction, which then proceeded to a foil and then to a more substantial sample such as a pin. In other situations, some other readily ignitable material was used as a fuse. A jet of fluorine produced the best ignitions in certain cases. With many materials, the reaction did not go to completion, so the unreacted portion had to be separated, characterized, and measured by weighing or other means. There were also problems with reactions of substances such as uranium, which have multiple oxidation states. A large amount of effort was required to obtain good data for UF_6 due to the formation of UF_3, UF_4, and ill-defined compounds between UF_4 and UF_6, but it was finally done successfully.

In the early fluorine-combustion work, the fluorine that was commercially available had significant impurity levels (1.14% O_2, N_2, CF_4, C_2F_6). Larry Stein of the Chemistry Division purified fluorine by fractional distillation and obtained a purity of 99.96%, which was considered satisfactory. The General Chemical Division of the Allied Chemical Co. built a similar still that produced 99.9% fluorine for commercial sale.

Some of the information on compounds was obtained by oxygen combustion methods. Enthalpies of formation were measured for UN, WS_2, ZrH_2, ZrD_2 (D is deuterium), MoS_2, SiC, Pu, PuC, Pu_2C_3, PuO_2, PuN, and as information on fluorides of the elements began to emerge, the fluorine combustion method was used to determine enthalpies of formation for BN, B_2O_3, HfB_2, NbB_2, SiC, SiO_2, TaB_2, UP, US, UB_2, ZrB_2, and Na_2C_2.

The combustion calorimetry work produced data for a standard temperature of 298 K (25°C). In practical applications such as process calculations, it is necessary to have enthalpy data over a wide temperature range. Ideally, if entropy values are available or can be calculated and the thermodynamics of pertinent phase transitions are known, free energies of formation can be obtained. This information would yield a complete set of thermochemical data for a compound, similar to those published in the widely used JANAF (Joint Army, Navy, Air Force) tables.

To provide the higher-temperature enthalpy data, a drop calorimeter capable of temperatures up to 1500°C was constructed (Fig. 3-10). This device consisted basically of a furnace in which a sample suspended by a wire was heated to a precisely known temperature and then dropped through a gate mechanism into a calorimeter where the heat evolution was measured as the sample cooled.

Although simple in principle, this equipment was highly complex in its design, construction, and operation, and several years of effort were required to work out all the problems and place it in operation in 1966. A second drop calorimeter of the same basic design, but with numerous modifications, was built to extend the temperature capability up to 2500°C. In the higher temperature calorimeter an electron-gun heater and a pyrometer for temperature measurement were used instead of the resistance furnace and thermocouple used in the 1500°C calorimeter. Later on, a third drop calorimeter having a temperature capability of about 3300°C was constructed. Induction heating was used to heat the specimen in that unit.

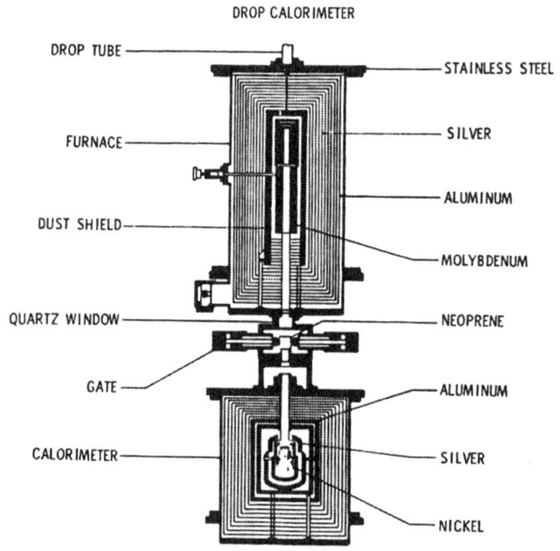

Fig. 3-10. Schematic of Drop Calorimeter

Ward Hubbard was the individual in charge of this work; others who participated include Milt Ader, Don Fredrickson, Elliott Greenberg, Jerry Johnson, Ralph Nuttall, Pat O'Hare, Howard Porte, Ed Rudzitis, Jack Settle, Rosemary Terry, Erv Van Deventer, Steve Wise, and Bob Kleb from Central Shops.

Energy Conversion and Storage

THERMOELECTRIC RESEARCH

In 1961, Hal Feder and Russ Edwards started a fundamental research program on thermoelectricity to contribute to the technological development of systems in which the thermoelectric effect is used to convert nuclear reactor heat directly into electrical power. Both liquid and refractory solid thermocouple materials were investigated. The criterion for evaluating such materials was a "figure of merit," Z, which was defined as follows:

$$Z = S^2\sigma/k$$

where S is the Seebeck coefficient, σ is the electrical conductivity, and k is the thermal conductivity. (The Seebeck coefficient is a measure of the effect of temperature difference between the two junctions of a thermocouple on the electrical potential that is generated.)

The work on liquid materials was directed mainly toward the indium-antimony (In-Sb) system. Seebeck coefficients were measured at compositions over the entire liquid range of the system at various temperatures from about 630 to 750°C. Temperature differences of 10 to 15°C were used for the two legs of the thermocouple. An addition of 20 at.% bismuth raised the Seebeck coefficient of indium by approximately a factor of two, and the values were of the order of 1 μV/°C.

Results were generally similar for the systems In-Bi and Sb-Bi.

The studies of solid materials were concerned primarily with uranium monosulfide (US), a compound known to be stable up to high temperatures. Uranium and thorium monosulfides were selected on the basis of their stabilities up to high temperatures and their electronic similarity to cerium sulfide, which was known to have a large Seebeck coefficient. Over the temperature range from 150 to 1200°C, Seebeck coefficients for the US were -40 to -90 µV/°C, which are somewhat lower than those for cerium sulfide. When converted to a figure of merit, the Z values for US were considered to be about a factor of ten too low to be useful for practical applications, and the program was discontinued in 1963.

During the course of these studies, a great deal of information was generated on the thermodynamic and physical properties of these systems in addition to their thermoelectric characteristics. Some of the individuals who contributed to this work were Russ Edwards, Phil Danielson, Clyde Metz, Frank Mrazek, and Marv Tetenbaum.

THERMALLY REGENERATIVE GALVANIC CELLS

At about the same time that the work on thermoelectric systems ended, a research study on thermally regenerative galvanic cells was initiated, largely through the efforts of Carl Crouthamel. The eventual goal of this effort, like that of the thermoelectric program, was to couple a nuclear reactor with a device that would convert the thermal energy directly into electricity. The idea was to dissociate a chemical compound by the reactor heat and then recombine the dissociation products to form that same compound in an electrochemical cell to produce the electricity. This was not a new concept created at ANL; work was already in progress at various laboratories, but it was felt that CEN's experience and capabilities for handling liquid metals and molten salts would be a major asset in mounting such a program.

Argonne researchers recognized that the efficiency of a thermally regenerative electrochemical system would be unlikely to exceed that of a steam turbine because it, too, would be subject to the Carnot cycle limitation. It might, however, have other advantages: (1) few, if any, moving parts, (2) adaptability to a wide range of plant sizes, (3) compactness and lower weight, and (4) ability to isolate the system rigorously through its useful life.

The first system to be investigated was the lithium/hydrogen cell, which consisted of a liquid lithium electrode, a hydrogen electrode in which the hydrogen was admitted by permeation through an iron diaphragm, and a molten LiCl-KCl electrolyte (m.p. 352°C). The overall cell reaction is

$$2\ Li + H_2 \rightarrow 2\ LiH$$

The lithium hydride product dissolved in the electrolyte. Because little was known about the phase relationships of LiH with the alkali metal halides, a considerable experimental effort was devoted to this subject. Some of the results are given in Table 3-2.

Table 3-2. Lithium Hydride Salt Systems

System	Minimum Melt. Pt., °C	Eutectic Comp., mol% LiH
LiH-LiCl	495.6	34.0
LiH-LiBr	453.3	29.7
LiH-LiI	390.8	23.5
LiH-LiF	684	Solid Solution

Information was also obtained on the systems LiH-KCl, LiH-NaCl, and LiH-LiCl-KCl. From the standpoint of phase relations, LiH behaved much like an alkali metal halide.

Initial electrical measurements on the Li/H_2 cell showed a high efficiency, but the current was limited by slow diffusion of the hydrogen through the Armco iron diffusion barrier. Other barrier designs and materials were tested, and some work was done on the regeneration of Li/H_2 cells. The final conclusion was that the overall efficiency that could be expected from the Li/H_2 cell was in the range of 8 to 12%, which is about the same as that for existing mercury-vapor-cycle and thermoelectric devices.

Work was also started on bimetallic concentration cells with liquid metal electrodes:

$$M_A \text{ / Electrolyte} + M_A \text{ Salt / } M_A \text{ in } M_B$$

where M_A is a volatile metal, and M_B is non-volatile. As the cell is discharged, M_A is transported through the electrolyte as ions, and its concentration increases in the M_B electrode. The cell is regenerated thermally by vaporizing M_A selectively from the M_A-M_B electrode and returning it to the pure M_A electrode. After an evaluation of the alkali metals, lithium was selected as the volatile metal, and the LiCl-KCl eutectic was chosen as the electrolyte. Polarization curves were measured for lithium electrodes in combination with other liquid metal electrodes, which included Te, Bi, Sn, Pb, Zn, and Cd. The potentials ranged from 1.93 V for the Li/Te couple down to 0.66 V for Li/Cd. On the basis of the potentials, vapor pressures, solubilities, and other factors, Li/Sn and Li/Bi were selected for further initial studies. Solid intermetallic compounds exist in both systems: Li_5Sn_2 and Li_2Sn in Li-Sn and Li_3Bi in Li-Bi. Thermodynamic functions of these systems were determined, and preliminary regeneration studies were started on Li-Bi, Li-Sn, and Na-Sn. The system Na-Bi was investigated by mass spectrometric and effusion measurements. Attention was turned to various other electrolyte compositions, including LiF-LiCl, LiF-LiCl-LiI, NaF-NaCl-NaI, LiF-LiI, and LiBr-LiI. The compound Li_3Bi was found to be soluble in molten LiF-LiCl. This discovery led to a series of solubility and absorption spectrum studies of lithium and sodium intermetallic compounds in molten salts. The solutions were of various colors.

Table 3-3 is a summary of the more promising regenerative galvanic cell systems that were operated. The only one that was run in a regenerative mode was the Na/Pb cell, with a regeneration temperature of 825-875°C. The cells all operated reversibly with no significant overvoltages.

Table 3-3. Regenerative Galvanic Cell Systems

System	Open-Circuit Potential, V	Current Density, A/cm^2	Temp., °C
Na/NaF-NaCl-NaI/Na in Pb	0.4	0.2	575
Na/NaF-NaCl-NaI/Na on Bi	0.6	1.1	530-615
Li/LiF-LiCl-LiI/Li in Bi	0.9	2.2	500
Li/LiF-LiCl-LiI/Li in Sn	0.6	1.0	500
$Li/LiF\text{-}LiCl\text{-}LiH/H_2$	0.3	0.2-1.0	525-570

Additional regeneration studies were conducted on a few of the candidate systems. In the Li-Bi system, the vapor pressures of the two components did not differ sufficiently for a practical operation. Engineering investigations were made on the regeneration of Na-Bi and Na-Pb cells. A cell operated at 543-600°C with regeneration temperatures of 775-875°C produced 7 A at 850°C. With the cell operating at 575°C, its open-circuit potential was 0.41 V. Vapor-liquid equilibrium measurements were made on the Na-Bi system, and the thermodynamics and phase equilibria for the system were characterized by emf measurements and thermal analysis studies, respectively. Other supporting information was obtained on the densities, viscosities, and surface tensions of various electrolyte salts and liquid metal solutions. Corrosion studies were also performed to determine the compatibilities of the liquid salts and metals with a number of metallic and ceramic materials that might be used as materials of construction.

During the operation of Na/Bi cells, the system proved to be electrochemically reversible (capable of both charging and discharging by an electric current) with high efficiencies, which suggested its possible use as a secondary (rechargeable) battery. A much higher voltage could be obtained, however, from the Li/Te system, which had been tested earlier. The Li/Te system was not suitable for a thermally regenerative cell because the high stability (large negative free energy of formation) of Li_2Te made thermal regeneration impossible, but that would be of no concern for a rechargeable storage battery. In a test of a Li/LiF-LiCl-LiI/Te cell, at 480°C, high current densities up to 5 A/cm^2 were achieved. By this time it was recognized that thermally regenerative cells could be made to work, but the low voltages and difficult engineering and materials problems in the regeneration cycle made it hard to justify further development work. The use of cells of this type in rechargeable batteries, however, seemed to show great promise, and the work began to head in that direction. This, of course, changed the potential practical application from the direct conversion of heat to electricity to one of electrical energy storage on utility systems and electric cars. For CEN, it was also a step away from nuclear work to a more politically acceptable area.

Numerous people were involved in the different phases of this program. The initial studies were done by Carl Crouthamel, Mel Foster, and Dave Anthers, who were soon joined by Dick Eppley, Bob Heinrich, Carl Johnson, Steve Banacek, John Allen, Al Fischer, and Stan Johnson. In 1964 Jim Hesson and Paul Shimotake joined the group and began looking at the engineering aspects. Elton Cairns appeared on the scene in 1967, and Jim Bingle, Ellen Hathaway, Gene Kucera, and Jim Peck joined the effort. Leadership in this work was provided mainly by Carl Crouthamel, Art Tevebaugh, and Elton Cairns.

LITHIUM-CHALCOGEN BATTERIES

The lithium/tellurium couple was the first to be investigated in the research and development program on rechargeable batteries. Several fundamental studies were made to gain a better understanding of the chemical and electrochemical characteristics of the system. Mel Foster and Gene Kucera made emf measurements to determine the effect of the changing cathode composition on the cell as it was discharged. Mel Foster, Carl Johnson, Kathleen Davis, Jim Peck, and Bob Schablaske established a phase diagram for the lithium-tellurium system, using thermal analysis and X-ray diffraction techniques. Carl Johnson and Mel Foster determined the phase diagram, which was of particular interest because it has a ternary eutectic at the unusually low temperature of

265°C. (In the case of the battery systems, the minimum operating temperature was usually the melting point of the electrolyte.) Vic Maroni and Ellen Hathaway conducted a series of Raman spectroscopic studies on the systems Li_2Se-LiF-LiCl, Li_2Te-LiF-LiCl, KCl-$MgCl_2$, $MgBr_2$-MgBr, and UO_2Cl_2 in NaCl-KCl-$MgCl_2$.

On the more applied side, Paul Shimotake, G. Rogers, and Jim Peck began operating cells of the type Li/LiF-LiCl-LiI/Se, in which the lithium and selenium were both liquids. These cells had an open-circuit potential of about 2.4 V and could produce short-circuit current densities of 11-13 A/cm^2. Work began on paste electrolytes, which served as mechanical spacers and electronic insulators between the electrodes while permitting the transport of lithium ions. These separators were formed by pressing a mixture of an inert powder such as $LiAlO_2$ with the molten salt electrolyte to form a structure that remained solid at the operating temperature of the cell. Porous metal structures were used in the liquid lithium and selenium electrodes to provide physical support and also to serve as current collectors. Paul Shimotake, Ramamritham Sridhar, and John Cassulo put a substantial effort into the development of these current collectors. Some additional initial investigations were conducted on the Li/S and Li/P couples as potentially useful battery cells.

Preliminary engineering design studies of batteries using cells of this type showed their potential for very high performance, both in energy capacity and power. In one progress report, the following practical applications were listed in order of what was considered to be higher-probability ones first: (1) spacecraft, (2) military communications, (3) materials-handling vehicles, (4) military vehicles, (5) boats and submarines, (6) remote locations, (7) buses and trucks, (8) urban automobiles, and (9) off-peak energy storage for central stations. The way things actually played out, the first funding was provided by the AEC for items (8) and (9) and by the U.S. Army for item (4). A rather surprising development was funding from the National Heart and Lung Institute to develop an implantable battery for an artificial heart, which produced some bad jokes about heartburn, but the idea appeared to be more feasible than one would think at first glance. The lithium/selenium system was selected for the heart battery; lithium/selenium and lithium/sulfur were both in contention for the other applications.

Elton Cairns, who came to ANL from the General Motors Research Lab, was in charge of this work in 1968-69. He was a widely recognized authority in the electrochemical field, which was a significant asset in obtaining funding from various organizations. He was highly competent both in the theoretical and experimental aspects of the work and a good leader. One quirk that bothered the staff at times was his proclivity to change the settings such as voltage or current on someone's experiment in their absence. He always had a good reason for making the change, but it was not always consistent with the experiment the investigator had in mind. Elton had another unusual practice; at personnel evaluation time, he would collect all the laboratory notebooks and read them in detail as a part of the evaluation. Probably not a bad idea; it was an objective measure and it encouraged good record keeping. One time Elton was asked what he would like to do eventually in his career, and he indicated that he would like to be on the faculty at the University of California with a few graduate students. That is what he is doing now.

FLUIDIZED BED COMBUSTION OF COAL

In 1969, the Division, as a part of the movement toward environmental research, began a program on the combustion of coal in a fluidized bed to which limestone was added as a means of reducing the emission of sulfur dioxide (SO_2) during combustion. A literature survey and preliminary laboratory experiments with fixed beds had shown that limestone (natural $CaCO_3$) was the most promising material for this purpose. An empirical model based on laboratory kinetic data was developed to determine the fluidized bed operating parameters such as bed height, superficial gas velocity, and optimum bed particle size.

Bench-scale experiments were conducted in a 6-in.-dia fluidized bed combustor to determine the SO_2 emissions under conditions prevailing in coal-fired power plants and industrial steam boilers. With a combustion temperature of 1600°F and a superficial gas velocity of 3 ft/sec, the following effects were observed:

1. The SO_2 concentration in the flue gas was reduced by 29 to 87% of that without the limestone additive.
2. The most favorable result was obtained with a calcitic limestone of 25-μm average particle size at 2.2 times the stoichiometric requirement.
3. For any particular limestone, the SO_2 removal increased with decreasing particle size.
4. The maximum utilization of CaO during the experiments was 38%.
5. The combustion efficiencies of carbon burned in two experiments were 96.7 and 97.1%.
6. Virtually all the ash from the coal was elutriated.
7. Bench-scale experiments at 1600°F showed that the source of NO_x compounds in the flue gas came from nitrogen compounds in the coal. In other words, little or no NO_x had been formed by fixation of atmospheric nitrogen in the combustion process.
8. The limestone addition produced a 30-40% reduction of nitric oxide in the flue gas.

Al Jonke led these initial studies, which were performed by Erv Carls, Roger Jarry, and Milt Haas.

Analytical Chemistry

By 1960, the scope of the Division's activities had expanded from the original work on solvent-extraction processes into a larger and much more diversified group of programs, including pyrometallurgical and fluoride volatility fuel reprocessing, reactor safety, calorimetry, and several other areas. Each of these depended heavily upon the Analytical Laboratory, which had to expand its range of capabilities to meet this need. The group then consisted of a large section for elemental analysis, plus other sections for the development of new analytical methods, X-ray diffraction and spectroscopy, gas chromatography, and radiochemistry. In 1963, a new section was formed to conduct studies with a newly acquired electron-probe microanalyzer. (The term "section" here refers only to the various organizations within the Analytical Laboratory and has no connection with the term used elsewhere for a major part of CEN as a whole.)

During the 1950s and 1960s, manufacturers of scientific instruments were capitalizing on advances in electronic technology to come out with a variety of new applications. One of these was the electron microprobe. During that period, the Division also acquired an atomic absorption spectrometer and a new 75-kV X-ray spectrometer. As these advances continued to be made, new designs of more mundane instruments such as analytical balances made the analytical work faster, easier, and more accurate.

The volume of work being handled by the Analytical Laboratory during the 1960s was extraordinary. In a typical year the elemental analysis section alone used over 100 methods to perform more than 13,000 analyses. The X-ray section determined several hundred diffraction patterns and made many spectrographic assays, identifications, and structural studies. The gas chromatography section analyzed numerous gas samples from glove boxes, furnaces, and calorimeters.

The analytical research was, for the most part, an outgrowth of the service work. As the Division's programs changed, existing analytical methods had to be modified or new methods had to be developed. In the majority of the samples, the element of interest was only a minor constituent, and the analytical method had to either tolerate or separate the matrix material from the element of interest. The matrix materials, which were often present in amounts as much as a million times greater than that of the element to be analyzed, were typically zinc, magnesium, salt fluxes, and aluminum oxide. Typical research problems included a method for dissolving PuO_2 and spectrophotometric determinations of Pu(IV) and ruthenium.

Mention was made earlier that the startup of EBR-II, which burned uranium-235 in a fast neutron flux, had a somewhat different spectrum of fission-product yields than those for thermal reactors. The burnup of spent fuel from a reactor can be determined from the concentration of a fission-product nuclide in the fuel if its fission yield is known. None of the 85 or so fission-product nuclides had all the nuclear and physical characteristics to be ideal for this purpose, but the Analytical Chemistry research group investigated a number of the most promising candidates for this purpose and obtained fission-yield data for them. Among these were Tc-99; La-139; Cs-133 and -137; Pr-141; Mo-95, -97, -98, and -100; Ru-101, -102, and -104; and total rare earths.

There was some concern as to whether the burnup monitors for the metallic EBR-II fuel would be applicable to oxide and carbide fuels. Tests that were conducted on irradiated uranium dioxide fuel showed that molybdenum and technetium were not suitable burnup monitors because of their tendency to migrate during irradiation, but that the rare earths were excellent candidates for this purpose. An X-ray spectrometric method for assaying the quantities of fission-product lanthanum, cerium, praseodymium, and neodymium was developed to measure the burnup of oxide fuels.

Bob Larsen, Carl Crouthamel, and Bob Meyer were in charge of various aspects of the analytical chemistry programs. Other members of the group included:

Murray Barsky	Carol Kosner	Bob Schablaske
Maureen Binelli	Gene Kucera	Chuck Seils
William Bloom	Verne Lemke	Gregory Smith
Eugene Bozisch	Russ Malewicki	Stan Siegel
Vera Drabek	Fred Martino	Nora Stalica
Florence Ferry	Bob Oldham	Ben Tani
Ruth Hanna	Al Panek	Ziggy Tomczuk
Bob Heinrich	Ray Popek	Florence Williams (Smith)
Myron Homa	Laury Ross	Jackie Williams
Ruth Juvinal	Dino Santelli	

Waste Processing and Gamma-Irradiation Facility

As of the end of 1962, routine disposal of liquid and solid radioactive wastes by CEN was discontinued, and the operation was transferred to the Plant Services Department. During the last year of operation, about 50,000 gallons of liquid waste was processed. The high-level gamma irradiation facility was also shut down after seven years of operation. Irradiations were conducted on 5,432 samples in 1961 and 2,443 in 1962, not counting the dosimetry samples used to measure the radiation fluxes.

Studies and Evaluations

Studies and evaluations were done from time to time on the feasibility and costs of projected applications of research and development programs of the Chemical Engineering Division. These might be best described as looking at the forest rather than the trees.

In 1964, Bill Mecham, Wally Seefeldt, Virgil Trice, and Milt Levenson reported on an interdivisional study that was being conducted for the AEC on the estimated costs of power generation by metal fueled reactors, based on a 1000 MW(e) plant. The following topics were covered.

1. A reference pyrochemical process
2. Preparation of fuel for processing
3. Removal of fission products
4. Separation of uranium and plutonium and their recovery as ingots
5. Composition of the product ingots
6. Estimated costs for the reactor fuel cycle

A second study was undertaken by this same group to compare the costs of reprocessing plutonium metal and oxide and carbide fuels by aqueous, fluoride volatility, and pyrochemical processes.

Bill Mecham, in 1967, conducted a study on the problems of tritium in power reactor fuel cycles. This study considered the sources of tritium in discharged fuel, its behavior both in aqueous and nonaqueous processes, and methods for its disposal. Because tritium cannot be separated economically from the large volumes of water used in aqueous processing, its disposal would have to be accomplished by dilution and release to the environment. In the case of fluoride volatility and pyrometallurgical processes, the tritium can be recovered in a small volume of gas, which can be stored or immobilized by further treatment.

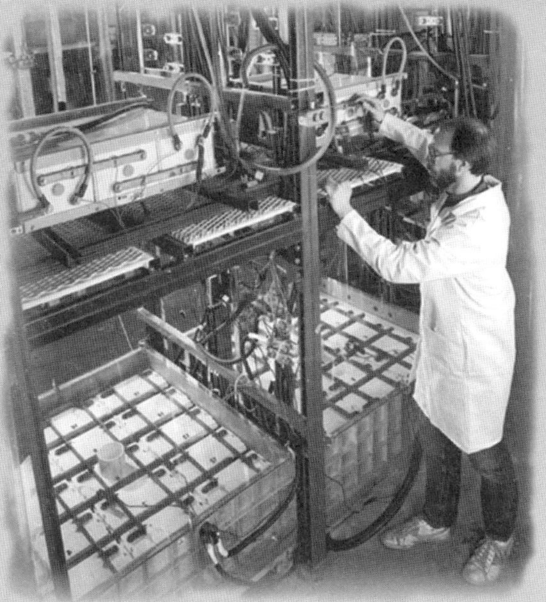

1970-1980: Transitions

1970-1980

(top, left) Calorimeter for determining basic thermochemical properties of a wide variety of materials.

(top, right) National Battery Test Laboratory used for standardized testing of cells, multicell modules, and batteries built by industrial firms.

(bottom) Solar energy testing facility in which arrays of different types of solar energy collectors were tested.

4

1970-1980: Transitions

The 1970s was a difficult period for the United States in that it seemed to present a continuous series of new problems. At the beginning of the decade, the foremost U.S. concern was to extricate itself from the extended Vietnam War. That was finally accomplished in 1975, but with little feeling of success. Richard Nixon defeated George McGovern by a landslide in the 1972 presidential election, but was forced to resign in favor of Vice President Gerald Ford in 1974 as a result of the Watergate incident. Meanwhile, in 1973, the Organization of Petroleum Exporting Countries (OPEC) suddenly stopped oil shipments to the U.S., causing a severe fuel shortage and an eightfold rise in the cost of crude oil, which, in turn, set off an inflationary spiral with interest rates rising to 15% or more. Jimmy Carter, by defeating Ford in the 1976 presidential election, inherited the problems of the continuing oil shortage, inflation, and hostility of some Mideastern nations. His last year in office became particularly difficult when Iranian militants seized 52 American embassy personnel as hostages.

In spite of increasing opposition by antinuclear and environmental groups, many commercial nuclear power plants were built and started up in the 1970s and construction of others extended into the 1980s. In 1975, the AEC was abolished. Its regulatory function was taken over by the Nuclear Regulatory Commission (NRC), and its other responsibilities were given to the Energy Research and Development Administration (ERDA). The ERDA was short-lived; in 1977 it was absorbed into a new cabinet-level Department of Energy (DOE). During this ten-year period, ANL had two new Laboratory Directors, Robert G. Sachs in 1973 and Walter E. Massey in 1979. In 1973, Leslie Burris replaced Richard Vogel as Director of the Chemical Engineering Division. Y-Wing, an office complex with an auditorium, and the National Battery Test Laboratory facility were added to Bldg. 205.

The large pyrometallurgical and fluoride volatility fuel reprocessing and reactor safety programs had all but disappeared by about 1970, and were supplanted largely by expanded programs on sodium technology, fluidized-bed combustion of coal, and high-temperature battery development. Work continued on materials chemistry and thermodynamics, analytical and reactor chemistry, and technologies employing liquid metal and molten salt solvents. These programmatic changes resulted in some layoffs and transfers to other divisions. Many of the staff members had to become familiar with new areas of work, and several people were hired to provide special expertise for the new programs. The work became more interdisciplinary and required increased capabilities in areas such as computer applications, materials science, electrochemistry, and electrical engineering. The Division's funding also shifted somewhat, with a larger proportion of the support coming in from non-nuclear branches of the DOE and from other federal agencies. In the mid-1970s, the DOE adopted a management policy of placing programmatic management "in the field," and CEN was given overall management responsibilities as the "lead laboratory" for the aqueous battery, advanced battery, and fuel-cell programs.

THE NATIONAL SCENE

In the early 1970s, the war in Vietnam was beginning to wind down, and, after a brief incursion into Cambodia, much of the American force in Vietnam was withdrawn. Peace talks were held in Paris in 1973, and the remaining U.S. troops were brought back home. Full-scale war erupted again in 1975 with a Communist victory; Marine guards and civilians had to be rescued from the U.S. Embassy and evacuated along with about 140,000 Vietnamese refugees. Many felt that the Vietnam War had ended in a U.S. defeat, and it was a bitter experience both for the veterans and the civilian population.

In 1972, President Nixon ran for a second term, defeating his Democratic opponent, George McGovern, by a large margin. At one point in the election campaign, the police nabbed five men who were attempting to break into the Democratic National Headquarters. The break-in, of itself, was of minor consequence and received little public attention until well after the election, but when the attempted cover-up that followed it was revealed some time later, several of President Nixon's advisors landed in jail, and Nixon finally resigned on August 8, 1974. Gerald R. Ford, the Vice President, assumed the presidency, and one of his first acts was to grant Nixon a full pardon. (The previous year, the Vice President, Spiro T. Agnew, had been convicted of income tax evasion and replaced by Ford.)

Jimmy Carter, a Democrat and former Governor of Georgia, defeated Ford in a race for the presidency in 1976. Carter, who had served in the nuclear navy as a submarine officer under Admiral Rickover, adopted a strong anti-nuclear stance, and played a major role in negotiating a nuclear non-proliferation pact, signed by 15 countries in 1977. A deeply religious man, devoted to peace, and perhaps idealistic, he was generally opposed to the idea of nuclear power because of its potential for nuclear proliferation. He called nuclear power "the option of last resort," withdrew funding for plutonium processing, and opposed the Clinch River Breeder Reactor proposed by Oak Ridge. In 1978, the U.S. signed a treaty to turn control of the Panama Canal over to the Panamanians by 2000, and the "Framework for Peace" in the Middle East was signed by Egypt's Anwar Sadat and Israel's Premier Menachem Begin after a 13-day conference at Camp David arranged by President Carter.

The decade ended on two particularly sour notes. The Soviets invaded Afghanistan, and Iranian militants seized the U.S. Embassy in Teheran, taking 52 American embassy personnel as hostages. They were held for 444 days.

NUCLEAR POWER

The nuclear power industry entered the 1970s with a highly optimistic view of its future. At the same time, the U.S. Navy was adopting nuclear power plants for its newer and larger vessels. Numerous contracts for the construction of civilian nuclear power plants were negotiated, and many were started up by the end of the decade. Other plants did not become operational until the 1980s and 1990s, due to the long construction times. The reactors that started producing power on a commercial scale in the 1970s are listed in Tables 4-1 and 4-2.

In 1973, the Arab oil-producing cartel, OPEC, suddenly slapped an embargo on their oil shipments, mainly in retaliation for U.S. shipments of arms to Israel during the Yom Kippur war. The price of crude jumped from $2 to $16/barrel. A gasoline and fuel oil shortage developed quickly in the United States. President Carter donned a sweater, admonished the public that they should do likewise and turn down their thermostats, and declared "the moral equivalent of war" on oil usage. Unnecessary use of automobiles was discouraged, and long lines

Table 4-1. U.S. Power Reactors in the 1970s

		Type	MW(e)	Start
Dresden 1 (Morris, IL) *	Commonwealth Edison	BWR	200	7/60
Indian Point 1 (Buchanan, NY)*	Consolidated Edison	PWR	257	1/63
Hanford N (Richland, WA)*		LGR	860	7/66
Haddam Neck (Haddam Neck, CT)	Conn. Yankee Atomic Power	PWR	590	1/68
San Onofre 1 (San Clemente, CA)*	So. Cal. Ed., San Diego G & E	PWR	436	1/68
Nine Mile Point 1 (Scriba, NY)	Niagara Mohawk Power	BWR	610	12/69
Oyster Creek (Forked River, NJ)	GPU Nuclear	BWR	619	12/69
Dresden 2 (Morris, IL)	Commonwealth Edison	BWR	794	6/70
R. E. Ginna (Ontario, NY)	Rochester Gas & Electric	PWR	470	7/70
Point Beach 1 (Two Rivers, WI)	Wisconsin Electric Power	PWR	485	12/70
Millstone 1 (Waterford, CT)	Northeast Utilities	BWR	660	3/71
Robinson 2 (Hartsville, SC)	Carolina Power & Light	PWR	683	3/71
Monticello (Monticello, MN)	Northern States Power	BWR	536	6/71
Dresden 3 (Morris, IL)	Commonwealth Edison	BWR	794	11/71
Palisades (South Haven, MI)	Consumers Power	PWR	780	12/71
Point Beach 2 (Two Rivers, WI)	Wisconsin Electric Power	PWR	445	10/72
Vermont Yankee (Vernon, VT)	Vermont Yankee Nucl. Power	BWR	504	11/72
Maine Yankee (Wiscasset, ME)	Maine Yankee Atomic Power	PWR	860	12/72
Pilgrim (Plymouth, MA)	Boston Edison	BWR	670	12/72
Turkey Point 3 (Florida City, FL)	Florida Power & Light	PWR	666	12/72
Surry 1 (Gravel Neck, VA)	Virginia Power	PWR	801	12/72
Quad Cities 1 (Cordova, IL)	Commonwealth Edison	BWR	789	2/73
Quad Cities 2 (Cordova, IL)	Commonwealth Edison	BWR	789	3/73
Surry 2 (Gravel Neck, VA)	Virginia Power	PWR	801	5/73
Oconee 1 (Seneca, SC)	Duke Power	PWR	846	7/73
Browns Ferry 1 (Decatur, AL)	Tennessee Valley Authority	BWR	1,065	8/74
Fort Calhoun (Fort Calhoun, NB)	Northern States Power	PWR	478	9/73
Turkey Point 4 (Florida City, FL)	Florida Power & Light	PWR	666	9/73
Prairie Island 1 (Red Wing, MN)	Northern States Power	PWR	503	12/73
Zion 1 (Zion, IL)	Commonwealth Edison	PWR	1,040	12/73
Kewaunee (Carlton, WI)	Wisconsin Public Service	PWR	503	6/74
Cooper (Brownville, NB)	Nebraska Public Power	BWR	764	7/74
Peach Bottom 2 (Delta, PA)	PECO Energy Co.	BWR	1,159	7/74
Indian Point 2 (Buchanan, NY)	Consolidated Edison	PWR	975	8/74
Oconee 2 (Seneca, SC)	Duke Power	PWR	846	9/74
3- Mile Island 1 (Londonderry Twp., PA)	GPU Nuclear	PWR	786	9/74
Zion 2 (Zion, IL)	Commonwealth Edison	PWR	1,040	9/74
Oconee 3 (Seneca, SC)	Duke Power	PWR	846	12/74
Arkansas Nuclear One (Russellville, AR)	Energy Operations Inc.	PWR	836	12/74
Peach Bottom 3 (Delta, PA)	PECO Energy Co.	BWR	1,035	12/74
Prairie Island 2 (Red Wing, MN)	Northern States Power	PWR	500	12/74
Duane Arnold (Palo, IA)	IES Utilities	BWR	538	2/75
Browns Ferry 2 (Decatur, AL)	Tennessee Valley Authority	BWR	1,065	3/75
Rancho Seco (Clay Station, CA)*		PWR	913	4/75
Calvert Cliffs 1 (Lusby, MD)	Baltimore G & E	PWR	825	5/75
James A. FitzPatrick (Scriba, NY)	New York Power Authority	BWR	780	7/75
Donald C. Cook 1 (Bridgman, MI)	Indiana/Michigan Power	PWR	1,020	8/75
Brunswick 2 (Southport, NC)	Carolina Power & Light	BWR	754	11/75
Edwin I. Hatch 1 (Baxley, GA)	Georgia Power	BWR	810	12/75
Trojan (Prescott, OR)*		PWR	1,095	

Indian Point 3 (Buchanan, NY)	New York Power Authority	PWR	965	8/76
Beaver Valley (Shippingport, PA)	Duquesne Light	PWR	810	10/76
St. Lucie 1 (Hutchinson Island, FL)	Florida Power & Light	PWR	839	12/76
Browns Ferry 3 (Decatur, AL)	Tennessee Valley Authority	BWR	1,065	3/77
Brunswick 1 (Southport, NC)	Carolina Power & Light	PWR	767	3/77
Crystal River 3 (Red Level, FL)	Florida Power Corp.	PWR	825	3/77
Calvert Cliffs 2 (Lusby, MD)	Baltimore G & E	PWR	825	4/77
Salem 1 (Salem, NJ)	Public Service Elec. & Gas	PWR	1,106	6/77
Joseph M. Farley 1 (Dothan, AL)	Southern Nuclear Operating	PWR	860	12/77
North Anna 1 (Mineral, VA)	Virginia Power	PWR	893	6/78
Davis-Besse (Oak Harbor, OH)	Toledo Edison	PWR	877	7/78
Donald C. Cook 2 (Bridgman, MI)	Indiana/Michigan Power	PWR	1,090	7/78
3 Mile Island 2 (Londonderry Twp., PA)	GPU Nuclear	PWR	792	12/78
Fort. St. Vrain (Platteville, CO)*		HTGR	330	1/79
Edwin I. Hatch 2 (Baxley, GA)	Georgia Power	BWR	820	9/79

* No longer in operation as of 1995.

Table 4-2. Foreign Power Reactors in 1970s

Country	No. of Reactors	Total Power, MW(e)
Armenia	1	400
Belgium	3	1,647
Bulgaria	2	800
Canada	8	5,136
Finland	2	1,155
France	5	4,693
Germany	8	7,358
India	3	390
Japan	21	14,288
Korea	1	556
Pakistan	1	125
Russia	8	5,292
Slovakia	1	408
Spain	1	438
Sweden	5	3,115
Switzerland	4	2,020
Taiwan	2	1,208
Ukraine	1	925
United Kingdom	22	5,434

developed at the filling stations, where the operators adopted various stratagems to deal with the situation, such as shorter hours or alternating days for odd and even license numbers. Christmas lights were frowned upon—there were some somber winter nights in those years. Many Argonne employees were glad that they had formed car pools. The oil shortage persisted through the 1970s, and then began to wane by about 1980 because of several factors, including conservation efforts, increasing availability of oil from other sources, and OPEC members' propensity to cheat on one another.

For the nuclear industry, the oil shortage was a double-edged sword. It made the American public aware of the nation's dependency on foreign energy sources and the need for alternatives to petroleum. At the same time, the steep increase in fuel prices threw the U.S. economy into an inflationary spiral with interest rates rising well into double-digit figures. As a result, the capital costs for nuclear plants rose sharply, and the price of yellowcake (uranium ore concentrate) rose from $7 to $35/pound. The Westinghouse Electric Corp. had to renege on its commitments to customers for uranium oxide at the agreed price. The AEC seemed to be losing its credibility, and the antinuclear forces began publicizing a variety of safety concerns. Permits for reactor construction slowed to a snail's pace due to intervenors in the licensing process and other roadblocks. The cost of capital during these delays was a major problem. In the 1950s, some had thought that nuclear power would be too cheap to meter; in the 1960s, the projected costs were mills per kilowatt-hour; and by the end of the 1970s they were in cents per kilowatt-hour. On top of all this, the projected use of electricity in the U.S. was revised significantly downward. The net effect of all these negative factors was to cast a dark shadow over the earlier optimistic outlook on the future of commercial nuclear power, and no construction of new power reactors has been undertaken in the U.S. since about 1979. There was talk of replacing existing reactors at the end of their lifetimes with a second generation of more fuel-efficient breeder reactors, but this, too, became a doubtful prospect in view of the ban on fuel reprocessing and concerns about safety, waste disposal, and potential proliferation of nuclear weapons.

The General Electric Co. and the Westinghouse Electric Corp. supplied most of the reactors for the nuclear power plants in the U.S. In most, but not all cases, the boiling water reactors were built by General Electric and the pressurized water reactors by Westinghouse. The Babcock and Wilcox Co. and Combustion Engineering Inc. also produced several reactors. Among the more prominent architect-engineering firms were the Bechtel Corp., Ebasco Dorsch Consultants, Sargent & Lundy Engineers, Stone & Webster Engineering Corp., Fluor Corp., and Gilbert Associates, Inc.

Two power reactor incidents of note occurred in the 1970s. The first was in 1975, when a fire broke out in the electrical cable systems of two 1,100-MW(e) Browns Ferry reactors operated by the Tennessee Valley Authority and located in Decatur, Alabama. The fire propagated through the entire cable system, shutting down both reactors and disabling the emergency core cooling system of one. The nuclear components of the plant were not affected, and there was no hazard to the public, but the antinuclear activists used the opportunity to further their cause.

A much more serious failure occurred on March 28, 1979, at Three Mile Island Reactor 2 near Harrisburg, Pennsylvania. The accident began as a small break in the loss-of-coolant system in which a valve stuck open, allowing coolant to escape from the vessel. The emergency core cooling system kicked in as designed, providing makeup water for the core. The operators, confused by the signals in the control room, shut off the emergency cooling system for several hours, whereupon the decay heat of the core vaporized the remaining cooling water in the vessel. Before the operators resumed the flow of emergency coolant, about one-third to one-half of the core melted down. The molten fuel and cladding dropped into the bottom of the vessel, where the molten metal from the cladding reacted with the water, forming a hydrogen "bubble." This situation prevailed for some time, giving the news media a field day with hour-by-hour reports, including much speculation about a possible explosion of the hydrogen. A small amount of radioactivity was released, but the situation was brought under control. The radioactive fallout from the incident was minimal, but the political fallout was enormous. People who claim to have been injured by fallout from Three Mile Island are still filing lawsuits. (One of the current programs in the Division includes an evaluation of the feasibility of using electrometallurgical methods to process the damaged core material and debris from the Three Mile Island reactor.)

The research and development work on nuclear fuel reprocessing by the Chemical Engineering Division at ANL, and at the other AEC laboratories as well, had become nearly dormant as of 1970. This change was in conformance with the AEC directives that the work at ANL should be shifted from nuclear to environmental programs. Although the research and development work on reprocessing had languished, there was some commercial activity in the 1970s. The first commercial plant, which was built by Nuclear Fuel Services, Inc., in West Valley, New York, and had a capacity of 1 ton/day, was started up in 1966. In 1972, it was shut down for retrofitting to meet increasingly stringent licensing requirements. Because of the high cost of retrofitting and the small capacity of the plant, further operation was deemed impractical, and the plant was shut down in 1976. A 1 ton/day plant for processing low-enrichment uranium oxide fuel was built by General Electric in Morris, Illinois. This process employed an aqueous Purex and ion-exchange first cycle to produce nearly fully decontaminated uranium oxide. The solid product was then to have been purified in a second cycle using fluoride volatility-fluidized bed technology to produce fully decontaminated UF_6. The process was never operated successfully, the main reason being that insufficient surge capacity had been inserted between the process steps to provide the necessary operational flexibility. Another problem was that the fluoride volatility portion of the plant was designed for "hands on" (non-remote) maintenance, which turned out to be impractical due to the radiation levels. This plant, however, is serving as a useful but expensive storage facility for discharged fuel subassemblies from commercial light water power reactors. Construction was begun on a larger (5-ton/day) central processing plant at Barnwell, South Carolina. Construction of this plant, which employed solvent-extraction technology, was completed in the late 1970s, but the ban on fuel reprocessing blocked its operation.

Public skepticism about reactor safety and concern regarding the potential proliferation of nuclear weapons had been growing, which was not surprising in view of the reactor accidents and the efforts of antinuclear and environmental groups. President Gerald Ford, feeling this pressure, suspended commercial processing, his main concern being the large amount of high-purity plutonium that might be diverted to nuclear weapons. President Jimmy Carter continued the ban on fuel reprocessing. Although the Unites States was attempting to set a good example, it was not taken very seriously by the rest of the world. Reprocessing continued to be pursued vigorously by several foreign countries, notably France, England, Japan, India, and Russia. France constructed a large 5-ton/day commercial plant at LaHague, where both

French fuel and that from other countries are processed. England, which has continued to recover fuel from its own nuclear power reactors, has recently built a 5-ton/day plant known as THORP (Thermal Oxide Reprocessing Plant), which replaced a former plant. When THORP was started in 1993, England began to sell its processing services to other countries. Japan has built a fuel reprocessing plant, which is scheduled to go into operation in 1998, and India has been using the Purex process since 1982 at the Tarapur Fuel Processing Plant, which has a 1/2-ton/day capacity. Fuel reprocessing is also suspected to be taking place in Russia, China, and possibly North Korea.

The AEC entered the decade of the 1970s with Milton Shaw continuing as the director of Reactor Development and Technology. Until 1973, Congress had supported the nuclear research programs with lump-sum funding covering both safety and breeder development, the allocation between the two being decided by the AEC. Shaw's general approach was to push the breeder development, thereby diverting funding away from research and safety programs such as the Loss of Fluid Test (LOFT). Safety became a contentious issue involving the AEC, Shaw, the reactor vendors, and the Advisory Committee on Reactor Safeguards. The friction increased to a point where the AEC prohibited its safety staff from communication with the regulatory staff on safety matters. This confusion contributed to a public loss of confidence in the nuclear regulatory process. The situation was not improved by the publicity attending the Karen Silkwood incident at Kerr-McGee in 1974. Hollywood did its part, too, by coming out with nuclear scare films such as *The China Syndrome*.

In 1973, Congress passed legislation abolishing the AEC, replacing it with the Nuclear Regulatory Commission (NRC) to perform the regulatory functions, and the Energy Research and Development Administration (ERDA) to handle the non-regulatory and promotional functions. In 1977, ERDA was absorbed by a new Cabinet-level entity, the Department of Energy (DOE), with James Schlessinger as the Secretary of Energy. Congress eliminated the Joint Committee on Atomic Energy (JCAE) in 1977. It is worthy of note here that, particularly in the earlier years, several senators and congressmen on this committee became quite knowledgeable about nuclear science and technology and played an important role both in promoting and guiding the programs. One effect of the demise of this committee was to put more policy-making power in the hands of the DOE managers.

When the DOE was created, its mission was quite different from that of its predecessors in that the nuclear component had been down-played sharply in favor of energy-conservation measures, environmental concerns, and alternative (preferably renewable) energy sources. It does, however, continue to have the responsibility for U.S. nuclear weapons programs.

THE LABORATORY

Argonne had two new laboratory directors in the 1970s. On April 6, 1973, Robert G. Sachs replaced Dr. Duffield, thereby becoming ANL's fifth director. Dr. Sachs, a physicist with a Ph.D. from Johns Hopkins, had taught at Purdue and the University of Wisconsin. He had also been at Argonne in 1946-47 and served as Associate Laboratory Director for High Energy Physics from 1964 to 1968. Dr. Sachs had the appearance and demeanor of a quiet, scholarly sort of person.

On July 2, 1979, Dr. Sachs was succeeded by Dr. Walter E. Massey. Dr. Massey had earned a Ph.D. in physics at Washington University and been on the faculties at the University of Illinois and Brown University. He gave the impression of being an energetic, ambitious, career-oriented individual. While serving as the Laboratory Director of Argonne, he also continued as professor of physics and in other activities at the University of Chicago. The Laboratory had two Deputy Directors at the time, Robert N. Laney, in charge of research and development, and Michael V. Nevitt, who was responsible for physical research. At the next lower level were several Associate Laboratory Directors, who had jurisdiction over general areas of related research and development work that encompassed parts or all of various divisions. Those concerned with CEN activities at the time were Jack Kyger (Engineering Research and Development), Gale Pewitt (Energy and Environment), and Mike Nevitt (Physical Research).

THE DIVISION

General Information

The management structure of CEN as of January 1970 is given in Table 4-3. At that time, Dr. Richard Vogel was the Director, Donald Webster was the Deputy Director and also Associate Director along with Les Burris and Arthur Tevebaugh. Everett Proud was the Assistant Director. Paul Nelson and Duane Barney became Associate Division Directors in the 1970s, and Paul then became the Deputy Director.

Dr. Vogel, in 1973, left Argonne to accept a position with the Exxon Corp. in Washington State. He subsequently moved on to the Electric Power Research Institute (EPRI) in California in 1983 and continued to keep in touch with his former colleagues at CEN through various technical contacts, including some contract work that the Division did for EPRI on post-irradiation fuel examinations. During his career, Dr. Vogel garnered numerous honors, including the AIChE Robert Wilson Award, and fellowships in the ANS and AIChE.

Leslie Burris was appointed to replace Vogel as CEN Division Director in 1973. As ANL entered the 1970s, many significant changes were coming about, not only in CEN, but also in the entire nuclear community. On the national scene, nuclear power was having problems both with economics and public acceptability. The antinuclear and environmental groups had gained enough support to have a major effect on the political decisions that were being made in Washington. These manifested themselves in the types of programs the AEC would support in the national laboratories, and it was clear that there would have to be a transition from nuclear to environmental research. There was still interest in the breeder reactors as a long-range energy source, and support continued in the areas of sodium chemistry and technology, reactor safety, and nuclear waste disposal. Nuclear fuel reprocessing as such was out. The Division explored new program areas, including cleaner methods of burning coal in power plants, solar collectors, fusion power, high-performance batteries for electric

Fig. 4-1. Leslie Burris

Table 4-3. CEN Organization as of January 1970

Division Director:	
Richard C. Vogel	
Deputy Division Director	
Donald S. Webster	
Associate Division Directors:	
Leslie Burris	Program Manager, Sodium Technology
Arthur D. Tevebaugh	Energy Conversion, High Temperature Materials, Calorimetry
Donald S. Webster	Coal Combustion, Fuel Reprocessing
Assistant Division Director:	
Everett R. Proud	
Section Heads:	
Elton J. Cairns	Energy Conversion & Physical Chemistry
Fred A. Cafasso	Sodium Chemistry
Carl E. Crouthamel	Analytical Chemistry, Neutron Cross Sections, Chemistry of Irradiated Fuels
Albert A. Jonke	Process Development, Fluidization Processes
Paul A. Nelson	Sodium Engineering
Group Leaders:	
Paul E. Blackburn	High Temperature Materials
Lester F. Coleman	Safety, Criticality & Special Studies
Norman E. Dudey	Nuclear Cross Section Measurements
John T. Holmes	Sodium Engineering
Ward N. Hubbard	Calorimetry
Carl E. Johnson	Chemistry of Irradiated Fuels
Robert P. Larsen	Analytical Chemistry & Burnup Analysis
Norman M. Levitz	Engineering Process Development
Robert J. Meyer	Sodium Analytical Methods
William E. Miller	Sodium Engineering
R. Dean Pierce	Pyrochemical Engineering
Joseph Royal	Publications Review Group
Martin J. Steindler	Chemical Process Development
Robert K. Steunenberg	Energy Conversion & Physical Chemistry

Harold M. Feder, Joseph E. Draley, and Charles E. Stevenson are listed as Senior Chemists or Engineers in a January 1970 description of the CEN staff.

vehicles and off-peak electrical energy storage, and fuel cells that were not limited to hydrogen as the fuel. The coal, battery, and fuel-cell programs developed into sizeable efforts in the 1970s.

When the AEC was disbanded into ERDA and NRC, and then DOE was formed, the interfaces between CEN and the funding offices became more fragmented. Each office wanted its own separate progress reports rather than the previous reports issued by the Division as a whole. The consequence was that well over 400 topical and progress reports were issued by CEN during the period 1970-1979. Because the funding and reporting functions were so diverse, much of the Division operated in a matrix-management type of structure in which the Division upper management remained cognizant of technical programs and maintained administrative control of the people and programs. However, many of the negotiations with the funding agencies as to the work to be done, and scheduling and funding requirements, were conducted at a lower (usually section-head) level under the general direction of upper management. Some of the CEN people involved in this arrangement were given the title "Program Manager." The individuals in these positions spent a lot of time and effort traveling to Washington, Germantown, or Morgantown visiting DOE managers in order to keep their programs healthy. There were jokes about taking along "groveling pads" to wear on their knees while begging for funding.

Matrix management had become a popular idea, and it was well suited to the Division's needs at the time, given its wide variety of programs and funding sources. Manpower and facilities could be organized into team efforts that often involved a variety of disciplines. This kind of flexibility tended to stabilize the Division as the individual programs waxed and waned.

One significant organizational change in CEN came about on July 1, 1971, when the analytical support groups from the Chemical Engineering and Chemistry Divisions were consolidated into a single cost center. This combined group was given the official name "Analytical Chemistry Laboratory" (ACL), a separate entity that would provide analytical services to the entire Laboratory, but would still remain a component of the Chemical Engineering Division. This organization was to be self-supporting by doing the work on a fee-for-service basis, at a cost of $25.50/ hour. During the initial stages of operation, Bob Larsen supervised the group in Bldg. 205, and Ralph Bane was responsible for the groups in Bldgs. 200 and 212. Norman Dudey was the first full-time manager of ACL until he left the Laboratory for another position after a year or so. Norm was replaced by Eugene Voiland, who had come to ANL from KAPL. After about three years, Voiland left for a position with General Electric at the Morris plant. Paul Cunningham, who was a Ph.D. chemist from the University of California at Berkeley and had been involved in CEN research programs for several years, became the manager of ACL in 1975. During his tenure in that position, he also served as Section Head for Analytical and Environmental Research in CEN. Since its inception, ACL has had to face a number of challenging problems, the foremost one being whether the new programs in the division would continue to generate a sufficient demand for analytical work to provide the necessary financial support. Analytical charges were highly visible on the cost reports, and in some instances the research groups felt they could save money by doing some of their own analyses. There were also instances in which a research group would have to purchase a specialized analytical instrument to obtain the data they needed and a question arose as to whether the research group or the ACL should have jurisdiction over that equipment. A means of funding some supporting development work on analytical methods also needed to be arranged. These problems were resolved eventually, and the ACL became well established in its present form.

When Cunningham left ANL in 1982 for a somewhat similar position at Los Alamos, David W. Green became the manager of the ACL and still is. Green is also a Ph.D. from Berkeley, who came to CEN from the Chemistry Division. He had worked previously as a Research Associate at the University of Chicago and taught at Albion College. In 1996, the ACL celebrated the 25th anniversary of its formation. Because of internal transfers among programs and other factors, it is difficult to establish exactly who the charter members of the ACL are, but they are believed to include those listed in Table 4-4.

The women who worked in CEN as secretaries, clerks, and other such positions were essential to the success of the Division's operations. Although their work was very much appreciated by the technical and administrative staff, it was usually not visible beyond initials on memos or internal reports. Many staff people probably don't realize how often the secretaries saved them from embarrassment or worse by correcting their spelling and grammar, retrieving some obscure piece of information from a file, handling phone calls adroitly, or working overtime to have material ready for the Review Committee or a trip. Many of them were instrumental in organizing the various parties, luncheons, and other social functions. A Secretaries' Day once a year falls far short of giving them the credit they deserve.

In the early 1970s, Alice Graczyk operated the Division Office under Dr. Vogel and then Les Burris. Anne Melton, Everett Proud's secretary, handled many of the administrative details of that office. Doreen Prucha was the principal secretary in Herb Brown's operation and Evelyn Rafacz continued to serve as Dr. Lawroski's secretary. Marie Driskell and Joy Swoboda were handling attendance records, purchase requisitions, equipment records, and a number of other miscellaneous items. Maria (Scaropoulos) Contos began working in the Technical Editing office. Some of the other secretaries and clerks in CEN at the time are listed in Table 4-5.

By the mid-1970s, the Division had once more outgrown the available space in Bldg. 205. A particularly pressing need was for an auditorium. Room A-059 was satisfactory for group meetings, but too small and not well laid out for Review Committee meetings, large program reviews, seminars, and other such functions. Office space was

Table 4-4. Analytical Chemistry Laboratory Personnel in the 1970s

Murray Barsky	Ruth Juvinall	Laurids Ross
Maureen Binelli	John Kartunnen	Dino Santelli
William Bloom	Gwendolyn Kesser	Robert Schablaske
Eugene Bogusch	Eugene Kucera	Charles Seils
Vera Drabek	Robert Larsen	Stanley Siegel
Florence Ferry	Vernon Lemke	Norbert Stalica
Irene Fox	Russell Malewicki	Benjamin Tani
Ruth Hanna	Fredrick Martino	Ralph Telford
Robert Heinrich	Robert Meyer	Zygmunt Tomczuk
Ben Holt	Robert Oldham	Florence Williams (Smith)
Myron Homa	Allen Panek	Jacqueline Williams
Kenneth Jensen	Ray Popek	

Table 4-5. Secretaries, Clerks, and Others in the 1970s

Karen Beres	Judith King	Darcell Richards
Alice Birmingham	Jackie Lehmann	Susan Rodighiero
Susan Butz	Sally Leonard	Susan (Cathimer) Roessler
Patricia (Wood) Canaday	Sheila Madson	Vita Shiffer
Ann Chaplin	Sharon Matlak	Maureen Sobczak
Di Ann Fager	Debra McCann	Jackie Stakkowski
Carol Gehrke	Marcella Moore	Janet Steinquist
Sherry Grisko	Kathy Mueller	Renee St. Germain
Paula Hataburda	Sofia (Gawenda) Napora	Katherine Wall
Carol (Richie) Hendricks	Bonnie Nolan	Candace Weiler
Barbara Jivery	Marge (Swanson) Panek	Melania Wilson
Mary Keigher	Julitta Pyle	

also in short supply again, due in part to the many industrial participants and other temporary people who were there at the time. Y-Wing was constructed in 1975-76 to fulfill these needs. Covering a total ground area of about 4,300 sq. ft, Y-Wing is a two-story structure consisting of an auditorium with a lobby area, 25 offices, a conference room, an open office area for secretaries and their equipment, and an exterior lobby, which also connects to the main part of Bldg. 205 on both levels. The auditorium and its lobby are attractively decorated and serve the purpose well, although there were a few start-up problems. There seemed to be a problem in finding projector lenses that were compatible with the distance and size of the screen. The other problem was more serious. For a time, the ventilation system was either inadequate or not functioning properly. On one hot day in August, during a major program review, the room became hot and essentially unventilated during a morning session. With the large crowd, the odor became unbearable and the meeting finally had to repair to the auditorium in Bldg. 200. That problem has long since been corrected, but it was an embarrassing experience for the people in the program who were trying to make a good impression on their reviewers and visitors.

Another addition was made to Bldg. 205 in 1978 to accommodate the growing need for laboratory space by the National Battery Test Laboratory. A particular concern was to have a separate facility where zinc/chlorine and other such batteries could be tested safely. The addition, designated J-156 and located at the end of J-Wing, has an area of about 1,550 sq. ft, and consists mostly of open laboratory space. It is currently used for battery and environmental work.

Although not a new addition, Room C-201, located upstairs at the north end of C-Wing, was used for a purpose some people were not aware of. When Dr. Vogel was faced with a task that required deep concentration, such as writing a journal article, he used that room as a hideaway with strict orders that he not be disturbed. If he "invited" you up into that room, you knew you were in for some very hard work.

The most serious accident in the history of the Division occurred late in the afternoon of October 17, 1972, when an accumulation of hydrogen and air exploded in a glove box located in Laboratory G-102. Two inter-connected glove boxes (Nos. 6 and 7) were being used to investigate a process for the preparation of uranium-plutonium oxide fuel pellets. The process involved the reduction of

UO_3 and PuO_2 powders to a UO_2-PuO_2 mixture by heating them to about 600°C in a stream of cracked ammonia (75% hydrogen-25% nitrogen) in equipment simulating a fluidized-bed reactor. The powder was then pressed and fired at 1650°C to form the ceramic pellets.

The explosion was more in the nature of a rapid deflagration than a detonation, but it made a loud noise that could be heard throughout the building. It had sufficient force to blow out the safety glass panels in the two glove boxes and the glass panels in doors and walls between the laboratory and the connecting corridor. The concrete block walls of the laboratory showed slight structural damage. Two employees who were in the laboratory suffered lacerations and bruises from the explosion and a third employee cut his hand on debris while evacuating the corridor. The laboratory had to be decontaminated to remove the plutonium activity. Several factors appeared to have been involved in the incident. Large quantities of hydrogen had apparently leaked into the box from a hydrogen exit line during an experiment three weeks earlier. A hydrogen-monitoring instrument gave no warning of a high level because it had been cross-connected after a calibration, and a high concentration of air resulted from inadequate ventilation of the glove boxes. The hydrogen-air mixture was most likely ignited by an electrostatic spark.

Division Administration

In CEN, responsibilities for management of the research and development programs pass from the Division Director down through the Associate Directors, and then to the various Section Heads, Group Leaders, and Program Managers. Management of the Division's administrative functions is the responsibility of the Assistant Division Director. Everett Proud had become the Assistant Division Director in 1966 and his secretary, Anne Melton, handled much of the routine work. A job description for the Assistant Division Director written by Everett at the time describes it as "being responsible for management of the Division to the level delegated to him by the Division Director. The responsibility falls into five categories: (1) personnel, (2) procurement, (3) budget management, (4) building facilities, and (5) the instrument shop." When Lee Mead departed from ANL in 1963, Ron Breyne had been given responsibility for the procurement function, and his role expanded while he served as an Administrative Assistant under Proud. In 1967, the Division was shocked by the premature death of John Schraidt, who had been handling the facilities and services operations plus a number of other responsibilities such as Division Safety Officer and had been intensely involved with the EBR-II Fuel Cycle Facility. He also performed some of the Associate Division Director's functions during the transition between Frank Masten and Everett Proud. Alice Graczyk had begun her career in CEN as Schraidt's secretary. John Natale was another major player in that area who handled many other projects. He was one of the original designers of the ubiquitous CENHAM glove boxes mentioned earlier.

In 1965, Herb Brown was placed in charge of the building facilities and services such as the machine shop, installation crews, reclamation crews, graphic arts, service requests and work projects. He also reported to Proud. Herb was assisted by Les Dorsey, Ernie Johnston, and Pleasant Kelsheimer. Doreen Prucha was the secretary. "Big Ernie" was particularly noted for his ability to install heavy safety-glass glove-box windows because of his size and strength. Les Dorsey, among other things, was given the job of tracking equipment inventories, which required checking the ANL serial number tag on every "sensitive" item in the Division, a job that required a lot of tact. Somehow, he always remained good-natured during the process.

After Schraidt's death, Les Coleman became the Division Safety Officer and handled the job with the same rigorous, no-nonsense manner that Schraidt had employed. Les also assumed responsibility for the instrument shop.

In 1974, Ron Breyne transferred to BIM (Biology and Medicine Division), where he served as Assistant Division Director, and Leo Morrissey took over the financial management and budgeting work. Later on, Paul Eident also joined in that effort. In 1978, Everett left CEN to work for Betsy Anker-Johnson, who had been appointed Associate Laboratory Director for Basic Energy Sciences. At that time, Ron Breyne and Herb Brown began to serve as joint Assistant Division Directors, with Ron in charge of the personnel, procurement, and budgeting operations and Herb handling the building and facility responsibilities.

Personnel

By 1970, the composition of the staff of CEN had changed significantly from that in the early days of the Division, when most of the people were in their 20s and 30s. Those individuals were now in their 40s and 50s, and they accounted for a peak in the age distribution curve, which has diminished gradually as new, younger people have joined the Division, and is currently disappearing due to retirements. Many of those retirees now work on a limited part-time basis as STAs (Special Term Appointees).

From the time the Division was formed until about 1990, the basic categories of full-time employees were as shown in Table 4-6. There were also functional titles, such as Group Leader, Section Head, Assistant Division Director, and Division Director. The technician/operator and secretary/clerk categories were paid hourly. The higher-level technical personnel were paid monthly, and the higher administrative levels were on a biweekly salary schedule. The claim is sometimes made that the qualifications of ANL engineers and scientists at the assistant, associate, and senior levels are equivalent to those of assistant, associate, and full professors at a major university. That is probably true in many cases, as evidenced by the fact that ANL staff people often teach courses, serve as thesis advisors, or conduct cooperative research with university professors, but whether it is a valid generalization is a moot point.

Table 4-6. Personnel Categories

Technician/Operator I	Secretary/Clerk I
Technician/Operator II	Secretary/ Clerk II
Technician/Operator III	Secretary/Clerk III
Senior Technician/Operator	Senior Secretary/Clerk
Engineering/Scientific Assistant	Administrative Assistant
Engineering/Scientific Associate	
Assistant Engineer/Scientist	
Associate Engineer/Scientist	
Senior Engineer/Scientist	

The Laboratory has always encouraged employees to avail themselves of educational opportunities, some in-house, and others on college and university campuses. In the 1970s, the MBA degree was beginning to be favored strongly in business organizations nationwide, and the University of Chicago offered an "Executive Program" to Argonne staff members. The purpose of this program, which resulted in an MBA degree, was to strengthen the managerial capabilities of individuals in positions of leadership. Several CEN people completed this training in the 1970s: Erv Carls, Eddie Gay, Dave Green, Martin Kyle, Paul Nelson, and Mike Yao. In addition to the regular staff, the Division has always had a large number of temporary personnel, including visiting scientists, consultants, industrial participants, trainees of various kinds, faculty appointments, graduate and post-doctoral students, co-op students, STAs, and others.

All personnel matters of the Division were administered by the Assistant Division Director. These included the recruiting and hiring of weekly-rated, salary, and engineering and scientific assistants. Final decisions in personnel selection were normally made with the concurrence of appropriate Group Leaders or Section Heads. Applications of more senior personnel were screened by the Assistant Division Director, who also administered the Division's Equal Opportunity Employment Program. Salary reviews and performance appraisal programs were also his responsibility. Other personnel matters included security clearances, travel, promotions, education and training, consulting and training contracts, and numerous other functions. He also served as the Division's representative on the ANL Personnel Committee B. Sofia (Gawenda) Napora was designated to maintain the CEN personnel records, and has continued in that capacity for years. She has an excellent memory for names of people who have been in the Division.

Personnel activities such as recruiting and hiring, salary increases, and promotion of the more senior staff members were handled by the Division Director and other upper management people. Applicants for employment were usually subjected to a schedule of interviews and asked to make a presentation on their thesis or whatever other research they had been doing. Promotion cases for those in this category were presented to ANL Personnel Committee A.

Procurement

In addition to office supplies and other such items that are purchased routinely by most business organizations, CEN had such a diversity of programs that procurement was often anything but routine. Frequently, the desired item did not exist and had to be fabricated, often from exotic metals or special ceramics. Many common items could be obtained from the CEN stockroom, but budgetary restraints had resulted in a marked decrease in the inventory by the 1970s, necessitating greater use of the Laboratory's central facilities. The Division maintained an extensive file of industrial catalogs that were very useful in tracking down some item that would fill the need. One intriguing catalog in that file was a listing of railroad locomotives with various options, including the desired color scheme. The reason for that catalog being in the file is uncertain, but it may have had something to do with the facilities at EBR-II.

The Assistant Division Director was responsible for all Divisional procurement. That covered all purchases, including equipment for the technical programs. Other items in this area were the establishment of work projects, outside contracts and consulting agreements, property-control records, equipment library, and equipment utilization program.

Budget Management

The Assistant Division Director's duties included all the financial management and budgeting activities in the Division. Funding for the technical programs required the preparation of 189a forms, which were basically program proposals to the funding agencies that outlined the work to be done, the schedule, and the costs. Technical personnel prepared the texts of the proposals, but manpower levels, costs, and cost codes were worked out with the budget people, who then tracked the expenditures to ensure that they remained within the budget. Materials and services and Division overhead had to be included in the costs. Taxes on programs were imposed in some cases to cover administrative costs of Program Directors and Associate Laboratory Directors. Manpower was also charged to programs for purposes such as quality assurance, technical editing, computer assistance, and other items. Budget management was not an easy task during the unstable funding conditions in the 1970s. Oftentimes a program would be part of the way through the fiscal year and suddenly hear from the sponsor that they could provide only a fraction of the funding level in the original projection at the beginning of that year. This created a big problem because of the multiplying effect. As a somewhat extreme example, if word was received midway through the fiscal year that the funding was 50% of the original amount, it meant the work had to stop immediately. The confusion was compounded in 1976, when Congress, unable to pass the budget legislation on time, changed the beginning of the fiscal year from July 1 to October 1. A separate set of program plans and 189s had to be prepared for this so-called "Transition Quarter."

In addition to budget management, the Assistant Division Director was responsible for handling time cards, staff attendance records, absentee reports, cost code assignments, special materials requests, and a number of other items, including arrangements for fund drives and Division social activities such as the annual picnic.

Building Facilities and Services

This aspect of the Assistant Division Director's responsibilities was handled by Herb Brown, who reported to Everett Proud and was assisted by Artie Freeman, Pleasant Kelsheimer, and Les Dorsey. Doreen Prucha was Herb's secretary and she was assisted by other secretaries or clerks as necessary. The primary responsibility of this office was the operation and maintenance of Bldg. 205 and any other areas under the jurisdiction of CEN. Installation work involved the Plant Service people, electricians, pipefitters, carpenters, laborers, and others. Some Plant Service personnel worked regularly in Bldg. 205. This office prepared service and work project requests and monitored performance of the work. A visitors' register was maintained in that office, and it held the keys to all storage spaces. Physical arrangements were made for special events such as program reviews, seminars, and open houses. Work in special facilities, including the caves and plutonium glove boxes, was overseen to ensure that all the safety equipment was functioning properly.

In addition to the multitude of complex systems that must be maintained to operate the building, the courtyard and nearby exterior areas were of concern. Ever since Bldg. 205 was built, it seems as if the "front yard" has been torn up at least 50% of the time for one reason or another. Jan Muller once quipped that "They should put a zipper in it."

The design of engineering and much of the laboratory equipment has always been an important aspect of the Division's activities. Schraidt's group did most of this work, although many engineers and a few chemists also did detailed design work. In some cases, design engineers, including Jim Kotora and

Israel Pollack, were brought in from other divisions for special projects. Johan Graae was involved in many different design efforts. Dick Malecha, although not a member of CEN originally, has been involved with the Division in various capacities for many years. Dick has a special expertise in the area of quality assurance and was involved in Laboratory-wide QA work for several years. Largely as a result of his influence, CEN was one of the first divisions to incorporate detailed QA-related specifications in shop drawings. It is difficult to make a sharp distinction between designing and drafting, the way it is done in CEN, but some of those who did the work were Harry Smith, Ernie Singleton, Bob Frank, Ray Stimak, Tony Keledus, and Norm Schoij. These people were assigned from Central Shops, but functioned as a part of CEN.

The machine shop was staffed by Central Shops personnel, but, like the designers and draftsmen, they were usually assigned to CEN for extended periods of time so they became acquainted with the CEN people whose work they were doing. Some of the longer-term machinists were Bill Voss, Tom Denst, Ed Lewandowski, Phil Brown, and Ted Wist.

There were others who worked in Bldg. 205 on a rather regular basis. Everybody from that period remembers the electrician, Pat Doolin, with his impish (and Irish) sense of humor. At one time two painters, both named "Wally," were working in the building, which sometimes caused a bit of confusion.

For many years, Bill Olsen operated the instrument shop. This facility had the capability to provide equipment design service, construction, maintenance, and repair of most of the instruments used by the Division. The principal function of the shop was to repair instruments being used in the research and development programs. Over the years enough equipment had accumulated from former programs to provide an inventory of items that could sometimes be used in new programs without the cost and lost time involved in ordering new ones. This office also maintained supplies of spare parts and chart paper.

Computers

By 1970, impressive progress had been made in the manufacture of miniaturized integrated circuits containing tens of millions of components on a single microchip, and it was inevitable that someone would soon exploit the technology in the form of a practical computer microprocessor. That occurred in 1971 when an American engineer, Marcian E. Hoff, combined the basic elements of a computer on a single silicon chip. That microprocessor was known as the Intel 4004, and it was followed immediately by frenetic development programs at many research laboratories to come out with their own versions. The first affordable desktop produced for personal use was the Altair 8800 and it was marketed by Micro Instrumentation Telemetry Systems in 1974. The Tandy Corporation, in 1977, was the first major electronics company to produce a personal computer by adding a keyboard and CRT monitor along with a cassette recorder to store programs. Soon thereafter, Stephen Wozniak and Steven Jobs formed Apple Computer Inc., and produced excellent machines that found wide acceptance. Apple became one of the outstanding success stories in the American business world. In the area of personal computers, International Business Machines, Inc. (IBM) was a sleeping giant; it didn't introduce a PC until 1981. It is amusing to recall at this point that *Popular Mechanics* magazine had predicted in 1949 that in the future a practical computer may weigh as little as 1.5 tons.

Mel Foster was coordinating the computer applications within CEN. At that time, essentially all of the computations were made on the large machines in the Applied Mathematics Division (AMD), which were connected to a station in CEN consisting of a card reader, paper tape reader, cardpunch, and printer. In 1974, Mel joined Dr. Vogel and Carl Crouthamel in their move to Exxon in Washington State and his job at CEN was taken over by Steve Gabelnick, who had been doing work on matrix-isolation spectroscopy and other studies that led to a familiarity with the ANL computer facilities.

In addition to the main-line computer facilities at AMD, the people in CEN began branching out into all sorts of applications to take advantage of the rapid advances in the technology that were occurring in the 1970s. To assist in the implementation of these applications, computer specialists were brought in from AMD and commercial organizations to contribute their expertise. Some of these people, including Glen Chapman, John Osudar, and Bob Land, became permanent members of the Division. Bob Kessie, a chemical engineer who had been involved in several earlier research and development programs and had been a long-time hobbyist in electronics, took to these new developments like a fish to water and did a lot of work with a Varian minicomputer on data acquisition and control. Some of the earliest programs to make extensive use of dedicated computer facilities were proof of breeding, calorimetry, examination of irradiated fuels, battery development and the National Battery Test Laboratory, physical properties research, and coal combustion.

Another important development in the mid-1970s was the introduction of the small hand-held calculators. When Bill Schertz arrived from Texas Instruments to join CEN he had one of those calculators, and also a digital watch, both of which were novel and attracted great interest. Irv Johnson purchased a Texas Instruments Model SR-51 in July 1975 and used it for many years. The hand-held calculators were a real time saver for those individuals who made many simple day-to-day calculations that did not justify the use of the large computers. These hand-held devices were in much demand by the staff, but ANL management was reluctant to provide them because of their cost (a few hundred dollars) and "volatility" (susceptibility to theft). Those who had them had to follow security procedures nearly as stringent as those for plutonium.

There were varied reactions, as one might expect, to the explosion of computer technology in the 1970s. A few would have preferred to stick with their slide rules and have as little as possible to do with it. Most realized that it was inevitable and began to take courses, read manuals, and learn to type. (This was the period when owner's manuals written by computer specialists were unintelligible to the average person—a practice that has not yet disappeared completely.) It was easy to be intimidated by the language. In those days most people had never heard of bits, bytes and baud, RAM and ROM, machine language, assembly language, floppy disks, hard disks, Fortran, Pascal, Cobol, Basic, hexadecimals, parity, and hundreds of other terms, many of which almost any 12-year-old could explain today. And then there were those for whom computers had become a compelling interest. Some people spent much of their time during evenings and weekends writing programs, which can become an obsessive activity. Home desktop computers as we know them were not available until the late 1970s, but a few hobbyists managed to assemble workable systems. Irv Johnson recalls his experience in developing a home computer capability: "My first step was to buy a KIM-1, a single-board basic computer. Input and output were by means of a keypad and four segmented video displays. A very primitive operating system was located in ROM on the board, and I had 1 KB of memory. Programming was by machine language. Later I added 4 KB of memory and was able to obtain a Tiny Basic that could be used. The addition of a homemade video terminal and keyboard resulted in a complete usable computer. Programs were saved on a regular small tape recorder. With this system, I learned a lot about programming, but could not do the complicated calculations I needed. I then built a Heathkit H11, a PC version of the Digital Equipment Corp. LSI-11. At first, input and output were by Teletype™ and storage was on paper tape. I then added a dual floppy disk (8 in.) and video terminal, giving me a real home computer for which I could write Basic, Fortran, or assembly language. By including a modem, I could then transfer programs between ANL and home. Later on, I went by stages to a 'PC,' an XT, an AT and finally a 486DX."

Application of computers in the technical programs has already been mentioned here and there. In 1978, the computer group started placing summaries of their activities in the Division reports. Their activities in 1978 and 1979 consisted of (1) computer modeling and simulation, (2) laboratory automation, (3) database development, and (4) general support (consulting, equipment specifications, *etc.*).

In the modeling area, a program was developed to calculate potential surfaces, equipotential contours, and current-density distributions for collector plates in lithium/metal sulfide battery cells. A model was constructed to calculate tritium concentrations during release and cleanup after a hypothetical accident in a fusion reactor. A NASA code was modified to calculate complex chemical equilibria (up to 12 components) in the formation of a non-ideal solution from condensed species.

Laboratory automation activities were supported in the National Battery Test Laboratory (NBTL), the ambient-temperature and lithium/metal sulfide battery programs, and the ion-microprobe facility. Three systems based on a Digital Equipment Corporation (DEC) PDP-11 family of processors were installed to perform the following functions:

1. Acquire data from the 50-cell lithium/metal sulfide cell lifetime testing facility and to acquire data and control cell tests for the development and engineering work, including the Mark 1A battery.
2. Automate experiments, including viscosity, calorimetry, density, and thermal diffusivity measurements of materials in the reactor-safety program.
3. Collect and process data from the ACL's new gas-chromatograph, mass-spectrometer system.
4. Calculate the results of coal-char combustion with sulfur retention by limestone in a fluidized bed (coal program).
5. Calculate the total flow of all fissile and fertile materials through a large number of nuclear reactors of three generic types (Pu-producing, Pu-consuming, and U-233-consuming).

Databases were assembled to assist CEN staff in reviewing DOE proposals, to centralize information collection, and to generate reports and statistical data. Databases were also set up to assist CEN administrative personnel in various activities, and general support was provided to Division members in the form of consulting, data-management services, a computer documentation library, CEN's time-sharing and batch-entry facilities, and audiovisual minicomputer course materials. In 1979, work was started on a database file for CEN publications.

The secretarial staff also saw a major change in the 1970s when the electric typewriters began to be replaced by "word processors." The word processor was a stand-alone system with a keyboard, video screen, floppy disk storage, and captive printer. This system had some of the features of the current computer word-processing programs in terms of editing, moving text, *etc.* Unlike most of the technical staff who were starting to use computers, the secretaries didn't have to learn to type to use these new devices, and they seemed to appreciate the versatility and capability to store their work on disks.

Miscellany

By the 1970s, the 20 years of the Division's existence was sufficient time for some traditions and lore to have developed. As mentioned before, the average age of the staff had increased. In the earlier days, many of the people were single, and there was a lively social life among the CEN members. One consequence of this, especially in the 1950s and 1960s, was a number of intradivisional marriages. Building 205 isn't exactly the Club Med type of atmosphere that one would expect to engender romantic relationships, but they did develop. By the 1970s, the following marriages had taken place between CEN members or between Division members and people closely associated with CEN:

Milt Blander - Marie-Louise Saboungi
Herb Brown - Norma Pinches
Carl Cushing - Beatrice Hjelte
Phil Danielson - Doris Andersen
Len Hanna - Ruth Peterson
Phil Fineman - Olga Giacchetti
Peter Kafalas - Eleanor Juco
Eugene Kucera - Gene McCloud
Ted Olszanski - Ruth Voiland
Ron Paul - Nola Joy Tasharski
Ed Peterson - Betty Reilly
Chuck Seils - Laureen Peterson
Jack Settle - Leila Buckner
Art Shor - Roberta Wagner
Carl Swoboda - Joy Holloway
Ronald Wingender - Doris Michalek

Three major social occasions are sponsored periodically by the Division: (1) a picnic, (2) a Thanksgiving season Turkey Raffle, and (3) an in-house buffet just prior to the Christmas-New Year's Day holiday break. One other affair that might also be mentioned is the Review Committee dinner, where the Division staff socializes with the Review Committee members and among themselves. The picnics, with food, drinks, games, and other activities, take place at Argonne Park and present an opportunity for the spouses and children of Division employees to become acquainted with one another. The Turkey Raffle usually involves some form of entertainment when the drawings are held in the Bldg. 205 auditorium. The annual Christmas parties, which were held originally in the large L-Wing lunchroom, were moved off-site for several years, and now take place in the A-059 Conference Room and adjoining tornado shelter under X-Wing. Secretaries' Day is generally observed off-site by small groups, as is Bosses' Day if the secretaries think they deserve it. Birthdays are sometimes observed by serving cake and coffee during the day. Program groups go out to lunch occasionally, sometimes to celebrate some special event, and individuals occasionally host group parties at their homes. At present, the age curve for the Division's population has reached the stage that there are numerous retirement parties, usually at the Flame restaurant.

For several years, the Laboratory has operated an after-hours "Exchange Club" in Bldg. 617 to encourage more informal contact among the technical staff members, who tend to become compartmentalized by geographic and administrative factors. It did seem to accomplish that purpose, and its use was expanded to include essentially all ANL personnel. Some used it as a "singles bar" when that idea was popular, and at closing time some of the groups would proceed on to a commercial establishment such as Ripples to continue the party. In general, the employees at ANL were probably not as close knit socially as those at sites such as Oak Ridge or Los Alamos because they were widely scattered over the Chicago metropolitan area and their social life tended to be more oriented toward their own communities.

Given the hundreds of people who have been in the Division over the years, it would be impossible to recount the wide variety of personal interests and activities that were represented. In general, they were probably rather typical of any Chicago suburban community. Possibly the longest standing group is the Division's Golf League. Their first annual tournament was in 1952. According to the *Argonne News,* the participants in the 1958 tournament were Jack Fischer, John Loeding, Don Fischer, Jim Bingle, Les Burris, Al Chilenskas, Ira Dillon, and George Bennett, who made the arrangements. More recently, regular members of the golf league have included Dave Green, Bill Miller, Les Burris, Vic Maroni, and Bob McBeth (from the Chemistry Division). For several years there was a volleyball group consisting mostly of members from the pyrochemical program, which included Dean Pierce, Bill Walsh, Les Burris, and Bill Murphy. This group didn't seem to recognize that volleyball is not a contact sport, and sometimes showed up at work somewhat battered and bruised. A CEN softball team was organized by Ron Breyne, Ziggy Tomczuk, and others. Dean Pierce has been a physical fitness enthusiast all his life and is an excellent athlete who has participated in triathlons and other such events. Now an STA, he still rides a bicycle to Argonne from his home in Naperville on the days he works. Paul Nelson became interested in running in the 1970s and participated in the Chicago Marathon. There are many others in the Division who are involved in athletics and sports, including a large number who coach or officiate in organized youth athletics.

Several people in the Division are boaters, including Leo Morrissey who has participated for years in the prestigious Mackinaw sailboat races. Powerboat owners include Jim Battles, whose enthusiasm for fishing has become legendary in the Division. There are a number of private pilots, two of whom, Ward Hubbard and John Ackerman, moved on to the more demanding sport of competitive gliding.

Another interesting group is the chess team, of which Dave Green, Ziggy Tomczuk, Ewald Veleckis, and Bill Walsh have been members. A surprising outgrowth of this activity came from Bill's experience in playing chess with prisoners at the Stateville Penitentiary. At an ANL seminar, a Dr. Carl Pfeiffer discussed the effects of chemical imbalances in the body on human behavior. Bill procured hair samples from some of the prison inmates and others and found strong correlations between abnormal behavior and the levels of trace elements and compounds such as zinc, copper, and amino acids. On the basis of these findings, he eventually established a clinic aimed primarily at curing behavioral disorders in children, using dietary changes and supplements, but no prescription drugs. He was the founder and is now president of the Carl Pfeiffer Clinic in Naperville, which has become highly successful in treating tens of thousands of patients per year, with many referrals from psychiatrists and other physicians.

A few musicians have surfaced in CEN. John McKee, Chuck McPheeters, and Jim Eberhart formed a guitar trio and sang numbers typical of the Sons of the Pioneers at the Turkey Raffles. On one occasion, Sandy Preto joined them, making a quartet. Sandy, who did electrochemical research studies in the battery program, had another talent. She took up belly dancing and, for a short time, moonlighted occasionally at an Aurora night club. She also presented her "Sasha" routine at one of the Turkey Raffles. Another musician who contributed much to some of the CEN social affairs was Cindy Wesolowski, who plays the accordion and also has an excellent singing voice. Many other Division personnel sing or play instruments, but have not been heard at CEN.

Some other performers who deserve mention are Bob Larsen and Ed Berrill, who could always get a lot of laughs, and Johan Graae, who was nearly always called upon to give an impromptu speech at parties. Pat Doolin, mentioned earlier as the electrician who was assigned to Bldg. 205 for years, was a talented magician. One of his best tricks was to pass a cigarette (lighted or not) through a borrowed coin; you could watch it from a distance of a foot or two and still not understand how he did it. Pat also served many years as a scoutmaster, as did Herb Brown with Martin Steindler as his assistant.

TECHNICAL PROGRAMS

As stated earlier, CEN's technical programs in the 1970s underwent a major shift from nuclear reactor technology toward environmental research and alternative energy options. Impetus was added to this trend by the oil embargoes. Another trend that affected the Division strongly was a decision by the AEC, ERDA, and especially DOE to place more of their management activities "in the field." By 1979, the total funding for CEN was about $31 million, with about a third of the amount being contracted out to industries, universities, and other ANL divisions. The distribution by program areas was roughly as follows: electrochemical (batteries and fuel cells), 58%; fission-related work, 16%; coal technology, 13%; basic energy sciences, 8%; other, 5%. By the end of 1979, about 284 people were employed by CEN, 222 of whom were staff. The increased management effort for the funding agencies required a significant part of the staff effort.

Lithium/Metal Sulfide Batteries

INTRODUCTION

In 1970, the research and development work on high-temperature batteries was still in an exploratory stage. A decision had been made to discontinue the work on thermally regenerative galvanic cells in favor of rechargeable batteries, which seemed to have more promise for commercial application. Although a large amount of information had been generated on active materials for the electrodes and electrolytes and numerous experimental cells had been built and tested, many questions remained as to their potential performance and lifetime as well as their suitability for battery configurations. The principal candidates for the positive electrode material were the chalcogens (sulfur, selenium, and tellurium), and lithium was favored for the negative electrode. Early on in the program, it became evident that the terms "anode" and "cathode" were going to lead to confusion in the case of rechargeable cells. The chemical definitions are that oxidation occurs at the anode (negative electrode) and reduction at the cathode (positive electrode) when a cell is discharged. On recharge, however, the situation is reversed so the negative electrode becomes the cathode and the positive electrode the anode. To avoid confusion, a decision was made to avoid the terms anode and cathode and simply call them negative and positive electrodes, which is true during both charge and discharge.

There was some uncertainty about the optimum composition for the molten salt electrolyte, which had to be capable of transporting lithium ions between the electrodes, have a sufficiently high decomposition potential to be stable at the cell voltage, and have a low enough melting point to permit a reasonable operating temperature. In addition to the electrodes and the electrolyte, suitable materials had to be available for components such as electrode separators, current collectors, electrical feedthroughs, and cell insulators and housings. Another uncertainty in the program at the time was the practical application envisioned for high-temperature batteries. Several funding agencies were supporting the work with different applications in mind, including electric-vehicle propulsion and load-leveling on electric utility systems.

The result of the above situation was that a concentrated effort was made during the first two or three years to obtain the additional information needed to select the most suitable combination of cell components. The effort was then focused on the development of specific cell and battery designs for defined applications. During this period, the program expanded greatly and included the addition of industrial participants to the overall effort. This expansion, together with other factors, produced several management changes in the 1970s. The first year or two, Art Tevebaugh was the cognizant Associate Division Director, and then Don Webster assumed this responsibility until Paul Nelson became the director of all the electrochemical programs in 1973. Elton Cairns was the Section Head in charge of the high-temperature battery work until he returned to the General Motors Research Laboratory in 1974, and then went on to Lawrence Berkeley Laboratory and the University of California in 1978. Managers of the High Temperature Battery Program were Dick Ivins (1974-76), Bob Steunenberg (1976-78), and Duane Barney (1978-84). Al Chilenskas was Manager of Advanced Battery Development. Section Heads at various stages of the program were Elton Cairns, Bob Steunenberg, and Eddie Gay. Group leaders at one time or another included Al Chilenskas, Bob Steunenberg, Eddie Gay, Bill Walsh, Jim Battles, Paul Shimotake, Don Vissers, Bill Miller, and Mike Roche.

INDUSTRIAL PARTICIPANTS, CONSULTANTS, AND CONTRACTORS

A major goal of this program was to involve industrial organizations in the effort so the technology could be transferred to industry as it was being developed. In 1973, battery firms were solicited for an assignment of one or more of their scientists or engineers to assist in the development work and assess the likelihood of future commercial feasibility of the technology. Approximately 20 industrial participants worked on the program for periods of a few weeks to a year on a cost-shared basis. Some of these participants were Ray Hudson and Ken Gentry from Eagle-Picher Industries, Inc., Dick Rubischko from Gould Inc., Jim Birk from the Atomics International Division of Rockwell International, Robert Hamilton from the Carborundum Co., and Doug Stakem from Catalyst Research Corp.

In 1975, cell development and fabrication programs were initiated at Gould, Eagle-Picher, and Catalyst Research. At about the same time, management of a program supported by DOE at Atomics International was transferred to ANL. Ray Hudson headed up the operation at Eagle-Picher, Dick Rubischko and later Barry Askew the one at Gould, Sy Sudar the one at Atomics International, and Doug Stakem the one at Catalyst Research. (Catalyst Research dropped out of the program after a couple of years, and the General Motors Research Laboratory joined the effort later with Elton Cairns as the principal investigator.) Procurement contracts were negotiated with Carborundum to develop and supply boron nitride (BN) felt separator materials. Dick Malecha was given the responsibility for handling routine management of these subcontractors.

Many consultants were used in the high-temperature battery program. Scott Wood of the Illinois Institute of Technology and Bruce Wagner of Arizona State University assisted the chemistry and materials groups. John Newman from the University of California, Berkeley, did extensive work on the mathematical modeling of the processes occurring in Li(Al)/iron sulfide cells, and one of his students, Richard Pollard, did a Ph.D. thesis on that subject. Some of the other consultants included Richard Alkire, University of Illinois; Theodore Beck, Electrochemical Technology Group in Seattle; Douglas Bennion, University of Utah; John Dunning, General Motors Research Laboratory; Robert Huggins, Stanford University; George Janz, Rensselaer Polytechnic Institute; Rob Selman, Illinois Institute of Technology; Warren Towle and William Tiedemann, Johnson Controls (formerly Globe-Union); and Ernest Yaeger and Boris Cahan, Case Western Reserve University. Certain members of CEN or program review committees who had special expertise in the area, including Ed Buzzelli, Westinghouse Electric Corp., Terry Cole, Ford Motor Co., Charles Tobias, University of California, Berkeley, and Wayne Worrell, University of Pennsylvania, also provided helpful technical advice.

In 1976, ANL and the Chicago Section of the Electrochemical Society co-sponsored a Symposium and Workshop on Advanced Battery Research and Design, which was attended by many of the above individuals and organizations, as well as a number of foreign participants. The proceedings are reported in ANL-76-8.

EXPLORATORY STUDIES

A large body of information had already been accumulated at ANL on the chemistry and technology of molten halide salts in the nuclear fuel reprocessing and regenerative galvanic cell programs and in the basic studies of molten metals and salts. For cells having lithium negative electrodes, it was clear that a lithium halide salt system would be the most suitable, and the LiCl-KCl eutectic was selected on the basis of its relatively low melting point (352°C), high decomposition potential, compatibility with construction materials, and low cost.

From the standpoint of theoretical specific energy (W-hr/kg), the lithium/sulfur cell (Li/LiCl-KCl/S) was the best of the systems under consideration, and a concerted effort was made for about three years to build and operate cells of this type. One of the several problems associated with this cell was the lithium electrode. The liquid lithium in the negative electrode caused several problems: (1) It was sufficiently soluble in the electrolyte to impart some electronic conductivity, which led to current inefficiency and self-discharge of the cell. (2) Its chemical activity was sufficient to cause some reduction of the potassium in the LiCl-KCl electrolyte,

$$Li + KCl \rightarrow LiCl + K\uparrow$$

which, over a period of time, resulted in metallic potassium deposits in the cell and a change in the LiCl:KCl ratio in the electrolyte. (3) The lithium metal, being a potent reducing agent, was not compatible with oxide ceramics such as magnesia (MgO) and alumina (Al_2O_3) used as electrical insulators. (4) At cell operating temperatures (generally 400-450°C), it was difficult to confine in the negative electrode structure. Confinement of lithium in the negative electrode was most likely related to two factors: competition between the lithium and the electrolyte to wet the electrode structure, and the fact that the cell housings were normally at the negative electrode potential, which meant the lithium could deposit on the housing wherever electrolyte was present. Many ingenious electrode designs using various metals in porous forms such as mesh, felts, and foams were designed and tested, but none was fully successful.

All of the above problems were finally resolved by going to lithium-aluminum alloy negative electrodes. The Li-Al phase diagram, which had been investigated earlier by Yao, Heredy, and Saunders [*J. Electrochem. Soc.* **115**, 1039 (1971)] and was refined by Jack Settle and Mike Myles in Jim Battles's group, is shown in Fig. 4-2. At 450°C, there is a composition range between 10 and 47 at.% Li, where two solid phases ($\alpha + \beta$) coexist in equilibrium and the electrical potential remains constant at 292 mV relative to elemental lithium. When cycled electrochemically, this material formed a porous, interconnected (skeletal) microstructure that is ideal for high-rate, efficient electrodes, and it did not change appreciably over thousands of cycles in well-designed electrodes. On recharge, the lithium all deposited in the electrode because of its decreased chemical activity in the Li-Al structure.

The price to be paid for these advantages was a significant, but acceptable decrease in the voltage and specific energy of the cells. The lithium-silicon system was also a candidate for the negative electrode. In this case, the electrode discharges through a series of voltage steps at 42, 150, 277, and 326 mV *vs.* elemental lithium, which correspond to the transitions

$$Li_{3.90}Si \rightarrow Li_{3.25}Si \rightarrow Li_{2.33}Si$$
$$\rightarrow Li_{1.71}Si \rightarrow Si$$

In practice, the Li-Si electrodes were generally operated between about 50 and 80 at.% lithium. A $Li_{3.25}Si$ electrode in combination with an FeS or FeS_2 positive electrode has a specific energy approximately 10% higher than that for Li-Al in a similar cell. Lithium-aluminum negative electrodes were used in most of the cells that were built and tested, but the Li-Si electrodes were also used successfully, particularly in Li-Si/FeS_2 cells at General Motors and in Li-Si/FeS cells at Gould. In some cases, Gould used mixtures of Li-Al and Li-Si. Some exploratory work was done on calcium negative electrodes at ANL, but it was not continued into full-scale cells. Subsequent studies have shown that a ternary Li-Fe-Al system is an excellent negative electrode.

A somewhat similar situation developed in the case of positive electrodes. On the basis of previous work reported in the literature and preliminary studies at ANL, the chalcogens (sulfur, selenium, and tellurium) were selected for the initial investigations. Of these three elements, sulfur had the advantage in terms of voltage and specific energy, and a large amount of effort went into research on the behavior of sulfur in electrodes, both in its basic chemical and physical properties and in the development of practical electrode structures. The high oxidation potentials of chalcogens required the use of refractory metals such as molybdenum or niobium for current collectors and other components of the positive electrode, and many different experimental electrode designs were evaluated. Selenium and tellurium, with additives such as

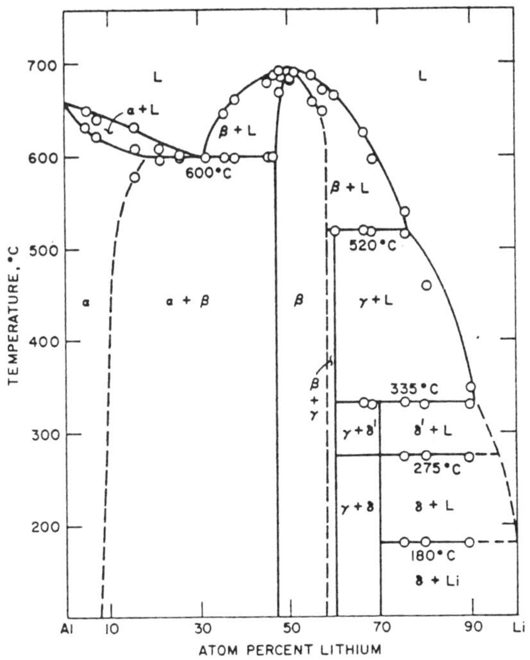

Fig. 4-2. Lithium-Aluminum Phase Diagram

thallium in some cases, were tested in many cell configurations, but were finally abandoned because of their lower voltages and specific energies and short cycle life caused by soluble species in the electrolyte. Much attention was given to sulfur electrodes, due in large part to the enthusiasm of Elton Cairns for lithium/sulfur cells because of their very high voltages and theoretical specific energies (W-hr/kg).

A major problem with sulfur was its near-zero electronic conductivity. Attempts to increase the conductivity by various additives such as carbon black and phosphorous sulfides were unsuccessful. Another serious problem was the tendency of sulfur to dissolve in the electrolyte, which caused a serious loss of capacity on extended charge-discharge cycling. It was clear at this point that a new approach had to be taken toward the positive electrode. Don Vissers, Ziggy Tomczuk, and Bob Steunenberg came up with the idea of using metal sulfides, which are solids and offer some degree of electronic conductivity. The iron sulfides (FeS and FeS_2) seemed to show the most promise, and they performed well in preliminary tests. Workers at the Atomics International Division of Rockwell International had the same idea at about the same time, but the ANL group managed to get a patent. The more promising metal sulfides for this application are given in Table 4-7.

From a technical standpoint, any of the above materials could be used successfully, but cobalt and nickel sulfides were deemed too expensive for large-scale practical application, which left FeS and FeS_2 as the materials of choice. (Later in the program, Laszlo Redey and Don Vissers developed a Ni_3S_2 reference electrode for use in molten salts that was highly successful and resulted in a patent and an IR-100 Award.) Titanium disulfide, TiS_2, undergoes a reversible reaction with lithium, and cells of this type were operated. The reaction differs from those of FeS and FeS_2, however, in that it involves a lithium insertion or intercalation mechanism that allows only one lithium atom to enter the TiS_2 lattice, which limits the specific energy.

As was the case with the Li-Al electrode, the iron sulfide electrodes produced lower voltages and specific energies than sulfur, but later testing showed that they performed very well electrochemically, with a long cycle life. At this time, the program became focused on the two systems:

Li(Al) (solid)/LiCl-KCl (liquid)/FeS (solid)

Li(Al) (solid)/LiCl-KCl (liquid)/FeS_2 (solid)

These systems, with solid electrodes and a liquid electrolyte, were of a more conventional configuration for a storage battery.

Table 4-7. Promising Positive Electrode Materials

Material	Potential *vs.* Li°, mV	Material	Potential *vs.* Li°, mV
FeS	1620	FeS_2	2066, 1620
NiS	1830, 1656	CoS_2	1997, 1900, 1663
Co_9S_8	1663	NiS_2	2034, 1830, 1656
Ni_3S_2	1656		

The overall discharge reactions for these cells are

$$2\ Li + FeS_2 \rightarrow Li_2S + FeS \quad (1.67,\ 1.33\ V)$$

$$2\ Li + FeS \rightarrow Li_2S + Fe \quad (1.33\ V)$$

The theoretical specific energy for the Li/FeS cell is 460 Wh/kg, and that for the Li/FeS_2 cell, if both voltage plateaus are used fully, is 680 Wh/kg. Both are respectable values in spite of the compromises made in both electrodes. At first glance, these discharge reactions appear to be straightforward, but that turned out not to be the case, especially for the FeS electrode. The discharge process in the FeS electrode proved to be very complex, and a large effort was made by the materials and chemistry groups to identify the reactions and the intermediate phases that were formed. As many as six electrochemical and four chemical reactions appeared to be involved. A variety of techniques were used to identify the phases and reaction paths, including X-ray diffraction and metallographic studies of partially charged and discharged cells. Two compounds, Li_2FeS_2 (W-Phase) and $LiK_6Fe_{24}S_{26}Cl$ (J-Phase, formed by reaction with the KCl in the electrolyte), were identified as intermediates in the discharge reactions. The J-Phase was shown to inhibit the kinetics of the discharge reaction. Studies of the reactions showed that alternative paths existed for the discharge process, one of which did not involve J-Phase. As shown in Table 4-8, the concentration of KCl in the electrolyte can be lowered to avoid the formation of J-Phase. The latter two salts improved the cell performance significantly, but at the price of higher operating temperatures.

Some cells were operated with FeS_2 positive electrodes, which produced two voltage plateaus, the first at about 1.6 V and the second at 1.33 V. Cells with FeS electrodes showed a single plateau at 1.33 V. Although the FeS_2 cells had a higher average voltage and specific energy, they had the disadvantages of a voltage change during discharge, corrosion of the metal cell parts, and long-term loss of capacity. Through the 1970s, most of the cells were of the Li(Al)/LiCl-KCl/FeS type, but at the end of the decade, the following options were still considered viable:

Negative Electrodes: Li-Al, Li-Al-Fe,
Li-Si, Li-Al-Si
Positive Electrodes: FeS, FeS_2
Electrolyte: LiCl-KCl, LiF-LiCl-LiBr

MATERIALS RESEARCH AND DEVELOPMENT

The high temperatures and corrosive materials used in the cells required an extensive effort on materials research and development. Martin Kyle handled much of this work until about 1973, when an expanded program was instituted with Jim Battles as Group Leader. John Smaga was a major contributor to the

Table 4-8. Electrolyte Systems

System	Composition, mol%	Melting Point, °C
LiCl-KCl (eutectic)	58.8-41.2	352
LiCl-KCl	66.7-33.3	425
LiF-LiCl-LiBr	22-31-47	430

materials work. A special laboratory facility was set up in J-101, where metallographic work could be done entirely within a controlled-atmosphere glove box. A unique feature of this facility was an arrangement in which samples inside the glove box could be observed from outside by a microscope mounted in the top of the box, either directly or by closed-circuit television. This arrangement was very useful in showing and explaining results to coworkers or visitors. This group also had a laboratory in Room A-141 that was used for corrosion testing and other purposes.

The materials program was far too broad to describe in detail. Extensive corrosion tests were conducted on all the metallic components of the cells. Fortunately, it turned out that low-carbon steel and stainless steel, which are relatively inexpensive and easily fabricated, could be used for cell housings and supporting structures for the lithium-aluminum electrodes. Stainless steel was preferred, mainly because it was not subject to oxidation by the air atmosphere outside the cells. The positive electrodes presented more of a problem, particularly with sulfur or FeS_2 as the active material. Refractory metals such as molybdenum could be used as current collectors, but they are expensive, heavy, and difficult to fabricate. For the FeS electrodes, iron or low-carbon steel was used as current collectors, but some corrosion occurred during long-term charge-discharge cycling. Adding iron powder to the electrode mix largely solved this problem. Alumina (Al_2O_3) was used successfully as electrical insulators in certain parts of the cells, although long-term exposure resulted in the formation of some $LiAlO_2$. The separator had to meet a severe set of requirements. It had to be resistant to reduction by the lithium electrode and to oxidation by the positive electrode. At the same time, it had to serve both as an electrical insulator and as a porous medium for the liquid electrolyte to permit the transport of lithium ions between the electrodes. A great deal of development work and testing resulted in two types of separators that gave excellent performance: boron nitride (BN) felt or woven cloth, and porous plaques pressed from magnesium oxide (MgO) and the electrolyte salt. The production of BN in these forms was a technical breakthrough by the Carborundum Co., which was making it for an undisclosed military application. It was formed by extruding boric oxide (B_2O_3) as fibers, which were converted to BN by treatment with ammonia, followed by a high-temperature exposure to nitrogen. Although Carborundum was most accommodating in developing and producing these materials, this effort demanded a great deal of guidance and attention from the Materials Group. Several options for electrode separators as of 1979 are listed in Table 4-9.

Another troublesome problem that was solved mainly by Mike Myles working with subcontractors was the development of a hermetic feedthrough from the positive electrode to an external cell terminal. (An extension of the cell housing served as the negative terminal.)

In addition to the traditional materials work, this group made very major contributions to two other areas of the program. One was in the identification of chemical phases and physical structures formed in the electrodes during different stages of charge and discharge. This work contributed to an understanding of the cell chemistry and provided guidance for future electrode designs. The other contribution was the identification of cell-failure mechanisms. The most common mode of cell failures was short-circuiting that developed over a period of time during charge-discharge cycling. This usually resulted from expansion of the negative electrode, which resulted in extrusion of the active material and breaching of the separator. Extensive studies of this phenomenon by the Materials and Chemistry Groups resulted in electrode design recommendations (loadings, geometry, *etc.*) that solved, or at least greatly alleviated this problem. Some of the other

Table 4-9. Status of Electrode Separator Materials

Form	Material	Comments
Woven Fabrics	BN	Used in early cells; satisfactory, but expensive.
Felt	BN	Well developed, satisfactory, expensive, but reasonable cost projected for mass production.
Powder	MgO, AlN, Si_3N_4	Pressed or vibratorily compacted; inexpensive, used satisfactorily in test cells.
Porous Plaque	MgO, Y_2O_3	In the experimental stage.

failures were caused by loss of separator integrity, escape of material from the positive electrode, and hardware problems. Extensive post-test examinations of weak and failed cells that were conducted by this group were the principal source of information needed to improve the cell designs. The excellence of this type of work by the Materials Group became well recognized and resulted in programs supported by other organizations.

CELL DEVELOPMENT

In the exploratory stages of the program, many different cell designs were used, most of them being cylindrical (pancake) assemblies held together by bolted flanges and ranging from about one to six inches in diameter. In a few cases, cells were stacked together in series to form small batteries. As the program became more focused on specific energy and power requirements and on configurations better suited to practical applications, cells of the types shown in Figs. 4-3 and 4-4 were developed. This development depended heavily on results of work by the Materials Group on electrode separators, electrical feedthroughs, and other components. Tom Kaun played a particularly major role in the development of new electrode and cell designs and obtained a large number of patents. The cylindrical and prismatic cells had a central positive electrode and two outer negative electrodes, which were grounded to the cell housing. Each of the negative electrodes normally had about one-half the capacity of the central positive electrode. These configurations were called "bicells" because they were, in effect, two cells connected in parallel with a common positive electrode. Although the earlier rectangular prismatic designs were also bicells, the more advanced designs, as shown in Fig. 4-4, had multiple negative and positive electrodes.

The electrodes were plaques of the active material that were usually made by pressing or pasting the active material into a metallic supporting structure, such as a mesh or foam. The function of the current collectors in the electrodes was to maintain a uniform current density and provide connections to a bus bar,

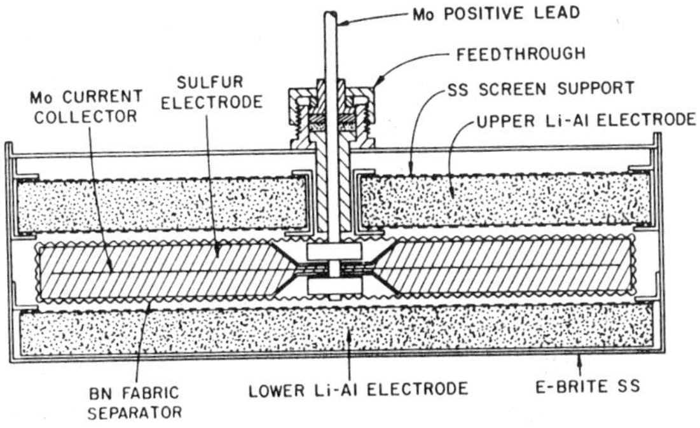

Fig. 4-3. Cylindrical Cell Design

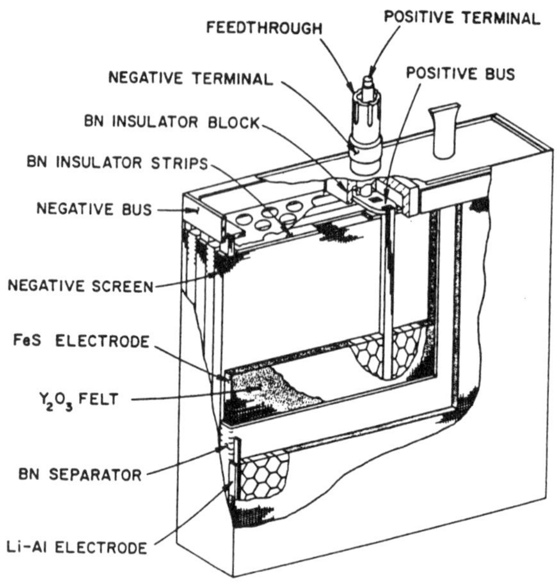

Fig. 4-4. Rectangular Prismatic Cell

which was connected through an insulating feedthrough to the outer terminal. Retainer screens or cloths were used in some cases on the electrode surface to prevent the escape of particulate material. The function of the electrode separator was to insulate the electrodes electrically from one another while at the same time containing the liquid electrolyte necessary for the transport of lithium ions between the electrodes. Cylindrical cells of the type shown in Fig. 4-3 were about 5 in. in diameter, and the prismatic cells (Fig. 4-4) ranged from about 5 × 7 × 1.5 in. up to 7 × 7 × 2 in.

In the late 1970s, the rectangular prismatic cell design was selected for continuing development because it was better suited to cells with multiple negative and positive electrodes. For example, the cell shown in Fig. 4-4 has five Li-Al negative electrodes and four FeS positive electrodes. The two outer negative electrodes are half as thick as the others. The positive electrodes are interconnected by a bus bar that leads to the positive terminal of the cell. The negative electrodes are grounded to the cell walls, which conduct the current to the negative terminal attached to the cell housing. The multiplate cell design permits the use of relatively thin electrodes with a high surface-to-volume ratio, which results in high specific power at acceptably low current densities, and high utilization of the active materials in the electrodes, which increases the specific energy.

"STATUS CELLS"

The program was coming under increased pressure from DOE to provide assurance that demonstrable progress was being made in the cell performance and cycle life. The response was to initiate an effort in which Eagle-Picher, Gould, and ANL each designed and built groups of identical state-of-the-art cells for standardized statistical testing. In general these "status cells" were of the prismatic multiplate design, about 5 to 7 in. wide, 7 in. tall, and 1 in. thick. The principal difference in the cells produced by the three organizations was that BN separators were used in the Eagle-Picher and ANL cells, while Gould preferred MgO powder.

There were also differences in the designs and materials used in the cell electrodes and hardware. The cells were tested at ANL by charge/discharge cycling until the cell capacity had decreased to 80% of its initial value, which usually occurred after several hundred and, in some cases, up to 1000 or so cycles. A 50-cell test facility (Fig. 4-5) was constructed in which the cells were cycled continuously at the operating temperature of 400-450°C. A computer was used to monitor the results.

Tests were conducted on other cells to determine the effects of various duty cycles, for example, discharges typical of a driving profile for an electric automobile. Effects of overcharge and overdischarge were examined.

In addition to the electrical performance testing, Al Chilenskas and his coworkers conducted safety tests on cells, which were subjected to vibration and shock with little if

Fig. 4-5. Fifty-Cell Test Facility for Li/Metal Sulfide Cells

any effect. Cells were also crushed to evaluate hazards that might result from a vehicle crash in which the molten salt electrolyte could be released from the cells at operating temperatures. These, too, were uneventful; most of the salt remained in the cell, and what little did escape solidified quickly.

BATTERIES

Al Chilenskas was in charge of the battery work in the program. All the potential applications of high-temperature batteries that were under consideration required voltages much higher than those of the individual cells, so a large number of cells had to be connected in series to form a useful battery. The battery then had to be enclosed in a housing with high-efficiency thermal insulation to maintain its operating temperature while minimizing the loss of thermal energy, which reduces the overall energy efficiency. Internal electric heaters were installed with the cells to heat the battery during startup and any time that additional heating was needed to supplement the resistive heating produced by normal charging and discharging. The development of heat-management strategies and high-efficiency thermal insulation was a major effort in the battery program. One quirk in the chemistry of the system that made the job a bit easier was an entropic effect in FeS cells that tended to level out the heating and cooling effects during the charge-discharge cycles.

Two basic types of high-efficiency thermal insulation were investigated: conventional and vacuum multifoil, with several different versions of each. The thermal conductivities of the better ones were in the range of $1\text{-}3 \times 10^{-3}$ W/m·K. Calculations for a small LiAl/FeS automobile battery with 1/2-in. peg-foil insulation produced by the Linde Division of Union Carbide would have a heat-loss rate of 66 W and energy densities of 87 W-hr/kg and 150 W-hr/L. Table 4-10 lists the three small experimental batteries were operated in the 1970s.

Another problem that had to be addressed was cell equalization during charging. Unlike nickel-cadmium and certain other batteries, lithium-alloy/iron sulfide cells have little tolerance for overcharging, so the individual cells had to be returned to the same state of charge. This required the development

Table 4-10. Experimental Batteries Operated in the 1970s

Year	No. of Cells	Voltage, V	Capacity, kW-hr	Thermal Insulation	Cycle Life
1977	6	8.8	1.0	Vac-Foil	11
1979	10	5.9	3.7	Conventional	34
1979	5	5.6	1.6	Vac-Foil	70

of a computer-controlled charge-equalization system, which was designed by Fred Hornstra and the Computer Applications Group. Once developed and placed in operation, the charge-equalization system performed well.

One of the "problems" with the battery development program was that the cells were beginning to exhibit long cycle life. The long cell life was desirable, but it resulted in tests that required months, or even up to a year, before a cell failed. Consequently, a great deal of time elapsed before the effect of a new design feature could be evaluated in a post-test examination. The DOE was becoming impatient with the rate of progress toward an in-vehicle battery demonstration, and, in 1978, mandated that one be done immediately. The CEN engineers and scientists felt that it was premature and so indicated, but the DOE insisted, so Eagle-Picher was given the job of assembling a battery (designated Mark IA). The Mark IA battery consisted of two modules, each containing 60 cells that were fabricated in the charged state, arranged in two rows of 30 each, and connected in series to provide the required voltage (see Fig. 4-6). Each module was enclosed in high-efficiency thermal insulation and had a capacity of 20 kW-hr.

During heatup, one of the modules developed a short circuit external to the cells, which propagated from cell to cell. The only external indications that something was awry

Fig. 4-6. Mark IA Battery

were fluctuations in the cell voltages and the internal temperature, which rose to over 1000°C. As is usually the case, the incident occurred in the middle of the night. A 24-hour watch had been set with Don Vissers and Sandy Preto on the night shift. They called the fire department as a precaution, but no action was required. The other module, which was alongside and connected in series, was unaffected by the failure.

Extensive examinations were conducted on the failed module, and electrolyte leakage was established as the primary cause of failure. The other module was not tested, but the cells were removed and used for other testing. A few days after the incident, Barry Askew, the Program Manager at Gould, concerned about the political aspects of the failure, asked a group of visitors from ANL, "Has there been much fallout over the Mark IA meltdown?" He was told that he could have found a better choice of words while talking to people recently involved in nuclear programs. The failure of the Mark IA battery was, in fact, a serious setback and may have been part of the reason for diminishing support of the program in the following years.

The program did continue, however, with plans for a Mark II battery in which Eagle-Picher and Gould would each produce a 10-cell module with emphasis on reliability.

By the mid-1970s, the potential applications being considered for high-temperature batteries had narrowed down to two: electric vehicles and load leveling. Al Chilenskas instigated design and cost studies on batteries for both of these applications. The design criteria for load-leveling batteries were (1) 100 MW-hr of electrical energy stored in about 7 hr and discharged over peak-load periods of 5 hr for five days per week, (2) a plant lifetime of 20 years, and (3) a cost (in 1978 dollars) of $45-55/kW-hr. Four conceptual designs were developed by ANL and outside contractors.

Throughout the high-temperature battery program, there were discussions concerning the most likely practical applications of electric vehicles. Delivery vans and commuters' cars seemed to be the best possibilities because they both involve relatively short, predictable routes that are compatible with daily, overnight recharging. In the early stages of battery development, at least, electric family cars capable of cross-country trips did not seem to be feasible. Electric vehicle design studies were conducted jointly by the Toyota Motor Company and ANL for a van and an automobile. The vehicle requirements and projected performance are shown in Table 4-11.

PERSONNEL

Because of the large scope of this program and the number of people involved in several aspects of the work, it is virtually impossible to give proper credit to each individual. The following is an attempt to list the contributors under the general areas into which they put most of their effort.

Chemical Research: John Ackerman, Milt Blander, Paul Cunningham, Jim Eberhart, Al Fischer, Ellen Hathaway, Myron Homa, Paul Hunt, Carl Johnson, Stan Johnson, Gene Kucera, Vic Maroni, Jane Marr, Al Martin, Carlos Melendres, Sandy Preto, Laszlo Redey, Mike Roche, Laury Ross, Marie-Louise Saboungi, Bob Schablaske, Rob Selman, Stan Siegel, Ramamritham Sridhar, Jim Sim, Bob Steunenberg, Condeocita (Cajigas) Sy, Ben Tani, Ziggy Tomczuk, Verne Trevorrow, Don Vissers, Bob Yonco.

Materials Research & Development: Jim Battles, Jim Bouquist, John Dusek (MSD), Martin Kyle, Jim Mathers, Frank Mrazek, Trent Latimer (MSD), John Mundy (MSD), Mike Myles, Ted Olszanski, Neil Otto, Jack Settle, John Smaga, Bob Swaroop, Wayne Tuohig (MSD), Dave Walker (MSD), Tim Wilson.

Cell Development & Testing: John Allen, Karl Anderson, Jack Arntzen, Lou Bartholme, Glen Chapman, Tom Cooper, Dennis DeNuccio, Ron Elliott, Bruce Feay, Bill Frost, Eddie Gay, Tom Kaun, Bob Kessie, Dennis Kilsdonk, John Kincinas, Nick Koura, Wally Kremsner, Fred Martino, Chuck McPheeters, Bill Miller, John Osudar, George Redding, Jim Riha, Bill Schertz, Paul Shimotake, Noel Vargo, Bill Walsh, Mike Yao.

Battery Development & Evaluations: George Bernstein, Ed Berrill, Steve Box, Al Chilenskas, Bill DeLuca, Johan Graae, Magdy Farahat, Ed Hayes, Fred Hornstra, Henry Kacinskas (ENG), Verne Kolba,

Table 4-11. Vehicle Requirements and Projected Performance

Vehicle	Van	Automobile
Curb Weight, lb	3150	2350
Battery Weight, lb	950	660
Payload, lb	795	330
Total Weight, lb	4900	3340
Acceleration Time (0-60 mph), sec	22	16
Maximum Speed, mph	95	85
Range (Overnight Charge), mi	100	100

Israel Pollack (ENG), Al Zielen (EIS), Sam Zivi (ENG).

Other: Duane Barney, Dick Beatty (QA), Elton Cairns, Jay Geller (QA), Dick Ivins, Dick Malecha, Paul Nelson.

Development of Aqueous Batteries for Electric Vehicles

The Electric and Hybrid Vehicle Research and Development and Demonstration Act, passed in 1976, called for the development of improved batteries for electric-vehicle propulsion. This activity, which was part of the Department of Energy's Electric Vehicle Project, was intended to reduce national petroleum consumption by the transportation sector. The Office of Electrochemical Project Management (OEPM) was established at ANL to manage the development of improved batteries suitable for electric-vehicle applications by the late 1980s. Three types of batteries were identified by OEPM as suitable candidates for this application: lead-acid, nickel/iron, and nickel/zinc. These names for the three systems are used universally in the trade, but may be a bit misleading in that the actual electrode materials are Pb/PbO_2, Fe/NiOOH, and Zn/NiOOH, respectively. The electrolyte in the lead-acid battery is sulfuric acid (H_2SO_4); an alkaline electrolyte such as a solution of potassium hydroxide (KOH) is used in the other two systems.

Neng-Ping (Mike) Yao, who had been instrumental in starting this program, was placed in charge as Program Manager and Director of OEPM. Mike had an interesting background, having been born into an aristocratic Chinese family and then receiving much of his education in Japan. He received his doctorate from the University of California at Los Angeles (UCLA), where he did his thesis work under Professor Douglas Bennion, a well-known electrochemist who later did some consulting work for the ANL battery program. Yao came to ANL from the Atomics International Division of Rockwell International, Inc., where he and Laszlo Heredy did the original work on the lithium-aluminum electrode. Mike was fluent in English, Chinese, and Japanese, a rather unusual combination. Prior to his assignment to OEPM, he worked a few years on the high-temperature battery program at ANL.

Specific managerial assignments in the OEPM program as of 1979 were as follows:

Mike Yao	Project Manager and Director, OEPM
Clint Christianson	Deputy Project Manager
Fred Hornstra	Manager, NBTL
Tien Lee	Technical Monitor, Lead/Acid Contracts
John Rajan	Technical Monitor, Nickel/Iron Contracts
Jim Miller	Technical Monitor, Nickel/Zinc Contracts
Glenn Cook	Group Leader, Battery Support Research
Bill DeLuca	Group Leader, Battery Components
Frank Foster	Contract Administrator
Jay Geller	Quality Assurance Engineer

Other members of the team included Mike Genge, Jiri Klinger, Dave Poa, Jodean Waters, and John Barghusen.

Industrial contractors on a cost-sharing basis performed most of the battery research and development work in this program. The main function of ANL, in addition to managing the contract work, was to monitor their technical progress by standardized testing of their cells, modules, and batteries in the NBTL (National Battery Test Laboratory), located in Bldg. 205. The workers at ANL conducted some research of a more basic nature on these systems.

INDUSTRIAL CONTRACTORS' PROGRAMS

Eight major industrial contracts for the battery technology development were initiated early in 1978: (1) lead-acid (Eltra Corp., ESB, Inc., Globe-Union, Inc.), (2) nickel/iron (Eagle-Picher Industries, Inc., Westinghouse Electric Corp.), and (3) nickel zinc (Yardney Electric Co., Energy Research Corp., Gould Inc.). The 1980 performance goals for the three types of batteries are listed in Table 4-12.

The NBTL (Fig. 4-7) was established to conduct independent testing and evaluation of various battery systems as they were developed. (At the time, suspicions were aired occasionally that battery manufacturers sometimes were not entirely candid about the true performance and lifetime of their products. These suspicions were real enough that they surfaced occasionally as humorous slides at professional meetings of electrochemists.) Cells and batteries developed within the DOE-sponsored programs and by private funding were tested in the NBTL (located in Laboratory H-102 of Bldg. 205 and expanded later into a new addition to the building, J-156). Those facilities as of 1980 were capable of the simultaneous testing of 27 cells or modules and one full-size (30-40-kW-hr) battery under simulated driving conditions as well as under normal cycling tests.

Graphical displays and tabulations of the results were directly available through a computer. The battery tests were normally in the following sequence:

1. Self-discharge (capacity loss in seven days).
2. Partial depth of discharge (to check "memory effects").
3. Energy density *vs.* power density (to provide performance and modeling data).
4. Simulated load profile tests (simulated electric-vehicle driving profile).
5. Peak power for sustained duration (30-sec at 80, 50, and 20% state of discharge).
6. Life-cycle tests (3-hr discharges until capacity declined to 75% of rated capacity).

In 1980, the capabilities of NBTL were expanded from 15 to 20 test stations, and work was started on further expansion to 40 test

Table 4-12. Performance Goals for Aqueous Batteries

Performance	Lead-Acid	Nickel/Iron	Nickel/Zinc
Specific Energy (W-hr/kg) at C/3 Rate[a]	42	54	64
Specific Power (W/kg), 30-sec avg at 50% Depth of Discharge	105	110	110
Cycle Life to 80% Depth of Discharge at C/3 Rate[a]	400	300	200

[a] The "C/3" rate corresponds to a full discharge in three hours.

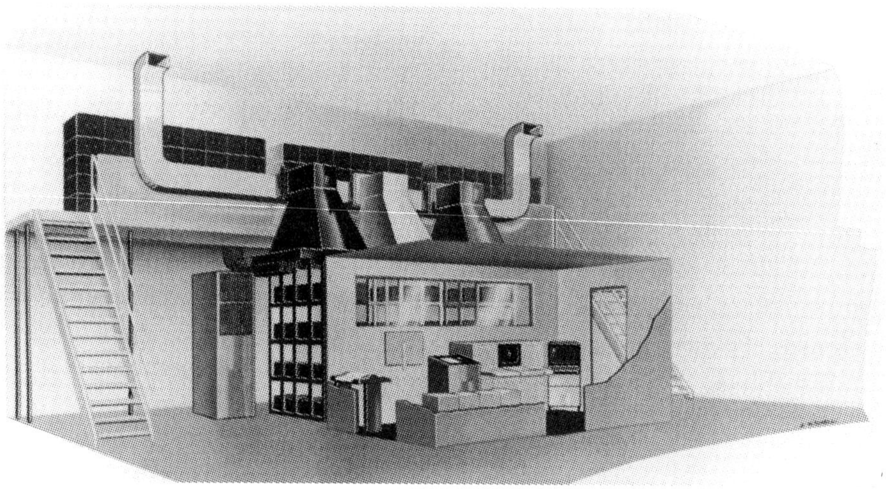

Fig. 4-7. National Battery Test Laboratory (NBTL)

stations. A typical cell/module test station had current and voltage capabilities of ±500 A and 0-12 V. A full-size battery test station had +300 A, -1200 A, and 10-200 V capacity, which required a large power supply. A system was constructed to measure sustained peak power from a battery. Three walk-in chambers were built to permit testing under various conditions of temperature and humidity. Other items included a computer-controlled simulation of driving profiles, instrumentation to measure individual cell voltages during battery tests, and an interactive control console to provide support for the individual control console terminals.

Most of the research and development in this program was performed by the battery contractors. Argonne performed some supporting basic and applied research work to complement the contractors' efforts. The following is a brief summary of the ANL activities.

ANL SUPPORTING RESEARCH

Zinc Electrode Studies. The loss of capacity of zinc electrodes on cycling is a problem that has plagued battery manufacturers for years. Several experimental and modeling studies were focused on mass transport, reaction mechanisms, and current distribution in zinc electrodes. Galvanostatic polarization experiments, combined with known information from the literature, resulted in a new theory involving passivation of the electrode caused by precipitation of zinc oxide (ZnO) at the surface. The effect of zincate supersaturation on the conductivity of the electrolyte was also investigated. This study demonstrated the effect of reduced electrolyte conductance upon zinc electrode performance.

Thermal Modeling of Ni/Zn Cells. A study to model heat transfer within Ni/Zn cells was undertaken under a contract with the University of Illinois. An experimental cell was constructed to measure temperature distributions over full-size electrodes during charge-discharge cycling.

Nickel Electrode Capacity Loss. During tests and evaluations of near-term alkaline batteries, three adverse effects had been noted: (1) an inability to obtain a full charge from some batteries after several charge/discharge cycles, (2) a lower charge acceptance of nickel electrodes in Ni/Fe cells as compared to Ni/Zn cells, and (3) effect of temperature on recharge efficiency. To investigate these effects, NBTL test data were reviewed, measurements were made on individual cell electrodes, and bench-scale experiments were performed. The first problem was quantified by NBTL data and individual tests, and some insight was gained into the problem, but the conclusion was that a detailed study would be required to reach an unambiguous understanding of the mechanisms involved. With regard to item (2), a small amount of soluble iron in the electrolyte resulted in more oxygen generation, which reduced the charge acceptance of the nickel electrode. Increasing temperature was shown to cause a marked capacity loss in the nickel electrodes.

Lead Positive Electrode Active Materials. In spite of the fact that the lead-acid battery is a well-established technology, the electrochemistry of the system is still not well understood. As an example, the utilization of active material in the positive (PbO_2) electrode is limited to 30-50%. *In-situ* laser Raman spectroscopic studies were conducted on the electrodes to identify the reaction products, and the study was extended to evaluate the merits of using $4PbO{\cdot}PbSO_4$, tetrabasicleadsulfate (TBLS), as the active material in these electrodes. The results showed that a significant amount of $PbSO_4$ and minor amounts of $3PbO{\cdot}PbSO_4{\cdot}H_2O$ and $PbO{\cdot}PbSO_4$ were formed. The relative electrochemical oxidation rates were ranked tentatively as follows:

$$TBLS > PbO{\cdot}PbSO_4 > PbSO_4$$

Upon discharge of the fully charged electrodes, only $PbSO_4$ was observed as a product. Thus, the only value of using TBLS in the fabrication of electrodes would lie in morphological effects produced in the first discharge. Another study was undertaken on lead positive active material, using neutron-scattering methods. The results showed that hydrogen (or deuterium) was incorporated into the β-PbO_2 lattice and that the lattice was lead-deficient.

Arsine-Stibine Emissions. Small amounts of arsine (AsH_3) and stibine (SbH_3) can be evolved from lead-acid batteries during charging. Both are highly toxic gases. Air samples were taken during the charging of lead-acid batteries both in workplace situations and in laboratory tests and analyzed for arsine and stibine. The results served as a basis for an engineering analysis that resulted in recommendations on the ventilation of battery compartments in electric vehicles.

Post-test Battery Analysis. Post-test analyses involving various analytical techniques and visual examinations were performed to determine causes of failure and to offer recommendations to the battery contractors that could be incorporated into their next generation of modules.

Battery Component Research and Development. Charging regimes and parameters have an effect on battery performance and lifetime. By 1980, NBTL had in place computer-control and data-acquisition systems capable of handling up to 15 concurrent battery charge/discharge cycling tests. Extensive studies were performed on the effects of constant-current, constant-voltage charging; equalization charging; pulsed currents; and other charging strategies.

Battery/Vehicle System Integration. The oil embargoes of the 1970s and the air pollution in large cities both generated pressure on the DOE to accelerate the development of electric vehicles, and their natural response was to push strongly for early demonstrations of the technology. Integration of battery prototypes into vehicles was an essential part of this endeavor. Most of the in-vehicle battery testing was to be done by JPL (Jet Propulsion Laboratory), and the batteries were to be supplied by the commercial manufacturers. The NBTL had the responsibility for providing information on the batteries to be tested, evaluating battery configurations suitable for specific vehicles, and evaluating demonstration results. Computer modeling of battery behavior was undertaken to provide a better understanding of the electrode and cell behavior. An equivalent circuit model was developed to predict the response of the battery to driving cycles, regenerative braking, charge/discharge strategies, and other such factors. Finally, materials availability/cost analyses were performed to evaluate the effects of near-term market penetration by electric vehicles on the cost and availability of key materials.

Fuel Cells

The major fuel-cell program in CEN, which began in about 1977, was directed toward cells that use molten carbonate as the electrolyte. Argonne became the lead laboratory with management responsibility for all the DOE work in that area, and CEN conducted an experimental research program of its own. Prior to that time, however, there were two small related programs that deserve mention. The first, which began in about 1971, did not involve fuel cells *per se* but was a close relative in that it employed an electrolytic cell with a molten carbonate electrolyte to replace CO_2 with oxygen in space capsules.

ELECTROLYTIC GENERATION OF OXYGEN FOR SPACE APPLICATIONS

The objective of this work, which was supported by the Ames Center of the National Aeronautics and Space Administration (NASA), was to recover oxygen from the CO_2 with high efficiency, and to recover an additional amount of oxygen from water to provide a total oxygen make-up stream of about 2.0 lb/astronaut-day. The carbon from the CO_2 was to be converted into a readily disposable or useful form. A particularly desirable reaction would be

$$4\ CO_2 + 2\ H_2O \rightarrow 2\ C_2H_2 + 5\ O_2$$

Although some progress was made, the research effort revealed several difficult technical problems, and it was terminated after about two years. Rob Selman did most of the work on that program. Sam Von Winbush, a visiting professor, also contributed to this work.

AQUEOUS ELECTROLYTES FOR HYDROCARBON FUEL CELLS

The second program, which began in about 1973 and was led by John Ackerman, was aimed at improved electrolytes for hydrocarbon/air fuel cells. The idea behind this work, which was supported by the U.S. Army Mobility Equipment Research and Development Command (MERDC), was to develop fuel cells that could use conventional fuels to propel military vehicles. Two potential advantages of such vehicles are that they would be quiet and that the fuel efficiency, in principle, could be high because fuel cells are not subject to the Carnot-cycle efficiency limitations of internal combustion engines. Conventional fuel cells with a phosphoric acid electrolyte had been used widely with hydrogen or alcohol as the fuel. The objective of this program was to use hydrocarbon fuels, and the effort was aimed mainly toward new electrolyte systems of much higher acidity. A number of organic and inorganic electrolyte systems with very high hydrogen-ion activities were investigated. John worked closely with the Army people at Ft. Belvoir during these investigations. This program was discontinued when funding was terminated.

MOLTEN CARBONATE FUEL CELLS

In 1977, CEN assumed its role as the lead laboratory for the DOE molten carbonate fuel cell programs with John Ackerman as the Program Manager. The overall objective of the DOE program was to develop fuel-cell power plants in which synthesis gas (hydrogen and CO) from a fuel processor such as a coal gasifier/clean-up system would be fed to fuel cells to generate electricity. The advantages of this type of plant are use of coal, very low pollutant emissions, and high efficiency (estimated at 60%, coal to bus bar). The increased use of coal, rather than oil, for power generation seemed like a particularly good idea in view of the oil embargo at the time.

The molten carbonate fuel cell program, like the battery programs, consisted of research and development work performed by a number of contractors under an ANL Program Office and supplemented by in-house work at ANL. The contractors included the Energy Research Corporation, Oak Ridge National Laboratory, the Institute of Gas Technology, and the General Electric Company. In addition to technical management, the CEN group assisted DOE with planning and evaluations. The molten carbonate fuel cell (see Fig. 4-8) consisted of a porous nickel alloy anode, a porous nickel oxide (NiO) cathode, an electrolyte structure that insulated the electrodes electronically but permitted passage of carbonate ions (CO_3^{2-}), and metal cell housings, which, in the case of cell stacks, were cell separator plates. The housings (or plates) bore upon the electrolyte structure to form a "wet seal" that separated the cell from its surroundings. The electrolyte structure, commonly called a "tile," was a composite

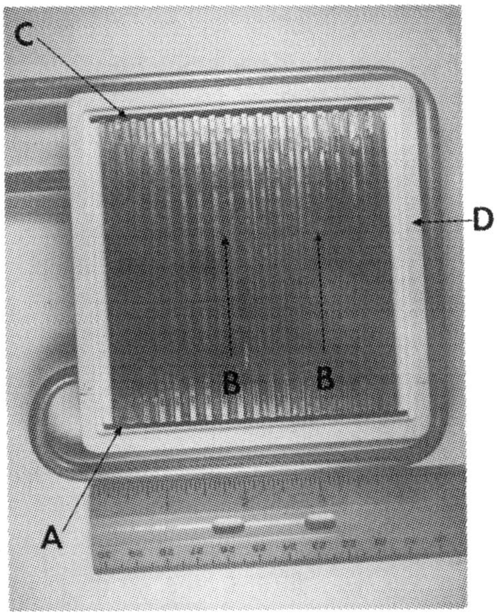

Fig. 4-8. Molten Carbonate Fuel Cell. Dimensions: 11 × 11 cm. A = gas entrance slot; B = gas channels; C = gas exit slot; D = wet seal area.

of very small (<1 μm) solid particles of lithium aluminate ($LiAlO_2$) and a mixture of lithium and potassium carbonates (Li_2CO_3 and K_2CO_3) that is liquid at the operating temperature of about 650°C.

Hydrogen and carbon monoxide in the fuel gas react at the anode with carbonate ion in the electrolyte to form carbon dioxide and water, giving up electrons to the external circuit. At the cathode, carbon dioxide and oxygen react and accept electrons from the external circuit to form carbonate ions, which are conducted through the electrolyte to the anode.

$$\text{Anode:} \quad H_2 + CO_3^{2-} \rightarrow H_2O + CO_2 + 2\,e^-$$
$$CO + CO_3^{2-} \rightarrow 2\,CO_2 + 2\,e^-$$

$$\text{Cathode:} \quad 2\,CO_2 + O_2 + 4\,e^- \rightarrow 2\,CO_3^{2-}$$

The ANL in-house effort was directed first toward improving the wet seal and then mainly toward improvements in the electrolyte tiles. The people involved in this work, in addition to Ackerman, were Pat Finn, Kim Kinoshita, Gene Kucera, and Jim Sim. In 1978, Dean Pierce was brought in as a Group Leader. One of the first problems to be tackled was corrosion of the stainless steel housing in the area of the wet seal where the two halves of the housing pressed directly against the electrolyte structure. The corrosion was due primarily to gradients in the chemical potential in these areas. The problem was alleviated greatly by flame-spraying aluminum onto the wet seal area and then heating the housing to form a diffusion layer between the aluminum and the Type 316 stainless steel housing. On exposure to the cell environment, the aluminum was oxidized, producing a graded structure, presumably of lithium aluminate ($LiAlO_2$), alumina (Al_2O_3), aluminum, and stainless steel. The protective coating of lithium aluminate and alumina was non-porous and adherent.

Most of the ANL work concerned the development of electrolyte tiles of adequate stability, strength, and performance. Above the liquidus temperature of the carbonate mixture (about 450°C), the tile was a very stiff plastic body held together by surface forces; below the liquidus temperature it was a hard, strong body resembling a ceramic. The carbonate content of the tile had to be high enough to transport carbonate ions at a rate sufficient for good cell performance without sacrificing the tile strength. The tiles typically had a carbonate content of about 65 wt%.

A key factor in the strength of the tile was the size and shape of the solid lithium aluminate particles. The compound exists in three different allotropic forms and can be prepared in various sizes and shapes (spherical "clumps," rods, bipyramids, and platelets). Methods were developed to prepare all three allotropic forms in high purity and to identify changes in the allotropic forms and particle morphology. The effects of these variables were evaluated in cell tests and post-test examinations. The initial tests were conducted with 7-cm-dia round cells, and in 1978 they were continued with 11-cm-square cells, which were more representative of a practical system.

The research and development work on electrolyte tiles continued in 1979, with more detailed studies of synthetic methods and the use of additional diagnostic techniques, including dilatometry to determine expansion, slumping and creep. New tile compositions were explored and some work was done on sintered electrolytes. By the end of 1979, the tile quality had improved significantly, and studies were started on long-term aging and degradation effects.

Coal Combustion

INTRODUCTION

In 1970, about 20% of the energy consumed in the United States was generated from coal. The rest was from oil (43%), natural gas (33%), and hydroelectric and nuclear (4%). Most of the oil was imported because domestic sources had been depleted to the point that their recovery costs were too high to compete with foreign supplies. The use of natural gas was increasing rapidly as pipelines were built, thereby displacing coal and oil for many applications, including the generation of electrical power. Estimates of U.S. gas reserves at the 1970 rate of consumption ranged from about 50 to 100 years. In contrast, U.S. coal reserves, if they could be used, would sustain the rate of energy consumption at that time for several hundred years. S. David Freeman, Director of the Ford Foundation Energy Policy Project in the early 1970s, described the situation very succinctly: "There are two things wrong with coal today. We can't mine it and we can't burn it."

The environmental and safety problems associated with the use of coal as an energy source had become a matter of considerable public concern. Underground mining was dangerous, and news report of numerous fatalities, injuries, and chronic illnesses were common. Environmental impacts from underground mining included acid mine drainage, erosion, leaching of pollutants from wastes, fires in mines and waste piles, silting of streams, and destabilization of land over the mines. Deep mining, which produced most of the higher-grade anthracite and bituminous coal, was giving way to strip mining, which is much safer and more benign environmentally, but has its own problems. The areas where strip mining is feasible tend to have lower-grade deposits such as high-sulfur bituminous coal, sub-bituminous coal, and lignite. The environmental effects consist mainly of piles of earth from the mantle over the deposits and the excavated areas where the coal was removed. Reclamation of strip-mined land is usually possible, but expensive. In Northern Illinois, old strip mines, which become filled with clear water, are common, and are sometimes used for recreational purposes such as swimming and fishing.

Once the coal is mined, burning it creates another set of environmental problems. The off-gases contain particulate matter, sulfur dioxide, nitrogen oxides, and carbon dioxide. Nearly all of the particulate matter can be removed by filters, which collect fly ash as a solid waste. The solid wastes from coal combustion contain a number of toxic substances that complicate their disposal. Liquid scrubbers can be used to treat the off-gases, but they are expensive and create liquid waste, which requires further processing.

The many problems associated with power generation from coal resulted in extensive environmental legislation to set standards and regulate that aspect of the industry. At the same time, there was an awareness that energy from coal was needed in the long run to conserve oil and gas. It was clear that new technologies were the only way to reconcile these two requirements, and the oil embargo of 1973 emphasized the need for timely development of these technologies.

In 1969, the Chemical Engineering Division had started a program on the combustion of coal in a fluidized bed to which particulate limestone ($CaCO_3$) was added to reduce the emission of sulfur dioxide (SO_2) during the combustion process. This program was funded initially by the National Air Pollution Control Administration (NAPCA) of the U.S. Department of Health, Education and Welfare. Spurred by the 1973 oil embargo, President Nixon, in 1974, recommended a $35 million expenditure for the direct (fluidized-bed) combustion of coal. Programs were sponsored by the Office of Coal Research of the U.S. Department of the Interior and the U.S. Environmental Protection Agency (EPA). There was pressure to move into the demonstration stage, and a 10-ton/day plant was being constructed by Pope, Evans and Robbins, Inc. Consideration was being given both to atmospheric-pressure fluidized-bed combustion (AFBC) and to pressurized fluidized-bed combustion (PFBC). An advanced high-pressure concept was pursued jointly by Argonne, Exxon Research and Engineering, and Westinghouse Electric Corporation. The equipment had progressed to a scale of about 500 lb of coal/hr and a conceptual design was in the works for a 3-MWe plant. The level of effort on the coal program in the Chemical Engineering Division was increased markedly in the 1970s. Extensive bench-scale engineering studies conducted throughout the decade were supported by more basic investigations that provided an understanding of the chemical and physical principles involved in the process. By the end of the 1970s, CEN, in addition to the in-house work, was providing administrative support in a number of areas to the Morgantown, WV, Energy Research Center in the form of program planning, project management, and program reviews. A substantial amount of the manpower in CEN was committed to these activities.

Al Jonke was in charge of the coal program until 1978, when Irv Johnson assumed that responsibility. Individual parts of the program at various times were headed up by Erv Carls, Irv Johnson, Mike Myles, Bill Swift, John (G. J.) Vogel, and John Young.

CLASSES OF COAL

Ideally, a fluidized-bed combustion unit should be able to accommodate coal from any source. Four different classes of coal, as shown in Table 4-13, are found in the United States. The values in this table, adapted from Bureau of Mines information (Bulletin 650), should be regarded only as typical, rather than specific data, for the purpose of illustration. Anthracite, although an excellent fuel, is very limited in supply, constituting only about 2% of the United States coal reserves.

Bituminous coal is plentiful and of good quality, except for the fact that much of it has a high sulfur content. This was the major reason for investigating new combustion techniques with a capability for *in situ* sulfur removal. It is interesting to note that Illinois has the largest bituminous coal reserves of any state in the nation, but Illinois coal is plagued with a high sulfur content (most of it 3 wt% or more) and, therefore, is useless for power generation without some means of sulfur control.

FLUIDIZED BED COMBUSTION OF COAL

Bench-Scale Engineering Experiments. The work that had been started in 1969 using a 6-in.-dia fluidized-bed combustion unit operating at atmospheric pressure was continued in 1970. The general features of this unit are shown in Fig. 4-9. The fluidized-bed combustor was 6 ft long, positioned vertically, and was equipped with a gas preheater for the fluidizing air and vibratory screw feeders for the coal and limestone. The flue gases passed through primary and secondary cyclone separators followed by a glass fiber filter. Infrared spectroscopy and other instrumental methods were used to monitor the composition of the flue gas. The principal variables in the operation were (1) Ca/S mole ratio in the feed, (2) type of limestone, (3) gas velocity, and (4) combustion temperature. Initial tests were conducted with pulverized limestone (25 μm), a superficial gas velocity of 3 ft/min, temperatures of about 800 to 900°C, and Ca/S ratios of 1.5-2.5. The temperature and gas velocity over these ranges had little effect, but raising the Ca/S ratio from 1.5 to 2.5 increased the SO_2 removal from 57% to 83%. Recycled limestone from the fly ash gave about the same results, and dolomite performed better than limestone (90 *vs*. 83%). The results from these runs produced sufficient information to develop a mathematical model from the data.

Table 4-13. Classes of Coal

Class	Anthracite	Bituminous	Sub-bituminous	Lignite
Quality	Excellent	Good	Fair	Poor
Availability	Poor	Excellent	Good	Fair
Fixed Carbon, wt%	86-98	50-86	40-60	<40
Moisture, wt%	<2	2-6	20-30	40
Ash, wt%	3-9	4-12	5-10	5
Sulfur, wt%	0-1	0-3 or more	0-1	0-1
Heating Value, Btu/lb	13,000	13,000	10,000	7,000

Fig. 4-9. Bench-Scale Fluidized-Bed Combustor

Work was then started on pressurized fluidized-bed combustion, using the 6-in.-dia bench-scale unit. This system was operated at 10 atm pressure with a compressed air feed, an air preheater, and an off-gas system made up of cyclone separators, filters, gas samplers, and pressure let-down valves. The unit was instrumented with a data-logging system. The principal coal used in the operation was a highly caking, high-volatile, bituminous type, but some work was also done with sub-bituminous coal and lignite. The coal feed was ground to a particle size of -14 mesh and dolomite was used as the sorbent for SO_2.

In the experiments with bituminous coal, the temperatures were about 800-900°C, the Ca/S mole ratios 1-3, and the gas velocities 2-5 ft/sec. The runs were all at 8 atm pressure with 3% oxygen in the flue gas and a bed height of 3 ft. Again, the effect of temperature was nil, and the sulfur retention increased with higher Ca/S ratios. The effluent gas met the

EPA standard of 1.2 lb $SO_2/10^6$ Btu of energy produced. The NO_x levels were also low, well below the EPA standard of 0.70 lb NO/10^6 Btu.

Experiments were conducted with the 6-in.-dia pressurized combustor to determine whether any difficulties would be encountered in processing lower-grade coals. One of these was a sub-bituminous coal with a high ash content of 17%, and the other was a lignite with a low heating value of 7,625 Btu/lb. The operating performance was excellent in both cases, which demonstrated the versatility of fluidized-bed combustion for processing coals of widely varying types and quality.

Some special tests were directed toward the effects of operating conditions on the concentrations of SO_2 and NO and other components in the flue gas. Sulfated dolomite was used in the bed. Changes in the temperature had no significant effect on the SO_2 and NO concentrations. A reduction of the operating pressure from 8 to 1 atm did not affect the SO_2 concentration, but did increase the NO level from about 200 ppm to approximately 400 ppm.

The continuing bench-scale work with the 6-in.-dia fluidized bed showed that dolomite was superior to limestone in experiments conducted at 950°C and 8 atm pressure with respect to sulfur retention, but NO concentrations in the flue gas were in about the same range (85-150 ppm, compared to the EPA standard of 500 ppm). The effects of excess combustion air were also investigated. This was of interest because combustion heat could be removed from the bed by generating steam in tubes immersed in the bed, by passing excess air through the bed, or both. With excess combustion air in the range of 17-50%, and with different Ca/S ratios, no consistent effect on sulfur retention by a dolomite bed was observed. The NO level in the flue gas, however, increased with the oxygen concentration, as expected (160, 200, and 220 ppm NO with 3, 6, and 9% oxygen, respectively). The combustion efficiency increased linearly from 86 to 98% when the excess air was increased from 17 to 75%. Studies of particle-size effects showed that the particle sizes of coal and dolomite had no effect on the NO level in the flue gas or the combustion efficiency. The sulfur retention, however, increased with decreasing particle size of the dolomite.

Scale-Up Studies. In 1975, the ANL program on fluidized-bed combustion of coal underwent a major expansion supported primarily by ERDA. In addition to continuing research on fluidized-bed combustion and regeneration of the limestone or dolomite, a proposal was made to construct at ANL a Components Test and Integration Facility (CTIF). This facility was a 3-MW (electric equivalent) coal-combustion unit with a pressurized fluidized bed. This unit was to be tied into ANL's steam plant to utilize its coal-handling and water-treatment equipment. The idea behind the CTIF was to provide support for a full-scale pilot plant about ten times that size.

In 1976 a conceptual design study was initiated on the Component and Test Integration Unit (CTIU). The objectives were to provide technical support for a pilot-plant-scale PFBC project; to provide a facility for materials evaluation, equipment-component testing, and instrument evaluation; and ultimately to investigate alternative and advanced concepts of PFBC systems. The CEN staff people put a great deal of thought and design effort into the CTIU project. The reference operating conditions for the combustor design are given in Table 4-14. The fluidized-bed combustor was contained inside a 12-ft-ID steel vessel with a pressure rating of 18 atm. The combustor itself was made up of a stack of flange-connected, refractory-lined modules. The modules were round, but the refractory liners in some cases were square. The hot gas removal system comprised three stages of particulate removal—two cyclone

Table 4-14. Operating Conditions for CTIU Combustor

Operating Pressure, atm	10	Excess Combustion Air, %	20
Bed Operating Temperature, °C	900	Superficial Gas Velocity, ft/sec	7
Bed Height, ft	8.4	Ca/S Mole Ratio	2
Freeboard Height, ft	23	Target Sulfur Removal, %	90
Fluidized Air, Dry, lb/sec	10	Combustion Efficiency, %	94

separators and a granular bed filter. Hot gases from the third stage cleanup flowed into a stationary turbine-test cascade. The system was thoroughly instrumented, including graphic displays, with overall control from a central module for process management and data acquisition and processing.

In August 1978, DOE decided to defer construction of the CTIU indefinitely and to use the existing facilities with comparable operating conditions and performance characteristics to achieve the program objectives.

Toward the end of the 1970s, DOE concluded that the industrial application of atmospheric-pressure, fluidized-bed combustion (AFBC) was ready for commercialization and was one of the technologies that could make a major impact quickly on the national energy problem. Four industries (chemical, petroleum, paper, and primary metals) accounted for about 70% of all fossil fuels used in boilers. The DOE recommended building one large (200,000 lb steam/hr) prototype AFBC for each of these industries. Also, several industrial AFBCs were being built, as were PFBC units. One of these was a 20-MWe PFBC test unit of the International Energy Agency (IEA) at Grimethorpe, England. After the design and engineering activities associated with the CTIU project had been brought to an orderly conclusion in late 1978, Erv Carls, who had been Technical Manager of that project, was appointed Experimental Program Manager of the Grimethorpe operation and spent two years in England in that position. During that time, he continued as an ANL employee on special assignment.

Limestone Studies. The term "limestone," which is calcium carbonate ($CaCO_3$), has sometimes been used generically in this program to include dolomite, a double compound consisting of calcium and magnesium carbonates ($CaCO_3 \cdot MgCO_3$). Limestone and dolomite occur naturally in large quantities, and both were investigated as sorbents for sulfur removal in the fluidized-bed combustion work. At the maximum operating temperature of 900°C normally used in these studies, calcium carbonate was stable, but magnesium carbonate in the dolomite decomposed (calcined) to the oxide, producing a $CaCO_3$-MgO combination. At 950°C, however, the $CaCO_3$ also decomposed to form CaO. This process apparently produced larger interstitial openings in the particles, and the SO_2 retention was better than values predicted by a mathematical model from the lower-temperature data. At a Ca/S ratio of 1.5, the sulfur retention was 96%, compared to a predicted value of 89%. The NO levels in the flue gas were very low in the two experiments—135 and 120 ppm. Tests at 950°C also showed that dolomite was more effective than limestone in retaining sulfur.

Studies were conducted on the kinetics of sulfation and regeneration reactions of dolomite, using thermogravimetric techniques.

The rate of the sulfation reaction

$$(CaCO_3 + MgO) + SO_2 + 1/2\ O_2 \rightarrow (CaSO_4 + MgO) + CO_2$$

was enhanced by the presence of water in the reactive gas stream, and the reaction was either first or three-quarters order in SO_2 concentration, depending, respectively, on the presence or absence of water vapor. The reaction was zero order with respect to water-vapor concentration, and it was one-fourth order in oxygen concentration with water vapor present. The apparent activation energy was 7.6 kcal/mol.

A rudimentary model for the sulfation reaction was developed on the following assumptions:

1. Diffusion through the macropores of the dolomite particles could be ignored as a rate-determining step.
2. The $CaCO_3$ crystallites were platelets.
3. Diffusion through the reaction layer was rapid relative to the movement of the reaction interface.
4. The effective diffusion coefficient was a function of the extent of the reaction.

The model fit the data and led to detailed structural studies of the material, using optical microscopy, X-ray diffraction, and scanning electron microscopy.

For economic reasons, the widespread use of limestone in fluidized-bed combustion of coal would require that sufficient quantities of suitable limestone be available at various locations across the country to avoid excessive shipping costs. Tests were therefore conducted to evaluate the effectiveness of representative limestones from different parts of the country for FBC use. Reaction rates of the limestone samples with a $SO_2/O_2/N_2$ mixture were determined by thermogravimetric analysis. The predicted values for the quantity of limestone required for a given quantity of coal, based on the test data, gave a range of 0.37 to 2.4 kg/kg coal. Sixty percent of the limestones were below 0.5 kg/kg coal. These results indicated that highly reactive limestones required to meet EPA sulfur emission standards are widely available in the U.S.

In the fluidized-bed combustion of coal, approximately one ton of limestone or dolomite would be sulfated for each 5 tons of coal (3 wt% S) combusted. The large volume of spent sorbent material would be a major waste-disposal problem unless some means were available for its recycle. A limestone regeneration process in which the sulfur could be recovered in a commercially useful form such as elemental sulfur, H_2S, or SO_2 might offset some of the processing cost. Various methods for regenerating limestone or dolomite were considered: (1) reductive decomposition of $CaSO_4$ by CO or hydrogen, (2) a two-step process in which $CaSO_4$ was reduced at 900°C to CaS, which was then converted to $CaCO_3$ and H_2S by reaction with a CO_2-H_2O mixture at about 600°C, and (3) a solid-solid reaction of CaS with $CaSO_4$ at about 1100°C to form CaO and SO_2. The reductive decomposition was believed to be the most promising of the three methods, and partial combustion of the coal was a potential *in situ* method for generating the reducing gases. Dolomite that had been sulfated in coal-combustion experiments was regenerated by using methane in a 3-in.-dia reactor. The effects of temperature, fluidizing velocity, fluidized-bed height, solid-bed residence time, and total reducing gas concentration in the flue gas on the regeneration of CaO were examined. The results, in brief, were as follows:

1. Deeper beds resulted in poorer regeneration.
2. Increased fluidizing gas velocity affected regeneration adversely.

3. Higher regeneration temperatures (>1040°C) increased the extent and rate of regeneration.
4. Establishment of two reaction zones in the reactor was very effective in preventing CaS buildup.
5. Decrepitation of the particles in the bed occurred during regeneration.

Additional components were installed in the bench-scale equipment, and the combustion and regeneration systems were separated. Also, the inside diameter of the regenerator was increased from 3 in. to 4 in., and a coal-feeding system was added. Other supporting regeneration studies continued in CEN, one of which involved the use of carbon to regenerate sulfated limestone in an externally fired rotary kiln. This approach was discontinued due to a lack of durable construction materials and the high cost of clean fuel required to heat the kiln.

Inorganic Salt Additives. The use of inorganic salts as additives to enhance the effectiveness of limestone beds was investigated. Initial studies were made with sodium chloride (Nalco), using synthetic flue gas at 850°C. Changes in the surfaces of the particles were determined by porosimetry and electron microscopy. Trace amounts (1 mol % or so) of the added salt apparently caused structural rearrangements and migration of pores to form larger pores, which increased permeation of the gases into the particles. A similar effect was noted with calcium chloride ($CaCl_2$) additions to limestone and dolomite. Additions of larger amounts of $CaCl_2$ caused the formation of a liquid capable of dissolving CaO. This effect could lead to complete sulfation of the material.

The success of this general approach suggested the possibility of using other salts such as sodium carbonate (Na_2CO_3) or sodium hydroxide (NaOH), which do not contain corrosive chloride ions. Because of concerns about the corrosion that might result from these salts, a laboratory-scale fluidized bed made of quartz was assembled to test various alloys in a simulated flue-gas environment at 850°C and 20-500 ppm SO_2 with NaCl or $CaCl_2$ added to the bed material. The alloy test specimens were located in and above the bed. Stainless steel and high nickel-alloy specimens in the bed showed higher corrosion rates with the salt additives, and specimens at 650 and 550°C above the bed exhibited extensive corrosive attack. In view of these results, a new, automated AFBC unit with a 6-in.-dia combustor was built with provisions for seven corrosion probes, each containing seven metal specimens (see Fig. 4-10). Several 100-hour tests were conducted, using coupons of nickel-, iron-, and cobalt-base alloys.

After the exposure, the coupons were examined by scanning electron microscopy and energy dispersive X-ray analyses. Additions of 0.3 mol % $CaCl_2$ to the sorbent had little effect on the corrosion rates, but additions of 0.5 mol % NaCl or 1.9 mol % Na_2CO_3 accelerated corrosion of most of the coupons. Samples near the top of the bed were attacked more severely than those near the fluidizing gas distributor. In general, most of the specimens were penetrated less than about 15 μm. The uncertainties in long-term corrosion rates calculated from these data were felt to be such that much longer test times would be required.

Experience from operation of gas turbines with hot flue gases from liquid fuel combustion units had indicated that alkali metal compounds such as chlorides and sulfates of sodium and potassium cause "hot corrosion" of gas-turbine hardware. This could also be a problem with PFBC combined-cycle power plants, where the flue gas is used to drive a turbine. Granular bed filters were being developed to remove these compounds, as well as particulate material. Diatomaceous earths and activated bauxites proved to be the best sorbents for these beds. Removals of 95-98% of the NaCl from the gas were readily achieved with these materials. Tests indicated that

Fig. 4-10. Corrosion Test Equipment

bauxite could be regenerated by washing with water. Particulate matter also had to be removed to prevent erosion of the turbine blades. A granular bed filter using sulfated limestone was investigated and found to remove 91 to 99% of the particulates from the gas. To reduce the loading of the bed, however, the use of a high-efficiency Donaldson TAN-JET® cyclone separator upstream from the bed was examined. The cyclone, by itself, was able to decrease the particulate concentration to about three times the acceptable level for use in a gas turbine.

Synthetic Sorbents. The possibility of developing synthetic sorbent materials as an alternative to limestone or dolomite was investigated. The thought was that it might be possible to synthesize a material having physical and chemical properties tailored to meet the specific needs of a fluidized-bed combustor.

Alkali or alkaline earth oxides deposited on a support such as α-alumina pellets were tested. Cylindrical α-Al_2O_3 pellets (1/8 × 1/8 in.) were impregnated with CaO, and their rates of sulfation were measured, using a synthetic combustion gas and a thermo-gravimetric analysis apparatus. The rate was 0.7 order with respect to SO_2 concentration and first order in the CaO concentration. The rate of sulfation was independent of water vapor content of the gas and increased with temperature from 750 to 900°C, above which it became independent of temperature. The sulfated additive ($CaSO_4$ in α-Al_2O_3) was regenerated with various reducing gases (H_2, CH_4, or CO), and the rate was 0.8 order in the reducing gas concentration. The rate was the same for CH_4 and H_2, but lower with CO. Other metal oxides (Na_2O, K_2O, SrO, and BaO) were also tested. The general conclusion was that the synthetic additives offered no significant advantages over limestone or dolomite.

Behavior of Trace Elements. Coal contains a large number of trace elements, which appear in the fly ash or the fluidized bed after the combustion process. Some of these elements, for example, mercury, lead, beryllium, and fluorine, are of particular environmental concern because of their toxicity. For this reason, studies were conducted to determine differences in the way trace elements behave in fluidized-bed combustors and in conventional coal-fired power plants. Preliminary data had indicated that the fluidized-bed system would reduce trace-element emissions.

Further studies of trace-element behavior in ten-cycle combustion-regeneration tests showed that nine elements (Cr, Tl, V, Mo, Zn, Ba, Fe, In, and Sr) were enriched in the sorbent as a result of coal-ash buildup on the surfaces of the dolomite particles. Eight elements that tend to be volatile in conventional coal-fired units were either enriched (Cr, Tl, and Zn) or remained stable (As, Cd, Pb, Ni, and Se) in the dolomite.

FOSSIL FUEL UTILIZATION AND CONVERSION

This program began with the title "Coal Conversion Technology Program" and was later renamed "Fossil Fuel Conversion and Utilization." Largely as a response to the oil crisis of the 1970s, a large national program on the conversion of coal into clean fuels was being sponsored by the U.S. Department of the Interior and other federal agencies, with support by the U.S. Bureau of Mines, private research organizations, and universities. The initial work at CEN was authorized by the AEC Division of Applied Technology in September 1974 and was aimed at problems related to an *in situ* (underground) coal-gasification process being developed at Lawrence Livermore National Laboratory (LLL).

When ERDA was formed, the work continued as part of the *in situ* gasification program under ERDA's Division of Petroleum, Gas, and *In Situ* Technology. In 1975, CEN was authorized to begin an additional project in support of coal liquefaction development called "Physical Parameters in Synthoil Process." This work was done in cooperation with the ERDA Pittsburgh Energy Research Center. With the advent of DOE, these programs were placed under its organization, and CEN began to provide technical support to DOE for its management of the overall alternative fuels development programs. This management support consisted of developing program plans, establishing objectives, reviewing contractors' proposals, providing contract placement documents for DOE, and presenting program reviews at contractors' annual meetings. This was much the same strategy that DOE was using in the battery, fuel-cell, and coal-combustion programs in CEN in the late 1970s, *i.e.*, the primary function of ANL was to assist DOE in program management, although a small in-house research and development effort was also allowed.

Char is produced as a by-product during the *in situ* gasification of coal. Two possible uses of the char were under consideration: (1) production of hydrogen by a steam-char reaction, primarily for use in the coal-conversion processes, and (2) combustion to generate heat or electricity. The gasification reactions to be studied were

$$CO + H_2O \leftrightarrow CO_2 + H_2$$
$$H_2O + C \leftrightarrow CO + H_2$$
$$CO_2 + C \leftrightarrow 2\,CO$$

A laboratory-scale unit to study the reaction of steam and oxygen with coal and char was originally designed to simulate *in situ* gasification conditions at a temperature of 800°C and a pressure of 40 atm, and it was later modified to operate at 1000°C and 136 atm. The variables to be studied were steam pressure, coal devolatilization rate, reaction temperature, and total pressure. During the construction of this unit, preliminary studies were conducted on char reactivity with oxygen and steam at atmospheric pressure and on the catalytic activity of coal ash. Chars were prepared from a number of coals and their reactivities with steam and oxygen at atmospheric pressure were measured, by using a thermobalance. The activation energy proved to be about one-half of that reported in the literature, which suggested that catalysis by inorganic matter had occurred or that the reactions had taken place inside small pores having a high length-to-diameter ratio.

Ashes of several coals were prepared, both by high-temperature oxidation and by low-temperature ashing techniques, and the surface areas and surface acidities were measured. Surface acidity is used in the petroleum industry as a measure of the activity of cracking catalysts. On a unit area basis, the coal ash had a higher concentration of surface acid sites than a commercial cracking catalyst, which indicated potential cracking activity by the coal ash. Studies were conducted on the reaction kinetics of the processes occurring in the gasification zone of the coal-gasification process, *i.e.*, steam-char, CO_2-char, hydrogen-char, water-gas shift, and methanation reactions. The coal samples were representative of coals where field tests were either in progress or were planned, and the type of coal was shown to have a significant impact on the steam-char reaction. The data from this work were intended for use by others in mathematical modeling of coal-gasification processes. In 1979, the experimental work on coal gasification was phased out due to budget constraints.

During 1976-77, CEN had a short-term experimental program on the Synthoil Process, a method for converting coal to a liquid fuel of low sulfur content that would be suitable for use in power production. The process consisted of liquefaction and hydrodesulfurization of the coal in a turbulent-flow, catalytic, packed-bed reactor. A slurry of the coal in recycled oil was treated with hydrogen at 450°C and 135-270 atm pressure in the presence of a Co-Mo/SiO_2-Al_2O_3 catalyst. The gross liquid products were centrifuged to remove the unreacted solids. The chemical reaction is exothermic, and a batch calorimeter was devised to measure the amount of heat evolved by the reaction. Plans were also made to determine heat-transfer characteristics to be used in engineering-design calculations, to develop improved methods for separating solids from the liquid product, and to test catalysts. Some preliminary work was done in these areas, but it was halted as a result of budget decreases and a shift of manpower to DOE management responsibilities. Several years later, after the immediate oil crisis was over, the national Synthoil effort in the 1970s was criticized severely by some politicians and the media as a "government boondoggle."

MAGNETOHYDRODYNAMICS

In the mid-1970s, a small CEN effort was initiated in support of the ANL program on magnetohydrodynamics (MHD). An MHD generator is basically an expansion engine that converts heat from combustion gases directly into electric power by passing the hot, partially ionized gases through a duct (or "channel") lined with electrodes and surrounded by magnetic coils that create a magnetic field across the duct. Movement of the electrically charged gas through the duct generates a current that is collected at the electrodes. The expanding high-pressure gas propels itself through the duct. Very high temperatures are required to ionize combustion gases, but an addition of small amounts of potassium or another alkali metal (seed particles) enhances ionization sufficiently to permit operation at temperatures of 2000-2500°C. Proposals for using MHD in generating plants usually involve passing the exhaust gases from a high-pressure combustion unit first through an MHD generator and then into a boiler to produce steam for a conventional steam-turbine generator. Plant efficiencies for this type of system are projected to be as high as 60%.

The downside of this technology lies mainly in the corrosive gases and particulate material, which, together with the high temperatures, present serious corrosion, erosion, and solids-deposition problems.

The coupling of high-pressure coal-combustion technology and high-efficiency (*e.g.*, superconducting) magnets was a natural for Argonne, which had been working in both areas. A small CEN effort to provide chemical and engineering support for the MHD program at ANL was started in about 1975. This became a sizeable program at ANL, in which CEN was responsible for the development of technology associated with the components downstream from the MHD channel-diffuser, and for processes to recover and recycle the potassium carbonate (K_2CO_3) seed material. The major focus was on interactions between the slags and seed materials in the gaseous and the solid or liquid phases, and development of a process for removing sulfur from the spent seed material. In 1979, the Division assumed an additional task of conducting tests of MHD steam-generating components in a 2-MW facility constructed at ANL. One of several possible arrangements for these downstream components in an MHD plant is shown in Fig. 4-11.

The overall objective of the MHD Heat and Seed Recovery Project at ANL was to provide technical data for designing and operating the downstream gas system and the seed-recycle process. An immediate project goal was to support a 20-MW unit of this type that was being built at the time by the University of Tennessee Space Institute. The longer-range goal was to provide the technical data needed for a 500-MW Engineering Test Facility (ETF) to demonstrate the feasibility of a coal-fired MHD power plant. The Argonne MHD Process Engineering Laboratory (AMPEL) was a facility constructed by the ANL Engineering Division to test MHD steam-generator components under prototypical MHD conditions. This facility could supply a mixture of combustion gas, potassium compounds, and coal ash at a rate of 1 kg/sec by burning a slurry of fuel oil, coal ash, and K_2SO_4 with air preheated to about 800°C. The Division was given the lead role in the experimental work with this unit. The equipment was operated in three different temperature regimes: 2000-1500°C, 1500-1200°C, and below 1200°C. Instrumented, temperature-controlled tubes were inserted in the system to measure heat-transfer characteristics and to obtain samples of condensates and other deposits. Preliminary experiments showed that entrained solids in the gas tended to block radiant heat transfer to the wall, and deposits of K_2SO_4 and a potassium-rich slag phase were found on the tube walls. These effects were not "show-stoppers," but they had to be taken into account in the equipment designs.

A computer model, called "MHDGAS," was developed to predict the equilibrium partial pressures of gaseous species and compositions of condensed solids and liquids in the various regions of the system. Later, the model was extended to include diffusion rates within the condensed phases, and to incorporate several other elements present at low concentrations in the coal ash (iron, sodium, chlorine, lead, phosphorus, and zinc) to predict their behavior in the process. This computer model proved to be highly effective, and it was used extensively in the engineering-design work.

Recovery of the potassium seed material and its desulfurization became an important part of this program because of environmental and economic factors. Two types of processes for removing sulfur from the K_2SO_4 were: (1) those that recover the sulfur in a commercially usable form (elemental S, H_2S, SO_2, and H_2SO_4), and (2) those that recover it in a disposable form ($CaSO_4{\cdot}2H_2O$ and $MgSO_4$). Both approaches were being explored as of the end of 1979.

PERSONNEL

In addition to those mentioned earlier, the following people worked in the coal

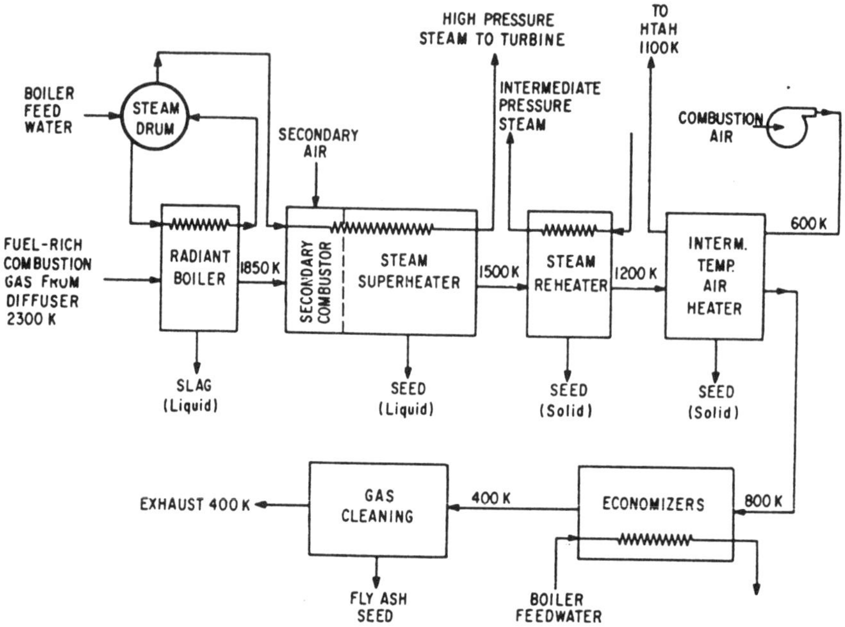

The combustion gas leaves the MHD channel-diffuser at a temperature of about 2000°C, a pressure of 1 atm, and a velocity of 100-200 m/sec. At this point, the gas contains liquid slag particles, many smaller than 1 μm in size, and about 1 wt% potassium vapor. In the first component, the radiant boiler, slag is separated from the combustion gas, while it is still at a high temperature, to minimize reactions with vaporized potassium species that occur at lower temperatures. The high concentration of NO_x produced in the combustor is allowed time to equilibrate to acceptably low levels by providing a relatively modest cooling rate of 200°/sec down to about 1600°C. At 1500°C or lower, additional air is injected into the fuel-rich gas to complete the oxidation of CO to CO_2. The oxidizing conditions also cause the potassium seed material to condense primarily as K_2SO_4, which is an efficient way of removing sulfur from the flue gas. The K_2SO_4 and small amounts of solid slag deposit as a liquid or solid on the tubes of the steam and air heaters. A gas-cleaning system is then needed to remove the remaining slag-seed particles before the flue gas is exhausted to the atmosphere.

Fig. 4-11. Downstream Gas System for an MHD Power Plant

combustion programs: John Ackerman, Lou Anastasia, Dale Bowyer, Ronald Brock, Tom Bump, L. Burnett, Tom Cannon, Stanley Che, Tsewei Chen, John Falkenburg, Don Fredrickson, Darrell Fee, Ray Green, Milt Haas, J. Hepperly, Chen-Chao Hsu, Hann-Sheng Huang, Bill Hubble, Roger Jarry, K. Javdani, Irv Johnson, Lowell Koppel (consultant), Sheldon Lee, John Lenc, Alain Lescarret, John Montagna, Dave Moulton, Russell Mowry, Tom Mulcahey, Bill Murphy, Satyendra Nandi, Frank Nuñes, John Pavlik, Walt Podolski, Devabhaktuni Ramaswami, Jim Riha, Satish Saxena (consultant), Charlie Schoffstall, John Shearer, Atul Sheth, Stan Siegel, Gregory Smith, Sherman Smith, Eugene Smyk, Bob Snyder, John Stockbar, Gale Teats, Clarence Turner, Noel Vargo, Elmo West, and Ira Wilson. Some of these individuals were from other ANL divisions.

The overall MHD Program at ANL was under the direction of Mike Petrick of the Engineering Division. The CEN work was led by Terry Johnson, who, along with Atul Sheth, did the computer modeling studies, which were a major part of the program. Others who were involved in various aspects of the MHD work in CEN were Paul Blackburn, Al Fischer, Carl Johnson, Sheldon Lee, and Bill Swift.

Sodium Technology

The basic chemistry program that had been started by Irv Johnson and Fred Cafasso in the early 1960s bore fruit in 1969 when Argonne was designated as the center for AEC's sodium analytical chemistry program. The major components of this program were development and standardization of improved analytical methods for impurity elements in sodium, and development of on-line meters for continuously monitoring the concentrations of impurity elements in flowing sodium. The objective of this program was to develop an advanced chemical-analysis and impurity-monitoring technology for the Fast Flux Test Facility (FFTF), a sodium-cooled fast reactor planned for construction at Hanford in the 1970s.

The impurity elements of greatest concern were oxygen, carbon, and hydrogen. In the core region of the reactor, cobalt-60 and manganese-54, which are intense gamma ray emitters, are produced by neutron activation of stainless steel. Oxygen reacts with stainless steel in the hot core regions to produce iron oxides. These, along with the Co-60 and Mn-54, are deposited in cooler regions outside the core (piping and the intermediate heat exchanger where heat is transferred to the secondary sodium system). The resulting high radiation levels that would prevail if oxygen corrosion of the stainless steel were extensive would make maintenance operations difficult. Carbon is transported from stainless steel components in the hot core region to cooler regions of the primary sodium system. The result is loss of strength of the stainless steel. A capability to monitor carbon concentrations in the sodium was needed to facilitate research on the kinetics of carbon transport. Monitoring of the hydrogen concentration in the secondary sodium system was needed for rapid detection of water-to-sodium leaks in the steam generator.

The basic research program was continued to provide a sound scientific foundation for understanding the behavior of impurity elements in sodium, especially their role in corrosion of stainless steel. Tom Kassner of the Materials Science Division headed up a Materials Performance Group that studied the roles of impurity elements in corrosion and loss of strength of stainless steel in the core region.

Because much of the sodium technology work at ANL was in CEN, we were assigned the responsibility of coordinating sodium chemistry programs among all the ANL divisions that were involved: besides CEN, Chemistry (CHM), Materials Science and Technology (MST), and EBR-II. Les Burris, assisted by Joe Draley, was the Sodium Technology Program Coordinator. Particularly important to this program was EBR-II because impurity meters could be tested under reactor operating conditions on the Radioactive Chemistry Loop, through which a small fraction of the primary sodium coolant could be circulated.

In the early 1970s, the Sodium Technology Program grew to an effort level of about 35 staff members. Fred Cafasso headed up the Sodium Chemistry and Analytical Standards programs. Paul Nelson was in charge of developing on-line impurity meters and administering a National Meter Program, which involved other national laboratories and commercial organizations.

ANALYTICAL STANDARDS PROGRAM

The objective of this program was to establish reference and standard methods for sodium analysis and to certify them for national use. Development of sampling methods and equipment was a considerable part of this effort. To help carry out these responsibilities, CEN organized, under Bob Meyer, a National Standards Working Group which involved representatives from several other laboratories in the United States. With the help of this group, a manual of analytical procedures was prepared for analyses of sodium and the reactor cover gas. The analytical methods were evaluated by a series of round-robin analyses by the various laboratories. After some difficulties with the procedures were resolved, an approved manual of sodium analytical methods was published. Researchers in the Analytical Standards Laboratory were Max Adams, Murray Barsky, Harry Edwards, Bob Haglund (MST), Al Panek, Mike Roche, Laury Ross, and Stan Skladzien.

One of the major advances in this program was the development of an improved method for determining the chemical activity of oxygen in sodium. Developed by Dale Smith of MST and Murray Barsky, the method involved equilibration of thin wires of vanadium with the liquid sodium at 750°C. The oxygen activity in the sodium could be derived from the oxygen content of the vanadium wire and data on the distribution coefficient of oxygen between vanadium and sodium.

NATIONAL METER PROGRAM

This program consisted of coordinating, as well as participating in, a national program to develop, test, and establish commercial availability for impurity meters to be used in FFTF and other sodium-cooled fast reactors. Meters were developed for oxygen, carbon, and hydrogen, as well as a detector of water leakage into sodium in a steam generator by monitoring the hydrogen concentration in the secondary sodium as it exited the steam generator.

Oxygen Meter. The oxygen meter chosen for testing and further development was an electrochemical device developed by the United Nuclear Corporation (UNC). It was a galvanic cell having a thorium oxide-yttrium oxide (ThO_2-Y_2O_3) solid electrolyte and a copper-copper oxide (Cu-CuO_2) reference electrode. The cell developed a voltage that was dependent on the oxygen activity in the sodium. The electrolyte was in the form of a thin-walled tube of about 3/4-in. diameter, which was produced by isostatic pressing. The reference electrode material was contained in the tube around which the sodium flowed. Several other reference electrode materials were tested, namely, Na-Na_2O, Sn-SnO_2, and oxygen gas. The oxygen-gas electrode was selected. Most of the development work on the oxygen meter was done by Lou Bartholme, Verne Kolba, Larry Marek, John McKee, Dick Schlueter (MST), and Don Vissers. Fred Cafasso and George Stavropoulos conducted some basic electrochemical studies of oxygen in sodium.

Twenty oxygen meters were tested to give a statistical measure of their performance, ten in CEN in a specially built loop, the Oxygen Meter Rig, and ten in Westinghouse Electric Corporation's Advanced Reactor Division at Hanford. Argonne people in this effort included John McKee, Paul Nelson, and Don Vissers. The meters were calibrated over an oxygen concentration range of 1 to 15 ppm, using the vanadium-wire equilibration procedure. The meters performed very well. Equipment used for oxygen meter calibrations is shown in Fig. 4-12.

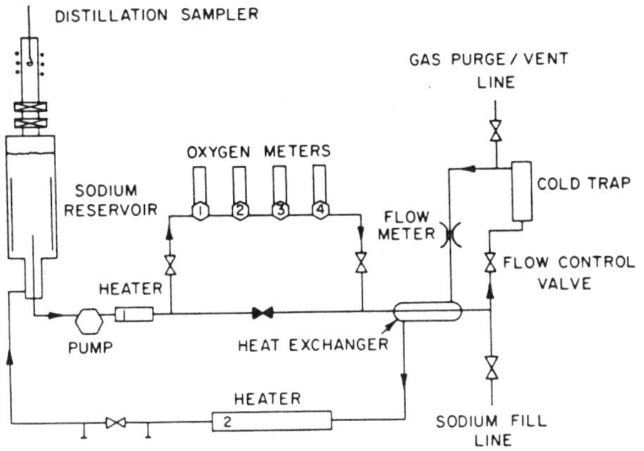

Fig. 4-12. Oxygen Meter Calibration Equipment

Carbon Meter. A carbon meter had also been developed by UNC. A diffusion-type meter operated at 750°C, it was based on the principle that the rate of carbon diffusion through an iron membrane, at constant sodium flow rate, depends on the concentration of carbon in the sodium. The carbon reaching the other side of the membrane reacted with a flowing moist hydrogen-oxygen mixture to form CO, which was catalytically converted to CH_4 (methane), whose concentration was measured in a flame-ionization detector. A meter was calibrated by equilibrating sodium of known carbon activity with a metal specimen (*e.g.*, Fe-8 wt% Ni). Figure 4-13 shows the equipment module that was used for the equilibration.

The response time of the carbon meter to a step change in carbon concentration was about five minutes—somewhat slow, but adequate for reactor use where rapid changes of the carbon concentration in sodium would not be expected. John Allen, Charles Luner, Paul Mack, Don Raue, Mike Roche, Westly Ruther (MST), and Stan Skladzien were involved in the development of the carbon meter.

Hydrogen Meter. The hydrogen meter was developed for rapid detection of minute leaks in a steam generator of a sodium-cooled fast breeder reactor. The meter was patterned after one used successfully by Électricité de France, which was based on the rapid diffusion of hydrogen through a nickel membrane and subsequent measurement of the hydrogen concentration with a mass spectrometer. A vacuum drawn on the back side of the membrane provided an activity gradient that caused the hydrogen to diffuse.

Two versions of the hydrogen meter were developed in CEN under the leadership of Don Vissers. Lou Bartholme and John Holmes were also involved in this work. One version was the equilibrium hydrogen meter. It measured the hydrogen pressure above the sodium, which could be related to the hydrogen pressure in the sodium by Sievert's Law:

$$S = KP^{1/2}$$

where S is the concentration of sodium in hydrogen, ppm; K is Sievert's constant, ppm-torr$^{1/2}$; and P is hydrogen pressure, torr. Figure 4-14 shows an in-line arrangement of a hydrogen detector. The remarkably precise

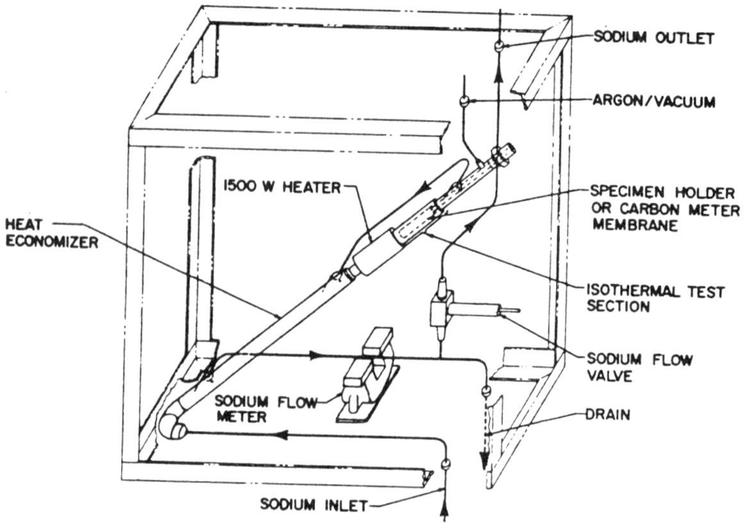

Fig. 4-13. Module for Specimen Equilibration Carbon Meter

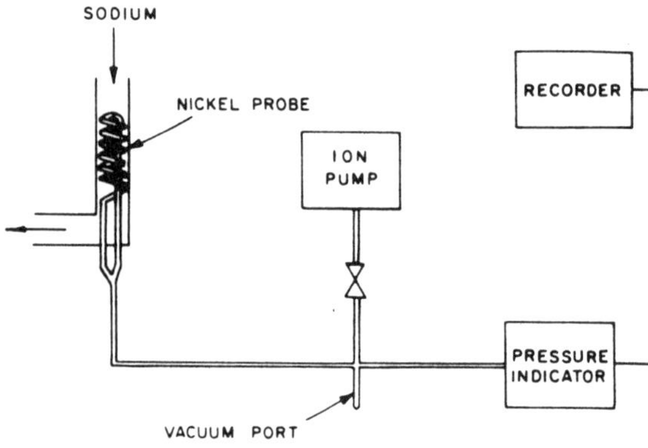

Fig. 4-14. Hydrogen Activity Meter

results from this meter are illustrated by the calibration curve in Fig. 4-15. The other version, the leak detector, was operated in the dynamic mode. The flux of hydrogen through the membrane was determined by monitoring the current to the ion pump, which maintained a vacuum on the back side of the membrane. When operated in the dynamic mode, the hydrogen meter was a highly effective leak detector. It satisfied very well the requirements for such a meter of high sensitivity, rapid response, reliability, and ruggedness. A 4% change from an initial hydrogen concentration of 0.1 ppm could be detected within ten seconds.

Impurity Meter Modules. Two modules containing impurity meters and provisions for equilibration and temperature and flow control of metal specimens were designed and built for testing by Don Cissel on the Radioactive

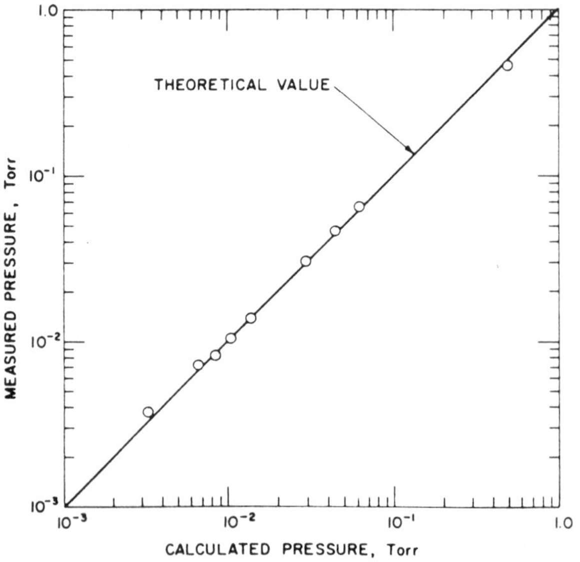

Fig. 4-15. Calibration of Hydrogen Activity Meter

Sodium Chemistry Loop at EBR-II (see Fig. 4-16). Such modules were slated for installation at FFTF.

Similar meter modules were also tested at CEN on a small pumped-sodium loop called the Apparatus for Monitoring and Purifying Sodium (AMPS). This apparatus was built to characterize impurity levels in cold-trapped sodium as a function of several operating variables. Cold trapping (precipitation of impurities on a cold finger) is the normal method of maintaining sodium purity in sodium-cooled fast breeder reactors. The modules performed very well, both at EBR-II and on AMPS. Both types of hydrogen meters were installed on the EBR-II secondary cooling system, where they proved to be very reliable. The principal CEN personnel in this project were Lou Bartholme, John Holmes, Verne Kolba, John McKee, Paul Mack, Chuck McPheeters, Paul Nelson, Don Raue, Stan Skladzien, Mike Slawecki, and Don Vissers.

At EBR-II, the oxygen concentration in the primary sodium was only 1 ppm, much lower than had been found by the previously used chemical distillation procedure. The hydrogen concentration obtained by the equilibrium hydrogen meter was only 60 ppb. The

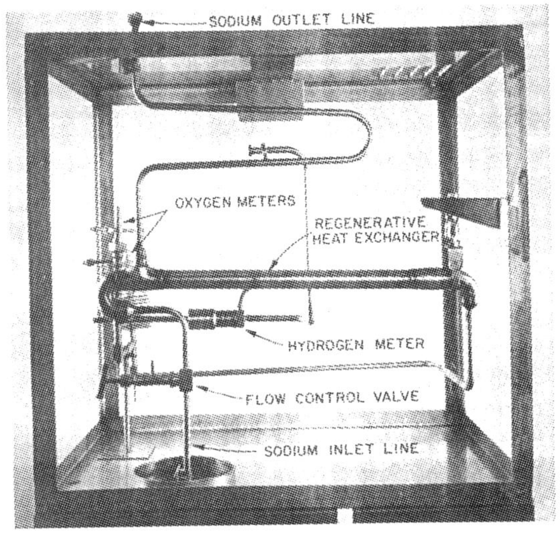

Fig. 4-16. Oxygen-Hydrogen Meter Module

concentrations remained essentially constant during the long time over which they were followed. This observation removed much of the concern about the level of impurities in cold-trapped sodium and on concentration fluctuations caused by system upsets. As a consequence, interest in installing impurity meters on the FFTF primary system waned. (FFTF has no secondary sodium system, the heat being exhausted to the atmosphere.) Installation, operation, and maintenance of the meters were eventually judged not to be cost effective, so the impurity meter program was phased out.

STEAM GENERATOR LEAK DETECTION

As part of the General Electric Company's program on steam generator systems development for a liquid metal fast breeder reactor (LMFBR), the Division was given responsibility for technical direction of work to evaluate leak-detection systems. Most of this work was conducted at the Sodium Components Test Installation (SCTI) at the Liquid Metal Engineering Center in Canoga Park, CA, where a large 30-MW steam generator was under test. The leak-detection system that was evaluated consisted of two ANL hydrogen detectors and two Westinghouse oxygen meters in the sodium, along with a gas chromatograph and a hydrogen meter in the cover gas. The cover-gas hydrogen meter consisted of a nickel diffusion tube surrounded by a heater that induced cover gas to flow past the diffusion tube by natural convection. The diffusion tube had to be heated to about 540-650°C to obtain performance comparable with that of in-sodium meters at 485-510°. Steady-state hydrogen and oxygen concentrations in the sodium were 60-80 ppb and 0.6 ppm, respectively. Steam-generator leaks were simulated by injecting hydrogen or water into the flowing sodium at rates ranging from 1×10^{-6} to 2×10^{-4} lb H_2O/sec.

Many tests were performed on the SCTI. At low sodium temperatures (about 200-300°C), the rate of solution of hydrogen in sodium is low, so bubbles of hydrogen entrained in the sodium reached the cover gas before dissolving. Under these conditions, a cover-gas hydrogen meter provided better sensitivity and better response than in-sodium hydrogen detectors. At temperatures above 370°C, the cover-gas monitor was unaffected by hydrogen injections that were easily detected by the in-sodium monitors. The oxygen meters also showed promise of good sensitivity, but alarm circuitry for them had not yet been developed when the program was terminated. From the information obtained, it was determined that water inleakage rates of as low as 2×10^{-5} lb H_2O/sec could be detected in less than one minute in the secondary loop of an LMFBR, well below leak rates that could cause significant damage in five minutes (approximately 10^{-3} lb/sec). The work at SCTI was terminated in October 1973. Chuck McPheeters and Don Vissers were involved in this work.

POST-1974 SODIUM TECHNOLOGY PROGRAM

The CEN sodium technology program underwent some significant changes in 1973 and 1974 as a result of budget cuts and shifting priorities by the AEC. During the 1974-1980 time span, the program consisted of three major areas of study: (1) tritium behavior and control, including development of in-sodium tritium monitors, (2) design and regeneration of cold traps used in LMFBRs for removal of impurities from sodium, and (3) processing of bulk sodium for use in LMFBRs or for disposal.

Tritium Behavior, Monitoring, and Control. A tritium monitor was developed to facilitate studies of tritium behavior in LMFBR systems. The tritium monitor, illustrated in Fig. 4-17, operated on the same principle as the that of the sodium meter, namely, the rapid diffusion of hydrogen isotopes through a nickel membrane. Sodium-containing tritium flowed on one side of the membrane while a sweep gas of argon-1% hydrogen flowed on the other. Tritium diffusing through the membrane was carried by the sweep gas into a proportional counter where the disintegration rate was monitored. The probe was designed to fit into an oxygen-meter housing such as the one that had been installed on many sodium systems and on the EBR-II primary and secondary sodium systems. Because the tritium count rate is affected by many variables, *e.g.*, membrane area and thickness, gas flow rate, counting tube volume, and counting efficiency, each meter had to be calibrated by analyzing sodium samples for tritium.

Tritium Removal by Cold Trapping. Tritium is generated in an LMFBR by three-particle fission and by activation of boron in the control rods and impurities in the core region. It migrates readily through the LMFBR system and, without effective control, would be released to the environment. Cold trapping

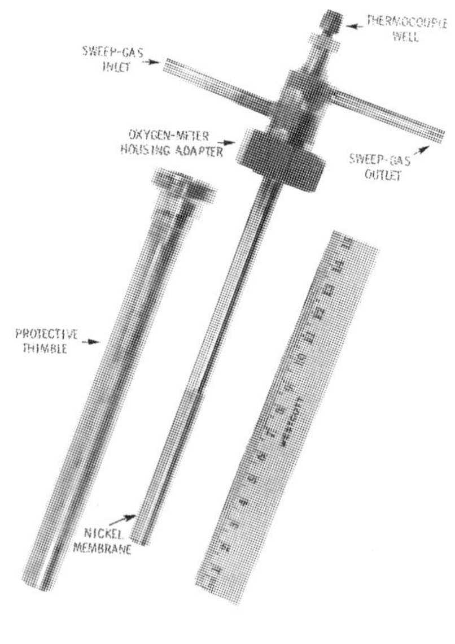

Fig. 4-17. Tritium Meter Probe

has been found to be an effective way of removing it from sodium and assuring minimum environmental release.

Hydrogen, one of the major impurities in sodium, is removed as its hydride (NaH) by cold trapping. A study of tritium removal showed that, of the two possible mechanisms, coprecipitation with sodium hydride and isotopic exchange with hydrogen in precipitated sodium hydride, coprecipitation was dominant (see Fig. 4-18).

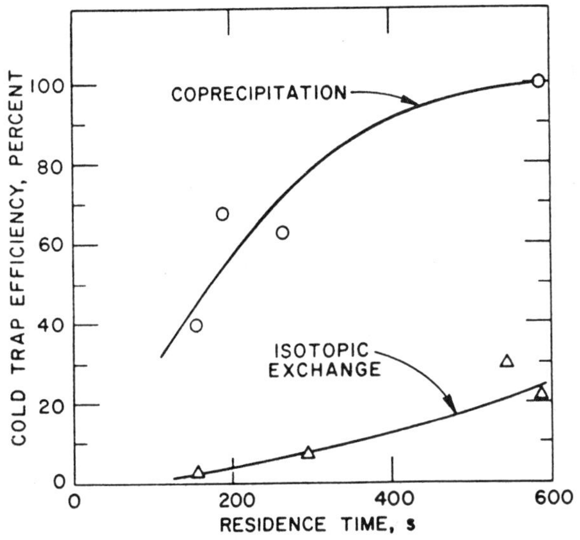

Fig. 4-18. Cold Trap Efficiency for Tritium Removal from Sodium

Tritium Behavior in EBR-II. Two tritium monitors were installed on the primary and secondary sodium systems at EBR-II and were used to study the behavior of tritium. In these studies, the cold traps on the primary and secondary sodium systems were bypassed to allow the tritium concentrations in these systems to build up. A computer model was developed to predict the changes in tritium and hydrogen concentrations. The measured and predicted concentrations, shown in Fig. 4-19, are in excellent agreement.

The development of the computer model required measurement of the permeation coefficient through the steam generator material (Croloy®: Fe-2.25 wt% Cr-1 wt% Mo). The

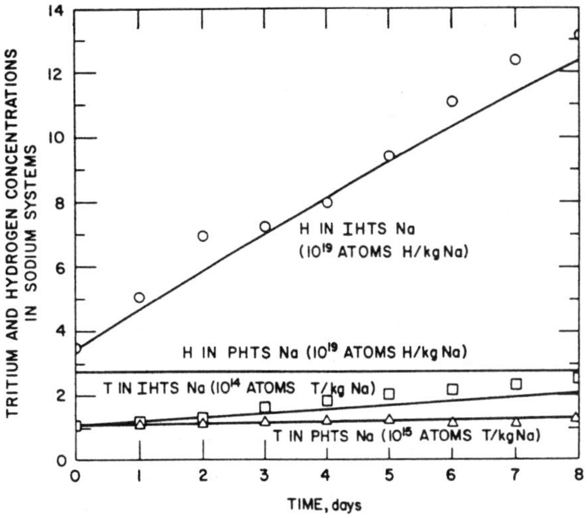

PHTS = Primary heat-transfer system
IHTS = Intermediate heat-transfer system
Symbols: Measured values
Solid lines: Computer-generated values

Fig. 4-19. Predicted and Measured Tritium and Hydrogen Concentrations

formation of iron oxide layers on the steam side of the Croloy reduced the tritium permeation coefficient by a factor of approximately 160 compared to that of unoxidized Croloy.

From this study and the study of tritium removal by cold trapping, it was concluded that, in a 1000 MW(e) fast breeder reactor, more than 99.8% of the tritium generated would be removed by the cold traps. Tritium release to the environment would be well below the limits set by Federal regulations.

Chuck McPheeters, Don Raue, and Tom Renner performed most of the work on tritium in the sodium program.

DESIGN AND REGENERATION OF COLD TRAPS

The major impurities in sodium, hydrogen and oxygen, are removed by precipitation in a cold trap where, at the low temperatures of the trap (about 125°C), the solubilities of sodium hydride (NaH) and sodium oxide (Na_2O) are very low.

Optimization of Cold Trap Design. Because of the reliance on cold traps to control impurities, particularly hydrogen and oxygen, optimization of the cold-trap design was studied in AMPS. Specially designed sacrificial cold traps were used to determine the distribution of deposits within the traps. A one-dimensional computer model was developed to predict the distribution of the precipitates. The deposition of sodium hydride was in excellent agreement with model predictions, but the agreement was not as good for sodium oxide. This was attributed to poor adherence of sodium oxide particles and to an inaccurate or drifting oxygen meter. Most of the precipitated material in an LMFBR cold trap is expected to be sodium hydride.

Cold Trap Regeneration. Cold-trap regeneration was investigated because of the likelihood that a secondary cold trap would plug in about one year as a result of diffusion, into the sodium, of hydrogen generated by water-side corrosion in steam generators. Replacement of a cold trap would be difficult and costly. A study was made of cold-trap regeneration by thermal decomposition of the sodium hydride under vacuum. The released hydrogen (and tritium) would be converted to water for subsequent disposal. Rapid release of hydrogen at temperatures between 300 and 400°C showed this to be a feasible procedure, but the program was terminated when the Clinch River Breeder Reactor (CRBR) was cancelled in 1983.

BULK SODIUM PROCESSING

This program consisted of two major activities: (1) reclaiming or upgrading sodium from various sources for use in the CRBR and (2) disposal of contaminated sodium. Sources of sodium for the CRBR included the two decommissioned reactors, Hallam and Fermi, which amounted to about one million pounds of the metal, and commercial-grade sodium, which contains too much calcium but is one-quarter as expensive as reactor-grade sodium. A simple, inexpensive method for calcium removal was therefore sought.

Reuse of Sodium from the Fermi Reactor. In experiments conducted by Don Raue and Art Youngdahl (MST), analysis of Fermi sodium showed that it was suitable for direct use provided that it was passed through a 20-μm filter to remove carbon-bearing particulates. The program was ended before Hallam sodium could be sampled and analyzed.

Removal of Calcium from Commercial-Grade Sodium. Sought in this program was a calcium-removal procedure that could be used at the reactor site so that tank-car quantities of commercial-grade sodium could be processed by passing it through a calcium-removal procedure and into sodium storage tanks. Several selective calcium-oxidation procedures were conceived, but the effort died when the CRBR was cancelled. These studies were made by Ralph Leonard, Chuck McPheeters, Don Raue, Tom Renner, Stan Skladzien, Ray Wolson, and Art Youngdahl (MST).

Recovery of Contaminated Sodium. A distillation process was developed for recovering sodium drained from used cold traps and other equipment components. Fission-product decontamination factors greater than 100 were achieved in a small (10.2-cm-dia) distillation column. The principal contaminant in the product was cesium. If desired, additional removal of cesium could be achieved by passing the recovered sodium through a reticulated vitreous carbon trap. Such a trap had been used successfully at EBR-II to remove cesium-137 from the primary sodium.

Disposal of Contaminated Sodium. A considerable quantity of contaminated sodium had accumulated in the U.S. as a result of LMFBR operations and reactor-development programs. In the CEN Division, Chuck McPheeters, Ray Wolson, Bali Misra, and Stan Skladzien investigated possible processes to convert the contaminated sodium and sodium-bearing waste to a disposable form.

A calcining process proved to be effective for converting sodium to sodium oxide, which could then be converted to a silicate glass for disposal. In the process, sodium was introduced into the upper end of a slanted rotary-drum reactor which had been preloaded with sodium oxide and was operated at a temperature of 200°C. Oxygen slowly introduced into the exit end of the reactor smoothly converted the sodium to sodium oxide. The process was also demonstrated, using a 55-gal drum as the reactor.

PERSONNEL

The following individuals were involved in the sodium technology program in the 1970s: Max Adams, John Allen, Murray Barsky, Lou Bartholme, Bob Blomquist, Les Burris, Fred Cafasso, Norm Chellew, Sanat Dhar, Joe Draley, Jim Eberhart, Harry Edwards, Magdy Farahat, Hal Feder, Erwin Filewicz (EBR), Joanne Fink, Al Fischer, Don Fischer, Sherman Greenberg (MST), Bob Haglund (MST), Dave Hebditch, John Heiberger, John Holmes, Chuck Honesty, Jimmy Humphrey, Les Jardine, Tom Kassner, Bob Kessie, Verne Kolba, Wally Kremsner, Romesh Kumar, Len Leibowitz, Ralph Leonard, Charles Luner, Paul Mack, Larry Marek, John McKee, Chuck McPheeters, Bill Mecham, Bob Meyer, Bill Miller, Bali Misra, Paul Nelson, Al Panek, John Pavlik, Don Raue, Tom Renner, Mike Roche, Laury Ross, Westly Ruther (MST), Atul Sheth, Stan Skladzien, Mike Slawecki, George Stavropoulos, Marv Tetenbaum, Ewald Veleckis, Pete Vilinskas, Don Vissers, Ray Wolson, and Art Youngdahl (MST).

Fuel Element Failure Detection

Considerable effort, mainly by Norm Chellew, Les Coleman, Bob Kessie, and Bill Miller, was devoted to developing methods for detecting fuel-element failures for application in FFTF. The methods under consideration were based on cover-gas analyses of one kind or another, for example, detection of xenon and krypton isotopes in the cover gas by gamma-ray spectrometry. Out of this work evolved a very effective gas-tagging technique in which different ratios of Xe-133 and Xe-135 were added to the plenums of the fuel elements. If a fuel element failed, not only could the failure be detected, but the particular subassembly could be identified by the isotopic ratio. This tagging technique was used with great success until EBR-II was shut down in 1995.

Solar Energy

Solar energy, at first glance, would seem to be an ideal way to meet many of the nation's energy needs—it is renewable, non-polluting, and available in large quantities. In practical application, however, it poses a number of difficult problems. One is its intermittent availability both on a daily and seasonal basis, and during long periods of overcast weather in many parts of the country. Various means of energy storage, including pumped hydro-electric systems, flywheels, storage batteries, and conversion to hydrogen have been proposed to solve this problem, but have not come to commercial fruition. The other major problem is that the low intensity of sunlight requires that its energy be concentrated in some way to provide the high-grade energy needed for most applications. For example, solar air-conditioning using the ammonia-water cycle needs a minimum temperature of about 120°C, and large central power plants require temperatures of at least 300-600°C. In the case of photovoltaic cells, each cell typically produces about 0.5 V, so many cells are needed to obtain a useful voltage, and a large area is required to achieve the necessary power. In the early 1970s, all of these problems were being addressed in projects at various research organizations.

In August 1975, the Solar Energy Project at ANL was transferred from the ANL Accelerator Research Facilities (ARF) Division to the Chemical Engineering Division. The program was aimed toward the development of advanced solar collectors based on a concept used originally by Roland Winston of the University of Chicago to collect Cherenkov radiation in research studies. This concept, known as the Compound Parabolic Concentrator (CPC), used two parabolic surfaces to reflect and concentrate solar radiation onto an absorbing medium. The parabolic surfaces were in the form of a trough oriented in an east-west direction, with the opening of the trough aimed south. To increase the effective area, multiple troughs were placed side-by-side to form an array.

The geometry of the collectors was designed to concentrate the radiation without focusing the light, which permitted the collectors to accept the radiation over a relatively wide angle. This feature eliminated the need for expensive sun-tracking equipment. Some CPC collectors with low concentration capability (2X) could be stationary throughout the year, but higher-performance units required periodic readjustments. The solar program was oriented initially toward high-temperature (150-250°C) collectors suitable for solar thermal power generation or for industrial process heat. For successful practical application, the life-cycle cost of delivered energy from solar systems would have to compete with that from fossil-fuel systems, so a major program objective was to develop low-cost solar collectors.

Flat-plate (non-concentrating) solar collectors were limited by heat losses to a maximum operating temperature of about 100°C, which is insufficient for absorption-cycle air conditioning. The required temperatures were readily achievable, however, with CPC collectors having modest concentration levels of 3-10X. With higher concentration levels, it appeared that more demanding applications such as industrial process heating and power generation could be satisfied.

Application of the CPC collector to the photovoltaic conversion of solar energy to electricity was also of interest. The use of reflective surfaces decreased the amount of silicon required for a given power output, which might allow a five- to eight-fold decrease in the cost per peak kilowatt (defined as a kilowatt at noon on a sunny day).

Initial collector arrays built and tested at ANL had demonstrated the capability to concentrate sunlight without tracking the sun, and the collection efficiency was higher than that for conventional flat-plate collectors. One of the major goals early in this program was to develop inexpensive materials and fabrication methods for the curved, reflective trough surfaces. Materials used in these studies included thermoformable plastics, foamed plastics, roll-formed metals with reflective surfaces, and silvered glass.

Several collector arrays were designed and constructed during the course of the program. A prototype thermal collector was designed, built, and operated at temperatures over 250°C. A typical design for a small, lightweight 1.5X thermal collector is shown in Fig. 4-20. This collector resembled a conventional flat-plate collector in its exterior appearance, but it required no sun-tracking or movement during use. The reflective troughs, fabricated from plastic or thin sheet metal, directed the incoming light onto an absorber located within an evacuated glass tube. A fluid flowing through tubes attached to the absorber plate recovered the thermal energy. A glass cover

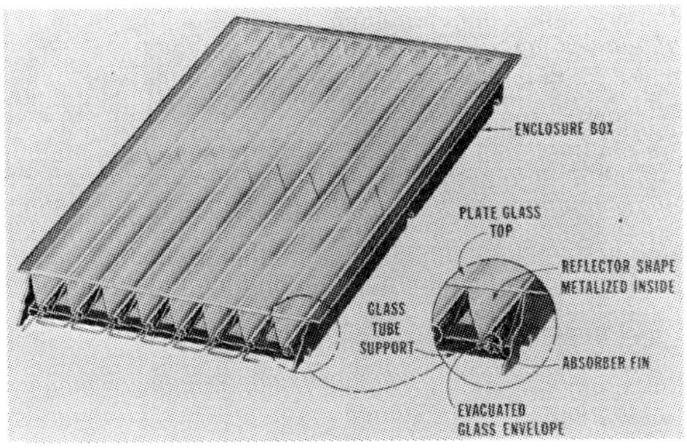

Fig. 4-20. Lightweight Solar Collector

over the assembly protected the troughs from dust, snow, *etc*. Collectors of this general type would be suitable for process-heat and air-conditioning applications.

Three prototype collectors were constructed: (1) a low-cost 3X concentrator having plastic mirrors and an area of about 6 sq ft, which would require repositioning four times per year, (2) a 1.5X concentrator with rolled aluminum mirrors and an area of about 8 sq ft, which would not require periodic adjustments, and (3) a 1.5X collector panel with an area of about 30 sq ft in a configuration envisioned for mass production. To expedite transfer of this technology to industry, ANL worked closely with several industrial firms, including General Electric Co., Chamberlain Mfg. Corp., Corning, Inc., and Owens-Illinois, Inc. In 1977, the 1.5X CPC cusp collector was selected by *Industrial Research* magazine as one of the nation's 100 most significant technological developments in that year. Two licensees of the DOE patent, Sunmaster Corp. and Energy Design Corp., began marketing commercial versions of this collector and half a dozen or so other firms also became licensees. By 1979, Sunmaster and Energy Design were selling units commercially, and a 200-ton air conditioning machine using a 13,400-sq ft collector array at the Frenchmen's Reef Hotel in St. Thomas, Virgin Islands, was started up as a demonstration project. Several CEN personnel were instrumental in making this demonstration a success. At this point, the CPC concept had been developed and demonstrated, so the CEN staff turned mainly to modeling studies for further design optimization and materials selection. Further attention was given to process heat for industrial use.

Research and development continued on the photovoltaic application of solar concentrators. A dielectric CPC panel was designed, built, and tested. This panel, which was four feet square, produced 125 W of electrical power, and it required only 20% of the silicon needed for a conventional flat-plate photovoltaic panel of equivalent power, which illustrated the potential for a major cost reduction.

As of 1978, the ANL solar energy program was an interdivisional effort in which CEN had the responsibilities for development of improved solar collectors, development of a passive cooling system based on seasonal storage of ice, and provision of technical assistance to the National Solar Energy Demonstration Program. Most of the development work had been subcontracted. As a part of the ANL program, a unique, fully instrumented facility was built for testing collectors and evaluating new designs. Figure 4-21 is a photograph of this facility, which was located in an area just south of the ANL Guest Facilities.

A new program was started in 1978 on thermal energy storage for cooling applications. This was an outgrowth of the CEN effort on the use of solar energy for cooling buildings. This concept involved a large, well-insulated tank of water buried underground near the building. Freezing units at the bottom of the tank formed and released sheets of ice, which floated upward. Refrigeration was achieved by vaporization of ammonia in the units, rising of the vapor to above-surface condensers, and gravity return of the condensate to the freezing units. Operation ceased when the air temperature rose

On the right is a 3X collector for process heat application that operates at 200°C; at the center, still covered, is an experimental 3X collector with low-cost plastic mirrors; to its left, is a 1.5X collector that operates at 150°C without tilt adjustments.

Fig. 4-21. Solar Energy Test Facility

above freezing, and the system was totally passive with no need for external power. Estimates indicated that a tank with a height and diameter of 15 ft or so could provide sufficient cooling capacity for a well-insulated family home.

As was the case with most of the other CEN programs in the late 1970s, the DOE felt that they needed more technical and management assistance in the national program, so a substantial part of the CEN manpower was diverted to those activities.

Bill Schertz was responsible for getting the solar program underway in CEN, and for the continuing management of the CEN and other ANL solar programs. Don Webster was the cognizant Associate Division Director. Other personnel who worked on the program included John Allen, Norm Levitz, Bill McIntire, Kent Reed, Bob Rush, and Alvin Wantroba.

Fuel-Cycle Studies

With the administrative reorganizations occurring during the changeover from the AEC to ERDA and then to DOE, the confusion of the financial transition quarter in 1976, and other political wind shifts in the 1970s, the engineers and scientists working on nuclear fuel-cycle programs sometimes felt as if they were living through a chapter of *The Perils of Pauline*. Programs seemed to be started and then terminated or redirected suddenly more in response to current political considerations than to the long-term potential benefits of the work. Nevertheless, a sizable amount of fuel-cycle technology was generated during this period.

DEVELOPMENT OF ADVANCED FUEL RECOVERY OPERATIONS

In mid-1969, development work on pyrochemical processes, in particular, salt-transport processes, was terminated. At a reduced level of effort, a program to support the reference process for recovery of fuel from liquid metal fast breeder reactors (LMFBRs) was instituted. The reference process was a modification of the Purex solvent-extraction process, mentioned in Chapter 1, that had been developed for recovery of spent light water reactor (LWR) fuels. It was basically a chop-leach process to segment and subsequently dissolve UO_2 and PuO_2 mixed oxide ("MOX") from fast reactor fuels. Concentrations of the dissolver solution constituents were then adjusted to provide a feed solution for the solvent-extraction operation. The major effort on the development of this process was at Oak Ridge National Laboratory. The objective of CEN's work was to investigate alternative steps that might improve the reference process. The major elements of the program were the following:

1. Liquid metal decladding.
2. Fluidized-bed conversion of dissolved uranium and plutonium nitrates to solid oxides.
3. Adaptation of centrifugal contactors for the solvent-extraction operation.
4. Electrolytic reduction of tetravalent plutonium in the solution to the trivalent form.
5. In-line analysis of mixed oxides during fabrication.

This program capitalized on CEN's extensive experience in liquid metal and molten salt technology, centrifugal contactors, and fluidized-bed technology. It was terminated, however, in April 1973, when major reductions were made countrywide in research and development funding by the AEC in order to concentrate financial resources on the Fast Flux Test Facility (FFTF), which was then under construction at Hanford, Washington.

Liquid Metal Decladding of Reactor Fuels. The most difficult and costly aspect of reprocessing LMFBR fuels by the Purex process was the head-end operation, which consisted of cladding removal, sodium and fission-product iodine removal, and introduction of fuel into the dissolver. The reference process for preparing fuel for dissolution called for shearing of whole fuel subassemblies to produce short pin segments, 1/4- to 1/2-in. long. Problems encountered at Oak Ridge with the shearing operation were expected to become more difficult with highly irradiated fuels. Stimulated by their extensive experience with liquid metals and molten salts, workers at CEN began an investigation of liquid metal decladding.

In liquid metal decladding, the cladding was removed by slowly immersing a subassembly, from which the end hardware had been removed, into molten zinc at 800°C. The zinc dissolved the stainless steel cladding. The uranium and plutonium oxides, which were not affected by the zinc, were collected in a basket at the bottom of the zinc pool. An overlying layer of molten salt served to collect the iodine, probably as potassium iodide, and also to restrict the rate of zinc vaporization. The amount of steel that could be removed far exceeded the solubility of iron in the zinc, indicating a corrosive, grain-boundary attack on the stainless steel. The zinc-steel slurry could be pumped to a zinc-distillation apparatus, where the zinc would be distilled for recycle, and the steel residue removed to waste.

Cladding from fuel elements could be removed in about two hours or less. Fission-product iodine was retained completely in the overlying salt layer. Centrifugal pumps were used successfully to transport liquid metal alloys between vessels.

In the proposed ANL head-end flowsheet, the uranium and plutonium oxides were reduced to metal by magnesium and calcium reductants dissolved in zinc. The product metals were recovered by distilling off the solvent metals or by converting them to nitrides by pumping the product solution through a bed of calcium nitride (Ca_3N_2). The uranium-plutonium product, either metal or nitride, was to be fed to a dissolver for preparing feed solution for the solvent-extraction steps.

Decladding by melting stainless steel cladding and allowing it to drain away from the oxide fuel columns at a temperature between 1490 and 1650°C was also investigated. Work ended at an early stage when the Advanced Fuel Processing Program was terminated in 1973.

Irv Winsch, Terry Johnson, Dean Pierce, and Bill Miller were responsible for most of the work on this program.

Fluidized Bed Conversion of U/Pu Nitrates to Oxides. Following recovery of uranium and plutonium from spent MOX fast reactor fuels, it is necessary to convert the uranium-plutonium nitrate solution from the solvent extraction process to oxides for refabrication of new fuel. The process in use at the time was a multipass batch procedure consisting of precipitation, filtration, and calcination. The presence of plutonium introduced several problems resulting from its toxicity, intrinsic gamma radiation, and need to avoid nuclear criticality. As a result, conversion and fuel-fabrication operations had to be conducted remotely in glove boxes. A need for an improved process amenable to continuous operation was apparent.

The Division investigated the use of fluidized beds, which may be operated in a continuous mode, for the nitrate-to-oxide conversion. Supporting the denitration effort was a laboratory program to characterize the properties of the oxides produced under various denitration conditions and to determine their dissolution rates in nitric acid.

A fluidized-bed pilot plant (see Fig. 4-22) was constructed and operated in the early 1970s. The major components were geometrically safe (4-in.-dia) denitrator and a fluidized-bed reduction unit (1.6-in. dia) for reduction of the UO_3-PuO_2 product of denitration to UO_2-PuO_2. These, plus auxiliary systems, were housed in a large glove box.

Denitration runs were performed over the temperature range of 300 to 450°C. About 25 kg of UO_2-2.2 wt% PuO_2 was prepared. (Had the program continued, products with higher plutonium concentrations would have been prepared.) A planned assessment of the sinterability of the mixed oxide product, important for the production of suitable oxide pellets, was not done because of program termination. However, oxide pellets produced in-house appeared to be satisfactory, having densities that were 84% of the theoretical value. The process was operable over the range of plutonium concentrations of interest and with plutonium nitrate alone.

This program, led by Norm Levitz, also involved Dave Grosvenor, Nick Quattropani, Gale Teats, and Sy Vogler.

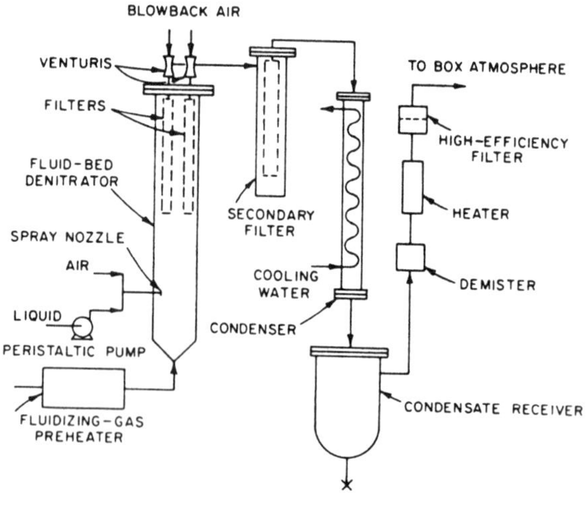

Fig. 4-22. Denitration Pilot Plant

Development of Centrifugal Contactors. Development of centrifugal contactors for use in the Purex-type solvent-extraction process for LMFBR fuels was begun in 1970. The objective was to guarantee nuclear criticality safety by developing a centrifugal contactor having a maximum diameter of four inches. Other advantages of the centrifugal contactor are high efficiency, low holdup of the organic and aqueous streams, and improved ease of operation and maintenance. Radiation damage to the organic solvent is minimized by the short time the solvent is in contact with the highly radioactive aqueous phase. Centrifugal contactors are so efficient that each unit in a bank of contactors can be counted as a theoretical stage.

Centrifugal contactors were first developed at the Savannah River Plant, where they were installed in one of the Purex lines for the extraction-scrubbing step. Their performance was outstanding. Responsible for their development was Don Webster, who was hired into the CEN Division as an Associate Director in 1967. He oversaw the CEN development of centrifugal contactors for LMFBR process applications. The principal investigator was George Bernstein, and others in the program included Dave Grosvenor, John Lenc, Dean Pierce, Nick Quattropani, and Noel Vargo. Others who became involved in applications of the centrifugal contactor were Ralph Leonard, Aldona Siczek, Janet Meisenhelder, Martin Steindler, and Tony Ziegler.

An annular mixing design was adopted for the ANL contactor (see Fig. 4-23). In this design, a large-diameter hollow rotor was employed within an outer casing. Mixing was effected by skin friction between the rotor and the casing. Rotor speeds ranged between 1,000 and 6,000 rpm; the larger the unit, the lower the speed. The mixed aqueous and organic phases flowed into the hollow rotor through a hole in the bottom and were separated efficiently by centrifugal action. The separated phases flowed through their respective ports, each going in an opposite direction to the next contactor.

Terminated in 1973, this program was resumed in 1977, and larger 25-cm (about 10-in.) diameter units were also constructed for application to LWR fuel processing. However, focus was soon restricted to the smaller units when commercial processing of LWR fuel in the United States was banned. A theoretical analysis of annular mixing in a centrifugal contactor resulted in development of a dimensionless dispersion number by which various combinations of variables (*e.g.*, rotor speed, width of annulus, residence time of fluids in the annulus) could be used to give satisfactory mixing.

Electrolytic Reduction of Plutonium in the Purex Process. Electrolytic reduction of plutonium from the tetravalent to the trivalent state, in which it is insoluble in the organic phase, was investigated in the early 1970s. This reduction is necessary to separate plutonium from uranium in the Purex process. Although chemical reductants, such as ferrous sulfamate [$Fe(H_2NHSO_3)_2$], are very effective, their ultimate appearance in the aqueous waste limits the waste-volume reduction by evaporation of the water. Successful electrolytic reduction of tetravalent plutonium was achieved in the presence of 0.005 *M* hydrazine (N_2H_4), but difficulties were encountered in maintaining electrode performance. As the program ended, titanium and gold electrodes were being tested. John Heiberger, Mike Krumpelt, and Martin Steindler performed the electrolytic reduction studies.

In-line Analysis during Fuel Fabrication. A program for conducting in-line analyses of U/Pu mixed oxide fuel during its fabrication was partially completed at program termination in 1973. An in-line X-ray fluorescence method was developed that was capable of measuring the U:Pu ratio to within

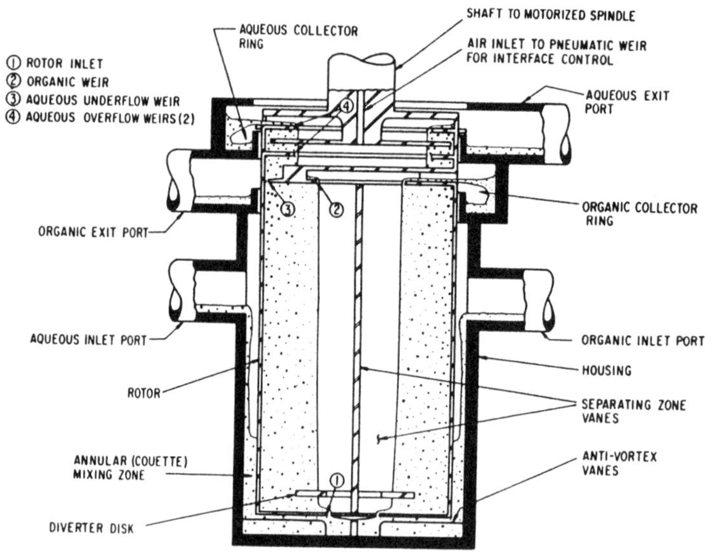

Fig. 4-23. Annular Centrifugal Contactor

the required accuracy of 0.5%. Work on an X-ray diffraction method to measure the ratio of oxygen to plutonium plus uranium, O:(U + Pu), was underway when the program was cancelled. This work was done by Martin Steindler and Glenn Schnizlein with support from the Analytical Chemistry Laboratory.

POST-1973 FUEL-CYCLE PROGRAMS

Following termination of the large fuel-cycle programs in 1973, the Division conducted a number of smaller activities in this area during the remainder of the decade. Most of this work was done under the direction of Martin Steindler. Because of its diversity, it is difficult to give due credit for each aspect of the work, but the following individuals were involved in one way or another. Tom Gerding, Les Jardine, Ruth Juvinall, Rudolf Keller, Bob Kessie, Bruce Kullen, Romesh Kumar, Norm Levitz, Ralph Loutfy, Bill Mecham, Bob Nelson, John Parks, Wally Seefeldt, Martin Seitz, Verne Trevorrow, Charlie Wach, and Irv Winsch.

Fuel Reprocessing. Beginning in October 1974, a one-year study was directed toward the reprocessing of carbide and nitride fuels. Reference flowsheets were evaluated with emphasis on environmental, safety, and materials-protection aspects of recovery operations. An integrated concept for reprocessing and remote fabrication of these advanced fuel types was developed.

Reliability of High-Level Waste Canisters. Also completed in 1975 was a one-year study to determine the reliability against failure and leakage of canisters used to contain high-level radioactive wastes in glass and calcined forms. The storage times considered for the canistered wastes were 10 years underwater at a reprocessing plant storage pool, followed by 100 years at a retrievable surface storage facility. Final emplacement would be in a geologic repository. In general, the study exposed a woeful lack of materials-property information needed for reliable analyses. Even so, canisters of 304L stainless steel showed promise of reliable storage of the wastes over the 100-year time span.

Encapsulation of Radioactive Wastes in Metal. The Division investigated incorporation of high-level waste in a metal matrix such as lead as an alternative to incorporation into glass monoliths. Considered was waste in the forms of calcined pellets and vitrified borosilicate glass beads. Dispersion of these wastes in lead produced a two-phase composite containing 60 to 70 vol% of the high-level waste. Advantages of the metal-matrix waste form compared to a glass monolith are higher thermal conductivity, mechanical strength, ductility, and impact resistance. The corrosion rate of lead in salt water is as low or lower than that of glass. Encapsulating the metal-waste composites would further delay the escape of fission products from the waste form. One centimeter of lead would be expected to reduce penetration by salt water for about 3,000 years.

Large cylindrical composites, 4-in. dia by 20 in. long, were made by pouring lead over glass beads contained in a low-carbon steel pot. Impact and leaching tests conducted on small cylindrical composites (2-in. OD by 2 in. long) revealed good impact resistance and low leach rates. A comparison of the cost of metal-matrix encapsulation with that of casting borosilicate glass monoliths showed the processing costs to be about the same within the sensitivity of the cost analyses. The program was terminated at the end of FY 1980.

Salvage of Alpha-Contaminated Structural Materials. Work was conducted to develop processes for cleanup of structural materials (*e.g.*, piping, discarded equipment, and discarded glove boxes) contaminated by transuranic elements. The objective was to reduce the contamination level to less than 10 nCi/g, thereby permitting disposal of the materials as low-level waste or, possibly, their return to beneficial use. Melting contaminated steel materials under a fluoride slag resulted in transfer of plutonium to the slag and a reduction of the plutonium concentration in the metal to below-detectable levels. The very low rate of accumulation of plutonium in the slag would allow many batches of metal to be processed before discard or treatment of the slag for recycle would be necessary.

Storage and Disposal of Tritium. A short study was made on the management of tritium formed in light water reactors (LWRs). Tritium in spent fuel reports to the low-level aqueous waste when the fuel is dissolved and recovered by the Purex process. A conclusion of the study was that the tritium-containing waste could be disposed satisfactorily by injection of the waste into deep underground aquifers.

Consolidation Techniques for Cladding Hulls. Under the leadership of Norm Levitz, work was performed to develop methods for safe and economic management of spent fuel-subassembly hardware components (Zircaloy cladding hulls plus other items such as spacers and fittings) that result from processing LWR fuels. The objective of the program was to reduce the large volume of this waste and eliminate the hazard due to the pyrophoricity of Zircaloy. Ancillary goals were separation of the hulls from long-lived radioactive contamination and, possibly, recovery of the zirconium. The study involved an assessment of the volumes of these wastes, characterization of their composition, and a review of current experience in handling them. This study culminated in publication of tentative criteria for managing cladding hulls and other subassembly hardware.

Studies of Fission-Product Migration in Geological Formations. In the late 1970s, CEN participated in the Waste Rock Interaction Technology Program managed by Pacific Northwest Laboratories. This program concerned the disposal of fission-product wastes in deep underground rock formations, *e.g.*, granite. The geochemistry group in CEN studied transport through the granite host rock of fission-product elements released from a breached waste canister. An experimental study of the migration of cerium, sodium, and cesium through granite (saprolite) under simulated groundwater flow showed that water intrusion did not adversely affect the effectiveness of the granite host rock as a barrier. A conclusion of this study was that site-specific studies are necessary for any host material being considered for a repository (*e.g.*, salt, basalt, or granite).

Other geochemical studies included (1) an applied program to develop well-logging techniques using a pulsed neutron source and gamma or neutron detectors to determine oil levels in the vicinity of a bore hole, and (2) a basic research program to investigate the migration of trace elements in the earth's crust on the assumption that these natural processes could be useful in the design of a repository.

Evaluation of Source Term for Ex-reactor Fuel Cycles. Work funded by the Electric Power Research Institute was performed to evaluate the plutonium source term for three reactor fuel cycles: LWRs fueled with uranium, LWRs fueled with plutonium, and LMFBRs (see Table 4-15).

The source term is the fraction of total plutonium that escapes containment and reaches the environment. For the technology then in use, the fraction was 1×10^{-8} for processing plants and 1×10^{-9} for fabrication plants. Improvements in that technology would reduce these fractions by 100 and 1000, respectively, and it was estimated that future technology could reduce them by another factor of 1000.

LWBR Proof-of-Breeding Analytical Support. In 1978, CEN became a major participant in a program to determine the breeding ratio in the Shippingport Light Water Breeder Reactor (LWBR), which was located in Shippingport, Pennsylvania. This reactor was fueled with a ThO_2-$^{233}UO_2$ fuel. Breeding, if any, would result from generation of U-233 by the reactions

$$
{}_{90}Th^{232} + {}_{0}n^{1} \rightarrow {}_{90}Th^{233}
$$
$$
{}_{90}Th^{233} \rightarrow {}_{91}Pa^{233} + \beta^{-}
$$
$$
{}_{91}Pa^{233} \rightarrow {}_{92}U^{233} + \beta^{-}
$$

in which the thorium-232 absorbs a neutron to form thorium-233, which then decays by beta emission first to protactinium-233 and then to the uranium-233, which is fissionable. Our role was to analyze precisely located segments

Table 4-15. Estimated Plutonium Releases, μg Pu/GW(e)-yr

	Currently Used	Currently Available		Anticipated		
Operation	LWR-U	LWR-U	LWR-Pu	LWR-U	LWR-Pu	LMFBR
Fabrication	-	-	0.7	-	0.0008	0.002
Processing	3450	34	72	0.003	0.008	0.027
Total	2350	34	73	0.003	0.009	0.03

of fuel rods. The results of these analyses were to be used by the Bettis Atomic Power Laboratory to calibrate an Irradiated Fuel Assay Gauge for measurement of uranium-isotope and selected fission-product concentrations in irradiated fuel rods. The fission-product concentrations were used to determine fuel burnup. The Irradiated Fuel Assay Gauge was to be used for making these measurements on hundreds of fuel rods after they had been irradiated for four years in the LWBR.

Very stringent requirements were set for analytical accuracy and precision. Process requirements included high-precision shearing of fuel rods to obtain segments at designated locations along the rod, complete dissolution of a segment, and minimal contamination on successive dissolution of the segments.

Designed and built was a shear capable of locating cuts to within 0.05 mm and of providing at least 99.7% recovery of a sheared segment. Total uranium and uranium isotopic abundance in samples of dissolver solution was determined by mass-spectrometric/isotopic-dilution procedures. Concentrations of the fission products, Cs-137, Ce-144, and Zr-95, were determined by radiometric techniques. Analyses of fuel-rod segments provided by Bettis showed that their requirements for high analytical precision and accuracy could be met.

Charles E. Stevenson handled the negotiations for this program. Norm Levitz was the Project Leader. Don Graczyk of the Analytical Chemistry Department played a key role in devising the exacting analytical procedures that were required. Other participants in this program are listed in the next chapter.

Nuclear Fusion Energy Research

It seems paradoxical that both nuclear fission and nuclear fusion release large amounts of energy although they are reverse processes. Fission, however, occurs with large atoms such as uranium and plutonium, which split into fission products, while fusion involves isotopes of hydrogen, which combine to form a larger atom (helium). Both nuclear reactions entail the conversion of a very small amount of mass into a very large amount of energy ($E = mc^2$). The three isotopes of hydrogen with atomic weights of 1, 2, and 3 (${}_1H^1$, ${}_1H^2$, and ${}_1H^3$) are called hydrogen (H), deuterium (D), and tritium (T). Natural hydrogen contains about 0.15% deuterium, which can be concentrated by fractional distillation or electrolysis of water, but many separation stages are required to produce essentially pure D_2O (heavy water). On the other hand, the supply is basically unlimited (sea water, for example). Tritium can be obtained only by nuclear reactions. It decays with a soft β^- emission and a half-life of about 12 years.

The two nuclear reactions that are usually considered for fusion reactors are D-D and D-T:

$${}_1D^2 + {}_1D^2 \rightarrow {}_2He^4$$
$${}_1D^2 + {}_1T^3 \rightarrow {}_2He^4 + {}_0n^1$$

To maintain the nuclear reactions at a "break-even" energy level requires a temperature of about 100 million degrees, and a product of plasma density and confinement time of 10^{14} sec/cm^3 for the D-T reaction. The requirements are even more stringent for the D-D reaction. If a lithium blanket to absorb neutrons from the D-T reaction surrounds the high-vacuum space around the plasma, the system has a capability for breeding new tritium *via* the reactions

$${}_3Li^6 + {}_0n^1 \rightarrow {}_2He^4 + {}_1T^3$$
$${}_3Li^7 + {}_0n^1 \rightarrow {}_2He^4 + {}_1T^3 + {}_0n^1$$

The bred T is recovered and used to fuel the plasma. Natural lithium consists of about 7.42% lithium-6 in lithium-7. Most of the tritium breeding occurs by the first (lithium-6) reaction, but the second one also contributes when the neutron energies are below about 3 MeV. The less severe requirements in operating conditions and the possibility of breeding strongly favor the D-T reaction over the D-D reaction for a practical fusion reactor. Lithium metal and its compounds are readily available from commercial sources.

Superconducting magnets are used to levitate and compress the plasma in the vacuum chamber. The best known fusion device is a toroidal or doughnut-shaped chamber, called a "Tokamak," which was used by the Russians, at Princeton University, and eventually by other laboratories. Other experimental devices included a "pinch" magnetic field (the Scyllac machine at Los Alamos) and magnetic mirrors at Livermore National Laboratory. In these devices, the plasma was introduced into the machine by a plasma gun. Another approach to fusion power involves the use of high-power lasers to heat pellets of fuel where the inertia of the material confines the plasma, but work in CEN was limited to the magnetic systems.

A highly simplified version of a fusion system is shown in Fig. 4-24, in which the vacuum chamber enclosing the plasma is called the "first wall." This wall is surrounded by a liquid lithium blanket, which serves as a coolant and neutron absorber, and the whole assembly is surrounded by superconducting magnets to confine the plasma.

Fusion power would have some compelling advantages if it could be developed into a practical technology: (1) The fuel resources, deuterium and lithium, are basically unlimited. (2) Fusion does not produce radioactive fission products, which create a serious waste-disposal problem for fission reactors. (3) There is no possibility of a catastrophic reactor excursion or explosion. The potential problems are a fire if hot liquid lithium should escape from the blanket, and confinement of tritium, which is not highly radioactive, but could be insidious and virtually impossible to recover were it to escape to the environment. The only other significant radioactivity would be that produced by absorption of neutrons by structural materials in the reactor; the levels of this activity would be orders of magnitude lower than those of wastes from a fission reactor. The principal difficulty with fusion is that maintaining a stable plasma is a formidable problem, which has not yet been solved. In the 1950s, practical fusion power was predicted for the 1990s; and now that we are in the 1990s, it appears that another 40 or 50 years may be needed for it to achieve fruition.

Although Argonne was not recognized as one of the leading players in the fusion program of the early 1970s, it made valuable contributions to the national effort. Later on, CEN became well known for its work on the first wall and other structural hardware, blanket materials, tritium recovery, and neutron dosimetry.

LITHIUM HYDRIDE SYSTEMS

When the fusion work started at CEN, Vic Maroni, Carl Johnson, and others had already been investigating molten salt systems containing lithium hydride (LiH), partly because of the previous interest in regenerative galvanic cells based on hydride systems. This work continued to be supported by the AEC Division of Physical Research, and it was a natural fit with the fusion program when it was started at ANL.

Thermochemical Properties and Solubilities. The thermochemistry of the Li-LiH system had been characterized earlier with respect to the standard free energy of formation of LiH, Sievert's constant for dilute solutions of LiH in lithium, and the equilibrium constant for the formation of LiH. A similar study was conducted on the Li-D system by measuring equilibrium pressures of D_2 at temperatures of 705-851°C, pressures of 0.5-760 torr, and compositions of 0.5-99% LiD. The results showed a small, but measurable difference in the LiH and LiD systems. Because nitrogen

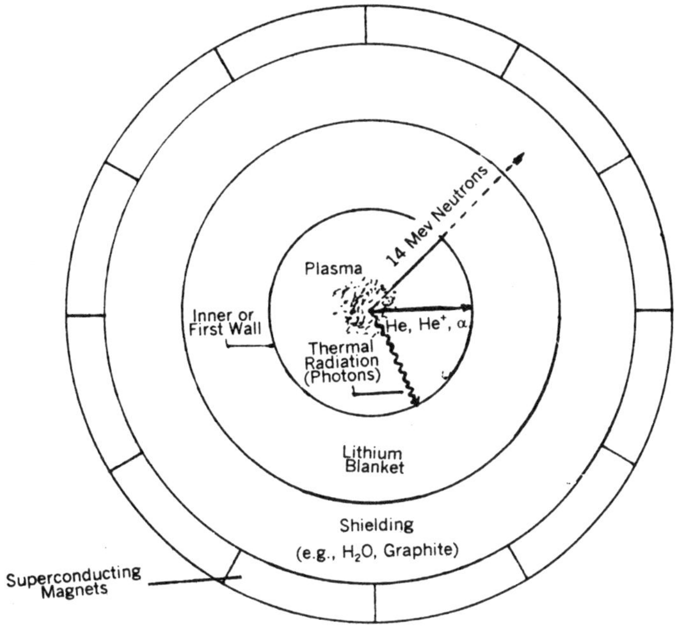

Deuterium and tritium are injected into the system by a plasma gun, where they are confined as a plasma by a magnetic field at a temperature around 100 million degrees. Nuclear fusion reactions occur, releasing large amounts of heat and high-energy neutrons, which are absorbed in a liquid lithium blanket. Helium atoms and ions, including alpha particles, are also generated. To produce useful electrical power, the lithium would be pumped through heat exchangers to convert water to steam to drive turbines. One of the greatest technical challenges, other than maintaining the fusion reaction, is to design a first wall that can withstand the high temperature differential, radiation damage from the neutrons, mechanical stresses, and corrosion by the lithium blanket, which may contain various impurities. Many of the engineering problems with fusion power stem from the enormous temperature gradients between the plasma at about 100 million degrees and the superconducting magnets, which must operate at temperatures near absolute zero.

Fig. 4-24. Simplified Representation of a Magnetic Fusion Reactor was a potential impurity in fusion-reactor blankets, measurements were also made on the Li-Li_3N system, and the results proved to be in good agreement with existing calorimetric data.

The possibility of extracting bred tritium from a lithium blanket in a fusion reactor was explored by measuring the distribution coefficients of LiT between lithium and molten halide salts at 400-600°C. The results are given in Table 4-16, where K_d is the volumetric ratio of tritium in the salt to that in the lithium. These results showed that the distribution coefficients are favorable for such an extraction.

The solubility of LiD in liquid lithium was measured because cold-trapping was under consideration as a method of removing hydrogen isotopes from lithium blankets. The solubilities (in mol%) were 0.50, 0.84, and 2.30, respectively, at 322, 375, and 451°C. Some studies were also conducted on the ternary system, Li-Al-H, to determine the effects of an additive on the activity coefficients of lithium and hydrogen. Aluminum was of practical interest because it and its alloys were potential low-activation structural materials in fusion-reactor blankets. At low hydrogen concentrations, the equilibrium hydrogen pressure increased monotonically with concentration, indicating a homogeneous condensed phase. At higher concentrations, however, the results suggested a miscibility gap at the lower temperatures.

Table 4-16. Distribution Coefficients of LiT between Lithium and Molten Salts

Molten Salt Composition	Temp., °C	K_d
LiCl-31 mol% LiF	550	4.1
LiCl-41 mol% KCl	600	2.6
LiBr-41 mol% RbBr	500	1.2

The solubility of oxygen, also a potential impurity in liquid lithium, was measured in a joint effort with Oak Ridge National Laboratory, and preliminary results showed that the oxygen solubility in mole percent, S, was represented by the equation

$$\ln S = 6.610 - 7140\ T^{-1}$$

where T is in degrees Kelvin.

Hydrogen Permeation Studies. Hydrogen and its isotopes are noted for their ability to diffuse through metals, especially at high temperatures. This Houdini-like behavior was of concern because of the possibility that tritium in the hot liquid lithium blanket would diffuse to the environment through the containment walls not only of the blanket itself but also the lithium pumps, the steam-generator tubing, and other auxiliary equipment. Studies were therefore initiated on methods of decreasing tritium permeability in various metals. Some of the initial concepts involved multilayer metals, impermeable impurity coatings on the metal surfaces, and inert gas purging with a tritium-trapping capability. Test specimens were normally in the form of sheet or tubing.

The permeation rate, Φ, in units of $[cm^3(STP)\cdot mm]/(cm^2\cdot hr\cdot atm^{0.5})$, can be expressed by the equation

$$\Phi = \Phi^{\circ} \exp(-Q_p/RT)$$

where Φ° is a pre-exponential term, and $-Q_p/RT$ represents the activation energy, which reflects the temperature dependence of the rate. Permeation rates were measured at 10-600 torr of upstream hydrogen pressure and temperatures of 400-800°C. For copper, Types 316 and 304 stainless steel, multilayered specimens of 304-SS/Nb, 316-SS/Cu/316-SS, and 304-SS/Cu/Nb, the values of Φ° ranged between about 1,000 and 12,000, with values of Q_p falling between about 15 and 27 kcal/mol.

The hydrogen-permeation characteristics of vanadium were investigated. The permeability varied greatly, depending on activation (hydrogen firing) and deactivation of the material. At the end of a sixty-day study, an approximately 1-µm-thick surface layer containing oxygen and carbon was observed.

In 1977, permeation studies were continued with emphasis on stainless steel/aluminum bronze combinations. The permeation rates fell from that expected for pure stainless steel to a value about 30 times lower after a few weeks. Examinations of the specimens suggested that the lower rates were due to delamination and oxidation of the aluminum bronze. A similar type of effect was found in the thermal conductivity of a 304-SS/Cu/304-SS tube. The effect was attributed to a buildup of oxide layers at the metal interfaces.

Hydrogen-permeation studies were continued on candidate austenitic (316-SS), nickel (Inconel-625 and -718), and refractory metal (Ti-base) alloys at temperatures between 100 and 800°C and hydrogen driving pressures from 10^{-1} to 10^4 Pa. The objective was to provide information on the probable rate and extent of tritium migration through structural materials in fusion-reactor systems. Surface oxidation of the small amounts of aluminum in the alloys by impurities in the gas reduced the hydrogen permeation rate by a factor of about ten at low hydrogen pressures. The values of Φ° for the stainless steel and Inconel alloys fell in the range of about 0.2-0.6 $cm^3/(mm \cdot s \cdot kPa^{0.5})$, with Q values of about 13-15 kcal/mol. Titanium-base alloys were of interest because of titanium's availability, structural strength, low induced radioactivity, and resistance to radiation damage. Titanium, however, reacts with hydrogen to form a hydride, so it might be expected to have a high hydrogen-permeation rate. A study of 5621 titanium alloy (Ti-5 wt% Al-6 wt% Sn-2 wt% Zr-1 wt% Mo-0.25 wt% Si) showed hydrogen-permeation rates about a factor of 1,000 higher than those of stainless steel. There was no obvious change in the visual appearance or the physical strength of a specimen after the tests.

LITHIUM BLANKET PROCESSING

The objectives of blanket processing in a fusion reactor are to prevent an ongoing buildup of tritium in the lithium and, in the case of a breeder, to recover the tritium for use as the reactor fuel. Two avenues were pursued for these purposes, chemical gas sparging and electrochemical evolution. In both cases, the tritium was first extracted as LiT into a molten salt such as LiF-LiCl-LiBr, which was saturated with metallic lithium from the blanket. In the experimental work, deuterium was used as a stand-in for tritium.

The gas-sparging method was investigated by circulating an argon-HCl mixture in a closed loop through molten LiF-LiCl-LiBr containing small quantities of lithium metal and LiD. The extent and rates of the HCl reactions with LiD and Li were followed by monitoring the off-gas stream for the reaction products, H_2, HD, and D_2, with a mass spectrometer. A typical experiment consisted of adding 5 mg LiD and 50 mg Li to about 900 g of LiF-LiCl-LiBr at 500°C while argon was bubbled through the melt. The HCl was then added to the argon to titrate out the Li and LiD. At completion of the reaction, the gas stream was passed through a hot (550°C) titanium bed, which served as a getter for the hydrogen isotopes. In all the experiments, the extraction was complete (within the experimental uncertainty) without excess HCl. The results of this work indicated that molten salt extraction with gas sparging would probably be capable of maintaining hydrogen levels at about 1 ppm by weight in lithium circuits.

The electrochemical method, which consisted simply of electrolyzing lithium hydride at a voltage below the decomposition potential of the molten salt, had the advantages that corrosive gases were not required and the dissolved lithium in the salt did not have to be oxidized. Experiments were performed in which a constant voltage was applied to the salt, and then 5 to 10 mg of LiD was added to the system. The current increased sharply as expected, and then decreased as the LiD was removed from the salt. Again, the removal of LiD was essentially complete, suggesting that this method could also maintain a hydrogen level of about 1 ppm in the lithium.

Lithium-Processing Test Loop. Work began in 1976 on the design and fabrication of a 50-gal (190-L) Lithium-Processing Test Loop (LPTL), which was started up in late 1977. The principal initial objectives were to (1) test the effectiveness of semicontinuous molten salt extraction using a single vessel as a pseudo-mixer-settler, (2) evaluate the efficiency of a high temperature zirconium getter trap and a regenerative cold trap in series for removal of non-metallic impurity elements, and (3) develop impurity control along with processing and monitoring methodology for large liquid lithium loop systems. Provisions were also made for attachment of various types of other experimental test equipment. The general layout of the LPTL, which was located in one of the shielded cells in Laboratory H-126, is shown in Fig. 4-25. A bench-scale mini loop was also constructed and operated to provide technical support for the LPTL.

By mid-1979, the LPTL had reached an operating time of 9,500 hr. At that time the initial objectives had been met, and the focus was shifted somewhat toward (1) demonstrating that the steady-state concentration of bred tritium in the blanket could be limited to about 1 ppm by a continuous tritium-recovery process, (2) developing on-line methods for monitoring tritium and impurities, (3) testing materials and equipment, and (4) providing technical input for other tritium-handling facilities.

Tests of the molten salt-extraction procedure with LiF-LiCl-LiBr showed that the salt was highly effective in removing hydrogen, deuterium, and carbon from the lithium. Cold trapping proved effective in removing entrained salt from lithium in the extraction unit from >2,000 ppm to <300 ppm. Analyses of zirconium wires exposed to flowing lithium for about 10,000 hr indicated that refractory metals could be used to getter O, N, H, and C. An in-line meter was devised for monitoring hydrogen (and its isotopes) by coupling a hydrogen-permeation probe with a mass spectrometer. Another useful monitoring device was a four-probe electrical resistance meter to determine total lithium impurities, including the hydrogen isotopes.

In October 1979, the electromagnetic pump channel of the loop developed a leak that released 10-20 gal (~ 40-80 L) of hot liquid lithium which caught fire. The fire was well contained, and there was no significant damage beyond the pump area. Although the operators were not pleased by this event, especially since it occurred during Fire Prevention Month, it tended to lay to rest some of the prevailing fears about the severity of lithium fires and difficulties in coping with them. The loop was repaired and placed back in operation.

NEUTRON DOSIMETRY

In 1975, a Dosimetry and Damage Analysis Center (DADAC) was established at ANL as a support group to provide standardized neutron-dosimetry data and damage-correlation parameters to programs determining irradiation effects. The overall program concerned two interrelated areas, one focused on characterization of irradiation facilities in term of neutron flux, fluence, and the energy spectrum, and the other on the development of computer codes to calculate damage parameters from observed irradiation effects. Two types of irradiation facilities were involved—reactors and accelerators. Among these were Argonne's EBR-II and CP-5 reactors and the

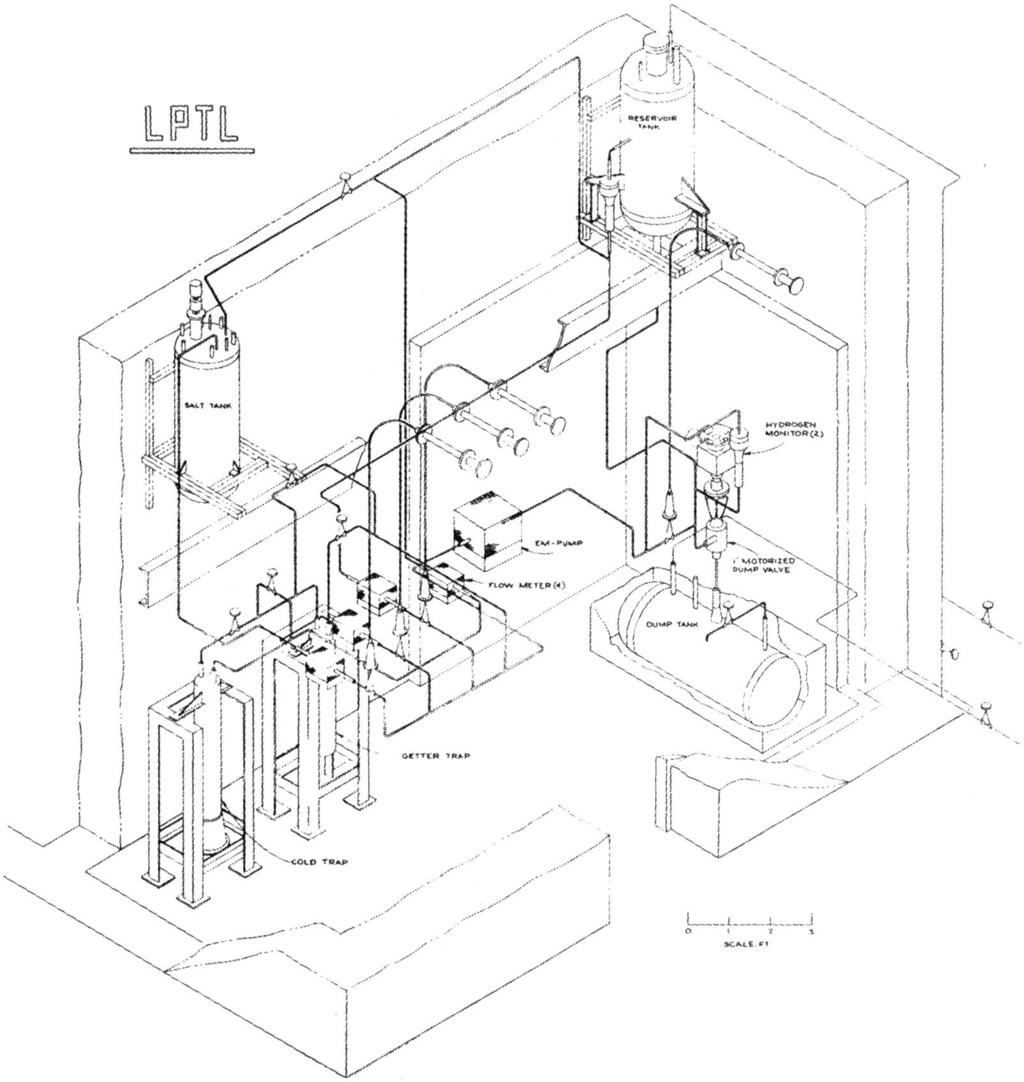

The Lithium Processing Test Loop, located in a shielded isolation cell in Laboratory H-126, contained about 50 gal (190 L) of liquid lithium metal at a temperature of about 500°C. Stainless steel was the material of construction. The molten lithium from a reservoir was pumped electromagnetically through flowmeters, a molten salt contactor, getter traps, and cold traps to study various methods of removing hydrogen isotopes and other impurities from the liquid metal phase. Instruments such as a conductivity meter and a mass spectrometer were used to monitor impurity levels. For reasons of cost, availability, and safety, deuterium was used as a stand-in for tritium, which would be the principal isotope in an operating fusion reactor.

Fig. 4-25. Lithium Processing Test Loop

High Beam Flux Reactor at Brookhaven. The accelerator facilities included the Rotating Target Neutron Source and the Davis Cyclograaff, both at Livermore, and a Van de Graaff and research reactor at Oak Ridge. Characterization of the accelerator sources was particularly difficult due to low neutron fluxes and sensitivity of the neutron distribution to the incident angle of the neutrons at close geometries. Damage effects from neutron irradiation include displacement per atom (DPA), primary knock-on-atom distribution, gas generation rates (H and Ne), and transmutation rates. By 1979, sufficient data were available to generate computer codes that could predict radiation-damage parameters with ±10% accuracy for most irradiations. Most of this work was being done to support the Fusion Materials Irradiation Test (FMIT) facility under construction at Hanford at that time. At the end of the 1970s, CEN was serving as a nationwide center for information on dosimetry and radiation damage in fusion systems.

TRITIUM HANDLING AND CONTAINMENT STUDIES

Argonne had played an important role in parametric design and analysis studies of the Tokamak Experimental Power Reactor (TEPR) and The Next Step (TNS) reactor, and in the late 1970s it was involved in similar studies of the engineering test facility (ETF), the commercial fusion reactor (STARFIRE), and the international tokamak reactor (INTOR). The Chemical Engineering Division was responsible for tritium handling and processing systems and, in some cases, for the entire blanket and shield portions of the reactor or facility. STARFIRE was managed by ANL, ETF by ORNL, and INTOR by an International Atomic Energy Authority (IAEA)-sponsored group involving the USA, European Community, Russia, and Japan. The Division, because of the preceding experimental and design work, was able to put together a library of global computer programs with the following capabilities.

1. Performing comprehensive analyses of vacuum pumping, tritium handling, system costing, *etc.*, for all types of fusion reactors.
2. Unfolding experimental air-detritiation test data to derive tritium reaction parameters, as well as adsorption and desorption rates.
3. Performing broad-based thermal-hydraulic calculations for a variety of first wall/blanket configurations with various coolants and structural materials.
4. Performing finite subsystem analyses of distillation cascades (fuel enrichment), magnetic diverter processing systems, *etc.*

The codes were updated continually as new information became available, and many were supplied to other USA and foreign fusion facilities at their request.

Vic Maroni spearheaded the fusion program of the 1970s. Wally Calaway, Ewald Veleckis, and Bob Yonco did most of the work on molten salt chemistry and blanket processing, and Erv Van Deventer conducted the hydrogen-permeation studies. Jim Weston was primarily responsible for the lithium test loop, and Larry Greenwood handled the neutron dosimetry effort. Rob Clemmer, Pat Finn, and Bali Misra performed the tritium handling and containment studies, which required extensive computer work. Norm Chellew, Bob Heinrich, Bob Land, and two students, Ralph Pelto and Ray Peterman, also contributed to this program.

Materials Chemistry and Thermodynamics

As mentioned in the previous chapter, the thermochemistry program consisted of two parts. One, which was based primarily on combustion calorimetry to measure standard enthalpies of formation ($\Delta H_f{}^\circ{}_{298}$), was under the direction of Ward Hubbard and Pat O'Hare. The other program, led by Paul Blackburn, was aimed primarily toward potential fast breeder reactor fuels and included phase studies together with determinations of higher-temperature heat capacities, enthalpies of transition, and entropies with the ultimate objective of providing a full set of thermodynamic functions for the materials of interest.

CALORIMETRY

As of 1970, fluorine combustion calorimetry had been used to obtain standard enthalpies of formation of 35 elemental fluorides and 21 other compounds. The more conventional oxygen bomb calorimetry had been used to study 15 reactions, and solution calorimetry was used for nine aqueous reactions. Empirical correlations were used to estimate enthalpies of formation of the rare earth fluorides, and work had begun on the use of theoretical calculations (extended Hückel molecular-orbital calculations and *ab initio* calculations based on the Hartree-Fock-Roothan approach), to predict thermodynamic data.

Standard enthalpies of formation were determined for the compounds shown in Table 4-17, using combustion and solution techniques.

Table 4-17. Compounds Investigated by Standard Enthalpy Measurements

As_2S_3	$Cs_2U_2O_7$	Pu_2C_3
As_4S_4	HF	Rb_2MoO_4
$BaMoO_4$	K_2UO_4	Rb_2UO_4
$BaZrO_3$	Li_2MoO_4	Sb_2S_3
$CaUO_4$	Li_3N	$SrUO_4$
Cs_2CrO_4	Li_2O	UF_6
Cs_3CrO_4	Li_2UO_4	UN
Cs_4CrO_4	$MgUO_4$	U_2N_3
$Cs_2Cr_2O_7$	MoS_2	$VC_{0.79,\ 0.85}$
Cs_2MoO_4	Na_2C_2	XeF_6
$Cs_2Mo_2O_7$	$NaHC_2$	
$CsNO_3$	α-Na_2UO_4	

The standard enthalpy of formation of plutonium sesquicarbide ($PuC_{1.466}$), measured by oxygen bomb calorimetry, was -19.4 ± 0.7 kcal/mol. A redetermination of the enthalpy of formation of liquid HF produced a value of -72.57 ± 0.09 kcal/mol. Because sodium acetylides were thought to have a role in carbon transport in liquid sodium reactor coolant, the standard enthalpies of formation of Na_2C_2 and $NaHC_2$ were determined from the enthalpies of their reactions with water. The results showed the enthalpies of formation to be 4.77 ± 0.40 kcal/mol for Na_2C_2 and 23.10 ± 0.27 kcal/mol for $NaHC_2$. These values, together with estimates of the entropy and incremental enthalpy functions, suggested that the free energy of formation of Na_2C_2 would make it thermodynamically unstable with respect to graphite and sodium. As information was becoming available on the chemistry of irradiated fuels, the thermochemistry of fission-product oxides became an important subject. The fission products, generated as elements, reacted with the oxide fuel to form mixtures of oxides. Standard enthalpies of formation of Cs_2MoO_4, Cs_2UO_4, Rb_2MoO_4, $BaMoO_4$, and Li_2MoO_4 were determined. Enthalpies of formation were also obtained on compounds of interest for oxide and carbide fuels, $VC_{0.85}$, $VC_{0.79}$, $NaUO_3$, and α-Na_2UO_4, and other compounds of interest to the non-nuclear energy programs, Li_2S, Li_3N, $CsAlO_2$, and $KAlO_2$. The enthalpy of formation of UF_6 was redetermined when a question arose about the previous value. In addition to the above, $LaNi_5$ was included because of its possible role as a hydrogen-storage medium, and, toward the end of the 1970s, measurements were started on organic compounds that constitute "building blocks" of coal constituents.

HIGH-TEMPERATURE THERMODYNAMIC STUDIES

The objective of this program was to obtain physical property and thermodynamic information on potential fast breeder reactor fuels. The work was predicated on the use of mixed uranium-plutonium oxide or possibly carbide fuel and liquid sodium coolant. Mass-spectrographic studies showed that the volatile species resulting from vaporization of $(U_{0.8}Pu_{0.2})O_{2-X}$, a typical reactor-fuel composition, were UO^+, UO_2^+, UO_3^+, PuO^+, and PuO_2^+. The partial enthalpies of sublimation of these species were determined and their partial pressures were derived for oxygen-to-metal (O/M) ratios of 1.92 to 2.0 in the solid phase at 1968°C. To calculate more accurate pressures from the measured ion intensities, one needs to know the ionization probabilities, fragmentation patterns, and partial relative ionization cross sections. Paul Blackburn and Phil Danielson obtained most of these data experimentally, and, with minor assumptions concerning the plutonium oxides, calculated the pressures of the species. The UO_2 and UO_3 pressures were then used to calculate the oxygen pressure. Oxygen pressures or potentials in the fuel were important for several practical reasons in that they affect (1) the extent of cladding attack by oxygen at the cladding/fuel interface, (2) the extent to which sodium coolant would interact with the fuel in case of cladding failure, and (3) redistribution of fuel and fission products by vapor transport. The oxygen potential could also affect diffusion and creep rates within the fuel. A computer model based on these results and other U-Pu-O phase studies was developed to permit oxygen-potential calculations over a range of variables. Calculated results from the model agreed well with experimental data in the literature that had been obtained under specific conditions by various investigators.

There was concern that interaction of sodium coolant with the oxide fuel in a reactor could produce swelling and cladding rupture in the event of a leak in the cladding. Earlier studies had indicated that the equilibrium phase between sodium and UO_2 in the temperature range of 600-1000°C is Na_3UO_4. Experiments conducted with mixed oxide (UO_2-19.8% PuO_2) gave an X-ray pattern similar to that of Na_2UO_4, indicating that the material was Na_2MO_4, where M is (U + Pu). When attempts were made to measure Na_2UO_4 with the mass spectrometer, sodium vaporized, leading to a three-phase region believed to be $NaUO_3$-Na_2UO_4-Na_2O_7, where both sodium and oxygen could be measured. Further studies were conducted in which UO_2-20% PuO_2 in the form of powders or sintered pellets was contacted with sodium at 900°C, yielding Na_3MO_4. X-ray diffraction showed this to be a single-phase material with a face-centered cubic structure and a lattice parameter of 4.77 Å. In another study, lattice-parameter measurements were used to determine the equilibrium O/M ratio of mixed UO_2-PuO_2 fuel as a function of temperature and the U/Pu ratio. To determine the kinetics of UO_2-20% PuO_2 reaction with sodium, sintered pellets of the material were exposed to sodium in sealed capsules at temperatures of 500-900°C, sectioned, polished, and examined under a microscope. The unreacted UO_2-PuO_2 and the Na_3MO_4 phases were readily distinguishable, so the extent of reaction could be measured. The product was observed as a surface layer that extended into cracks near the surface, which were somewhat enlarged because the product had a lower density than the original material.

A technique called "matrix isolation spectroscopy" was used to investigate thermodynamic properties of materials at temperatures beyond the range of direct experimental measurements. For example, thermodynamic data obtained by this technique were important in determining equation-of-state relationships used to project the outcome of design-basis accidents in a fast breeder reactor. In using this technique, a solid sample of the material of interest was heated in a Knudsen effusion cell, and the resulting molecular beam in a stream of inert gas impinged on an infrared transmitting window in a liquid helium cryostat, freezing the molecules in an inert gas lattice. These low-temperature conditions greatly simplify interpretation of visible and infrared spectroscopic data. Construction and testing of this equipment were completed in 1970. These studies began with uranium oxides and were extended later to include oxides of cerium, praseodymium, europium, terbium, and thorium.

REACTOR FUEL STUDIES

***Thermodynamic Properties*.** Although the high-temperature thermodynamic studies provided much of the basic information needed to evaluate the safety of fast breeder reactors and possible consequences of an accident, additional investigations were focused more specifically on this subject. For example, Len Leibowitz, Martin Chasanov, and Bob Blomquist measured the speed of sound in liquid sodium to provide data for analyses of shock-wave propagation. Further studies by these individuals plus Don Fredrickson and Don Fischer extended the results and provided additional thermodynamic data.

A variational calculation method based on thermodynamic considerations was developed to determine the effects of burnup, void volume, and starting fuel composition on the phase composition and pressure of an oxide fuel-fission product mixture produced when the fuel is heated to 3225°C. The vapor pressure was important in evaluating design-base accidents, and the phase composition was necessary for post-accident heat-removal analysis. The thermal diffusivity of UO_2 was required to evaluate safe methods of cooling an LMFBR core in case of a meltdown.

Arc-melting experiments showed that liquid phases of iron and UO_2 separated cleanly on solidification, and that the zirconium favored the oxide phase, as expected. Because a concrete base might be located under the core of a reactor such as FFTF, high-temperature experiments were conducted with liquid UO_2-concrete and molten steel. The concrete forms a liquid at about 1100°C, and at some point the liquid phase of UO_2, PuO_2, fission products, and concrete constituents (SiO_2, CaO, and Al_2O_3) will rise to the surface and float on the molten steel. Information on the fission-product distribution between these two liquid phases is needed to estimate the heat distribution between the phases. This study was then extended to a situation in which basalt would be used as a sacrificial material that would, in case of a meltdown, dissolve the molten fuel, thereby reducing the heat flux. Experiments were performed on the distribution of representative fission products (lanthanum, cerium, niobium, molybdenum, and ruthenium) between molten iron and a 6 wt% solution of UO_2 in molten basalt. As expected from thermodynamic predictions, rare earths and niobium were found in the basalt phase, and the noble metals (ruthenium and molybdenum) were in the iron phase. During these studies, the solubility of UO_2 in basalt was found to be 5 to 7 wt% at 1550°C and 50 to 55 wt% at 2200°C.

The viscosity of liquid basalt was measured over the range 1350-1950°C, using a rotating-cup viscometer, and similar measurements were made on molten alumina over the range 2121-2469°C. Thermal-diffusivity measurements on UO_2 were also made by Bob

Blomquist, Jim Haley, Jack Fischer, and Martin Chasanov.

In 1975, a study was begun on various methods of retaining an ex-vessel core after an accident. The function of the core retainer would be to prevent uncontrolled downward travel of the core debris following a breach of the reactor vessel ("China syndrome") and to remove fission-product decay heat. Although a variety of solutions had been suggested and most of them were considered in the study, the major effort was aimed at sacrificial barriers. Barrier materials under consideration included basalt, alumina, magnesia, urania, and thoria.

The post-accident heat-removal (PAHR) studies consisted of two parts. One was a comprehensive computer modeling study that incorporated the extensive physical property and thermodynamic information that was already available; the other was an experimental effort to provide additional key data. The matrix-isolation method was used to generate thermodynamic functions of fuel-vapor species at high temperatures. Data on UO_2 were obtained by Steve Gabelnick, Gerry Reedy, and Martin Chasanov. Dave Green and Gerry Reedy produced data on uranium nitrides, plutonium oxides, and plutonium nitrides. Joanne Fink, Martin Chasanov, and Len Leibowitz prepared a preliminary set of property values for thorium and uranium systems.

In addition to the fission-product heating in a meltdown situation, consideration must be given to heat produced by chemical reactions of the materials. One such situation was found in the modeling of post-accident heat removal in gas-cooled fast reactors, where the high-temperature (~2100°C) reaction of UO_2, graphite, and stainless steel is exothermic. Heat evolution from this reaction must be included in the development of procedures for containing failed core material in gas-cooled fast reactors.

The Division participated in various international efforts to assess the state of thermodynamic data for reactor safety analysis, to identify what further information was needed, and to determine the most effective ways of promulgating the existing data.

Effects of Irradiation on Reactor Fuels. Nuclear fuels in a reactor undergo a variety of changes due to radiation exposure, heat, contact with the coolant, and mechanical stresses. The fissile material is converted to fission products, which may react chemically with one another, with the fuel, and with the cladding. Fission gases can pressurize the fuel elements, and cladding can be embrittled or affected otherwise by the high temperatures and by coolant corrosion. The principal variables affecting the fuel are the radiation level (mainly of neutrons), the time of exposure, and the temperature. The burnup of the fuel (% of fissile material consumed) depends upon the flux and energy distribution of the neutrons and the time of exposure. The objective of these studies was to evaluate the chemical and physical effects that occurred in the fuel during irradiation.

Determination of Burnup in Fast Reactor Fuels. In the late 1960s, a program for measuring burnup in fast reactor fuels was based on developing analytical methods for fission products that could be used as burnup monitors, and on establishing accurate values of their fission yields. An ideal burnup monitor would be a fission product whose fission yield is insensitive to neutron energy. The primary candidates lie on the two peaks of the fission-yield curve. On the heavy element peak are the four major rare earth elements (lanthanum, cerium, praseodymium, and neodymium), plus iodine, tellurium, and barium. On the light element peak are zirconium, ruthenium, and molybdenum. Radioactive nuclides of these elements were investigated as potential burnup monitors.

The fission yield of a candidate burnup monitor was determined from the amount of that fission-product nuclide produced from a known number of fissions. A new technique for quickly determining the number of fissions consisted of using solid-state track recorders (SSTRs) and foil activation. The technique involves placing a thin, uniform deposit (nanograms) of a fissile material in contact with a suitable dielectric such as mica. As fissions occur, the recoil fission fragments produce damage tracks in the dielectric, which, after chemical etching, are visible under a microscope. Simultaneous irradiation of an adjacent foil (approximately 300 mg) of the fissile material produces enough fission-product nuclides for radiochemical analysis or direct counting with a Ge(Li) detector.

Using the above technique, fission-yield determinations were made for Ce-144, Zr-95, Zr-97, Mo-99, Ru-103, I-131, Te-132, and Ba-140. Distinguishing features of these nuclides were short half-lives (1 to 65 days), fission yields greater than three percent, and well-established photon emissions (gamma rays) from disintegration. Fission yields are different for Pu-239 and U-235, but for a particular fissile material the yields vary little with neutron energy. As a result of this work, Zr-95 and Ba-140 were frequently chosen as burnup monitors.

Fast Reactor Neutron Reactions. With the cooperation of the EBR-II and the Materials Science Divisions, the CEN Division measured the capture-to-fission ratios of the principal heavy elements in fast reactor fuels, determined the extent of helium production and its possible relationship to radiation damage of cladding and structural materials, and measured the fission yield of tritium.

Capture-to-fission ratios for Pu-239, U-235, U-238, and U-233 were determined as functions of radial and axial positions in the EBR-II core. A study was made of helium production in sections of the stainless steel control and safety rod thimbles in EBR-II. Nitrogen in stainless steel diffuses from colder to hotter zones in stainless steel, where it undergoes (n,α) reactions to produce helium. Because helium acts to stabilize voids in stainless steel, it can accelerate loss of ductility of the stainless steel under irradiation. This study showed that helium generation rates could be calculated with an accuracy of 25-30%.

Knowledge of how much tritium is generated in fuel during irradiation is important in the design of reactors and processing plants. Accurate measurements of the tritium fission yield were difficult because of its low yield relative to those of other fission products, its long half-life (11 years), and the possibility that reactions other than fission could produce tritium. Experiments in CEN revealed a tritium fission yield of 2.25×10^{-4} (±20%) per fission of U-235. This value is two to three times that for the thermal yield of 0.8 to 0.9×10^{-4} tritons per fission.

A program to measure the tritium yields from fast fission of Pu-239 was begun in 1972. Early measurements showed a considerably higher yield than that for fast fission of U-235. The irradiation experiments were redesigned to accommodate the higher yield and were performed in EBR-II, but the program was canceled before the samples could be analyzed—part of the general cutback in LMFBR research and development to provide additional funding for construction of the Fast Flux Test Facility.

Dosimetry and Damage Analysis. On July 1, 1971, a dosimetry program was established in the CEN Division. The program comprised three major tasks: (1) development and application of experimental and analytical techniques for characterizing fast-neutron irradiation environments in terms of neutron flux, fluence, and energy spectrum; (2) establishment of methods for computing irradiation correlation parameters (e.g., displacements per atom) from dosimetry data; and (3) service dosimetry.

Development of Dosimetry Methods. The principal means by which dosimetry methods were developed and demonstrated was by participation in the Interlaboratory Reaction Rate (ILRR) program. The participating laboratories utilized different methods, and the results were compared through the ILRR program. The participating laboratories and their principal methods were as follows: (1) Argonne, SSTRs and foil activation, (2) Hanford Engineering Development Laboratory, foil activation, (3) National Bureau of Standards, ion chamber, (4) Aerojet Nuclear Corporation, proton recoil and neutronics calculations, (5) Los Alamos National Laboratory, neutronics calculations.

Neutron reactions comprise both fission and nonfission reactions (*e.g.*, capture). Rates for both types of reaction were needed. The required accuracy for nonfission reactions (±5%) was considered to be in hand. Additional work, however, was required to achieve a higher accuracy (±3%) for fission reactions because of their use for calculating reactor power and fuel burnup, and for developing fuel behavior correlations.

Investigators at ANL exposed SSTRs and adjacent massive foils (300 mg) of U-235 in a low-power facility (the Coupled Fast Reactor Measurement Facility, at Aerojet Nuclear Corp.) to establish a factor, K, for the relationship between the counts per minute for a particular fission product (*e.g.*, Zr-95) counted in a particular Ge(Li) detector and geometry to the number of fissions per gram established by the SSTRs. Because SSTRs could not be used in a high-flux reactor such as EBR-II, K factors were used in conjunction with foil activation in EBR-II to determine fission rates. From work in the low-power facility, K factors for four different radionuclides were shown to be constant over a wide range of neutron energies. With the foil technique, fission rates in EBR-II could be measured with an accuracy of ±2.5%.

Radiation Damage Correlations. Correlating radiation damage effects with the neutron environments was just getting underway when the dosimetry program became a casualty of cutbacks in LMFBR research and development funding to provide additional support for construction of the Fast Flux Test Facility. The focus of the dosimetry then shifted to the Controlled Thermal Reactor (CTR) program.

Service Dosimetry. A service dosimetry program was organized to establish a capability for performing routine dosimetry measurements at ANL and EBR-II, and for coordinating ANL's overall dosimetry development program. This dosimetry program was responsible for designing the dosimetry aspects of all ANL experiments conducted in EBR-II. In the course of this and other work at EBR-II, over 600 dosimetry measurements were made.

IRRADIATED MIXED OXIDE FUELS

Studies of the chemistry of irradiated mixed oxide (UO_2-PuO_2) fuels were initiated in the late 1960s when the AEC selected mixed oxides as the fuel of choice for LMFBRs. These studies were continued throughout the 1970s. The objective was to gain an understanding of the complex chemical reactions and processes that take place in the fuels during irradiation. Knowledge of these phenomena was expected to lead to the development of methods for increasing fuel lifetime in the reactor.

Sampling and Analysis. As oxide fuels are irradiated, the temperature across the 3-mm radius varies from about 2500°C at the center to about 700°C at the cladding. This large temperature gradient produces restructuring of the fuel, an oxygen potential gradient, and a redistribution of the fission products (Fig. 4-26). In the hot central region, the

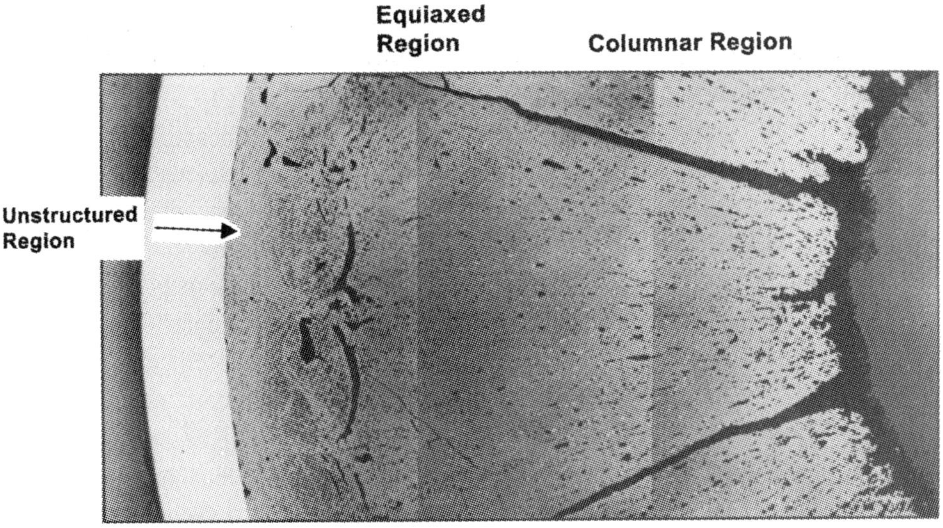

Fig. 4-26. Cross Section of Irradiated UO_2-PuO_2 Fuel Pin. Courtesy of Oak Ridge National Laboratory.

mixed oxides crystallize into columnar grains that radiate outward from the center. The resulting densification of the fuel causes a central void to form. Beyond the columnar grains is a region of equiaxed grains, and beyond that and near the cladding is an unstructured region of the original fuel.

In early work, an electron-probe microanalyzer, which accomplished microsampling and micro-analysis directly, was used to determine point-to-point concentrations of the elements across a fuel pin. This instrument cannot detect elements in the low mass range from hydrogen to carbon and has low sensitivity for elements carbon through oxygen. The capability of the electron-probe microanalyzer was supplemented in 1972 by acquisition of an ion-probe microanalyzer (IMMA), which provided a capability for determining the distribution of the whole range of elements with good sensitivity (ppm range). With this instrument, a small sample area (down to 3 μm) was ionized by a monoisotopic beam impinging on the sample surface. The secondary ion beam, comprising the sputtered sample ions and some of the scattered primary ion beam, is collected in a double-focusing mass spectrometer where element masses can be identified and their concentrations determined.

Oxygen Distribution in Irradiated Fuel. The solid-state diffusion of oxygen resulting from the thermal gradient drives oxygen toward the outer edge of the fuel, thereby establishing an oxygen gradient that is a function of the total oxygen content and the temperature gradient. The most important single property of a mixed-oxide fuel is the oxygen potential, which governs the chemical form of the fission products, their chemical activity (and, hence, the extent of fuel-cladding interactions), and in the event of a cladding rupture, the extent of sodium-fuel reaction.

A new method, based on the distribution of molybdenum in the fuel, was developed for calculating the oxygen potential at points in the fuel, mainly along a radius. The basic concept of the method was that at each point in the fuel (along the temperature gradient), a local equilibrium exists between the oxidized and reduced forms of a fission product, which functions as a "redox" indicator. For a variety of reasons, the Mo-MoO_2 system was chosen, resulting in the following equation:

$$\Delta G(O_2) = \Delta G^{\circ}_f(MoO_2) + RT \ln [a(MoO_2)/a(Mo)]$$

where $\Delta G(O_2)$ represents the oxygen potential, $\Delta G^{\circ}_f(MoO_2)$ is the standard free energy of formation of MoO_2, a(Mo) is the activity of molybdenum in a noble metal inclusion, and a(MoO_2) is the activity of MoO_2 in the oxide matrix. This equation was found to work well in the columnar grain region, but in the cooler fuel regions, cesium molybdate (Cs_2MoO_4) forms, and the MoO_2 activity is no longer proportional to its concentration in the oxide matrix.

Behavior of Cesium. Considerable research was devoted to the behavior of cesium because of its high volatility and resultant mobility in oxide fuels. The transport and reactions of fission-product cesium in a fuel element may affect the performance of fuel pins in several ways, one of which is restriction of the axial flow of fission gases (xenon and krypton) to the plenum of the fuel element. In-pile experiments indicated that gas-flow restrictions develop in the region of the interface between fuel and blanket sections. This gas-flow restriction was attributed to the deposition of cesium uranate (Cs_2UO_4). The prevention of fission-product gas flow could cause the pressure to build up in the fuel region and eventually rupture the cladding.

The reaction of cesium to form cesium molybdate competes with that to form cesium uranate. Thermodynamic calculations led to the conclusion that, if the oxygen-to-metal (O/M) ratio is much less than 2.00 (hypostoichiometric, *e.g.*, 1.97), there will be a tendency for the cesium to migrate to the fuel-blanket interface. If the O/M ratio is greater than 2.00 (hyperstoichiometric, *e.g.*, 2.10), formation of cesium uranate in the fuel can be expected. This could lead to swelling of the cladding. If the O/M ratio is close to stoichiometric, the cesium will tend to remain in the fuel as cesium molybdate.

The chemical reaction of gaseous cesium with uranium oxide to form Cs_2UO_4 is the following:

$$2\ Cs(g) + O_2(g) + UO_2(s) \rightarrow Cs_2UO_4(s)$$

where the oxygen pressure and uranium oxide activities are for hyperstoichiometric UO_2. The amount of Cs_2UO_4 formed is limited by the amount of oxygen available above that needed to form stoichiometric UO_2. The rate of reaction of cesium with hyperstoichiometric UO_2 as a function of temperature and the initial O/U ratio is shown in Fig. 4-27.

Studies of irradiated cladding showed that oxygen-containing fission products such as cesium uranate can accelerate intergranular attack of the stainless steel cladding. Although the level of attack was not too serious, understanding and controlling the formation of such compounds were important. Darrell Fee developed a model of cladding attack based on the reaction of cesium with the protective layer of Cr_2O_3 on the cladding surface to form cesium chromate (Cs_2CrO_4), which is non-protective. Once the protective Cr_2O_3 layer on the stainless steel is breached, which occurs slowly and is temperature-dependent, the subsequent rate of attack depends only on the rate of generation and release of oxygen formed as the fuel is fissioned. Cesium chromate forms on the cladding when the

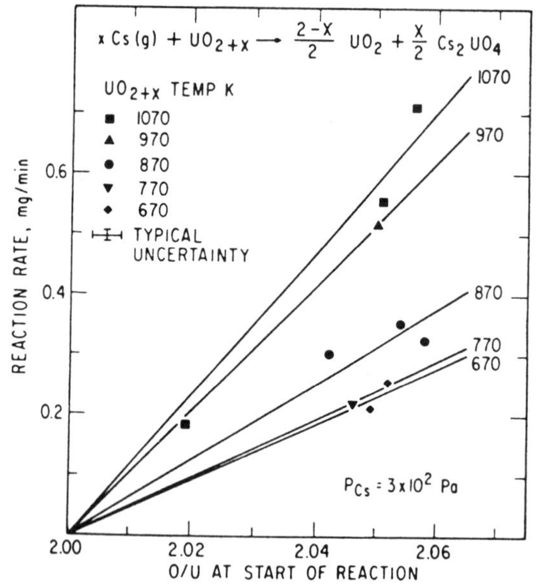

Fig. 4-27. Effect of Temperature and O/U Ratio on Reaction Rate of Cesium with UO_2

cesium equilibrium pressure of the fuel exceeds the equilibrium cesium pressure on the cladding. Thus the distance between the two curves shown as the shaded area in Fig. 4-28 represents the driving force for cesium chromate formation and, thus, cladding attack. The model accurately predicted the in-reactor attack of LMFBR fuel cladding.

Chemistry of GCFR Fuels. When the Clinch River Breeder Reactor was cancelled in the late 1970s, attention turned to the behavior of mixed oxide fuels in a Gas-Cooled Fast Reactor (GCFR). This work relied heavily on the LMFBR base technology, but there are significant differences between an LMFBR and a GCFR. Unique to the GCFR are the use of vented fuels, roughened cladding to promote heat transfer, a helium coolant, and direct contact of the primary coolant with the steam generator.

Again, cesium chemistry occupied center stage because of concern about deposition of

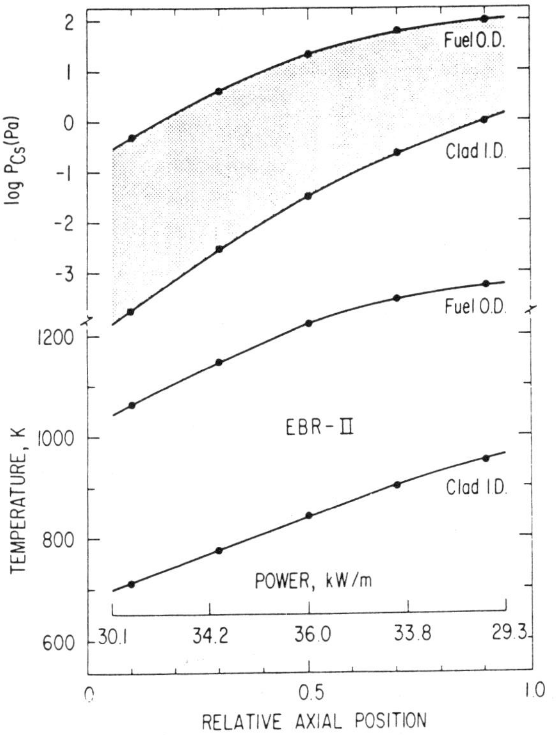

Fig. 4-28. Temperatures and Equilibrium Cesium Pressures in EBR-II Fuel Pin

cesium uranate at the fuel-blanket interface, restricting axial flow of the fission-product gases and their release to the helium coolant. A major accomplishment was the elucidation of the Cs-U-O phase diagram (Fig. 4-29) by Darrell Fee and Carl Johnson. Of the ten possible cesium uranates, only three, Cs_2UO_4, $Cs_2U_2O_7$, and $Cs_2U_4O_{12}$, can exist in equilibrium with hyperstoichiometric UO_2 and liquid/gaseous cesium in the temperature range of 600-1000°C.

Molten Salt Chemistry

Throughout almost the entire history of the Chemical Engineering, and, later, the Chemical Technology Division, the research and development programs have involved the use of molten salts. Among these programs were pyrochemical and fluoride volatility fuel reprocessing, high-temperature batteries, fuel cells, fusion blanket processing, the coal-combustion and magnetohydrodynamics work, and the current programs on electrometallurgical processes. In addition to these applications, molten salts were often used as solvents in basic research studies where more conventional aqueous or organic solvents could not be used for one reason or another. In

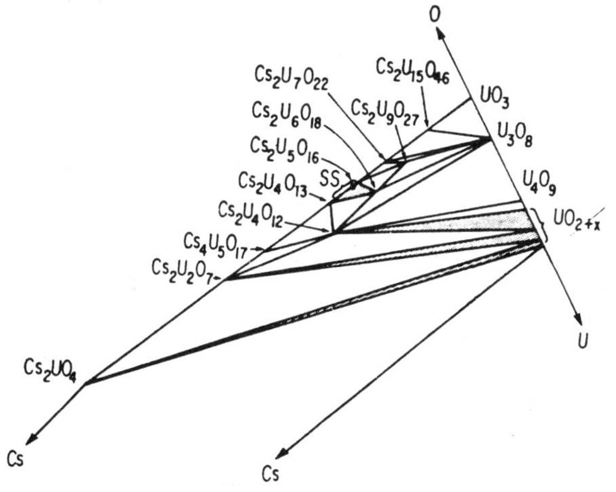

Fig. 4-29. Cesium-Uranium-Oxygen Phase Diagram

the early 1970s, the focus was largely on molten-salt electrolytes for high-temperature batteries, but the scope then expanded into the basic chemistry of salts involved in the fuel cell, coal, and fusion programs.

PHASE EQUILIBRIA

In 1970, researchers in the battery program were investigating candidate systems suitable for use in lithium/chalcogen cells of the general type Li/MX/C, where Li was lithium or a lithium alloy, MX was a molten alkali metal halide salt containing lithium ions, and C was one of the chalcogens (sulfur, selenium, or tellurium). Theoretically, a lithium/sulfur cell would have provided the highest voltage and capacity, but attempts to operate such cells were unsuccessful for several reasons. Elemental lithium presented the problems of solubility in the electrolyte, which resulted in self-discharge due to electronic conductivity of the salt, a tendency to plate out on any metal surfaces at negative potential in addition to the electrode during recharge, and corrosion by chemical reduction of ceramic insulators. As pointed out previously, alloying the lithium with aluminum solved these problems. In evaluating the chalcogens for the positive electrode, consideration had to be given to the phase relationships among the molten halide electrolyte, the elemental chalcogens (S, Se, or Te), and the reaction product (Li_2S, Li_2Se, or Li_2Te). Multicomponent electrolytes, such as LiF-LiCl-LiI, had to be used because they were liquid at reasonable temperatures. (Ternary phase diagrams of the systems LiF-LiCl-LiI and NaF-NaCl-NaI had been developed previously by Carl Johnson and Ellen Hathaway.) A phase diagram, to represent the positive electrode in a Li/LiF-LiCl-LiI/S cell, would have five components, LiF, LiCl, LiI, S, and Li_2S. Because the electrolyte composition is invariant in composition, however, a binary or ternary salt can be regarded as a single phase, making possible pseudoternary systems comprising the salt, Li_2S, and S.

Implicit in the phase-diagram studies are mutual solubilities of the components. Solubilities of Li_2Se were measured in LiCl-KCl, LiF-LiCl-LiI, and LiBr-RbBr eutectics as a function of temperature. All showed the usual straight-line log (solubility) *vs.* 1/T plots, and the solubilities were the same for salts having all-lithium cations. With LiCl-RbBr, the solubility was lower. The solubility of selenium in LiBr-RbBr was about the same as that of Li_2Se (0.03 mol% at 365°C), but the selenium solubility from mixtures of Se and Li_2Se was considerably higher (1.5 mol%), probably due to the formation of selenides. Reverse experiments on the solubility of molten halide salt in Se and Li_2Se-Se mixtures showed very low values, a favorable result from a practical viewpoint. Companion studies of sulfur and Li_2S in LiBr-RbBr showed low solubilities of 0.3 and 1.5 mol%, respectively, but Li_2S-S mixtures resulted in values of 8-10 mol%—a bad omen for the use of sulfur as a positive electrode material.

Paul Cunningham, Ellen Hathaway, Stan Johnson, and Vic Maroni conducted studies of the pseudoternary phase diagrams employing emf, differential thermal analysis (DTA), and chemical analysis techniques. Isothermal sections of the systems Li_2Se-Se-(LiBr-RbBr) and Li_2Se-Se-(LiF-LiCl-LiI) are shown in Figs. 4-30 and 4-31. Further investigations were conducted on the solubilities of the molten salts and Li-Li_2S in each other, using emf measurements. The results elucidated some details of the pseudoternary phase diagrams and provided thermochemical data that resulted in a standard free energy of formation of -94.0 kcal/mol for solid Li_2Se at 360°C.

Several other phase-equilibrium studies that related to various applied programs were conducted by different individuals: CaI_2-$CaCl_2$, CaI_2-CaF_2, and CaI_2-$MgCl_2$ by Ram Sharma; LiI-KI and LiI-RbI by R. Sridhar and

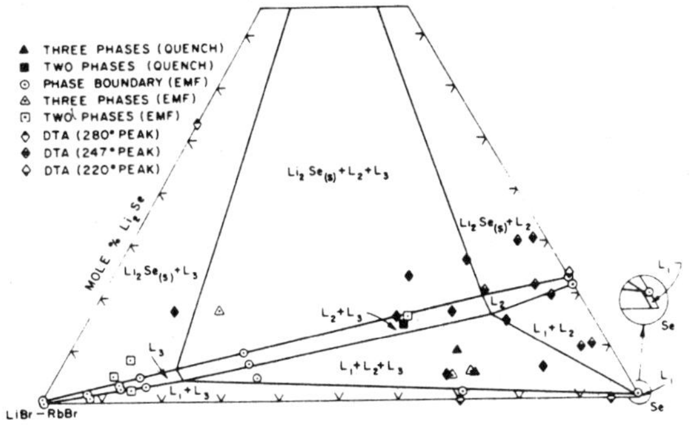

Fig. 4-30. Phase Diagram for Li_2Se-Se-(LiBr-RbBr) System at 360°C

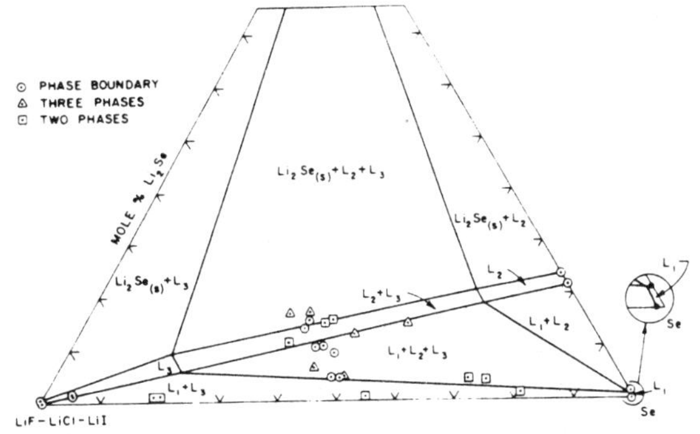

Fig. 4-31. Phase Diagram for Li_2Se-Se-(LiF-LiCl-LiI) System at 390°C

Carl Johnson; LiF-AlF_3, NaF-AlF_3, and Li_3AlF_6-Na_3AlF_6 by Marie-Louise Saboungi; the reciprocal ternary system Na-NaOH-Na_2O-NaH by Mike Myles and Fred Cafasso; and Li-LiH, Li-LiD, and Li-LiT by Ewald Veleckis, Vic Maroni, and Bob Yonco. Marie-Louise Saboungi, Jane Marr, and Milt Blander determined the solubility of FeS in LiCl-KCl, and Marie-Louise Saboungi and Al Martin measured the stability of "J-phase" ($LiKFe_{24}S_{26}Cl$) as a function of molten salt composition. Gene Kucera and Marie-Louise determined the solubility of MgO in CaO-$CaCl_2$.

MOLTEN SALT SPECTROSCOPY

Two phosphorus compounds, P_4S_7 and P_4Se_3, were of interest as positive electrode materials for electrochemical cells. Vic Maroni and Bob Schablaske investigated the structures of liquid and solid P_4S_7. Five of the sulfur atoms were present in P-S-P bridges, and two were in the terminal P-S bonds. Bands for the two stretching modes expected for the terminal P-S bonds were found in the solid, but not in the liquid. This observation suggested that the terminal sulfur atoms might have been dissociated in the molten state or might have become cross-linked with terminal atoms on adjacent molecules, a behavior proposed earlier for P_4S_{10} molecules. Apparently, the "P_4S_5" internal cage structure remained intact in the liquid state. Raman studies of solid P_4Se_3 produced results consistent with previous studies in the literature.

The Raman spectrum of the iodate ion, IO_3^-, in molten nitrates was examined as part of a program to explore possible methods of removing fission-product iodine from off-gases in scrub towers. Iodate is one of the major reaction products from the reaction of iodine with nitrates. Melts containing the iodate were prepared by heating 15 mol% I_2O_5-85 mol% MNO_3 mixtures, where M was Li, Na, Rb, Cs, Ag, or Tl. Three Raman bands attributable to IO_3^- were observed in all cases. Changes of the cation in the above sequence produced downward shifts in the symmetric stretching frequencies, which indicated that the cation polarization effects on the IO_3^- and NO_3^- ions were similar. Vic Maroni and Ellen Hathaway did that investigation.

Vic Maroni studied the structures of divalent metal ions in molten halide salts, including the systems LiCl-$MgCl_2$, LiCl-KCl-5 mol% ZnI_2, and LiCl-KCl-5 mol% $ZnCl_2$. The zinc iodide and chloride produced identical spectra consistent with the presence of $ZnCl_4^{2-}$, which indicated that the zinc was tied up as chloride. This result explained the high removal efficiency of iodine by LiCl-KCl cover salt used in zinc decladding and the low vapor pressure of ZnI_2 over the melt. In the case of magnesium, $MgCl_4^{2-}$ was shown to be the predominant monomeric species, and the complete Raman spectrum of $MgCl_4^{2-}$ was characterized.

Several other materials were examined by Raman infrared, emission-spectrum, and/or X-ray methods: tungstates, molybdates, $NaNH_2$, KCl-$SnCl_2$, KX-PbX_2 (where X= Cl and Br), and molten carbonates. Vic Maroni, John Bates, Paul Cunningham, and Ellen Hathaway were responsible for most of this work.

Although not strictly molten-salt chemistry, another area that received attention was the characterization of vapor species related to the above systems. George Papatheodorou played the major role in this work, but several others were involved. An interesting phenomenon that had been observed earlier was that trivalent metal halides such as the chlorides and bromides of aluminum, gallium, indium, and iron form volatile complexes with normally non-volatile halides, thereby enhancing their volatility. This effect was primarily of theoretical interest, but could have practical implications for chemical separations, crystal growth, or high-power gas lasers. The objectives of research in this area were to find new vapor complexes and to determine their structure, thermodynamic properties, and mode of formation.

Spectrophotometric and vapor-density measurements were made on the pairs $CoCl_2$-$InCl_3$, $PdCl_2$-$InCl_3$, $PdBr_2$-$AlBr_3$, and $SmCl_3$-$AlCl_3$ at temperatures and pressures of 180-825°C and 10-1200 kPa. Equilibrium constants were determined for the reactions, which were of the type

$$CoCl_2(s) + In_2Cl_6(g) \rightarrow CoIn_2Cl_8(g)$$

Raman and resonance Raman spectra of several gaseous species, including $CuIn_2Cl_8$, $CuAl_2Cl_8$, $CuCl_2$, and $InAlBr_4$, were determined. Another interesting and potentially useful aspect of these reactions is that they can be used to chlorinate transition metal and lanthanide oxides to the chlorides by addition of $AlCl_3$. The complex chloride vapor species can then be used to recover high-purity chlorides or to effect chemical separations. Solid uranium dioxide was chlorinated by this technique, using Al_2Cl_6 vapor and chlorine at 300°C to produce solid UCl_5, which was then transported as a UCl_5-Al_2Cl_6 vapor complex from which high-purity U_2Cl_{10} crystals were deposited. This work was then continued, using $FeCl_3$, $AlCl_3$, and $POCl_3$ as "carrier" gases, and research was aimed at a systematic characterization of the thermodynamics of the complexing reactions. Mixtures of UO_2 and ThO_2 were chlorinated, and the products were separated as UCl_3 and $ThCl_4$. Enhancement of the volatility of $ErCl_3$ by $AlCl_3$, $GaCl_3$, and $InCl_3$ was also examined.

THEORETICAL STUDIES

By the mid-1970s, a large body of data on phase equilibria in molten salts had been generated both in the basic and applied programs of the Division, and still more information was available from the literature. Nevertheless, much was still missing, especially for ternary and higher systems, in terms of phase diagrams and especially thermodynamic properties. Milt Blander and Marie-Louise Saboungi made a significant contribution toward improving the situation by invoking theoretical concepts to correlate the existing data and to predict the behavior when experimental data were missing or incomplete.

Their first effort, in 1974, was to use a thermodynamically self-consistent empirical extension of conformal ionic solution (CIS) theory to calculate liquidus phase diagrams for charge-asymmetric ternary reciprocal systems. Eight such systems were calculated, and the results agreed well with the limited experimental data that were available. Their conclusion was that the theory could be a significant aid in calculating phase diagrams *a priori*, in judging and correlating experimental data, and in minimizing the amount of experimental data needed to construct ternary phase diagrams. The conformal ionic solution theory was used to derive an expression for the total excess free energy of mixing for additive ternary systems. In some cases, the solution properties of ternary systems could be predicted solely from measurements of the subsidiary binaries.

The studies were then extended to incorporate molecular dynamics (MD) computer simulations for deriving the macroscopic and microscopic properties of classical charged particles from intermolecular potentials. Equations derived by this approach resulted in predictions of thermodynamic properties of molten-salt solutions that agreed closely with experimental measurements.

Marie-Louise Saboungi, Harold Schnyders, Mel Foster, and Milt Blander applied the CIS theory to the prediction of liquid-liquid miscibility gaps in reciprocal molten salt systems and found excellent agreement with experimental data for the system Na, Ag||NO_3, Br obtained by Blander and Gene Kucera. Thermodynamic calculations were used to determine the solubility products for sulfides of Ag, Cu, Cr, Fe, Mg, Mn, Na, Pb, Ti, and Zn, which were of interest to the battery program, and a computer-assisted analysis was conducted on the phase diagrams in the NaF-AlF_3-LiF-CaF_2-MgF_2-Al_2O_3 system, which is important in the aluminum industry.

Liquid Metal Chemistry

Most of the extensive studies that involved the chemistry of liquid metals in the 1970s were closely connected with the battery, fuel-reprocessing, sodium technology, and fusion programs and have been mentioned previously. The systematic investigations of solubilities and thermodynamic properties of various solutes in liquid metals that had been going on in Irv Johnson's group were concluded. Data were obtained by Irv, Bob Yonco, and John Heiberger on dilute solutions of lanthanum, cerium, and praseodymium in liquid zinc and in liquid cadmium. Also in Irv's group, Bob Yonco, John Heiberger, and Jack Fischer investigated solutions of neptunium in liquid magnesium, and Mike Krumpelt studied the thermodynamics of neptunium solutions in liquid cadmium. Ram Sharma made measurements on the systems magnesium-lead and magnesium-tin, and Mike Myles determined the phase diagram of the system copper-magnesium-calcium. These investigations, along with those performed earlier, provided a large, systematic data base on the solubilities and thermodynamic properties of liquid metal solutions, primarily those of zinc and cadmium.

Marie-Louise Saboungi and Jane Marr computed isothermal cross-sections of the ternary phase diagrams for the systems Al-Li-Mg and Ca-Li-Mg, using solution theory based on statistical mechanics and thermodynamics. Although some uncertainties about the subsidiary binaries and other input information precluded entirely unequivocal results, this type of analysis proved to be a powerful tool for generating complex phase diagrams with a minimum amount of experimental effort.

Further experiments and theoretical studies were directed to the solubilities of gases in liquid metals. This subject was of practical interest to the applied programs, particularly fusion energy and sodium technology, but was also intriguing from a basic viewpoint in that a better understanding of the unusual solvency properties of liquid metals might be gained by looking at the simplest of solutes, the noble gases. Ewald Veleckis, Bob Blomquist, Wally Kremsner, George Redding, and Bob Yonco had obtained experimental data on the solubilities of helium, argon, and xenon in liquid sodium. Their results were used in a theoretical study based on bond energies and surface-tension considerations. The general conclusion was that the first nearest neighbor interaction in a liquid metal is attractive (positive bond energy), and the second neighbor interaction is repulsive (negative bond energy).

Other Basic Research

NUCLEATION STUDIES

A small study was undertaken by Milt Blander on bubble nucleation in liquids to gain a better understanding of one mechanism for contact vapor explosions that occur in industry and could be hazards in nuclear reactors and the transmission of liquefied natural gas. The limits of superheat for 17 cryogenic hydrocarbons ranging in volatility from ethane (b.p., -88.2°C to cis 2-butene (b.p., 3.7°C) were measured and were about 0.885 of the critical temperature. The limits of superheat for mixtures were linear functions of the mole fractions.

An analysis of contact vapor explosions of hydrocarbons showed a definite correspondence between the explosiveness of hydrocarbons poured on a hot substrate and the conditions based on nucleation theory that closely approach the limit of superheat of the hydrocarbons. Jim Eberhart and Wally Kremsner were also involved in these studies.

METEORITIC STUDIES

In the early 1970s, Milt Blander instigated a combination of theoretical and experimental studies on the chemical processes involved in the origin of chronditic meteorites with the objective of developing clues as to the early history of the solar system and possible precursors leading to the formation of planets. This work in the Chemical Engineering Division was performed by Milt Blander and Gene Kucera. Also involved in the program were Lou Fuchs of the ANL Chemistry Division, H. N. Planner and Klaus Keil of the University of New Mexico, and L. S. Nelson and N. L. Richardson from Sandia Laboratories. The National Aeronautics and Space Administration (NASA) supported the work.

The concepts under study were based on the constrained equilibrium theory, which leads to predictions of the chemistry of meteorites and planet precursors. Chrondules are millimeter-sized spherical bodies found in the matrices of chrondites, the largest class of meteorites. They appear to be silicate droplets that crystallized rapidly within the chrondites, which crystallized 4.6 billion years ago. To investigate the way in which the spherulites were formed, experiments were performed on the crystallization of metastable, supercooled Mg_2SiO_4-SiO_2 mixtures.

Small beads of $MgSiO_3$ were formed by melting the material on loops of iridium wire, using a CO_2 laser in a furnace. Each bead was then held at a temperature ranging from 970 to 1346°C and then remelted by the focused laser beam. During recooling in the furnace, the metastable, supercooled droplets crystallized with a characteristic brightening caused by the heat of fusion. Some of the chrondules that formed a glass were devitrified by heating.

Thin, polished sections of the spherules were examined by optical microscopy and an electron microprobe. The microprobe analyses indicated that the SiO_2 content of the material had decreased from the original value of 60 wt% to values between 44 and 56 wt%. The only crystalline material detected by X-ray was forsterite (Mg_2SiO_4); the residual amorphous material was undoubtedly a glass. Based on the size of the spherules and the cooling conditions, the onset of crystallization was estimated to have been between 985 and 1346°C.

The micrographic examinations of the thin chrondule sections showed a variety of crystal morphologies. By combining this information with other data on furnace temperatures, nucleation temperatures, and the phase diagram of the $MgSiO_4$-SiO_2 system, Blander and his co-workers were able to correlate the crystal morphology, size, and texture with the degree of subcooling. A comparison of these spherulites with those in the chondrites showed many similarities, and indicated that those in the chrondites had crystallized at temperatures in excess of 600°C and, most likely, above 900°C.

These observations place limitations on current theories of the origin of chrondules. For example, if they were formed by direct condensation of metastable supercooled liquids, they could have formed in a nebula of solar composition at pressures as low as 10^{-3}-10^{-4} atm. If formed by impact melting during a collision of a "meteorite" with another body, the chrondules must have been in a thermal blanket at high temperatures after the collision.

A wealth of information was also obtained from a study of the carbonaceous Allenda chrondite. The mineralogy and textures of 20 calcium-aluminum-rich inclusions from this chrondite indicated that they had crystallized from a liquid formed either by impact or direct condensation.

Another group of researchers had found anomalous oxygen isotope compositions in the calcium-aluminum-rich inclusions in the Allende meteorite, and concluded that the inclusions could never have been liquid, contrary to their textural appearance and to the constrained equilibrium theory. When their isotopic data were normalized relative to a consistent standard, the data did not support their conclusions but were consistent with the liquid droplet model and constrained equilibrium theory.

THERMAL CONDUCTIVITY OF ASSOCIATING GASES

Associating gases were known to have enhanced thermal conductivities and heat capacities, which could be useful in heat-transfer or power cycles. Thermal conductivity measurements were performed on a number of associating vapors to (1) characterize the types (size) of polymers present and gain a fundamental understanding of the factors that influence bond strength, (2) determine the thermodynamics of the individual association reactions and, hence, the extent of such reactions, and (3) determine the magnitude of the thermal conductivity augmentation. In addition to the experimental work, *ab initio* molecular orbital calculations were performed on various species to gain a better understanding of the process and provide a basis for comparisons of the experimental results.

The initial thermal conductivity measurements were made on H_2O, D_2O, acetone, 2,2,2-trifluoroethanol, and acetic acid vapors, using a thick hot-wire cell in the ranges of 0.07-2.0 atm and 47-143°C. The pressure dependence of the thermal conductivities for all five substances indicated that dimers were the only significant polymeric species. The enthalpies of association ranged from -3.36 to -4.75 kcal/mol for the first four species, but the value for acetic acid was -15.55 kcal/mol. The strong association of acetic acid suggests that it, or a derivative such as CF_3COOH, might be a good candidate for a heat-transfer or power-cycle working fluid.

A new thermal conductivity apparatus was constructed for use at pressures up to 500 kPa and temperatures up to 250°C, and additional measurements were made on various fluoropropanols, a fluorobutanol, pyridine, trifluoroacetric acid, and a methanol-water mixture. In the methanol-water mixtures the results indicated the presence of a methanol-water dimer and trimer, methanol tetramer, and water dimer. Trifluoroacetic acid associated very strongly, with a high thermal conductivity similar to that of acetic acid. The methanol-water system with a 2:1 trimer species also showed strong association (an enthalpy of -10.43 kcal/mol). The work on associating gases was performed primarily by Dave Frurip and Larry Curtiss.

FLUID CATALYSIS

Fluid catalysis involves the catalysis of a desired reaction in a system of two immiscible liquids. Hal Feder and Jerry Rathke began a program on this subject in 1977. The reactions that were being investigated have potential application in fuel conversion, energy storage, or feedstock manufacture. The objective was to elucidate the kinetics whereby homogeneous organometallic catalysts activate small molecules such as CO and direct their subsequent reactions into specific paths. According to one of the progress reports, "intimate details connecting the molecular structures of the catalysts, the nature of the intermediate species, and the thermochemical kinetic parameters of each reaction step are sought by a combination of techniques including chemical, spectral, and isotopic studies at high-temperature and high-pressure conditions." Due to space considerations, the intimate details are left to the imagination of the reader. Particular attention was directed toward synthesis gas, which consists mainly of hydrogen and carbon monoxide and is a principal reactant in the industrial production of numerous organic chemicals. Because it is potentially available from any carbonaceous material (*e.g.*, coal, wastes, biomass), it could become a key chemical intermediate for production of fuels or organic chemicals from new sources. One of the areas of work in this program concerned the use of a soluble mononuclear metal complex to catalyze the reaction of carbon monoxide and hydrogen. A particularly exciting development was the discovery of a new system for homogeneous conversion of methanol to ethanol. This system has the potential practical advantages that the ethanol itself is not homologated to higher alcohols, and that the main byproduct is carbon dioxide rather than water, so the ethanol product is anhydrous.

One branch of this research effort was to explore new ways of using homogeneous catalysts that would enhance the separability of the catalysts and the reaction products. Three approaches were identified: (1) partition catalysts (two immiscible liquid phases with controlled partition of the catalyst between them), (2) supported catalysts (catalyst anchored by strong bonds to solid oxides or porous polymers), and (3) surfactant catalysis (catalytic centers attached by covalent bonds to lipophilic or lipophobic entities in such a way as to produce micelles or bilayer membranes).

ENVIRONMENTAL CHEMISTRY

In 1973, Paul Cunningham and Stan Johnson undertook a program of environmental studies related to atmospheric pollution, and, later on, Ben Holt and Romesh Kumar joined the effort. The Analytical Chemistry Laboratory was also a major contributor. The objectives were to gain an understanding of the chemistry of airborne particulate matter, with emphasis on the mechanisms and kinetics for the formation of primary (at the source) and secondary (in the atmosphere) sulfates, nitrates, and other pollutants. This effort involved the development of air-sampling devices, instruments, and analytical techniques to characterize atmospheric aerosols, in addition to laboratory experiments to investigate the formation of pollutants.

Infrared (IR) spectroscopy was selected as the primary method of analysis, and, in the initial studies, air samples were taken at ANL and in Chicago. The results indicated that the smaller particles (<1 μm dia) were composed mainly of ammonium sulfate [$(NH_4)_2SO_4$], whereas the larger particles consisted of carbonates, silicates, and surface nitrates with almost no ammonium sulfate. The sulfate was believed to have come from a heterogeneous oxidation of SO_2 to SO_3, which then reacted with ammonia.

Two possibilities were proposed for the oxidation of sulfur dioxide: (1) a homogeneous oxidation of SO_2 to SO_3, and (2) a heterogeneous reaction in which the SO_2 was first sorbed on a cloud droplet or aerosol particle and then oxidized to sulfate. Because the isotopic ratio of O-18/O-16 in the product was significantly different for these two oxidation mechanisms, it was possible to identify which mechanism was predominant in a particular case.

Air samples were taken by two methods, inertial impaction, in which the particles impinged on a collection surface, and filtration. Introduced in 1977 was IR Fourier-transform analysis, which greatly increased the sensitivity and accuracy of the analyses and permitted determinations of nitrate, acidic and neutral sulfate, ammonium ions, and total hydrocarbons on particle samples collected by impactors or filters. A multiple internal reflection technique was then adopted, which greatly increased the sensitivity and made possible very short sampling times for polluted industrial and urban air as well as power plant plumes. These advances in sampling and analytical techniques made it possible to analyze specimens directly on the collectors with no intermediate handling.

Laboratory experiments were conducted in a flow reactor system in which seed aerosol, reactive gases (O_3, SO_2, NO, NO_2, and NH_3), and ultraviolet light could be introduced in a controlled way to simulate typical or extreme conditions for particulate formation.

Much of the environmental work was done in cooperation with other organizations. In 1975, ANL field studies were partially supported by the Midwest Interstate Sulfur Transport and Transformation (MISTT) study. In the summer of 1975, samples were collected in St. Louis, MO, and in two rural locations in central Illinois. The total sulfate was lower in the rural areas, but the acidity was higher. In 1976, ERDA implemented the Multistate Atmospheric Power Production Pollution Study (MAP3S), which leads one to wonder if acronyms are raging out of control when their letters have to be exponentiated. Samples were collected at College Town, PA; Charlottesville, VA; Rockport, IN; and Uptown and Roquette, NY. In general, there were wide variations in sulfate activity, and the nitrate levels were very low. The aerosol acidities were higher in the Eastern states than in the Midwest.

Two items that were beginning to be addressed at the end of the 1970s were (1) a need for data on liquid-vapor phase equilibria in the system NO_x-SO_x-H_2O-NH_3 to gain a better understanding of nitrogen behavior in the atmosphere, and (2) the role of SiO_2, which is believed to be produced by the reoxidation of SiO resulting from the reduction of SiO_2 by hot carbon char during coal combustion. Over 25% of the submicron particulate in samples from the Southwest proved to be SiO_2.

The use of dolomite to control sulfur dioxide emissions in coal combustion plants was mentioned earlier. Some supporting basic chemistry studies in that area were conducted by the Environmental Chemistry Group. In the regeneration of spent dolomite, two competing reactions occur in the reduction of calcium sulfate by hydrogen, one producing calcium sulfide and the other, calcium oxide. The sulfide was the predominant product, but the oxide was more desirable from a process standpoint. Introduction of as little as 1% water vapor caused the reaction to produce only calcium oxide.

Studies of the dolomite-SO_2 reaction were undertaken to evaluate the effect of structural disorder in the dolomite on the reaction rate. Experimental methods for defining the magnitude of morphological disorder (porosity, surface area, *etc.*) were well established, but a quantitative determination of structural disorder of a reacting solid in terms of defects, dislocations, stacking disorder, cleavage, *etc.*, as compared to an ideal, well-ordered crystal was a difficult problem. X-ray diffraction and high-voltage electron microscopy were selected as possible investigative techniques. In studies of the conversion of dolomite to the half-calcined product, the conversion yield at 640°C was about 15% after several hours, but at 800°C the yield was close to 100% within 20 min. This result suggested that the higher temperature might have caused additional structural disorder, but X-ray could not confirm this due to the complexity of the system. The electron-diffraction and electron-microscopic studies of half-calcined dolomite and 50% sulfated material showed numerous disorder features. Further detailed basic studies on dolomite and its sulfated products were not planned because the complexity of the systems seemed to preclude definitive results.

Analytical Chemistry Laboratory

The creation of the Analytical Chemistry Laboratory (ACL) as a separate entity within CEN in 1971 was mentioned earlier in the discussion of the Division's organization. The stated functions of ACL at the time were the following:

1. To provide assistance to scientists and engineers in solving analytical problems that were outside their area of expertise or that required specialized equipment.
2. To perform routine analyses, thereby using the Laboratory's resources efficiently.
3. To conduct research and development studies, as required, to provide the ACL with the capabilities to meet the needs of the Laboratory's programs.

In the late 1970s, about two-thirds of the work by ACL was for CEN programs; the remainder was scattered widely among other ANL divisions, the main ones being the Chemistry and Materials Science Divisions. The workload was approximately 1,500 jobs per year, which required more than 16,000 separate analyses. The staff amounted to about 30 people.

In 1977, ACL received direct support from DOE for a development program on instruments to identify and measure organic environmental contaminants. The instrument that was developed consisted of a gas chromatograph, time-of-flight mass spectrometer, and computer system. In 1979, a plutonium analytical facility, which had been under construction for several years, became operational. This facility was used primarily for work involving plutonium-containing wastes.

The ACL had a major role in nearly all of the research and development programs in CEN, not only in performing analyses, but also serving in an advisory capacity, and, in some cases, lending personnel to work within particular programs. It is probably no exaggeration to say that few, if any of the major research and development programs in CEN could have been conducted successfully without the participation of the ACL.

Nuclear Safeguards

In June 1970, at the request of AEC's Office of Safeguards and Materials Management, the Division began work aimed at developing general design criteria for safeguarding Special Nuclear Materials (SNM) in nuclear facilities. Heading the group organized for this effort was Dr. Stephen Lawroski, the former Director of CEN. Members of the group included Leonard Link, John Loeding, Charles Bean, and Wally Seefeldt. The materials covered by SNM were plutonium, uranium-233, and uranium highly enriched with U-235. Some attention was also given to protection of facilities handling low-enrichment uranium.

In the mid-1970s, this work came under the aegis of the newly formed Nuclear Regulatory Commission (NRC), and the group began to assist the Office of Standards Development of NRC with the preparation of safeguards measures. These measures were designed to deter, prevent, or respond to (1) the unauthorized possession or use of significant quantities of nuclear materials through theft or diversion and (2) sabotage of nuclear facilities. The two basic safeguards measures were physical protection and materials control and accounting. The Division's program provided background information, rationale, and technical data for NRC criteria, guides, and reports in support of these safeguards objectives. The criteria were the basis for NRC's rules and regulations; the guides defined the rules and identified acceptable ways of complying with them; the reports helped the licensee to understand the regulations and guides.

The output of the group was a set of draft criteria, guides, and reports for use by NRC. The NRC prepared all final versions for publication. Included in the output of this group were the following:

1. Draft criteria for design and startup of processing plants.
2. Three guides on security clearance requirements.
3. Three guides on holdup of SNM in equipment, design and placement of physical barriers, and materials-protection contingency measures at processing and fabrication plants.
4. Reports on factors affecting theft vulnerability and on protection of hazardous materials such as by-product fission products and small quantities of SNM.
5. A study of the feasibility of dynamic inventory in fuel fabrication and processing plants and of potential methods for accomplishing such an inventory.
6. A report on bullet-resistant materials to aid licensees in design and construction of control rooms and central alarm stations for nuclear facilities, *e.g.*, reactors and processing plants.

This program came to an end in 1977 when the contract for performing it expired.

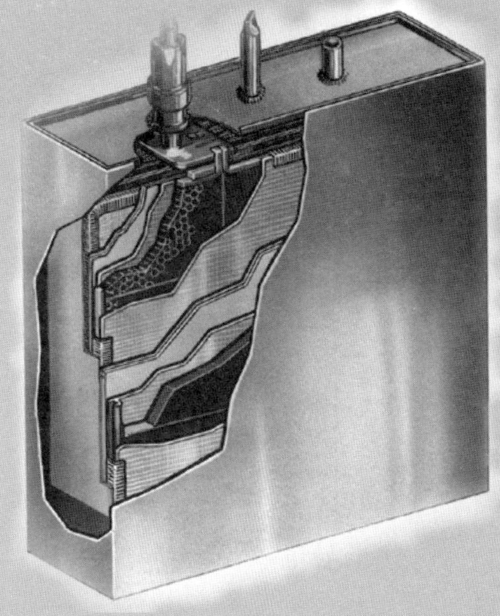

1980-1990: CMT – New Name, New Missions

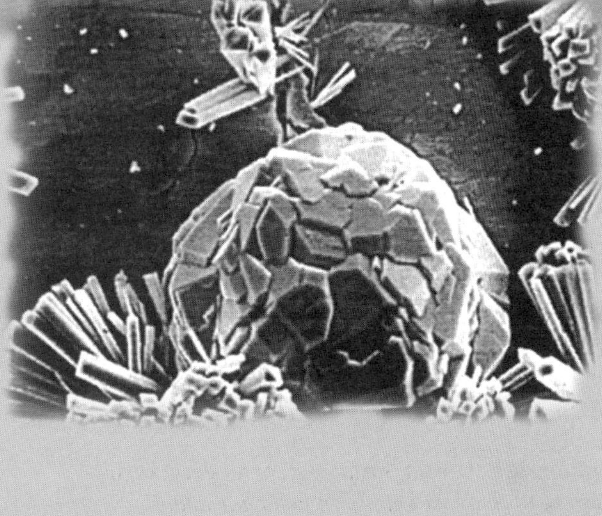

1980-1990

(top, left) Molecular models used in basic science studies of materials properties.

(top, right) Prismatic design developed for lithium/iron sulfide cell intended for electric-vehicle application.

(bottom) Bank of centrifugal contactors, which are used to remove transuranic elements from a waste stream through differences in the distribution of the waste components between immiscible aqueous and organic phases.

(bottom, right) Micrograph showing surface of waste glass sample after corrosion testing to study its long-term behavior under conditions relevant to geologic disposal.

5

1980-1990: CMT—New Name, New Missions

At the beginning of 1980, the morale of the nation was depressed by the continuing hostage situation in Iran and an unfavorable economy. President Jimmy Carter was soundly defeated by Ronald Reagan in the 1980 election. Reagan and his Vice President, George Bush, served two terms, and in 1988 Bush defeated Michael Dukakis in a race for the presidency.

Although the United States was not technically involved in a war in the 1980s, neither was it entirely at peace. Unrest was rampant throughout the world with assassinations, hostage taking, and hijacking, and the U.S. was involved in several limited military actions. The Cold War continued with Reagan taking a strongly anti-Communist stance (the "Evil Empire") until 1985 when he met with Mikhail Gorbachev, which relaxed the tension somewhat, and the Berlin Wall was opened in 1989, about seven months after a disastrous explosion and fire at the Chernobyl nuclear power plant. The Communist regime in the Soviet Union was beginning to crumble.

In the U.S., nuclear utilities were still constructing power plants and putting them in operation, but no orders were placed for new plants. Public confidence in nuclear power had been shaken by the accidents at Three Mile Island and Chernobyl. In addition to the perceived safety problems, the disposal of nuclear wastes became a major issue that was at least as much political as it was technical.

During the 1980s, scientific research and development was subjected to severe budget cuts, not only by the U.S. government, but also by industries attempting to improve their profitability. The decline in financial support was accompanied by decreasing enrollments of university students in science and engineering courses.

At Argonne, Dr. Alan Schriesheim replaced Dr. Massey as Laboratory Director on May 10, 1984. The Laboratory, by this time, had shifted its emphasis strongly toward environmental and alternative energy research in accordance with changes in the DOE priorities. Several new divisions or groups had been formed at Argonne to handle the new programs. Within the Division, the level of effort on high-temperature lithium alloy/metal sulfide batteries decreased significantly, and a greater emphasis was placed on nuclear technology. A major nuclear initiative was the Integral Fast Reactor (IFR), which combined an EBR-II-type metal-fueled reactor with an on-site electrometallurgical fuel recovery and recycle system.

Les Burris, who continued as Director of the Chemical Engineering Division until 1984, was succeeded by Dr. Martin Steindler, a chemist and long-term member of the organization. On July 19, 1982, the Chemical Engineering Division was renamed the "Chemical Technology Division" (CMT) to reflect more accurately the nature of its work. Another long-term ANL tradition ended in the 1980s when the emergency telephone number was changed from "Dial 13" to "911," which had become a national standard.

The 1980s was a difficult period for the CMT Division, as well as many other research organizations, because of shifting national priorities, funding uncertainties, and other pressures. During the course of the decade, the number of full-time personnel in the Division decreased by about 40%.

NATIONAL AFFAIRS

The people of the United States entered the 1980s with a feeling of despair about the hostages in Iran and worries over an economy gone awry. The inflation rate was 12%, unemployment in "rust belt" cities had reached 15-20%, and the prime lending rate was 15%. The country was sliding into a recession that persisted well into 1982. Morale was diminished further by Operation "Eagle Claw," a clandestine military attempt to rescue the hostages from the embassy in Teheran and remove them from Iran by airlift. Not only was the feasibility of the operation questionable, but poor planning and execution led to its complete failure with eight American fatalities and five injuries. The Iranians responded by dispersing the hostages to ensure that any further such attempts would be futile. The American public was heartsick over the situation, and people tied yellow ribbons around trees to express their frustration.

The faltering economy, the hostage situation, and several other factors set the stage for a decisive defeat of Jimmy Carter and Walter Mondale by Ronald Reagan and George Bush in the 1980 presidential election. Reagan came on with an optimistic attitude, confidence about the future, and a strong sense of patriotism. The hostages were released two days before his inauguration, which further enhanced his popularity. But the euphoria was short-lived. On March 30, only two months after his inauguration, Reagan was shot in an assassination attempt by John W. Hinckley, who was pronounced not guilty by reason of insanity. Reagan's press secretary, James Brady, and two secret service officers were also wounded. On August 3, the U.S. air controllers went on an illegal strike and were fired summarily on August 11 by Reagan's orders. This event had a chilling effect on the union movement throughout the U.S.

Reagan and Bush were reelected by a landslide vote in the 1984 contest against Walter Mondale and Geraldine Ferraro. In the 1988 election, George Bush and Dan Quayle defeated Michael Dukakis and Lloyd Bentson, giving the Republicans a 12-year tenancy in the White House.

In the early 1980s, the "cold war" with the Soviet Union was still the primary concern in American foreign policy. Reagan was strongly anti-Communist and felt that the U.S. should deal from a position of strength. One of his top priorities in office was to increase the size and power of the U.S. military forces. Terrorist activities and limited military actions seemed to have replaced all-out wars around the world. Argentina, in 1982, attacked the Falkland Islands and was defeated by the British, with the U.S. providing only logistic support. In 1983, a truck bomb killed 237 U.S. Marines, who were assigned to a peacekeeping mission in Beirut, Lebanon. Just six months earlier, 50 people had lost their lives in an attack on the American Embassy in Lebanon. In October of that same year, the United States mounted an attack on the island of Grenada over a concern that Cubans were building an airstrip there for shipment of communist arms to Nicaragua and El Salvador. In spite of overwhelming firepower, 133 Americans were killed or wounded. Reagan was almost obsessive about the leftist Sandinistas who had taken control of Nicaragua in 1985, and strongly supported the Contra guerrillas who were attempting to overthrow the Sandinista regime. A covert operation in which proceeds from arms sales to Iran were diverted to aid the Contras resulted in the so-called "Iran-Contra" affair, which surfaced in 1987 and culminated in congressional hearings that featured Marine Lt. Col. Oliver North, Jr. North and arms dealer Richard Secord were convicted, but North's conviction was later overturned in an Appeals Court. Reagan denied any knowledge of the operation and escaped unscathed, which was one of the reasons he was dubbed the "Teflon President." Also in 1987, Iraqi missiles killed 17 Americans on the

U.S.S. *Stark*. Iraq issued an apology. The Iranians, in an attempt to block oil shipments through the Gulf of Iran, laid mines and built gun emplacements, which provoked skirmishes with the U.S. Navy that erupted into a crisis when the cruiser, U.S.S. *Vincennes*, mistakenly shot down a civilian Iranian airliner with a loss of 290 lives. The last U.S. military action in the 1980s was an attack on Panama to free it from dictatorial control by strongman Gen. Manuel Noriega, who had been indicted by the U.S. on drug-dealing charges.

Numerous other instances of terrorism, such as plane hijackings and hostage taking, occurred throughout the 1980s. Three nuns and a lay worker were shot in El Salvador in 1981, and a South Korean Boeing 747 was shot down by the Russians, killing 269 passengers. Indira Gandhi, Prime Minister of India, was assassinated by Sikh terrorists, and anti-Sikh riots ensued with an estimated 1,000 deaths in 1984. In 1985, a TWA airliner was hijacked by two Shiite Moslems, the cruise ship *Achille Lauro* was hijacked with one American fatality, terrorists seized an Egyptian Boeing 737 with 59 killed in a rescue attempt, and nine were killed at airports in Rome and Vienna. In 1988, Soviet and Kuwaiti planes and an Aegean cruise ship were hijacked with further loss of life. The biggest blow to the U.S. was the bombing of a Pan Am Boeing 747 over Lockerbie, Scotland, which killed 259 passengers and nine on the ground. Besides the terrorism, concern was growing about the escalation of chemical and nuclear weapons. Israel bombed an Iraqi nuclear facility in 1981, Iraq bombed an Iranian nuclear plant in 1985, Libya used chemical weapons on Chad in 1987, and Iran and Iraq used chemical weapons on each other in their border war of 1988.

The 1980s were marred by several major accidents. In 1984, a toxic gas leak from a Union Carbide plant in Bhopal, India, resulted in 2,000 deaths and 150,000 injuries. A chartered plane carrying U.S. military personnel crashed in Newfoundland in 1985, killing 248 U.S. soldiers. In 1989, the Exxon tanker *Valdez* ran aground in Prince William Sound, Alaska, releasing 11 million barrels of crude oil. The decade was topped off by the Chernobyl disaster in 1989.

The Soviet Union underwent some wrenching changes, starting with the successive deaths of several leaders: Leonid Brezhnev in 1982, Yuri Andropov in 1984, and Konstantin Chernenko in 1985, who was replaced by Mikhail Gorbachev. Reagan, who had been adamant about any negotiations with the Communists up to that time, met with Gorbachev on November 19, 1985, and the tension began to thaw. It was during this period that Reagan said, "Mr. Gorbachev, tear down that wall!" Relations started to improve gradually, and the Berlin Wall was finally opened to the West in 1989.

The U.S. continued its space program, with emphasis on the Space Shuttles. Original launchings of the four shuttles were *Columbia* (4/12/81), *Challenger* (4/4/83), *Discovery* (8/30/84), and *Atlantis* (10/3/85). The shuttle flights became almost routine until January 28, 1986, when the *Challenger* exploded 72 seconds after lift-off, killing all seven aboard. This was a severe shock to the American public, compounded by the fact that one of the casualties was a school teacher, Christa McAuliffe, whose participation had been highly publicized, and many children witnessed the accident on television. The explosion was attributed to an O-ring failure in the propulsion system, and further shuttle flights were postponed until the *Discovery* was launched on September 9, 1988. An unmanned space probe, *Magellan*, which was launched on May 4, 1989, to explore Venus, used radar to generate a geological map of the planet.

Reagan and Bush both seemed to be basically neutral on the issue of nuclear power, which Carter had opposed. Carter had been more concerned about future energy sources for the U.S., and had encouraged research and development work on conservation and alternative technologies to meet the need. The national laboratories were funded to conduct research and development in these areas, with an emphasis on the use of subcontracts and other working agreements to transfer the new technologies to industry as they became available. Reagan had an opposing viewpoint on this matter. He felt that private industries should fulfill these needs themselves, using their own resources. An example of this approach was a reduction of annual government funding from about 155 to 35 million dollars for solar energy research. Reagan, intent on decreasing federal expenditures, brought in a Budget Director, David Stockman, who achieved notoriety by studying the budget in great detail and coming up with specific funding cuts for nearly every activity in the government. The one area of research and development funding that remained healthy was national defense, but about 80% of that money was diverted toward weaponry procurement rather than technology development.

Significant changes were also occurring in the economic and business institutions of the country. Strong emphasis was placed on bottom-line profits for the next quarter or two with little thought given to long-range strategies. Corporate takeovers and severe downsizing became prevalent. Once the oil crisis of 1979, the recession of 1982, and the high interest rates of the 1970s had subsided, the U.S. underwent a period of rapid economic expansion. The only major concerns were the ongoing rapid increases in the national debt and trade deficits. The emphasis on short-range profits at the expense of long-range planning by industry caused the downsizing or complete elimination of many corporate research laboratories. This was a particularly serious problem in view of foreign competition. In 1960 about 17% of the 47,000 U.S. patents were issued to foreign holders. By 1987, the figure was about 49% of the 88,000 U.S. patents. Although these figures are not necessarily a quantitative measure of the problem, they illustrate a trend. The fates of many corporate research laboratories were exemplified by a subcontract that the CMT high-temperature battery group had with the Carborundum Company's research organization to develop and supply boron nitride cloth separators for lithium alloy/iron sulfide cells. The Carborundum research organization had an excellent reputation and lived up to it by developing an innovative process for boron nitride fiber production and forming the fibers into a variety of felts and woven cloths on the scale required by the battery program. At this point, the Carborundum Company was taken over by the Kennecott Copper Corp., which absorbed most of the Carborundum research activities into the Kennecott Labs. Within a matter of months, Kennecott was taken over by Sohio and the process was repeated. Even the large, highly respected corporate research laboratories, where major discoveries such as the transistor were made and Nobel-prize winning basic research was done, became the Rodney Dangerfields of the corporate world. One exception to this trend was in the computer industry, where much highly competitive research was done, but much of that consisted of short-range efforts toward incremental extensions of the existing technology.

Similar observations could be made about U.S. universities and colleges. The public perception of science and technology seemed less favorable than in the past. The proportion of American university and college students majoring in the sciences had dropped by about 40% in the 1980s. The reason for that is not obvious, but self-interest would suggest to a student entering college that exiting with an MBA would be more likely to lead to a higher salary and better employment opportunities than would a Ph.D. in one of the sciences, and in considerably less time. The prevailing conventional wisdom that a person entering the work force will have to make six or seven career changes during his or her lifetime is a disincentive to spending the ten years or so beyond high school that are required to become a research scientist in some specialty. In the graduate schools, foreign students who constituted about 50% of the total in science and engineering took up much of the slack. Some of the foreign students returned to their native countries, while others remained permanently in the U.S. and have now become a large proportion of the nation's practicing scientists and engineers. The universities also felt the brunt of reduced federal funding in the 1980s. Research grants and contracts were reduced or eliminated in many cases, and funding for facilities was hard to find.

Thus, all three of the nation's major research and development institutions—industry, the universities, and the national laboratories—were feeling the financial crunch of the 1980s. Two major reductions in force, one near the end of 1981 and the other in mid-1982, decreased the manpower at Argonne from about 5200 to 4200.

Although the 1980s were not a particularly sanguine period in general, they had their bright spots, among which were "firsts" that were achieved by several women. Sandra Day O'Conner was the first woman appointed to the Supreme Court (1981), Dr. Sally K. Ride was the first woman astronaut in space (1983), Dr. Kathryn D. Sullivan was the first to walk in space (1984), Geraldine Ferraro was the first woman to run on the national ticket for Vice President of the United States (1984), and Hanna H. Gray was the first female president of the University of Chicago. Apparently President Gray's first introduction to the Argonne site was when she was to be honored at a reception on a hot, humid summer day in the cafeteria building. A tornado alert sounded, and everybody was herded into the basement where they sweltered while standing under water pipes dripping with condensation. Apparently, she found the situation somewhat amusing, at least in retrospect.

THE NUCLEAR INDUSTRY

By 1980, nuclear power had become a controversial issue, partly because of concerns such as safety, nuclear proliferation, and radioactive waste storage, and partly because of uncertainty about its economic competitiveness. People on both sides of the nuclear issue raised many complex questions, for which there were no unequivocal answers. Few if any plans were being made for new nuclear power plants after 1979, but many plants were still under construction and scheduled for startup in the 1980s and 1990s. The more significant ones are listed in Table 5-1.

The waning interest in nuclear power was reflected by the fact that several major universities in the U.S. were phasing out their nuclear engineering departments, which, in some cases, included research reactors.

THE DIVISION

Dr. Walter Massey continued in the position of Laboratory Director at Argonne until May 10, 1984, when he was succeeded by Dr. Alan Schriesheim. Schriesheim had earned a Ph.D. in physical organic chemistry at Pennsylvania State University and had been General Manager of the Technical Department in the Corporate Research Laboratories of the Exxon Research and Engineering Company. Dr. Schriesheim was the first Laboratory Director with a background primarily in industrial research, which was no doubt a major asset in the competitive environment of the 1980s. He had a purposeful, thoughtful manner and seemed very interested in receiving input from the staff. He maintained a schedule of breakfasts with senior technical personnel with no particular agenda, just to talk over whatever was on their minds. Another aspect of his administration that requires mention was

Table 5-1. Power Reactors Started up in the 1980s and 1990s

		Type	MW(e)	Start
Arkansas Nuclear One 2 (Russellville, AR)	Energy Operations, Inc.	PWR	836	3/80
North Anna 2 (Mineral, VA)	Virginia Power	PWR	897	12/80
Joseph M. Farley 2 (Dothan, AL)	Southern Nuclear Operating Co.	PWR	860	7/81
Sequoyah 1 (Soddy-Daisy, TN)	Tennessee Valley Authority	PWR	1148	7/81
Salem 2 (Salem, NJ)	Pub. Serv. Electric & Gas Co.	PWR	1106	10/81
McGuire 1 (Cornelius, NC)	Duke Power Co.	PWR	1129	12/81
Sequoyah 2 (Soddy-Daisy, TN)	Tennessee Valley Authority	PWR	1148	6/82
Susquehanna 1 (Berwick, PA)	Penn. Power & Light Co.	BWR	1082	6/83
San Onofre 2 (San Clemente, CA)	So. Cal, Ed. & San Diego G. & E.	PWR	1070	8/83
St. Lucie 2 (Hutchinson Island, FL)	Florida Power & Light Co.	PWR	839	8/83
LaSalle County 1 (Seneca, IL)	Commonwealth Edison Co.	BWR	1078	1/84
Virgil C. Summer (Parr, SC)	So. Carolina Electric & Gas Co.	PWR	885	1/84
McGuire 2 (Cornelius, NC)	Duke Power Co.	PWR	1129	3/84
San Onofre 3 (San Clemente, CA)	So. Cal. Ed. & San Diego G. & E.	PWR	1080	4/84
LaSalle County 2 (Seneca, IL)	Commonwealth Edison Co.	BWR	1078	10/84
WNP-2 (Richland, WA)	Wash. Pub. Power Supply Syst.	BWR	1157	12/84
Susquehanna 2 (Berwick, PA)	Penn. Power & Light Co.	BWR	1091	2/85
Callaway (Fulton, MO)	Union Electric Co.	PWR	1171	4/85
Diablo Canyon 1 (Avila Beach, CA)	Pacific Gas & Electric Co.	PWR	1073	5/85
Catawba 1 (Clover, SC)	Duke Power Co.	PWR	1129	6/85
Grand Gulf (Port Gibson, MS)	Energy Operations, Inc.	BWR	1173	7/85
Byron 1 (Byron, IL)	Commonwealth Edison Co.	PWR	1105	9/85
Waterford 3 (Taft, LA)	Energy Operations, Inc.	PWR	1075	9/85
Wolf Creek (Burlington, KS)	Wolf Creek Nucl. Oper. Corp.	PWR	1160	9/85
Palos Verde 1 (Wintersberg, AZ)	Arizona Public Service Co.	PWR	1221	1/86
Limerick 1 (Pottstown, PA)	PECO Energy Co.	BWR	1055	2/86
Diablo Canyon 2 (Avila Beach, CA)	Pacific Gas & Electric Co.	PWR	1087	3/86
Millstone 3 (Waterford, CT)	Northeast Utilities	PWR	1149	4/86
River Bend (St. Francisville, LA)	Energy Operations, Inc.	BWR	936	6/86
Catawba 2 (Clover, SC)	Duke Power Co.	PWR	1129	8/86
Palos Verde 2 (Wintersberg, AZ)	Arizona Public Service Co.	PWR	1221	9/86
Hope Creek (Salem, NJ)	Pub. Serv. Electric & Gas Co.	BWR	1031	12/86
Clinton (Clinton, IL)	Illinois Power Co.	BWR	930	4/87
Alvin W. Vogtle 1 (Waynesboro, GA)	Georgia Power Co.	PWR	1162	5/87
Shearon Harris (New Hill, NC)	Carolina Power & Light Co.	PWR	860	5/87
Byron 2 (Byron, IL)	Commonwealth Edison Co.	PWR	1105	8/87
Beaver Valley 2 (Shippingport, PA)	Duquesne Light Co.	PWR	833	11/87
Perry 1 (North Perry, OH)	Cleveland Electric Illum, Co.	BWR	1205	11/87
Palos Verde 3 (Wintersberg, AZ)	Arizona Public Service Co.	PWR	1221	1/88
Fermi 2 (Newport, MI)	Detroit Edison Co.	BWR	810	1/88
Nine Mile Point 2 (Scriba, NY)	Niagara Mohawk Power Co.	BWR	1080	4/88
Braidwood 1 (Braidwood, IL)	Commonwealth Edison Co.	PWR	1120	7/88

South Texas Project 1 (Palacios, TX)	Houston Lighting & Power Co.	PWR	1250	8/88
Braidwood 2 (Braidwood, IL)	Commonwealth Edison Co.	PWR	1120	10/88
Alvin W. Vogtle 2 (Waynesboro, GA)	Georgia Power Co.	PWR	1162	5/89
South Texas Project 2 (Palacios, TX)	Houston Lighting & Power Co.	PWR	1250	6/89
Limerick 2 (Pottstown, (PA)	PECO Energy Co.	BWR	1055	1/90
Seabrook (Seabrook, NH)	No. Atlantic Energy Serv. Co.	PWR	1150	7/90
Comanche Peak 1 (Glen Rose, TX)	Texas Utilities Co.	PWR	1150	8/90
Comanche Peak 2 (Glen Peak, TX)	Texas Utilities Co.	PWR	1150	8/93

the activities of his wife Beatrice, who was responsible for many improvements in the overall appearance of the Laboratory. The cafeteria was remodeled with various enhancements, including trees among the tables, and the lobbies and other public areas of many of the buildings were redecorated. By the time the numerous improvements in the appearance of the buildings and grounds were completed, just about all of the last vestiges of the "military base" appearance of Argonne were gone, and the site had more the appearance of a university or corporate research campus.

The Associate Laboratory Directors most concerned with the programs in the Chemical Engineering Division were Dr. John J. Roberts (Energy and Environmental Technology) and Dr. Charles E. Till (Engineering Research and Development).

As of 1980, the upper management of the Chemical Engineering Division comprised the following:

Division Director:	Leslie Burris
Deputy Division Director:	Donald S. Webster
Associate Division Directors:	Duane L. Barney Fred A. Cafasso Martin J. Steindler
Assistant Division Directors:	Ronald L. Breyne Herbert L. Brown, Jr.

Paul Nelson and Al Jonke do not appear on this list because they were in Associate Laboratory Directors' organizations at the time, but both returned to CMT, whereupon Paul replaced Don Webster, who had retired, as the Deputy Division Director, and Al returned to his position as an Associate Division Director. Because of the many programmatic changes in the 1980s and the numerous reorganizations to accommodate these changes, Section Heads, Group Leaders, Program Managers, *etc*., are not listed here, but are mentioned in the section on technical programs.

When the Chemical Engineering Division was formed in 1948, its work consisted largely of engineering research and development work on solvent-extraction processes for the recovery of spent reactor fuels. By 1980, the Division had undergone two major stages of diversification from that original mission. The first stage was a move into other nuclear areas such as nuclear safety, materials research, and basic research on the properties of reactor materials. The second stage involved non-nuclear research and development of all types of energy sources, a mission that ERDA had assigned to all the multipurpose national laboratories. The Division's activities at the time fell into three general categories: (1) process and equipment development through engineering-scale demonstrations, (2) applied chemistry, and (3) basic research in a variety of areas, some of which were related to the applied technology programs.

The distribution of funding in 1981 was approximately as follows: electrochemical technology (batteries and fuel cells), 62%; fusion power technology, 15%; coal combustion technology, 15%; basic energy sciences, 8%.

In view of this diversification, the time was ripe for the Division to adopt a more descriptive name, and, after a period of considerable agonizing by Les Burris and others, the name "Chemical Technology Division" was selected. There was some concern over the fact that a long-standing and highly respected group at Oak Ridge also had that name, but that did not seem to create a problem. Consequently, on July 19, 1982, CEN became CMT. The change was announced in a memorandum from John Roberts, which is reproduced in Fig. 5-2. A few people grumbled that the new name was not compatible with basic research, but it stuck with no apparent ill effects.

In 1984, Les Burris, after eleven years as Division Director, decided it was time to step down, and he was succeeded by Dr. Martin J. Steindler, who had started to work in the Division in the early 1950s after receiving a Ph.D. from the University of Chicago. Martin, after being initiated into the Division by working for Hal Feder, had been the lead player in much of the work on fluoride volatility processes and was responsible later for most of the nuclear work after the fuel-reprocessing programs had ended. It is an interesting fact that, even up to the present time, each engineer who has been the Division Director was succeeded by a chemist, and each chemist has been succeeded by an engineer. (Jim Battles is a Metallurgical Engineer.)

Fig. 5-1. Martin Steindler

As indicated earlier, budget cuts required a significant reduction in force at CMT from about 280 in 1980 to approximately 170 in 1989. One factor that alleviated the situation somewhat was that most of the staff members who had joined the Division in the late 1940s and early 1950s were reaching, or at least approaching, retirement age. The Laboratory offered attractive early retirement packages to those who were in the appropriate age brackets. There was an exodus of several upper management people from the Division. Don Webster retired in 1982 and Al Jonke in 1985. Duane Barney transferred to Washington, DC, on special assignment as an ANL employee. Fred Cafasso, in 1984, left the Division to accept an upper management position in the Chemistry Division. Numerous other people in the Division either retired or found work elsewhere, some in other divisions at Argonne. The tight funding situation in the Division was made worse by the fixed costs such as building maintenance, accounting, library, and utilities, which could not be reduced significantly.

The management situation in the Division tended to be fluid, with the various programs starting and stopping and several key people leaving. After the dust had settled, the upper management of the Division ended up as follows in 1989:

Division Director:	Martin J. Steindler
Deputy Division Director:	Paul A. Nelson

ARGONNE NATIONAL LABORATORY **ANNOUNCEMENT**

July 19, 1982

To: All Employees

From: Walter E. Massey

Subject: Chemical Technology Division - Name Change

Effective immediately, the Chemical Engineering Division name will be changed to the Chemical Technology Division. The management appointments, the organizational structure, and the scope will remain the same.

There are several reasons for changing the division's name. The first is that the division's major activities, which are process and equipment development through engineering scale demonstrations, applied chemistry, and basic chemical research in areas related to the division's applied technology programs, are better characterized by the new division name than the old name. The name "Chemical Engineering" implies that the division's activity is limited to process development and demonstration, and that is not the case. Finally, the staff includes physical chemists as well as chemical engineers, and the change of name will help in recruiting the former, without hindering recruiting of the latter.

WEM:pam

Fig. 5-2. A New Name for the Division

Associate Division Directors: David W. Green
Carl E. Johnson

Assistant Division Directors:
Ronald L. Breyne
Herbert L. Brown, Jr.

Responsibility for the Technical Editing Group, which had been under Gwen Kesser, was given to Joe Harmon in 1980. Those who served under Joe as technical editors at various times during the 1980s include Susan Barr, Dave Hamrin, Richard Keener, Penelope Raptis, and John Simmons. Maria Contos and Valerie Gaines were secretaries. Maria and Joe, in cooperation with the Computer Applications Group, instituted a system for computerized record keeping of all the Division's publications. Without that system, it would have been very difficult, if not impossible, to have written this volume. As of

1989, the Technical Editing Group consisted only of Joe and Maria. Unfortunately, the volume of work did not decrease in proportion to the size of the group.

The other supporting groups in the Division remained intact for the most part, albeit at reduced levels. As with the research and development groups, the workload was eased somewhat by the use of part-time retirees or other employees hired on Special Term Appointments (STAs).

Because the secretaries and clerks have an essential role in the operation of the Division, but do not normally receive personal recognition in publications, we have done our best to list in Table 5-2 all those who were in the Division at some time in the 1980s.

Helen Hill managed the Division office throughout the decade as the Director's secretary, first under Les Burris, and then under Martin Steindler. Marilyn Osborn was the Deputy Director's secretary, under Don Webster, and then Paul Nelson. Stefanie Chapman worked for Fred Cafasso, Lucille Jensen for Martin Steindler, Sherry Grisko for Duane Barney, Doris Worthington for Al Jonke, Judy Popik for John Ackerman, Cindy Wesolowski for Carl Johnson, Jan Steinquist for Dave Green, and Loretta Cescato for Jim Battles. Jan Muller was given additional administrative responsibilities in Herb Brown's office. Anne Melton, who had worked for Everett Proud and then Ron Breyne, retired in 1981. Others in Ron's office were Sofia Napora, Maria Colunga, Laurie Malak, Cindy Mau, Debbie Morgan, Teresa Clark, Debbie O'Rourke, Guadalupe Franchini, Heidi Rudolph, and Laurie Ammerpohl.

Table 5-2. Secretaries and Clerks in the CMT Division in the 1980s

Lauren Ambrose	Guadalupe Franchini	Marilyn Osborn
Laurie Ammerpohl	Dorothea Gabrovic	Peggy (McBride) Parks
Margaret Anderson	Valerie Gaines	Judith Popik
Vanessa Arzate	Sherry Grisko	Roberta Riel
Lilia (Mojica) Barbosa	Guadalupe Guzman	Heidi Rudolph
Jacqueline Bertoletti	Marian (Gardner) Harkins	Kathleen Shields
Alice Birmingham	Helen Hill	Janet Steinquist
Marlene Bukowski	Lucille Jensen	Carol Stogsdill
Mary Burke	Kerri Kilian	Virginia Strezo
Susan Bykowski	Wendy (Donnelly) Lamb	Wendy (Zanelli) Strle
Laurie (Malak) Carbaugh	Dawn Landis	Audrey Taylor
Janet Carothers	Teresa Lang	Donna Tipton
Loretta Cescato	Lee Legerski	Sandra Tummillo
Stefanie Chapman	Charlotte LeGrand	J. Van Dahm
Teresa Clark	Stephanie (Malm) Malak	Denise Voss
Maria Colunga	Cindy (Mau) Meurer	Coleen Weeks
Maria Contos	Deborah Morgan	Cynthia Wesolowski
Valerie Crudup	Janice (Lewin) Muller	Cynthia Wilkinson
Nancy Deloria	Sofia (Gawenda) Napora	Kathy Willis
Linda Ellis	Deborah O'Rourke	Doris (Michalek) Worthington
Elaine Estand		

The Division did not escape Mrs. Schriesheim's aggressive beautification program, and it was probably a good thing. Building 205, after 50 years of hard use, was ready for a cosmetic upgrade, and the Division management was agreeable to the idea. In the public areas, lighting was modernized, floors were retiled, walls were painted, and broken plaster was repaired. Some of those things had been done before, but not all at once. The overall effect was impressive; one had the feeling of being in a brand new building. A major item in the upgrading was the Division library, which had been relocated to the east end of L-Wing in the 1960s when Dr. Vogel moved the Division Office to A-Wing. The library, which occupies about 2,000 square feet, and had undergone only a few changes since that time, was completely revamped in 1985-86 to satisfy Mrs. Schriesheim's decorating requirements and Martin Steindler's insistence on functionality; it is now an efficient, attractive place to work.

The morale of the personnel in the 1980s was obviously affected by the budget cuts and layoffs, but perhaps even more so by increasing micromanagement of the programs, especially by DOE. There were more and more program reviews, on-site inspections, and reporting requirements, as well as hundreds of new regulations relating to safety, quality assurance, personnel matters, environmental concerns, and other areas. At the same time, DOE was requesting many technical studies and evaluations that had to be done with existing personnel, often on short notice. Few would argue with the good intent of most of the individual items, but, taken as a whole, they required large amounts of additional paperwork and time. The increasing tendency toward micromanagement of the national laboratories eventually reached a level where it became a national issue.

One rather plaintive comment heard frequently from scientists and engineers during this period was that "research just isn't fun anymore." The threat of program funding cuts and layoffs, along with the heavy monitoring, and often, criticism of their work, seemed to suggest that their efforts were unappreciated and probably unworthy of support. The Laboratory Directors and Division Management worked hard to attract new programs and were successful enough that a bad situation did not become a lot worse. By the end of the 1980s, things had stabilized somewhat with the Division reduced in size.

With respect to computers, the 1980s was an exciting period for members of the Division, as well as the rest of the world. IBM (International Business Machines Corp.) came out with the first desktop personal computer ("PC") in 1981, which was immediately cloned by several other manufacturers. The PC machines were based on MS-DOS (the Microsoft Disk Operating System) up until the mid-1990s. The Apple Computer Company developed a competitive line called Macintosh, which used a graphical interface with icons, which many people found easier to use than the PC machines. Apple computers were more popular for use in schools due the their ease of operation, and Apple dominated that niche of the market for many years. On the other hand, more software was becoming available for the PCs owing largely to Microsoft's aggressive marketing. The two systems were not compatible, and each had its proponents, sometimes with a nearly religious fervor. It was a bit reminiscent of the battle between "Beta" and "VHS" when videocassette tape recorders first came on the market. Both PCs and Macs found their place in the Division as more desktop computers were acquired.

By 1980, the Division had experienced a major growth in laboratory automation, data acquisition, instrument control, and data reduction. Much of the commercial equipment that was being used was packaged with internal computers. Most of the individual systems used in the experimental work were of the PDP-11 family. To centralize these activities, a VAX 11/780 was procured in December 1980. The VAX and PDP units are manufactured by the Digital Equipment Corp. A PDP-11/44 served as a "front end" for connection of computers in the laboratory programs to the VAX, and the VAX provided for direct connection of individual terminals to the ANL Central Computing Facility. The VAX and associated equipment were installed in a climate-controlled enclosure on the service floor under A-Wing.

In 1981, the Computer Group began to explore office automation for administrative functions such as personnel, budget, and procurement, and in 1982 the system became available for almost everybody's use. By the end of the year, 132 terminals were in use in the laboratories and offices. Software packages became available, including (1) MASS-11 for word processing, (2) DIGICALC for spreadsheet work, (3) T_EX for typesetting purposes, (4) BASIC and PL-1 language compilers, and (5) electronic mail.

At this point, nearly everybody had access to terminals and user-friendly software. MASS-11 on the computer gave the secretaries many more options than had been available on the single-purpose word processors. Some staff members began writing report drafts, memos, *etc.*, on the computer and giving them to a typist in electronic form for proper formatting and corrections. The final step had been taken in the transition from typewriter and carbon paper to computer.

By the end of 1983, the Computer Group consisted of seven CMT staff members and three full-time consultants, and the VAX was being used by more than 250 people. That same year, the publications file program was completed. To meet the needs of the Nuclear Fusion Group, the VAX was connected to a Cray mainframe at Lawrence Livermore National Laboratory. As the system continued to grow, a divisional local area Ethernet system was installed to provide more bandwidth. Connections were extended to thousands of other organizations and the Internet.

Programs such as Proof of Breeding, the National Battery Test Laboratory, TRUEX work, and the high-temperature thermodynamic studies of materials were heavy computer users, and a MicroVAX-II was acquired in 1985 for the high-temperature work. In 1988, the 7-year-old VAX was replaced with a VAX 6220, which had nearly three times the capacity of the older system. Toward the end of the 1980s, the computer terminals began to be replaced by desktop personal computers, which use the VAX as a disk, file, and/or print server. A block diagram of the CMT computing facilities as of 1989 is shown in Fig. 5-3. From about 1987 on, laser printers were used almost exclusively, along with Adobe Post-Script software. Very high-resolution output could be obtained with typesetting software.

Throughout the decade, the Computer Group, under Steve Gabelnick's leadership, was involved in a wide variety of activities:

1. Laboratory data acquisition and control
2. Computer modeling and simulation
3. Post-analysis of experiments
4. Graphic interpretation
5. Division information management and database development
6. Operating system maintenance
7. Computer networking
8. Equipment procurement
9. Advisory, educational, and consulting services

Members of the Computer Group included the following: Steve Gabelnick, Glen Chapman, Jacqueline Copple, Msiri Kazadi, Bob Kessie, Joe Kulaga, Bob Land, Carl Meyer, and John Osudar.

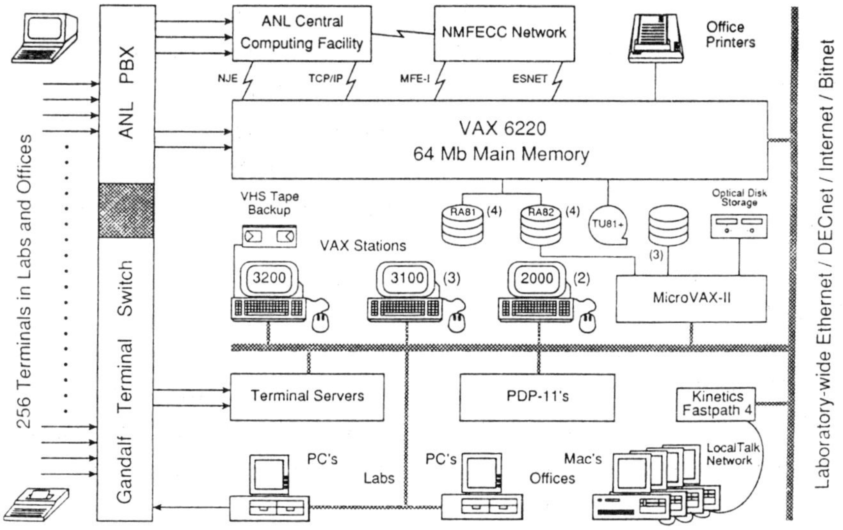

Fig. 5-3. Block Diagram of CMT Computing Facilities

TECHNICAL PROGRAMS

As the Division entered the new decade, the technical programs remained essentially unchanged for a year or two until the Reagan budget cuts and new DOE priorities began to take their toll. In 1981, the DOE made a decision to transfer overall management responsibility for the high-temperature battery program from CMT to Sandia National Laboratories and Lawrence Berkeley Laboratory and to reduce the funding severely for the Li-Al/FeS battery research and development at ANL. At the same time, there was a more gradual, but steady reduction of effort in the ANL and contractors' work on aqueous battery systems. The fuel-cell program remained relatively healthy, and the major emphasis was shifted from molten carbonate to solid oxide fuel cells. The remaining solar energy work was transferred out of the Division in 1982. The fossil fuel program and magnetohydrodynamic studies continued throughout the 1980s, as did the applied physical chemistry and basic research work.

Several programs were initiated on hazardous waste management and various aspects of nuclear technology, including separations of reactor fuel constituents, high-level active waste repositories, an electrometallurgical process for recycling fuel from the proposed Integral Fast Reactor (IFR), and methods for recovering plutonium residues from a diversity of other materials.

In the 1980s, several major trends were evident in the Division's programs. The overall budget and personnel were reduced drastically. There was a significant shift of effort from electrochemical programs to nuclear technology. The individual programs tended to be smaller, with less emphasis on industrial contractors, outside management activities, and end-use demonstrations, and more emphasis on research and development.

Lithium-Alloy/Iron Sulfide Batteries

The primary effort in the advanced battery work was on the lithium-aluminum/iron monosulfide (Li-Al/FeS) cell because of its performance and reliability. Cells using iron disulfide (FeS_2) instead of the monosulfide were also of interest due to their potential for higher performance, and the emphasis had begun to shift toward that system by the end of the decade.

The failure of the 20-kWh Mark I battery module in 1979 brought about a revised strategy and schedule for the development of lithium alloy/iron sulfide cells, as shown in Table 5-3.

Because a major objective of this program was to transfer the technology to industry, approximately 60% of the funding was expended through industrial subcontracts. Two major cell and battery development contracts were with battery manufacturing firms, Eagle-Picher Industries, Inc., and Gould Inc. Other related contracts were negotiated with the Carborundum Division of the Kennecott Copper Corp., General Motors Research Laboratories, Rockwell International, the Institute of Gas Technology, and the University of Florida.

ELECTRODE CHEMISTRY

As of 1980, previous work had shown that iron monosulfide (FeS) and iron disulfide (FeS_2) were both promising materials for positive electrodes, and a lithium-aluminum alloy containing about 50 at.% lithium (LiAl) was the favored negative electrode material, although there was also some interest in lithium-silicon alloys. A LiAl/FeS cell undergoes the following overall reaction:

$$2\ LiAl + FeS \rightarrow Li_2S + Fe + 2\ Al$$

which exhibits a voltage plateau at about 1.33 V and a theoretical specific energy of about 460 Wh/kg of active materials. A LiAl/FeS_2 cell discharges by the overall reactions:

$$2\ LiAl + FeS_2 \rightarrow Li_2S + FeS + 2\ Al$$
$$2\ LiAl + FeS \rightarrow Li_2S + Fe + 2\ Al$$

with voltage plateaus at 1.67 and 1.33 V, respectively, and a theoretical specific energy of approximately 680 Wh/kg. In the case of the lithium-silicon electrode, which is normally started at a composition corresponding to $Li_{3.25}Si$, a series of compounds

Table 5-3. Development Schedule for Lithium/Iron Sulfide Batteries

Year	1983	1986	Post 1990
System	Li-Alloy/FeS	Li-Alloy/FeS	Li-Alloy/FeS_2
Specific Energy,[a] Wh/kg			
Cell	100	100-125	160-225
Battery		80-100	130-180
Peak Specific Power,[b] W/kg			
Cell	125	150-250	240-350
Battery		120-160	190-360
Lifetime, Deep Discharges[c]			
Cell	500	1000	1200
Battery		800	1000
Battery Heat Loss, W		75-150	50-100

[a] At a 4-hr discharge rate.

[b] Peak power sustainable for 20 s at 0 to 50% state of discharge.

[c] End of life defined as 20% capacity loss or a coulombic efficiency below 95%.

is formed as lithium is removed during discharge:

$$Li_{3.25}Si \rightarrow Li_{2.33}Si \rightarrow Li_{1.71}Si \rightarrow Si$$

Cells with FeS or FeS_2 positive electrodes and $Li_{3.25}Si$ negative electrodes have specific energies about 10% higher than those with LiAl negative electrodes. Based on these considerations, the negative electrode work was broadened to include Li-Si and Li-Si-Al as well as Li-Al as the active material.

A special test cell was designed so as to apply current perpendicular to the electrode face, causing the current distribution to be one-dimensional, varying only through the thickness of the electrode under test. This one-dimensional-electrode (ODE) cell was used for studies of galvanostatic (constant-current) cycling, current interruption, electrode-potential relaxation, and high-intensity current pulsing. The current-interruption data provided values for the area specific resistance (ASR) of the electrode, which is in units of ohm-cm^2, and is a useful value in the engineering design of practical cells.

Studies were continued on FeS and FeS_2 positive electrodes. The discharge process for the FeS electrode, which appears simple at first glance,

$$2\ Li + FeS \rightarrow Li_2S + Fe$$

turned out to be much more complex when it was examined in detail. Several chemical and electrochemical reactions are involved, but these are the overall reaction paths:

$$FeS \rightarrow J\ phase\ (LiK_6Fe_{24}S_{26}Cl)$$
$$J\ phase \rightarrow X\ phase\ (Li_2FeS_2)$$
$$J\ phase \rightarrow Li_2S$$
$$FeS \rightarrow Li_2S$$
$$X\ phase \rightarrow Li_2S$$

The "J" and "X" phase designations, which were acquired through some arcane process during their characterization by X-ray diffraction, are used extensively in the literature on lithium/iron sulfide batteries. Emf studies of these reactions resulted in a fairly good thermodynamic characterization of the FeS discharge process, and, together with literature data, produced values of -102.5 and -99.9 kcal/mol at 700 and 800 K, respectively, for the free energy of formation of Li_2S.

Current-interruption techniques were used to investigate voltage-loss processes (mainly electrode polarization) that were limiting cell performance. The results proved valuable in mathematical modeling work and in cell design optimization.

A common problem in conducting electrochemical studies on systems involving molten salt electrolytes is the lack of a suitable reference electrode. Laszlo Redey and Don Vissers found that the Ni/Ni_2S_3 couple makes an excellent reference electrode for molten halide salt systems. Tests of the electrode showed a linear response of the emf to the logarithm of the sulfide ion concentration, and this Nernstian response was in excellent agreement with that predicted theoretically for a two-electron reaction. The usefulness of this electrode was improved further by incorporating a thermocouple to provide corrections to the emf during temperature changes. The potential versatility of this electrode for various applications was recognized nationally through an IR-100 Award by *Industrial Research Magazine.*

Studies were also conducted on FeS_2 positive electrodes, which discharge by the following sequence of phase changes:

$$FeS_2 \rightarrow Li_3Fe_2S_4 \rightarrow Li_{2+x}Fe_{1-x}S_2$$
$$\rightarrow Fe_{1-x}S \rightarrow Li_2FeS_2 \rightarrow Li_2S + Fe$$

Emf measurements were made on the first reaction ($FeS_2 \rightarrow Li_3Fe_2S_4$), using a specially designed cell, with the following result:

$$E\ (V) = 1.5558 + 0.0004785\ T\ (°C)$$

An interesting quirk of this reaction is the positive temperature coefficient, which indicates an entropic heat absorption during discharge that could simplify heat management of a practical battery. Additional solubility data were obtained on the Li-Fe-S phases formed in the FeS_2 electrode.

The ODE cell was used to conduct extensive measurements on various types of electrodes (Li-Si, Li-Al, Li-Al-Si, FeS, FeS_2, NiS, and others) and several different electrolyte compositions. Nearly all of the early work had been done with LiCl-41.2 mol% KCl molten salt electrolyte, which is a eutectic composition that melts at 352°C. This electrolyte has two serious shortcomings. (1) Because of its high melting temperature, it requires high cell operating temperatures approaching 400°C or higher. (2) Its KCl content is sufficiently high to cause the formation of J-phase, which inhibits the kinetics of the FeS discharge reaction in a cell. Formation of J-phase can be suppressed in two ways. One way is to increase the temperature, and the other is to decrease the potassium ion content of the electrolyte. Cells were operated with electrolytes of various compositions in an effort to improve their electrochemical performance and/or decrease the operating temperature. An experimental investigation to seek molten salt systems having lower melting temperatures revealed several combinations of alkali metal halides (*e.g.*, LiF, LiCl, LiBr, KCl, KBr, CsCl, CsBr) that had adequately high lithium ion concentrations along with melting temperatures in the general range of about 235 to 350°C. Consideration had to be given, however, to other factors such as solubilities of the active materials, stability toward reduction by the Li-Al electrode, and corrosion or other interactions with the cell hardware.

MATERIALS ENGINEERING

In a high-performance electrochemical cell, the electrodes contain highly aggressive compounds, which place severe limitations on the choice of materials for cell components such as the separator, current collectors, housing, insulators, and other components. Chemical compatibility and mechanical stability were evaluated by two methods: static immersion tests in mixtures that simulated the positive and negative electrodes, *i.e.*, FeS/LiCl-KCl and LiAl/LiCl-KCl, and tests in 50-Ah cells. Nickel proved to be very corrosion resistant in the FeS static tests, but was attacked in the cell tests. The attack, however, could be reduced greatly by adding iron powder to the electrode. The same situation prevailed with low-carbon steel. The beneficial effect of iron powder additions was quantified later by accelerated static immersion tests.

Some work was done on protective coatings applied to low-cost substrate materials, most often, low-carbon steel. Coatings of Ni, Ni-Mo, Cr, and TiN-TiC appeared promising for FeS electrodes, and no corrosion was observed with LiAl electrodes coated with TiC, TiN, or TiC-TiN. Tests of low-carbon steel electrodes, electroless and electroplated with nickel, showed a reduction of the corrosion rate in FeS electrodes, but their long-term integrity appeared questionable.

At the Li-Al electrode, deposition of aluminum on low-carbon steel resulted in a brittle, intermetallic reaction layer, which limited the lifetime of the component. A galvanic deposition study was made by coupling a low-carbon steel electrically to an aluminum electrode in LiCl-KCl electrolyte and measuring the current at temperatures of 470-530°C. The rate showed an Arrhenius temperature dependence, *i.e.*, $G = G_o \exp(-Q/RT)$, where the activation energy, Q, was 21.4 kcal/mol, and the galvanic deposition rate constant was 1.65 × 10^{-2} g/(cm^2 sec). Diffusion-rate measurements of aluminum in nickel and low-carbon steel at 475°C indicated diffusion coefficients of 3.4 × 10^{-11} and 2.4 × 10^{-11} cm^2/sec, respectively. The predominant phase for the aluminum/steel couple was $FeAl_3$, and in the aluminum/nickel couple there was an inner zone of Ni_2Al_3 and an outer zone of $NiAl_3$. When the rate data for deposition and diffusion were combined, the results showed that the overall aluminum reaction rate was controlled by the initially slower galvanic deposition and was linear, but once a certain transition time had been reached, the rate was limited by diffusion and was parabolic.

By 1983, materials work in the lithium-aluminum/iron sulfide program had broadened to include other areas such as sodium/sulfur and glass electrolyte cells, so further work is discussed under those topics.

CELL DEVELOPMENT

Li-Al/FeS Cells. Engineering modeling studies were conducted to generate mathematical correlations of performance and lifetime with cell design parameters and operating variables. These correlations were obtained through multiple regression analyses of test data from previously operated cells. Specific energies were predicted for up to 800 cycles of 84 Li-Al/FeS multiplate cell designs. The results were used to identify cell designs most likely to meet the Mark II cycle-life goals, *i.e.,* less than 20% decline in specific energy over 500 cycles.

The Mark IA cells had shown a significant capacity decline during cycling (typically 20% within 80 cycles). An effort was undertaken to provide design recommendations that would reduce the rate of capacity loss in Li-Al/FeS cells. By inserting a reference electrode into an engineering-scale bicell, the charge/discharge characteristics of the positive and negative electrodes could be monitored individually. The results indicated that the capacity fading problem had been caused by a capacity decline of the negative (Li-Al) electrode, and that the problem could be nearly eliminated by a high lithium content of more than 60 at.% in the electrode. On the basis of this and previous work, more stable capacity could be achieved by using LiCl-rich LiCl-KCl electrolyte, a lithium content greater than 46 at.% in the negative electrode, and an operating temperature above 460°C. As a result of these changes, the capacity-loss rate of ANL's and contractors' cells was reduced significantly to less than 0.03% per cycle.

Post-test analyses had shown that swelling of the electrodes was one of the causes of cell failure on cycling. Studies were undertaken to determine the magnitude of the swelling and the forces involved. Measurements were made both with a small-scale special apparatus at Argonne and with a device specified by Argonne for tests of two full-scale multiplate cells at Eagle-Picher, one with boron nitride (BN) felt separators and the other with BN fabric separators. A number of factors were found to have an effect on swelling: (1) loading density of the electrodes, (2) percent utilization of the active materials, (3) compressibility of the separator, and (4) degree of mechanical restraint during cycling. Temperature had no significant effect. As additional cells were constructed and tested, electrode and current-collector hardware designs were improved to control swelling.

In the operation of a practical battery, situations could arise in which the system would have to be cooled to room temperature, and then reheated to the operating temperature. There was some concern about the effects of alternate freezing and thawing of the molten salt electrolyte. Tests showed, however, that as many as 60 freeze-thaw cycles had no serious effects on well-designed cells at 25, 50, and 75% state of charge.

In 1983, computer modeling was used to correlate cell designs with the performance of

Li-Al/FeS cells. The validity of the computer model was verified by experimental results, and the model was then extended to Li-Al/FeS_2 cells.

At this point, the Li-Al/FeS system had reached a stage of development where the cell designs had been established, and further work was directed toward the design, fabrication, and testing of battery modules to meet the schedule for the Mark II phase of the program. A comprehensive report, entitled "Li Alloy/FeS Cell Design and Analysis Report" (ANL-84-93), was compiled by Eddie Gay with the intention of providing the technical information that would be needed by any organization that might have a future interest in the technology.

Li-Al/FeS$_2$ Cells. The program goals for the Mark III battery cells were a specific power of 200 W/kg, a specific energy of 300 Wh/kg, and a lifetime of 1,000 charge-discharge cycles. Cells having iron disulfide (FeS_2) positive electrodes appeared to offer the best shot at achieving these goals. Iron disulfide, also called "pyrite," in its crystalline form has a yellow metallic luster similar to that of gold. Because it has often been mistaken for gold by panners, it has been named "fools' gold," a term that the cell developers preferred to avoid. During the battery program, Duane Barney displayed a very large crystal of the material on his desk.

For reasons mentioned earlier, cells having FeS_2 electrodes were operated only on the upper (first) voltage plateau, which is at about 1.65 V *vs.* the Li-Al negative electrode. Investigations of Li-Al, Li-Si, and Li-Al-Si electrodes showed that their capacity densities were generally similar. A Li-Al alloy of 53 at.%, rather than the usual 48 at.% lithium, was selected to increase the lithium activity when the cell was near the end of discharge. A new electrolyte composition, 25 LiCl-37 LiBr-38 KBr (mol%), with a eutectic melting temperature of 310°C, was selected because of its higher lithium content and lower melting temperature than the LiCl-KCl used previously. The lower operating temperature tended to slow the loss of sulfur from the FeS_2, which had been a source of capacity decline.

Figure 5-4 shows a comparison between a Li-Al/LiCl-KCl/FeS_2 cell operated on both voltage plateaus and a higher-performance Li-Al/LiCl-LiBr-KBr/FeS_2 cell of the same capacity (24 Ah) discharged only on the upper voltage plateau. The higher performance of the upper-plateau cell was attributed to two factors: (1) improved electronic conductivity of the electrode and (2) the increased liquidus range of the LiCl-LiBr-KBr electrolyte. Utilization of the active material was about 89% of the theoretical capacity, and the cell was operated for more than 300 cycles with no change of capacity. This, along with other results, led to a decision to use the upper-plateau cells in future work.

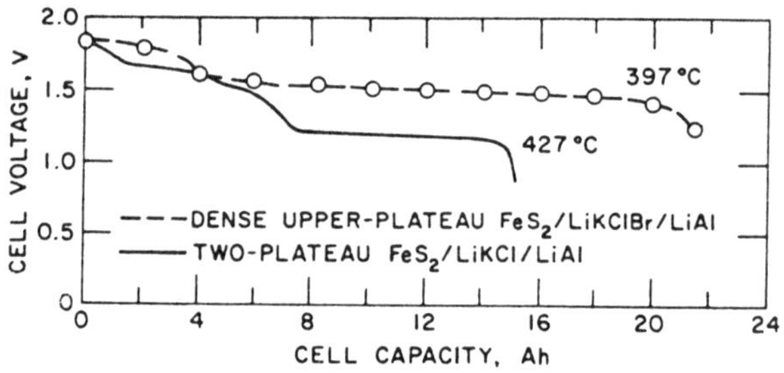

Fig. 5-4. Discharge Curves for Two-Plateau and Upper-Plateau-Only FeS_2 Cells

The cell design effort then continued, with the following activities:

1. Experiments to determine the influence of physical electrode parameters on cell performance.
2. Examinations of electrochemical electrode processes with candidate active materials that provide effective overcharge protection for lithium/metal sulfide cells.

3. Demonstration of long life and ultra-high pulse power capability for a Li-Al/FeS_2 cell.
4. Modeling studies of cylindrical electrodes, which are simpler and less costly than flat-plate designs.
5. Measurement of potential changes during different kinds of electrochemical testing by use of voltage sensors inserted at various locations within a Li-Si/FeS_2 cell.

High-intensity current-pulse techniques were used to relate performance to cell-design parameters. The self-cooling effect during high-power discharge of the FeS_2 electrode permits very compact packaging of the cells in a battery. At this point in the development work, a specific power of 150-300 kW/kg in 1- to 1000-ms pulses appeared feasible. The very high power-pulse capability of the system could lead to several special applications such as powering electromagnetic launchers.

Lifetime tests were performed on a 24-Ah Li-Al/LiCl-LiBr-KBr/FeS_2 bicell. Discharge and charge rates were 4 and 8 h, and cutoff voltages were 1.25 and 2.05 V, respectively, which kept the cell within the upper-plateau range, and the operating temperatures were between 388 and 427°C. The cell capacity remained essentially constant over 400 charge-discharge cycles and 5,400 h. The results of a post-test analysis of this cell indicated that the design had a potential for 1,000 cycles of high-performance operation.

A number of advances were made in the various components of the cells. The composition of the LiCl-LiBr-KBr electrolyte was changed from 25-37-38 mol% to 34-32.5-33.5 mol%, thereby increasing the conductivity about 25%. The possibility of replacing boron nitride (BN) felt separators flooded with the molten salt electrolyte by electrolyte-starved MgO powder separators was investigated. The powder separators are rigid bodies held together by capillary forces in the pores. The advantages are increased resistance to chemical attack, especially in overcharge-protected cells where metallic lithium is present, and the ease of confining a solid rather than a liquid material in the desired locations. A modification in the positive electrode composition to 50-50 mol% FeS_2-CoS_2 plus 5 mol% Li_2S was also studied.

In a battery configuration, many cells must be connected in series to achieve a useful voltage. Because the cell capacities are not exactly the same for various reasons, passing the correct amount of current through one cell to effect a full charge may overcharge the next one. An electronic charge-equalization device was developed to avoid the problem, but it was cumbersome in that it required an individual electrical lead to each cell. The ideal situation, which occurs in certain other battery systems, is to have a cyclic chemical process that begins to take place when a cell is fully charged, so additional current has no effect. A breakthrough for the Li-Al/FeS_2 system occurred in 1986 when it became evident that chemical overcharge protection was a possibility. This was done by limiting the capacity of a lithium electrode of a specific composition such that lithium metal was produced when the voltage rose at the end of charge. The lithium metal dissolved in the electrolyte and diffused to the positive electrode, where it was oxidized to the sulfide, thereby creating a "chemical short circuit." This so-called "lithium-shuttle" mechanism proved to be a practical method for avoiding overcharge damage to the cell. In the engineering development of overcharge-tolerant cells, a Li-Al_5-Fe_2 alloy, which has a potential of -260 mV *vs.* the usual Li-Al electrode, was used.

Up to about 1986, essentially all of the cell development work had been on cells having rectangular, flat-plate electrodes. In that year a model was developed for a design in which seven rod-shaped positive electrodes having an outer annulus of MgO powder separator material were contained in a cylindrical cell.

Although the anticipated electrical performance for that cell design was comparable to that of a flat-plate cell (a specific energy of 190 Wh/kg and specific power of 185 W/kg), the fabrication cost should be lower.

Electrochemical investigations were conducted in 1987 on thin-electrode primary Li-Si/FeS_2 cells under investigation at Sandia National Laboratories. The main objective was to determine the effect of compressive force on cell performance and the optimum ratio of electrolyte to active material. The work at Sandia, adapted from the earlier studies at Argonne, was aimed toward military applications, which is a classic case of research finding use in an area far afield from the original purpose. Cells were later produced commercially by the U.S. subsidiary of the French company, SAFT (Société des Accumulateurs Fixes et de Traction).

In 1988, a major emphasis was started on bipolar battery designs. In a bipolar battery, the cells were stacked in a pancake fashion so that the positive and negative electrodes had a common current collector, the bipolar plate (see Fig. 5-5). This type of arrangement greatly reduced the cell weight by eliminating much of the hardware and also resulted in a uniform current density, which improved the electrochemical performance. Calculations projected outstanding performance for a bipolar electric car battery—a 640-km range and acceleration similar to that achieved with a high-performance gasoline engine.

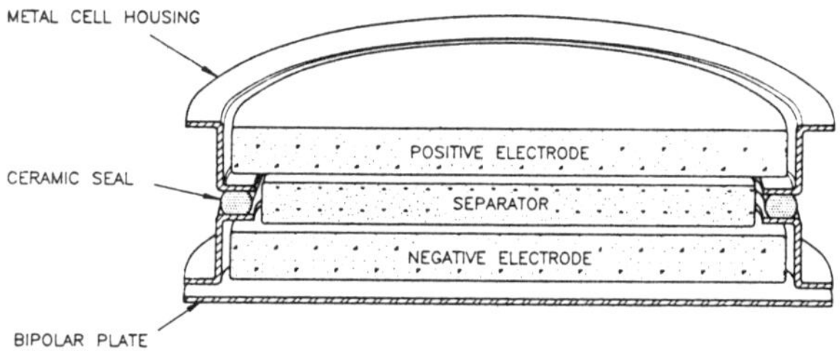

Fig. 5-5. Bipolar Li-Al/FeS_2 Cell Design

The major problem to be faced in the development of bipolar cells was the peripheral seal. This seal had to provide electrical insulation, form a gas-tight bond between the metal components (usually iron and molybdenum), have a coefficient of expansion compatible with those of the metal components, and be resistant to corrosion by the molten salt electrolyte and the active materials in the electrodes. Early attempts to use gaskets or other arrangements had not been satisfactory. Remarkably, Tom Kaun and the others in the group were successful in developing a seal that met all these requirements. Tests of the hermetic seal were conducted by constructing and operating a 3-cm-dia bipolar cell with a LiAl-10 mol% $Li_5Al_5Fe_2$ negative electrode and an upper-plateau FeS_2 positive electrode. The cell showed very good electrical performance, and a post-test analysis after 400 h of operation confirmed the integrity of the hermetic seal.

CELL TESTING

Eagle-Picher Industries, Gould Inc., and Argonne continued to build groups of state-of-the-art "status cells" and occasional modules consisting of several cells to measure progress toward the program goals. Sixty Li-Al/FeS bicells were tested in 1980. Capacity fading, which had been a concern at the time, was reduced to 0.02%, 0.03%, and 0.08% per cycle in the ANL, Eagle-Picher, and Gould cells, respectively, which was a significant improvement. Specific energies for the Eagle-Picher and Gould cells were 80 and 93 Wh/kg, respectively. The Gould cells had a short lifetime of about 80 cycles due to problems with the boron nitride felt separators in their particular design, so they decided at that time to switch to MgO powder separators. In 1981, Eagle-Picher and Gould each fabricated over 100 cells. The specific energies of these cells ranged from about 70 to 90 Wh/kg, and specific powers were between approximately 55 and 75 W/kg. It did not appear at this point that the Li-Al/FeS system was going to be able to meet the 1996 performance goals, despite the fact that a number of marginal design improvements looked feasible. The results for the last sets of status cells are listed in Table 5-4. The following year, the two contractors each built four "best effort" cells in which the specific power was increased markedly, to 102 and 132 W/kg, respectively.

Post-test examinations yielded valuable information on cell failure mechanisms, in-cell corrosion reactions, and electrode morphology (*i.e.,* microstructure, active material distribution and utilization, reaction uniformity, impurities, and cross-contamination of electrodes). The two major causes of cell failure were extrusion of material from the iron sulfide electrode and localized protrusions of Li-Al alloy through the separator. Almost all the other problems were short circuits caused by electrical feedthrough failure or metallic deposits in the separators. The results of post-test analyses of failed cells were an invaluable guide in the development of improved electrode beds and hardware components. The techniques that Jim Battles, John Smaga, and others developed for the post-test examinations have subsequently found much wider application than in this program alone.

MODULES, BATTERIES, AND AUXILIARY COMPONENTS

Table 5-5 lists the small batteries that were constructed and tested in 1980.

A detailed analysis was made of the heat generated in Li-Al/FeS cells during constant-current charge-discharge cycling. The heat-generation rate proved to be substantially higher during discharge than during charge, and it increased significantly at the end of charge and the end of discharge. Calculated internal temperatures of a five-cell module agreed well with experimental values. Work also continued on the development of charging equipment and on commercialization studies in which various market factors were assessed.

Efforts were continued on establishing working relationships with potential users of the Li-Al/FeS battery, in both institutional and private sectors. A joint program with Eagle-Picher Industries was under way to develop a 30-kWh fork-lift battery for the U.S. Army, an economic analysis for a bus battery was proposed to the U.S. Department of Transportation, and technical discussions were held with the Chrysler Motors Corp., General Motors, Inc., and the Ford Motor Co. In 1982, a study of battery-vehicle interactions was performed jointly by ANL and the Toyota Motor Co.

Table 5-4. Performance Data for Status Cells

Performance Characteristic	Eagle-Picher	Gould
Ave. Peak Capacity, Ah	157	340
Ave. Specific Energy, Wh/kg	80	90
Ave. Peak Specific Power, W/kg	70	80
Cycle Life	330	345
Capacity Loss per Cycle, %	0.06	0.02

Table 5-5. Small Batteries Tested in 1980

No. of Cells	Battery Voltage, V	Battery Capacity, kWh	Thermal Insulation	Cycle Life
9	10.6	3.0	Conventional	72
10	12	2.5	Vac-foil	71
10	12	2.5	Vac-foil	79
10	11.6	3.8	Conventional	270
10	11.8	4.0	Conventional	150

Also in 1982, the development and testing activities on Li-Al/FeS batteries were split off from the rest of the program and incorporated into the Industrial Electrochemical Projects programs under Mike Yao. In 1983, a study supported by DOE was directed toward a full-scale Li-Al/FeS battery design for a Chrysler T-115 van. An artist's conception of this 162-cell battery is shown in Fig. 5-6. A complementary program was funded by the Electric Power Research Institute (EPRI) for the development of a 10-cell Li-Al/FeS module to be built by Gould Inc. and tested at ANL.

By 1984, a wide interest had developed in the Li-Al/FeS battery system, both in the U.S. and internationally. Contractors to the U.S. Department of Defense were using the technology to build advanced, nonrechargeable, thermal batteries. The U.K. and Canada both had aggressive programs on potential commercial and defense applications. Varta, A.G., in Germany and SAFT in France were also involved. The foreign discussions included a visit from a Soviet delegation to ANL at Gould's request in the early 1980s. The ANL people attending the meeting were told that they were to limit their discussions to material that had been published.

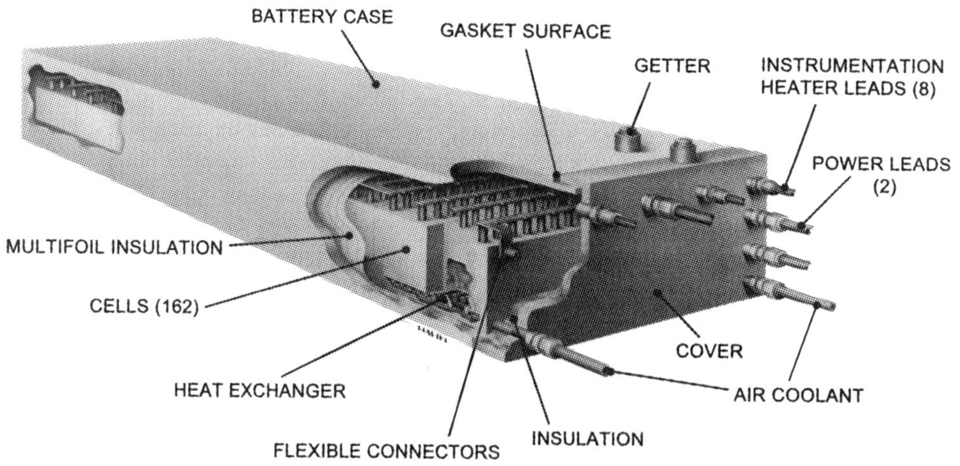

Fig. 5-6. Design for a Li-Alloy/FeS Van Battery

The Soviet group of about ten people included a female interpreter, and it was understood that they would speak only Russian. One member of the group very definitely had the appearance and demeanor of a policeman, and he said nothing during the technical discussions, leading to a suspicion that he was a KGB man. It was apparent that the Soviet scientists had read the ANL literature and were familiar with the names of the ANL staff people they were meeting. That evening Gould hosted a dinner at their headquarters in Rolling Meadows, and, after several of the traditional toasts, some of the Soviets became not only jovial, but also bilingual. The consensus of the ANL people after the meeting was that the Soviets were doing very little, if any research on lithium-alloy/metal sulfide batteries, although they claimed to have a program.

In 1985, Gould delivered two nine-cell, 12-V Li-Al/FeS modules to ANL for testing in a program funded jointly by EPRI, DOE, and the Tennessee Valley Authority. Under a power regime specified by the standard J-227a/D simulated driving profile, one of the modules delivered a specific energy of 104 Wh/kg (on a cell-weight basis), which was considered acceptable for electric van propulsion. Gould then constructed two 36-V batteries, which were tested in 1987. The first battery, designated Mark I, was based on the design used in the 12-V modules. Its power and its specific energy (49-110 Wh/kg of cell at a 40-A constant-current discharge and 77-112 Wh/kg of cell on a simulated driving profile) were considered satisfactory, but its lifetime was inadequate (113 cycles). The failure was caused by iron deposition in the separators. The thermal insulation and thermal control auxiliaries limited the specific energy of the Mark II battery.

An effort was made to improve thermal control of the battery by developing a high-efficiency, lightweight, low-cost insulated jacket in a collaborative program between ANL and the Westinghouse Oceanic Division (formerly Gould Ocean Systems Division) of Cleveland, Ohio. A part of this work was done jointly between ANL and Meyer Tool and Manufacturing Co. of Oak Lawn, Illinois. If developed successfully, this technology could also be useful in several other applications, *e.g.*, sodium/sulfur batteries, sodium/chloride batteries, monolithic fuel cells. Two full-scale insulated cases were fabricated and tested. The heat loss was 208 W, which was considerably higher than the 100-W goal, but it appeared that this figure could be decreased by using a chemical getter to lower the residual gas pressure and by optimizing the types and thicknesses of the insulation layers. One of the cases was used to house a Gould 36-V electric van battery module, which achieved a lifetime of about 103 cycles. The specific power for the module was 71 W/kg and the specific energy for a driving profile was 70 W/kg. The corresponding values for the cells themselves were >95 W/kg and >110 Wh/kg.

By the end of the 1980s, the basic technology of Li-Al/FeS cells and batteries had been developed, but additional work was still needed to improve their performance and lifetime before they could be considered ready for commercial application.

PERSONNEL

The lithium-alloy/iron sulfide program was hit hard by the budget cuts in the early 1980s. The effort dropped from about 35 full-time people in 1980 to approximately 10-12 by 1984 and to about 4 or 5 full-time people by the end of the decade. Different individuals were in charge of the overall program as it underwent the resulting organizational changes: Duane Barney, John Ackerman, Jim Battles, and Paul Nelson. Others who participated in the program at one time or another during the 1980s include Karl Anderson, Jack Arntzen, Lou Bartholme, Milt Blander, Ira Bloom, Alan Brown, Al Chilenskas, Clint Christianson,

Bill DeLuca, Mike Duoba, Magdy Farahat, Al Fischer, Don Fredrickson, Eddie Gay, Kenneth Gillie, Ed Hayes, Roger Hogrefe, Tim Holifield, Fred Hornstra, Tom Kaun, John Kelley, Dennis Kilsdonk, Verne Kolba, Wally Kremsner, Gene Kucera, Dick Malecha, Jane Marr, Fred Martino, Bill Miller, Jim Miller, Bali Misra, Frank Mrazek, Mike Myles, Neil Otto, Sandy Preto, George Redding, Laszlo Redey, Mike Roche, Laury Ross, Paul Shimotake, John Smaga, Bob Steunenberg, Bob Swaroop, Ziggy Tomczuk, Tony Tummillo, Ravi Varma, Don Vissers, John Weber, and Carl Webster. The secretaries were Sue Bykowski, Marian (Gardner) Harkins, and Sherry Grisko.

Other High-Temperature Battery Research

SODIUM/SULFUR BATTERIES

Sodium/sulfur battery cells employ metallic sodium as the negative electrode and elemental sulfur as the positive electrode. The sodium and sulfur react electrochemically, at first forming polysulfides (Na_2S_5, Na_2S_4, Na_2S_3, Na_2S_2), and finally Na_2S if the reaction is carried to completion:

$$2\,Na + S \rightarrow Na_2S$$

The open-circuit voltage of the Na/S cell is about 2.0 V. The cell design usually consists of a ceramic tube containing the sodium, which is surrounded by the sulfur. The ceramic separator between the reactants is made of "beta alumina," which is capable of sodium ion conduction. A minimum operating temperature around 300°C is required to maintain liquid phases in the electrodes and to assure adequate ionic conductivity of the beta alumina.

The term "beta alumina" is actually a misnomer because it implies a phase of pure aluminum oxide, which is not the case, but it has come into general use. The material actually consists of Na_2O-Al_2O_3, Na_2O-MgO-Al_2O_3 and similar compositions, in which aluminum and oxygen atoms form spinel-like blocks that make up hexagonal layers spaced apart by Al-O-Al columns. Sodium ions occupy the spaces between the planes and are relatively free to move about in directions perpendicular to the c-axis. For use in cells, the tubes are fabricated by pressing and sintering the powder, which provides lower, but isotropic conductivity (the same in all directions). In most of the work at ANL on Na/S cells, the solid electrolyte was β''-alumina, which has the approximate composition $Na_2O \cdot MgO \cdot 5Al_2O_3$. The addition of MgO was shown to improve the sodium ion conductivity of the material.

Joe Kummer and Neil Weber of the Ford Motor Co. had reported work on the Na/S system, and interest in the system developed rapidly. By the mid-1970s significant research efforts were in progress at the Ford Motor Co., Dow Chemical Co., General Electric Co., General Motors Corp., and TRW in the United States. Foreign programs included the British Rail Technical Centre, Chloride Silent Power, Ltd., as well as UKAE, Harwell in Great Britain, Brown Boveri & Cie and VARTA Batterie AG in Germany, Laboratoires de Marcoussis in France, and Yuasa Battery Co., Ltd., in Japan.

Sodium/sulfur and lithium-alloy/iron sulfide batteries had a lot in common in that they were both novel electrochemical systems, they evolved at about the same time, they both involved high-temperature operation, and they were being considered for the same practical applications—electric vehicles and load leveling on utility systems. In the development work, they tended to have the same general types of problems, such as corrosion, internal stresses, cell leakage, heat management, and capacity matching of cells during charging. The two systems were definitely competitors with respect to anticipated performance, cost, and lifetime.

The battery groups at Argonne were not deeply involved in Na/S cell and battery development, but they were able to make useful contributions to the technology due to the facilities and expertise available at ANL for extensive performance testing and post-test diagnostics.

In 1983, post-test examinations were started on Na/S cells from the Ford Aerospace and Communications Corp. to identify life-limiting mechanisms, long-term degradation of critical components, and characterization of the sulfur-electrode morphology. Detailed examinations showed that the morphology of the sulfur electrode was complex and involved the phases Na_2S_2, Na_2S_4, Na_2S_5, and sulfur. Phases resulting from corrosion of metallic components were $NaCrS_2$ and $NaAlS_2$. When the cells were cooled down from the operating temperature, the composition gradients persisted, so it was possible to observe the reaction fronts. It appeared that several metastable phases were formed in the sulfur electrode. There was still concern about the integrity of the β''-Al_2O_3 electrolyte in spite of advances in the cell design, and freeze-thaw cycling produced fractures. In at least one case, fracture was caused by inadvertent cooling during electrical testing. In 1986, work was completed on testing of four load-leveling cells, five electric-vehicle cells, and one cell for a satellite application from Ford. Department of Energy support for Na/S cell work at ANL was terminated at that time.

Tests were then initiated on Na/S cells from Chloride Silent Power, Ltd., in Great Britain. Test results from four 20-Wh cells translated into a vehicle range of 218 km and a peak power of 108 Wh/kg. One failed at the 388th cycle; the others ran for 524-807 cycles. Several other cells and small modules were tested and examined. One of the modules was a one-third size electric vehicle battery containing 960 cells (300 Ah at 64 V), which came complete with charging and thermal management systems. Although one cell failed after 155 cycles, the module was still delivering 80% of its capacity after 225 cycles.

By the end of the decade, the Na/S and Li-alloy/iron sulfide batteries were in similar positions. Tests had shown that performance and lifetime goals could be met on occasion, but not consistently. It was also becoming more difficult to justify large development programs in an era of tight budgets and a general lack of enthusiasm on the part of the public for electric cars and energy conservation.

GLASS ELECTROLYTE SYSTEMS

Although nearly all the work on sodium/sulfur cells was done with the β''-alumina solid electrolyte, an interest developed in glasses capable of sodium-ion conduction as a possible alternative. Glass electrolyte had several properties that could be advantageous: (1) ease of fabrication into different shapes, (2) high strength in thin sections, (3) an absence of grain boundaries, and (4) isotropic ionic conductivity. The Dow Chemical Company had developed a sodium borate glass for use in sodium/sulfur cells, and a joint effort between the Materials Science and Technology and CMT Divisions at Argonne had produced a sodium-ion conducting glass called NASICON, which was composed of sodium, zirconium, and silicon oxides. The Dow glass had a high resistivity of 24,000 ohm-cm at 300°C, while the values for NASICON and the β''-alumina were about 500 and 5 ohm-cm, respectively. The resistivity of the Dow glass was too high for use in practical cells and that of NASICON was at best marginal. A value of about 200 ohm-cm was considered to be the maximum for cells using thin NASICON electrolytes.

In 1982, a systematic study of NASICON-type glasses was undertaken to determine optimum compositions in the Na_2O-Al_2O_3-ZrO_2-SiO_2 system for glass electrolytes to be used in Na/S cells. Many factors had to be considered, including the composition regions that form glasses, the temperatures required, ionic conductivity, glass transition temperatures, coefficients of thermal expansion, and chemical stability toward sodium and mixtures of sulfur and sodium sulfides. A large number of glass compositions were prepared and characterized with respect to physical properties and corrosion resistance. The composition, $42Na_2O$-$8Al_2O_3$-$5ZrO_2$-$45SiO_2$, was selected and given the name "ANL glass." The resistivity of the material was 221 ohm-cm at 300°C, and its corrosion resistance in the Na/S cell environment proved similar to that of the β''-Al_2O_3.

During the development work on the glass electrolyte, various design studies were conducted on cells. The most promising design concept appeared to be one in which the sodium was contained in many small (1.5-mm OD) glass electrolyte tubes that were immersed in the sulfur. One specific conceptual design was a 150-Ah cell consisting of 4,000 such tubes, which had a projected specific power of 2,000 W/kg and specific energy of 200 Wh/kg.

A great deal of effort was put into techniques for fabricating glass electrolyte tubes having small diameters and thin walls. Production of uniform 500- to 1,000-g glass billets from the high-purity reagents was straightforward, but drawing the tubes was not so easy. Several small-scale Na/S cells were tested and achieved 36-40 cycles at 4-h charge and 8-h discharge rates, and the cell failures were not necessarily related to the electrolyte. These and other small-scale experiments revealed no inherent problems with the glass electrolyte, although it was still a long way from practical application in full-size batteries.

Because of its ability to conduct sodium ions, the glass electrolyte can be used in a reference electrode in which the reference material is completely isolated from the rest of the system. A miniature reference electrode was developed and used to characterize the sodium-sulfur system. This characterization was done by means of emf measurements and techniques such as coulometric titrations.

COMPOSITE SOLID ELECTROLYTE SYSTEMS

On the basis of some preliminary work reported earlier by Hitachi, Ltd., and the Electric Power Research Institute (EPRI), a small program was started in the late 1980s on composite solid electrolytes consisting of about 50 vol% each of sodium beta alumina and glass. The ANL glass appeared to be ideal for this purpose. The beta-alumina is expensive and difficult to fabricate into shapes such as flat plates for use in high-performance cells, but it is an excellent conductor of sodium ions. The glass is more amenable to fabrication in various configurations, but its conductivity is mediocre. The thought was that it might be possible to capitalize on the better features of both by forming them into a glass-ceramic mixture. Preliminary attempts to prepare a composite electrolyte resulted in a material having a resistivity of 1700 ohm-cm. The initial results were encouraging, but it was clear that more work would be required to develop a viable composite electrolyte. After a significant additional effort, which included a change in the glass composition to Na_2O-Al_2O_3-SiO_2, composites with excellent resistivity values (about 30 ohm-cm at 250°C) were obtained.

ALTERNATIVE CELL SYSTEMS

One can easily think of many combinations of elements and/or compounds that have the thermodynamic properties to generate a voltage when placed in an electrochemical cell. Very few make the grade as useful systems, however, because of a multitude of other conditions that must be satisfied, such as kinetic factors, electrochemical reversibility, compatibility with materials, and the availability of a suitable electrolyte, to name a few. The following systems were selected for exploration largely on the basis of information that had been developed in the existing battery programs.

Calcium/Metal Sulfide Cells. This work was predicated on the use of calcium as a low-cost substitute for lithium in the lithium-alloy/metal sulfide cells. The cell components were negative electrodes of $CaAl_{1.2}Si_{0.4}$, positive electrodes of $Fe_{0.93}Co_{0.07}S_2$, boron nitride felt separators, iron (negative) and molybdenum (positive) current collectors, and LiCl-NaCl-$CaCl_2$-$BaCl_2$ electrolyte (m.p., 383°C). A 154-Ah bicell of this type, which was operated for 102 cycles (3500 h), had a specific energy of 47 Wh/kg and a specific power of 50 W/kg. (The theoretical specific energy of this cell was 610 Wh/kg.) Multiplate cells were projected to have higher specific energy and power (>120 Wh/kg and 170 W/kg, respectively). A problem was encountered with declining coulombic efficiency, however, which limited the cell life to about 300 cycles. The cell life was limited by contamination of the felt separator by finely divided particles from the electrodes.

Sodium/Alkali Nitrate Cells. The sodium/alkali nitrate cells consisted of a molten sodium negative electrode, β"-alumina solid electrolyte, and a molten alkali nitrate positive electrode. The overall cell reaction is

$$2\ Na + MNO_3 \rightarrow Na_2O + MNO_2$$

where MNO_3 is the alkali nitrate, and MNO_2 is the corresponding nitrite. Three alkali nitrates were used: $NaNO_3$, $LiNO_3$, and $NaNO_3$-$LiNO_3$ eutectic. A 1.62-V cell with the eutectic electrode was operated at 206°C for 14 cycles. Potentiometric titrations and X-ray data showed the following reactions of $NaNO_3$ at 345°C and $LiNO_3$ at 275°C:

$$2\ Na + NaNO_3 \leftrightarrow Na_3NO_3$$
$$4\ Na + 2\ LiNO_3 \rightarrow Li_2O + Na_3NO_3 + NaNO_2$$

Additional cell tests were conducted reasonably successfully, but it appeared that an operating temperature of 330-350°C would be required, giving them no significant advantage over the existing sodium/sulfur cells, which operate at about 330°C.

Sodium/Metal Chloride Cells. Investigations were conducted on Na/MCl_x cells in which a metal chloride was substituted for the sulfur electrode in sodium/sulfur cells. Nickel dichloride ($NiCl_2$) was used for the positive electrode, and a secondary electrolyte of liquid $NaAlCl_4$ was required. Thus, the basic cell was

$$Na/Na\beta''\text{-}Al_2O_3/NiCl_2\ (NaAlCl_4)$$

Variations of this system included the use of a composite electrolyte instead of beta alumina, iron chloride instead of nickel chloride, and additions of sulfur to the positive electrode. Conceptual cell-design studies for this system, using high-surface-area cells, yielded specific-power and specific-energy projections of about 300-500 W/kg and 200 Wh/kg, respectively.

The experimental studies were focused on means of increasing the performance of the $NiCl_2$ electrode. In the initial investigations, an 18 vol% nickel electrode had an area-specific impedance of 1.2 ohm-cm^2 at 300°C. By fabricating the electrode with larger and more uniform porosity, the area-specific impedance was reduced to 0.80 ohm-cm at 260°C. These impedance values were better than those for state-of-the-art cells, and these results pointed the way for further studies in the 1990s.

Aqueous Battery Research and Development

DEVELOPMENT OF ELECTRIC-VEHICLE BATTERIES BY CONTRACTORS

The program on aqueous-electrolyte batteries continued with the major effort consisting of research and development by industrial contractors on three types of batteries:

Lead-Acid:

Eltra Corp., Plymouth Meeting, PA
Exide Management and Technology Co., Yardley, PA
Globe Battery Division of Johnson Controls, Inc., Milwaukee, WI

Nickel/Iron:

Eagle-Picher Industries, Inc., Joplin, MO
Westinghouse Electric Corp., Pittsburgh, PA

Nickel/Zinc:

Energy Research Corp., Danbury, CT
Exide Management and Technology Co., Yardley, PA
Gould Inc., Rolling Meadows, IL

A milestone in 1981, which called for recommendations to DOE on a reduction in the number of contracts, resulted in three remaining contracts: lead-acid by Globe and nickel/iron by Eagle-Picher and Westinghouse. Work on the nickel/zinc system was discontinued in 1982 because of cycle-life problems caused by the zinc electrode, which is notorious for its poor behavior. (Joe Consiglio, who served as a consultant to CMT, once observed wryly that research on shape changes and dendrite formation in zinc electrodes has provided lifetime employment for a lot of electrochemists over the years.) In about 1985, the Globe name was shortened to "Johnson Controls." From 1984 to the end of the decade, essentially all of the contract work on aqueous batteries consisted of the Johnson Controls effort on lead-acid batteries and the Eagle-Picher effort on nickel/iron batteries. Significant progress was made on both systems.

Lead-Acid Batteries. This technology had been around for a long time, so the principal research activities dealt mostly with reducing the weight, increasing the power and energy, and extending lifetime during deep-discharge cycling. Eltra worked on lightweight, expanded-mesh electrode grids. Exide investigated electrode separator materials, changes in sulfuric acid electrolyte concentration, and cell design variations, including tubular positive plates. Johnson Controls was concerned mostly with an innovative design in which the electrolyte was pumped through porous electrodes. This approach increased the utilization of active material from about 30% to 55%. In 1983, Johnson Controls delivered full-size (108-V) electric-vehicle batteries to the National Battery Test Laboratory (NBTL) at Argonne for verification tests and to the Jet Propulsion Laboratory, Tennessee Valley Authority, and Detroit Edison for in-vehicle testing. This was considered to be intermediate state-of-the-art (ISOA) technology. Johnson Controls continued to provide cell modules and small batteries for NBTL and in-vehicle testing throughout the 1980s.

Nickel/Iron Batteries. Like lead-acid batteries, nickel/iron batteries (invented by Thomas Edison) were a well-developed technology that had been used in various industrial applications, but not for electric-vehicle propulsion, which demands high specific power and energy together with long cycle life and low cost. Experience had shown that most of the problems associated with the nickel/iron cell were at the nickel oxide (positive) electrode. Westinghouse, in an effort to reduce weight and cost, came up with a sintered steel wool electrode plated with nickel. They also investigated electrolyte and gas-management systems. Although reasonably good progress was made in achieving higher specific energy and power, and an extended lifetime of >900 cycles, the Westinghouse contract was concluded in 1983.

Eagle-Picher took the approach of thick, sintered electrodes in the nickel/iron cell. In 1982, the Northrup Corp. purchased and placed in operation 12 electric vehicles with Eagle-Picher batteries. Full-size batteries were delivered to the Jet Propulsion Laboratory, the Tennessee Valley Authority, and the U.S. Postal Service in 1984. In 1985, a multicell module achieved an equivalent of >55,000 miles of simulated electric car driving in NBTL, and long lifetime (>30,000 miles) was obtained in tests at the Tennessee Valley Authority. Although specific power and energy were still marginal for electric-vehicle use, the nickel/iron system appeared to have a potential for commercialization, and Eagle-Picher focused their effort on cost reduction.

NATIONAL BATTERY TEST LABORATORY (NBTL)

Electrical Performance Testing. Entering the 1980s, NBTL continued its function of testing contractors' and other cells, modules, and batteries (Fig. 5-7). In 1981, the testing capability was increased to 50 test stations (45 for cells and modules and 5 for full-size batteries). The following year, three high-current modules and two battery cyclers were added, and the environmental testing annex was commissioned. In this facility, three chambers were provided in which batteries could be tested at temperatures from -30 to +50°C, and safety measures, including special ventilation, were taken to permit testing of zinc/chlorine batteries.

Fig. 5-7. Nickel-Zinc Module Being Prepared for Testing in National Battery Test Laboratory

Computer hardware and software were expanded, improved, and modified to handle many specialized tasks such as a library of driving cycles, data acquisition from individual cells in a 256-cell battery, pulsed-current charging and discharging, and thermal testing.

By 1985, over 1700 cells had been tested since 1978, and a capability had been added for testing high-temperature Li-FeS and Na/S cells. At that time the laboratory had assumed a somewhat different role with less emphasis on ANL contractors' cells and batteries and increased activity in general-purpose testing of a wide variety of cells and batteries. Ragone plots were developed for a variety of battery types. These are logarithmic plots of specific energy *vs.* specific power (see Fig. 5-8). Ragone plots are a convenient way of displaying data to be used in estimating battery discharge times for almost any load profile.

Post-Test Analysis and Diagnosis Laboratory (ADL). When electrical performance testing was completed, the cells, modules, or batteries were subjected to

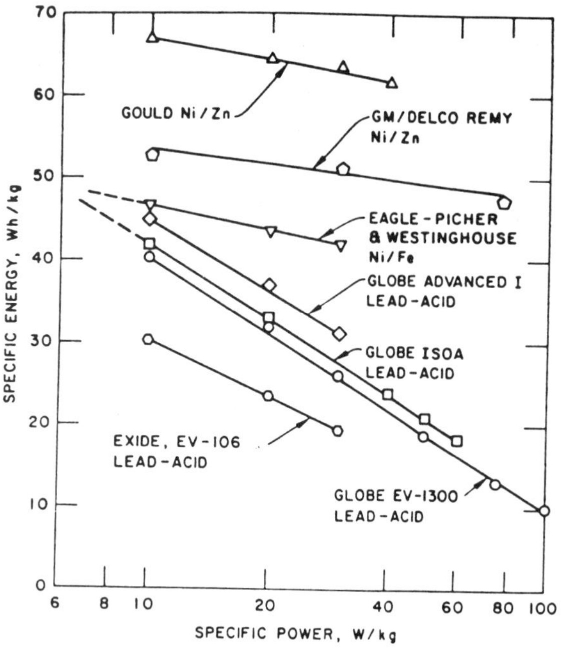

Fig. 5-8. Ragone Plot for Various Batteries Tested at NBTL

post-test examinations, which usually showed the cause of failure or performance loss. Other valuable information was obtained, such as corrosion, chemical phases in the electrodes, distortion of cell hardware, and many other effects.

The analysis normally proceeded through three stages: (1) evaluation of NBTL test results, non-destructive testing, dissection, and visual examination, (2) detailed analyses of component materials, individual electrode cycling, examinations by scanning electron microscopy (SEM), X-ray fluorescence, and other appropriate instrumental methods, and (3) identification of cell-failure mechanisms.

Failures of lead-acid cells usually occurred in the positive (PbO_2) electrode due to deterioration and/or shedding of active material from the electrode, or to current-collector corrosion. In the nickel/iron cells, failures were most often a result of contamination of the nickel oxide (positive) electrode by iron or mechanical failures of the cell hardware. As mentioned earlier, the nickel/zinc cells were plagued by shape changes and dendrite formation in the zinc (negative) electrode. Although these effects could be ameliorated somewhat by resorting to measures such as vibrating electrodes, further development work on this system was halted.

When a post-test examination was completed, the contractor was provided with a report of the findings and recommendations for corrective action. On many occasions, contractors' representatives observed the examinations as they were being performed.

SUPPORTING BATTERY RESEARCH

An in-house program was conducted by CMT to provide supporting research backing up the more applied work being done by the contractors and ANL. Much of the work under this task involved mathematical modeling of cell and battery behavior and heat effects produced by battery operation. The modeling effort also included studies of battery-vehicle integration in electric vehicles, and an analysis was made in 1980 of all the battery performance and vehicle data available at the time. Models were generated for the thermal effects in batteries (heat-generation rates, heat removal by convection and radiation, and temperature distribution). Models were developed for the equivalent electrical circuit of a battery to determine how it would behave as a component in an integrated system such as an electric vehicle. A mathematical model was constructed for vibrating electrodes (of particular interest to the Ni/Zn battery). Up until about 1983, mathematical models were based largely on statistical experimental data. A new approach was taken at that time, in which fundamental electrochemical principles, and engineering considerations, were integrated into a physical-mathematical model applicable to any battery system with minor modifications. Results from this model and those from experimental data were in excellent agreement both for lead-acid and nickel/iron batteries.

Several supporting research studies were conducted on lead-acid cells. Overcharging can result in the production of arsine (AsH_3) and stibine (SbH_3), which come from arsenic and antimony present as alloying elements or impurities and are highly toxic. The production rates of these gases under different conditions were quantified by experimental measurements and used to make recommendations for ventilation in charging facilities. Laser-Raman and neutron-scattering techniques, in cooperation with the Solid State Science (SSS) Division, were used to study the structures of electrodes in lead-acid cells, and these were followed by an in-depth, neutron-diffraction study of PbO_2 on Pb, Pb-Sb, and Pb-Ca-Sn grids. These studies revealed important information on the behavior of PbO_2 electrodes in lead-acid cells.

Current distributions in zinc electrodes and ionic conductivities of zincate-potassium hydroxide electrolytes were investigated. Zinc is transported between the electrodes of a Ni/Zn cell in the form of zincate (ZnO_2^{2-}) ions. Laser-Raman and nuclear magnetic resonance techniques were used to investigate the structure of the zincate ions. Only one dominant Zn^{2+} species with tetrahedral coordination was found. Resistance factors in the separators and the effects of electrolyte agitation in Ni/Zn cells were evaluated. Penetration of the separator material by growth of zinc deposits was a continuing problem in these cells. In the case of nickel/iron cells, about the only problem unique to that system was iron contamination of nickel electrodes, which was examined in some detail.

Most of the other supporting work was of a general nature, such as thermal management, charging and discharging regimes, self-discharge rates, overcharge effects (gas evolution and water loss), and effects of temperature.

OTHER BATTERY APPLICATIONS

Although the primary emphasis of the aqueous battery program was on electric-vehicle propulsion, two other applications were given some attention: (1) load leveling on utility networks and (2) hybrid fuel-cell/battery vehicles.

Load Leveling. The requirements for load-leveling batteries differ considerably from those for vehicle propulsion in that the major consideration is low cost, rather than low weight. Implicit in the cost are manufacturing cost, cycle life, and volumetric energy density (kWh/m^3). A cost and design study of lead-acid batteries in 1980 indicated that they might be marginally acceptable for bulk electrical energy storage, and goals were set for an acceptable system: a lifetime >4,000 cycles, a maximum initial cost of $50-75/kWh, and a minimum energy density of 75 kWh/m^3.

Contracts were awarded to C&D Batteries, Inc., and Exide for three tasks: (1) research and development on advanced lead-acid cells and modules, (2) demonstration of pilot-line production of advanced modules, and (3) construction of a 5-MWh battery for testing in the Battery Energy Storage Test (BEST) facility. The C&D contract was dropped for lack of funding. Exide, in 1982, exceeded 6,000 cycles with 1,800-Ah cells and 15,000 shallow cycles with 200-Ah cells. In 1983, three modules delivered to ANL for parametric cycle-life tests exceeded the design goals.

Hybrid Vehicle. Studies were performed for DOE on a hybrid vehicle concept that combined a methane-fueled phosphoric-acid fuel cell with a battery for an electric bus. The idea was to operate the vehicle primarily on the fuel cell with a small amount of surplus power to charge the battery. When extra power was needed for acceleration or hill climbing, the additional energy would be retrieved from the battery. This application would require a smaller, but higher-power battery than the all-electric vehicle.

INDUSTRIAL ELECTROCHEMICAL TECHNOLOGY

In the early 1980s, the Chemical Technology Division undertook a program aimed at reducing the amount of energy consumed nationally in electrolytic processes for the production of various materials. Although this work was not battery-oriented, it is mentioned here because it was done as an adjunct to the aqueous battery program. Preliminary surveys indicated potential process improvements for the commercial production of chlorate, hydrogen peroxide, ozone, chlorine and sodium hydroxide (chloralkali process), aluminum, and zinc. Most of the ANL in-house work was performed on processes for the conversion of Al_2O_3 to Al_2S_3, followed by electrolysis of the Al_2S_3 in a molten salt solvent, *e.g.*, NaCl-KCl-$MgCl_2$. The proposed process ran into several difficulties, including an inability to convert the aluminum oxide completely to sulfide and incomplete electrolytic reduction of the Al_2S_3, so the effort was terminated.

Studies of the chloralkali process indicated that the NaOH concentration in the cells could be increased from the usual 29% to about 50% without serious polarization problems in membrane cells. The conclusion was that the membrane, rather than polarization or conductivity effects, was the limiting factor on energy savings and technical feasibility in the production of 50% NaOH in membrane cells.

In 1983, the Electrochemical Industrial Associates (IEA) program was proposed. This program, sponsored by ANL, was intended to provide consulting and some experimental support to dues-paying industrial members in the technology of batteries, fuel cells, and industrial processes. The program was unsuccessful; the industries preferred to work through special research and development contracts. As DOE funding of the electrochemical programs had decreased steadily through the 1980s, the Division did pick up some "work for others," but the overall program continued to diminish in scope.

PERSONNEL

In 1980, this was the largest program in the Division, and it was organized as the Office of Electrochemical Program Management under the direction of Mike Yao. Group Leaders were Clint Christianson, Fred Hornstra, Glenn Cook, and Ralph Loutfy. Financial support both for contract work and for in-house research diminished throughout the decade, causing certain reorganizations. The engineering development work on lithium-alloy/metal sulfide batteries was moved into this organization in the mid-1980s, and all the battery work had been consolidated into one entity by 1990. Paul Nelson, who had become the Deputy Division Director in 1982, had jurisdiction over the electrochemical programs. After Mike Yao's departure in 1985, John Ackerman was placed in charge of the battery and fuel-cell work, and Clint Christianson became a Section Head responsible for the aqueous battery program. In 1989, Mike Myles was given overall responsibility for the electrochemical programs. The following is an attempt to list those associated with this program in the 1980s:

John Ackerman, John Barghusen, Joe Becker, Ed Berrill, Dick Biwer, Ira Bloom, Alan Brown, Louis Burkel (MST), Earl Carothers, Al Chilenskas, Clint Christianson, Glenn Cook, Dave Corp, E. Creamer, Bill DeLuca, Eva Folke, Don Fredrickson, Eddie Gay, Mike Genge, Mark Hash, Ed Hayes, Roger Hogrefe, Fred Hornstra, Chen-Chao Hsu, Tom Kaun, Jiri Klinger, Dick Kmiec, Marvin Kronenberg, Johnsee Lee, Tien Lee, Ralph Loutfy, Jane Marr, Janet Meisenhelder,

John Miller, Nguyen Mingh, Frank Mrazek, Tom Mulcahey, Mike Myles, Ron Paul, Dave Poa, John Rajan, Laszlo Redey, Lewis Ruppert, Paul Shimotake, Merril Singleterry, John Smaga, Greg Smith, Carl Swoboda, Tony Tummillo, Ravi Varma, George Vasilopulos, Don Vissers, John Weber, Carl Weber, and Mike Yao. Secretaries included Marlene Bukowski, Wendy Strle, Elaine Estand, Sherry Grisko, Kerry Kilian, Wendy Lamb, Dawn Landis, and Charlotte LeGrand, who was the head secretary until she left the Division in 1985.

Fuel Cells

The Division's fuel-cell program had become well established by 1980, and, like several other DOE programs, consisted of two responsibilities: (1) technical management of DOE's programs on advanced fuel-cell development and (2) supporting research and development on molten carbonate fuel cells (MCFCs). This responsibility was expanded in 1981 to include solid oxide fuel cells, which were being developed at the Westinghouse Electric Corp. In 1982, DOE management of its fuel-cell programs was shifted from Washington Headquarters to the Morgantown Energy Technology Center in Morgantown, West Virginia.

In the mid-1980s, workers at ANL came up with a novel concept known as the monolithic solid oxide fuel cell, which had a potential for very high performance and a capability to run on hydrocarbons, as well as the hydrogen, carbon monoxide, natural gas, or alcohol fuels normally used in fuel cells. Initial funding for this work was provided by the Defense Advanced Research Project Agency (DARPA) of the Department of Defense (DOD). The CMT fuel-cell programs experienced the same budget squeeze as the battery programs and others, but managed to survive through the 1980s.

MOLTEN CARBONATE FUEL CELLS

The molten carbonate fuel cell (MCFC), as outlined in the previous chapter, consisted of a lithiated nickel oxide cathode and a porous nickel anode separated by a rigid electrolyte tile composed of molten Li_2CO_3-K_2CO_3 in a solid $LiAlO_2$ matrix. The operating temperature was about 1000°C, and each cell generated about 0.8 V. The fuel of greatest interest was "synthesis gas" (hydrogen and carbon monoxide) produced from coal in a fuel processor. Advantages claimed for this system were (1) use of coal, (2) very low pollution, and (3) high efficiency (50% from coal to bus bar).

Two major contractors were involved in the MCFC program, the General Electric Co. and United Technologies Corp. Their objective was to develop and fabricate full-size (0.5- to 1.0-MW) cell stacks and provide DOE with information needed to proceed with stack development. Supporting work was performed by the Energy Research Corp., mainly on possible incorporation of a reforming catalyst into the fuel cell, and by the Institute of Gas Technology on methods of handling sulfur impurities in the fuel.

Electrolyte Studies. Work continued, in cooperation with members of the Materials Science Division (MSD), on the preparation of tile electrolytes from $LiAlO_2$ powder and Li_2CO_3-K_2CO_3 salt by pressing and sintering. The sintered electrolytes needed to be thin and porous (about 60%). Information from the literature suggested that tape casting might be an alternative to pressing and sintering for electrolyte preparation. The tape-casting procedure involves preparation of a thick slurry (slip) of the powders in a mixture of plasticizer, binder, deflocculating agent, and solvent. The mixture is distributed on a plate in a specified thickness, dried to form a pliable sheet, and fired to form a thin ceramic. Tape casting did not prove to be suitable for molten carbonate fuel cells, but experience with the technology became very useful in later work on monolithic solid oxide fuel cells. Work continued on the pressed and sintered tiles to optimize conditions for their preparation and to determine their behavior (swelling and other effects) during in-cell tests.

Cathode Development. Efforts were also continued on the preparation of cathodes, which were basically nickel oxide (NiO) doped with lithium oxide (Li_2O) to impart electrical conductivity. The cathodes were made initially by in-cell oxidation and lithiation of a porous, sintered nickel plate in the molten carbonate electrolyte under an air-carbon dioxide atmosphere. A more controlled procedure was developed in which electrode plates of the composition $Li_{0.05}Ni_{0.95}O$ were prepared for later assembly into the cells.

Although the lithiated nickel oxide cathode performed reasonably well, it underwent a very slow dissolution process that limited its lifetime to about 4,000 hours, which was far short of the 40,000-hour goal. This was something of a setback in the program, which resulted in work on alternative cathode compositions. Certain oxides such as Li_2MnO_3, $LiFeO_2$, and ZnO were expected to be chemically stable and to exhibit electronic conductivity when doped with appropriate additives. Detailed studies were conducted on Li_2MnO_3 and $LiFeO_2$ doped with Ca^{2+}, Fe^{3+}, Al^{3+}, and Nb^{5+} with some encouraging results. As of 1989, NiO continued to be the cathode of choice, but the alternatives were still of interest as possible replacements.

Anode Development. The anodes in MCFCs were a porous nickel alloy. The idea of using electrically conductive oxides was explored. Potential candidates for this purpose were oxides that might have dual valence states, *i.e.*, $LiFeO_2$, Li_2TiO_3, Li_3TaO_4, Li_3VO_4, Li_3NbO_4, Li_2ZrO_3, MnO, and CeO_2. Good conductivity requires electronic defects, which means that the material must be non-stoichiometric. Detailed studies of these compounds showed undoped $LiFeO_2$ to be a promising possibility, and an investigation was started on MnO. As of 1989, however, this work had not progressed to the stage where these materials were ready to replace porous nickel as the anode.

Methods were investigated for introducing controlled porosity into oxide electrodes during their fabrication. Pore-size distribution, shape, and degree of interconnection are all factors in achieving a successful electrode. A technique that proved particularly successful was to incorporate small ceramic fibers produced by a spray process into the electrode mix. The procedure worked well both for pressed and sintered and tape-cast electrodes.

Contract Work and Commercialization. Argonne's role as technical manager of the DOE Fuel Cell Program prohibited any design or development work at the Laboratory; its activities were limited to basic research and technical support. Stated briefly, the goals set in 1981 for the contract work on MCFCs were as follows.

1. Demonstration of 20- to 100-cell stacks of full-size cells.
2. Design of facilities for testing full-scale stacks.
3. Operation of single cells under a variety of conditions.
4. A reference design for a coal-fueled fuel-cell plant.
5. Experimental operation using the product of a coal gasifier.

In 1982, DOE decided to place most of the development work on MCFCs at United Technologies Corp., but General Electric, Energy Research Corp., the Institute of Gas Technology, and ANL remained in the program. The DOE plan was to develop a pilot plant on a schedule consistent with an expectation that commercialization would begin to accelerate after 1987.

Three system-design studies were made. These studies employed an MCFC model from Physical Sciences, Inc. (PSI) on reforming methane into CO and H_2, based on the premise that natural gas would be the best fuel for entering the market. The study included three cases: (1) a conventional external reformer, (2) a reformer heated by the sensible heat in recycled fuel, and (3) a reformer integral with the anode of the fuel cell (internal reforming). Energy-conversion efficiencies were estimated to be 50-60%. One conclusion was that operation at pressures of 6-10 atm would be economically advantageous.

The problem mentioned earlier about limited lifetime of MCFCs came to light in 1982, and the original goal of building and operating a full-scale stack by the mid-1980s was delayed by 3-4 years. In 1983, a small plant analysis was made for phosphoric acid fuel cells operating on natural gas fuel, and this was followed by a wider market application analysis conducted by the ANL Energy and Environmental Sciences Division (EES) with participation by the CMT and Engineering (ENG) Divisions.

In 1984, the United Technologies work was transferred to an organization called International Fuel Cells (IFC), which had previously been a part of United Technologies. At about the same time, the DOE set a goal of a full-scale stack of cells having a lifetime of 40,000 hours (roughly five years) in the 1990s. By 1985, IFC had operated 20-cell stacks that gave excellent performance at first, but then developed problems with the NiO cathodes. At that time, several other ANL divisions were participating in the fuel-cell program, namely, Materials and Components Technology (MCT), Engineering (ENG), Energy and Environmental Sciences (EES), and Materials Science and Technology (MST). With the increasing competition for the diminishing funding and the proliferation of divisions at ANL, there was a general tendency toward fragmentation of the larger applied programs into specialized groups in different divisions.

Development of MCFCs by contractors continued in the 1980s. International Fuel Cells continued operation of a 20-cell, full-area (0.7 m^2), 25-kW stack for 1,700 hours, and ERC ran a four-cell, 0.4-kWh stack with internal reforming of the fuel.

SOLID OXIDE FUEL CELLS

Tubular Cell Design. The development of solid oxide fuel cells (SOFCs) began several years later than that on other advanced fuel cells. The principal difference between SOFCs and the previous types was the use of a solid, rather than liquid electrolyte to transport ions between the electrodes. In the early 1980s, the Westinghouse Electric Corp. was working on a solid oxide cell that consisted of a metal oxide (lanthanum manganite) cathode, a cermet (Ni or Co in ZrO_2) anode, and a Y_2O_3-stabilized ZrO_2 solid electrolyte capable of transporting oxygen ions. Electrical connectors between the cells were lanthanum chromite, which is electronically conductive and served the same function as the metallic bipolar intercell connectors in the MCFCs.

The Westinghouse cells had a tubular configuration and operated at about 1000°C. At the anode, hydrogen and carbon monoxide in the fuel gas react with O^{2-} ions from the electrolyte, giving up electrons to an external circuit and forming water and carbon dioxide as the products. At the cathode, oxygen accepts electrons from the external circuit, forming O^{2-} ions in the electrolyte. The electrodes were very thin, typically about 25 μm, while the electrolyte and interconnectors had thicknesses of several hundred micrometers. In a power-plant application, 100 or so cells would be connected electrically in series with parallel gas flows to form a module. The SOFCs were of a simpler design than the MCFCs, but the thermodynamic efficiency was 10-20% lower. The thought was that the SOFCs would be used in smaller, lower-cost plants with a pressurized gas and steam turbine bottoming cycle. An overall efficiency of at least 60% was anticipated for this type of system.

The program goals for 1985 were to demonstrate (1) acceptable cell lifetime and performance, (2) a thermal management system in a submodule of 15-25 cells, and (3) a 5-kW generator. Argonne had the same role as in the other programs, that is, to provide technical management and support. The support work included independent cell testing to assess the technology, modeling to gain insight into the cell behavior, and other general technical support.

By 1984, Westinghouse had obtained stable cell performance for more than a year and had operated a 300-W unit. The following year, over 9,000 h of stable performance was achieved, and plans were made to scale up the system to 5 kW in 1986. Argonne management of the Westinghouse program was terminated because of a possible conflict of interest with Argonne's development work on their monolithic solid oxide fuel cell (MSOFC).

Monolithic Solid Oxide Fuel Cell. In 1981, workers in CMT came up with a novel design for a solid oxide fuel cell capable of outstanding performance. Because the cell components were all thin solids, it appeared that they could be fabricated into a compact, monolithic structure that would achieve a power-to-volume ratio about 100 times that for molten carbonate cells. Development work on MSOFCs began at ANL in 1983 as a joint effort of the CMT and MST Divisions supported by DOE and the Defense Advanced Research Projects Agency (DARPA). The DARPA support derived from the fact that a compact, very high-power fuel cell could have a number of military uses.

A monolithic fuel cell stack, along with the chemical reactions and current-flow pattern, is shown in Fig. 5-9. The electrode materials and chemical reactions are basically the same as those in conventional tubular SOFCs. The two electrodes and intervening electrolyte, each having a thickness of about 0.025 to 1.0 mm, were fabricated in one piece as a thin, three-layer sheet formed into a honeycomb-like cell with a structure much like corrugated paperboard. The cells were stacked with sheets of electrically conductive material between them to provide a current path from cell to cell, similar to the bipolar plates used in MCFCs. Fuel and air were passed through the cells in alternating passages through the honeycomb. At the operating temperature of about 800-1000°C, hydrocarbon fuels were reformed within the cell, so the cells could be operated with hydrocarbon, as well as hydrogen or other fuels. The cells could be arranged for parallel flow of the reactants, or stacked with

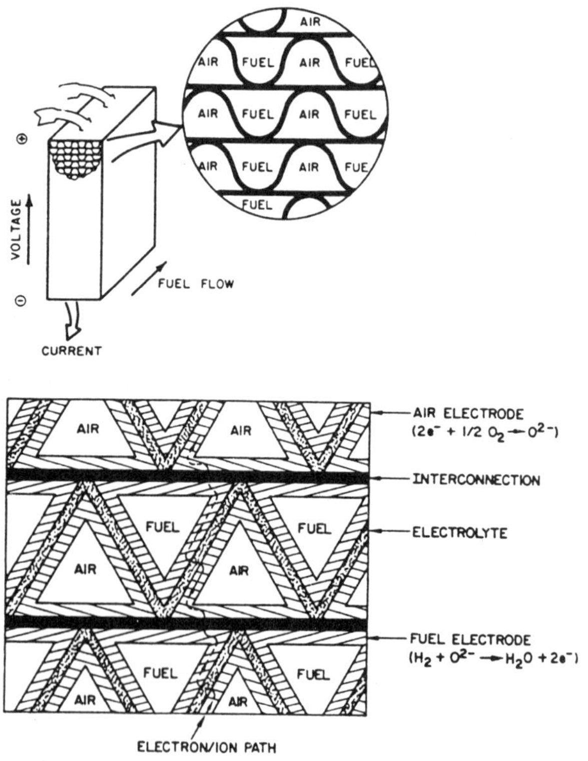

Fig. 5-9. Monolithic Solid Oxide Fuel Cell

each cell oriented 90° from the adjacent ones to establish a cross-flow pattern. One of the advantages of the monolithic design was that manifolding of the gaseous reactants and products was relatively simple. The cross-sectional areas of the gas passages in the honeycomb were equivalent to a diameter of 1-2 mm.

The cathodes (on the air side) consisted of strontium-doped lanthanum manganite ($LaMnO_3$), and the anodes (on the fuel side) were a nickel-yttria (Y_2O_3)-stabilized zirconia (YSZ) cermet. The solid electrolyte was Y_2O_3-stabilized zirconia ZrO_2, which conducted oxygen ions between the electrodes. The material connecting the cells in a stack was strontium-doped lanthanum manganite. It is interesting to note here that the research leading to the selection, synthesis, and testing of these materials was highly dependent on advances in solid-state science that were occurring very rapidly at that time.

Development of the electrodes was tricky. The cathode needed sufficient porosity to permit a reasonable rate of oxygen diffusion to the cathode/electrolyte interface where the oxygen molecules react. The electrolyte had to be very thin so its electrical resistivity (for oxygen ions) would not limit the current density (and power) too much, while at the same time it had to be leak-tight with no cracks or open porosity. In the anode, the electrical conductivity depended on particle-to-particle contact, which was lost if the nickel content was too low. If it was too high, however, thermal expansion of the electrode was too great to match it with the other layers.

Fabrication of the three-layer composites was an equally challenging task. Sequential tape casting, coextrusion, and other procedures were investigated with some success. A critical factor during fabrication was to match not only the coefficients of thermal expansion of the finished components, but also their shrinkage patterns during firing in the fabrication process. Any gas leakage between the electrodes would cause a loss of efficiency, and a great deal of effort went into the elimination of cracks or open porosity. By 1986, this effort was largely successful, and two cells were operated for more than 700 hours at a temperature of 1000°C and a current density of 50 mA/cm^2, using hydrogen fuel and air. Successful operation was also achieved with hydrocarbon fuels (methane, ethane, propane, octane), ethanol, and natural gas containing 5 ppm sulfur. By 1987, essentially all the defects had been eliminated, and the cell performance was approaching the theoretical level.

Although the basic technology had been demonstrated on a small scale, much work was going to be needed in scaling the system up to practical power level of at least a few kilowatts. A decision was made that a team consisting of ANL, Allied Signal Aerospace/AiResearch, and Combustion Engineering should do this development work. Allied Signal was to be the prime contractor and ceramic fabricator, Argonne was to develop and improve materials technology, and Combustion Engineering was to develop and test fuel-cell systems. Table 5-6 gives fuel-cell performance projections for different types of cells powering electric vehicles.

Table 5-6. Fuel Cell Performance Projections for Electric Vehicles

Fuel Cell Types	Power Density, W/L
Conventional Phosphoric Acid	302
Proton Exchange Membrane	1,575
Molten Carbonate	700
Monolithic Solid Oxide	2,779

PERSONNEL

The MCFC program continued into the 1980s, and work on the MSOFCs began in 1983. The funding agencies at the time felt very strongly that Argonne's role in the programs had to be limited to program management and technical assistance to the agency, with only a limited experimental effort supporting the technology development. In both fuel-cell programs, the basic technological concepts had been demonstrated at ANL by the mid-1980s, and remaining development, scale-up, and demonstration efforts were relegated to industry. In the Division, the result was that the sizable research effort in the early 1980s dwindled down to a small group of four or five people by the end of the decade. Up until about 1987, Dean Pierce and Darrell Fee were the group leaders for the molten carbonate and solid oxide programs, respectively, both of them reporting to John Ackerman, who was also the technical manager for all DOE fuel-cell programs. In 1987, Dean and Darrell both left the fuel-cell program, and Mike Myles became responsible for all the CMT electrochemical programs. Jim Smith assumed responsibility for the molten carbonate work, and Chuck McPheeters was given the solid oxide program. Later on, Mike Krumpelt took on the solid oxide work, and became particularly recognized for his assessments of potential fuel-cell applications.

Other individuals who worked on fuel cells in the 1980s include Paul Blackburn, Alan Brown, Terry Claar (MST), Glenn Cook, Dennis Dees, Tony Fraioli, John Heiberger, Hann-Sheng Huang, Tom Kaun, Kim Kinoshita, Verne Kolba, Ted Kraft, Gene Kucera, Frank Mrazek, Roger Poeppel (MST), John Rajan, Jim Sim, Joe Stapay, Bob Steunenberg, John Young, and Stan Zwick. Judy Popik was the principal secretary for these groups; Marlene Bukowski and Alice Birmingham also provided secretarial services.

Coal Utilization

Fluidized bed combustion of coal in a limestone bed continued to be the Division's major coal technology program in 1980.

ARGONNE PREMIUM COAL SAMPLE PROGRAM

In 1984, Argonne was awarded the Premium Coal Sample Program, the objective of which was to provide the coal research community with well characterized and stabilized samples of coal for research purposes, *e.g.*, development of a general understanding of the basic structures and the fundamental reaction processes of coal. The program was a joint effort between the Chemistry and Chemical Technology Divisions. The responsibility of CMT was design, construction, and operation of the coal-processing facility. The Chemistry Division was responsible for characterizing and administering the coal sample bank.

Two separate, sealed enclosures were constructed in Building 211 (formerly the cyclotron building) between Buildings 200 and 205. Each enclosure had a volume of approximately 1,500 cubic feet. A nitrogen atmosphere containing less than 100 ppm by volume of oxygen was maintained in each enclosure. In the first enclosure, about 3/4 metric ton of coal could be crushed and ground to the desired particle size. The crushed material was blended in a large blender located between the two sealed enclosures and then transferred to the second enclosure, where it was apportioned by means of a spinning riffler either to 5-gallon glass carboys or to a filler-sealer machine. This automatic machine metered the coal in either 5- or 20-g amounts into amber borosilicate glass ampoules, which were then flame sealed. About 5,000 of the 20-g ampoules and 10,000 of the 5-g ampoules were produced in each coal-sample run. The balance was stored in the 5-gallon carboys for packaging later and distribution to the research community. A variety of coals were processed in the mid-1980s.

SEPARATION OF COAL MACERALS

An outgrowth of the Premium Coal Sampling Program was a research program in the Chemistry Division on coal macerals. Macerals are microscopically distinguishable organic remains, which are present in all coals. The separation of coal macerals from the mineral matter and sulfur content of coal could make possible the tailoring of feedstocks for optimum utilization of coal and for more efficient coal-conversion technologies.

The Chemistry Division used density-gradient centrifugation for separation of macerals. By this procedure, only milligram to gram amounts of macerals could be separated in a single run. Therefore, the CMT Division was asked to develop a continuous process for production of large quantities of macerals. Two process concepts were investigated, both based on the fact that when a paramagnetic fluid in placed in a magnetic field gradient, the apparent density of the fluid increases. The magnetic field could, in theory, be "tuned" to produce a density at which the coal macerals could be separated. The first concept was a simple flow system in which separations were attempted at the very high magnetic fields produced by a superconducting quadrupole magnet, using manganese chloride as the paramagnetic fluid. In the second concept, centrifugation was carried out using a colloidal suspension of magnetite, which has a magnetic susceptibility 1,000 times that of manganese chloride. This allowed use of a simple electromagnet. Unfortunately, neither system worked very well, so the program was terminated.

FLUIDIZED-BED COMBUSTION OF COAL

At the end of the 1970s, the DOE concluded that atmospheric fluidized-bed combustion (AFBC) of coal was ready for commercialization. The two major applications of this technology were generation of steam for industrial purposes (boiler sizes of 50,000 to 200,000 lb of steam per hour), and for generation of electricity in large utility boilers. An AFBC steam boiler at Georgetown University had been operated without major problems, and its builder offered an entire series of AFBC boilers for commercial use. The Tennessee Valley Authority (TVA) was proceeding with construction of a 20-MW AFBC pilot plant and design of a 200-MW power plant.

In 1980, programmatic control of CMT's fossil energy program was transferred from DOE Headquarters to the Morgantown Energy Technology Center (METC) in Morgantown, West Virginia. Additional refinements of AFBC technology by CMT were authorized. The CMT program was aimed at solving problems, filling information gaps, improving process operability and economics, and reducing environmental impacts. In the early 1980s, work was concentrated in two areas: (1) development of a methodology for predicting the amount of a given limestone or dolomite sorbent required to retain effectively the SO_2 produced during combustion of coal, and (2) development of a hydration process for enhancing the extent to which a sorbent could be sulfated.

Prediction Methodology for Sorbent Utilization. A key to the widespread use of the AFBC technology was the ability to predict accurately the sorbent feed rate required for different limestones and dolomites that keep SO_2 emissions within environmental standards. To provide a prediction methodology, Darrell Fee undertook the development of two models, one to predict in-bed sulfation and the other to predict particle attrition within the fluidized bed. The background information for development of these models was published in a sourcebook on the utilization of limestones and dolomites (ANL/CEN-80-10). This sourcebook contained the chemical, physical, and morphological properties of 29 limestones and dolomites, as well as their distribution in the United States.

A CMT in-bed sulfation model was developed, for which knowledge of sorbent-specific sulfation behavior was required. This was best obtained by thermogravimetric analysis (TGA). For this purpose a standard TGA was defined. Comparison of model predictions with actual combustor experience showed that the model underpredicted the sulfur retention for a given Ca/S molar ratio by 20 to 30%. Two reasons were advanced for this discrepancy: (1) the extent of sulfation in the freeboard region above the bed was not taken into account, and (2) the standard TGA did not adequately show the in-bed sulfation behavior of a sorbent. In a large Babcock and Wilcox boiler, the amount of sulfur captured in the freeboard region was 9 to 10% of the total, which is a significant amount. Experiments also showed that the maximum amount of sulfation found by the TGA was generally slightly lower than that in an actual AFBC. The reasons for this were not pinned down.

In 1982, the model was revised to include sulfation in the freeboard area and to take into account the variation in sorbent particle size in the fluidized bed. (Equations for the model appear in ANL-83-55, pp. 64-65.) Better agreement with operating experience was achieved with this model. The model was something of a *tour de force*, but it was never fully perfected. (Darrell Fee, who had been the leader of this effort, became group leader of the SOFC group at that time.) Other people who contributed to development of the model were Liang-Shih Fan, Irv Johnson, John Lenc, Mike Myles, John Shearer, Greg Smith, Ira Wilson, and S. Wong.

A stochastic model was developed to predict the change in the size distribution of sorbent particles as a result of attrition. Pneumatic-transport attrition data were used to verify the model. The model predicted the experimentally measured particle-size distribution to within half an order of magnitude.

Hydration-Enhanced Limestone Utilization. In a search to increase the sulfation of limestone, hydration and subsequent in-bed dehydration of partially spent sorbents greatly enhanced their sulfur-retention capacity. Sulfation of limestone is impeded by buildup of a surface layer of $CaSO_4$. Treatment with water hydrates residual CaO to form $Ca(OH)_2$, with a resultant expansion that cracks open the surface layer. Subsequent dehydration of the $Ca(OH)_2$ yields a highly reactive form of CaO for reaction with additional SO_2.

A demonstration of the hydration process in CMT showed that nearly 90% of the CaO could be utilized by employing five hydration-sulfation cycles conducted alternately in the same fluidized-bed unit. Similar results were obtained at the Grand Forks Energy Technology Center. However, elutriation losses of fine particles were high because of decrepitation of the sorbent particles caused by the hydration-dehydration cycles. Make-up of the sorbent losses resulted in a high sorbent-to-coal ratio for achievement of high calcium utilization (see Fig. 5-10).

Sorbent hydration in a fluidized-bed unit was also investigated. With residence times of only 10 to 20 minutes, the fluidized-bed hydrator would be only 2.5 to 5% of the size of the fluidized-bed combustor. Economic assessments projected operational savings of $214,000 per year for an industrial boiler producing 100,000 lb of steam per hour. The projected period for payback of the additional

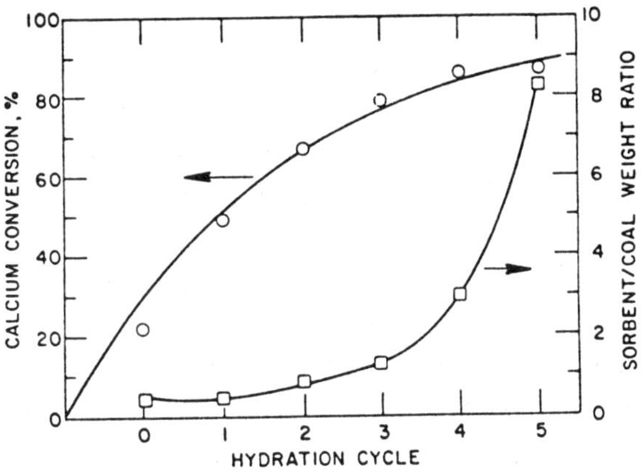

Fig. 5-10. Results of Limestone Hydration Experiments

capital costs required to install the hydration process was 1.5 yr for sulfation and hydration in the same fluidized-bed unit and 1.1 yr for separate sulfation and hydration units. The DOE terminated financial support of the hydration process despite industrial interest in it because of the increase in sorbent utilization.

Because atmospheric fluidized-bed combustion was considered ready for commercial use, emphasis in CMT was shifted to development of pressurized fluidized-bed combustors (PFBCs) for utilities and to advanced fluidized-bed concepts such as multistage combustion and circulating beds.

CLEANUP OF HOT GASES FROM PRESSURIZED FLUIDIZED-BED COMBUSTORS

Work continued in the 1980s on hot gas cleanup technology for PFBCs. The high-temperature, high-pressure off-gas generated from a PFBC is expanded through a gas turbine to recover energy. Alkali metal sulfates are formed by chemical reactions of alkali metal chlorides in the off-gas by the reaction

$$2\,NaCl + H_2O + SO_2 + 1/2\,O_2 \rightarrow Na_2SO_4 + 2\,HCl$$

Atmospheric-pressure studies in the late 1970s showed that the alkali metal chlorides could be removed effectively by beds of diatomaceous earth and activated bauxite. In the early 1980s, an experimental apparatus that allowed tests to be made at 10 atm pressure was fabricated and placed in operation (see Fig. 5-11). The hot flue gas was first passed through a primary (hot) cyclone to remove the bulk of the particulate matter. A small fraction of the gas was then diverted through the bauxite sorber unit to test its effectiveness for removal of the alkali compounds. The main flow of the flue gas continued through a cold cyclone, after which both gas streams were combined and filtered through a metallic filter for final knockdown of the fine particulate material. This system removed essentially all the particulate matter from the flue gas. The activated bauxite was regenerated by water leaching. The capture efficiency of the activated bauxite for alkali metal compounds was not decreased with repeated regeneration.

Tests were conducted to determine the extent to which activated bauxite and diatomaceous earth were themselves sources of alkali. Under conditions closely resembling those of actual PFBC operation, the amount of sodium and potassium evolved from activated bauxite was very small. On the other hand, evolution of alkali metal compounds from a diatomaceous earth (Emathlite) continued at a

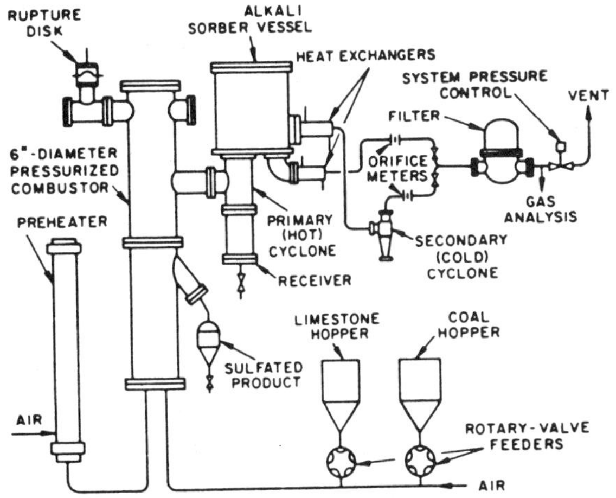

Fig. 5-11. Pressurized Fluidized Bed Facility

very high rate over the 50-hour duration of the experiment. Diatomaceous earth was therefore dropped from further consideration as an adsorbent.

The concentration of alkalis entering the absorber bed was analyzed by a computer-controlled, flame-emission spectrometer supplied by Ames Laboratory. A batch-type alkali and particulate sampling train (APST) was also installed as a backup for analysis of alkali metal content of the off-gases. Results obtained with the Ames emission spectrometer were often one or two orders of magnitude lower than those obtained with the APST. This difference was traced to capture or absorption of both NaCl and KCl vapors on the walls of the Type 304 stainless steel sampler to the Ames analyzer. As a satisfactory substitute for the Type 304 stainless steel was not readily available, use of the Ames sampler was abandoned. To obtain the alkali content of the PFBC off-gas, small alkali absorbers were installed in the off-gas stream exiting from the PFBC. (No stainless steel sampling line was used.) After a given time for absorption of alkali metal compounds, analyses of the analytical beds gave a time-averaged alkali metal concentration. Excellent results were obtained by this analytical method. The sodium vapor concentration resulting from combustion of a high-sodium lignite was 1.42 and 1.50 ppm as measured by two activated bauxite beds and 1.30 ppm by a diatomaceous earth bed. Since more than 99.8% of the NaCl was removed in a bauxite sorber unit, the alkali metal specification for an industrial gas turbine of 0.024 ppm was concluded to be readily achievable.

In about 1988, the PFBC was fitted with silicon carbide candle filters. All of the off-gas from the PFBC was passed through these candle filters. These, in conjunction with a hot cyclone through which the exiting flue gas was passed, removed over 99.8% of the particulates from the flue gas, which is adequate removal of particulate from gas to be expanded through a turbine.

The hot off-gases from fluidized-bed combustion contain organic matter that consists partly of carcinogenic or mutagenic polycyclic aromatic hydrocarbons (PAHs). The only analytical method then available for analyzing these gases for those organic compounds was gas chromatography (GC) combined with a mass spectrometer (MS) and a computer spectra-reduction system. The CG separates the complex mixture of components, and the MS identifies the mass of each compound. Before analysis by this procedure, the PAHs had to be concentrated by a factor of a million. This was done by adsorbing them on a resin and then desorbing them by volatilization or by solvent extraction. Nonuniform behavior of the PAHs in stack gas cast doubt on the veracity of the analyses.

On-line monitoring systems were then investigated because they would eliminate the problems of sampling and sample concentration and also provide real-time data. Double (or triple) mass spectrometry, multiphoton laser ionization mass spectrometry, and induced fluorescence in a molecular beam were evaluated as possible analytical methods, but proved inadequate for in-line monitoring of PAHs in stack gas, and the program was discontinued.

MANAGEMENT ACTIVITIES

A variety of management activities were performed for the Morgantown Energy Technology Center. They comprised the following:

1. A PFBC technology assessment.
2. An evaluation of cyclone performance in PFBC applications.
3. An assessment of methods for the discharge and handling of solids from PFBC combustors.

4. A study of heat-exchanger design and operation for PFBCs.
5. A study of the technology status of particulate removal from high-temperature, high-pressure combustion gases.
6. A survey of instrument and measurement needs for FBCs.
7. A review of the hot gas cleanup system proposed for the Curtis-Wright 13-MWe pilot plant.
8. An assessment of international fossil energy programs based on reviews of the literature, foreign-travel trip reports, and interviews of foreign visitors and U.S. travelers.
9. Development of a Materials-Process-Product-Model computer code to obtain cost and process information for various energy technologies.
10. A conceptual design for an AFBC for installation in the Argonne power plant.
11. An up-to-date assessment of fluidized-bed technology, both in the U.S. and elsewhere.
12. Design of advanced fossil energy systems as a partial update of a 1976 Energy Conversion Alternatives Study performed by NASA (NASA TM-73871).
13. Technical assessments for environmental control for coal utilization. These were directed mainly to NO_x from various coal-related energy technologies.
14. Management of the Atmospheric Fluidized-Bed Cogeneration Air Heater Experiment for DOE, the objective of which was to assess materials and process performance of in-bed air heaters for co-generation of electricity and process steam in an AFBC.

Walt Podolski and several other individuals associated with the coal program contributed to these various efforts.

PERSONNEL

Irv Johnson remained in charge of the coal program until 1981, when Al Jonke returned to the Division and assumed that responsibility. Other CMT personnel associated with the program in one way or another in the 1980s were: Rajiv Agrawal, Roy Benson, Milt Blander, Amrit Boparai, Alan Brown, Erv Carls, Martin Clemens, Gary Dyrkacz, Tony Engelkemeier, Magdy Farahat, Darrell Fee, Jack Fischer, Bob Heinrich, Jim Helt, Richard Henry, Hann-Sheng Huang, Terry Johnson, Sheldon Lee, John Lenc, Peter Lindahl, Shelby Miller, Russell Mowry, Mike Myles, Robert Nelson, Frank Nuñes, Arthur Pelton, Walt Podolski, Leo Raphaelian, Carmen Sabau, Satish Saxena (Consultant), John Schneider, Charlie Schoffstall, John Shearer, Gregory Smith, Sherman Smith, John Stockbar, Bob Swaroop, Bill Swift, Gale Teats, George Vandegrift, Ravi Varma, Don Vissers, John Vogel, John Weber, Florence Williams, Ira Wilson, John Young, and Tony Ziegler.

Magnetohydrodynamic Heat and Seed Recovery Technology

Throughout the 1980s, the CMT Division continued as the lead division in a multi-divisional effort called the MHD Heat and Seed Recovery Project. This project, headed up by Terry Johnson, was organized to develop components downstream from the MHD generator (also called the "channel-diffuser"). These components comprise the MHD bottoming cycle, which is described in Chapter 4. This work was done in cooperation with other DOE contractors, including Babcock and Wilcox, the University of Tennessee Space Institute, and Mississippi State University. The work performed by CMT was directed to (1) control of NO_x, and SO_x, (2) measurements of the fouling of convective steam-generating tubes in the radiant boiler, (3) thermodynamic support studies, (4) recovery of the K_2CO_3 seed material, and (5) evaluation of candidate materials of construction.

The Fossil Energy Users Laboratory (FEUL), initially called Argonne Magnetohydrodynamics Process Engineering Laboratory (AMPEL), was constructed to accommodate a wide range of experiments under conditions simulating those in the radiant heat boiler where steam is produced from hot gases flowing from the MHD generator. Shown in Fig. 5-12, the facility had two test legs. One leg had an oil combustor capable of burning liquid fuel in various types of oil-solid mixtures and coal slurries; the other leg had a two-stage cyclone coal combustor. These combustion units were housed in two metal buildings (approximately 25 ft wide by 55 ft long by 30 ft high) erected near the Argonne power plant. Instrumentation and facility-control systems were located in an adjacent trailer.

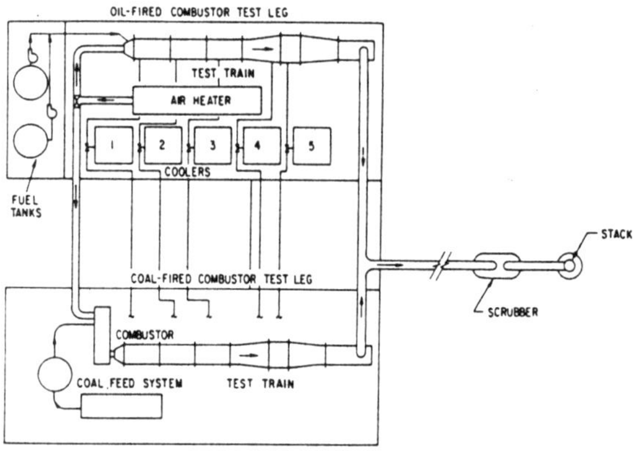

Fig. 5-12. Major Components of Fossil Energy Users Laboratory (FEUL)

CONTROL OF NO_x AND SO_x EMISSIONS

Most of the work in this area was on reducing the concentration of nitric oxide (NO) in the gas entering the radiant boiler, which contained NO at a concentration of 3,000 to 10,000 ppm, to below the emission standard of about 500 ppm in the gas exiting the boiler. Tests were run in FEUL to verify a chemical kinetics model that was based on reaction-rate constants for 29 reactions involving oxygen, carbon, hydrogen, and nitrogen. The model showed that the required NO concentration in the exit gas could be achieved by cooling the gas from its entrance temperature of 2000°C to the exit temperature of 1500°C at a rate no higher than 200 to 300 degrees per second. The tests showed that the model could be used to predict reliably the behavior of NO in an MHD power plant.

The SO_x was efficiently removed from the combustion gas by reaction with the K_2CO_3 seed material to produce K_2SO_4.

FOULING OF STEAM GENERATOR TUBES IN THE RADIANT BOILER

The work conducted in FEUL consisted mainly of fouling tests designed to investigate the nature and thickness of deposits formed on the steam tubes in the radiant boiler. Such information was needed for the design of the radiant heat boiler. The fouling conditions in this boiler were especially severe because the gas coming from the MHD generator contained about 2 wt% K_2SO_4 and 0.2 wt% coal ash. In fouling tests conducted through 1987, tube banks were exposed to combustion gas at temperatures up to 1500°C. The test train for the fouling tests is shown in Fig. 5-13. These tests demonstrated that the character of the tube deposits depended primarily on whether the gas temperature was above or below 1063°C, the melting point of K_2SO_4. At gas temperatures below 1063°C, entrained seed and ash formed a low-density, loosely adhering deposit that was readily removed by soot blowing. At gas temperatures above about 1300°C, the outer surface of the growing deposit reached the melting temperature, and the depositing seed and ash material flowed off the tubes, resulting in a stable deposit thickness and a steady-state heat-transfer

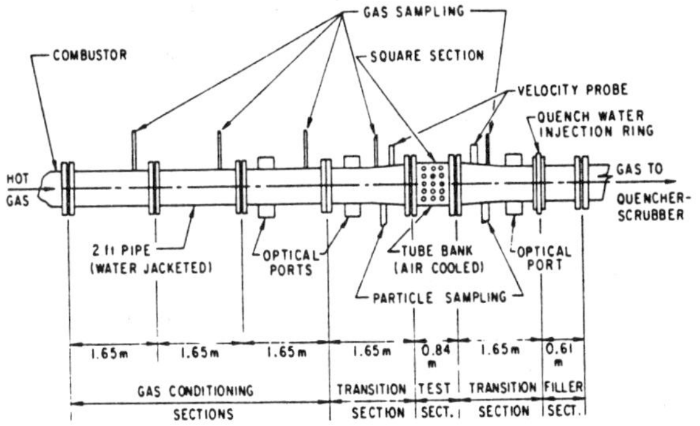

Fig. 5-13. Test Train for Fouling Tests in Fossil Energy Users Facility

coefficient. Gas temperatures between 1100 and 1300°C produced dense, adherent deposits that were difficult to remove by soot blowing

Argonne developed a model for calculating the fouling efficiency (gas-side, heat-transfer coefficient of fouled tube divided by that of a clean tube) of a convective heat exchanger operating at temperatures below about 1,300°C. As shown in Table 5-7, calculated and measured fouling efficiencies were in good agreement.

Table 5-7. Measured and Calculated Fouling Efficiencies

	Fouling Efficiency		
Test	Measured	Calculated	Difference, %
1	0.80	0.82	2.5
2	0.85	0.75	12
3	0.73	0.77	5.5
4	0.69	0.69	0

FOLLOW-UP APPLICATIONS OF FEUL

When MHD studies using FEUL waned in the late 1980s, FEUL was used to investigate (1) SO_x-NO_x control in slagging combustors, which were being developed for potential application to coal-fired gas turbine systems and for retrofitting existing gas- and oil-fired boilers, and (2) char combustion.

In the slagging combustion studies, an investigation of SO_x-NO_x was conducted in a staged cyclone coal combustor with limestone injection. The tests were performed in the FEUL coal-fired test leg, which consisted of a primary combustion stage into which coal and primary air were injected, and a secondary combustion stage. Funding was provided by the Coal Technology Corp. and the Illinois Coal Industrial Committee. Limestone and a pressure-hydrated limestone were used to getter SO_x. Best results (80% reduction of SO_2 emission) were realized when the first-stage combustor was operated in an oxygen-deficient mode (60 to 70% of stoichiometric air), and the second stage was operated with excess air (160 to 170% of stoichiometric). This work also showed that satisfactorily low NO_x emissions could be achieved in slagging combustors.

The pyrolysis (or mild gasification) of coal yields volatiles that are recovered for their chemical value and a clean char that can be utilized in industrial or utility boilers. In FEUL, very stable combustion of a low-volatile char was realized. Combustion efficiencies ranged between 91 and 96%, which is lower than the 99.5% required for boiler application. The work was halted before conditions that might have increased the combustion efficiency could be explored.

RECOVERY OF POTASSIUM CARBONATE SEED MATERIAL

The primary function of the MHD seed-reprocessing system was conversion of K_2SO_4 discharged from the radiant boiler to K_2CO_3 for recycle to the coal combustor. The most promising chemical process was high-temperature reduction of K_2SO_4 to K_2S and conversion of the K_2S to K_2CO_3. With minimal funding in 1980 and 1981, CMT limited its efforts to managing seed-reprocessing studies by industrial firms and to developing a computer code for single- and three-stage contactors. A sharp reduction of total MHD funding in 1982 forced termination of this work. Seed-recovery work was resumed again in 1988 for one year when CMT investigated the reduction of K_2SO_4 in a molten pool of sulfates and sulfides contained in a ceramic-lined reactor. A process flowsheet based on thermodynamic studies was developed and provided to Babcock and Wilcox for testing by them.

SUPPORTING STUDIES

Thermodynamic Support Studies. Thermodynamic support studies were conducted in 1980 and 1981 to investigate reactions between slag constituents and K_2SO_4 and to evaluate the characteristics of the seed-slag deposits in the radiation heat boiler. At temperatures from about 2000 down to 1400°C, the K_2SO_4 reacted with the slag to form potassium silicates, aluminates, and aluminosilicates. Recovery of potassium from these materials was judged to be very difficult, and probably uneconomical. Because of these reactions, it was estimated that 6-10% of the potassium would be unrecoverable.

Other support studies included electron-probe analyses of the slag-seed deposits, thermal diffusion of potassium in the slag, and thermodynamic modeling of coal-seed-slag interactions. This work was also a casualty of the sharp reduction of MHD funding in 1982.

Materials Studies. Corrosion of heat-exchanger tube materials by MHD combustion gases was investigated in 1987 and 1988. Austenitic and ferritic steels were exposed to MHD conditions for up to 2,000 hours in autoclave tests. Austenitic steels performed satisfactorily under steam superheat conditions; corrosion rates of ferritic steels were unacceptably high. Neither austenitic nor ferritic steels performed acceptably at the temperatures of an intermediate air heater. The work was terminated before tests of other candidate alloys could be done.

PERSONNEL

Terry Johnson was in charge of this work. Because of its highly interdisciplinary nature, many individuals from other divisions were involved. Those from CMT included Paul Blackburn, Carl Johnson, Stan Johnson, Jiri Klinger, Bali Misra, Mike Myles, Charlie Schoffstall, Atul Sheth, Eugene Smyk, Bill Swift, Doug Warinner, and Stan Zwick.

Energy and Municipal Waste Technology

ENERGY FROM MUNICIPAL WASTE

In 1982, Argonne was designated as the lead laboratory for DOE's Energy from Municipal Waste program. The purpose of the program was to undertake research that would enhance the ability of the private sector to complete the development, demonstration, and commercialization of technologies for deriving energy from municipal solid waste (MSW). The principal CMT tasks were to (1) investigate the basic thermodynamic mechanisms associated with the pyrolysis of MSW, (2) evaluate the catalytic upgrading of pyrolytic liquids, and (3) analyze the external combustion of MSW as a supplemental source of energy for a conventional steam boiler.

Pyrolysis of MSW. Although pyrolysis of MSW had not previously been technically or commercially successful, the possibility that MSW could be degraded to produce stable, transportable gaseous and liquid fuels continued to be considered seriously. The basic mechanisms of pyrolysis were not understood. In CMT, a variety of cellulosic materials—paper, newsprint, packing materials, wood wastes, grass clippings, *etc.*—were pyrolyzed in a bench-scale reactor. Gases were analyzed by mass spectrometry. Liquids were analyzed by gas chromatography and by gas chromatography/mass spectrometry. So complex were the liquids that complete identification of molecular species was not possible, but they appeared to be substituted furfurals and furfurols. Various amounts of tars and chars were also produced in the pyrolysis experiments. The reactor experiments were supplemented by determination of activation energies obtained by thermogravimetric analysis.

Two subcontractors, the Solar Energy Research Institute (SERI) and the University of Arizona, were also involved in this effort. Using a flame pyrolyzer with a molecular-beam, mass-spectrometer analyzing device, SERI studied the influence of reactor conditions on the compositions of solid and gaseous products at low pyrolysis temperatures. Investigators at the University of Arizona performed pyrolysis tests in a high-pressure autoclave reactor. Studies of MSW pyrolysis were phased out in 1988.

External Combustion of MSW. In a large-scale combustion demonstration of the use of MSW as a supplemental fuel in existing steam boilers, a refuse-derived fuel (RDF) has been cofired with coal. Preparation of the RDF resulted in difficult storage and transportation problems with high capital and maintenance costs. Co-firing of RDF with coal was, therefore, not economically beneficial. The CMT Division investigated an alternative—burning the coal and MSW in separate units. The gaseous combustion products would be combined with those from combustion of coal for generation of steam. Economic analyses of MSW combustion in a dedicated unit showed that the cost of MSW-derived power would be competitive with that from coal.

In 1988, a computer model was developed to simulate combustion of MSW in a combustor shown schematically in Fig. 5-14. The MSW pyrolyzes and burns as it is transported horizontally along the grate from left to right. Part of the heat from burning the waste is recovered in water walls that line the flame zone. The remainder of the heat in the flue gases is used for generation of steam. The model provided valuable information on the effects of dimensional and operating variables.

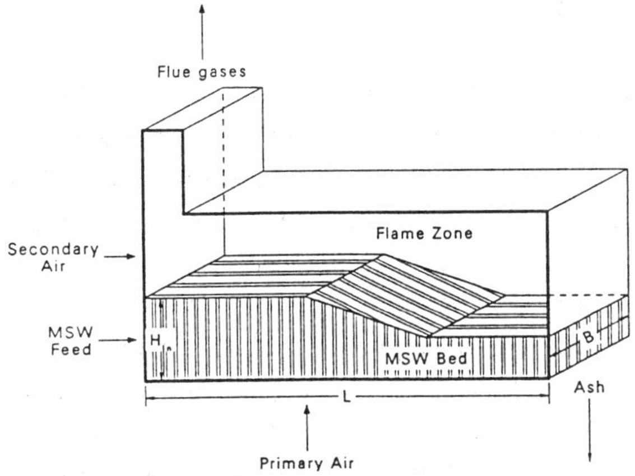

Fig. 5-14. Municipal Solid Waste Combustor

HAZARDOUS ORGANIC WASTES

The technical feasibility of a microwave-assisted chemical process for detoxification (conversion to harmless materials) of hazardous organic wastes, such as trichloroethylene (TCE), polychlorobiphenyls, dichlorobenzene, and ethyl ketones, was investigated in the late 1980s. Initial results showed that more than 80% of the TCE adsorbed on active carbon (loaded with copper and chromium catalysts) was detoxified in one pass when heated by microwave radiation to temperatures approaching 400°C. At 500°C, more than 98% was detoxified.

The work was extended to testing detoxification in a microwave-induced plasma reactor. A stable plasma was first established at atmospheric pressure, after which a chlorohydrocarbon-bearing vapor was introduced. In a single pass, 95.9 to 98.3% conversion of C_2H_3Cl and C_2HCl_3 was achieved.

Shabir Ahmed, Amrit Boparai, Jack Demirgian, Jim Helt, Richard Henry, Romesh Kumar, Mike Myles, John Schneider, Ravi Varma, Ronald Wingender, and John Young worked on this program.

TREATMENT/DISPOSAL OF REACTIVE WASTES

Work was performed in the late 1980s on development of an integrated process for conversion of waste sodium, including sodium-containing radionuclides, to a glass form. In the process, shown in Fig. 5-15, sodium was injected into the reactor, using nitrogen as an aspirator and atomizer. An air stream containing a suspension of glass-forming materials was introduced to oxidize the sodium. Molten glass was formed directly in the reactor.

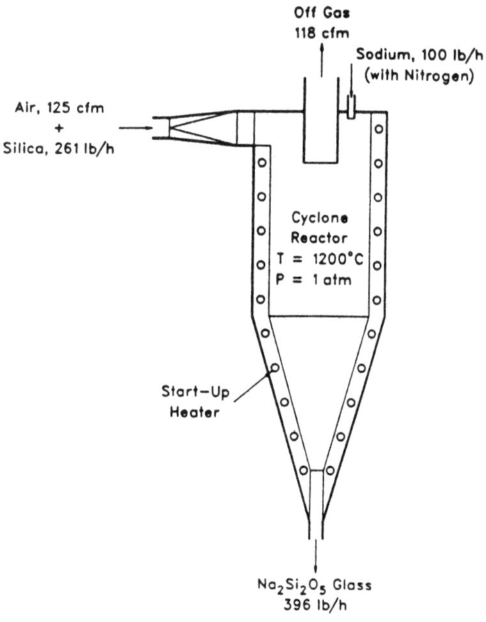

Fig. 5-15. Cyclone Reactor for Converting Sodium to Glass

If no radioactive contaminants were present, a simple soda-silica glass was a suitable waste form. If radioactive contaminants were present, other constituents, such as B_2O_3, Al_2O_3, and CaO, had to be added to provide a glass waste form having high resistance to leaching of the radioactive contaminants by water.

Argonne provided technical assistance to commercial demonstration of the process. Chemical Technology personnel in this program were Alan Brown, Jim Helt, Romesh Kumar, John Rajan, Ravi Varma, and Don Vissers.

Solar Energy

The ANL solar energy research program, which was a joint effort among several divisions, continued in the 1980s. This program supported the presidential goal in 1978 of achieving 20 quads of energy production by the year 2000 from solar energy. (A quad is 10^{15} Btu.) The Chemical Technology Division's activities included solar collector development, ice storage, and thermal storage. The solar energy program was somewhat unusual in that the results of the development studies were adopted quickly by industrial firms, some of which had been industrial participants, and placed on the market as commercial hardware in many cases. Funding for the solar-collector work, which was a large part of the CMT program, was being withdrawn gradually, and in 1983 the CMT program was transferred to the EES Division along with Bill Schertz, who had spearheaded the effort.

SOLAR COLLECTOR DEVELOPMENT

Three major types of solar collectors were under consideration: (1) flat plate, (2) evacuated tube, and (3) parabolic trough. Each had its problems. The flat-plate collectors had inadequate performance characteristics, the initial cost of the evacuated-tube collectors was too high, and those parabolic reflectors that tracked the sun had operational and maintenance problems. The non-imaging compound parabolic concentrator developed by CMT appeared to show the most promise, and the Solar Energy Group in CMT, having become the acknowledged leader in this technology, was the focus for commercialization activities, hardware tests and evaluations, and support for establishing standards. The group served as a clearinghouse for manufacturers and other solar researchers in universities and other government laboratories.

A study of seven first-generation collector designs showed optical losses of about 5-20% because of a necessary gap between the absorber and the collector. Most of the designs also had a low solar absorbance of approximately 0.8. Second-generation design studies at ANL resulted in two concepts—a four-facet "gaplossless" collector and a generalization of it, the multifacet gaplossless collector, both of which were modifications of the classic compound parabolic concentrator cusp. The concept was developed at ANL and patented by DOE. A third design overcame the problem of low solar absorbance by making an absorber-reflector system an optical cavity with increased absorbance through multiple-plate reflections. The final design to be studied employed a multifacet gaplossless reflector as a secondary concentrator coupled to a conventional focusing parabolic trough. This combination permitted a relaxation of the accuracy requirements for the primary mirror shape, with a resulting cost reduction.

Tests were conducted on nine collector modules: (1) three first-generation ANL prototype units, (2) three first-generation industrial units, and (3) three second-generation designs by ANL and the General Electric Company. The results showed that the ANL computer programs gave accurate predictions of the performance of all the designs, and that the optical losses were indeed recouped by the gaplossless and enhanced absorption designs.

By 1982, the nonimaging concentrating collectors developed at ANL had been introduced to the market and were being sold commercially, so the research program was phased out. A small follow-on effort continued, however, on the computer modeling, and evaluations were made on advanced coating materials that showed promise of increasing the absorptivity from about 0.78 to 0.92. The mode of operation was also being changed to allow steam generation in the collector instead of an organic heat-transfer fluid. Tests of a prototype steam-generating collector demonstrated 50% conversion of sunlight to steam at 135°C.

ICE STORAGE

The objectives of the ice storage program were outlined in the preceding chapter. The general idea was to develop passive refrigeration devices for heating and cooling buildings, using solar power. The near-term program objective was to determine the feasibility of the concept through theoretical and experimental studies. The longer-range goal was to develop the technology to the research-prototype stage, at which point it could be taken over by the industrial sector.

This work continued in the early 1980s, primarily on materials and hardware design. Cylindrical heat pipes were replaced by Roll-bond® panels (a proprietary product in commercial use), which are basically copper or aluminum sheets with integral fluid channels. The immediate objective was to increase the surface area available for deposition of ice and for heat transfer, and the new material was used both for the evaporators (underground) and the condensers (above ground). Several full-scale heat pipes were fabricated for outdoor use during the winter. The evaporator consisted of four Roll-bond heat pipes attached to a plastic cylinder. Each evaporator unit had about 3 m^2 (32 ft^2) of surface area. The aboveground condenser units consisted of Roll-bond panels having an area of about 15 m^2 (160 ft^2).

Initial attempts to operate the facility were unsuccessful in that the "freeze and release" procedure used on the laboratory scale did not perform properly in the scaled-up equipment. After suitable modifications, however, the system did function satisfactorily, and about 18 tons of ice was produced in 1982. Mathematical models of the freezing process were developed to permit extrapolation of the experimental results to other system designs.

THERMAL STORAGE

Three projects on thermal energy storage were in progress in 1980: (1) Stratified Storage and Analysis, (2) Testing of Latent Heat Storage Units, and (3) Development of Thermal Energy Storage Devices Using Cross-Linked, High-Density Polyethylene.

Estimates by others had indicated that thermal stratification in storage tanks could improve the performance by 5-15% over fully mixed systems. Typical tanks behave as mixed systems, and special designs were required to achieve stratification. Semiempirical modeling studies were conducted to predict accurately temperature-time and -space behavior within tanks under a wide variety of operating conditions.

Approximately 12 manufacturers had latent heat-storage units on the market or nearly ready to be marketed in 1980. Because the manufacturers' test procedures were not uniform, the performance results were inconsistent. This project consisted of testing available units in cooperation with Oak Ridge National Laboratory (ORNL) to assess the industry-wide performance of heat-storage units and to develop testing standards. Oak Ridge was then to do full-scale testing, using the final test procedure, and the results were to be published jointly by ORNL and ANL.

A cross-linked, high-density polyethylene (HDPE) developed jointly by Monsanto Corp. and the University of Dayton was of interest as a latent heat storage medium due to a phase change at 130°C involving a latent heat of about 40 cal/g. The material does not change shape or fuse at temperatures up to 150°C. A heat-storage device using this material could be used for solar-powered air conditioners ("chillers") and industrial processes requiring heat at temperatures up to 150°C. A prototype unit was designed for a 3-ton chiller. Ethylene glycol, water, or a combination of the two would be used as the heat-transfer fluid.

PERSONNEL

The solar programs were headed up by Bill Schertz, and CMT personnel involved in those programs in the 1980s included John Allen, Anthony Gorski, Arthur McGarity, and Alvin Wantroba. About 10 people from other divisions were involved.

DESICCANT STUDIES

A small program was started in 1982 on the use of solid and liquid desiccants for dehumidification in heating, ventilating, and air-conditioning systems. The DOE at that time was stressing the development of novel desiccants that would remove moisture with a minimum expenditure of energy. The work began in the solar energy group headed by Bill Schertz and was performed by Sidney Bourne, Jack Demirgian, Tony Fraioli, Jiri Klinger, Wally Kremsner, and Jack Parks.

In the initial studies, a promising family of solid desiccants, two manganese oxides, ε-MnO_2 and Mn_3O_4, were identified, and it was found that the rectangularly tunneled microfiber structure of these compounds results in minimum energies for their water-adsorption characteristics. There appeared to be a possibility that the weak adsorbate-adsorbent interactions of the microfibers might be used to provide easy mass transport of water into the interior of silica gel structures, which are limited in capacity by pore blocking during the water adsorption cycle. Composites of silica gel particles were prepared with MnO_2 and metal fibers as additives to determine their mass-transport and heat-transfer characteristics.

The concept behind the liquid desiccants was to develop a system that could be regenerated by a liquid-liquid separation, thereby avoiding the energy cost of vaporizing water. A search was initiated for organic liquids whose mixtures with water have the requisite phase-separation properties, and a computer model was developed to aid in predicting the performance of such systems. The organic compound N-cyclohexyl-2-pyrrolidone (CHP) had the necessary water-adsorption and liquid-phase separation properties when mixed with water, but there was a need to improve the degree of CHP-water separation. Work was done on pertinent phase diagrams and other properties to determine the effects of various additives. The water-miscibility properties of six amines and seven amides were determined. The phase diagram for water-diisopropylamine indicated that an excellent liquid-liquid separation occurred on heating to 65°C .

Nuclear Fusion Energy Research

Unlike many other CMT programs, the work on fusion energy, under the leadership of Vic Maroni, remained at a relatively constant level of effort (approximately 10 people) throughout the 1980s. This activity fell under the purview of Fred Cafasso, an associate division director, who transferred to the Chemistry Division in 1983. Dieter Gruen of the Chemistry Division was then given overall responsibility for Vic's group and the basic research group headed by Milt Blander.

The program continued to consist of (1) systems analysis and engineering experimentation, (2) materials research and development, and (3) neutron dosimetry and damage analysis. Some significant changes of emphasis, however, were in the wind. There was an interest in replacing the liquid lithium in the tritium breeder blanket with a solid lithium-bearing material. The Division was being asked to play a larger role in national and international fusion reactor designs, and joint projects with other ANL divisions, other laboratories, and industry. Ongoing work on neutron dosimetry and damage analysis was to be expanded. Work was to be discontinued on liquid lithium blanket processing and hydrogen isotope permeation, and operation of the lithium processing loop was curtailed.

SYSTEMS ANALYSIS STUDIES

Studies of fusion reactor systems, which had been in progress since 1969, had established the Division as an internationally recognized leader in the design and analysis of tritium-processing systems. In 1980, CMT personnel participated in several major design studies for tokamak fusion reactors, including the STARFIRE commercial reactor, the Engineering Test Facility (ETF), and the International Tokamak Reactor (INOR). In this international project, STARFIRE was managed by Argonne, ETF by Oak Ridge, and INOR by the International Atomic Energy Authority (IAEA), which involved the U.S., the European Community, the U.S.S.R., and Japan. These studies included the establishment of tritium-supply guidelines for fusion-reactor startup projections, design of a complete tritium fuel-processing facility, a parametric analysis of the STARFIRE vacuum system, design of a solid tritium breeder blanket, and evaluation of the effects of physical and chemical parameters on tritium (spill) cleanup systems.

A special study was conducted to consider mechanisms involved in the recovery of tritium from a solid breeder blanket. What seemed to be needed was a low-density blanket material having a fine grain size (about 1 μm) and a bimodal pore distribution, that is, small grains and fine porosity within fairly coarse particles (about 1 mm) with a much coarser porosity between the particles. This design study identified critically needed data such as the solubility of T_2O (H_2O) in Li_2O; compatibility of breeder material, coolant, and structural alloys; thermal conductivity and radiation stability of the breeder material; and the kinetics of tritium release. Properties databases were developed for the candidate materials.

Several major design studies were made in 1982. One study was aimed toward achieving a world-wide consensus on tokamak fusion reactor designs. Another study was focused on blanket design and involved both solid blankets (Li_2O) and liquid metal blankets (lead-83% lithium). In addition, the Division participated in a study sponsored by the Electric Power Research Institute (EPRI) to assess the risk factors involved in the tritium fuel cycle. The main CMT effort in that study was to define the tritium systems that would be needed in a fusion reactor plant.

In 1983, CMT participated in a blanket comparison and selection study and tritium systems work in support of mirror and tokamak designs. Both liquid metal and solid blankets were included. Because the blanket must generate tritium and convert the energy from the lithium transformation reaction into sensible heat, areas of concern were tritium recovery and inventory in the blanket, and thermal-hydraulic effects. The tritium systems work was concerned with processing the tritium and deuterium to fuel the plasma, the water coolant, and the atmosphere in the tritium areas. A flowsheet was developed for a continuous tritium-processing system for the upgraded Mirror Fusion Test Facility (MFTF-B) at Livermore National Laboratory.

It was recognized that an order of magnitude improvement in performance was necessary to make fusion power an attractive commercial energy source. Two design assessments were made in 1985 on (1) the technical and economic benefits of using polarized fuels in tritium-deuterium reactors and (2) the benefits of using high-temperature, high-pressure helium as a heat-transfer medium. Neither concept appeared to offer a good possibility of achieving the desired performance. The economic aspects of fusion power were explored further in a 1986 study based on three fusion reactor designs. The bottom line was that the bus-bar cost of electricity could be about 32 mill/kWh in 1986 dollars, assuming a developed fusion technology. This value compares with a cost of about 52 mill/kWh in 1985 for Illinois electricity produced by coal and nuclear plants.

The Tritium Systems Test Assembly (TSTA) at Los Alamos National Laboratory was a full-scale test of tritium handling systems to be used in fusion reactors. A joint team involving ANL, Los Alamos, and the Japan Atomic Energy Research Institute (JAERI) was formed to study the addition of a breeder blanket interface (BBI) to the TSTA. Three types of blankets were considered: (1) liquid lithium metal, (2) solid Li_2O, and (3) an aqueous salt solution containing lithium. By 1989, BBI reference designs had been developed for items (2) and (3), along with development schedules and estimated costs. Funding was not available for ongoing development work on all the options, and at the end of the year, strategies were being planned to determine the most cost-effective paths to pursue.

Although fusion reactors would not produce the large quantities of highly radioactive wastes characteristic of fission reactors, waste disposal is still an important factor and a matter of public concern.

Therefore, a study was conducted on the reprocessing and disposal requirements for prototype fusion-reactor modules using liquid metal or solid breeder blankets. Reprocessing of the structure would require remote handling methods and, for some systems (depending on the details of the reactor design), residual activity may be too high for shallow land burial. Reprocessing of a lithium metal blanket would result in no major costs or disposal problems.

In 1981, as part of a U.S./Japan exchange program, Patricia Finn of CMT spent six months in Japan, mostly at JAERI. Pat is very knowledgeable about high-temperature, refractory oxide systems and other pertinent technology, having been a major contributor to CEN's programs on the thermodynamics and other physical properties of such systems, and also to applied programs such as the fuel-cell work. The experimental program at JAERI was directed toward Li_2ZrO_3 and Li_2TiO_3 as solid breeder blanket candidates. The work included studies of fabrication methods, thermal diffusivity and conductivity, compatibility with structural materials, and tritium-release behavior after irradiation.

Chemical Technology personnel who were involved in these systems-analysis studies were Rob Clemmer, Pat Finn, Carl Johnson, Ralph Leonard, and Bali Misra. Many people from other divisions also contributed.

MATERIALS RESEARCH AND DEVELOPMENT

Liquid Breeder Blanket Processing. In 1980, the work on processing of liquid lithium breeder blankets and the corrosion studies continued, but the interest had turned more toward solid breeder blankets, so the lithium-processing and hydrogen-permeation studies were de-emphasized. An investigation of possible causes of the LPTL leak that caused the fire mentioned in Chapter 4 revealed that iron crystallites and other solids had built up in the area of the electromagnetic pump. These deposits might have restricted the liquid metal flow, thereby producing cavitation and erosion of the channel walls. The location was in a region of high residual stress from cold working during fabrication. Crystals of the compound Li_xCrN_y were found, which verified a suspicion that nitrogen had a role in the mass-transfer corrosion of stainless steel in lithium loops. The LPTL was repaired, but further operation was curtailed due to other priorities.

Liquid lithium-lead-bismuth (Li-Pb-Bi) alloys were considered as a possible alternative to lithium as a blanket material. Potential advantages were a relatively high breeding ratio (>1.2), a thinner breeding zone, a melting temperature 30 to 40°C lower than that of lithium, and less susceptibility to contamination by air or water vapor. Compositions of 10 and 15 at.% Li in Pb/Bi at a mole ratio of 0.773 were found to be totally liquid at 131 and 136°C, respectively.

Hydrogen Permeation Studies. An extensive survey of hydrogen permeation data for pure iron, ferritic alloys, and martensitic alloys revealed that alloys containing small amounts of aluminum had lower permeabilities than aluminum-free alloys. Experiments on iron-chromium (Fe-Cr) alloys showed that small aluminum additions of 2 wt% reduced the hydrogen permeation rates of the alloys by as much as a factor of 200 in an oxidizing atmosphere.

SOLID BREEDER BLANKETS

Although liquid lithium metal seems at first glance to be the obvious choice for a fusion reactor breeder blanket, it has several serious drawbacks: (1) safety, due to its flammability, (2) corrosion, (3) electrical conductivity, which makes it difficult to pump through magnetic fields, and (4) high chemical reactivity, which attracts impurities such as water vapor, oxygen, and nitrogen from cover gases. Most of these problems can be overcome (and a few new ones introduced) by the use of a solid breeder blanket. Lithium-bearing ceramic materials such as Li_2O, $LiAlO_2$, Li_2ZrO_3, Li_4SiO_4, Li_2SiO_3, and Li_2TiO_3 were considered for this purpose. Two key engineering areas were involved in research on processing solid blankets: *in situ* tritium recovery and thermal-hydraulic performance in the breeding zone. In processing the solid blanket, the breeding medium remains stationary, and tritium is recovered by sweeping a purge gas (helium or nitrogen) through channels in the solid bed. The temperature must be kept high enough to achieve acceptable diffusion rates, but also low enough to avoid sintering or other microstructural changes in the bed.

The TRIO Experiment. Early work on solid breeder blanket processing was focused on comprehensive in-pile tests of a miniaturized blanket assembly, which were assigned the somewhat strained acronym, TRIO (Tritium Recovery in-Pile), at the Oak Ridge Research Center (ORR). Figure 5-16 shows the experimental configuration of the TRIO-01 experiment. This experiment was designed to simulate conditions in an operational solid fusion reactor blanket on a miniaturized scale. A lot of development work and verification testing had to be done in preparation for the experiment. An "engineered microstructure" with the desired bimodal pore distribution was developed for the γ-$LiAlO_2$ blanket material, and extensive chemical analyses were performed to identify trace contaminants. A gas loop was used to verify performance of the gas-analysis system. In addition, neutron fluxes and energy spectra, as well as tritium production rates, were determined in a sample of γ-$LiAlO_2$.

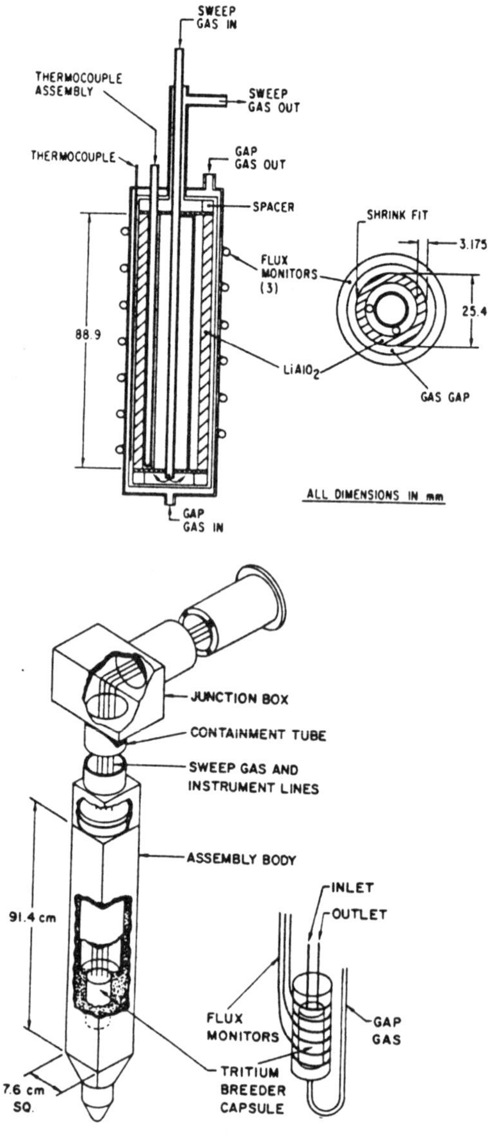

The solid breeder material (γ-$LiAlO_2$, about 70% dense, bimodal pore structure) is in the form of a hollow cylinder, and the purge gas is swept past the inside surface. Temperatures, monitored continuously, are varied over a range of 400-1000°C with less than a 150°C gradient across the solid. Neutron flux is monitored continuously by self-powered neutron detectors and by dosimetry wires. The assembled capsule fits into a cylindrical hole in an aluminum core piece. Sweep-gas lines pass upward out of the reactor to a gas-analysis package where tritium levels and the chemical composition of the sweep gas are measured.

Fig. 5-16. Tritium Breeder Blanket Experiment (TRIO-01) for Fusion Reactor

The irradiation phase of the TRIO experiment was conducted at Oak Ridge in 1983. Thirty-three tests were made at various temperatures and sweep-gas compositions. With a helium-0.1% hydrogen sweep gas, tritium released from the assembly was easily recovered, and tritium retention in the blanket was extremely low. Nearly all of the tritium was recovered in the reduced form as T_2 or HT. (From an overall tritium handling standpoint, it would be desirable to have the tritium in an oxidized form, namely, T_2O or HTO.)

The experiment also tested the heat-transfer performance of the assembly. The temperature profiles across the solid bed were within 10°C of those predicted by heat-transfer computations. Post-test examination of the solid bed indicated no significant changes in the porosity or microstructure of the γ-$LiAlO_2$. Between the irradiations and analytical work involved in the TRIO experiments, quite a few CMT people were involved in this phase of the fusion program. Pat Finn and Rob Clemmer were group leaders. Others included Del Bowers, Al Fischer, Larry Greenwood, Myron Homa, Rob Klemmer, Ralph Leonard, Dick Malecha, Bali Misra, Gerry Reedy, Mike Slawecki, Shiu-Wing Tam, and Sy Vogler. Tony Scandora from Electronics Technologies (ELT) was assigned to CMT.

Thermodynamics and Kinetics of Solid Breeding-Blanket Materials. It was becoming apparent that removal of tritium from an oxide breeder blanket was a complex process that would require thermodynamic studies of equilibrium reactions, solubilities, phase relationships, *etc.*, and investigations of the kinetics involved in diffusion, adsorption-desorption, and other rate-limiting processes. In the case of tritium produced in a Li_2O bed by the neutronic reaction $^6Li(n,\alpha)$, the tritium was expected to be released from the bed by dehydration of LiOT and to pass into the helium purge stream. An experimental study was undertaken to determine the solubility of tritium in Li_2O and to determine the phase relationships and vaporization behavior of the Li-O-T system. Because of the difficulties involved in handling tritium, the studies were actually conducted on the Li-O-H system, which is thermodynamically very similar to the Li-O-T system. The reaction of water vapor with Li_2O produces lithium hydroxide (LiOH), which dissolves in the Li_2O, causing a lower activity of LiOH and reduced partial pressures of gaseous LiOH and H_2O over the system. This effect is an important consideration in establishing conditions for retention or release of tritium in the Li_2O blanket. The solubility of LiOH in Li_2O was, therefore, measured as a function of water-vapor partial pressure and temperature. The solubility of LiOH (or H) in Li_2O proved to be proportional to the concentration of water vapor in the helium carrier gas.

Measurements were also made on the vaporization of Li_2O (as LiOH) in the presence of water vapor: $Li_2O + H_2O(g) \leftrightarrow 2\,LiOH(g)$. The experimental results can be expressed as follows: $\log p(LiOH) = 9210/T + \log p(H_2O)$, where p is in atmospheres and T is in degrees Kelvin.

The SOLGASMIX computer code was used to study the use of a H_2-H_2O mixture to exploit isotopic exchange as a means of enhancing tritium recovery from blanket material at temperatures of about 600 and 1000°C. The addition of small amounts of hydrogen and/or moisture to the tritium purge gas strongly affected both the chemical form of the released species and the residual tritium in the solid material. For γ-$LiAlO_2$ breeder material and an oxygen activity of about 10^{-15}, a temperature of 600°C was more favorable for tritium release than 1000°C, and most of the tritium was in an oxidized form. Higher tritium release, although in the reduced form, could be achieved at both temperatures by using low oxygen activities ($<10^{-25}$). This study was then extended to evaluate the relative performance of three candidate blanket materials with respect to tritium release, which turned out to be $LiAlO_2 > Li_2O > Li_4SiO_4$. On the basis of these results, further work was initiated on the thermodynamics and kinetics of tritium release from γ-$LiAlO_2$.

Both calculational and experimental methods were used to characterize further the mechanisms involved in the recovery of tritium from γ-$LiAlO_2$. One study was focused on the dissolution, adsorption, and desorption of gaseous H_2O. Isotherms were developed for surface adsorption of H_2O and the solubility of OH^- in $LiAlO_2$ as a function of H_2O(g) partial pressure. The nature of the adsorption curves indicated that more than one type of surface site was involved in the process. Further work showed that physisorption occurred at low temperatures, but chemisorption became dominant at higher temperatures. Computer models were devised to predict tritium release from a ceramic breeder, and more detailed studies of adsorption-desorption and other phenomena involved in the process were continued through the end of the 1980s.

The thermal conductivity of a ceramic breeder material is important because it determines the temperature difference between the interior of the solid breeder material and the coolant/breeder interface. The thermal conductivity of 85% dense γ-$LiAlO_2$ at temperatures of 827-1127°C was about 2.5 W/(m·°C). The effects of porosity and temperature proved to be complex, however, because of different heat-conduction mechanisms involved. This finding suggested that caution be exercised in extrapolating γ-$LiAlO_2$ thermal conductivity data to other temperatures and/or porosities.

The effects of the blanket configuration on the thermal conductivity were investigated. The use of beryllium or its oxide in the form of microspheres as a neutron multiplier in "sphere-pac" $LiAlO_2$ blanket beds was evaluated by a computer study. Both mixed-sphere (γ-$LiAlO_2$-BeO) and coated-sphere (γ-$LiAlO_2$ coated with BeO) systems were studied. The coated-sphere arrangement resulted in a higher thermal conductivity.

This was a rather wide-ranging effort that included many individuals: Bob Blomquist, Wally Calaway, Norm Chellew, Rob Clemmer, Larry Curtiss, Pat Finn, Al Fischer, Carl Johnson, Bob Land, Len Leibowitz, Vic Maroni, Bali Misra, John Shearer, Shiu-Wing Tam, Marv Tetenbaum, Erv Van Deventer, Ewald Veleckis, and Bob Yonco.

AQUEOUS BREEDER BLANKET

The International Thermonuclear Experimental Reactor (ITER) project included various efforts to produce conceptual designs of a tokamak reactor that could be used to test components for a prototype fusion reactor. As a part of this study, CMT researchers examined the technical requirements for an aqueous lithium salt blanket (2 *M* LiOH or $LiNO_3$) in which tritium fuel could be bred. The study included corrosion effects in the blanket and a breeding blanket interface (BBI) for recovering the tritium product. The BBI was a complex system which required a sizeable design effort, but no experimental work was done in this area.

DOSIMETRY AND DAMAGE ANALYSIS

Work continued on the effects of the intense radiation fields in fusion reactors on component materials. The primary goal continued to be a characterization of (1) the neutron fluence and energy spectrum of neutrons produced in reactor and accelerator test facilities and (2) the damage effects on materials, including nuclear displacements, internal gas production, and transmutation rates. The group in the Chemical Technology Division, led by Larry Greenwood, had a lead role in collecting and correlating this type of information, which was essential because the irradiations were being conducted at several different facilities, including those at ANL.

The principal reactors used as irradiation facilities at the time were the Oak Ridge Research Reactor (ORR) and the High Flux Isotopes Reactor (HFIR) at Oak Ridge, the Experimental Breeder Reactor (EBR-II) at ANL-W, and the Omega West Reactor at Los Alamos National Laboratory. Accelerators can also produce high fluxes of 14-MeV neutrons similar to those in a fusion reactor, so irradiations were conducted in several accelerators—the Rotating Target Neutron Source (RTNS) at Lawrence Livermore National Laboratory (LLL), the cyclotron at the University of California at Davis, and the newly commissioned Intense Pulsed Neutron Source (IPNS) at ANL. As information was accumulated in the extensive data files at CMT, it was forwarded to the National Magnetic Fusion Energy Computer Center (NMFECC) at LLL.

Each facility had its own unique neutron flux, energy spectrum, geometry, *etc*., so dosimetry measurements were required in order to put the irradiation effects on a common basis. To perform the dosimetry measurements, small wires or foils were included with the materials to be irradiated, and the radioactive products were measured in each sample by gamma-ray spectroscopy. The activation rates for up to 38 different nuclear reactions were then used to adjust the neutron flux spectrum, which was obtained by a neutronics calculation. This was the same technique used to characterize the neutron environment in the TRIO experiments. The cross-section data are also essential for fusion reactor diagnostics and the handling of long-lived wastes.

By 1981, extensive data files of damage parameters had been generated for 36 isotopes, including six key elements for which damage parameters up to 50 MeV had been obtained from accelerator irradiations. Prediction of damage in fusion reactor materials required prior calculation of neutron-energy-dependent cross sections, which were used to determine the number of atomic displacements and transmutations at each neutron energy. In 1982, new theoretical treatments were developed to include (n,γ) capture reactions and β-decay, and all the files were updated accordingly. The Chemical Technology Division pioneered these dosimetry and damage techniques at higher neutron energies (50 MeV) for accelerator neutron sources.

Two new activities were undertaken in 1984. One involved a joint program with the Princeton Plasma Physics Laboratory in which cross sections for the reactions $^{54}Fe(n,2n)^{53}Fe$ and $^{27}Al(n,2n)^{26}Al$ were measured near 14 MeV. Since both reactions have thresholds near the fusion neutron energy of 14 MeV, their yields are highly sensitive to the plasma temperature, which determines the neutron spectrum near 14 MeV, and the yields of the reactions can be used for plasma diagnostics. They are also useful in calculating the radioactivity levels of activation products in fusion-reactor wastes. The other new program was the measurement of spallation yields of specific isotopes to determine the neutron spectrum. The utility of this technique was demonstrated in collaborative experiments between IPNS and the Los Alamos Meson Physics Facility (LAMPF). Work then continued on cross-section determinations for a variety of isotopes of interest to the fusion program. Vanadium, iron, copper, nickel, and molybdenum were of particular interest.

The formation of helium by (n,α) reactions in structural materials was studied by experiments with nickel, copper, and zinc. As an example, helium concentrations up to about 100 ppm were generated in copper in a neutron fluence of 6×10^{22} n/cm^2. Subsequent work showed that more helium than was expected was produced in iron samples.

By the end of the 1980s, the fusion research group in CMT had become well known nationwide and internationally for its comprehensive experimental studies and computations relating to radiation damage from the neutrons in fusion systems. Their SPECTER computer code was used to determine atomic displacements, gas production, and other transmutations in a wide variety of materials. Near the end of the decade, a part of the program effort took an interesting twist when it was realized that their work had possible applications relating to the Strategic Defense Initiative, and several avenues of investigation were proposed in the area of nuclear arms treaty verification.

The people who were active in this program were Rob Clemmer, Don Graczyk, Larry Greenwood, Ray Popek, Chuck Seils, and Bob Smither.

"COLD FUSION"

Early in 1989, two electrochemists, Stanley Pons of the University of Utah and Martin Fleischmann of the University of Southampton, England, startled the scientific community with an announcement that they had achieved "cold" (room-temperature) nuclear fusion by electrolyzing heavy water (D_2O), using platinum and palladium electrodes. They claimed that the cell produced more heat than could be accounted for by the chemical reactions alone, and they thought they observed gamma rays at energies typical of fusion, although that result was disputed. Because of the important scientific and technological implications of these results, laboratories all around the country started trying to reproduce the experiments, generally with negative or ambiguous results.

At CMT, Laszlo Redey, Mike Myles, Dennis Dees, Mike Krumpelt, and Don Vissers became interested in the subject and conducted two series of experiments: (1) differential temperature analysis of identical light- and heavy-water electrochemical cells and (2) calorimetric measurements on operating heavy-water cells. Wrought palladium cathodes and platinum anodes were used in both cases. The first series of experiments was unsuccessful because identical electrochemical conditions could not be maintained in the two cells. In the second series of experiments, cells were operated in a calorimeter sensitive enough to detect the levels of excess heat reported by Pons and Fleischmann. The results of the energy-balance measurements and calculations showed no significant excess heat.

Efforts to reproduce the cold fusion experiments at various laboratories gave mixed results, but generally tended to be unconvincing (for example, no neutrons were detected). There were (and still are) some scientists, however, who believe that the phenomenon is real. Toward the end of 1989, the CMT workers proposed a joint research effort with Case Western Reserve University, where they had seen some evidence of excess heat in their experiments, to investigate the matter more thoroughly, but the proposed program had to be dropped for lack of funding.

Steel Research Initiative

In 1987, a program that resulted from a joint steel industry/federal laboratory research initiative was aimed at reducing equipment, operating, and energy costs in the steel industry through the use of new, innovative technologies. The purpose of the ANL program was twofold: (1) the development of an electromagnetic process for the continuous casting of various grades of steel in large-aspect-ratio shapes such as steel strip and (2) the development of an economical process for the purification of ferrous scrap. Although this program was under the general direction of Jim Battles, essentially all the work was done in the Energy and Environmental Systems, Electromagnetic Technology, and Materials and Components Technology Divisions. Bali Misra of CMT also contributed some effort to this program.

ELECTROMAGNETIC CASTING OF STEEL

In the production of steel strips, sheets, and other shapes, liquid steel from the production furnace is either cast into large ingots or slabs, or fed into a continuous casting process, and then subjected to rolling and other operations to produce the final product. Electromagnetic continuous casting involves the use of magnetic fields to form a liquid sheet of the metal, which is levitated until it is solidified by cooling. The electromagnetic continuous casting studies were focused on levitation experiments, development of a caster design, fluid-flow analysis and experimentation, heat-transfer analysis, and sensors for on-line monitoring and process control.

Levitation Experiments. The experimental work was directed initially toward the development of a combination of high-frequency magnetic fields and eddy currents. Tin, which has a melting point of about 250°C, was used as a stand-in for steel to permit a more convenient operating temperature. A test magnet, 20 cm wide and 13 cm high with a 30-cm-long ferrite core, was designed and built to study the magnetohydrodynamic stability of the liquid metal. The experiment was started by inserting a sheet of tin into the magnet and melting it by eddy-current heating. The liquid tin sheet was disrupted before levitation, apparently due to a mismatch between the electrical conductivities of the stainless steel window and the liquid tin. Somewhat better results were obtained when the stainless steel was replaced by copper or Monel metal, but the results indicated that the low frequency (10 kHz) and power output (30 kW) of the equipment would both have to be increased.

Electromagnetic Design. A computer code (PE2D) was used to evaluate various magnet geometries for a horizontal electromagnetic caster. A cross section of a conceptual design for a bench-scale unit is shown schematically in Fig. 5-17. The window-frame design was expected to yield a very uniform horizontal field (± 0.2%). Special design features were the copper shields and the side guards. The frequency of the AC field was such that it penetrated less than one-third the thickness of the steel sheet. The steel levitated due to longitudinal eddy currents normal to the direction of the AC field, and the magnetic force generated by interaction of the eddy currents with the AC field.

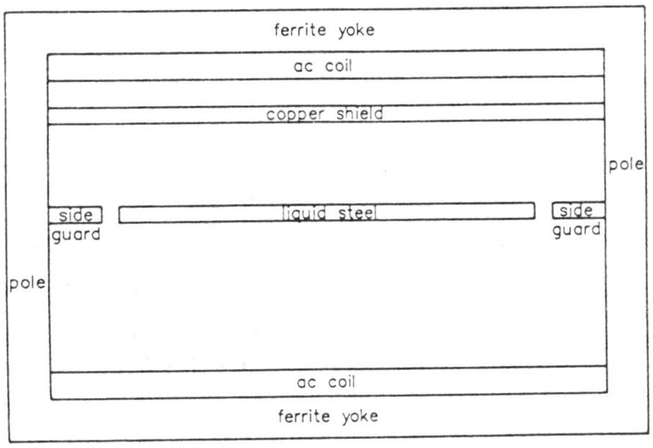

Fig. 5-17. Cross Section of Conceptual Horizontal Caster

Feed System Development. It was assumed initially that a conventional liquid metal feed system would be suitable for a bench-scale caster. Preliminary analysis, however, showed that low turbulence levels are required to achieve good stream quality and steady-state flow, and means are necessary to remove inclusions and minimize heat loss. A parametric study using a three-dimensional computer code, COMMIX, provided design guidance for the feed system.

Test results from a series of experiments on wave formation at a gas/liquid metal interface showed that a magnetic field could damp out waves formed on a liquid metal surface by a nitrogen gas stream. A second series of experiments, using a water model, indicated that the geometric characteristics of the nozzle affecting the shape and velocity distribution of the emerging liquid stream were (1) the aspect ratio of the nozzle cross section, (2) the area contraction ratio from inlet to exit, (3) the length of the straight section, and (4) the corner curvatures of the exit cross section. To evaluate the effects of these geometrical factors, eleven Plexiglass® nozzles of various geometries were built and tested. To obtain further design information, an experimental feed system for recirculating liquid tin was designed and fabricated. That system was used to investigate the flow of molten metal through different types of nozzles.

Heat-Transfer Analysis. A key element in the casting process was the necessity to cool the molten metal quickly as it left the caster so it maintained the desired dimensions and shape as it solidified. Radiative cooling was insufficient, so convective cooling had to be provided. That had to be done in such a way, however, that it did not disturb the flow of the liquid steel. To proceed with a parametric investigation of the problem, a set of reference conditions had to be established. These included items such as the outlet temperature, thickness of the steel strip, casting speed, coolant pressure, electromagnetic heat flux, emissivity of the steel, coolant gas, coolant velocity, magnet length, and coolant channel height. A literature review revealed a variety of cooling-enhancement techniques, including the addition of fine water droplets to a coolant gas, which was highly effective.

Sensors and Process Control. The initial effort in this area was focused on sensors and control systems for on-line process monitoring and automatic control of the casting process. A wave detector based on capacitance changes was developed for quantifying and characterizing surface-wave formation caused by the coolant gas. A spectrophotometer was modified and fitted with fiberoptic light guides for measurements of tracers and water-jet thickness variation. A review of sensor and control needs indicated that certain components were available commercially, but that further research and development would be required in some cases.

STEEL SCRAP BENEFICIATION

The purpose of this work was to develop processes that will produce high-quality, deep-drawing steels from low-grade ferrous scrap material. The Institute of Scrap Iron and Steel at that time estimated a ferrous scrap inventory of about 750 million tons, and that inventory was growing in spite of the fact that more units of steel were being produced from scrap than from ore in the domestic production of steel. A major impediment to greater use of scrap was the presence of undesirable residual elements such as copper and tin, which degrade the quality of steel. This part of the steel research was done by Milt Blander and his coworkers and might be more appropriately placed in the basic research studies, but it is germane to the general area of steel production.

Removal of Copper from Scrap Steel. In the common types of ferrous scrap material, copper is present in concentrations of about 0.2 or 0.6 wt%. For deep-drawing steel, the allowable copper content is 0.1 wt% or less; for steel tubing or pipe, it is 0.2 wt%; the lower grade rebar steel can contain 3 wt% or more. Thus, to produce high-grade steel from low-grade scrap, the copper content must be reduced by a factor of six or more.

With the common steel-making slags in an electric-arc furnace, all of the copper remains in the molten metal. Thermodynamic data indicate that the copper could be extracted into the slag by the addition of sulfides. The reaction $FeS + 2\ Cu \leftrightarrow Cu_2S + Fe$ can be driven to the right by using a slag in which Cu_2S has a low activity coefficient, such as a polyvalent sulfide, *e.g.*, an Al_2S_3-FeS slag. With this type of system, distribution coefficients (wt % Cu in slag/wt % Cu in metal) as high as 30 were obtained. Even with this large distribution coefficient, however, a slag volume about equal to the metal volume would be needed to decrease the copper content by a factor of ten. This copper-extraction procedure leaves some sulfur in the steel. This sulfur is removed by exposing the steel to metallic calcium or aluminum.

Preliminary economic analyses of the above procedure indicated a potential saving of about 20% for the production of liquid steel, and savings up to 30% may be achievable if the spent slag is reprocessed and reused.

Removal of Tin and Other Tramp Elements. Tin is present in ferrous scrap material at concentrations of 0.05 or 0.02 wt%, depending on the type of scrap. Permissible levels for tin are 0.01 wt% in deep-drawing steel, 0.02 wt% in pipe and tubing steel, and 0.03 wt% or more in rebar steel. Calcium reacts with more electronegative metals such as tin, *i.e.*, Ca + Sn → CaSn, and the literature reports that calcium can be used to remove tin and sulfur from molten steel. Preliminary experiments with *in situ* electrochemical generation of calcium for this purpose were promising.

Sodium Technology

From the late 1960s through the middle 1970s, the major efforts in sodium technology at Argonne were on development of continuous on-line impurity monitors and improved chemical analytical procedures for the major impurities of concern in sodium, namely, oxygen, hydrogen, and carbon (as Na_2O, NaH, and Na_2C_2). These were to be applied at the FFTF (Fast Flux Test Facility), then under construction at Hanford, Washington. When prototype monitors and the improved analytical procedures were tested at EBR-II, it was found that the concentrations of these impurity elements were much lower than had previously been measured by much less sensitive analytical methods. Moreover, the cold traps on the primary and secondary sodium systems maintained essentially constant concentrations of oxygen and hydrogen. This finding eliminated the need for the expensive on-line monitors and costly facilities to house them.[1] In the Chemical Technology Division, work was shifted to optimization of cold-trap design, *in situ* regeneration of plugged cold traps, purification of sodium to reduce the level of radioactive contamination, and disposal of radioactive sodium and sodium-containing wastes. This program was managed by Chuck McPheeters.

OPTIMIZATION OF COLD TRAP DESIGN

Sodium impurities, particularly sodium oxide and sodium hydride, are normally controlled by low-temperature precipitation and deposition of the precipitates in cold traps. In a cold trap, the sodium enters at the top of an annular section created by a circular divider that extends to near the bottom of the cold trap, and then flows upward through the central section. Both the central and annular sections are packed with wire mesh, which provides nucleation sites for the impurity precipitates. In the cold trap the sodium is cooled to a temperature of about 125°C. When a cold trap becomes filled with precipitates, a sharp increase in the pressure develops across the unit. Only a small fraction of the main sodium flow (2% or less) is diverted through a cold trap for removal of oxygen and hydrogen

[1] Work did continue on the development of hydrogen leak detectors in the Components Technology Division for long-range applications to liquid metal fast breeder reactors, but the hydrogen leak detector was not required on FFTF because it was not equipped with a steam generator.

impurities. In large LMFBRs, hydrogen produced by water-side corrosion of steam-generator tubes diffuses into the secondary sodium system at a high rate. Although the hydrogen is removed continuously by the secondary system cold trap, much of it diffuses through the intermediate heat exchanger into the primary system, where it is removed by the primary system cold trap. The major impurity deposited in the primary and secondary system cold traps is sodium hydride (NaH). Cold-trap lifetime is limited by the deposition of sodium hydride in the primary and secondary system cold traps.

In the CMT Division, work was undertaken to increase the cold-trap capacity, and hence, its useful lifetime by optimization of the design. A one-dimensional computer model was developed with which distribution of impurity deposits within a cold trap was calculated from temperature profiles, impurity concentrations, wire-mesh packing density, and geometrical dimensions. This model was tested using small, simple cold traps installed on AMPS (Apparatus for Monitoring and Purifying Sodium). The cold traps consisted of 19-mm-dia tubes, 560 mm long, with thermocouples along the centerline and air cooling along the exterior to establish a temperature gradient along the trap's length. The cold traps were packed with wire mesh to different densities to provide extended surface area for impurity nucleation and growth. The amount of impurity deposited in a trap was determined by continuously measuring inlet and outlet impurity concentrations. These measurements were made using on-line oxygen and/or hydrogen meters.

After an impurity-deposit experiment, the cold trap was cut into segments to determine mass distribution within the trap. The distribution of NaH precipitates was in excellent agreement with model predictions, but that for Na_2O was not. The difference was attributed mainly to poor adherence of the Na_2O crystals on the stainless steel surfaces. Inaccurate or drifting oxygen-meter calibration may have been a contributing factor.

A two-dimensional model was then developed to aid in optimizing cold-trap design for maximum capacity. This model, named MASCOT (Model for Analyzing Sodium Cold Traps), was used to determine, by means of a parametric study, which cold-trap design variables have the greatest impact on capacity. The base case was the cold-trap design for the Clinch River Breeder Reactor (CRBR). The results of this study are shown in Fig. 5-18, where capacity, defined as percent of mesh volume filled with impurity, is plotted as a function of the L/D (length/diameter) ratio and the ratio of annular downflow area to the central upflow area (ANN/CEN). The other curves are cases of different mesh packing densities. The most important factor is the ratio of the annular area to the central area. The capacity can be increased about threefold by increasing the ratio from 1.0 to 80 (large annular area, small central sodium return pipe).

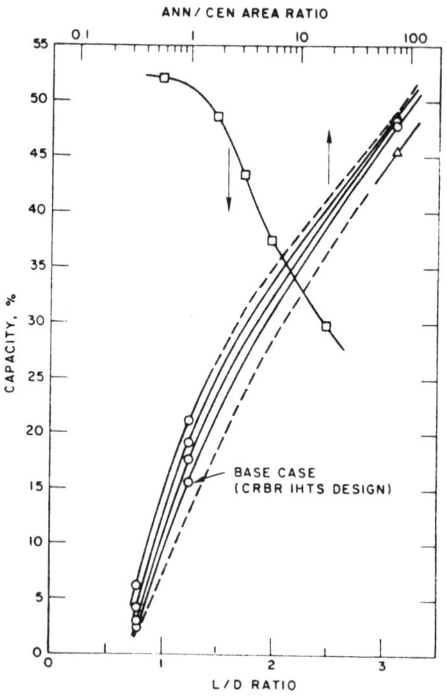

Fig. 5-18. Cold Trap Capacity as a Function of Area and L/D Ratios

The L/D ratio is also important and should not exceed about 1.5. The CRBR cold-trap design was revised to reflect the results of this study.

To verify the accuracy of the MASCOT computer model, calculated and measured sodium hydride distributions were compared for cold traps from actual operating systems. These included the primary system cold trap of the Fermi reactor, two industrially fabricated cold traps used on experimental systems by the General Electric Co. and Mine Safety and Appliance Co., and two experimental cold traps operated on an ANL system. In all cases, the measured and calculated sodium hydride distributions were in excellent agreement.

The MASCOT was subsequently used to design cold traps having volumes of 3,000 to 7,000 L to satisfy the needs of liquid metal fast breeder reactors ranging in size from 1,000 to 3,500 MW(t).

IN SITU REGENERATION OF COLD TRAPS

A method of increasing cold-trap life is *in situ* regeneration, which is achieved by thermal decomposition of NaH, the bulk constituent, under a vacuum. The feasibility of *in situ* regeneration was demonstrated in small-scale tests in 1981, so plans were made to demonstrate the procedure on a full-size, plugged cold trap that had been removed from the EBR-II secondary system in 1970 and placed in storage.

Chemical Technology Division staff designed, fabricated, and installed the regeneration system (see Fig. 5-19). In the regeneration test, approximately 13,200 L of gas, predominantly hydrogen, was removed in 97 h. This represented 97-98% of the original quantity of hydrogen in the trap.

Subsequent efforts were directed to selecting a viable, safe, and economical method for treating the gas, which also contained some tritium, for disposal. A variety of processes were reviewed for

Fig. 5-19. Cold-Trap Regeneration System

separation, concentration, conversion, and stabilization of the tritium-contaminated hydrogen. The most direct, cost-effective approach was to consider the hydrogen and tritium gases as one entity, convert them to water (H_2O and T_2O), and incorporate the water into concrete formed in a steel drum. Chemical Technology Division staff recommended that the gases be processed as they exited the regeneration unit. The Sodium Technology Program was terminated before gas conversion and immobilization of the tritium-contaminated water in concrete could be demonstrated.

SODIUM PURIFICATION BY DISTILLATION

Operation of LMFBRs, particularly with some fraction of breached fuel elements, results in contamination of sodium with radioactive fission products. Decontamination of the sodium should aid in the rapid detection of additional fuel-cladding failures and also allow easier future reuse of sodium after reactor decommissioning. The radioactive species of major concern are cesium-137, barium-133, fuel particles, and various fuel-coolant reaction products.

The CMT Division explored several methods for decontaminating sodium. These included soluble gettering, reaction with various solids (including graphite), and low-pressure distillation. A reticulated vitreous carbon trap had been used successfully at EBR-II for removing ^{137}Cs, but it is specific only for cesium and risks introduction of carbon-bearing species into the sodium. The Division selected low-pressure distillation for development.

The feasibility of separating ^{137}Cs and ^{133}Ba was first demonstrated in a small, total-reflux distillation column 100 mm in diameter and 610 mm long. This small column was packed with wire mesh, and the reboiler was filled with 7.7 kg of sodium spiked with ^{137}Cs and ^{133}Ba. The decontamination factor for ^{137}Cs (concentration ratio of that in top product to that in the reboiler) was 200. The ^{133}Ba reacted with small amounts of Na_2O to form BaO, which plated on the column surfaces.

These encouraging results led to construction of a larger (204-mm ID) column consisting of three 0.61-m-high packed sections with a reboiler at the bottom and condenser at the top. Contaminated sodium was fed between the bottom and middle sections; and product sodium was withdrawn from between the middle and top sections. When the column was operated, ^{137}Cs in the product stream was below the background level, so an accurate decontamination factor could not be determined. Again, BaO plated out on the column surfaces.

A series of runs was made with $Na^{131}I$ in an attempt to determine the HETP (height equivalent to a theoretical stage). As nearly as could be determined, the HETP was in the range of 100 to 300 mm.

Development of the distillation process was concluded with a test using primary sodium from the Fermi reactor. The ^{137}Cs level in the product was again very low, 3×10^{-6} ppm, which is much lower than the concentration of approximately 52 ppm permitted for shallow land burial.

SODIUM WASTE TECHNOLOGY

In 1980, the Division provided technical support to ANL-West on development of a melt/drain, evaporation, calcination (MEDEC) process for disposal of radioactive waste containing elemental sodium. The MEDEC process was to be incorporated into the Sodium Process Demonstration plant planned for construction at ANL-West to demonstrate the safe treatment of wastes containing metallic sodium. The process consists of (1) melting and draining sodium from waste and scrap, *e.g.*, a cold trap, (2) evaporation to remove residual sodium, and (3) calcination to convert sodium into Na_2O. Incorporation of Na_2O into a high-sodium glass was subcontracted to Battelle Pacific Northwest Laboratory. Development work by CMT on the calcination step was completed in about 1980 and is described in Chapter 4.

The sodium evaporation step required work in three areas: (1) treatment of tritium gas released during removal of sodium from a cold trap,[2] (2) development of sodium-vapor-detection instrumentation for determining the end point of an evaporation cycle, and (3) determination of the effects of organic waste items (plastic bags, paper wipes, *etc.*) inadvertently mixed with the waste. The Argonne Chemistry Division conducted work on the treatment of tritium-containing gases.

For the sodium-evaporation step, four methods were tested for detecting the end point of the evaporation step: (1) measurement with ordinary vacuum gauges, (2) use of atomic absorption, (3) determination of breakdown voltage, and (4) use of a quadrupole mass spectrometer. The quadrupole mass spectrometer proved to be the most reliable and

---

[2] This reprepresents a situation in which *in situ* regeneration is not practiced, and the cold trap is slated for disposal. It is assumed, however, that the bulk of the hydrogen and tritium was removed by thermal decomposition.

effective method. The design and operating procedure for it were prepared for ANL-West.

Organic materials added to the sodium did not cause rapid or violent reactions at temperatures up to 450°C and under an argon atmosphere. Nor did they interfere with the detection of the evaporation end point. However, a carbonaceous residue accumulated in the boiler, and it was recommended that organics be removed to the extent possible before the processing of sodium-containing wastes.

SODIUM-CONCRETE REACTIONS

In some of the modeling scenarios of core-meltdown accidents, liquid sodium containing heat-generating debris comes in contact with the concrete basement of the containment vessel. In such a case, water emanating from the concrete would react with the sodium, thereby producing NaOH and other reaction products. The resulting phase relationships among the non-gaseous species are complicated, but literature data indicated that they had been well characterized at temperatures below about 550°C. No experimental work, however, had been reported above that temperature, so an experimental investigation of the Na-Na_2O-NaOH-NaH system was undertaken.

Equilibrium hydrogen pressures were determined as functions of temperature and hydrogen content above mixtures with fixed Na:O ratios. Four isotherms obtained from a sample having a Na:O ratio of 1.983 provided information on the boundaries between the various phase fields. Due to other priorities, no further work was done in this area.

PERSONNEL

The sodium program was in Fred Cafasso's part of the division, and then under Martin Steindler. Chuck McPheeters and Verne Kolba had Group Leader responsibilities. Others in the program were Joanne Fink, Vic Kremesec, Len Leibowitz, Bali Misra, Don Raue, Stan Skladzien, Charles Smith, Marv Tetenbaum, and Ray Wolson.

Nuclear Separations Technology

ANALYTICAL SUPPORT FOR LWBR PROOF-OF-BREEDING

The program to provide analytical support for determination of the extent of breeding in the Light Water Breeder Reactor (LWBR) at Shippingport, PA, continued into the 1980s (see Chapter 4). The role of CMT was to analyze segments of fuel rods for total uranium, individual uranium isotopes, and certain fission products, the latter to enable a determination of burnup. The fuel segments were cut from precisely located positions along a fuel rod. The work was carried out under the direction of Bettis Atomic Power Laboratory (BAPL), which had the responsibility for determination of breeding.

Delayed neutron profiles of fuel rods selected for analysis by CMT were obtained by BAPL, using a Production Irradiated-Fuel Assay Gauge (PIFAG). The CMT analyses enabled Bettis to calibrate the PIFAG so that it could be used later to obtain total uranium and the amounts of uranium isotopes (U-232 and U-233) in 500 fuel rods statistically selected from a total of 17,192 fuel rods discharged from the reactor after four years of irradiation (end-of-life). From these measurements, Bettis could obtain the fissionable uranium content of the entire core.

Analyses of the fuel rods started with precision shearing of the rod into segments by a specially designed shear that simultaneously pulverized the fuel and freed it from the cladding during the shear stroke. Segment lengths ranged from 17.5 to 53.3 cm. From the initial boundary of the segment, small lengths of the fuel rod (about 1/4 in.) were sheared until the final boundary of the segment was reached. All of the material from an original segment was collected in an aluminum sample can, and the entire can and contents were dissolved in a 13.6 *M* HNO_3-0.06 *M* HF solution at 195°C and a pressure of 125 psig in a tantalum-alloy dissolver. The dissolver solutions were analyzed for total uranium and uranium isotope abundances by a combination of mass spectrometric isotopic dilution (MSID) and alpha spectrometry. High-precision, high-accuracy analyses were required.

Following tests of a prototype dissolver, two complete dissolver systems were installed in one of the manipulator-equipped M-Wing caves in Building 200. Each dissolver system consisted of a tantalum alloy vessel mounted in a stainless steel shell, associated liquid and gas handling equipment, and electronic balances for weighing the fuel charge. Tantalum alloy vessels were required to provide corrosion resistance to the highly corrosive nitric-hydrofluoric acid dissolver solution.

Computer programs were developed for automation of the dissolving operation. After dissolution of a fuel segment, a dissolver was thoroughly cleaned before dissolution of the next segment. Cross-contamination was less than 0.005%, well within the 0.01% stipulated by BAPL.

The campaign for destructive analysis of seventeen fuel rods of various types (core, blanket, and reflector) began in August 1985 and, with 24-hour operation during weekdays, was concluded in November 1986. The fuel recovery on shearing of the fuel rods averaged well above the minimum of 99.75% required by BAPL. The fuel dissolution procedure consistently provided 99.99% recovery of each fuel segment. Performance of the analytical operations was also outstanding. Analyses of two separate dissolver samples from dissolution of 135 fuel-bearing segments by the MSID method showed an average difference of only 0.01%. Measurements of precision and accuracy for cesium-137 and cerium-134 also exceeded requirements. (Analysis for zirconium-95 was dropped.)

The BAPL staff stated that they were very pleased with the high quality results achieved in this program, which made possible a "fine-tuned" calibration of the PIFAG. Later, BAPL reported breeding ratio values in the LWBR that were slightly greater than one (1.013 and 1.015), but too low to be of practical interest.

One task remained—disposal of the 700 L of dissolver solution produced in the end-of-life campaign. The dissolver solution was mixed with cement in 5.7-L cans to produce a monolithic solid. Excess lime was used to ensure acid neutralization. Two primary cans were placed in a secondary can, which, in turn, was loaded into a 55-gal drum containing a specially designed shielded cask. Suitable for ultimate disposal as transuranic (TRU) waste, they were first shipped for interim storage (up to 25 yr) at a site operated by the Rockwell International Corp. at Richland, WA.

This program required a sizeable group that included Roy Benson, Ron Brock, Tom Cannon, Phil Deeken, Jim Fagan, Don Graczyk, Joe Hoh, John Kincinas, Ralph Leonard, Fred Martino, Bob Meyer, Bob Nelson, Jack Parks, Don Raue, Lewis Ruppert, Glen Tomlinson, Verne Trevorrow, Charlie Wach, Prince Walker, Irv Winsch, and Tony Ziegler. Norm Levitz had the overall responsibility for the program.

TRUEX PROCESS DEVELOPMENT

Process Description. In 1984, the CMT Division began a cooperative program, under the leadership of George Vandegrift, with the Chemical Separations Group of the ANL Chemistry Division (CHM) to develop the TRUEX (Transuranic Extraction) process. This process, which was invented by members of the Chemical Separations Group led by Phil Horwitz, uses an extractant, abbreviated CMPO, in combination with tributyl phosphate (TBP), the conventional Purex process extractant, and a diluent. The diluent is typically a normal paraffinic hydrocarbon, such as dodecane, which is used as the TBP diluent in the Purex process, or a nonflammable chlorocarbon, such as carbon tetrachloride or tetrachloroethylene.

The chemical structure of CMPO is shown in Fig. 5-20. The name for it is octyl(phenyl)-N,N-diisobutyl carbamoyl phosphine oxide. Its discovery was the result of basic research studies in the Chemistry Division on a series of neutral, phosphorus-based bifunctional extractants.

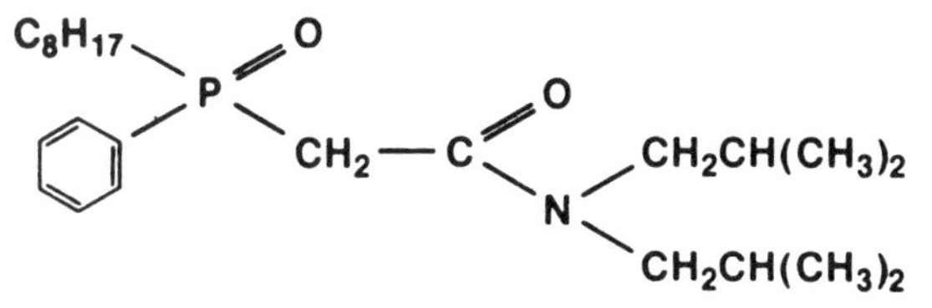

Fig. 5-20. Chemical Structure of CMPO Extractant Used in Purex Process

The distinguishing feature of CMPO is its ability to extract trivalent elements from nitric acid. The intended application of the TRUEX process was for processing a variety of Hanford wastes to remove the transuranic (TRU) elements, specifically neptunium, plutonium, and americium. In nitric acid solutions, these elements exist predominantly in the trivalent state or can easily be put in the trivalent state. In the Purex process, neptunium and americium (and a small fraction of the original plutonium) report to the first-cycle aqueous waste. Transuranic wastes, because of their long half-lives, must be disposed of in a deep geologic repository. Removal of the TRU content of a waste to below 100 nCi/g of solid makes it a "nonTRU waste" eligible for less expensive, near-surface burial. Application of the TRUEX process to nuclear wastes in the United States would (1) greatly lessen the amount of waste that must be buried at the Waste Isolation Pilot Plant (WIPP), (2) recover substantial amounts of TRU elements, *e.g.*, plutonium, that otherwise would be lost, and (3) alleviate long- and short-term problems with waste storage that threaten to curtail the production of plutonium.

In concept, the TRUEX process is the same as the Purex process. The solvent-extraction step consists of the conventional extraction, scrubbing, and stripping operations.

Because of their small size and high throughput, centrifugal contactors were an integral part of implementing the TRUEX process. One advantage of centrifugal contactors is that they can fit into conventional glove boxes. Another advantage is that the number of stages allocated to the extraction, scrubbing, and stripping operations, all of which are conducted within the same unit, can easily be changed to achieve desired process objectives and to accommodate new process flowsheets.

Plutonium Finishing Plant Waste. In 1985, a flowsheet was developed and tested for removal of americium and plutonium from Plutonium Finishing Plant (PFP) waste, one of the wastes at the Hanford site. This waste, which contains americium and plutonium in low concentrations, is generated when plutonium-bearing scrap is converted into concentrated plutonium nitrate solutions. The flowsheet was demonstrated in a 14-stage centrifugal contactor (five extraction, two scrub, four americium stripping, and three plutonium stripping stages). Results were excellent. The aqueous waste contained <10 nCi of TRU per gram of solid. The americium and plutonium product streams contained >99.9% of that in the feed solution.

In 1987, CMT personnel assisted Hanford personnel in making the first TRUEX process tests, using actual PFP wastes. No problems were encountered in operating the centrifugal contactors and in implementing the PFP flowsheet.

Processing of HCl/Brine Solutions. In 1985, work was started on the development of a solvent-extraction process to remove plutonium and americium from chloride salts generated at Los Alamos in pyrometallurgical processes for recovering and refining plutonium. A two-cycle flowsheet (Fig. 5-21) was devised in which plutonium, oxidized and maintained in the tetravalent state by sodium chlorite, was extracted from a water solution of the salt waste into TBP with 25 vol% tetrachloroethylene (TCE) solvent in the first cycle. In the second cycle, americium was removed from the brine solution into a 0.5 *M* CMPO-TCE solvent. In each cycle, the solvent extracts were scrubbed with 6 *M* HCl, and the products were stripped into a dilute HCl solution.

The CMT Division had the responsibility for developing the first cycle, which is a Purex-type cycle. The Chemistry Division developed the TRUEX second cycle. During 1987, the Purex-TCE cycle was installed in a glove box at Los Alamos. The separation was made in a 16-stage contactor fabricated largely of Hastelloy C276 (see Fig. 5-22). Very good plutonium recoveries were achieved, and the Purex-TCE cycle was adopted for production use. Some corrosion of the Hastelloy C276 occurred, and an ANL-type contactor made of Kynar® was fabricated for Los Alamos by a French firm, Robatel.

The TRUEX-TCE cycle was not put into operation, primarily because americium recovery was of secondary importance, and because glove-box space was not available for its installation.

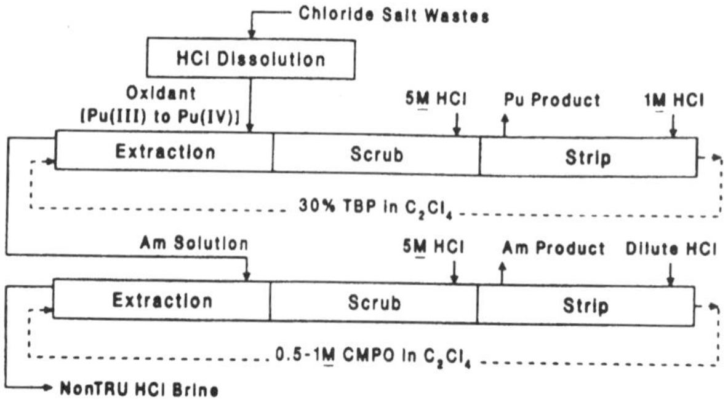

This flowsheet exemplifies the manner in which a Purex solvent-extraction cycle can be coupled with a similar TRUEX cycle to achieve a three-way separation of uranium, plutonium, and TRU elements. A separation of this type could be useful in developing processes for treating nuclear wastes.

Fig. 5-21. Process for Recovery of Americium and Plutonium from Chloride Salt

Fig. 5-22. Sixteen-Stage Centrifugal Contactor Used in TRUEX Process

Generic TRUEX Model. Because of the wide variety of actinide wastes in the United States, a large effort was devoted to developing a generic TRUEX model by which distribution coefficients of the key elements could be determined from a knowledge of the waste composition. Flowsheets based on the distribution coefficients could then be developed. The model enabled determination of stagewise composition changes in counter-current multistage contactors.

Considerable success was achieved with the model. Curves A, B, and C in Fig. 5-23 show the ultimate agreement between the model and experimental data as successive improvements in the model were made. Figure 5-24 shows the close agreement between experimental and calculated values for stage-to-stage concentrations of nitric acid in the aqueous phases of a 14-stage centrifugal contactor.

Radiolytic and Hydrolytic Degradation of the TRUEX Solvent. If the TRUEX process were used to extract TRU elements from the first-cycle waste of the Purex process, the TRUEX solvent would be degraded by radiolysis and hydrolysis. The

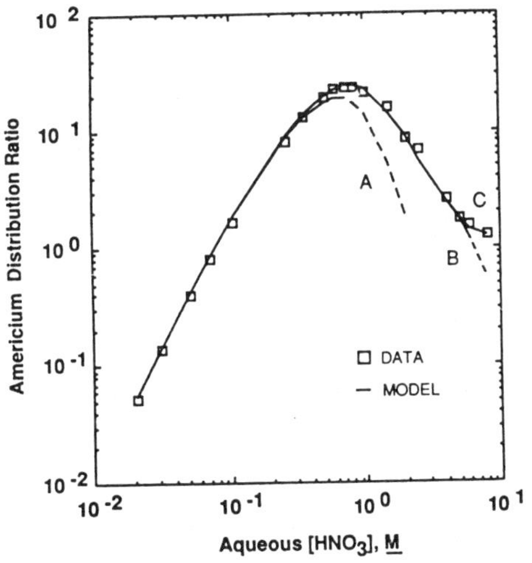

Fig. 5-23. Effect of Nitric Acid Concentration on Extraction of Americium by TRUEX Process (A, B, and C show successive model refinements.)

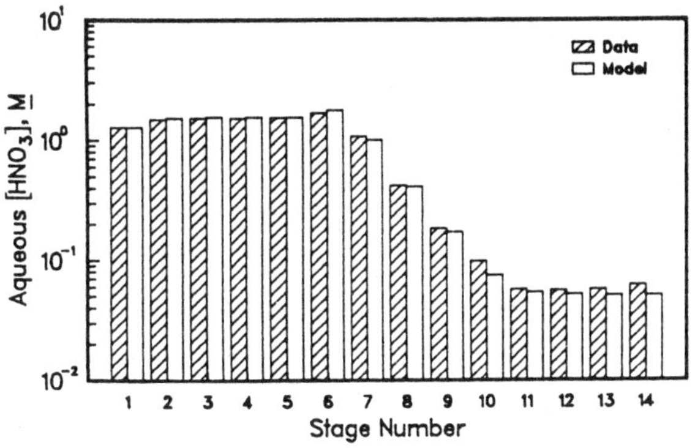

Fig. 5-24. Experimental and Calculated Values for Nitric Acid Concentrations in 14-Stage Centrifugal Contactor

extent of solvent degradation by these mechanisms was determined by changes in the americium distribution coefficient between 2.0 *M* nitric acid and the TRUEX solvent (CMPO diluted with a paraffinic hydrocarbon). The extent of degradation is similar to that for TBP in the Purex process. Degradation by hydrolysis was approximately one-thirtieth of that by radiolysis.

Extensive studies of radiolytic and hydrolytic damage to CMPO were also made at nitric acid concentrations of 0.01 and 0.05 *M* and at temperatures of 70 and 95°C. From these, G values (molecules of CMPO destroyed per 100 eV of absorbed radiation) were obtained, and rate constants for hydrolysis of CMPO at any temperature could be derived. Algorithms expressing the amount of decomposition of CMPO by radiolysis and hydrolysis were incorporated into the Generic TRUEX model. Washing a degraded TRUEX solvent with a carbonate solution effectively removes the degradation products.

Centrifugal Contactor Development. Because of the importance of centrifugal contactors in the TRUEX process, continued improvements were made in their design and operation. One of these was the design of a remotely operable contactor for operation within a glove box or a shielded cell facility.

Provisions were made for easy installation and removal of major components, such as the motor-rotor assembly. Several contactors were designed and fabricated at ANL for installation at other DOE sites, including Hanford, Y-12 at Oak Ridge, and Los Alamos.

To permit design modifications to accommodate different flowsheets, models were developed for calculating aqueous and organic flow through the contactor and vibrational parameters of spinning rotor-motor combinations.

Personnel. George Vandegrift was in charge of the CMT work on TRUEX. Other members of the group were Natalie Blake, David Chaiko, David Chamberlain, Lorac Chow, Clifford Conner, Jacqueline (Leddin) Copple, Don Fredrickson, Eddie Gay, Ray Jaskot, Jane Kwok, Ralph Leonard, Frank Mrazek, John Rajan, Monica Regalbuto, Lucinda Reichley-Yinger, Carmen Sabau, Wally Seefeldt, Ian Tasker, Verne Trevorrow, Pui-Kwan Tse, Erv Van Deventer, and Jake Sedlet.

Ralph Leonard was in charge of the centrifugal contactor effort, which also included George Bernstein, Bob Frank (TD), Ralph Pelto, Ron Wigeland, and Tony Ziegler.

RECOVERY AND SEPARATION OF MOLYBDENUM-99

In 1985, the Division began a small program, under the leadership of George Vandegrift, to assess the feasibility of substituting low-enriched uranium (LEU, 20% ^{235}U) for high-enriched uranium (HEU, 93% ^{235}U) for production of the fission product ^{99}Mo. This substitution was required by the United States government to prohibit export and internal use of HEU because of its potential use in nuclear weapons. Most of the world's supply of ^{99m}Tc for medical-diagnosis applications is produced from ^{99}Mo.

Molybdenum-99 decays with a half-life of 66 h to metastable technetium-99, which has a half-life of 6.5 h. Targets containing ^{235}U need to be irradiated for five or six days to reach a steady-state concentration of ^{99}Mo. The targets must then be processed rapidly to produce a ^{99}Mo source from which ^{99m}Tc may be stripped periodically. For example, an alumina column on which ^{99}Mo has been deposited could be "milked" several times to remove ^{99m}Tc before most of the ^{99}Mo has decayed.

Targets of HEU consisted of uranium aluminide in curved plates, uranium-aluminum alloys in modified fuel rods, and electrodeposited films of UO_2 on the inside surfaces of cylinders. Substitution of LEU for HEU would require dissolving and separating ^{99}Mo from nearly five times more uranium. The greater amount of uranium would also likely affect the removal of other impurities from the ^{99}Mo. To overcome partially the low concentration of ^{235}U in LEU while maintaining the same target geometries, denser uranium targets were sought. The objective of the Division's work was to evaluate alternative target materials and to develop procedures for dissolving them, and for separating ^{99}Mo from the bulk of the uranium and other target material constituents.

Early consideration was given to U_3Si_2 dispersed in aluminum and clad in aluminum. Although dissolution procedures for the target material constituents were developed, they were cumbersome and unattractive, and the presence of large amounts of silicon caused downstream processing problems. Another candidate was electrodeposited uranium to form freestanding hemispherical shells. Don Vissers was achieving some success in producing dense deposits, but lack of sufficient funding precluded further pursuit of that approach. As the decade ended, the search for suitable target materials was continuing.

The individuals involved in this work included Del Bowers, Ed Huff, Samson

Marshall, Laszlo Redey, George Vandegrift, and Don Vissers.

Integral Fast Reactor Electrometallurgical Process

THE INTEGRAL FAST REACTOR CONCEPT

The Integral Fast Reactor Concept evolved from a growing awareness of the potential of metal fuels in fast reactors. This was based on the following two discoveries at the Experimental Breeder Reactor II (EBR-II), a sodium-cooled, pool-type fast reactor in which a uranium-fissium[3] metal fuel alloy was employed.

1. High fuel burnups (in excess of 10%) could be achieved by providing annular space in the fuel element to accommodate fuel expansion during irradiation.
2. Metallic fuels provide a high degree of inherent safety. Sharp temperature rises in the fuel resulting from loss of flow in either the primary or secondary sodium systems cause the metal fuel assemblies to expand and the fuel elements to move apart. This slows down the chain reaction, and the reactor shuts itself down without human intervention.
3. Although initially proposed as a breeder, an IFR could also be designed instead to "break even," *i.e.*, generate an amount of plutonium equal to that burned, or to consume more plutonium than is generated (a converter reactor).

[3] Fissium is a mixture of noble metal fission products, molybdenum, ruthenium, rhodium, and palladium, which was present in EBR-II driver fuel at a total concentration of 5 wt% in ratios corresponding to their fission yields. Fissium produces an equiaxed metal grain structure in uranium, which, under irradiation, expands uniformly in all directions.

The success achieved with metal fuels in EBR-II spurred Chuck Till to inquire of Les Burris and Bill Miller about the possibility of reviving the melt-refining process, which had been successfully employed in the early years of EBR-II operation in the adjacent Fuel Cycle Facility to recover discharged fuel. Burris noted that the melt-refining process would not be practical for recovery of high-burnup fuel because a large fraction of the fuel (30 to 50%) would have to be removed for separate processing to remove noble metal fission products. In a subsequent search for a process suitable for IFR fuel, Bill Miller came up with the idea of using electrorefining, which had been investigated by the French in the 1960s for the purification of unirradiated uranium.

Electrorefining soon became a major focus of research and development work in CMT. The Division was given the responsibility for developing, under the direction of Les Burris, an on-site process for recovering plutonium and uranium from core and blanket materials, removing fission products from them, reenriching the core alloy with plutonium bred in the blanket, and disposing of wastes. Integrating the process into a closed cycle for recycling fuel from an IFR became a major feature of the IFR concept. The close coupling of the recovery process with the reactor accounts in large measure for the name, Integral Fast Reactor.

Because the fission products would be completely removed in the electrorefining process, a new alloying metal was sought that would confer the same benefits to the fuel as does fissium. Zirconium was found to be equally effective, and a fuel alloy consisting of uranium, plutonium, and zirconium in the proportions of 70/20/10 wt% was chosen as the IFR driver fuel. Zirconium has the additional advantage of raising the melting point of the fuel. With this fuel, burnups greater than 18.5% were achieved in EBR-II, far exceeding the design burnup of 10%. The composition selected for the blanket material was uranium-10 wt% zirconium.

PROCESS FLOWSHEET DEVELOPMENT

Two basic steps were incorporated in the original process flowsheet: (1) electrorefining for removal of fission products from core and blanket material and (2) halide slagging for concentrating plutonium in the blanket material from a concentration of about 5 wt% or less to a value sufficient to enrich the driver fuel (30 to 40 wt%). In this flowsheet, metallic fuel-element segments were dissolved in liquid cadmium at 500°C, leaving behind the cladding hulls. The resulting solution of uranium in cadmium was made the anode of an electrolytic cell. Under an applied voltage, uranium was transferred through a molten chloride salt electrolyte to a solid metal cathode. The molten salt electrolyte was the LiCl-KCl eutectic (m.p., 352°C) containing a few percent UCl_3 to facilitate uranium transport. The uranium deposit was recovered mechanically and melted to form an ingot for preparation of new fuel and blanket pins. In halide slagging, the blanket material was melted under a molten salt consisting of barium and calcium chlorides and a small amount of UCl_3, which oxidizes plutonium preferentially into the salt phase. The salt was then added to the electrorefining electrolyte for plutonium enrichment of the core material. The operating temperatures for the electrorefining and halide slagging steps were 500 and 1250°C, respectively.

The following changes and improvements in this flowsheet were made to overcome some difficulties with it and to incorporate more efficient process arrangements:

1. The solution rate of uranium in cadmium was found to be intolerably low. To overcome this problem, the anode basket containing fuel or blanket segments is moved into the salt phase. Under an applied voltage, uranium and plutonium are oxidized to their chlorides as, simultaneously, an equivalent amount of uranium is deposited on the cathode.
2. Plutonium accumulates in the salt phase and is not transportable to a solid metal cathode when uranium is present because $PuCl_3$ is substantially more stable than UCl_3 (by about 20 kcal/mol). Electrotransport of plutonium requires use of a liquid cadmium cathode. In cadmium, plutonium has a low activity coefficient of 10^{-4}, which makes its transport to cadmium possible.
3. Halide slagging, which was shown to work in accord with theoretical predictions, was found to be unnecessary. Blanket fuel could also be processed by electrorefining. Plutonium accumulates in the salt phase as $PuCl_3$ while uranium is deposited on the cathode. Plutonium is removed later from the salt to enrich core material. The simplified process is shown in Fig. 5-25.

PROCESS FREASIBILITY AND DEMONSTRATION

Two inert-atmosphere glove boxes with instrumentation and argon purification systems were installed in Laboratory G-118 for conducting electrorefining and the previously mentioned halide-slagging experiments with plutonium-bearing materials. The principal investigators were Bill Miller and Ziggy Tomczuk. Ray Wolson was responsible for maintaining the facility and keeping plutonium accountability records.

In the early electrorefining experiments, uranium was transported from a cadmium solution to a solid iron mandrel. Important

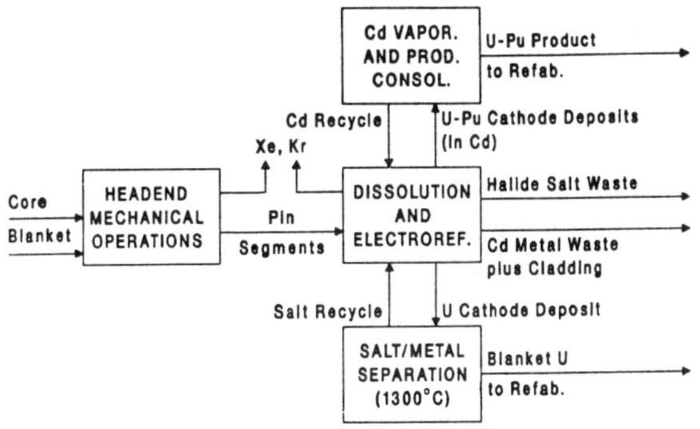

Pin segments of core or blanket material are contained in a basket suspended in the electrolyte salt. Under an applied voltage, uranium and plutonium are dissolved anodically, with uranium transporting to a solid metal cathode and the plutonium reporting to the salt. Subsequently, plutonium, along with some uranium, is transported to a liquid cadmium cathode.

In this flowsheet, the role of the cadmium pool in the bottom of the electrorefiner changes from that of an anode to a collector of any noble metal fission products and zirconium that might escape from the anode basket. It would also collect any uranium that might fall from the dendritic deposit of uranium on the solid cathode. The electropositive fission-product elements (alkalis, alkaline earths, rare earths, and yttrium) collect in the salt, as does sodium used as a thermal bond in the fuel elements. Iodine also reports to the salt.

The product metals are recovered by vaporizing cadmium and any adhering salt, and melting them into ingots for manufacture of new fuel pins.

Fig. 5-25. Electrorefining Process for Driver and Blanket Fuel of the Integral Fast Reactor

results of these early studies were the following:

1. High uranium recoveries (greater than 99%) were demonstrated.
2. Uranium and plutonium were distributed between salt and metal phases in accord with theoretical predictions. Their total concentrations in the salt and metal phases could easily be adjusted by using cadmium chloride ($CdCl_2$) as an oxidizing agent or lithium as a reducing agent.
3. Fission products behaved as predicted. The noble metal fission products remained in the cadmium. The electropositive elements transferred to the salt. A uranium decontamination factor of 240 was realized in a run spiked with a small amount of irradiated uranium.
4. The low-carbon steel vessel used to contain the molten salt and cadmium phases showed no evidence of corrosion after one year at 500°C.

Because plutonium reports to the salt electrolyte, consideration was given to using a cup containing liquid cadmium as the cathode for deposition of both uranium and plutonium. On codeposition of uranium and plutonium, uranium dendrites grew out of the cadmium surface when the cadmium became saturated with uranium. The activity coefficient of uranium at its saturation concentration in cadmium is one, the same as that for uranium metal, so the pronounced tendency for dendrite formation is understandable. These dendrites had a fern-like structure and, on growing out of the cadmium cup, they short-circuited the cell when one of them came in contact with the cradle holding the cathode cup. Various methods for preventing dendrite growth (stirring, cutter blades, *etc.*) were unsuccessful.

As the process evolved, uranium was removed on a solid metal cathode, and plutonium was removed by transport to a cadmium cathode. It was found later that by maintaining a sufficiently high ratio of plutonium to uranium in the salt, saturation of the cadmium with uranium would not occur and, hence, dendrite formation would not occur. If the activity of plutonium in the salt was greater than that of plutonium in the plutonium-saturated cadmium, plutonium could be "pumped" into the cadmium, with the intermetallic compound, $PuCd_6$, precipitating as the additional plutonium was transferred.

A small amount of rare earth elements will be deposited along with the plutonium in a cadmium cathode. The decontamination factors for cerium and neodymium in two experiments are shown in Table 5-8. These decontamination factors are low, but ample—in fact, desirable. The high radiation level resulting from contamination of the plutonium product with rare earth fission products is a strong deterrent to diversion of plutonium to weapons production.

The TRU elements, americium, curium, and neptunium, were found to accompany the plutonium quantitatively. They can, therefore, be recycled with plutonium to the reactor where they are consumed by fission and contribute to the energy generated by the reactor. The absence of these elements in the electrorefining waste streams is a major advantage of the electrorefining process.

LARGE-SCALE DEMONSTRATION OF ELECTROREFINING

Justified by the successful electrodeposition of uranium on a 300-gram scale, a large-scale electrorefiner with a capacity of 20 kg of uranium was installed in a large, inert-atmosphere glove box in Laboratory J-118. Only the deposition of uranium was investigated. Work with plutonium was prohibited because the amounts of plutonium that would be required far exceeded the laboratory and building security limits. Eddie Gay was in charge of operating this facility.

Cross-section diagrams of the large electrorefiner are shown in Fig. 5-26. A close replica of this unit was designed and built for installation in the EBR-II Fuel Cycle Facility. The purposes of the large electrorefiner were to (1) demonstrate fuel dissolution, deposition of 10-kg amounts of uranium on an iron-mandrel cathode, and product handling, (2) establish the duty cycle for the electrorefiner, and (3) evaluate equipment performance.

Dissolution in cadmium had been attempted, but, as noted previously, the rate of uranium solution was impracticably low, so anodic dissolution became the norm. Successful demonstration of the deposition of 10 kg of uranium was carried out many times. Figure 5-27 is a photograph showing the highly dendritic deposit of 10 kg of electrodeposited uranium.

Several other uranium-deposition options were examined. Approximately 4 kg of uranium was successfully transported into liquid cadmium, but this was not an efficient method for uranium transport. Chemical deposition into liquid cadmium was also tried with limited success. Neither of these options was considered further for uranium collection.

The deposition of 10 kg of uranium required 40 to 50 h. That is much too long for practical application of electrorefining. Reducing the uranium deposition time was identified as a challenge to the researchers as the process development moved into the 1990s.

Table 5-8. Decontamination Factors for IFR Electrorefining Process

Experiment	Ratio	Discharged IFR Fuel	Cathode Product Ingot	Decontamination Factor
ENG 20	Nd/Pu	0.069	0.012	5.8
ENG 21	Nd/Pu	0.069	0.022	3.1
ENG 20	Ce/Pu	0.043	Not obtained	--
ENG 21	Ce/Pu	0.045	0.0085	5.3

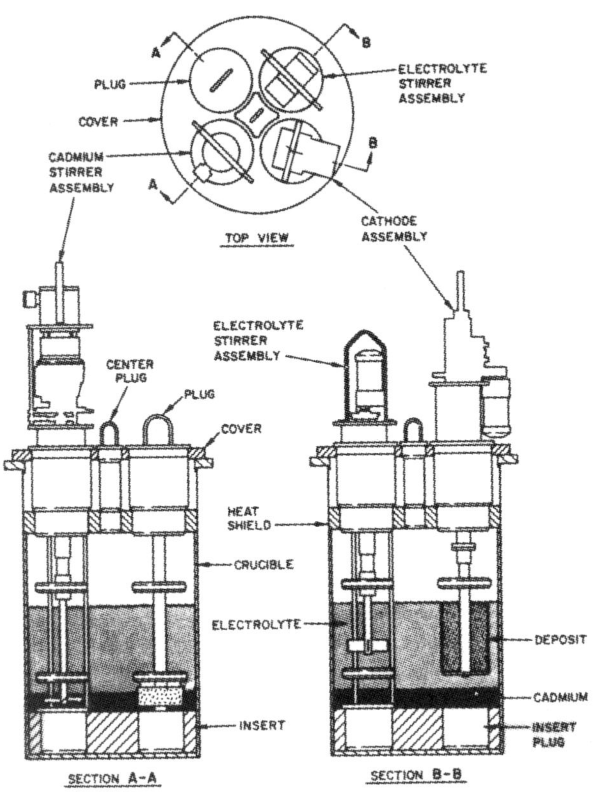

Fig. 5-26. Large-Scale Electrorefiner for Spent Nuclear Fuel

Fig. 5-27. Dendritic Uranium Deposit (~10 kg) from Electrorefining

IFR WASTE MANAGEMENT

Nearly all the fission products in the electrorefining vessel are in the anode basket and the electrolyte salt. These become part of the two major waste streams resulting from the electrorefining—a metal waste stream and a salt waste stream.

Metal Waste. The anode basket contains most of the noble metal fission products, zirconium (the alloying element in core and blanket fuels), and stainless steel cladding hulls. To this are added fuel assembly hardware, waste ceramic materials from preparation of new fuel pins and from operations to consolidate the uranium and plutonium products produced in electrorefining, and cadmium waste streams. The change from fuel dissolution in cadmium to anodic dissolution did not significantly alter the composition of the metal waste stream. One of the cadmium waste streams is produced in treating salt wastes to remove the TRU elements. It contains transuranic and rare earth elements and yttrium. Also, even with anodic dissolution, some cadmium must be removed occasionally from the cadmium cathode pool in the electrorefiner. Cadmium, however, is only a temporary constituent of the metal waste, since it is vaporized and condensed for recycle.

Initially, the metal waste was encapsulated in a heavy-walled steel container for disposal in a repository. In a later innovation, copper powder was added to the metal waste in a copper canister, and the mixture was compacted by hot isostatic pressing before being placed in a steel canister for final disposal.

Salt Waste. In the initial waste flowsheet, actinide elements in the salt removed to waste were extracted into cadmium with a Cd-1 mol% Li alloy. Most of the electropositive fission products also transferred to the cadmium. Only cesium, strontium, and iodine remained in the salt phase. The cadmium was combined with other metal wastes.

Figure 5-28 shows the distribution coefficients of several metals as a function of lithium concentration in cadmium. The distribution coefficient is defined as $D(M) = Y_M/X_M$, where Y_M is the mole fraction of MCl_3 in the salt, and X_M is the mole fraction of M in the liquid cadmium solution. These data, and other data for transuranic elements, indicate that a single contact with a Cd-1.0 mol% Li alloy should produce an actinide concentration of less than 1 ppb in the salt. In practice, however, such low values could not be attained. This was attributed to the presence of small amounts of uranium and plutonium oxide or nitride impurities in the salt. Although the actinide content of the treated salt could be low enough (less than 100 nanocuries per gram) to classify the salt as non-TRU, the presence of cesium and strontium would still require its burial in a deep geologic repository.

The major difficulty with the above treatment of salt waste was the loss of several percent plutonium to waste. Work was therefore begun on separating the actinides from the rare earths to enable recovery of the actinides. The large differences between the actinides and the rare earths (see Fig. 5-28) show that they can be separated by a multistage extraction process that employs cadmium

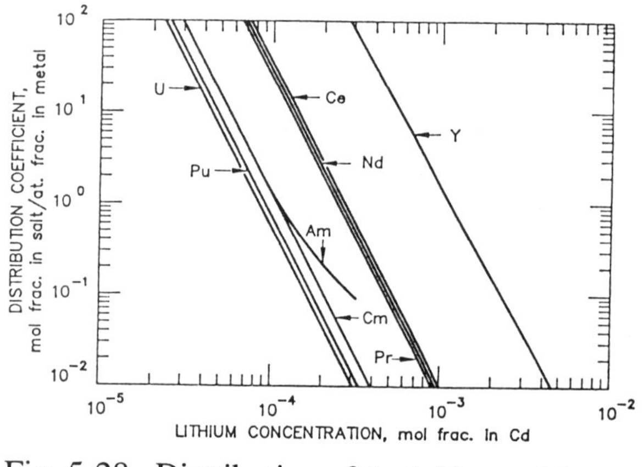

Fig. 5-28. Distribution of Actinides and Rare Earths between Salt and Cd-Li Alloys

containing a small amount of uranium as the extractant. Calculations showed that more than 99% of the actinides accompanied by less than 20% of the rare earths could be achieved in five to six equilibrium stages of countercurrent contact. Work was therefore initiated on development of suitable centrifugal contactors for this operation.

After actinides and rare earths had been removed from the waste electrorefiner salt, the treated salt was dispersed in a Portland cement mortar (salt concentration about 10 wt%) and cast into a corrosion-resistant container for disposal. Although the resistance of the cement mortars to water leaching was satisfactory, hydrogen and chlorine were produced by electrolysis. The resultant pressurization of the container was a major concern. Therefore, disposal of the salt in cement mortars was abandoned, and attention was turned to investigating other media, such as zeolites, for immobilization of the salt waste.

PERSONNEL

Les Burris was the leader of the IFR effort until his retirement, when Jim Battles assumed that responsibility. Others in the program were John Ackerman, Jack Arntzen, John Basco, George Bernstein, Del Bowers, Erv Carls, Don Fischer, John Gabor, Eddie Gay, John Heiberger, Bob Heinrich, Joe Herceg, Ed Huff, Ken Jensen, Jerry Johnson, Irv Johnson, Terry Johnson, Bill Kann, Bob Kessie, Michele Lewis, Len Leibowitz, Dick Malecha, Bill Miller, Bali Misra, Tom Mulcahey, Dave Poa, Nick Quattropani, Don Raue, Jack Settle, Mike Slawecki, Bob Steunenberg, Ziggy Tomczuk, Dave Warren, and Florence (Williams) Smith.

PLUTONIUM RESIDUE RECOVERY (PuRR) PROCESS

The Plutonium Residue Recovery Process (PuRR), although not a part of the IFR

Program, had an indirect relationship to it in that it utilized pyrochemical procedures and was concerned with plutonium. It involved a joint effort between Lawrence Livermore National Laboratory (LLNL) and ANL to develop an effective process for the recovery of plutonium from intractable residues, and, later on, to remove transuranic elements from wastes to a non-TRU level to simplify their disposal. An adjunct effort between Savannah River Laboratory (SRL) and ANL was directed toward the use of a pyrochemical head-end treatment to convert intractable residues into a form suitable for subsequent aqueous processing.

The experimental program, which began in 1988, concentrated on the reduction of PuO_2 in the residues by a liquid metal-molten salt system to produce a solution of metallic plutonium in the liquid metal, which is a more convenient material for plutonium recovery either by pyrochemical or by aqueous processes. Reductions of PuO_2 by this technique had been well established in earlier pyrochemical programs using combinations such as zinc-magnesium as the liquid metal and molten LiCl-KCl-$MgCl_2$ as the salt.

Potential feed materials for the PuRR process included any of 22 categories of weapons-production residues. The most intractable of these materials, acid-leached LECO® analytical crucibles (ground) from SRL and incinerator ash heels from Rocky Flats (calcined and ground), were selected for the studies. The plutonium contents of these materials ranged from >1 wt% to about 20 wt%.

Several solvent/reduction systems were considered; initial experiments were conducted with Cu-Mg-Ca and Zn-Mg reductants for the LLNL effort (Rocky Flats material), and Al and Al-Mg for the SRL material. The salt used for the Cu-Mg-Ca system was $CaCl_2$-15 wt% CaF_2, and for the Zn-Mg and Al-Mg systems it was $MgCl_2$-22 wt% NaCl-19 wt% KCl-4 wt% MgF_2. Good reductions of the PuO_2 (98% or higher), which were achieved in nearly all cases, far exceeded recoveries by aqueous processes. Additional reduction studies resulted in plutonium recoveries averaging 98.0% after one reduction stage and 99.4% after a second stage, based on analyses of the salt, but these results could not be confirmed by analysis of the metal ingot products, which was difficult because of the heterogeneous nature of the product.

Savannah River then requested a determination of whether plutonium remaining in the slag from bomb reductions of PuO_2/PuF_4 by calcium could be extracted by pyrochemical procedures, and also asked ANL to design and fabricate a heated transfer line for transferring molten salts and metals. (Such lines had been developed earlier at ANL in pyrochemical research and development programs.)

Much of the effort in 1989 was concerned with the CaO byproduct resulting from the use of calcium as a reductant. To eliminate the CaO waste and simultaneously recover the calcium for recycle, work was begun on incorporation of a calcium electrowinning step in the process. Regeneration of the calcium metal was performed successfully. The use of carbon anodes resulted in the production of CO or CO_2, rather than oxygen or chlorine, as the by-product of the electrolysis and also reduced the operating voltage.

In spite of the promising results, this program was phased out around 1990 because of a political realization that large quantities of plutonium would become available from reduction of the nuclear weapons stockpile, and there was little incentive to recover more of it. The effort was diverted to the use of this technology to convert wastes into a non-TRU form for disposal, as mentioned earlier.

The individuals working on this program included Jerry Johnson, Irv Johnson, Tom Mulcahey, Dean Pierce, and Dave Poa.

Nuclear Waste Management

YUCCA MOUNTAIN PROJECT

The proposed site for the first high-level waste repository in the United States is in the tuff beds of Yucca Mountain, Nevada. The site is unique among those being characterized internationally because it lies several hundred meters above the water table in a zone unsaturated with water. Investigators in the CMT Division studied the reactions of glass and spent irradiated reactor fuel under anticipated repository conditions. The general focus was on the post-containment period, *i.e.*, that period following canister deterioration (estimated to be 300 to 1000 years after emplacement of the waste), during which time the waste glass monolith is exposed to the repository environment.

Waste Glass Studies—Unsaturated Tests. Although large amounts of water are not expected to contact the waste during the repository lifetime, water vapor or small amounts of transient water may contact the waste at any time after emplacement. Unsaturated tests were designed to study glass/steel/fluid interaction under those conditions. Groundwater from the Yucca Mountain site was dripped onto a model waste package composed of a glass cylinder (SRL 165) sandwiched between two perforated stainless steel disks at 90°C. The rate of water introduction was varied from one drop every 3.5 days to one drop every 14 days. The glass was doped with actinide elements (uranium, plutonium, neptunium, and americium). The nature and degree of reaction were assessed from solution analysis and inspection of the waste package surface by various analytical techniques, such as scanning electron microscopy.

The test apparatus is shown schematically by components in Fig. 5-29. Both batch- and continuous-mode testing were done. In the batch mode, the tests were terminated at selected 13-week intervals. In the continuous mode, the test package, along with any adhering water, was transferred to a new test vessel, and the test was continued in that manner until little additional information could be learned from further exposure.

Forming on the glass surface was a layer of clay, which spalled off, allowing further reaction. This ongoing, cyclic process plays a major role in glass erosion. The release of lithium, present in the matrix glass at a concentration of about 1 wt%, was a good monitor of the extent of glass reaction because little or no lithium is incorporated into secondary phases that form on the glass surface.

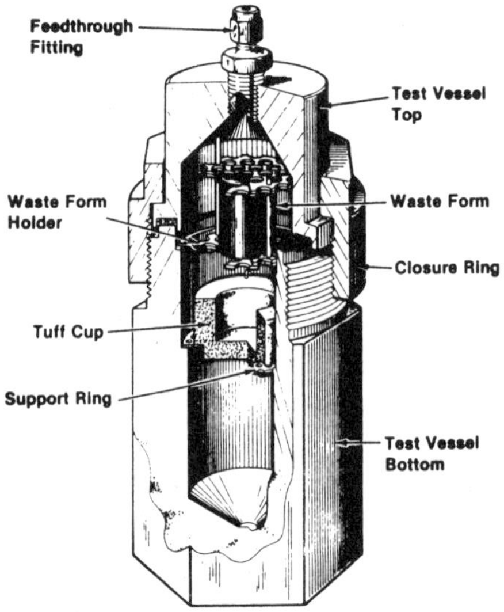

Fig. 5-29. Test Apparatus for Unsaturated Zone Testing

Hydration of Glass by Water Vapor. Before high-level nuclear waste is contacted by liquid water, it will be contacted by water vapor. Some water will be sorbed on the glass surface. Hydration will occur between the glass and this small amount of sorbed water, which will release alkali metals into the sorbed film, decreasing the vapor pressure and allowing further sorption of water. By this mechanism, significant quantities of water may condense on, and subsequently react with the glass. As the glass reacts, released species quickly approach their solubility limits, and precipitates form in the solution. These are incorporated into a potpourri of mineral phases—predominantly zeolites and silicates that form on the on the outer surface of the glass (see Fig. 5-30). The vapor-hydration experiments were extremely useful in projecting performance of glass because they disclosed the mineral phases that control the reaction rate.

Vapor/Hydration Leach Experiments. The extent of vapor hydration or aging of glass is a function of the glass composition, relative humidity, reaction time, and temperature. Of interest was whether aged glass reacted differently to liquid water than fresh glass. The difference in reaction rates was striking. Aged glass reacted significantly faster with liquid water than did fresh glass. Even more striking, however, was the behavior of plutonium and americium. Very little plutonium or americium appeared in solution in leaching tests using fresh glass. In contrast, with aged glass, considerable amounts of plutonium and americium were found in the solution as colloidal precipitates. This effect is illustrated by the data in Table 5-9. The two types of glass are a Westinghouse Savannah River Co. (WSRC) glass and a West Valley Demonstration Plant (WVDP) glass. The importance of this finding is that groundwater transport of colloidal suspensions may differ greatly from that of dissolved materials.

Static Leach Experiments. After high-level wastes have been enclosed in a repository, the most likely scenario for release of waste constituents is through leaching and transport in groundwater. Two sets of static leach-rate tests were conducted by the CMT Division in the 1980s: (1) general leach-rate characterization studies in the early 1980s and (2) studies in the late 1980s that were pertinent to the Yucca Mountain repository. Two techniques were used to measure the small amounts of radioactive materials leached from the wastes: (1) neutron-activation analysis (NAA), and (2) solution analysis after spiking the waste form with radioactive tracers. These methods complemented each other because there are instances when NAA is not

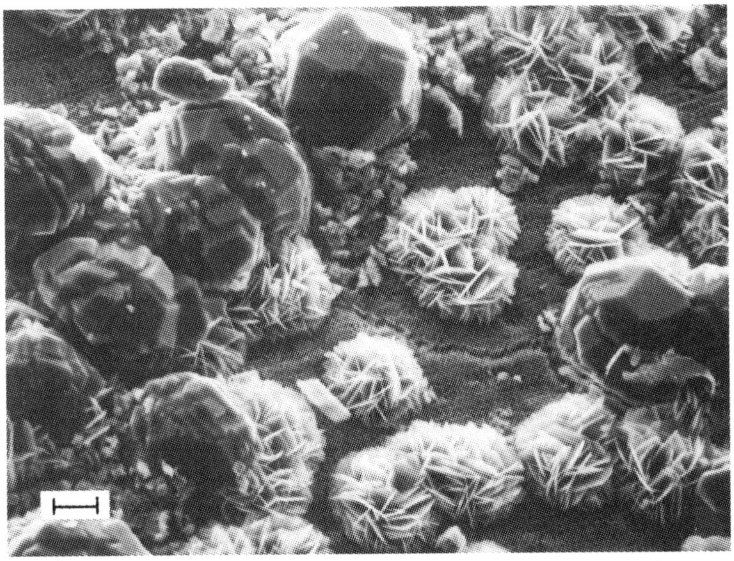

Fig. 5-30. Scanning Electron Image of Hydrated Glass Surface Showing Complex Silicate Precipitates

Table 5-9. Release Rates of Elements from Fresh and Hydrated Glasses

Elements	Normalized Release, g/m^2			
	WSRC Glass		WVDP Glass	
	Fresh	Hydrated	Fresh	Hydrated
B, Li, Na	1.5	30	10	25
Am, Pu	0.1	2	0.001	0.2
Np, U	0.6	15	2	10

applicable. The close agreement between NAA and spiking for amounts of selected constituents from a Savannah River Laboratory (SRL) glass, SRL-211B, is illustrated in Fig. 5-31.

Leach tests were conducted on three Savannah River glasses, 131A, 211B, and 211C. Glass samples prepared for leaching were spiked with radiotracers from the four major classes of fission products (alkali, alkaline earth, rare earth, and noble metal), as well as the actinide elements (uranium, plutonium, neptunium, and americium). Static leach tests were conducted, following procedures specified by the Materials Test Center, namely, contact with deionized water at 90°C for 3, 7, 14, 28, and 42 days. Some tests, however, were run for up to 546 days to gain insight into the long-term nature of glass leaching. In one of the 546-day tests of an SRL-131 glass, six distinct layers were found by analytical electron microscopy.

After about one year, there was little further accumulation of any element in the leachate. Effective leachate saturation was reached by formation of a "protective" gel layer on the glass surface. This occurred because of saturation of some elements in solution and formation of precipitates. For each glass, the matrix elements of the glass (Si, Na, Li, B, Al, Ca, and Fe) leached at about the same rate, but the nature of the gel layer, the composition of the precipitates, and the behavior of the waste elements differed greatly among the glass compositions. It was concluded that the behavior of waste elements must be determined in tests using glass having the same composition as that of the actual waste. Uranium and neptunium continued to leach from the glass and were not found on the glass surface, whereas plutonium and americium accumulated slowly in the leachate, their concentrations stabilizing after 28 days.

The Yucca Mountain leach tests were performed with leachants characteristic of the tuff repository. These tests showed that the ratio of the glass surface area to the leachant volume is a critical parameter in static leach experiments because it affects the rate at which the solution becomes saturated or approaches saturation by various glass components and,

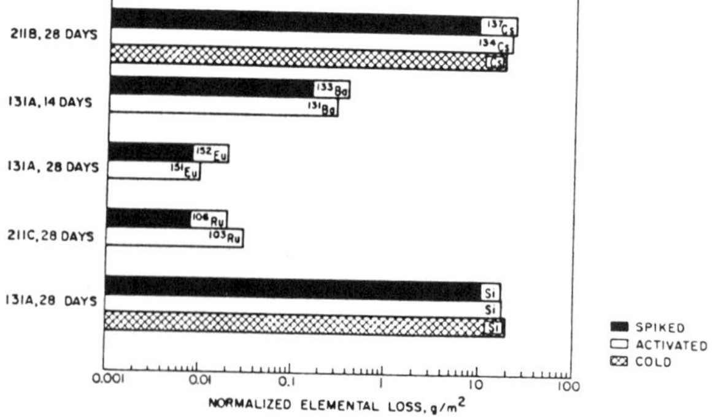

Fig. 5-31. Losses of Representative Elements from Simulated Glass Wastes

therefore, the eventual composition of precipitated phases. Release rates of elements were controlled more by solubility than by the existence of a protective layer.

Natural Analog Studies. Naturally occurring glasses are useful for validating both predictive computer codes for long-term waste performance and laboratory experiments used to accelerate glass reactions. The experimental alteration by water vapor of basalt and obsidian glasses, whose compositions are similar to synthetic waste glasses, was investigated and compared with the alteration of synthetic glasses under the same conditions. The amount of water sorbed and, hence, the extent of glass alteration increased with relative humidity at all temperatures. Reactions occurred in a thin film of condensed water, and precipitates formed in only a few days due to rapid saturation of the small amount of water by dissolving constituents of the glasses.

Spent Fuel Studies. Although most of the work on storage of nuclear wastes in the 1980s was concerned with the storage of high-level radioactive wastes immobilized in glass, a study was initiated in 1988 to determine the release of radionuclides from spent fuel samples (from light water reactors) exposed to varying amounts of water following breach of containment in a repository. The high level of ionizing gamma radiation causes a reaction of oxygen and nitrogen in the air to form nitric oxide, which then combines with water to form nitric acid. The nitric acid causes corrosion of the fuel storage canister. Relative rates of corrosion of candidate cannister materials were Cu>Cu-Ni>Al-bronze>> Inconel 825. Isotopic exchange of U-235 and U-238 could be determined with high precision by mass spectrometric isotope dilution. This program was continued into the 1990s to determine the behavior of fission products and actinide elements.

ANALOG EXPERIMENTS FOR PROPOSED HANFORD REPOSITORY

In the early 1980s, one of the candidate repository sites was at Hanford, Washington, in underlying Pomona basalt. Borosilicate was the solid waste form, and bentonite plus basalt was the backfill around a canistered waste form. Analog experiments were performed in which the migration of elements from the glass waste by a water stream was followed through a sequence of materials, *i.e.*, waste form, bentonite clay, and the host basalt rock. The laboratory experiments are illustrated in Fig. 5-32, and Fig. 5-33 shows the distribution of neptunium from water at 90°C flowing at a rate of 250 m/yr, which is higher than that expected in a repository. At the water flow rates expected in a repository, neptunium would migrate about 200 m per 1,000 years. The Hanford site for a repository was eliminated from further consideration in 1984, so the analog experiments were stopped. This work was part of the geoscience program conducted by Martin Seitz and Neil Sturchio with Rex Couture and George Vandegrift.

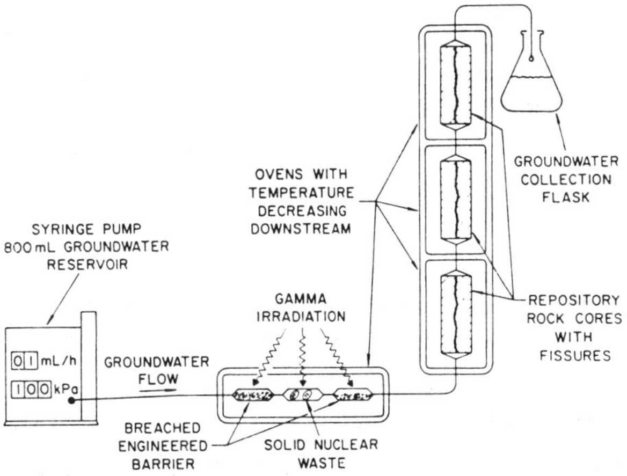

Fig. 5-32. Design of Groundwater Stream Simulation Experiment

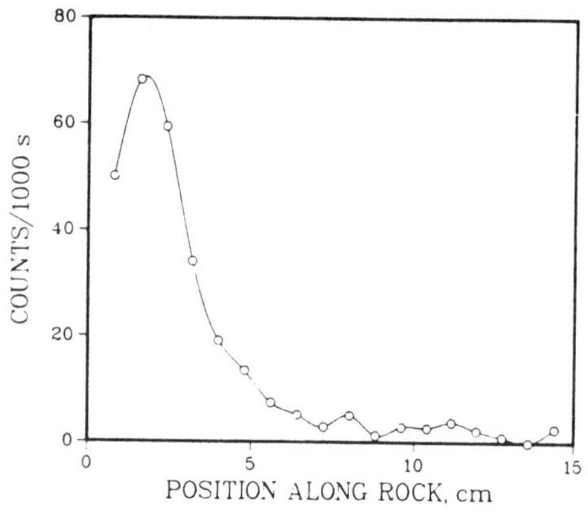

Fig. 5-33. Distribution of Neptunium-237 on a Basalt Core Surface

BRITTLE FRACTURE STUDIES OF SOLID WASTES

Because mechanical impacts to solid wastes might occur during processing, handling, storage, and transportation, the CMT Division began the development of a generalized methodology for characterizing the impact fracture of brittle materials for evaluation of material properties and use in accident analysis of full-size waste forms. Fracture of a solid waste would increase the surface area available for leaching of fission products by groundwater and produce a quantity of respirable particulates (<10 μm).

A preliminary model was developed on the basis of a particle-size distribution resulting from impacting small Pyrex® cylinders (38-mm dia by 100-mm length) by a drop weight. The particle-size distribution was lognormal. Subsequently, simulated waste glasses (SRL-131 and PNL-76-68), several Synroc formulations, and conglomerate brittle materials such as sandstone were tested. For the materials tested, all were found to follow a lognormal particle-size distribution. A similar lognormal distribution was found at Pacific Northwest Laboratories (PNL) for free-fall impact tests of full-scale canisters (0.61-m dia by 2.3-m length), indicating that the CMT model may be applicable to full-size waste forms. In the PNL tests, the respirable fraction was found to be 0.15 wt%. This CMT study was terminated in 1982. These studies were done by Bill Mecham, Wally Seefeldt, and Martin Steindler.

NUCLEAR METHODS OF OIL WELL LOGGING

Workers in CMT investigated the application of nuclear oil well logging methods to determine the amount of oil remaining in an underground formation after removal of the easily available oil. In the proposed method of logging an underground reservoir, the radioactivity of the reservoir is first determined by a gamma-sensitive probe. Then, water containing a gamma-emitting radionuclide is injected, and gamma logging is repeated. The difference in measured radioactivity can be related to the residual oil content. The difference in radioactivity between an oil-saturated core and an oil-free core on injection of water containing Na-22 is shown in Fig. 5-34. This work was not continued beyond demonstration of the feasibility of the method.

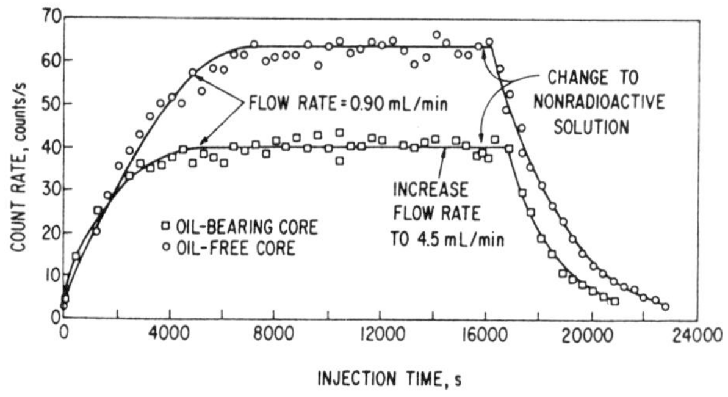

Fig. 5-34. Gamma Signals from Cores Injected with Radioactive ^{23}Na-Bearing Solutions

PERSONNEL

The glass studies were performed under the leadership of John Bates. Other members of the group included Teofilo Abrajano, Don Fischer, Tom Gerding, Les Jardine, Jim Maser, Bill Mecham, Gerald Reedy, and Ewald Veleckis. Two postdocs, Charles Byers and Bill Ebert, also made substantial contributions, as did several members of the Analytical Chemistry Laboratory (ACL) and the Chemistry Division (CHM).

Martin Seitz led the other work, which was performed by Rex Couture, Tom Gerding, Neil Sturchio, George Vandegrift, and Don Vissers. Again, several members of the ACL and CHM were involved significantly in the effort.

Applied Physical Research

In the early 1980s, the Applied Physical Research Program was concerned mainly with the physical and thermodynamic properties of oxide reactor fuels. By 1984, a sufficient amount of information had been accumulated in handbooks and computer databases to permit useful modeling of reactor accidents. About that same time, it became apparent that metallic reactor fuels could achieve much higher burnups than previously thought possible by redesigning the fuel pins to accommodate the swelling of the fuel. This result was a major breakthrough that led to the IFR concept, which kindled a renewed interest in metallic fuels in the Applied Physical Research Program.

OXIDE REACTOR FUELS

Gas-Cooled Fast Reactor Fuel. In support of gas-cooled fast-reactor (GCFR) technology, a brief investigation was conducted on the transport and reaction of oxygen and fission products in mixed uranium-plutonium oxide fuels. Although the GCFRs are similar to LMFBRs in that they both use oxide fuels, they differ in that the GCFR fuel pins are vented, the cladding is roughened, helium is used as the coolant, and there is direct contact between the coolant and the steam generator. Large axial temperature gradients that occur within the GCFR fuel pins lead to complex chemical and transport phenomena, which are related primarily to the chemistry of oxygen and cesium in the fuel pins. Under these conditions, solid-state thermomigration results in redistribution of oxygen (as well as actinides) in the fuel.

A mathematical model used a defect structure for UO_2-PuO_2 fuel involving oxygen vacancies and electronic disorder. The model relates a thermodynamic factor, $1 + \delta\ln\gamma_V/\delta\ln C_V$ (where γ_V is the oxygen-vacancy activity coefficient, and C_V is the vacancy concentration), to the oxygen potential of the fuel. Extensive data were available and accurate models already existed for the oxygen potentials, so only one quantity, the heat of transport of oxygen vacancies, Q_V, was required to obtain profiles of the steady-state oxygen redistribution under a given set of reactor operating conditions. The value of Q_V depends on the stoichiometry of the system. Experimental data showed that the value of Q (the heat of transport) decreases to zero as the mean valence of the plutonium goes from 4.0 to 3.0, which is consistent with the model.

High-Temperature Physical Properties. In modeling reactor accidents, one needs information about the thermophysical properties of reactor materials, such as fuels, coolants, and structural materials, from room temperature to very high temperatures (about 5700°C). The purpose of this work was to provide reliable values of these properties for reactor safety analysts. Available experimental data were assessed and calculations were made by using thermodynamic relationships as well as approximation and extrapolation techniques when necessary. Uncertainties in the recommended values were also estimated from experimental deviations, data-fitting errors, and approximations or extrapolations.

Recommended values for the properties of sodium were documented by Joanne Fink and Len Leibowitz, and Joanne published the computer codes used in the calculations in a sister document. Both documents were sent to the United Kingdom for comparison with recommended values from scientists in the United Kingdom, France, and the Federal Republic of Germany. A magnetic tape of the property values and the computer codes was sent to a database at Oak Ridge, which is entitled Safety Analysis Computerized Reactor Data (SACRD).

Equations used to fit experimental data on the enthalpy and heat capacity of UO_2 were evaluated by Fink and Leibowitz, taking into account factors such as contributions from phonons and thermal expansion and an anomalous electronic contribution. The best fit was obtained with a two-term, nonlinear equation below the transition temperature (2397°C), and a linear equation above that temperature. The two terms of the nonlinear equation represent contributions from the photons and electrons. Previous assessments of the thermal conductivity of UO_2 had resulted in a three-term nonlinear equation representing contributions from phonons, electrons, and a conduction decrease due to dislocations. A statistical study showed that the data did not justify the inclusion of the dislocation term. A two-equation formulation fit both the thermal conductivity and the thermal diffusivity data better, and it was consistent with the theoretical basis of the equations used for enthalpy and heat capacity.

Enthalpy measurements were performed on thoria-urania (ThO_2-UO_2) fuels for the Westinghouse Electric Corporation's Bettis Atomic Power Laboratory. The equipment, which was capable of operation up to about 3200°C, was an induction-heated drop calorimeter with a platinum crucible. This type of calorimeter was described earlier. Enthalpy measurements were made on ThO_2 between 2227 and 2948°C, and on 70 mol% ThO_2-30 mol% UO_2 between about 2000 and 2900°C. As of 1981, the enthalpies and heat capacities of UO_2, PuO_2, ThO_2, $(U,Pu)O_2$, and $(Th,U)O_2$, as well as $(Th\text{-}U)O_2$ containing 30, 15, and 8 wt% UO_2, had been determined by CMT workers.

Thermodynamic functions are needed for vapor species of the fuel in the analysis of hypothetical reactor accidents. Experimental studies were conducted in which the matrix-isolation, Fourier-transform infrared spectroscopic techniques mentioned earlier and calculational methods were used to estimate data that were not experimentally available. These studies yielded improved values for the thermodynamic functions used in the calculation of vapor pressures. Preliminary results indicated that the vapor composition was very sensitive to both the temperature and the composition of the condensed phase, whereas the total pressure was relatively insensitive to both. Further investigations were conducted on vapor species in equilibrium with hypostoichiometric (oxygen-deficient) UO_2. The vapor in equilibrium with UO_{2-x} was a complex mixture consisting of O, O_2, U, UO, UO_2, and UO_3 molecules. Figure 5-35 shows the composition of vapor in equilibrium with $UO_{1.96}$ as a function of temperature.

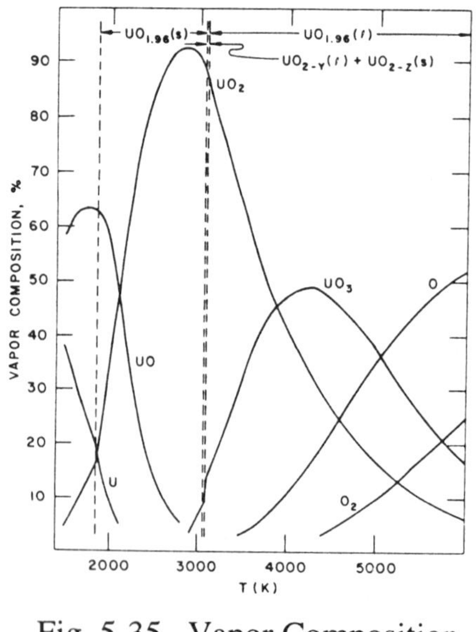

Fig. 5-35. Vapor Composition in Equilibrium with $UO_{1.96}$

Thermal Conductivity of UO_2. The thermal conductivity of reactor fuel is one of the more important pieces of information needed in reactor safety analysis. Investigators at three other laboratories had published values for the thermal conductivity of UO_2 at its melting point. These results were evaluated carefully with a proprietary heat-transfer computer code, and the discrepancies were largely resolved by identifying errors in the workers' analyses of their results. This study produced a value of 5.6 W/m·K. There was still some uncertainty in the value, but no measurements were made by CMT workers because further studies were in progress at other laboratories.

METALLIC REACTOR FUELS

A review was conducted on the available thermodynamic and transport properties of U-Zr and U-Pu-Zr alloys, which had been selected as potential blanket and driver fuels in the IFR. The existing data were found to cover only a relatively small part of the composition and temperature ranges of interest. To understand fuel behavior under both normal and accident conditions, a wide composition range is needed because zones of widely different composition are generated within the fuel by irradiation and thermal gradients.

Thermal Conductivity. Equipment was set up to measure the thermal conductivity of metal fuels by a comparative (axial rod) method, which had a precision of about 5%. In this method, a sample stack was prepared with the alloy of interest placed between two samples of a reference material of known thermal conductivity. Two thermocouples were placed a known distance apart in each of the reference materials, and heaters were used to establish a thermal gradient across the stack. The heat flux through the test sample could then be derived from the measured heat fluxes through the reference materials. To test the method, the thermal conductivity of a National Bureau of Standards (NBS) stainless steel standard was measured at temperatures from 325 to 825°C. Inconel was used as the reference material. The results indicated an accuracy of about 2%. Because stainless steel and Inconel interact with the IFR alloys, the thermal conductivity of tantalum was measured, and it was then used as the reference material to determine the conductivity of a U-11.36 wt% Zr alloy. The results agreed with estimated values based on literature data.

Conductivity measurements were then made on a single U-11.4 wt% Zr sample at 50 temperatures between about 300 and 900°C. The data were fitted to a second-degree polynomial equation: $\lambda = -1.763 + 4.871 \times 10^{-2}T - 1.330 \times 10^{-5}T^2$, where λ is the thermal conductivity in W m^{-1} K^{-1}, and T is the temperature in degrees Kelvin. A relatively large scatter in the data suggested that insufficiently long equilibration times were used for stabilization of numerous transition phases at the higher temperatures, but the data were in general agreement with results from earlier studies.

Next were thermal conductivity measurements on U-8.4 wt% Zr and two cladding alloys, D9 (an austenitic-type steel similar to Type 316 stainless steel) and HT9 (a ferritic-type steel similar to the 400-series steels). Some difficulty in obtaining reproducible results was experienced with the HT9 because of a slow phase transition. The variation in thermal conductivity, however, was fairly small over the temperature range of the measurements. In general, the thermal conductivity data for the cladding materials were in reasonable agreement with the literature values for steels of comparable compositions.

Thermal Expansion. Linear expansion measurements were conducted on IFR fuel alloys, using a dilatometer in a helium-atmosphere glove box. Two alloys were investigated in the initial experiments: U-8.4 wt% Zr and U-20 wt% Pu-1.3 wt% Zr. In both cases, the results were in reasonable accord with literature data, although some phase transitions were observed in greater detail. Measurements were then made on two additional fuel alloys (U-19 wt% Pu-10 wt% Zr and U-26 wt% Pu-10 wt% Zr) and the two cladding alloys (D9 and HT9). In the tests with U-20 wt% Pu-1.3 wt% Zr alloy, the thermal expansion values on heating differed significantly from those on cooling. A separate differential thermal analysis (DTA) indicated that the difference was probably due to a delayed solid-state phase transition at 347°C.

Thermal expansion of the cladding alloy, HT9, was measured at temperatures up to about 900°C. Data from the heating and cooling cycles were in agreement, and a phase transition was found at about 800°C. The D9 alloy did not show such a transition.

Fuel-Cladding Interactions. Previous studies had indicated that plutonium-rich layers can segregate on the surfaces of U-Pu-Zr fuel alloys when they are exposed to cladding, NaK, or a helium atmosphere. This behavior was ascribed to the availability of nitrogen at the alloy surface. Therefore, tests were performed in which selected IFR alloys were heated in nitrogen-helium mixtures that simulated the nitrogen activity estimated for the cladding. A thermogravimetric analysis (TGA) technique was used to follow the sample weight during the reaction. In a typical test, exposure of the U-20 wt% Pu-1.3 wt% Zr alloy to a flowing stream of 50 ppm nitrogen in helium for 12 hr at 700°C produced a distinguishable 10-μm-thick surface layer. Scanning electron microscopy/energy dispersive spectrometry (SEM/EDS) showed that the layer had undergone major compositional changes involving a considerable enrichment of zirconium and a large depletion of uranium. This discovery suggested that the reaction products of nitrogen with zirconium and plutonium were more stable than those with uranium, which resulted in migration of zirconium and plutonium to the surface.

Because low-melting eutectics occur in the U-Fe and Pu-Fe systems, there was concern that cladding integrity might be compromised by the formation of liquid phases during irradiation of the fuel. Experiments were performed in which mixtures of the cladding alloy, HT9, and either a binary or a ternary fuel alloy (U-Pu or U-Pu-Zr) were heated in a DTA apparatus at 10°C per minute to about 1550°C in a yttria crucible, held for two hours, and then cooled. No unusual effects were observed until the binary fuel reached a temperature of 1155°C and the ternary fuel reached 1105°C, where large exothermal peaks occurred, indicating an energetic reaction. Scanning electron microscope examination of the residues from these tests indicated that the iron in the cladding had reacted with zirconium in the fuel to produce Fe_2Zr or Fe_3Zr. The DTA results showed that the zirconium did not change the eutectic temperature. Improved compatibility of U-Pu-Zr fuels over U-Pu fuels appeared to result from the presence of zirconium to react with iron, making the iron less available for melting. Zirconium appears to offer an additional benefit because of its strong affinity for elements such as carbon, oxygen, and iron, which form surface layers and thereby retard iron diffusion into the fuel.

The DTA results indicated eutectic freezing temperatures for the U-Fe and U-Fe-Pu systems that were somewhat lower than those obtained from phase diagrams or calculated. The difference was about 25°C for the binary fuel and 35°C for the ternary fuel. At the end of the 1980s, work was just starting on a more detailed study of the U-Fe system.

Phase Studies. A review of previous data on the U-Pu-Zr system had revealed large uncertainties in the solidus and liquidus temperatures, estimated to be about ±125 and ±75°C, and other inconsistencies in the ternary phase diagrams. Calculations of a ternary phase diagram require thermodynamic functions for all phases existing in the three binary subsystems. An analysis was completed on the U-Zr system. In the U-Pu system, however, the published solidus-liquidus values were inconsistent with the accepted enthalpy of fusion of uranium. The U-Pu system is complex because of the large number of phase transitions in the constituent elements. Further work was in progress on the development of an approach for a thermodynamic analysis that could accommodate the numerous phase transitions.

Some experimental work was done on a U-26 wt% Pu-10 wt% Zr alloy, a typical IFR fuel composition. Differential thermal analysis techniques were utilized. The results of this work provided guidance in evaluating a conflict in the literature data between the U-Pu phase diagram and the enthalpy of fusion of uranium. It appeared that the U-Pu phase diagram should be redetermined. The calculated diagram for U-Zr was in good general agreement with literature data, although there were some minor differences, which were resolved. By 1989, a more reliable ternary U-Pu-Zr phase diagram based on experimental data and calculations was nearing completion.

FISSION-PRODUCT BEHAVIOR

Fission-Gas Behavior in LWR Fuels. In a light water reactor (LWR), gaseous fission products, krypton and xenon, are released during the irradiation, with the result that internal pressure develops with burnup. Escape of these gases into the gap between the fuel and the cladding decreases the heat transfer because the thermal conductivities of these gases are lower than that of helium, which was used to fill the gap initially. The electric utilities, for economic reasons, wished to increase the burnup by about a factor of two. At the time, a great deal of information was available on fission-product gas release from UO_2 fuels, but it was not generally applicable to commercial LWR fuels. The Electric Power Research Institute (EPRI) sponsored a Multiple Cycle Plutonium Utilization (MCPU) program under which a detailed post-irradiation study of LWR fuel rods was initiated to determine the behavior of fission products under prototypical operating conditions. The study was aimed originally toward mixed oxide (UO_2-PuO_2) fuels, but it was reoriented toward UO_2 fuels alone because of federal restraints on fuel reprocessing and use of plutonium in commercial reactors.

The quantities of released fission gases and helium from irradiated fuel elements obtained from the Consumers Power Big Rock Reactor were measured. Fission-gas release was about 5-8% from three of the five fuel rods, but only about 0.2-0.5% from the other two. This discrepancy was difficult to rationalize but may have resulted from differences in the peak power and the temperatures within the rods. The relationships among power density, thermal conductivity as the gases are released, and the temperature are complex. Because of these factors, modeling of the system to predict fission-gas release is difficult. The investigators concluded that a statistical modeling approach might be more appropriate.

Model Development. The potential public risk associated with an accident at a commercial nuclear power plant results from the release of radioactive fission products produced in the fuel during the reactor operation. Under normal circumstances, the fission products remain within the fuel. Under accident conditions, however, some of the fission products may be released from the fuel and find their way into the environment. The amount of a specific fission product released to the environment is called the "source term" for that species.

A program was initiated to identify and characterize the radioactive fission products that might be released to the environment by an LWR under a variety of accident conditions. The program consisted of four components:

1. Assessment of thermochemical data for compounds formed from fission products, fuel, cladding, and structural materials.
2. Development of a core chemistry model to predict the behavior of fission products in the fuel, cladding, coolant system, and containment building.
3. Measurement of additional thermochemical data needed to complete the model development.
4. Development of a transport model to predict fission-product release rates.

The initial assessment indicated that pertinent thermodynamic data were lacking for the compounds Cs_2Te, $ZrTe_2$, and Cs_2ZrO_3, and specimens were prepared for enthalpy of formation and heat capacity measurements.

A core chemistry (fuel and fission product) computer model was developed to calculate equilibrium vapor pressures of fission-product gases and condensed (liquid and solid) phases formed by interaction of the fission products with one another, with oxygen in the fuel, and with uranium. The main emphasis was on the partial pressures of the more volatile fission products of Xe, Kr, I, Cs, Te, Mo, Sr, Ba, and Ru.

A scaling model, based on information from the core chemistry model, was developed to evaluate the movement of fission products from the fuel through the cladding, reactor vessel/coolant system, and containment building. The scaling model showed good agreement with experimental measurements.

Further experimental investigations showed that the fractional release rate of the fission gases was proportional to that of the other volatile fission products. The mechanism of fission-gas release, however, was not well understood but was believed to involve migration of the gas bubbles within the fuel.

Release of Fission Products from Breached LWR Fuel Pins. An experimental study was conducted on the release rate and chemical form of fission products from an irradiated LWR fuel pin into which a defect was introduced. A 10-cm-long section of the fuel pin, sealed at both ends, was inductively heated in a vacuum, and the gaseous species escaping through a hole drilled in the cladding were identified and measured by means of a quadrupole mass spectrometer. In the mass spectrometer, the ion current is related to the partial pressure of a species by the equation, $p = k\,(IT)$, where p is the partial pressure, I is the ion current, T is the temperature, and k is a constant. The defected pin could be regarded as a Knudsen cell, and the data were analyzed accordingly. Values of k were determined for gaseous CsI, I_2, Te_2, Kr, and Xe.

To develop the methodology for the work with irradiated fuel pins, a simulated pin containing UO_2 and inactive fission-product elements was prepared. A 1-mm-dia hole was drilled in the cladding, and the pin was heated to 600-900°C in a vacuum. Both CsI and Cs were found in the emitted gas, but no tellurium was evident. Further tests suggested that the tellurium had reacted with zirconium to form solid $ZrTe_2$. The rate of cesium release decreased with time, probably due to the development of a CsI-depleted layer on the top of the sample, but all of the cesium eventually escaped from the sample.

After extensive testing and calibration, the equipment was moved into a shielded cave facility. As a final test, a mockup fuel-pin section was made and filled with a powdered mixture of UO_2 and inactive fission-product compounds. Below 1290°C , the only species found were Cs and CsI. At 1290°C , there was evidence of SnTe, probably from a reaction of tin in the Zircaloy cladding with $ZrTe_2$. On continued heating up to 1590°C, Cs, Te, and I were observed.

An irradiated fuel sample was then heated stepwise from 450 to 1500°C over a 28.5-hr period. Over this temperature range, Xe, Kr, Cs, and Te were found. The cesium came off first, followed by the tellurium. The release rates of Xe and Kr were about the same and were relatively complex, with different activation energies in two temperature regimes and a dependence on the previous history of the specimen. A second test made with another sample produced somewhat similar results; krypton and xenon were released at all temperatures. Cesium was first noted at 990°C, tellurium began to appear at 1160°C, and elemental iodine was observed at 1260°C.

In a reactor accident situation, rupture of the cladding would most likely expose the fuel to oxygen. An unclad, irradiated fuel-pin section was heated at temperatures up to about 1140°C in a nickel Knudsen diffusion cell equipped with an outside manifold to permit controlled additions of oxygen. No effect of oxygen addition on the xenon release rate was observed in six temperature cycles, but on the seventh cycle it increased about 100-fold. When the oxygen flow was stopped, the xenon release rate returned to its original value. The lack of an effect in the first six cycles was attributed to a reaction of the oxygen with the nickel cell until a layer of nickel oxide was formed. After that point, the release of xenon was proportional to the amount of additional oxygen retained in the cell.

The behavior of fission-product iodine was studied by modifying the diffusion cell so that condensable gaseous species could be collected on a cold plate. Condensed specimens were collected from fuel samples that had been heated to 914, 1290, and 1395°C. Only the first specimen contained a significant amount of iodine, which was present as iodide. This was a significant result in that it indicated that the release rates of cesium and iodine were not the same, as had been assumed previously in degraded core accident analyses.

Downstream Behavior of Volatile Fission Products. After fission products are released from a defected fuel pin, they are likely to interact with a steam-hydrogen mixture flowing upward in the reactor vessel. The apparatus, shown schematically in Fig. 5-36, was constructed to mix gaseous, nonradioactive fission-product species with a flowing gas stream containing various proportions of air, steam, hydrogen, and/or aerosol particles. As the temperature of the flowing gas stream decreased, the fission products were expected to condense as aerosol particles and deposit on the duct wall. Chemical reactions of the fission-product species with the gas-stream components were also a possibility. The major component of the experimental unit used in these studies was a 3.7-m-long stainless steel pipe (duct). Volatile fission products and superheated steam entered the duct at temperatures up to 1000°C, pressures up to 600 kPa, and flow rates up to 60 kg/h. The tube was fitted with liner sections which could be removed for analysis of the deposits.

Among the possible volatile fission products present in a reactor accident, cesium, tellurium, and iodine were selected for study because of their harmful biological effects if released to the air. In the presence of steam, the cesium and iodine would be expected to

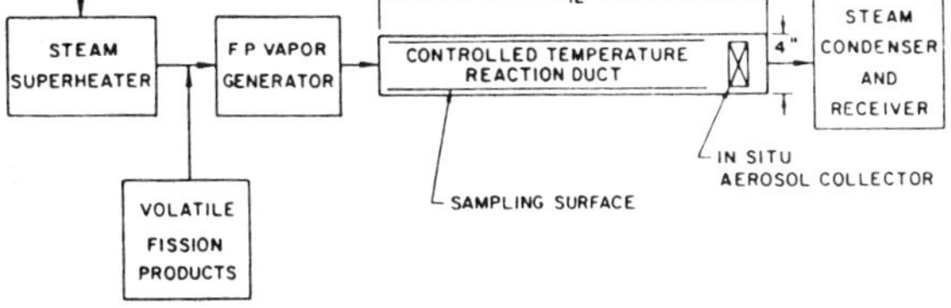

Fig. 5-36. Experimental Setup for Study of Downstream Behavior of Volatile Fission Products

react to form CsI, which is volatile at the high temperature, but would condense as it reached the cooler downstream part of the duct. Condensation of the CsI began where the steam temperature and the duct wall temperatures had decreased to about 700°C and increased with a further decrease in temperature downstream. Scanning electron microscopy indicated that the CsI had condensed as liquid particles or drops.

In 1985, five experimental runs were made—two with CsI, one with CsOH, one with CsI-CsOH, and one with Te-CsI-CsOH. During this work, scanning electron microscopy was used to view the deposits; the results are illustrated in Fig. 5-37. With CsI and CsOH alone, deposition occurred where the wall temperature had decreased to 680 and 670°C, respectively. With the CsOH-CsI mixture (molar ratio 8.4), the CsOH started to deposit at the same location as CsOH alone, but the CsI deposited farther downstream. This behavior was attributed to an equilibrium reaction

$$CsOH + CsI \leftrightarrow CsOH \cdot CsI$$

The idea was that the large excess of CsOH molecules prevented CsI condensation until the CsOH concentration had been decreased by condensation, thereby freeing the CsI. In the case of the Te-CsI-CsOH mixture, all three elements, Cs, I, and Te, were found at the downstream end of the duct. In two other experiments, one with CsI and the other with CsOH, aerosol particles formed in the gas stream were collected on stainless steel wires and strips, which were examined by scanning electron microscopy to determine the particle-size distribution and calculate the collection efficiency.

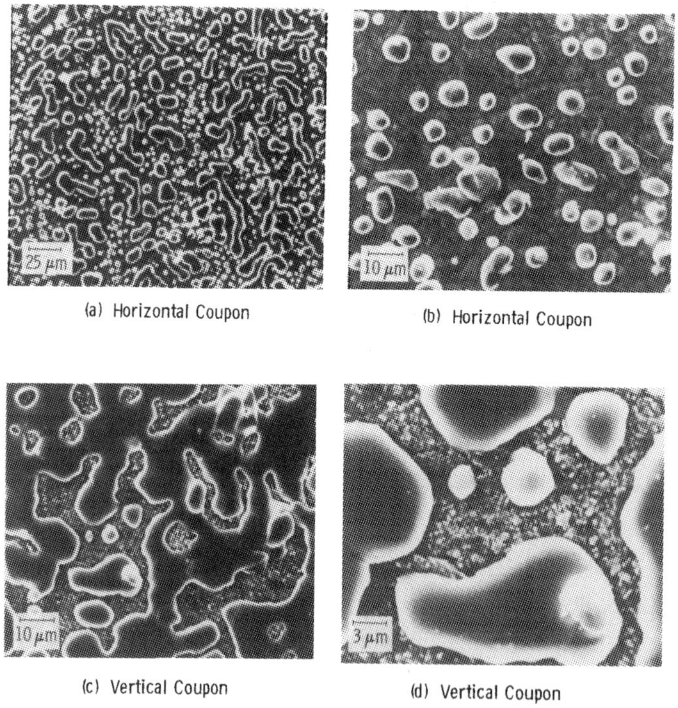

Fig. 5-37. Scanning Electron Microscopy of CsI Deposition in Transport Duct

Attention was then directed toward the revaporization of material from deposition sites and movement further downstream that could be important in loss-of-coolant accidents. Revolatilization experiments were conducted with the same duct arrangement. The results suggested that the quantity of volatile fission products converted to aerosols depends on the cooling rate of the carrier gas. The deposits could heat up by radioactive decay, and a flow of gas over the heated deposits could revaporize the materials, possibly leading to their escape into the containment vessel.

A study was made of the interactions of various fission products with one another, with steam and hydrogen, and with materials of construction. The system is complex. For example, the behavior of iodine depends on its reactions with cesium and steam, and the behavior of tellurium depends on its reactions with cesium.

A separate follow-on investigation was conducted on the CsI-CsOH system. A quadrupole mass spectrometer was used to measure partial pressures of vapor species over liquid CsOH, CsI, and solutions of the two salts in a silver Knudsen cell over various temperature ranges between 349 and 762°C. The partial pressures of CsOH, $Cs_2(OH)_2$, CsI, Cs_2I_2 were determined. The existence of the compound CsI·CsOH was shown by derivation from the experimental results.

In 1984, separate, but similar experiments were performed in which a mixture of nonradioactive compounds and metals composed of CsI, CsOH, TeO_2, SrO, Ag, In, Cd, Sn, and Mn was loaded into a stainless steel boat, which was then placed inside a stainless steel pipe. A steam-hydrogen mixture was passed over the mixture as the temperature was increased rapidly and then held for one hour at 500, 750, and 1000°C. The tube was then disassembled, and chemical analyses were performed on the residue in the boat and the downstream deposits. At all three temperatures, I, Cs, Te, and Cd had been transported from the boat; in this case, cadmium was transported only at the highest temperature. The quantities of cadmium transported were consistent with calculated values when it was assumed that the pure metals had formed a liquid alloy. The mode of transport of iodine was not known, but it was probably through formation of the CsI·CsOH species mentioned earlier. The behavior of tellurium was not completely understood, but was thought to have been related to interactions with the indium.

Fission-Product Release from Core-Concrete Interactions. The objective of this effort, which began in 1985, was to predict the release of certain fission products that would occur if the molten core mixture penetrated the bottom of the reactor vessel in a severe accident. Three important fission products, La, Ba, and Sr, whose oxides can form solid solutions with the UO_2 fuel material, were of particular interest. In a laboratory-scale simulation of a reactor accident, UO_2 fuel granules containing La_2O_3, BaO, and SrO, separated from concrete by stainless steel and zirconium chips, were heated to about 2110°C for 132 min in a helium stream saturated with water. This charge was supported on a perforated tungsten plate. Upon completion of the run, a tungsten condenser tube was removed, and the deposits were dissolved in nitric acid and analyzed for La, Sr, Ba, Fe, Cr, Ni, Mn, Ca, and Mg. The amounts of these species that were vaporized ranged from 0.01 to about 0.15%, except for manganese, for which the value was 33%. This high value was attributed to the relatively low boiling point of the metal (2041°C). The behavior of the other species was in reasonable agreement with values calculated from existing thermodynamic data. The following year, five more experiments were completed. For the elements La, Sr, and Ba, the amounts released were in the order Ba>Sr>La, and most of the loss was by vaporization, with particulate entrainment making only a minor contribution (<10%). At this stage of the work, the investigators summarized their conclusions as follows:

1. Core-concrete melts containing metals give lower oxygen-potential values than do core-concrete melts containing no metals.
2. The release of stainless steel components follows the vaporization behavior of the elements in the steel.
3. The trend in the release of Ba and Sr is in accord with thermodynamic calculations.

By 1987, the experimental conditions had been modified somewhat. The materials tested were stainless steel, concrete (basaltic or limestone), and UO_2 doped with La_2O_3, SrO, BaO, and ZrO_2. The materials, alone or as mixtures, were vaporized from a ZrO_2 crucible into a He-6% H_2 gas stream to which up to 600 ppm water was added to maintain the desired partial molar free energy of oxygen in the desired range (-420 to -550 kJ). The fraction of the sample that had been vaporized was determined by weight change and by chemical analysis of the condensates. The experimental results were compared with the SOLGASMIX computer code. The results were difficult to interpret due to their complexity, but it was clear that the transport of stainless steel constituents, Fe, Cr, and Ni, was considerably lower than in runs with the steel alone.

Detailed post-test examinations using electron-probe microanalysis were done on the ZrO_2 crucible and its contents. The solidified mass of concrete, UO_2, and steel in the bottom third of the crucible appeared to have been at least partially liquid during the experiment, and the thickness of the crucible had been reduced by about 40% in that area. The steel phase, which was approximately spherical, was embedded in the concrete-UO_2 mixture, which probably accounted for the reduced transport rate of Fe, Cr, and Ni. The major phases in the mixture were UO_2 containing CaO and ZrO_2, and silicates and zirconates of calcium and magnesium. When molybdenum, rather than ZrO_2, was used as the crucible material, the transpiration results were nearly identical.

The status of information from this work after two more years of similar experimental and calculational studies is summarized briefly as follows:

1. The releases of Sr, Ba, La, and U in a severe reactor accident as described here will be less than 1% of their inventory. This result is as low, or lower than source-term values in the Reactor Safety Study sponsored by the U.S. Nuclear Regulatory Commission (NUREG-75/104). The aerosol release will be less than about 0.5% of the total core-concrete mass.
2. The release of refractory fission products is suppressed by shifts in the partial molar free energy of oxygen in the gas phase toward more positive values.
3. In a severe reactor accident, an oxide phase (concrete, UO_2, and ZrO_2) will lie over a metallic phase containing zirconium and stainless steel. The gas emanating from this concrete-core mixture will have a partial molar free energy of oxygen that is in equilibrium with the overlying oxide phase. The calculational procedure assumes that H_2O and CO_2 formed by decomposition of the concrete will first be reduced as they contact the metal phase and then reoxidized as they contact the oxide phase.
4. Accurate source-term information is not yet possible, given the existing thermodynamic database.

Fission-Product Chemistry in Source-Term Experiments. In 1984-85, the Source-Term Experiments Program (STEP) consisted of four reactor-accident simulations run in the Transient Reactor Test (TREAT) facility. The behavior of fission products released from irradiated UO_2 fuel rods (3-4.5 at.% burnup) overheated in flowing steam to the point of catastrophic failure of their Zircaloy cladding (about 2200°C) was investigated. Materials released from the fuel rods were collected on various coupons and wires in three types of aerosol-characterization devices. A CMT team consisting of Joanne Fink, Mike Roche, Chuck Seils, Dave Steidl, Carl Johnson, and Bob Ritzman was involved in identifying and characterizing the deposits and developing plausible mechanisms for their formation.

In the STEP-1 experiment, no CsI was found in the deposits, but droplet-shaped deposits containing nickel and iodine were observed at three locations. The deposits were not identified positively, but they appeared to be an amorphous form of nickel iodate. X-ray diffraction showed the presence of SnO_2, which apparently came from the Zircaloy cladding, which contained 1.5 wt% Sn. Molybdenum was found with Cs (and the Sn), which suggested the formation of Cs_2MoO_4, which is known to be a stable gas-phase molecule with a significant pressure over the fuel. Tellurium, usually along with cesium, was detected on a few isolated samples, but other laboratory data had shown that tellurium deposition occurs at a higher temperature than that of the collectors. In the STEP-3 and -4 experiments, the initial flow velocities and the peak temperatures were much lower. The predominant constituent in the STEP-3 aerosol deposits was Fe, and in STEP-4 it was Si, as opposed to STEP-1 and -2 where the main constituents were Cs, Mo, and Sn.

PERSONNEL

Carl Johnson was in charge of the Applied Physical Research Program in the 1980s. Others involved in this effort were Jack Arntzen, Paul Blackburn, Bob Blomquist, Martin Chasanov, Magdy Farahat, Darrell Fee, Joanne Fink, Dave Green, Bill Gunther, Irv Johnson, Len Leibowitz, Yung Liu, Art Pelton, Bob Ritzman, Mike Roche, Chuck Seils, Jack Settle, Bill Shinn, Bruce Spencer (RAS), Shiu-Wing Tam, Marvin Tetenbaum, and Ewald Veleckis.

Basic Research

The CMT basic research programs in the 1980s continued much as before. In 1982, the programs on environmental chemistry and on thermodynamics and electrochemistry were incorporated formally into the Materials Science and Technology Division (MST), which had been formed at that time to show the Laboratory's commitment to materials research. These programs, however, remained in CMT administratively and physically. Fred Cafasso led this work through the early 1980s, Vic Maroni after that.

HOMOGENEOUS CATALYSIS

Research in this area, which began in the 1970s and continued through the 1980s, involved the use of *in situ* catalysts to promote desired chemical reactions in organic systems. In 1980, work was in progress on the use of a cobalt carbonyl catalyst for (1) the catalytic hydrogenation of carbon monoxide to form methanol and other products and (2) homologation of methanol to form ethanol, which is commercially important as a solvent, a chemical feedstock, and a fuel extender. Ethanol is currently produced from ethylene obtained from petroleum and by fermentation of feed grains. The research to date, which had proved highly successful, was not based on unprecedented chemical reactions, but rather on unprecedented combinations of known reactions.

Hydrogenation of Carbon Monoxide. The primary reaction products from the hydrogenation of CO were methanol, methyl formate, and ethylene glycol. Although these were the only products formed under special conditions, several others were usually formed by branching reactions. Variations of the ratios of primary and other reactions were determined at different carbon monoxide and hydrogen pressures. The amount of methanol in the product ranged from about 43 to 90%, depending on the conditions.

Further studies of catalytic carbon monoxide hydrogenation were based on a new concept and observation. Methanol can be formed by the disproportionation of formate ion. By this reaction, methanol can be formed from water and carbon monoxide:

$$3\ CO + 2\ H_2O \rightarrow CH_3OH + 2\ CO_2$$

This reaction is highly desirable in terms of raw material costs because it uses water, rather than elemental hydrogen, as the hydrogen source. It is also thermodynamically favored over the commercial process, which uses hydrogen, by a free energy change of -19.5 *vs.* -5.9 kcal/mol.

$$CO + 2\ H_2 \rightarrow CH_3OH$$

The products of the reaction with water were not always limited to methanol. Acetone and other organic materials were also obtained in some cases.

Although the larger negative free energy change for the water reaction is highly desirable, the reaction must also have an acceptably high rate, which requires the use of an effective catalyst. The reaction is promoted by a variety of metal complexes containing oxide or oxo ligands, *e.g.*, PbO, PbO_2, $Pb(O_2CH)_2$, and $PbTiO_3$. The reaction occurs at 200-350°C when these materials are slurried or dissolved in molten formate salts. Not all of these catalyst materials are stable in the presence of the CO and H_2O; for example, PbO is rapidly reduced to metallic lead. A search was conducted for lead-oxide-like compounds in which the lead is coordinated strongly to oxo ligands. Materials of this type include $PbTiO_3$, $PbMoO_4$, $PbWO_4$, and PbV_2O_6. A flow reactor containing a slurry of $PbTiO_2$ in molten KO_2CH was operated for 19 days at CO/steam pressures from 1.7 to 11.2 atm with no loss of catalytic activity or lead formation. With the $PbTiO_2$ catalyst, the product was about two-thirds methanol, one-third ethanol, and a small amount of propanol. These higher alcohols are desirable if the product is to be used as a fuel extender. A mechanism was developed for the catalytic reactions involved in the formation of the alcohols.

A research study was then focused on the mechanism involved in the disproportionation of formate ion to produce methanol:

$$4\ O_2CH^- \rightarrow CH_3OH + CO + 2\ CO_3^{2-}$$

This reaction occurred at a temperatures around 230°C in slurries of lead compounds such as the oxide, titanate, or phthalocyanine in potassium formate salts. The reaction mechanism appeared to involve the intermediate species, formaldehyde and methyl formate. Details of the mechanism were explored in several separate experiments.

A mechanism was identified for the formation of methanol from a soluble organometallic formate. The mechanism includes a pre-equilibrium process in which trimethylsilyl formate was equilibrated with hexamethyldisiloxane, carbon dioxide, and methyl formate at 250°C. Methanol was then produced by decarboxylation of methyl formate. Isotope labeling studies, using $(CH_3)_3SiO_2CD$, showed that the methanol and methyl formate came solely from formate on the silicon, and not from the appended methyl groups. The decarboxylation reaction turned out to be the rate-limiting step. Zinc phthalocyanine was identified as a catalyst for the formate disproportionation and a related carbon monoxide hydrogenation.

This research continued along the same lines in 1986, although it was described a bit differently as new catalytic chemistry involving small molecules such as CO, CO_2, and O_2. It had been hypothesized that the mechanisms involved in the hydrogenation of carbon monoxide described here bore some significant similarities to the use of a metal oxide catalyst system (supported CuO-ZnO) in the commercial production of methanol, and that the mechanisms in this study might lead to improvements in the commercial process. A rather detailed mechanism was proposed for the catalytic effect of hexamethyldisiloxane, and activation of hydrogen by the oxide was identified as the rate-determining step. Based on the premise that the activation step involves nucleophilic attack on the hydrogen molecule, nucleophilic siloxide catalysts, exemplified by $NaOSi(CH_3)_3$, were tested and found to increase markedly both the rates of hydrogen activation and carbon monoxide hydrogenation.

Ethanol Synthesis. Homogeneous catalytic systems for cobalt-catalyzed methanol homologation to ethanol by carbon monoxide and hydrogen had been reported extensively in the literature. The economic viability of such processes was in doubt, however, because of problems with inadequate conversion, selectivity, and catalyst recovery. The investigators in CMT discovered a new catalyst, iron pentacarbonyl, $Fe(CO)_5$, which catalyzed the homologation of methanol to ethanol in the presence of Syngas and a tertiary amine at temperatures above 180°C, through the reaction

$$CH_3OH + 2\,CO + H_2 \rightarrow CH_3CH_2OH + CO_2$$

The only significant side-reaction product was methane. This reaction uses hydrogen efficiently in that carbon dioxide, rather than water, is the oxygenated by-product. The roles of the iron carbonyl, $Fe(CO)_5$, and tertiary methyl amine, $N(CH_3)_3$, are included in a detailed mechanism that was worked out for the above reaction. A conversion rate of 13% per hour with 70% selectivity was achieved in a batch reactor.

At this juncture, the researchers found that carbonyl complexes of metals other than cobalt were effective catalysts when used with the amine. For example, the manganese compound, $Mn_2(CO)_{10}$, was very similar to iron in its effectiveness. Further work showed that $RhCl_3$, $Mn_2(CO)_{10}$, and $Ru_3(CO)_{12}$, separate or combined with $Fe(CO)_5$, were all effective catalysts. For the $Fe(CO)_5$ and $Mn_2(CO)_{10}$ systems, a bimolecular methyl transfer from a tetramethylammonium cation to the transition metal carbonyl anion appeared to be the rate-limiting step.

An experimental study was made of the kinetics of the several individual reactions which, when added up, correspond to the one shown above. Kinetics of the individual reactions were determined by measuring the rate of disappearance of pertinent species by reagent-specific analytical methods.

As the studies progressed, additional transition metal complexes were found to be active catalysts for selective homologation of methanol in methanolamine solutions *i.e.*, $RhCl_3{\cdot}3H_2O$, RhI_3, $Ru_3(CO)_{12}$, and $Cr(CO)_6$. These compounds were tested individually as catalysts, and also as co-catalysts with $Fe(CO)_5$. As individual catalysts, RhI_3 had the highest catalytic activity; $RhCl_3{\cdot}3H_2O$ was less active, but more selective for ethanol production. The $Ru_3(CO)_{12}$ was similar to $Fe(CO)_5$ in its catalytic activity as was $Mn_2(CO)_{10}$, but the selectivity for methanol production was better with the manganese compound. The $Cr(CO)_6$ showed no catalytic activity under these conditions. In the tests of mixed catalysts, the $Mn_2(CO)_{10}$-$Fe(CO)_5$ combination was the most promising. A methanol conversion rate of 14% per hour at 220°C was achieved.

The thrust of these studies was then turned toward replacing the tertiary amines with inorganic basic compounds that are less expensive, more stable, and more easily separated from the product. Alkali metal formates, *e.g.*, KO_2CH, function as an inorganic base and also as a methyl carrier. The only metal carbonyl complex that was found to be active in the presence of KO_2CH was $Mn_2(CO)_{10}$. The KO_2CH-$Mn_2(CO)_{10}$ system was reported to work well. This completed the homologation studies, which were successful in achieving their objective.

Catalytic Oxidation. This area of work, which was started in 1984, involved the development of catalysts for controlled air oxidations of substrates. Substrate oxidations for which the new strategies were being sought were (1) oxidation of methane to methanol or formaldehyde, (2) oxidation of higher alkanes to alcohols, aldehydes, or carboxylic acids, (3) oxidation of higher olefins to epoxides or, in the presence of water, to glycols, and (4) oxidation of methanol to formaldehyde.

A study was directed toward the use of organic bases, such as an amine or sulfide (designated here as "B"), to bind oxygen reversibly. This reaction forms an organic oxide adduct (B-O), which may be an amine oxide or a sulfoxide:

$$2\,B + O_2 \xleftrightarrow{\text{(catalyst)}} 2\,B\text{-}O$$

One potential application of this idea is the recovery of pure oxygen from air by cycling this equilibrium reaction back and forth through pressure and temperature changes. A thermodynamic estimate indicated that a B-O bond enthalpy of about 68 kcal/mol would be needed to achieve the desired result. A collaboration between the CMT researchers and the National Institute for Petroleum Energy Research resulted in a value of 70 kcal/mol for the N-O bond enthalpy in pyridine-N-oxide and a value close to that for trimethylamine-N-oxide. Autoclave experiments, however, showed that under conditions where oxidation was rapid and selective, the equilibrium constant for the reaction with molecular oxygen was too large to demonstrate reversibility.

Another application of interest was the use of organic base oxygen carriers with catalysts to control the aerobic oxidation of certain substrates. The concept was demonstrated by experiments in which cyclohexene was converted to cyclohexanediol by trimethylamine oxidation and by osmium tetroxide-catalyzed transfer of oxygen from trimethylamine-N-oxide. These preliminary results were highly promising.

On further investigation, mechanistic studies were conducted on the oxidation of tertiary amines containing α-hydrogen atoms. A rate-limiting electron transfer from the amine to the oxygen molecule, including the sacrificial loss of an α-proton, appeared to be a characteristic of this type of system. Future work was directed to amine oxidations with amines that do not contain α-protons or that contain α-protons not susceptible to abstraction.

Soluble Oxide Chemistry. By 1987, the work described so far had undergone some changes of emphasis and evolved into a situation where it could be categorized better as two topics: (1) soluble oxide chemistry and (2) phthalocyanine chemistry. Basic research on the nucleophilic activation of hydrogen had shown that alkali metal formates, hydroxides, and carbonates catalyze the water-gas shift reaction in triethylene glycol solution:

$$\underset{\text{(catalyst)}}{CO + H_2O \leftrightarrow CO_2 + H_2}$$

These were the first examples of homogeneous catalysts for this reaction, which is catalyzed by solid transition metal oxides in the existing commercial process. Homogeneous catalysts of this type might be expected to be more tolerant of impurities such as sulfur than are the solid oxides.

Attention was directed toward soluble oxide catalysts that promote hydrogenation and water-gas shifting by catalytic mechanisms that are different from those of other homogeneous catalysts and appear to be more relevant to suspected metal oxide surface chemistry. A study of the effect of formate on the rate of the water-gas shift reaction indicated that hydrogen activation occurs at the oxygen center of a formate or hydroxide, and a metal center is not needed.

The lack of sufficiently robust solubilizing ligands had limited the application of homogeneous metal oxide complexes in synthetic gas catalysis. The tristrimethylsilylmethyl ligand, a relatively new species at the time, had been used in other syntheses, and appeared to have good thermal stability. Synthetic procedures were developed for the preparation of tristrimethylsilylmethyl zinc formate and the corresponding hydroxide complex, both of which have solubilities in excess of 0.1 *M* in hydrocarbons. The initial studies were focused on soluble complexes of zinc because zinc oxide is an important component of catalysts used in the commercial production of methanol. The results of this research were highly significant in that they opened the door to a possibility that a variety of polar metal oxide and hydroxide complexes could be used in organic solvents at high temperatures and gas pressures.

Phthalocyanine Chemistry. Interest in metallophthalocyanines was sparked by the observation that ZnPc (Pc = phthalocyanine) catalyzed many reactions, even at the low concentrations (about 10^{-5} *M*) achievable with its limited solubility. It had been shown to catalyze the disproportionation of trimethylsilylformate and the addition of carbon dioxide to metal hydrides. Phthalocyanines are also highly stable. A sample of ZnPc heated for two weeks in a mixture of H_2 and CO_2 (300 atm) at 250°C showed no evidence of decomposition.

The investigators discovered that ring carbons in phthalocyanines can be metallated efficiently with metals such as lithium and magnesium, which opened up the possibility of introducing a variety of structural features into phthalocyanines. One motivation for such a substitution was to increase the solubility of the phthalocyanine, thereby increasing its effectiveness as a catalyst. Success was achieved when trialkylsilyl groups were incorporated into phthalocyanine rings to produce a mixture of trimethylsilyl-substituted phthalocyanines, *i.e.*, $H_2Pc(SiMe_3)_x$, where Pc is phthalocyanine, Me is the methyl group, and x is 2, 3, or 4. Nuclear magnetic resonance (NMR) spectra were determined to provide structural information on the molecule.

Although the trimethylsilyl complexes showed promise, their thermal stability was limited to temperatures below about 150°C. Therefore, the emphasis was shifted to alkyl-substituted phthalocyanines, which have higher thermal stabilities. A synthetic procedure that was developed for very stable metallophthalocyanines was used to produce the following complexes:

$Co(R_8Pc)$,
$Rh(R_8Pc)(CO)Cl$,
$Rh(R_8Pc)(PBu_3)Cl$,

$Rh(P_8Pc)(PhCN)Cl$, and
$Rh(R_8Pc)(PBu_3)$-Cl-$Rh(R_8Pc)Cl$,

where $R = C_5H_{11}$, Pc = phthalocyanine, and $Bu = C_4H_9$.

High-Pressure NMR Studies. In 1986, work began on the design and construction of a high-pressure, multinuclear NMR probe suitable for *in situ* kinetic studies. The existing apparatus was not capable of handling the extreme pressures required (*e.g.*, 300 atm) and the corrosivity of some of the materials under investigation. The following year was devoted to working out the design details. The new nuclear resonance detector (shown in Fig. 5-38) was developed in the shape of an elongated toroid. This design provided greatly improved sensitivity for ^{13}C and other insensitive nuclei associated with organometallic systems, and because of its confined magnetic field, the detector was ideal for use in high-pressure metal probes. This equipment was a major asset in expanding the scope of the research effort.

In 1989, investigations were started in which real-time kinetic and spectroscopic techniques at high temperatures and pressures were used to determine the chemistry of small molecules derived from coal, natural gas, and petroleum. Maximum concentrations of reactive gases were achieved by making use of their complete miscibility with supercritical fluids. A detailed NMR study, using the new probe, was initiated on the hydrogenation of cobalt octacarbonyl

$$Co_2(CO)_8 + H_2 \leftrightarrow 2\ HCo(CO)_4$$

Data were generated on enthalpies, entropies, and rate constants of the system.

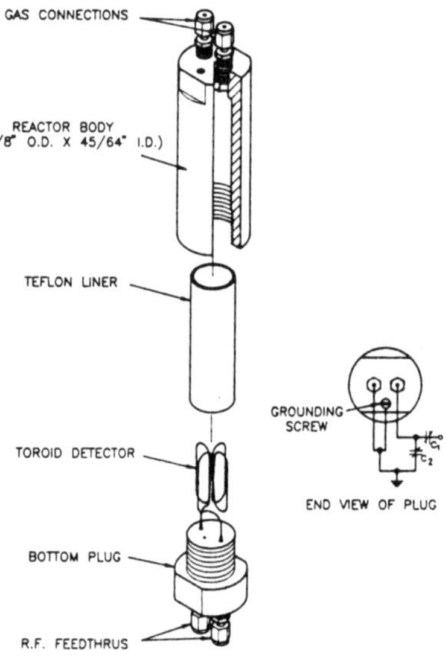

Fig. 5-38. High-Pressure Nuclear Magnetic Resonance Probe

Personnel. This area of research was initiated originally by Hal Feder, who kept in close touch with the work while at CMT and then during his retirement up to the date of his death in January 1998. Jerry Rathke is the Group Leader of this effort; others in the group in the 1980s were Ira Bloom (a postdoc at the time), Michael Chen, John Heiberger, Bob Klingler, and Theodore Krause, plus several postdocs and other temporary people. Larry Curtiss, although not a member of the group, contributed to a special aspect of the research.

THERMODYNAMICS

Molten Salt Systems. An effort was underway to develop fundamental solution theories to understand better the properties of solutions, to expand the ability to predict solution properties, *a priori*, and to facilitate the use of theories in technological applications. The derivation of thermodynamic properties of multicomponent ionic reciprocal molten salt systems by use of the conformal ionic solution (CIS) theory was mentioned in the previous chapter. The applicability of this theory was demonstrated by computations of the ternary phase diagrams of AlF_3-CaF_2-NaF, AlF_3-CaF_2-LiF, and CaF_2-LiF-NaF that are consistent with thermodynamic information.

The coordination cluster (CCT) theory is a statistical mechanical theory involving the activity coefficients of dilute solvents in multicomponent systems. The CCT theory was used not only for molten salt mixtures, but also for metallurgical slags and aqueous solutions. For the molten salts, it provided a rationalization of the energetics of association of cations and anions in dilute solutions. The CCT theory was extended to perform calculations for a multicomponent solvent based solely on data from the subsidiary binary systems.

Molecular dynamic studies of $MX-AX_3$ molten salts (where X is a halogen, M is an alkali metal, and A is a trivalent metal) at 500 K (227°C) allowed the six radial distribution functions and the near-neighbor and next-nearest neighbor coordination number to be deduced. The long-range order (up to 9.2 Å), due to X^- sharing, may exist in melts with large radius ratios. Theoretical and experimental studies of the complexing of S^{2-} and O^{2-} ions by Al^{3+} in LiCl-KCl melts suggested that large screening occurs in the formation of AlO^+ and AlS^+, and that the coulomb complexing model, which was used to treat this phenomenon, had a potential for predicting the effects of solvent and solute radii on complex formation.

Further studies of the energetics of bonding and association of Al^{3+} with divalent ions in molten salts were performed in which *ab initio* quantum mechanical calculations were made on the reactions of gaseous polyatomic molecules containing Al-F and Al-Cl bonds with oxygen- or sulfur-containing species. Evidence was found for the stability of Al-O and Al-S bonds in molten halides, which meant that it might be possible to create simple ionic pair potentials from which those bonds in a molten salt could be simulated by a molecular dynamics calculation.

Additional *ab initio* quantum mechanical calculations were made to study Al_2F_7 species, which were expected to occur in ordered AlF_3 melts. The calculations indicated a single Al-F-Al bridge that is bent in the presence of lithium ion. The bent bridge structure was in agreement with experimental observations on fluoride melts. Structural details were obtained by molecular dynamics calculations of melts that simulated $NaCl-AlCl_3$ and $KF-YF_3$ mixtures. Studies of melts simulating $KF-YF_3$ revealed a new type of intermediate-to-long-range order, a triangular triplet of Y^{3+} ions, and very few free YF_6^{3-} species at the K_3YF_6 composition.

In a separate study, emf measurements from an $Al/AlCl_3-NaCl(s)//NaCl-AlCl_3/Al$ cell were used to obtain total excess free energies of mixing per mole. The results showed an unusual degree of ordering near the 50-50 $NaCl-AlCl_3$ composition. Near these compositions, the activity coefficients of many halide solutes increase drastically, with a resulting decrease in their solubilities.

Molecular orbital calculations were continued, in which vibrational frequencies were predicted for comparison with observed infrared and Raman spectra. The results showed that frequencies for $AlCl_4^-$ ions in melts were similar to those for the gas-phase ion. These studies led to a continuation of some earlier work on Mg_2O^{2+} and MgO in molten fluoride melts. In comparing the energetics of these two ions, consideration was given to reactions involving $Li_2Mg_2OF_4$ and Li_2MgOF_2. Structure and bond-energy determinations for these species indicated that coordination of the oxygen anion by either one or two magnesium ions in the LiF melt was of nearly equal stability in the LiF melt.

Further calculations were performed on the stabilities of MgO and Mg_2O^{2+} in four alkali halide melts (LiF, LiCl, NaF, and NaCl), and the results were consistent with experimental data from the literature. Vibrational frequencies were predicted for complexes of magnesium oxide (F_2MgO^{2-}, Cl_2MgO^{2-}, F_4MgO^{2-}, and F_3MgO^{3-}) and aluminum oxide (F_3AlO^{2-} and Cl_3AlO^{2-}). Vibrational frequencies were also calculated for the five- and six-coordinated Al^{3+} ions in a fluoride melt (AlF_5^{2-} and AlF_6^{3-}). Of the six calculated vibrational frequencies, the three that had been observed experimentally agreed with the calculations. Calculations for one sulfide ion complex, S_3^-, were also in agreement with frequencies obtained by Raman spectra.

Molecular dynamics calculations were performed on MAX_4 melts (M is an alkali metal, A is a trivalent metal, and X is a halogen), in which the X^- anions are polarizable. Such melts are analogous to the chloroaluminates. With polarization of the anions, A_2X_6 molecules as well as the AX_4^- ions are present. This result is in contrast with the observation of AX_4^-, $A_2X_7^-$, and $A_3X_{10}^-$ anions in an ionic melt with nonpolarizable anions. All the species postulated to exist in chloroaluminate melts were observed in the calculations, but their relative stabilities differed from those in real melts.

In 1988, a semiempirical quantum mechanical method, called Modified Neglect of Differential Overlap, was used to investigate the structures of $Al_2X_7^-$ and $Al_3X_{10}^-$ ions. The $Al_2Br_7^-$ anionic species was found to have a bent bridge structure similar to that found previously for $Al_2Cl_7^-$ ion in the *ab initio* studies. For the $Al_3X_{10}^-$ species, chain structures were favored over ring structures with bent bridges for the chlorides and bromides, and linear bridges were retained in the fluorides.

The solubility of $CoCl_2$ in NaCl-$AlCl_3$ melts exhibited a sharp minimum at the most ordered composition of equimolar NaCl and $AlCl_3$. The concentration dependence of solubilities was accurately described by the coordination cluster theory and the solubility product principle. The results provided the information necessary to predict solubilities in all other acid chloroaluminates.

Neutron-diffraction measurements were made on a series of trivalent halides at temperatures slightly above their melting points. The cation-anion coordination numbers were 4.0 for $AlBr_3$, 3.8 for $GaBr_3$, and 3.7 for GaI_3. This finding indicated that $AlBr_3$ melts consist of Al_2Br_6 molecules, but the two gallium salts were about 70-80% dimerized. The $NdCl_3$ structure factor was used to obtain a model structure by the Reverse Monte Carlo method. The resultant structure bore a close similarity to a randomized structure of solid $NdCl_3$ for features up to about 6 Å and no remnant of the solid structure at about 8 Å.

These studies were a valuable contribution in gaining an understanding of the thermodynamics and structural characteristics of molten salt systems. They provided data that were useful not only in adding to the general basic information on molten salts in the literature, but also in selecting and optimizing molten salt systems that were being used in the Division's applied programs.

Alloys. Activity coefficients for dilute solutions (about 0.1 mol%) of lithium in Bi-Pb, Pb-Mg, Bi-Tl, and Bi-Al alloys were determined as a function of solvent composition and temperature to test the range of validity of the coordination cluster theory (CCT). The most extreme case of non-ideality was the Li-Bi-Al system, where the Bi-Al solvent exhibits a miscibility gap and a large positive deviation from ideality. In all the cases, the calculations derived from the CCT were in agreement with precisely measured emf data.

The scope of this study was extended to include sodium, as well as lithium, as the solute in Bi and Sn solvents. Limiting excess potentials of the alkali metals were obtained from Nernstian emf plots at very low concentrations. Data from these experiments and the literature were used to evaluate the limiting slopes of ten binary systems involving lithium or sodium. The limiting slopes are related to correlation functions that provide information on the distribution of atoms around the solute. Electrical resistivity measurements were also made on the Na-Bi system over a temperature range of 400 to 500°C within which several liquid and solid phases exist in the phase diagram. A peak in the resistivity occurred at 20 at.% Pb, which suggested a boundary between localized and nonlocalized states.

Electromotive force studies were made on the Na-Sb system, which differed from most of the alloys studied previously in that it is a semiconductor near the composition of 50 at.% sodium. Moreover, it was believed that the antimony atoms tend to form "covalent" bonds. Activity coefficients were measured by means of a β''-Al_2O_3 solid electrolyte, which conducts sodium ions. The initial results indicated a large deviation from ideality. In addition, two inflections were observed in plots of the concentration dependence of the logarithm of the activity coefficients of sodium. These inflections appeared to relate to valence states of -3 and -1 for the antimony. The limiting slope of the activity coefficient for very dilute solutions of Na in Sb was very small and positive, which parallels the behavior of other "covalently" bonded systems.

These studies continued with measurements of the thermodynamic properties of liquid sodium-lead alloys at temperatures of 380-450°C. Coulometric titrations were used to measure the sodium activity as a function of temperature and alloy composition. Information derived from the resulting data included excess Gibbs free energies of mixing, excess stabilities and concentration fluctuation functions, and entropies of mixing. Plots of the logarithm of the sodium activity *vs.* the sodium concentration showed two peaks corresponding to NaPb and Na_4Pb.

In the 1985 CMT Annual Report, the investigators who had been working on the above alloy systems made the following statement: "Despite the recent strong interest in alloys undergoing a metal-to-nonmetal (MNM) transition, our basic understanding of these solutions remains rather primitive." There was no comprehensive theoretical treatment in which the various electronic, magnetic, electrical, and thermodynamic properties of these alloys were included. The current effort had been focused on the activity coefficients of alkali metals (Li, Na, and K) alloyed with metals having very different electronegativities (Pb, Sb, and Bi). Such alloys were considered to be "liquid semiconductors." A goal of the work was to determine whether the continuous change from metallic to semiconductor-like behavior, alloy ordering and structure, and transitions between ionic and covalent-like solutions could be related to thermodynamic properties.

All the emf measurements to date had employed β-Al_2O_3 or β''-Al_2O_3 solid electrolyte except for the system Li-Bi, where molten LiCl-KCl was used. For the first time, the potassium form of β-Al_2O_3 was used successfully. The K-Pb alloy showed unexpected behavior in that the MNM transition occurred at a potassium concentration of 50 at.% and not at a composition corresponding to the ratio of the chemical valences (80 at.% potassium). Coulometric titrations were performed on the K-Pb system by means of the potassium β-Al_2O_3 solid electrolyte, which could contain the volatile potassium at pressures up to 2 atm in the emf cell. Data on the emf *vs.* temperature at 300 to 600°C at constant compositions were accurate enough to obtain reliable data on the partial molar heat capacities. The concentration dependence of the heat capacities showed a large anomalous peak at about the equiatomic K-Pb composition. Enthalpies of the equiatomic alloy were obtained by calorimetric methods from the melting point at 589°C to 851°C. The heat capacity was very large at low temperatures, but decreased to values typical of "normal" liquids as the temperature was raised. This effect is consistent with order-disorder transitions and could be related to long-range ordering or to the formation of ordered clusters such as Pb_4.

In collaboration with personnel from the Materials Science Division and the Intense

Pulsed Neutron Source, neutron-diffraction experiments were undertaken on the equiatomic K-Pb liquid alloys. The results led to model calculations based on known structural units in the KPb crystal (nearly regular Pb_4 tetrahedra surrounded by K_4 tetrahedra oriented in the opposite direction). Results of the calculation were in excellent agreement with experimental data. Further investigation showed that all the equiatomic alkali metal-lead compounds had the same crystalline structure, known as a Zintl compound. This structure involves a Pb_4 tetrahedron interlocked with a larger M_4 tetrahedron, where M is an alkali metal other than lithium.

In other work, a hydrogen-titration method was used to measure the thermodynamic properties of hydrogen solutions in metals and alloys. Due to hydride formation, certain binary alloys (Ca-Al, Ca-Si, and Mg-Si) were of interest as potential media for hydrogen storage. Studies were initiated on the Ca-Al system, which has two intermetallic compounds, $CaAl_2$ and $CaAl_4$. Equilibrium hydrogen pressures over mixtures of these compounds were measured over a temperature range of about 400-650°C. The results showed that the Ca-Al system could serve as a "high-temperature" hydrogen-storage system with an absorptive capacity of 2.1 wt% H, which compares favorably with 1.9 wt% for FeTi and 1.4 wt% for $LaNi_5$, which are well-known "low-temperature" hydrogen-storage media.

The hydrogen-titration technique was extended to the Ca-Si system, which was complicated by uncertainties in the phase diagram. In addition to three well-established silicides, Ca_2Si, CaSi, and $CaSi_2$, evidence had been reported for two additional intermediate phases, an "E" phase and Ca_3Si_4. Hydrogen-titration isotherms for the calcium-rich portion of the phase diagram provided unambiguous evidence for the existence of the "E" phase, proof that the reaction of silicides with hydrogen is reversible, and corroboration of the validity of the theory underlying the HTM method by correct predictions for the stoichiometries of Ca_2Si and CaSi.

Personnel. In most of the basic programs, including this one, much of the work was collaborative with other ANL divisions or outside organizations such as universities. Much use was made of students, postdocs, visiting professors, and other temporary personnel. Milt Blander had the primary responsibility for most of this work. Other CMT members working in this area included Bob Blomquist, Ira Bloom, Yat Ching Chen, Larry Curtiss, Dave Frurip, Jerry Johnson, Jane Marr, Zoltan Nagy, Art Pelton, Marie-Louise Saboungi, Jack Settle, Marv Tetenbaum, Ewald Veleckis, and Bob Yonco.

THERMOCHEMISTRY

The thermochemistry studies were closely related to the thermodynamics work, but differed in that they generally involved calorimetric measurements to obtain basic thermochemical properties such as enthalpies, free energies, entropies, and heat capacities of specific substances with high accuracy and precision. (Because thermochemical data relate naturally to absolute temperature, all the temperatures in this section are expressed in degrees Kelvin. Energies are expressed in kilojoules, kJ.)

Zeolites. Work on zeolites was undertaken because these minerals are of considerable importance in various energy and environmental applications. Zeolites are involved in the extraction of geothermal energy, nuclear waste storage, isolation of radioisotopes, transport of radioactive species in groundwater, and numerous industrial applications such as water treatment. Although the chemical and physical properties of natural and synthetic zeolites had been studied extensively, almost no work had been done to obtain accurate, reliable data on their thermochemical properties.

Samples of five naturally occurring minerals—analcime, natrolite, mesolite, scolectite, and thompsonite—were obtained from Professor William Wise of the University of California at Santa Barbara. Enthalpies of formation for analcime of the composition $Na_{0.96}Al_{0.96}Si_{2.04}O_6 \cdot 1.02\ H_2O$ were determined by hydrofluoric acid reaction calorimetry, and the enthalpy functions to 625 K were measured by drop calorimetry. These data, together with results from low-temperature heat capacity measurements, provided the data for a complete table of the thermochemical properies of analcime between 0 and 625 K. The enthalpies of formation of dehydrated analcime and natrolite were also determined. There was a concern that, in a hot-dry-rock geothermal energy facility, native feldspar and water might react at low temperatures to produce analcime and quartz, a high-volume mixture that could plug the inlet and outlet bore holes through which the water must circulate. Based on the thermochemical values that were determined for analcime and literature data for the other substances involved in the reaction, the equilibrium temperature for the reaction is 380 K, which indicated that plugging should not be a problem.

There was concern, however, about a reaction of native albite with water to form voluminous analcime, which could cause bore-hole plugging. To evaluate that possibility, reliable thermochemical data were needed for albite, water, analcime, and aqueous quartz. Reasonably good data were available for all of these except albite.

Calorimetric measurements were made on several materials related to analcime (natrolite, scolecite, mesolite, and heulandite), and the resulting data were used to compute enthalpies, heat capacities, entropies, and free energies of formation. Information from this work indicated that the structure of water in these zeolites is close to that of ice. This finding contradicted results of earlier work which had indicated that the structure of zeolitic water in analcime is more characteristic of liquid water.

As experimental data became available on thermochemical values for various zeolites, it was possible to develop rudimentary methods for estimating thermochemical properties of materials that were too time-consuming or difficult to measure. Such estimates were made of the heat capacities and entropies of the hypothetical anhydrides of natrolite, mesolite, and scolecite, as well as the standard entropies of laumontite and Na-wairakite and the standard enthalpy of formation of gonnardite.

In 1984, an evaluation was made of all the earlier experimental results on zeolites and silicates. Recently determined thermodynamic functions for analcime showed that the values could not be used to predict stability ranges for naturally occurring combinations of analcime + quartz and analcime + natrolite because analcimes from various sources had different Si/Al ratios. These ratios were not necessarily the same as that of the analcime used in the calorimetric measurements. Estimates were therefore made on the effect of analcime composition on its Gibbs free energy of formation. A phase diagram that was constructed was fully compatible with natural occurrences and the experimentally determined stability of analcime.

A collaborative research effort was started with Professor Joseph Smith of the University of Chicago on the thermochemical properties of certain molecular sieves related to zeolites. Aluminophosphates were of interest as industrial absorbents, catalysts, and catalyst supports. Another molecular sieve, synthesized by the Union Carbide Corporation, was silicalite, which is essentially pure silica and can be considered to be the end member of a series of silica-rich zeolites. It is used for removal of organic contaminants from water. The standard molar enthalpy of formation of this material was determined to be -904.6 ± 1.1 kJ/mol, which indicates that its stability is comparable to that of α-quartz. Unlike α-quartz, silicalite can be dissolved readily in hydrofluoric acid, which makes it a convenient reference substitute for α-quartz in calorimetric studies. The following year (1985), low-temperature heat-capacity and high-temperature enthalpy measurements on silicalite provided the additional information needed to establish its thermodynamic properties at temperatures from 5 to 1500 K. In addition, this work showed that the material began to decompose at about 1300 K, contrary to previous reports in the literature.

Calorimetric data on hydrated and dehydrated mordenite, one of the few zeolites that can be dehydrated without altering the framework structure of the material, yielded data on the binding enthalpy of water.

In modeling studies of rock-water interactions, which are important to geothermal energy and to high-level nuclear waste repositories, reliable thermochemical data on zeolites are essential. The natural zeolites presented a problem in gathering valid thermochemical data because of the variability in their compositions. The variabilities are not only in the silicon-to-aluminum ratios, but also in differences in the substituted cations, *i.e.*, Ca^{2+} for $2Na^+$. Studies were initiated on the effect of Si/Al ratio on the standard molar enthalpies and entropies of formation. The mineral, faujasite, was used because it is one of the few zeolites that can be dehydrated without affecting the basic aluminosilicate structure. Faujasites having Si/Al ratios of 1.23 to 2.84 were investigated. The enthalpy of formation of gibbsite, $Al(OH)_3$, was also determined because it was needed as auxiliary data in deriving values for the zeolites.

In continuing studies, enthalpies of formation of four faujasites in the hydrated and dehydrated forms were measured. The binding energy of the water of hydration was highly variable, even within this family of structurally similar zeolites. Apparently, there was no way to predict with certainty the effect of water of hydration on the enthalpy of formation. Further study of the effect of Si/Al ratio indicated that a sufficient correlation exists between that ratio and the enthalpies of formation to predict the effect of only small changes in the ratio.

An effort was directed toward certain other minerals of importance to nuclear waste storage, particularly at the Nevada site. The thermochemical properties of clinoptilolite, dehydrated clinoptilolite, and tobermorite were determined.

Uranium Compounds. Among the numerous thermochemical studies of binary uranium fluorides reported in the literature, the enthalpy of formation of only one of them, UF_6, appeared to have been established with any degree of certainty. A cooperative program to establish firm enthalpy of formation values for UF_3, UF_4, α- and β-UF_5, and, if possible, U_2F_9 and U_4F_{17}, was initiated among the CMT thermochemistry group and other groups at the Netherlands Research Foundation (ECN) in Petten, Los Alamos National Laboratory, and the Chemistry Division at Argonne. In 1980, results were obtained for α- and β-UF_5, UF_4, and UF_3. Data obtained for UF_3 and UF_4 by solution calorimetric measurements at ECN and by fluorine bomb measurements at ANL were not in agreement.

In irradiated nuclear fuels, the noble metal fission products, ruthenium, rhodium, and palladium, form exceptionally stable intermetallic compounds with the uranium: URu_3, URh_3, and UPd_3. Thermochemical data were desired on these compounds both for evaluating their role in practical applications such as fuel reprocessing and in the testing of theories of chemical bonding. An experimental determination of the enthalpy of formation of UPd_3, which indicated that it is, indeed, very stable, lent credence to the Brewer-Engel theory of chemical bonding.

Several thermochemical cycles that had UO_2 in common were yielding data for various uranium compounds that were inconsistent with results obtained from other cycles, which led to concern about the currently accepted value for the enthalpy of formation of UO_2. A redetermination of the enthalpy of UO_2 by fluorine combustion calorimetry gave a result in exact agreement with the previously accepted value.

The hydrolysis of UF_6 in excess water produces uranyl fluoride (UO_2F_2), but in small amounts of water other compounds, including UOF_4, $U_2O_3F_6$, and $U_3O_5F_8$, can be produced. Apart from UO_2F_2, no thermochemical data were available for these compounds. A research effort was started to determine the enthalpies of formation for UO_2F_2 and UOF_4 and to estimate values for the other two compounds. The experimental value for UO_2F_2 was in excellent agreement with a critical assessment of the literature data. The standard enthalpies of formation for UO_2F_2 and UOF_4 were -1654.8 ± 2.1 kJ/mol and -1924.6 ± 2.4 kJ/mol, respectively. An interesting result was that UOF_4 begins to decompose to UO_2F_2 and UF_6 at about 500 K.

The double fluoride, $AgUF_6$, was studied for various reasons, one of which was to investigate the relationship of thermochemical results to the oxidation of silver from Ag^+ to Ag^{2+}. The thermochemical values led to the rather surprising result that, thermodynamically, UF_6 is capable of oxidizing Ag^+ to Ag^{2+}:

$$UF_6 + 2\,AgF \rightarrow AgUF_6 + AgF_2$$

Normally, UF_6 is not thought of as a fluorinating agent. Potent fluorinating agents such as ClF_3, BrF_3, and XeF_2 do not fluorinate AgF to AgF_2 due to kinetic constraints. Apparently, the same is true of UF_6.

Some work was done to provide thermochemical information on uranium compounds, which could be used to predict the behavior of uranium in the Nevada Nuclear Waste Site. Because the question being addressed was the behavior of dissolved uranium in contact with concrete at ambient temperatures, the compound, $UO_3{\cdot}2H_2O$ (schoepite), which is the most hydrated form of UO_3, was chosen for the study. The standard enthalpy of formation at 298.15 K, the low-temperature heat capacity, the standard entropy, and the enthalpy increments up to the limit of stability of the material were determined. Information was then derived for the thermochemical values for the decomposition of UO_3 hydrates and the solubilities in acid of $UO_3{\cdot}2H_2O$ and $UO_3{\cdot}0.9H_2O$ as a function of temperature.

Chalcogen (S, Se, Te) Compounds. In spite of the fact that sulfides are used extensively in industry and have been the subject of a great deal of research, much of the existing thermochemical information on inorganic sulfides was seriously in error. In fact, it could not be used for practical applications in some cases. Fluorine combustion calorimetry was well suited to thermochemical measurements on sulfides, which were difficult to perform by more conventional calorimetric methods. The CMT group had been able to obtain reliable data on a variety of sulfides (As_4S_4, As_2S_3, Sb_2S_3, MoS_2, US, US_2, $CuFeS_2$, and K_2S). Plans were made to begin investigations of sulfides of aluminum and tungsten because of their technological importance, and pure and mixed sulfides of cobalt, chromium, manganese, and nickel, which are strategically important materials. The CMT Division was probably the only group that had a capability for measuring thermochemical values for these materials by fluorine combustion calorimetry.

Fluorine-combustion calorimetry was used to measure the enthalpy of formation of GeS_2, which is used in optical and electrical applications. Research on $ZrTe_2$ showed that the compound is quite stable, suggesting that zirconium cladding might act as a getter for fission-product tellurium in a reactor accident.

Because of its use as a desulfurization catalyst and intercalation compound, TiS_2 was of interest, and there was some controversy regarding the existence of a low-temperature transition at about 370 K. In collaboration with workers at the Exxon Corporation and the U.S. Bureau of Mines, the standard molar enthalpy of formation, ΔH^0, and high-temperature enthalpy increments, $H^o(T) - H^0(298.15\ K)$, were determined. No evidence was found for the transition at 370 K.

Thermochemical data were also obtained on K_2SO_3, $NaLiSO_4$, and $KHSO_5$, which were of commercial or research interest.

In the mid-1980s, the technical community was becoming excited about the subject of "high-tech" materials. A three-day symposium at the American Chemical Society meeting in Chicago in 1985 was devoted to discussions of newly synthesized inorganic compounds, and an entire issue of *Scientific American* (October 1986) concerned these materials and their potential applications. The term "high tech" probably defies an exact definition, but it did include a number of chalcogenide compounds of metals (sulfides, selenides, and tellurides). These compounds have a wide variety of potential applications: (1) high-temperature lubricants, (2) uses in electronics such as semiconductors, thin films, glasses, photoresists, photomicrography, (3) active materials in rechargeable batteries, (4) catalysts, (5) materials in solar-energy conversion, (6) ceramics, (7) materials in extractive metallurgy, and (8) corrosion inhibitors in sulfur-containing atmospheres. In many cases, thermochemical data on these compounds were unreliable or nonexistent, and fluorine-combustion calorimetry was uniquely suited to the purpose of obtaining this information.

In cooperation with Bell Communications Research and with Kenneth Volin and Sherman Susman of ANL's Materials Science Division, the enthalpies of formation of vitreous and crystalline $GeSe_2$ were determined. Molybdenum selenides were of interest because of their use as solid lubricants, electrode materials, and materials for solar-energy conversion. Enthalpies of formation were determined for $MoSe_2$ and Mo_6Se_6.

Vanadium monosulfide was of interest because of its use as a catalyst and its role in the mechanical and other properties of stainless steels. The standard enthalpy of formation of $VS_{1.043}$ was found to be -230.3 ± 2.2 kJ/mol. Tungsten diselenide (WSe_2) was of interest, owing to its use as a high-temperature lubricant. The value obtained for the standard enthalpy of formation of WSe_2 indicated that high-temperature vaporization of the selenide does not lead to the formation of $Se_2(g)$ or $Se(g)$, as others had suggested, and that further studies of the system were needed. The energy of combustion in fluorine of the semiconductor, black phosphorus, together with earlier data from white phosphorus, gave a value of -21.2 ± 2.1 kJ/mol for the transition of α-white to black allotropes of phosphorus.

Theoretical Studies. Larry Curtiss and Dave Chaiko of CMT and John Pople from Carnegie-Mellon University used quantum-mechanical calculations to derive thermochemical data for the N_2H radical, which had not been detected experimentally, but had been postulated as an intermediate species in atmospheric and combustion processes that involved nitric oxide (NO). The calculations showed that N_2H is unstable with respect to dissociation into N_2. Also determined were shifts in the vibrational frequencies of N_2H from those in N_2H_2 and N_2, which should be useful in any experimental attempt to detect the species. The enthalpy and entropy of dissociation were determined from the calculated vibrational frequencies and energy of formation.

Calculations were also made on a complex of water and carbon dioxide that is suspected of being a factor in global warming. *Ab initio* calculations were performed to determine the binding energy, structure, and vibrational frequencies of the $CO_2 \cdot H_2O$ complex, and a set of thermodynamic functions for this entity was generated.

In a cooperative effort with the Physics Division, a new set of thermodynamic functions for dimeric boric oxide (B_2O_2) was generated by a combination of *ab initio* calculations and photoelectron spectroscopy data. The results provided strong evidence that the molecular structure was O=B–B=O, rather than alternative configurations such as B-O-B-O. This work provided thermodynamic functions of improved accuracy for B_2O_2.

Personnel. Pat O'Hare was in charge of the calorimetric work. Other CMT people in this effort were Ward Hubbard, Howard Flotow, Ian Tasker, and Larry Curtiss. The CMT Calorimetry Group decreased in size until 1988, when it disappeared with the departure of Pat O'Hare. Some of the equipment, however, is still being used by other groups.

ELECTROCHEMISTRY

The basic electrochemical studies were concerned with experimental and theoretical investigations of processes occurring at electrodes and in electrolytes. Phenomena such as electrocatalysis, electrochemical corrosion, deposition and dissolution, and complexation in molten salt electrolytes were of interest. The experimental techniques included electrochemical measurements (relaxation and cyclic voltammetry methods), surface analysis (electron spectroscopy, atomic emission spectroscopy, scanning electron microscopy), and spectroscopic measurements (Mössbauer, Raman, UV-visible).

Electrode Kinetics and Electrocatalysis. A study was directed toward those factors that determine the rate and mechanism of the electroreduction of water on carbon-supported iron phthalocyanine (FePc) electrodes. Spectroscopic and electrochemical methods, together with quantum theoretical calculations, were used to correlate electronic structure and catalytic activity. Cyclic voltammetry showed that FePc films on glassy carbon electrodes in 0.05 *M* sulfuric acid solution catalyzed oxygen reduction at first, but became inactive on continued scanning, and then reactivated after about an hour on open circuit. The deactivation was attributed to slow transport of charge carriers within the p-type semiconductor film electrode. Films prepared from solutions of FePc in an organic solvent had a significantly higher activity than those formed by vapor deposition, apparently because the two methods produced different polymorphic forms of the films.

Quantum-mechanical calculations were performed on FePc and its analog, iron porphyrin (FeP), as well as their oxygen adducts. The charge of the iron is about the same in the two molecules, but the electrons are distributed differently. Calculations on the structures of the oxygen complexes of FePc and FeP showed that oxygen in the FePc-O_2 complex binds side-on to the iron with the O-O axis parallel to the FePc plane, whereas the oxygen in the FeP-O_2 complex is bound end-on to the Fe. These differences may determine what makes for a good electrocatalyst (FePc) or a mere oxygen carrier (FeP).

Continued work in this area using rotating ring-disc electrodes indicated that oxygen reduction in the FePc system proceeds by the formation of hydrogen peroxide (H_2O_2), and the loss of catalytic activity on cycling appears to be due to adsorption of peroxide intermediates. *In situ* laser Raman studies of the FePc films on gold and copper electrodes confirmed an earlier conclusion that the FePc was coordinated to water molecules in the solution. *Ab initio* calculations to study the solvation of O_2 by water molecules in solutions indicated that the energy of interaction between H_2O and O_2 molecules is only a few kilocalories, which accounts for the low solubility of oxygen in water. Ensuing studies showed that hydrogen peroxide could be anodically oxidized on the FePc films by a dissolution process that takes place under charge-transfer control.

Meanwhile, theoretical studies continued on the solvation of O_2 and O_2^- in water. These solvation effects were of practical importance because of their role in electrocatalysis. The most probable structure was one that includes both hydrogen-bonded O_2-HOH interactions and electron-donor O_2-OH_2 interactions, according to a newly developed molecular orbital method of calculation.

In continuing investigations of electrocatalysis, laser Raman and cyclic voltammetry techniques were used to study the structure and properties of iron protoporphyrin IX (FePP) as they relate to its catalytic activity in the electrochemical reduction of oxygen. This work was done in collaboration with John McMahon of Williams College. The FePP showed weaker catalytic properties compared to the FePc studies earlier. The conclusion from this work was that the catalytic activity and stability are functions of the molecular electronic structure of the compounds and perturbations caused by interactions with surrounding atoms, molecules, and surfaces.

Metal Dissolution and Deposition. Spectroscopic and electrochemical studies were conducted on anodic corrosion and passivation of metals in aqueous and molten salt environments. The anodic polarization of Fe, Co, Ni, Cu, Ti, Ta, and Mo in molten LiCl-KCl was measured at temperatures of 375 and 450°C. The corrosion-passivation behavior was similar to that found in aqueous systems, which was somewhat surprising. The oxides of Fe, Co, and Cu were reversibly reproducible, but those of Ta, Ti, Ni, and Mo were not. Consequently, Fe, Co, and Ni were suggested as possible cathodes for secondary batteries having molten salt electrolytes.

Many of the reactions involved in electrode kinetics studies are so fast that special measuring techniques are necessary. Several improvements were made in the equipment and the data-gathering methods. A galvanostatic double-pulse technique was used to investigate the kinetics of nickel in LiCl-KCl, LiCl-RbCl-$CaCl_2$, and LiBr-KBr-CsBr eutectic melts. The exchange current density of the reaction ranged between 0.1 and 1.7 A/cm^2.

Spectrochemical methods were applied to the identification of anodic corrosion films on silver in 0.05 *M* sulfuric acid. The dominant phase was Ag_2SO_4, but a new phase, $Ag_7O_8SO_3$, which had not been reported previously, was also observed. The conversion of Ag_2SO_4 to higher oxides *i.e.*, Ag_2O and AgO, was slow from an electrokinetic viewpoint. Anodic films on silver in 0.05 *M* Na_2SO_4 were basically the same as those found in H_2SO_4.

The formation of anodic corrosion films on nickel in 0.05 *M* NaOH solution at room temperature was studied by cyclic voltammetry and *in situ* laser Raman spectroscopy. An anodic film that formed on the nickel corresponded most closely to "hydrated" Ni_2O_3. Studies were also made of nickel in aqueous borate media. Infrared studies were conducted on anodic corrosion films formed on iron. Laser Raman spectroscopic and X-ray diffraction techniques were used to identify coatings found on Type 304 stainless steel that had been subjected to tensile tests in the Materials Science and Technology Division. Raman bands that could be attributed to Fe_2O_3 and Fe_3O_4 were observed in test samples that had been exposed to solutions containing dissolved oxygen as an impurity, and this result was confirmed by X-ray diffraction.

In collaboration with the Materials Science and Technology Division, a cell and electrolyte-recirculation system was constructed to conduct simultaneous spectroscopic and electrochemical studies of aqueous corrosion at high temperature and pressure (300°C and 10 MPa). Cyclic voltammetry studies of nickel in 100 ppm Na_2SO_4 solution at 25 to 290°C indicated that passivity of the nickel disappeared above about 100°C and pitting corrosion occurred above 250°C. At potentials near the region of oxygen evolution, the presence of Ni_2O_3 was noted. This study also included an evaluation of the use of AC impedance measurements to obtain further electrochemical information.

A study was made of lead in dilute Na_2SO_4 solution. Cyclic voltammograms showed typical anodic dissolution/passivation behavior. The passive film at low and high temperatures was $PbSO_4$, and at intermediate temperatures it was $PbO \cdot PbSO_4$. Laser Raman spectroscopy and X-ray diffraction were used to characterize the types of compounds that could conceivably form as lead corrosion products. The various forms (β- and γ-NiOOH, $Ni_2O_3 \cdot 2H_2O$, $Ni_3O_4 \cdot xH_2O$, and NiO_2) were thought to differ only in the degree of structural disorder and the nature of intercalated molecules, ions, *etc.*

An effort was devoted to the electrode kinetics and mechanistic aspects of metallic corrosion in aqueous media. The processes are basically electrochemical and involve anodic oxidation of a metal accompanied by reduction of a component in the solution, and the anodic oxidation is often the rate-controlling step. Stress-corrosion cracking is a manifestation of this type of process that has been a problem in light water reactors. A capability was developed for investigating high-temperature, high-pressure electrode kinetics. Among various reference electrode designs that were tested, a palladium/hydrogen pH-sensitive (pseudoreference) electrode gave experimental results close to the theoretical values at 90 atm pressure and a temperature of 250°C. Deviations were observed at lower temperatures, probably due to decreased hydrogen permeability of the palladium.

In 1985, a high-temperature, high-pressure test loop that incorporated a built-in, fast-pulse electrochemical cell was built and placed in operation, and several working-electrode, counter-electrode combinations and reference electrodes were designed, built, and tested. The palladium/hydrogen electrode was tested further and found to be a stable reference electrode suitable for electrochemical relaxation measurements at high temperature and pressure. A cooperative program was initiated with John Halley of the University of Minnesota in which he was to perform molecular-dynamics calculations and the CMT workers were to provide experimental data. The initial experiments were made on relatively simple systems, including Fe^{2+}/Fe^{3+} and Cr^{2+}/Cr^{3+} redox reactions on a gold electrode.

A cyclic-voltammetry investigation of the Fe^{2+}/Fe^{3+} couple showed that the overall rate of the reaction increased fourfold between 25 and 150°C, an increase typical of diffusion-controlled reactions. In these measurements, ultra-clean media were required to avoid interferences, and a computer model was developed to sort out the diffusional effects from the surface reaction kinetics. Also, *ab initio* molecular orbital calculations were used to develop potential energy functions for hydrated transition metal cations, which can be used to provide calculations of electron-transfer rates.

These studies also included the development of computer models to estimate the effect of the structure of the interfacial solution layer on the transient electrode kinetics measurements, and a model to determine the effect of mass transport on the calculation of corrosion rates from electrochemical polarization measurements.

Because of the role of adsorbed molecules on surfaces in corrosion and other electrochemical reactions, some *ab initio* studies were conducted on this phenomenon. The studies emphasized computation of the energetics of surface processes and dealt with shifts in vibrational frequencies, binding energies, binding sites, ionization energies, and charge-transfer effects of the adsorbed molecules. Collaborative work was performed with

Professor John A. Pople of Carnegie-Mellon University, in which interactions of lithium and beryllium with water were investigated. The beryllium atoms were in clusters of one to five. A dramatic increase in the interaction energy occurred when the cluster size reached three beryllium atoms, with no further major changes as it went from three to five. The water molecules preferred corner binding sites over edges or faces of the clusters. An investigation of water interactions with Mg and Na atoms showed that they, like Li and Be, form complexes with H_2O that have metal-oxygen bonds with charge transfer to the metal atoms.

Personnel. Among the personnel in the electrochemical studies were Ira Bloom, Alan Brown, Dennis Dees, Mike Krumpelt, Ralph Loutfy, Carlos Melendres, Zoltan Nagy, Laszlo Redey, Jack Settle, and Bob Yonco.

OTHER SPECTROSCOPIC AND STRUCTURAL STUDIES

In the previous chapter, mention was made that many metal halides form gaseous complexes with "acidic gases" such as $AlCl_3$ or $FeCl_3$, which enhance their volatility by factors as large as 10^{14}. Such complexes may have practical utility for chemical separations, metallurgical processes, high-efficiency lamps, and lasers. This program consisted of spectroscopic and thermodynamic measurements to find new vapor-phase complexing agents and vapor species and to determine their electronic, vibrational, and structural properties. George Papatheodorou played a leading role in much of this work.

In 1980, spectrophotometric investigations were conducted on gas-solid reactions of $FeCl_2$ and $ErCl_3$ with $AlCl_3$, $GaCl_3$, and $InCl_3$ as the carrier gases, and *ab initio* molecular orbital calculations were made on the complexes $BeF_2{\cdot}AlF_3$, $MgF_2{\cdot}AlF_3$, and $AlF_3{\cdot}BeF_2{\cdot}AlF_3$. The calculations indicated that the cation MF^+ can move relatively easily about the nearly spherical anion AlF_4^- to form corner-, edge-, or face-bridged structures. The most stable structure for the complex $AlF_3{\cdot}BeF_2{\cdot}AlF_3$ was two edge-type bridges. In general, there was a preference for edge-type, two-fluorine bridging in these systems.

A statistical mechanical model was developed to calculate the entropies and free energy functions of vapor molecules. Only the atomic masses and a characteristic interatomic distance are needed for the calculation, which was tested successfully for the gaseous MX_n species where n = 1-6, and homonuclear clusters of 2-7 atoms.

Ab initio molecular orbital calculations were made on the bonding energies of $LiOH{\cdot}H_2O$, which was suspected of being an important species in the vapor of high-pressure molten carbonate fuel cells.

INTERFACIAL CHEMISTRY

Ab initio studies were initiated in 1984 on the properties of molecules adsorbed on surfaces. The idea behind these studies was to add a theoretical component to the various experimental efforts on aqueous corrosion and interfacial chemistry. Shifts in the vibrational frequencies, binding energies, binding sites, ionization energies, and charge-transfer effects of adsorbed molecules were investigated.

Zeolite Catalysis. In 1984, work was started on the surface chemistry of zeolites, which are aluminosilicate minerals having an open structure. Due to this open structure, they can function effectively as catalysts. Spectra of several types of zeolites were obtained by dusting or impacting zeolite particles on the surfaces of ATR (attenuated total internal reflection) elements. This technique permitted *in situ* measurements of gas adsorption/desorption processes and chemical surface reactions. *Ab initio* studies were also begun on molecules adsorbed on surfaces with emphasis on the energetics of surface processes.

Diffuse reflectance studies were concentrated on the reactivities of two iron-modified versions of ZSM-5 and offretite, which are synthetic zeolites that have potential utility as molecular-shape-selective catalysts. Protonated sites of differing stability were examined by means of deuterated probe molecules (D_2O and CH_3OD). Analysis of the infrared band positions of reactants and products in the zeolite structure provided new information on the role of the various protonated sites in hydrogenation and dehydrogenation reactions. The O-H and C-H stretching vibrations revealed further information on the reactions involved in hydrocarbon reforming within the zeolite.

By 1986, significant progress had been made in the use of Fourier transform infrared (FTIR) diffuse reflectance spectroscopy to investigate the reactivity of synthetic zeolites toward light hydrocarbons. These studies were augmented by adding a gas chromatograph to the diffuse-reflectance cell so that the reaction products could be identified simultaneously with the FTIR spectra. This arrangement shed new light on the effect of temperature on carbon-hydrogen bond scission in the catalytic conversion of alcohols to gasoline-type hydrocarbons. The catalyst used for this purpose was a synthetic zeolite, H-ZSM-5.

Investigations continued on the vibrational dynamics of zeolite framework structures. Twelve different structures were by then analyzed fully, providing the theoretical groundwork for detailed lattice dynamics calculations. Potential energy functions were then determined on several well-studied model systems having a rutile structure. A valence force field model involving only three parameters resulted in very good fits to observed vibrational data for several dioxides and fluorides having the rutile structure.

Ab initio molecular orbital calculations were used to investigate rotational barriers and vibration frequencies of organic templating cations used in the crystallization of synthetic zeolites. The use of organic bases such as tetraalkylammonium hydroxides greatly increased the number of gel/solution synthesis possibilities for producing microporous materials such as zeolites. The role of these organic materials, however, was not well understood, and the subject was investigated by researchers at the Materials Science Division and the Intense Pulsed Neutron Source along with those in CMT. Molecular orbital calculations were performed on the energies of the conformational rearrangements required for tetrapropylammonium ion to serve as a template for the zeolite ZSM-5.

Continuing work with the combined infrared spectroscopic and gas chromatographic techniques was directed toward the catalytic reactions of several light hydrocarbon reactions on aluminosilicate and aluminophosphate framework systems, the interactions of light alcohols and Brönsted acid sites of H-ZSM-5, and the catalytic properties of silica- and cobalt-substituted zeolites.

In 1988, a systematic study was conducted on the catalytic properties of a series of aluminum phosphate ($AlPO_4$) molecular sieve structures. A group of low-molecular-weight organic oxygenates (light alcohols, ethers, aldehydes, and esters) was used to observe reactant selectivity effects and any homologation behavior. The key factors affecting homologation proved to be pore directionality, pore connectivity, the nature of the functional groups of the reactant, and the presence of pre-existing carbon-carbon bonds in the reactant.

Further structural studies of the effects of molecular sieve frameworks on the vibrational frequencies and rotation of diffusion of occluded template molecules were continued. Results from *ab initio* molecular orbital and inelastic neutron scattering techniques indicated that cobalt-substituted aluminophosphate molecular sieves possess a strong oxidizing capability, converting methanol to aldehyde, and that cobalt/silicon substitutions can activate methane, which is homologated to C_2 and C_3 alkanes and alkenes.

Personnel. The CMT people involved in this work included Vic Maroni, who spearheaded the effort, Larry Curtiss, Torben Brun, Bob Kleb, Jim Newsam, Bob Beyerlein, and D. E. W. Waughan.

SPECIAL BASIC RESEARCH INVESTIGATIONS

The following are several basic research studies that were related to areas of practical interest. In some cases, basic investigations were necessary to support certain aspects of the applied programs. Other studies were exploratory studies of subjects that might lead to future applied programs.

Gas-Condensed Phase Equilibria. Equilibria between gases and liquids or solids in complex multicomponent systems are important in several energy technologies, including magnetohydrodynamics and coal combustion, desulfurization, and gasification. The objective of this program was to generate computer programs and database information to examine the chemistry of these complex systems.

In the condensation of alkali sulfates in combustion gases, Na_2SO_4-K_2SO_4 solutions were found to condense at somewhat higher temperatures than Na_2SO_4 alone, but to form a liquid at much lower temperatures. Thus, under typical burner conditions, a much larger fraction of the alkali sulfate condenses to form corrosive liquids rather than the less corrosive solids. A second conclusion was that the presence of dolomite as an entrained solid greatly increases the possibility that corrosive sulfate liquids, rather than solids, could be formed in turbines and condensers. (As mentioned elsewhere, dolomite is used in fluidized-bed combustion systems to remove SO_2 from the off-gases.)

This work was then expanded to incorporate reciprocal molten-salt solution theories to handle more complex systems such as mixtures of Na^+, K^+, Ca^{2+}, Mg^{2+}/SO_4^{2-}, Cl^-, and CO_3^{2-}. In 1987, a computer program was used in a collaborative study with the ANL Materials and Components Technology Division to calculate the condensed and gaseous products in the combustion of high-sulfur coal. Under both oxidizing and reducing conditions, the condensed phases were saturated with silica (SiO_2), and the major gaseous carriers of sodium and chlorine were NaCl and HCl, respectively. Silicates were the major phase of the condensates at temperatures above about 1327°C, and the concentrations of sulfates (under oxidizing conditions) and sulfides (under reducing conditions) increased with decreasing temperature. These results are considerably different from the assumptions that had usually been made in the past about the compositions of corrosive deposits from coal combustion or conversion.

Impurity Interactions in Liquid Metals. A study was directed toward the reactions of nitrogen with selected metallic elements in liquid lithium because nitrogen was known to play a key role in corrosion and mass transfer in chromium-containing steels used for containment of liquid lithium blankets in the magnetic fusion program. The systems under study were Li-Cr-N and Li-Ca-N.

Scanning electron microscopy and ion-microprobe mass analyses, together with sampling and wet chemical analyses, were completed, and the results were combined with earlier resistometric data to gain a better understanding of the mechanism of the reaction of chromium with nitrogen in liquid lithium. The data supported the proposed formation of Li_9CrN_5 in the solutions. The solubility of this compound is significantly higher than that of chromium metal, which could account for the enhanced corrosion and mass transfer observed in lithium-stainless steel systems.

Precipitation techniques, based on the formation of nitrides that are less soluble than Li_3N, were also investigated as a means of reducing nitrogen levels in liquid lithium. Experiments with calcium additions indicated a finite solubility for the compound LaCaN and an equilibrium between that compound and Ca_3N_2. These results indicated that the nitrogen levels could be controlled in a range of 100-1000 ppm by weight through careful adjustment of the calcium activity in the lithium.

Sulfide Species in Molten Salt Solvents. The properties of sulfide species formed in molten salt electrolytes were germane to the chemistry of high-temperature electrochemical systems, including the metal sulfide-based batteries. Electronic-absorption, resonance-Raman, and electron-spin-resonance spectra were obtained for species formed in molten CsCl-$AlCl_3$ solutions containing different S/Al ratios. The results showed that the S_3^- radical was the predominant sulfur species, and that this radical tends to form adduct-type molecular ions such as $AlCl_3{\cdot}S_3^-$.

Other investigations were concerned with the structural and chemical behavior of polyhalide and polysulfide anions in molten alkali halides of the type used in batteries with chalcogen positive electrodes. Identification of the charge-bearing species and elucidation of mechanisms for capacity loss and self-discharge were important to the applied battery program. A spectrophotometric study of solutions of Li_2S-FeS_x, where x is between 1 and 2, depending on the state of charge of a LiAl/FeS_2 battery, provided firm evidence for the existence of at least two sulfide-containing species.

Physical Chemistry of Minerals and Materials. The aim of this research was to elucidate the dissolution and transport characteristics of certain 3d transition metals such as Cr, Mn, Ni, and Cu in nonaqueous media. This area of research had a potential for eventual applications, including the recovery of these strategically important metals from naturally occurring deposits or industrial by-products.

In a solvent-extraction type of separation, Co^{2+} dissolved with equal concentrations of Fe^{3+}, Ni^{2+}, and Cr^{3+} in a low-melting (150°C) carboxylate eutectic phase and was extracted selectively into a dodecane phase containing bis-(2-ethylhexyl) phosphinic acid. This result demonstrated the possibility of using ligand-exchange reactions in solvent-extraction separations.

Experiments were conducted with molten LiCl-KCl eutectic as a solvent for two polysulfide ore concentrates from iron-cobalt-copper-nickel mineral deposits in Missouri and Idaho. Additions of acidic post-transition metal halides (*e.g.*, $BiCl_3$, $HgCl_2$, $SnCl_2$, and $ZnCl_2$) were effective in dissolving most of the ore-bound transition metals.

Continuing studies were performed on anion/cation coordination chemistry in low-melting molten salts containing organic anions. The ligand field spectra of 3d transition metal ions were studied in formate, acetate, and thiocyanate eutectics. Octahedral coordination was seen in all cases. The tendency toward octahedral coordination was attributed to crystal-field effects, while formation of complexes with low negative charge was believed to result primarily from electrostatic effects. A highlight of this work was the synthesis and characterization of a unique halocarboxylate salt that had metastable fluid properties at room temperature. Further studies led to a glassy alkali metal trifluoroacetate (TFAc) that formed a metastable liquid at temperatures as low as 22°C. This salt, of the composition CsTFAc-NaTFAc-KTFAc (50-25-25 mol%), remained fluid for long periods of time at room temperature in an anhydrous condition, but crystallized if traces of water or other impurities were present. Electronic absorption spectroscopy was used to examine the structural properties of the TFAc anion and to investigate the coordination chemistry of 3d transition metals (V^{3+}, Cr^{3+}, Fe^{2+}, Co^{2+}, Ni^{2+}, and Cu^{2+}) in TFAc solutions. The ligand field absorption spectra were similar to spectra of the same cations in aqueous solutions, but the molar absorptivities were 2 to 20 times higher.

Lixiviation studies were conducted on naturally occurring oxides and sulfides of the transition metals. Molten LiCl-KCl alone was not effective for this purpose, but successful lixiviation by LiCl-KCl occurred when acidic metal halides such as $BiCl_3$, $HgCl_2$, $SnCl_2$, and $ZnCl_2$ were added. Molten LiCl-KCl by itself, however, was successful in recovering metals from oxide mattes produced by roasting sulfide ore.

An electrolytic reduction was performed on LiCl-KCl solutions of transition-metal ions leached from ore samples. Complete electrodeposition of cobalt was achieved with a current efficiency near 100%, using graphite electrodes.

These studies were then directed toward sea-based minerals. Pacific sea nodules containing about 30 wt% Mo, 7 wt% Fe, and smaller amounts of Ni, Co, and Cu, all in oxide form, were treated with two candidate extraction salts, LiCl-KCl eutectic and NaCl-KCl-$MgCl_2$ eutectic. The latter extracted close to 100% of the Mn, Fe, Ni, Cu, and Co from the nodules, while the LiCl-KCl extracted only about 50% of the Mn, Cu, and Co. Cyclic voltammetry showed that cobalt can be electrodeposited from molten alkali halides with nearly 100% coulombic efficiency.

Taken together, the results of this work suggest the possibility of a compact pyrochemical processing plant to recover these metals at the site of the deep sea mining operation.

Further work was done in which equilibrations of nodule material and molten NaCl-KCl-$MgCl_2$ were sparged. Chlorine gas and hydrogen chloride (HCl) were produced by the lixiviation reaction. Manganese, cobalt, nickel, and copper were all extracted as divalent ions, and the iron was trivalent. The sequences of reactions involved were established by absorption spectra, X-ray diffraction, and chemical analysis.

In larger-scale experiments (50 g of nodule, 250 g of salt), iron and copper were recovered as $FeCl_3$ and $CuCl_2$ by volatilization at 450-500°C. (A complex vapor species, $FeCuCl_5$, appeared to be involved in the volatilization.) Nickel, and then cobalt, were recovered as the metals by electrolysis, leaving only manganese in the salt. Preliminary tests indicated that the manganese could be precipitated as an oxide by a controlled reaction with oxygen. This processing scheme was the subject of one of the *Industrial Research Magazine* IR-100 awards for 1985.

These studies, initiated and directed by Vic Maroni, were done by Bob Yonco and some temporary personnel.

Electrochemical Reactions in Submerged Arc Welding. Underwater arc welding of steel plate entails high currents and DC voltages (*e.g.*, 500 A and 30 V) which result in electrochemical reactions at the surfaces of the weld pool and the weld wire that is fed in. The weld wire is usually anodic while the weld pool and base plate are cathodic. The purpose of these studies was to determine mechanisms for alteration of the weld metal chemistry that would improve the quality of the welds.

After arc welding with SiO_2-MnO-FeO fluxes, the weld wire would be exposed to a relatively high oxygen partial pressure, and a half-cell reaction can be written as

$$nO^{2-} + M \rightarrow MO_n + 2\,ne^-$$

where M is a metal at the weld wire-slag interface, and n is related to the valence of M in the oxide. (For steel, M would be largely iron.) This reaction produces an oxide which finds its way into the slag and the weld metal. Experimental results for the oxygen content of the weld metal were consistent with the proposed mechanism.

The cathodic reactions result in reduction and deposition of metal (*e.g.*, Fe, Si) at the interface. The relative amounts of the species in the deposits depend upon the possible oxidation-reduction reactions between the various metals and oxides. An analysis of experimental results indicated that both the electrochemical and back reactions were important.

The underwater welding studies were initiated by Milt Blander. Art Pelton was involved in the work, but most of the effort was supplied by temporary personnel or people outside the Division.

SOLID OXIDE SUPERCONDUCTORS

In the 1980s, a discovery was made that a new class of materials consisting of copper oxide and oxides of certain Group II and III metals such as barium and the rare earths exhibited superconductivity at temperatures approaching 100 K (-173°C). This was an exciting development in that superconductivity had been limited previously to metals at temperatures below about 10 K (-263°C), which required liquid helium or hydrogen as a coolant. Because of the much higher T_C value (the temperature below which the material is superconductive) of the new oxide systems, sufficient cooling can be achieved with liquid nitrogen, which is much less expensive and more readily available.

This development stimulated much interest and sparked a variety of new research initiatives at Argonne. The Chemical Technology, Materials Science, and Components Technology Divisions entered into a collaborative effort on the following activities: (1) measurements and analysis of the vibrational densities of states of the high-T_C materials; (2) *ab initio* and semiempirical molecular orbital studies of $(Cu\text{-}O)_n$ clusters similar to those found in the high-T_C structures; (3) synthesis of high-T_C phases by oxidation of metallic precursors; (4) X-ray diffraction studies of single-crystal high-T_C phases; and (5) bonding of high-T_C ceramics to metals.

Infrared, Raman, and theoretical group methods were used to look at the vibrational densities of a variety of perovskite-type structures, including some with high T_C values. The predicted number of phonons was compared with the experimental data. Several phonons of $La_{1.85}Sr_{0.15}CuO_4$ and $YBa_2Cu_3O_{7-x}$ were observed and assigned unambiguously to the structure.

Molecular orbital calculations indicated the likelihood of mixed valence states of the different Cu sites in $YBa_2Cu_3O_{7-x}$ which fall in the ranges of 2.5-3.0 and 2.0-2.5 and may cause structural features that result in superconductivity. The superconducting phase for T_C = 94 K appeared to have an oxygen stoichiometry near 7.0.

Studies were continued on the connection (if any) between phonon behavior and high-temperature superconductivity. Results from $La_{2-x}M_xCuO_4$ (where M = Sr, Ba) showed an excellent fit between a valence-bond-type force field model and the four Raman-active phonons and two of the seven infrared-active phonons.

Attempts were made to fabricate superconducting ceramic materials in bulk and coating forms by oxidizing mixtures of metals corresponding to the stoichiometry of the desired oxide product. The initial results suggested that superconducting films could be fabricated on metal precursors if the conditions were controlled carefully. The most successful bonding procedure involved the use of indium foils with $YBa_2Cu_3O_7$ disks that had been sputter-coated with silver. A composite of the same ceramic material + 10 vol% silver sandwiched between two copper disks had an interfacial electrical resistivity of about 10^{-6} ohm-cm^2. Several additional assemblies of this general type were fabricated with good results. In 1989, preparations were started for a research effort on the preparation of high-T_C superconducting ceramic films by the oxidation of liquid alloy precursors.

Individuals who worked on various aspects of this program were Vic Maroni, Milt Blander, Ira Bloom, Alan Brown, Wally Calaway, Larry Curtiss, Dieter Gruen (Chemistry Division), Mark Hash, Stan Johnson, Carlos Melendres, Shiu-Wing Tam, Ben Tani, and Marv Tetenbaum.

ENVIRONMENTAL CHEMISTRY

The Environmental Chemistry Program continued in the 1980s with the same basic objectives: that is, determine the mechanisms involved in the formation of atmospheric sulfate and nitrogen-bearing aerosols, and improve the instrumental capability to conduct the studies.

Characterization of Atmospheric Aerosols. Although chemical analysis of aerosol particles can be accomplished by analytical techniques such as neutron activation, X-ray fluorescence, and atomic absorption, identification of the actual chemical species is not so straightforward. Methods had been developed for collecting time- and size-classified atmospheric particles, which provided some information on sources and mechanisms of formation.

The Attenuated Total Internal Reflection (ATR) instrument described in the last chapter was field-tested and found effective even with very short sampling times, *i.e.*, about one minute. The instrument combined inertial impaction for sample collection with an infrared detector that used multiple internal reflection to achieve very high sensitivity. Tests were conducted both at ground level and aboard aircraft. A typical sample covered only about 10% of the impactor surface, so the individual aerosol particles did not interact with one another. The capability of the device was evaluated in a weeklong series of airborne field experiments conducted around a power plant near Centralia, Washington. During traverses of the aircraft through the plume from the stack, a single traverse six miles from the stack produced a sample sufficient for analysis even though the plume itself was very clean.

The sensitivity of the Fourier-transform infrared (FTIR) spectroscopic method for condensed-phase nitrate was extended to the sub-microgram level. A band at 1384 cm^{-1} was useful for nitrate levels down to about 0.1 μg. A less sensitive band at 2430 cm^{-1} was better suited to higher levels (>40 μg).

A study was also conducted on determinations of the bisulfate ion, which has major absorption bands at 1205 and 600 cm^{-1}. A procedure was developed whereby the bisulfate ion could be measured quantitatively, which made possible a quantitative measure of the acidity of acid sulfates present in the aerosols.

In 1981, work was started on the use of X-ray diffraction to identify the crystalline phases in the fine-particle fraction of the aerosols. X-ray diffraction patterns were obtained and interpreted for 50-μg samples of aerosols in the 0.3-1.0 μm size range. Samples taken at Argonne and at State College, PA, showed that $(NH_4)_2SO_4$, $(NH_4)_3H(SO_4)_2$, $3(NH_4NO_3)\cdot(NH_4)_2SO_4$, and $2(NH_4NO_3)\cdot(NH_4)_2SO_4$ were frequently observed species. Numerous other double salts of ammonium and sulfate that contained metallic ions such as lead and calcium were also found. Some complexing of natural species and those generated by human activities may have occurred. Atmospheric scientists had not considered this possibility previously.

Work continued on the effects of collection time and particle size on the nature of aerosol particles. Particles of different sizes had very different chemistries, different sources and mechanisms of formation, and different behavior in the atmosphere. In general, the submicrometer particles were formed by gas-to-particle conversion, contained carbon and compounds of ammonia and sulfuric and nitric acids, had a relatively long residence time in the atmosphere, and tended to be regional in their occurrence. Larger particles, on the other hand, were formed mostly by mechanical processes, contained self-derived species such as carbonates and silicates, had fairly short residence times, and tended to be of local origin.

Sulfate Formation. Atmospheric sulfates, which are present in varying amounts in suspended particulate matter and in rain or snow, are composed of (1) primary sulfates formed within sources of emission to the atmosphere and (2) sulfates formed in the atmosphere, primarily by the oxidation of SO_2. Oxygen isotopy was used to investigate mechanisms by which these sulfates may be formed. The resulting information was expected to be useful in formulation of pollution-control strategies.

There are numerous pathways by which primary and secondary sulfates could be formed, which complicates research on the subject. Initial studies were aimed at the oxygen isotopy of sulfur dioxide to sulfate reactions that were amenable to laboratory simulation. The early results indicated that primary sulfates resulting from the formation of SO_3 at elevated temperatures, followed by hydrolysis, could be distinguished from other atmospheric sulfates.

Mechanisms of primary sulfate formation were studied by comparing the oxygen isotopies of sulfates formed from SO_2-air-water vapor mixtures at temperatures of 450-600°C in the presence of three different catalysts: platinum, Fe_2O_3, and charcoal. The results indicated that, of all the sulfate formed in this temperature zone in the exhaust chambers of power plants and other combustion systems, only the sulfates that are produced through the intermediate formation of gaseous SO_3 have a relatively high ^{18}O content. Application of the oxygen isotopy technique to samples of precipitated water and suspended aerosols at Argonne in 1976-1978 produced data on the relative percentages of primary sulfates in the two types of samples. In general, the percentage of primary sulfate in the precipitated water did not show a strong seasonal variation, but in the aerosol samples the level was about 10% in the summer and 30-40% in the winter. This work showed that the oxygen isotopic technique could be very useful in assessing the relative contributions of "local" *vs.* "distant" sources of sulfur emissions at a receptor site.

When the oxygen-isotopic technique was used to determine the scavenging effect of rain on sulfate in a power-plant plume, the results showed that rain was about 300 times more effective in removing primary sulfates than in removing the SO_2. In the power-generating units, about 48% of the scavenged sulfates originated as SO_3. In the exhaust from a diesel engine, about 60% of the total sulfate was SO_3-derived primary sulfate.

Oxygen isotopic studies were conducted on the effect that diurnal variations in the oxygen isotope ratios ($^{18}O/^{16}O$) in water vapor has on the ratios in atmospheric sulfates. Field experiments, contrary to earlier results, indicated no consistent differences in the day and night samples. Also, no excessive enrichment of ^{18}O was found in water vapor samples collected on charcoal.

One of the questions involved in the atmospheric-pollution studies was the degree to which hydrogen peroxide (H_2O_2) might be responsible for the oxidation of SO_2 to sulfate, and whether this oxidation limits the aqueous-phase formation of sulfuric acid. This subject was of interest to the U.S. Environmental Protection Agency because it might show whether efforts to reduce atmospheric formation of sulfuric acid and its subsequent deposition should focus on the reduction of H_2O_2 rather than SO_2 in the atmosphere. A determination was made of the dependence of the $^{18}O/^{16}O$ ratio in the sulfate on that in the H_2O_2 by which sulfate was formed by aqueous oxidation of SO_2. A relationship was derived whereby the relative importance of the H_2O_2 oxidation in the atmosphere could be assessed by measuring the isotopic ratios in the sulfate, water, and H_2O_2 of rainwater. Some preliminary results were obtained on rainwater.

Nitrogen Chemistry. The initial objective of this study was to determine the vapor-liquid equilibria for the system NO_x-H_2O-SO_x-NH_3 at compositions that could be present in the ambient environment, with special attention to compositions involved in acid rain. The program included the development of instruments for measuring individual nitrogen oxides at low concentrations in a mixture. One such instrument was a chemiluminescent NO_x analyzer, which is based on a chemiluminescent reaction between NO (nitric oxide) and O_3 (ozone). The instrument was developed to a stage where it could be used in air at atmospheric pressure, but further work was needed to adapt it to the vapor-liquid equilibrium studies.

An interesting application of this device was its use to investigate the production of NO and NO_2 by lightning in a thunderstorm. By correlating the NO_x measurements with other chemical and meteorological data, the amount of NO_x generated in a single lightning flash was estimated to be 1×10^{26} molecules/flash.

The catalytic effect of nitrogen oxides on the conversion of sulfur dioxide to sulfuric acid was studied by means of a computer model that supposed a sequence of reaction steps with known or estimated rate constants. The parameters in the model were evaluated using part of the experimental data, and then the model successfully calculated the kinetic data from the rest of the experiments.

Ammonia is a ubiquitous species in the atmosphere that has a role in the neutralization of acid gases and particles formed by industrial and automotive pollutants. It may also be a source of nitrate in the atmosphere through oxidation by various mechanisms. Preliminary results indicated that NO is produced by the photochemical oxidation of ammonia.

One of the reaction systems of interest was the conversion of SO_2 to sulfate in the presence of H_2O and O_2, and possibly the influence of NO_2 and $HONO_2$. When gas mixtures of NO_x, H_2O, and O_2 were prepared for this work, the NO_2 concentration decreased much more rapidly than would be expected on the basis of rate constants published for the homogeneous NO-NO_2-HONO-$HONO_2$ system. When surface reactions of the nitrous and nitric acids were taken into account, however, the calculated results agreed well with the experimental data. These experiments indicate that heterogeneous reactions may have an important effect on the NO_x-SO_x interactions under certain conditions.

Carbon Studies. Carbon was of major interest as a component in atmospheric aerosols, and also in other areas of materials science. Therefore, an investigation was begun on the reactions of carbon with a variety of gases, including SO_2 and NO_2. The carbon was deposited on the internal reflection units in the FTIR spectrometer either by sputtering or by deposition as carbon black from a methane flame. Unlike the flame-deposited carbon, the sputtered carbon was quite reactive when exposed to SO_2 or NO_2. After exposure to air, to a CO_2-O_2-SO_2 mixture, and then to air again, absorption bands corresponding to CO, sulfite ion, sulfate ion, and NO_3 were measured. This technique permitted detection of gases such as SO_2, CO_2, CO, and H_2O

vapor simultaneously with the solid-phase material.

Personnel. In the 1980s, Romesh Kumar became the Group Leader of this program, replacing Paul Cunningham who had also been the Section Head in charge of the Analytical Chemistry Laboratory. Members of the group included Dave Drapcho, Ben Holt, Bill Hubble, Ron Isaacson, Stan Johnson, Wally Kremsner, Teresa Lang, Eugene Nielson, and Stan Siegel. Tony Engelkemeier, Don Graczyk, and Ben Tani of the Analytical Chemistry Laboratory, and Louis Fuchs of the Chemistry Division, also contributed significantly to this effort.

GEOCHEMISTRY OF THERMAL SYSTEMS

This was a new basic research program that originated in 1985. Its objective was to investigate specific chemical phenomena in geothermal systems through detailed chemical and isotopic analyses of rock, minerals, water, and gases sampled from appropriate geological sites. The resulting data were used to characterize the dynamic processes that occur with time in these natural systems. This type of information could prove useful in practical areas such as nuclear waste isolation, geothermal energy exploration and development, and exploration for minerals, oil, and natural gas. In some respects, this work was similar to the environmental and zeolite studies in that the compositions of the materials and/or the physical conditions for their formation often show considerable variability in natural systems. These factors create a challenge in the interpretation of data and mathematical modeling, but do not preclude an ability to generate an understanding of these systems and to make certain predictions concerning their future behavior. The arsenal of analytical methods used in these studies included neutron activation, inductively coupled plasma/atomic emission spectrometry, atomic absorption spectrometry, electron microprobe analysis, mass spectrometry, and alpha spectrometry.

An interesting analytical sampling technique was developed whereby fluid inclusions in a polished mineral specimen could be released by a laser and analyzed in a noble gas isotope ratio mass spectrometer. Before analysis, the samples were subjected to neutron irradiation to produce isotopes of Ar, Kr, and Xe from selected elements such as K, Ca, Cl, Ba, Br, U, and I, which can then be analyzed together with the naturally occurring gases in the same inclusions.

Uranium and Thorium Series Disequilibrium Studies. The first studies were made on samples of drill cores from the Biscuit Basin Rhyolite Flow in Yellowstone National Park. Data from one core were consistent with a model of single-stage uranium addition involving water/rock mass ratios as high as 10^5 or more. Results from the other core showed much less uranium mobility. Preliminary dating studies of ^{238}U hydrothermal precipitates, which were conducted in collaboration with Carl Binz of Loras College, Dubuque, Iowa, suggested that none of the samples was older than 200,000 years. The ^{238}U disequilibrium among its decay series members was used to determine the time scale of hydrothermal activity as part of the Continental Scientific Drilling Program (CSDP). The sites of interest were at Yellowstone, Wyoming; Valles, New Mexico; and Long Valley, California. Samples were collected in 1986 from Yellowstone, Long Valley, and the caldera at Valles. Most of the veins were older than 300,000 years, which was the resolution of the technique. Analyses of ^{226}Ra in Yellowstone samples suggested that a redistribution of radium and barium had occurred.

In 1987, large-volume thermal water samples were collected from 13 hot springs and geysers in Yellowstone and analyzed for radium, thorium, and uranium to determine the geochemical behavior of these elements, as well as their potential application as tracers and indicators of groundwater flow rates in hydrothermal systems. Studies (still under the CSDP) continued in 1988 with investigations of the disequilibrium between ^{226}Ra and ^{230}Th in drill core samples of hydrothermally altered rhyolite from Yellowstone. The $^{226}Ra/^{230}Th$ ratios showed that the cation compositions of samples were consistent with an ion-exchange equilibrium between zeolites and coexisting thermal waters. Data were obtained on equilibrium constants and diffusivities. The zeolite-water exchange appeared to have been operating for at least 8,000 years.

The $^{230}Th/^{234}U$ method was well established for determining the age of travertines from 10,000 to 350,000 years. The use of ^{226}Ra for determining lower travertine ages was investigated to determine the potential usefulness of this method. These studies, conducted on thermal waters and travertines from Mammoth Hot Springs in Yellowstone Park, indicated that ^{226}Ra activities at depths to 70 m below the surface were much lower than expected. Nevertheless, the rate of travertine deposition (about 1 cm/yr) was accurately estimated from the slope of the best-fit curve in plots of the ^{226}Ra activity *vs.* depth.

Oxygen and Carbon Isotopic Studies. Studies of oxygen isotope distribution were completed in a collaborative effort with K. Muehlenbachs of the University of Alberta. In the initial investigations, some disequilibrium was observed between thermal water and most solid phases. Further work with silicate and carbonate samples from Yellowstone showed that the $^{18}O/^{16}O$ ratio in minerals decreased with depth, reflecting the thermal gradient. Most SiO_2 minerals had $^{18}O/^{16}O$ ratios higher than equilibrium values predicted from the temperatures and the thermal water compositions. This effect was attributed either to deposition of the minerals at higher temperatures or to water with a high ratio. The $^{18}O/^{16}O$ ratio decreased with increasing sample age. The investigators speculated that the initial deposition of amorphous SiO_2 occurs relatively rapidly after a new fracture opens and allows a flow of SiO_2-supersaturated water. When the flow decreases or stops, equilibration takes place with SiO_2 deposition.

Hydrothermal Energy and Volcanism. A catastrophic eruption of the Nevado del Ruiz volcano in Colombia occurred on November 13, 1985. In a cooperative effort with Prof. S. Williams of Louisiana State University and Dr. Y. Sano of the University of Tokyo, thermal water and gas samples were taken from the Nevado del Ruiz volcano and several other Central American volcanoes. In water samples taken from hot springs at the Nevada del Ruiz volcano, a significant increase in the SO_4/Cl ratio occurred about 13 months after the eruption, and then decreased to the normal value within a month. The change may have resulted from the input of a pulse of magmatic SO_2, but there was no change in the $^{34}S/^{32}S$ ratio.

Deep Carbon Geochemistry. The introduction of magmatic heat into potential source rocks for oil and gas may cause significant hydrocarbon deposits in situations where such deposits would not be produced by conventional oil- and gas-forming processes. The compositions, quantities, and mobilities of hydrocarbons released by magmatic heat are not well known and cannot be predicted from currently available data.

Analyses were initiated on samples from approximately 1.6-billion-year-old sedimentary rocks of the Animikie Basin that were intruded by magmas of the Duluth complex in northern Minnesota. The analytical studies were performed on carbon and carbon isotopes, including characterization of kerogen fractions. The results suggested that large-scale mobilization of hydrocarbons might have occurred in this formation. In another study, step-combustion experiments at temperatures between 200 and 550°C generated data that would be useful in determining the degree of fractionation of carbon isotopes that may occur in high-temperature geological environments.

Personnel. In general, this was a highly collaborative effort that involved many outside individuals and organizations. The Principal Investigator was Neil Sturchio. Martin Seitz had a part in the initial stage of the program. Teofilo Abrajano, John Bohlke, a postdoctoral appointee, and Ben Holt (STA) also contributed to the effort. Much of the effort was supplied by the Analytical Chemistry Laboratory.

Analytical Chemistry Laboratory

The Analytical Chemistry Laboratory (ACL) entered the 1980s with a staff of about 25 people and Paul Cunningham as Manager. In 1982, as mentioned earlier, Paul departed for Los Alamos and Dave Green took over the operation. At that time, about two-thirds of their work was for the Division, notwithstanding the fact that the ACL had been designated as a central facility for the entire Laboratory. Like many of the other programs in the Division, ACL was subject to a financial squeeze in 1982, which resulted in the loss of two staff members. Because the work load was somewhat variable, however, the ACL was able to use the services of well-qualified, part-time analytical chemists who had retired or were available for other reasons to even out the load at a relatively low cost. After that time, the group grew steadily, reaching a manpower level of about 50 by the end of the decade.

In 1980, an organic analysis facility in Building 221 had just been completed, in which a primary consideration was for safety in handling carcinogenic, toxic, or flammable materials. The facility was equipped with a computerized chromatographic-mass spectrometry system.

The capabilities of ACL for various types of work were increased in the early 1980s by the acquisition of several new pieces of equipment, including a thermal ionization mass spectrometer; an inert gas fusion system for hydrogen, oxygen, and nitrogen analyses; a pulsed nitrogen laser unit for determinations of uranium in soil and groundwater; an inductively coupled plasma/atomic emission spectrometer; and an energy-dispersive X-ray analyzer to be used with the scanning electron microscope. Various other equipment upgrades and acquisitions continued throughout the 1980s, with emphasis on increased use of computers for automation and data processing.

The ACL consisted of three groups as of 1982:

1. A Chemical Analysis Group that was involved primarily with wet chemical and instrumental analysis.
2. An Instrumental Analysis Group that dealt mostly with nuclear decay counting, mass spectrometry, gas chromatography, neutron activation, inert gas fusion, and isotopic analysis.
3. An Organic Analysis Group that conducted analyses of complex organic mixtures, including fossil fuels, toxic materials, pesticides, and drugs.

Later on, in about 1988, a fourth group, called the Environmental Analysis Group, was added to respond to the growing amount of environmental research in the Division.

The capabilities of ACL were heavily publicized in the early 1980s to attract more work both from other ANL divisions and from outside sources. At the time, the two major efforts were on the proof-of-breeding program in CMT and work for the Energy and Environmental Systems (EES) Division. Large efforts were also required on the TRIO experiments in the nuclear fusion program, analysis of Integral Fast Reactor fuel for oxygen and hydrogen, and support for the TRUEX program, all discussed earlier.

The ACL began doing a substantial amount of work for outside (non-ANL) organizations that included the Environmental Protection Agency (EPA), the Nuclear Regulatory Commission (NRC), General Public Utilities (GPU), Fermilab, the Commonwealth Edison Company, the University of Illinois, the Gas Research Institute, the National Park Service (NPS), and others.

The following is a recap of most of the work performed by ACL in the 1980s. Many of these items were mentioned in the previous technical discussions; other items were not reported previously because they were funded directly from sources other than CMT.

- Proof of Breeding Project
- Preparation of Isotopic Standards for the Applied Physics Division
- Workshops for Division of Educational Programs
- DOE Environmental Survey (of about 40 DOE sites)
- Detector Upgrade of Thermal Ionization Mass Spectrometer
- Accelerator Mass Spectrometer Dating with Calcium-41
- Development of Masked Multichannel Scaler (MMS)—subject of an IR-100 Award from *Industrial Research Magazine*
- Analytical Support for Region V EPA Projects
- Multiple Analyses of Uranium Alloys for the ANL Program on Reduced Enrichment for Research and Test Reactors
- Gas Analysis for Argonne Premium Coal Sample Program
- TRIO Program (Fusion Reactor Blanket Studies)
- New Method of Isolating and Purifying Uranium for Isotopic Analysis
- Analysis of High-Temperature Semiconductors
- Analytical Method for Organic Carbon in Lake and River Sludges
- Groundwater Analysis for Geosciences Programs
- Enriched Uranium Booster Target Project
- Loss of Cooling Accident Project
- Molten Corium-Concrete Interaction Program
- Development of a Dioxin Cleanup Method
- Pyrochemical Process for Integral Fast Reactor
- High-Sulfur Dry Scrubber Test
- National Acid Precipitation Assessment Program (National Park Service)
- Spent Fuel Leaching Tests
- Americium Certified Reference Samples for New Brunswick Laboratory
- TRUEX Process Flowsheet Development
- Energy from Municipal Waste Program
- Performance of Fuel in a High-Burnup LWR Reactor
- Polyurethane Foam Pyrolysis Studies
- Quantitation PCB/Pesticides in Environmental Samples
- Fourier Transform Infrared Microscopy
- Advanced Instrumental Methods for Analyzing Organics in Solid Waste
- Analysis of Dioxins from Municipal Waste Incineration
- Characterization of Organics in Solid and Sludge Wastes from Emerging Energy Technologies
- Plastic Pipe Program
- Rocky Flats Plutonium Metal Exchange Program
- Determination of ^{226}Ra and ^{228}Ra in Water
- Determination of Protactinium in Environmental Samples
- X-Ray Diffraction of Superconducting Materials
- Environmental Radiochemistry
- Plutonium Residue Recovery Program
- Advanced Mass Spectrometry Capability
- Analysis of Dioxins from Municipal Waste Incineration
- Chemistry of Bottom Sediments of the Chicago River
- Technology for Verification of Treaties Limiting Chemical Weapons

This list, although not all-inclusive, illustrates the variety and scope of the ACL programs. The organization of ACL at the end of the 1980s is shown in Table 5-10.

Table 5-10. Organization of the Analytical Chemistry Laboratory in 1989

Dave Green, Manager
Fred Martino, Assistant

Amrit Boparai		Bob Heinrich
Peter Aznavoorian		Del Bowers
Gerald Baudino		Harvey Goodspeed (STA)
Susan Bussey (STA)		Francis Markun
Jack Demirgian		Carmen Sabau
Laura Lamoureux		Stan Siegel (STA)
Gerry Reedy (STA)		Elane Streets
John Schneider		Ben Tani
Christine Snyder		Tony TenKate
Stephen Swanson		Lesa Wetter
Yifen Tsai (STA)		
Ronald Wingender (STA)		Pete Lindahl
		Ralph Bane (STA)
Don Graczyk		Ken Jensen (STA)
Tony Engelkemeier		Jane Marr
Alice Essling		Steven Newnam
Howard Flotow (STA)		Lynn TenKate
Irene Fox		
Doris Huff (STA)	Admin. Asst.:	Vanessa Arzate
Ed Huff	Secretaries:	Lilia Barbosa
Everett Rauh		Alice Birmingham
Laury Ross (STA)		Virginia Strezo
Florence Smith		(Janet Steinquist was the
Charles Stevens (R.A.)		lead secretary up to 1989.)

1990-1998: INTO THE FUTURE

1990-1998

(top) Samples being prepared for atomic absorption analysis.

(bottom, left) Inert atmosphere glove-box facility for investigating molten metal/molten salt processes and for preparing radioactive wastes for disposal.

(bottom, right) Apparatus developed for treating a wide variety of spent nuclear fuels.

6

1990-1998: Into the Future

The year 1998 marks the 50th anniversary of the Division, and we have regarded that year as the cutoff date for this history. (The work in the technical programs is generally reported up to the end of 1997.)

As the United States entered the 1990s, the economy was healthy, and it appeared that there might be some respite from military conflicts, but, once again, that was too good to be true. On August 2, 1990, Iraq invaded Kuwait, and the U.S., with the approval of its allies, responded with a military action that sent the Iraqis back home. Other foreign military involvements included an effort to protect the delivery of relief supplies to Somalia, and peacekeeping duties in Haiti and Bosnia. By 1991, the Soviet Union had broken up into the Commonwealth of Independent States, and Boris Yeltsin was elected president of the Russian Republic on July 10 of that year. On February 1, 1992, Presidents Bush and Yeltsin issued a joint statement that the Cold War was over.

In the 1992 U.S. presidential election, Bill Clinton and Al Gore defeated incumbents George Bush and Dan Quayle, which gave the Democrats control of the administration and both houses of Congress. In 1994, however, the Republicans captured majorities in the House and Senate. Clinton and Gore were reelected over Bob Dole and Jack Kemp in the 1996 election, while the Republicans maintained control of Congress. During the mid-1990s, the political atmosphere became highly charged, and some of the activities of the Clinton administration (referred to as "Whitewater") were under investigation by Special Prosecutor Kenneth Starr. The 1998 election resulted in a small gain in House seats for the Democrats, but both houses remained under Republican control.

The economy flourished in the 1990s, with a skyrocketing stock market. Although some nuclear reactors had been shut down, total nuclear power production was increased due to higher outputs and plant factors. In the computer market, IBM (International Business Machines, Inc.) cut 25,000 jobs as a result of a $6 billion loss. By the 1990s, IBM had lost its dominance of the computer industry, and Microsoft, Inc., was rapidly taking its place.

At Argonne, Laboratory Director Alan Schriesheim decided to resign in June 1996. He was succeeded by Dr. Dean Eastman, who resigned in July 1998. As of December 1998, Associate Laboratory Director Frank Fradin was serving as acting Laboratory Director. In the Chemical Technology Division, Martin Steindler resigned as Division Director in 1992, and was replaced by Jim Battles, who served in that capacity until 1994, when Jim Laidler assumed that responsibility.

The experimental work in the Division in the 1990s was mostly a continuation of existing projects, although the recycling of IFR fuel was phased out, and more emphasis was placed on environmental problems, such as the treatment of nuclear and other hazardous waste materials.

NATIONAL AFFAIRS

In the 1990s, the U.S. economy was in good shape, with reasonable interest and inflation rates and a low unemployment level of about 5.5%. The stock market was healthy, and underwent a boom in which the Dow Jones industrial index exceeded 9,000 in 1998. The market showed signs of faltering in 1998, however, due to world-wide economic problems that had arisen primarily in Asia. One matter of growing concern was the rapidly increasing national debt (about 5.5 trillion dollars in 1997). Both political parties espoused a balanced budget, but their opinions differed over priorities. The decade was marked by anomalous weather conditions, including severe flooding in several areas, which was attributed to the El Niño heating effect in the Pacific Ocean.

As the nation entered the 1990s, its foreign affairs appeared to be going well. General Noriega of Panama surrendered on January 3, and the United States signed an arms control agreement with the Soviet Union. On August 2, 1990, however, Iraq suddenly invaded and occupied the monarchy of Kuwait, a tiny, but oil-rich country. The U.S. immediately launched "Operation Desert Shield," in which over 230,000 troops were sent to Saudi Arabia and were followed by another 200,000 when President Saddam Hussein continued the Iraqi military buildup. The U.S. action was sanctioned by a U.N. Security Council resolution. A ground war called "Desert Storm" began on February 24. The Iraqi forces were soundly defeated within 100 hours. Bush elected not to extend the fighting to Baghdad, leaving Hussein, as it turned out, to continue his mischief. In 1992, the U.S. sent troops to Somalia to protect the delivery of relief supplies, but the operation was limited and relatively ineffective, with a final evacuation of the troops in 1995. A U.S.-led multinational military force was sent to Haiti in 1993 in a peacekeeping mission to reinstate President Aristide after he had been ousted by a military coup. In 1995, the U.N. sent 60,000 troops, including 20,000 Americans, to Bosnia in a peacekeeping effort. In Russia, both the economy and the government were in bad shape, and the people had difficulty in adapting to a capitalist system after generations of totalitarian rule. It seems likely that the present regime will remain in power, but the situation remains tenuous.

At the beginning of the decade, the White House was occupied by George Bush, his popular wife, Barbara, and his dog, Millie, who was credited with writing a book. Dan Quayle was Vice President, and the Democrats controlled both houses of Congress. The 1992 election was a contest among Bush, Bill Clinton, and third-party candidate Ross Perot. Some pundits attributed Clinton's victory to Perot, whose campaign apparently siphoned a substantial number of conservative voters away from Bush. The Democrats retained control of Congress. Although Clinton was a popular president, the 1994 mid-term election was a landslide victory for the Republicans, who captured both houses of Congress and many state governorships. Clinton and Vice President Albert Gore were reelected in 1996, and there was little change in the makeup of Congress. During these power shifts in Washington, relationships between the Democrats and Republicans became more acrimonious as both parties struggled to keep or gain power and to promote their different ideologies. Clinton and his administration were accused of a series of wrongdoings known collectively as "Whitewater," and Kenneth Starr was appointed as a Special Prosecutor to investigate the charges. At the present time, the matter has not been resolved. Both houses of Congress remained under Republican control in the 1998 election, but the Democrats gained five or so seats in the House of Representatives.

Although the U.S. space program also felt the effect of federal belt tightening, NASA made a number of remarkable achievements in the 1990s. The Hubble Space Telescope, which could make astronomical observations free from atmospheric effects, was launched in 1990. After the main lens was found to be defective, crew members of the space shuttle *Endeavor* made the necessary repairs in 1993.

During the period 1990-1997, the U.S. space shuttles *Columbia, Atlantis, Discovery,* and *Endeavor* completed a total of nearly fifty missions that involved military applications, satellite launchings, a wide variety of scientific experiments, and other activities. The Russians had a similar program involving seventeen space shuttles designated *Soyuz TM-9* through *TM-25*. In the mid-1990s, the U.S. teamed up with the Russians in manning the Russian Space Station MIR, which was about nine years old at the time and has been showing its age by a series of scary, but not fatal, mishaps. In addition to the manned space missions, the U.S. and other countries deployed several unmanned space probes. Some of the satellites were used for military and communications purposes; most were used to obtain all kinds of information on space phenomena, including the poles of the sun, short wavelength and magnetic effects, ultraviolet radiation, moon mapping, weather effects, X-ray sources in the Milky Way (stars, pulsars, black holes), asteroids, comets, auroras, solar winds, and radio signals. Two probes of particular interest to the public were used to explore the surface of Mars. The result was a series of fascinating pictures and chemical analyses of some of the rocks. A somewhat startling revelation toward the end of the 1990s was evidence that large quantities of water (as ice) may be present on the back side of the moon.

ENERGY AND THE ENVIRONMENT

The total consumption of energy in the United States in 1995, and the amounts used for electric power generation are shown by source in Table 6-1. Coal is the source of about 55% of the energy used for power generation, and the tonnage per year continues to increase. Petroleum and natural gas account for a larger percentage of the total energy use because of industrial and domestic heating and transportation consumption. Somewhat surprisingly, nuclear, in spite of a decreasing number of reactors in operation, increased by about 17% in the period 1990-1995 due to upgraded performance and higher plant factors. Despite their popular appeal, solar and wind contributions are negligible.

Perceptions about energy and the environment both by the public and by many scientists were changing significantly during the 1990s. According to some experts (*ANS Fuel Cycle and Waste Management Division Newsletter,* Spring 1998), sources of fossil energy (coal, oil, gas) and nuclear energy (uranium and thorium) are being found to exceed greatly the amounts previously believed to be available. Their conclusion was that development of better technologies to produce electricity cannot be justified on the basis of resource depletion. At the same time, a White House panel (*Chemical & Engineering News,* May 5, 1998) stated that the Nation has no coherent energy policy and may be losing out in a multibillion dollar global energy market. Both groups felt that U.S. research and development on energy, including nuclear, is seriously underfunded. Energy research funding in the U.S. decreased in real dollars by about 75% between 1978 and 1996, and is less than half of that in Japan. Concerns about global warming by greenhouse gases (mainly CO_2) seem to have softened slightly the antinuclear attitude by some environmental groups, although they still raise questions about nuclear proliferation and waste disposal.

If resource depletion is not a problem, there appears to be less justification for improved breeder reactors such as the IFR, although they may have significant advantages in safety, reduction or elimination of long-lived transuranic elements, and low waste volumes. An ongoing debate continues as to whether existing LWR designs, advanced LWRs, LMFRs , LMFBRs, or some other type should be the reactors of the future.

Table 6-1. U.S. Energy Consumption in 1995 by Source

Source	Total		Power Generation	
	10^{15} Btu	%	10^{9} kWh	%
Petroleum	34.6	38.3	61	2.0
Natural Gas	22.2	24.6	307	10.2
Coal	19.6	21.7	1,653	55.2
Nuclear	7.2	8.0	673	22.5
Hydroelectric	3.4	3.8	296	9.9
Biofuels	2.8	3.1	-	-
Other[a]	0.5	0.5	6	0.2
Total	90.3	100.0	2,995	100.0

[a]Other renewable sources (geothermal, solar, wind, etc.).

Dr. Lawroski may have put his finger on the problems of the IFR when he once said it was probably 25 years ahead of its time. Nuclear fusion power is an attractive concept, and its principles are well understood, but its practical application remains an elusive goal. Whichever of these reactor systems may be selected for future use, the Chemical Technology Division is well positioned to do development work on its fuel cycle—it has had experience with all of them.

With respect to the environment, a great deal of progress has been made on the reduction of air, water, and land pollution. Although much remains to be done, many of the problems have been identified, and solutions are being implemented. For several years, loss of ozone in the stratosphere has received a great deal of attention, and chlorofluorocarbons have been banned from general use in the U.S. and most other countries. The chemistry of interactions between chlorine compounds and atmospheric ozone is complex and poorly understood, so it is difficult to determine what measures, if any, should be taken. The emphasis now, however, has shifted toward global warming due to CO_2 in the atmosphere. It is not entirely clear that global warming exists, and if it does, whether it is a result of CO_2 produced by human activities. The official position of the U.S. government is that global warming is, indeed, a threat, and that immediate action must be taken to reduce CO_2 emissions.

Nuclear waste disposal has been an ongoing national problem for both technical and political reasons. The Federal Government has not yet been able to make good on its promise to accept and dispose of commercial nuclear wastes in exchange for a tax that was levied for this purpose on all the nuclear power that has been produced to date. These wastes are currently being stored underwater as spent fuel assemblies at various locations. Yucca Mountain in Nevada was selected as a site for permanent burial of nuclear wastes, but it has become controversial on both political and technical grounds, and no alternatives have been deemed acceptable at the present time. There are also differing points of view about the acceptability of various physical and chemical forms of nuclear wastes for permanent storage. Nuclear waste forms and the suitability of geological formations for this purpose are a major program in the Chemical Technology Division. Other studies in the Division have concerned the separation of transuranic elements, which have very long half-lives (*i.e.*, 10^5 years or so), and fission-product elements, which decay to acceptable levels in a few hundred years. Such a separation might simplify the nuclear waste disposal problem.

The Division has been involved in a wide variety of other environmental research activities such as air pollution, municipal waste incineration, and disposal of non-nuclear hazardous wastes. The coal combustion, fuel cell, and battery programs in CMT have all been aimed primarily at providing technologies for cleaner air.

THE LABORATORY

In 1996, Argonne National Laboratory celebrated its 50th anniversary. One of the events associated with this anniversary was the publication of a history of the Laboratory, *Argonne National Laboratory, 1946-96,* by Jack M. Holl, University of Illinois Press, Urbana and Chicago, 1997. Any ANL employee who reads this scholarly, yet very readable, history is bound to develop a fuller appreciation of the extraordinary efforts of the Laboratory Directors and other upper management people to keep the organization afloat through crisis after crisis. Argonne was not alone in facing these problems of budget cuts, repeated changes of perceived national needs, political wind shifts, and new technological challenges; the other national laboratories had similar problems. Another readily available source of information about Argonne's history, as well as its current status, is an ANL home page on the Internet at www.anl.gov.

The Reagan budget cuts of the 1980s hit the Laboratory hard. The number of full-time ANL employees dropped from over 5,000 in 1980 to less than 3,700 during that period, but then increased again to nearly 5,300 by 1994. The big research plum that nearly everybody wanted a part of was the Superconducting Super Collider (SSC), which was to have been a huge project. Argonne's lack of success in getting in on a significant piece of this action was probably a blessing in disguise. The SSC project, first to have been located in Illinois, and then in Texas (President Bush's home state), finally collapsed due to political controversy along with technical and financial difficulties. Meanwhile, Drs. Schriesheim, Fradin, and others began developing a very broad basis of support from universities, industries, and the DOE for the Advanced Photon Source (APS) to be located at ANL. This synchrotron had a wide appeal because it would be primarily a users' facility and could be used for a remarkably wide range of research areas covering the gamut from the most basic studies to immediate applied problems. The APS has become a reality and now operates as an independent entity, under the leadership of Dr. David Moncton, within the ANL organization. Another initiative that was pushed hard by Drs. Charles Till and Yoon Chang was the Integral Fast Reactor (IFR), which was basically an advanced version of EBR-II technology. This program was being pursued quite successfully until 1993, when it was halted abruptly by the new Clinton administration. At the beginning of his presidency, Clinton announced that "We are eliminating programs that are no longer needed, such as nuclear power research and development." He did compromise, however, in that continuing work was allowed on the actinide recycling program and the processing of fuel from EBR-II, which was to be decommissioned. This, in effect, ended Argonne's long history of reactor research and development.

In June 1996, Dr. Schriesheim resigned after 12 years as the Laboratory Director. He was replaced by Dr. Dean Eastman, who had received a Ph.D. from the Massachusetts Institute of Technology in the field of electrical engineering. Dr. Eastman had been in various research and R&D management positions at IBM and had held concurrent positions as a visiting professor at MIT and the University of Chicago. In July 1998, Dr. Eastman resigned, and, as of this writing, Dr. Frank Fradin, who had served for many years as an Associate Laboratory Director for Physical Research at ANL, is serving temporarily as the Laboratory Director.

THE DIVISION

In 1995, as a part of the improvements in the general appearance of Building 205, and, in particular, the exhibit in the lobby, a new logo, shown below at the left, was adopted for the Division. This design, which was selected from several proposed by Media Services, has more visual impact than the previous triangular logo (below right) designed by Martin Steindler in about 1986. In the new logo, the "atom" represents the nuclear programs being conducted in the Division, the lightning bolt suggests the various electrochemical programs, and the flask symbolizes the Analytical Chemistry Laboratory. Another assertion of the Division's identity appears on the Internet as the home page **www.cmt.anl.gov**. The Analytical Chemistry Laboratory also has a home page: **www.cmt.anl.gov/acl/acl.htm**.

CHEMICAL TECHNOLOGY DIVISION

CHEMICAL TECHNOLOGY DIVISION APPLIED SCIENCE BASIC SCIENCE PROCESSES DEVICES

Organization

A number of changes occurred in the upper management of the Chemical Technology Division, as indicated in Table 6-2. The terms of service in this table are only approximate because changes in the assignments did not normally take place at the beginning or end of a year.

Martin Steindler announced his resignation as Division Director in 1992, and was replaced by Jim Battles, a metallurgical engineer with a Ph.D. from Ohio State University. Jim had been involved in various CEN/CMT research and development programs over many years and had served in several managerial positions, including Associate and Deputy Division Director. In 1994, Jim stepped down with retirement in mind and was succeeded by Jim Laidler, who had earned a doctorate in Materials Science at the University of Virginia. He then held managerial positions at Hanford in the area of

Fig. 6-1. James Battles

Fig. 6-2. James Laidler

nuclear fuel reprocessing before and after a stint in the Army, followed by a position at ANL in Dr. Till's office before he joined the CMT Division.

The position of Deputy Division Director was eliminated when Jim Battles became the Division Director. During the course of the 1990s, the Division had to be rather nimble in reorganizing its structure to accommodate major changes in the technical programs that were dictated by the funding agencies. By 1998, the structure of the Division seemed to have stabilized, so it now consists of the Administration and four technical departments: (1) Nuclear Technology, (2) Electrochemical Technology and Basic Science, (3) Waste Management, and (4) the Analytical Chemistry Laboratory.

Table 6-2. CMT Upper Management Changes in the 1990s

Year	1990	1991	1992	1993	1994	1995	1996	1997	1998
Division Director									
Martin Steindler	X	X	X						
Jim Battles				X	X				
Jim Laidler						X	X	X	X
Deputy Division Dir.									
Paul Nelson	X	X							
Jim Battles			X						
Associate Division Dir.									
Jim Battles	X	X							
Dave Green	X	X	X	X	X				
Jim Laidler				X	X				
Chuck McPheeters						X	X	X	X
Mike Myles			X	X	X	X	X	X	X
Assistant Division Dir.									
Ron Breyne	X	X	X	X					
Karen Larsen					X	X			
Mike Boxberger							X	X	X

As of 1998, the administrative functions of the Division continue to be the responsibility of Mike Boxberger, assisted by Sofia Napora, Patricia Halerz, and Martha Teitlus. Leo Morrissey handles the Financial and Procurement Services with the help of Lauren Ambrose, Vanessa Mendez, Judy Popik, and Heidi Terdic. Mike Slawecki is the Manager and Janice Muller the Deputy Manager of the Facilities and Support Services Group, which includes Kurt Alford, Ronald Tollner, and Irma Zepeda. Jan is also the Division Environmental Compliance Officer. Ray Wolson is the Safety Officer, assisted by John Rajan and Gale Teats. Dick Malecha and Art Frigo manage the Engineering Design Group, which includes Dan Preuss and Yolantha Siegfried. Roberta Riel is the Quality Assurance Representative. The Computer Group, headed by Steve Gabelnick, consists of Glen Chapman, Jacqueline Copple, Joe Kulaga, and John Osudar. Joe Harmon, assisted by Maria Contos, continues as the Manager of the Editorial Services Group.

Personnel

During the period 1990-1998, the personnel level of CMT (all categories) decreased from about 260 to 235. During this same period, the percentage of Special Term Appointees (STAs) increased from about 14 to 20, which means the level of effort decreased somewhat more than the total figures would suggest, because STAs are generally part-time employees. Extensive use was made of temporary personnel such as the STAs, postdoctoral appointees, students, visiting professors, and other scientists. Several programs were staffed almost entirely by temporary people. There was also much collaborative work with people from other ANL divisions and outside organizations.

When one considers the loss or reduction in funding for some of the major CMT programs in the 1990s, such as the IFR, batteries, fuel cells, fossil fuel research, and fusion power, the Division appears to have survived remarkably well. The age distribution of people in the Division has been such that retirements have been peaking in the 1990s—it seems as if another retirement party is announced almost weekly.

The women in the Division who perform the secretarial and administrative tasks tend to be unsung heroines in that their names are not mentioned in the publications, which are the final products of the Division's work. Table 6-3 lists those who were in the Division through 1998.

Dr. Laidler instituted a loose-leaf *Members of the Chemical Technology Division* book, which was an inspired idea. This has proved helpful to everybody, but especially to new employees and to part-time people who do not have daily contacts with other members of the Division. Oftentimes one recognizes names that appear on reports and elsewhere and recognizes faces in the hallways, but has no other easy way of connecting them. One of the items to be listed on these sheets is outside interests. Perhaps one can gain a bit of insight into the types of people in CMT from this information. Because of some differences in the ways of categorizing such activities, an accurate count is difficult, but Table 6-4 gives a general idea of what many CMT people like to do in their spare time. Other activities that received five responses each were boating & canoeing, flying, softball, dancing, photography, and hunting. Many respondents listed home repair and maintenance, but that is often neither optional nor recreational.

In Retrospect

It may be appropriate at this point to reflect on some of the changes that have transpired over the past 50 years. Several long-term members of the Division were asked to offer their views on this subject. One of the most striking things about the Division when it was

Table 6-3. Secretaries and Other Administrative Assistants

Lauren Ambrose	Sherry Grisko	Debbie Rausch
Jody Bailey	Patricia Halerz	Roberta Riel
Lilia Barbosa	Marian Harkins	Lysbeth Rizzi
Norma Barrett	Mary Ann Hejny	Sue Rura
Alice Birmingham	Helen Hill	Kay Shields
Joyce Bringle	Christine Johnson	Gina Schierbeek
Mary Burke	Linda Lane	Carol Stogsdill
J. Butler	Stephanie Malak	Marietta Strachan
Laurie Carbaugh	Susan McKinney	Virginia Strezo
Janet Carothers	M. McLaughlin	Wendy Strle
Loretta Cescato	Vanessa Mendez	Martha Teitlus
Dianne Cheek	Jan Muller	Heidi Terdic
Maria Contos	Sofia Napora	Donna Tipton
Marilyn Elison	Marilyn Osborn	Sandra Tummillo
Elaine Estand	Maria Pacholok	Cindy Wesolowski
Anita Franchini	Judy Popik	Mary Lou White
Guadaluipe Franchini	Mary Powell	Irma Zepeda

Table 6-4. Outside Interests of CMT People

Reading	40	Volunteer Work	20	Basketball	9
Music (general)	36	Tennis/Raquetball	18	Theater/Movies	9
Traveling	33	Computers	16	Music (classical)	9
Bicycling	30	Swimming/Scuba	15	Sailing	8
Hiking/Walking	28	Golf	14	Bowling	8
Gardening	27	Woodworking	14	Volleyball	8
Camping	24	Cooking	14	Drawing/Painting	7
Church/Bible	24	Sewing/Crafts	14	Ultimate Frisbee	7
Skiing	24	Running	12	Soccer	6
Fishing	22	Cars	10		

first formed and in the early fifties is the ages of the employees. The people in upper management were in their 20s and 30s, as were most of the other staff. Those people represented a large peak in the age distribution curve that persisted through the years until it was finally diminished by retirements and other factors. One item of interest that arose in the interviews with the CMT staff was that, particularly in the earlier days, there were certain "movers and shakers" who had a major effect on the work of the Division. Among those who were mentioned specifically were Hal Feder, Milt Levenson, Carl Crouthamel, and Elton Cairns.

When the Division was formed, it had a single mission—to develop solvent-extraction processes for the decontamination and recovery of uranium and plutonium from spent reactor fuels. The scope of the mission was then expanded to include fluoride volatility and pyrometallurgical processes, and at about the same time, to processes for feed materials and waste handling. Diversity continued in a variety of directions, including reactor safety, materials chemistry, and new analytical methods. The EBR-II project was a large Laboratory-wide undertaking in which the Division was heavily involved in the fuel recycling. Sodium chemistry and technology and reactor safety became major programs. At about that time, pressure developed to redirect the Division's efforts toward environmental programs, which resulted in the work on batteries, fuel cells, fusion power, and coal combustion. The Analytical Chemistry Laboratory became a separate entity within the Division. The nuclear work that continued was directed largely toward the processing of fuel from EBR-II, which was being decommissioned, and waste processing; some work continued on separations chemistry and contactor development. Basic research has continued at a modest level through most of the history of the Division. Computers and other solid-state electronics have had a profound effect on the way work is done, both in the scientific and administrative parts of the Division. At first, computers were used almost exclusively for mathematical calculations, and, as languages such as Basic, Fortran, Cobol, Pascal, and others were developed, many people began writing their own programs. Further development of hardware and software greatly expanded the utility of computers to include word processing, spreadsheets, graphics, design, and a host of other applications up to the present stage, which includes rather easy-to-use features such as e-mail and the Internet, as well as very powerful, highly sophisticated systems that can be interconnected with supercomputers at other sites. Mathematical modeling is a useful tool for describing and understanding all kinds of phenomena and operations, and it has been used extensively in CMT programs. The complex Gaussian molecular orbital calculations that have been made almost routinely by Larry Curtiss and others in the Division would have been impossible just a few years ago. Another area of great importance to the work in CMT is the development of new instruments and other equipment based on computer technology, especially for analytical chemistry.

Complaints are often heard about the increasing degree of micromanagement, particularly by the funding agencies. When management systems such as PERT (Program Evaluation and Review Technique) were introduced, many felt that it was fine for building a submarine or a bridge, where all the tasks were predictable and could be placed in a schedule network. The research people complained that in their work, such management systems required "inventing on schedule," which was not possible. Nevertheless, milestone schedules became a part of the promises one had to make to procure funding, and there was a tendency to be overoptimistic in order to sell the program. Over the years, management by the funding agencies became more heavy-handed, with many requirements concerning safety, training, and a host of other items, and inspections have become more common. All of this tends to make experimental work slower and more difficult, and probably discourages spontaneity.

But for those who yearn for the "good old days," there were other unpleasantries in getting the work done. Nearly all of the work was classified, which meant keeping notebooks and other records in locked files. Because many people throughout the building were working with radioactive materials, everybody had to wear special shoes or shoe covers, and nearly every lab had hand and foot counters. Food and coffee were not allowed in the offices. Group meetings and reports were weekly, and attendance at weekly seminars was compulsory. Associate Division Directors, and often the Division Director, attended the group meetings, and they were not shy about criticizing your work. There was no air conditioning except as needed for sensitive equipment.

TECHNICAL PROGRAMS

Electrochemical Technology

In the 1990s, CMT continued its investigations of various types of batteries and fuel cells for electric vehicles, load-leveling, and other energy-storage applications. This work consisted of research and development on promising systems, performance and lifetime testing, post-test cell examinations and diagnostics, modeling studies, technology transfer to industrial firms, and technical management functions for the Department of Energy (DOE).

LITHIUM ALLOY/METAL SULFIDE BATTERIES

By 1990, the technology of lithium alloy/metal sulfide batteries had matured to the point that the DOE felt it was time to push them into practical application through a variety of joint agreements with industrial firms that would expedite technology transfer into the private sector. The Li/FeS technology had developed to the small battery level, while that of the higher-performance Li/FeS_2 system was ready for scale-up to full-size cells for electric vehicle batteries.

Industrial R&D Contracts. A new industrial R&D contract was negotiated between DOE and SAFT America whereby SAFT would develop lithium/iron sulfide batteries for electric vehicle propulsion. In support of this three-year, cost-shared effort, CMT was to provide technology transfer to SAFT, technical management for DOE, and the necessary R&D to solve any critical technical problems. The contract called for delivery of three full-size Li/FeS electric van batteries for laboratory and in-vehicle testing. In addition, SAFT was to design, fabricate, and test full-scale Li/FeS_2 cells, using the CMT technology.

In 1991, the General Motors Corporation, Ford Motor Company, and Chrysler Corporation formed a partnership designated as the U.S. Advanced Battery Consortium (USABC), to accelerate the development of advanced batteries for electric vehicles. This consortium was then extended to the electric utility industry through an agreement between USABC and the Electric Power Research Institute (EPRI); this partnership pooled its resources with those of DOE. These events resulted in a major restructuring of the DOE battery R&D program, which, in turn, caused a realignment of the CMT efforts so as to support the USABC objectives. The technologies that CMT discussed with USABC included lithium/sulfide, sodium/sulfur, nickel/metal hydride, and lithium/polymer battery systems. The sodium/nickel chloride system was added to the list due to a recent technical breakthrough by CMT that had markedly increased the performance of the nickel chloride electrode.

Through most of the 1990s, the Division continued to provide technical management and support to the USABC. In 1993, CMT was awarded two contracts, in the form of Cooperative Research and Development Agreements (CRADAs) by the USABC—one for developing bipolar Li/FeS_2 batteries and the other for testing advanced batteries in the Analysis and Diagnostics Laboratory. Performance targets for the Li/FeS_2 batteries were an energy density of 300 Wh/L, a power density of 600 W/L, a lifetime >1000 charge-discharge cycles, and a cost <$100/kWh.

In-House Research and Development. The in-house development work on Li/metal sulfide batteries in the 1990s was devoted almost entirely to the bipolar Li-alloy/FeS_2 cells and batteries for electric vehicle applications. A conceptual battery design is shown in Fig. 6-3, and Fig. 6-4 is an exploded view of the internal components of the system. The bipolar cells have a uniform

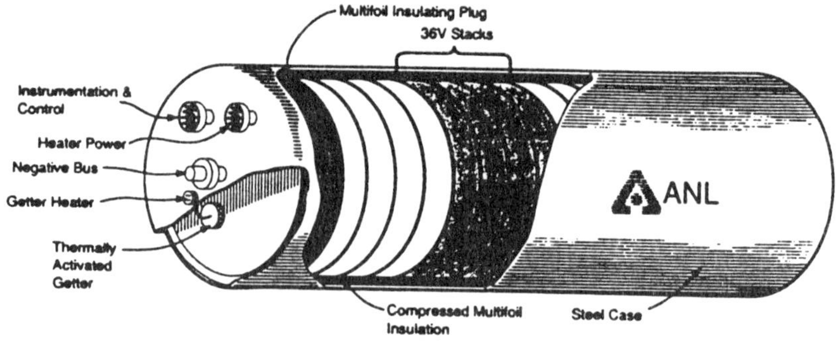

Fig. 6-3. Design of Li/FeS_2 Battery

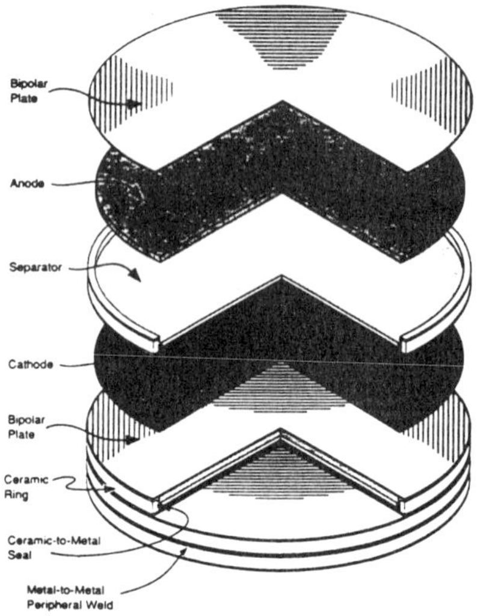

Fig. 6-4. Design of Li-Alloy/ FeS_2 Bipolar Stack

current distribution across the electrode surfaces, and this type of battery configuration has a low impedance. Calculations showed that the lower hardware weight of bipolar cells would increase the specific energy by 50% and the specific power by more than 100%. The extensive experience that had been gained previously on electrode materials, separators, electrolytes, construction materials, thermal insulation for batteries, and other items in the prismatic cells was directly applicable to the bipolar cells. One major challenge remained, however, in developing electrically insulating, hermetic seals to join the rims of the circular cells used in the bipolar system. A concentrated effort on this problem by Tom Kaun and his colleagues finally met with remarkable success. The remaining in-house work on bipolar batteries consisted mainly of scaling up the peripheral seals to full-size cells and various incremental improvements in the other cell components to increase performance and lifetime and to lower the cost.

Optimization studies were made on the electrode separators, which consist of MgO powder in a matrix of the molten salt electrolyte. Physical and electrical properties of the powder separators were investigated as a function of MgO content, electrolyte composition (LiCl-KCl, LiF-LiCl-LiBr, LiCl-LiBr-KBr, or LiF-LiBr-KBr), and temperature. Following these tests, the LiCl-LiBr-KBr (25-37-38 mol%) electrolyte was used in most of the cells at an operating temperature of 425°C. Because the long-term durability of MgO powder separators was of some concern, evaluations were conducted on a composite rigid separator consisting of MgO and chalcogenide pressed powders sandwiched between porous metal plates. These separators performed well and showed good chemical and physical stability.

The major effort was on development of the peripheral seals for the bipolar cells. To produce the seal, a ceramic ring was sealed to molybdenum on one side to form the FeS_2 electrode housing, and a steel assembly was sealed to the other side of the ring to form the lithium-electrode alloy housing. New sealant materials were developed, which were electronic insulators, bonded tenaciously to the metals and ceramics, and were resistant to chemical attack by the materials in the cell. The seals also had to have a graded coefficient of expansion to match those of the steel and molybdenum. A sealed bipolar Li/FeS_2 cell (3-cm dia) operated over 450 cycles and 2,000 h while retaining >90% of its original capacity.

Four of these cells were combined into a stack to form a module. The cells performed well, and the "lithium-shuttle" mechanism prevented overcharging of cells with no external charge equalization. In a bipolar stack, cells normally fail eventually by short circuiting, in which case the short-circuited cell ceases to function, but simply becomes an electronic conductor between the adjacent cells. The result is a small voltage loss, but the battery can continue to operate.

The next major step was to scale up the bipolar cells. A new project, funded jointly by the State of Illinois and DOE, was initiated with the objective of fabricating and testing a prototype bipolar Li/FeS_2 battery made up of 15-cm-dia cells having capacities of 50 to 75 Ah. A stack of 22 cells would constitute a 36-V module, and modules would be series-connected to form a 220-V electric vehicle battery. The calculated performance of a General Motors Impact, in which a bipolar Li/FeS_2 battery was substituted for the advanced lead-acid battery then used in the vehicle, could provide a range of 250 miles and accelerate from 0 to 60 mph in 8 seconds.

The CMT program on lithium-alloy/metal sulfide batteries was phased out in the late 1990s because the technology had reached a stage where commercial development by industry was appropriate.

Personnel. The development work on lithium/metal sulfide batteries was under the general direction of Don Vissers. Most of the modeling and evaluation studies were done by Paul Nelson, and Al Chilenskas continued to work on battery engineering and design as an STA after he retired.

Others who contributed to this program were Vincent Battaglia, Ira Bloom, Dennis Dees, Mark Hash, Gary Henriksen, Andrew Jansen, Tom Kaun, Frank Mrazek, and Laszlo Redey.

SODIUM/METAL CHLORIDE BATTERIES

The sodium/metal chloride battery system uses a molten sodium negative electrode, a β''-alumina solid electrolyte, and a solid metal chloride such as $NiCl_2$ or $FeCl_2$ as the positive electrode. The positive electrode has a secondary electrolyte (catholyte) of molten $NaAlCl_4$. This type of cell can be represented as:

$$Na//\beta''\text{-}Al_2O_3//NaAlCl_4/NiCl_2/Ni$$

The operating temperature is around 260°C. Although the $Na/NiCl_2$ system has a high theoretical specific energy of 790 Wh/ kg, the performance of existing single-tube cells was rather anemic, with a specific energy of 100 Wh/kg and a specific power of 100 W/kg. Preliminary modeling studies of the system showed that the surface area of the solid electrolyte needed to be increased to reduce the current density, and that the performance of the metal chloride should be increased.

A concerted effort was focused on ways to increase the performance of the $NiCl_2$ cathode. The following cell was used for these studies:

$$Al/NaAlCl_4\text{, additives}/Ni/NiCl_2$$

The aluminum electrode served both as a counter and reference electrode because of its well-defined chemistry and fast kinetics. The results indicated that the soluble nickel species was present, and it also made the cell construction simpler. By fabricating the nickel electrode with larger and more uniform porosity, its specific impedance was decreased to about one-half of that in state-of-the-art cells. The performance properties of porous electrodes were also investigated with a complete Na/β″-alumina, $NaAlCl_4/NiCl_2/Ni$ cell.

During charge, the nickel electrode developed a poorly conductive surface layer of the $NiCl_2$ product. An addition of 2 wt% sulfur (based on the weight of the chloroaluminate salt) to nonporous nickel electrodes resulted in a marked improvement in the capacity, impedance, and coulombic efficiency.

The electrochemical studies, which included techniques such as current-interrupted galvanic cycling and cyclic voltammetry, were complemented by direct microscopic examinations of the morphological changes that occurred in the electrodes. The data from these studies were used to develop a mathematical model of the electrochemical processes that were involved. Finite-element calculations were made on 22 cases involving a variety of design variables.

The Chemical Technology Division provided technical management and support for a contract that DOE had negotiated with Beta Power, Inc., to develop conceptual designs for high-performance Na/MCl_2 batteries. Two such designs had the potential of achieving 150 Wh/kg at the 3-h discharge rate and 150 W/kg at 80% depth of discharge, and a new contract was placed to develop the batteries for electric vehicle applications.

In 1991, a significant breakthrough was made in the form of a "synergistic" $NiCl_2$ electrode in which the usable capacity was increased by a factor of five, and the impedance was reduced to one-third of that of a baseline $Ni/NiCl_2$ electrode. The tests also showed that the operating temperature could be reduced as low as 153°C with good performance. The improved performance was effected through a combination of additives and morphological modifications. Calculated battery performance data based on the new CMT cell design indicated that the $Na/NiCl_2$ batteries could meet the mid-term USABC objectives for electric vehicles, and that bipolar cells, if developed successfully, would come close to meeting the extremely stringent long-term goals (specific energy, 200 Wh/kg; specific power, 400 W/kg).

A detailed study of the solubility of $NiCl_2$ in the chloroaluminate electrolyte indicated that the solubility can be decreased by adjusting the basicity of the system and through the use of additives. By applying this information to cell designs, research cells exceeded 500 cycles with no significant performance degradation. In addition, static, cycling, and stress-corrosion tests indicated that Ni-200 could be used safely for cell housings and current collectors. These improvements in the cell designs also permitted much faster recharging times (1-2.5 h *vs.* 10-15 h). At this juncture, the major improvements in $Na/NiCl_2$ battery technology attracted the interest of the Hughes Aircraft Co. and Eagle-Picher Industries, Inc., who fabricated some sealed cells.

By 1994, CMT had brought the $Na/NiCl_2$ technology to the stage where it could be transferred to industry. For electric cars, this system offers several advantages:

1. It has a wide range of operating temperatures (150-400°C).
2. It stores more energy per unit weight of nickel than other nickel battery systems (<2 kg/kWh *vs.* about 6.3 kg/kWh).
3. It has a high degree of overcharge protection and long cycle life.
4. The cells can be hermetically sealed and are environmentally safe.
5. The cells fail by short circuits, permitting continued operation of series-connected stacks.

6. The raw materials are abundant and relatively inexpensive.

Laszlo Redey, Jai Prakash, and Andy Jansen conducted these research studies.

OTHER ADVANCED BATTERIES

Lithium-Polymer Electrolyte Batteries. In 1992, a research program was started on the lithium-polymer electrolyte (LIPE) battery. In a CRADA arrangement with the 3M Company and Hydro-Québec, the CMT Division was to provide technical support of the development of LIPE batteries. These batteries are all-solid-state electrochemical devices that consist of ultra-thin (<100-μm) layers that are laminated together, and they operate at moderate temperatures (typically 60-80°C). The negative electrode in usually lithium but may be a lithium alloy or lithium-carbon. The electrolyte is generally a polyether such as polyethylene oxide (PEO) containing a dissolved lithium salt. The positive electrode is a composite of electrolyte (to provide conductivity) and some active material, most often a metal oxide or sulfide (*e.g.*, VO_2, MnO_2, or TiS_2).

The objective of this research was to develop a suitable reference electrode for cells of this type. A reference electrode is essential in analyzing the processes occurring within an electrochemical cell. It carries no current, but provides a constant potential against which the potentials of operating cell electrodes can be measured individually. This is a powerful diagnostic tool for studying the electrochemical reactions occurring at the individual electrodes. Development of a reference electrode for the LIPE cells was a challenging task because it had to satisfy several requirements—it had to be reversible with fast kinetics, chemically compatible with the system, and long-lived. It also required a geometry that was suitable for the thin electrodes of the cell. To pursue these objectives, an inert atmosphere glove-box facility was set up for fabrication and testing of the LIPE cells.

Lithium was an obvious candidate for the negative electrode, and several types of lithium or lithium-alloy reference electrodes were evaluated. Screening tests were conducted, in which lithium-tin alloy showed the best performance. These reference electrodes had to be deposited on a stainless steel substrate as ultra-thin (50-1000 Å) tin layers. Electrochemical tests were then performed in a cell in which the electrolyte material was PEO that contained dissolved lithium trifluoromethane sulfonate ($LiCF_3SO_3$) salt. The cell can be represented as

$$Li/PEO_{10}\text{-}LiCF_3SO_3/Li_xSn$$

where x is 8 or 10. During a coulometric titration in which lithium was transferred into the tin electrode, three voltage plateaus were observed. The third plateau, with a potential of 0.53 V *vs.* lithium, showed excellent stability over two weeks of monitoring.

Development of alternative cathodes was also a part of the LIPE project. In the operation of a lithium-polymer electrolyte cell, lithium ions are repeatedly inserted into and withdrawn from the transition metal oxide structure of the positive electrode. This process tends to produce structural changes in the metal oxide matrix, particularly under conditions of overcharge or overdischarge, so the cell operating conditions must be carefully controlled during cycling. Post-test analyses and other detailed studies of the cells were made, including modeling. Performance data based on tests of state-of-the-art materials, including α-MnO_2 positive electrodes, translated into a projected specific energy of 180 Wh/kg. Any further information on the design and performance of these batteries is considered proprietary.

Dennis Dees was the leader of the group working on lithium-polymer electrolyte batteries, which also included Khalil Amine, Vincent Battaglia, Rex Gerald, Gary Henriksen, Andrew Jansen, Christopher Johnson, Arthur Kahaian, Keith Kepler, Robert Klingler, Gene Nielson, Jai Prakash, Laszlo Redey, Mike Thackeray, Robert Turner, and John Vaughey.

Lithium-Ion System. Rechargeable "lithium-ion" batteries are becoming popular as power supplies for various consumer electronic devices such as laptop computers, camcorders, and cellular phones, and are also believed to have potential application to electric vehicles. The electrochemical process that occurs in these batteries, which operate at room temperature and generate about 4 V, involves the shuttling of lithium ions between two electrodes during charge and discharge. The purpose of this research was to determine if Li-Mn-O spinels can be used in place of the present $LiCoO_2$ cathodes, which are costly and considered hazardous.

The focus of the work was on (1) identification and characterization of Li-Mn-O phases that would be formed as lithium is inserted into, and extracted from, the cathode during discharge and charge and (2) methods of preparing cathode material of the desired composition. A portion of the Li-Mn-O phase diagram in the composition region of interest is shown in Fig. 6-5. The compositions that an electrode would have during charge and discharge lie on the tie line between $LiMn_2O_4$ and $Li_4Mn_5O_{12}$, and can be represented by the general formula $Li_{1+\delta}Mn_{2-\delta}O_4$ (δ between 0 and 0.33). Lithium can be extracted from, and inserted into, $Li_xMn_2O_4$ at approximately 4 V *vs*. metallic lithium when x is between 0 and 1.

Synthesis of single-phase $LiMn_2O_4$ spinel is not a simple process. When cubic $LiMn_2O_4$ is heated to 1200°C, it forms a number of different phases and crystal structures. The pertinent reactions were studied systematically by X-ray diffraction, thermogravimetric analysis, and differential thermal analysis, and a suitable temperature sequence for preparing single-phase, stoichiometric $LiMn_2O_4$ was developed.

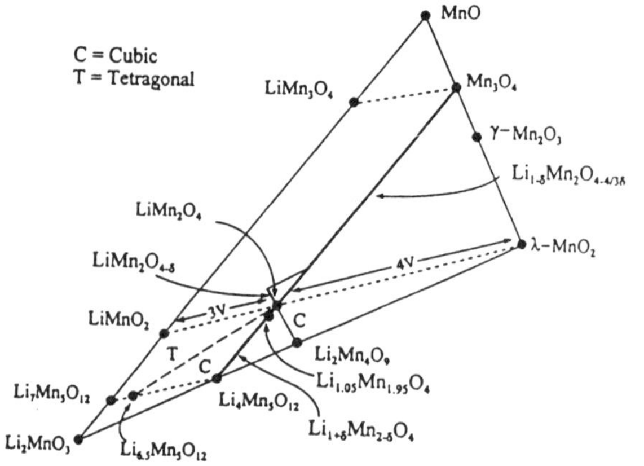

Fig. 6-5. Partial Phase Diagram of the Li-Mn-O System

Detailed structural studies were made on α-MnO_2, which is a potentially attractive positive electrode material for 3-V lithium secondary batteries. The CMT researchers substituted the water component in $MnO_2{\cdot}H_2O$ with Li_2O to fabricate "lithia-stabilized" electrodes, which have significantly superior performance over the parent α-MnO_2. The structures of hydrated, dehydrated, and lithiated α-MnO_2 were determined by the Rietveld profile refinement technique and time-of-flight neutron diffraction data. Structural studies were initiated on the Li-V-O system, which also shows promise as an insertion-type electrode. Initial results were promising and this work is continuing.

Contributors to the lithium-ion battery research were Khalil Amine, Roy Benedek, Dennis Dees, Gary Henriksen, Andrew Jansen, Christopher Johnson, Keith Kepler, Michael Manimelto, Michael Thackeray, and John Vaughey. Dennis Dees was the leader of this effort.

High-Power Energy Storage System. In 1997, a new CMT program was started on a high-power energy-storage device that could be coupled with a fuel cell in a hybrid vehicle. Integration of the galvanic properties of lithium battery "insertion" electrodes and the high surface areas of supercapacitors to form a lithium-ion based galvanic stack may be a method to achieve this end. Testing of a $Li_4Ti_5O_{12}/LiCoO_2$ cell couple gave promising results, and a conceptual design for 400-V galvanic stack based on a $Li_4Ti_5O_{12}/LiNi_{0.82}Co_{0.18}O_2$ system was developed. The design comprised a stack of 144 spirally wound cells divided into 18 modules enclosed in a thermally controlled jacket. The $Li_4Co_5O_{12}/LiCoO_2$ cells have shown excellent lifetimes in excess of 112,000 cycles. Work on this system is continuing with emphasis on modeling studies and further optimization of the electrode materials.

Khalil Amine was Project Manager for the high-power lithium ion battery program. Others working on the project were Vincent Battaglia, Andrew Johnson, Mike Thackeray, and John Vaughey.

Nickel/Metal Hydride System. The nickel/metal hydride battery is a promising contender to replace the popular nickel/cadmium batteries because of its superior performance, low impedance, and absence of toxic cadmium. It also avoids the shape-change vagaries of the cadmium electrode. The cell reaction for a nickel/metal hydride battery can be represented as

$$MH_x + xNiO(OH) \leftrightarrow M + xNi(OH)_2$$

where M is a rare earth alloy such as $LaNi_5$, with partial substitution of the La or Ni by a small amount of Sn, Mn, Co, Al, or possibly other metals. The present versions of Ni/MH batteries are afflicted by hydrogen-management problems associated with charge-discharge cycling and self-discharge, active material deterioration, and long-term capacity loss due to corrosion of the MH electrode.

A combination of electrochemical and neutron-diffraction studies was used for *in situ* investigations of the cells. Neutron diffraction revealed the locations of deuterium (a hydrogen isotope) in the molecular structure. A special quartz tube cell was designed to minimize background scatter caused by the construction materials, electrolyte, and the presence of remaining protons. A commercial $NiO(OH)/Ni(OH)_2$ electrode was deuterated and used as the positive electrode. The electrolyte was 15% KOD/D_2O. The electrochemical properties of the MH electrodes and Ni/MH cells were measured by interrupted galvanic cycling.

Three alloys were investigated: $LaNi_5$, $LaNi_{4.88}Al_{0.12}$, and $LaNi_{4.4}Al_{0.6}$. The $LaNi_5$ electrode had a low, but stable capacity. The $LaNi_{4.4}Al_{0.6}$ had the highest capacity and a small capacity drop. The $LaNi_{4.88}Al_{0.12}$ had a high initial capacity, but poor stability. The capacity-loss mechanism proved to be corrosion of the aluminum by the caustic electrolyte.

The *in situ* neutron diffraction measurements employed the General Purpose Powder Diffractometer at the ANL Intense Pulsed Neutron Source (IPNS). The data indicated that $LaNi_5$ and its derivatives have a hexagonal space group structure containing two crystallographically distinct nickel sites, Ni(1) and Ni(2). In aluminum-substituted $LaNi_5$, the aluminum substituted selectively on the Ni(2) sites.

As of 1997, the CMT work on Ni/MH batteries had become limited to testing in the Electrochemical Analysis and Diagnostic Laboratory (EADL).

The limited amount of research that ANL was allowed to do on the Ni/MH cells was performed by Laszlo Redey, along with Vincent Battaglia, Andrew Jansen, Dan Simon, and temporary and part-time personnel.

ELECTROCHEMICAL ANALYSIS AND DIAGNOSTICS LABORATORY

The EADL continued its mission of testing and evaluating advanced battery systems for applications such as electric vehicles and utility load-leveling. During the 1990s, evaluations were performed for the Department of Energy (DOE), the Electric Power Research Institute (EPRI), the U.S. Advanced Battery Consortium (USABC), and others. The resulting information was used by these organizations to identify the most promising avenues for further research and development. Test results were also useful in generating a database for modeling of projected battery performance.

Performance and Life Evaluations. A very brief summary of test results from six types of electric vehicle battery in 1990 is shown in Table 6-5. Where two sets of values are listed, the data are for batteries from different manufacturers and/or of different designs. The specific energies translate into ranges of 47 to 154 miles for a van having a 695-kg battery.

Although some of these batteries are suitable for specialized practical applications such as fork lifts and golf carts, none of them even come close to matching the power, range, and price of a typical family car powered by a gasoline engine. It was clear that a major technological breakthrough would be required to achieve widespread acceptance of electric automobiles.

Tests were continued on sodium/sulfur batteries from Powerplex Technologies, Inc., Chloride Silent Power, Ltd., and ASEA Brown Boveri. Operating temperatures for the sodium/sulfur system were in the range of 310 to 340°C. By 1992, the cycle life of these batteries was in a range of about 600-800.

The zinc/bromine batteries were supplied by the Studiengesellschaft fur Energie-speicher und Antriebssysteme (SEA) Research Group for Energy Storage and Propulsion Systems in Austria, which had licensed the technology from the Exxon Corp. and improved the original Exxon design.

Table 6-5. Summary of 1990 Test Results from Electric Vehicle Batteries

Technology	Specific Energy,[a] Wh/kg	Peak Power,[b] W/kg	Battery Life,[c] cycles
Sodium/Sulfur	81 79	150 90	>200 >200
Zinc/Bromine	75	53	>300
Nickel/Metal Hydride	54	182	289
Nickel/Cadmium	55	191	>250
Nickel/Iron	56 51	92 111	370 >200
Lead-Acid	36 33	125 87	370 715

[a] Measured at 3-h discharge rate.
[b] From driving profile discharge at 50% depth of discharge.
[c] For 100% depth of discharge cycles (">" indicates ongoing test as of December 1990).

The zinc/bromine system exhibited reasonably good specific energy (75 Wh/kg), but low specific power (53 W/kg). Two disadvantages were a need to conduct an electrochemical maintenance procedure after every six or so cycles, and certain safety issues.

Performance tests were conducted on nickel/cadmium battery modules manufactured by SAFT (Industrial Storage Battery Division) in France. This battery had the highest peak-power capability of any of the systems (191 W/kg) and completed more than 1,018 driving cycle discharges. It was then sent to EPRI for further evaluations. Two other batteries from SAFT having capacities of 200 and 80 Ah were tested for the fuel cell/battery bus project discussed later. These units chalked up 9,084 and 4,104 cycles, respectively, which correspond to >75,000 and 35,000 bus miles.

Two types of advanced nickel/iron batteries from Eagle-Picher Industries with capacities of 220 and 179 Ah were evaluated. The larger module had a lifetime of 370 cycles. Several additional nickel/iron batteries from Eagle-Picher were then placed on test. The longest operating module completed 918 cycles.

Tests continued on lead-acid batteries from several sources: Sonnenschein Battery Co.; Chloride Motive Power; Johnson Controls, Inc.; Bell Computer, Inc.; Electrosource, Inc.; GNB Industrial Battery Co.; Yuasa/Exide, Inc.; Optima Batteries, Inc.; and Hawker Energy Products, Inc. Of the lead-acid batteries tested, Electrosource modules achieved the highest specific energies (43 and 36 Wh/kg), an Electrosource module had the highest peak power (171 W/kg), and a Yuasa/Exide module showed the longest cycle life (494 cycles).

Performance data were obtained in 1990 on 3.6-Ah (C-size) nickel/metal hydride cells from the Ovonic Battery Co. The thought was that these cells might be scaled up to an electric vehicle battery. The cells exhibited the highest volumetric energy density (186 W/L) of any battery system examined to that time. The specific energy and peak power were 54 Wh/kg and 182 W/kg, respectively, but their longest lifetime was 289 cycles. The following year, tests of cells scaled up to 25 Ah showed an impressive peak power of 200 W/kg, but the lifetimes of these cells were limited to about 500 cycles. In 1997, a new CRADA with the Advanced Battery Consortium was started to evaluate Ni/MH cells and modules from USABC battery cost reduction contracts with GM-Ovonics and SAFT.

Tests were also conducted on Li/FeS cells from SAFT of America, Inc. The specific energy and peak power were 66 Wh/kg and 64 W/kg, and the lifetime was 163 cycles.

Post-Test Analysis. A joint failure analysis was conducted on a 19-kW Na/S battery that had been fabricated by Chloride Silent Power and tested at ANL for 241 cycles. Examination of 17 failed cells showed that most of the failures resulted from cracks in the glass seal between the β''-alumina solid electrolyte and an α-alumina top cap. The examinations also showed that the use of sodium safety cans alleviated, but did not prevent entirely, a potential for failure propagation due to loss of the sodium seal. A cell that had suffered from discharge polarization during cycling was examined by radiography, which indicated that the polarization was caused by a reduced sodium level in the electrolyte annulus. A maldistribution of sodium on the electrolyte surface could result in high localized current densities. Post-test examinations of Na/S batteries from Chloride Silent Power under a CRADA with USABC were completed in 1994.

Examinations were performed on failed cells from a 36-V Li/FeS module fabricated by Westinghouse Naval Systems Division and tested at ANL for 141 cycles. The cause of failure proved to be short circuits due to deposition of metallic iron in the separators. Difficulties with thermal management of the module had exacerbated the problem; most of the failures were in the central portion of the module, where the temperature had risen more than 50° over the normal operating temperature of 450°C.

Thermal Li/FeS_2 cells from Sandia National Laboratories were also examined to investigate failures resulting from Li_2S deposits within the separators of these primary cells. The Sandia thermal cells consisted of an FeS_2 cathode, a MgO powder separator, molten LiCl-KCl electrolyte, and a Li-Si anode. Microscopic examinations showed that Li_2S deposition was affected by a variety of factors mostly related to the compositions of the cell constituents. Further basic studies of the system were recommended.

Post-test evaluations were made on a prismatic Li/FeS_2 cell built and operated through 434 cycles by SAFT, America. Declining capacity and coulombic efficiency were attributed to expansion and extrusion of the positive electrode material into the MgO powder separator, which caused a partial short circuit. Other findings, however, were positive—there was little or no corrosion of the molybdenum components, and formation of aluminum intermetallic compounds with nickel and stainless steel components in the negative electrode was minimal.

A Ni/Fe module from a battery pack built by Eagle-Picher for in-vehicle testing in 1983 was removed from testing after 66,000 actual and equivalent miles of operation. Capacity decline of the module was totally attributable to voltage reversal of one cell caused by degradation of the iron plates through local overcharge.

Much of the effort on post-test examinations was directed to lead-acid batteries from various manufacturers. A module from Chloride EV Systems, Ltd., that had undergone 715 cycles was subjected to post-test analysis, which showed that its declining performance was caused by high levels of antimony and poor adhesion between the grids and the active materials. Two valve-regulated lead-acid (VRLA) technologies for utility cycling applications were evaluated: (1) gelled-electrolyte modules from Johnson Controls, Inc., and (2) absorbed-electrolyte modules from GNB Battery Co. Failure of the Johnson Controls unit resulted from a manufacturing defect that caused corrosion of a strap for the positive plate lugs. The GNB batteries had a longer lifetime and failed through corrosion of the positive electrode grids. A study of stibine (SbH_3) emissions during operation of the GNB batteries showed that only minor ventilation was needed to avoid a toxicity problem. Other post-test examinations of various lead-acid batteries showed that positive grid corrosion was the most common mode of failure. Overcharge and variations in the oxygen-recombination efficiency were factors in the corrosion process.

The post-test analytical and diagnostic work by CMT achieved growing recognition for its excellence, and, in 1993, the USABC chartered ANL to draft a procedural outline for the standardization of post-test analyses. In 1997, the ANL post-test analysis activities had declined and that laboratory was closed. At the time of this writing, however, a facility for the testing of advanced fuel cells was about to go into operation.

Evaluations and Studies. In 1997, evaluations were conducted on an electric vehicle battery pack from Chrysler Corp., which was constructed from Ovonic Ni/MH modules. A study was conducted for DOE on the technical status of foreign electric vehicle batteries. In addition to the performance and lifetime evaluations, Argonne has supported the Advanced Battery Consortium by having representatives on the Ovonic Nickel/Metal

Hydride Working Group and the Battery Testing Procedures Tiger Team, as well as various other USABC planning and evaluation groups. In 1998, evaluations were continued on the cost reduction programs and the lithium-polymer battery development program.

Personnel. The battery testing and evaluation activities were managed by Bill DeLuca. Others involved in the effort were Brandon Cole, Kenneth Gillie, Roger Hogrefe, John Smaga, Tony Tummillo, and Carl Webster.

FUEL CELL RESEARCH AND DEVELOPMENT

The DOE continued to insist that Argonne's primary role in the fuel-cell area be the technical management and evaluation of industrial research and development contracts, with small in-house programs on supporting research. This in-house work consisted mainly of materials studies and mathematical modeling of fuel-cell systems. By 1990, the Argonne scientists and engineers had established the basic designs and feasibility of molten carbonate and monolithic solid oxide fuel cells, but further work was needed to improve performance and lifetime and to scale up the systems for practical application. As in the battery program, the advent of polymer electrolytes opened the possibility of a new type of fuel cell, and work in this area was started in 1992.

Monolithic Solid Oxide Fuel Cells. As indicated in the previous chapter, the monolithic solid oxide fuel cell (MSOFC) is made up of two multilayer composites, one being the anode/electrolyte/cathode and the other being the cathode/interconnect/anode. The anode is nickel, the cathode is strontium-doped lanthanum manganite, and the electrolyte is yttria-stabilized zirconium oxide (YSZ). The electronically conductive intercell connector material is strontium-doped lanthanum chromite. The nickel anode is in the form of a Ni-YSZ cermet.

In 1988, Allied Signal Aerospace/AiResearch began development work on the MSOFC with technical support from ANL, and this support continued into the 1990s. Long-term effects on the cell components were evaluated by accelerated tests at elevated temperatures (1000-1200°C) in air for 1000 h. A small increase in the cathode resistance was observed, but not considered to be a serious problem. Since the MSOFC would be expected to operate with coal gas or natural hydrocarbon fuels that contain low-ppm concentrations of sulfur, the effect of sulfur on anode performance was evaluated. The interfacial resistance increased significantly with H_2S concentrations in the low-ppm range. Follow-up studies on the free energies of H_2S reactions with ZrO_2 and Y_2O_3 indicated that such reactions were thermodynamically possible, but thermogravimetric experiments showed that they did not occur, probably because of kinetic factors.

A major disadvantage of conventional fuel cells is that they require hydrogen as the fuel. Attempts to use methanol or ethanol fuel by adding special catalysts have met with only limited success. One scientific wag put it succinctly: "All they have to do is to solve three problems with the methanol fuel cell—the anode, the cathode, and the electrolyte." The situation may be different, however, for solid oxide fuel cells, which operate at temperatures of 800-1000°C. At these temperatures, potential fuels such as hydrocarbons, alcohols, and natural gas can reform on the nickel cermet anode to form electrochemically active species such as hydrogen and carbon monoxide. Exploratory experiments were conducted in which a polarization curve (voltage *vs.* current density) was measured for a solid oxide fuel cell using hexadecane as the fuel. The curve was similar to those for hydrogen fuel but showed a voltage reduction that was attributed to a high water content in the fuel stream. The water was added to support the internal reform reaction and to suppress the formation of tars and coke through cracking reactions. Subsequent studies were made on the operation of lower-temperature (500-700°C) solid oxide fuel cells with methanol as the fuel. The performance was nearly the same as that with hydrogen. Thus, it was established that solid oxide fuel cells can operate with fuels other than hydrogen but may have problems with sulfur in natural fuels and with the formation of carbonaceous deposits in the case of hydrocarbons. Work that was done on sulfur-tolerant anodes indicated that $La_{0.7}Mg_{0.3}CeO_{3.35}$ and $La_{0.8}Nb_{0.2}CeO_{3.7}$ show promise for this purpose.

There was an interest in decreasing the operating temperature of MSOFCs to 500-800°C so that metal components could be used in the cells, thereby increasing design flexibility. Theoretical considerations suggested that certain perovskites might have sufficient ionic conductivity to serve as electrolytes at 600°C, and the conductivities and ionic transference numbers were determined for six compositions of Sr-Zr-O or Sr-Zr-Sn-O. Although these materials have the oxygen substoichiometry needed for a fuel cell, the conductivities were still too low to be practical. In a preliminary study of ortho-phosphates, tests of strontium-doped $LaPO_4$ showed that materials of this type might be promising candidates. Studies were then conducted on Bi_2O_3-Al_2O_3 compounds, along with incorporation of Bi_2O_3 into a refractory host lattice ($LaAlO_3$). Conductivities of $Bi_2Al_4O_9$ and La-Al-Bi-O mixtures were determined, using platinum electrodes and humidified H_2/O_2 gas at 500-800°C. The conductivities were of the order of $10^{-2}\ \Omega^{-1}cm^{-1}$ with ionic transference numbers of about 0.65. This conductivity, although below the goal, is about the same as that of yttria-stabilized zirconia, the present electrolyte material. Some preliminary studies were performed on ceria-based and bismuth aluminate systems as potential solid electrolytes.

The sealants in a solid oxide fuel stack, in addition to forming gas-tight seals along the edges of each electrode and between the manifold and stack, must be chemically and thermally stable at 1000°C and compatible with H_2, O_2, water vapor, and the cell materials. Initial studies of two commercial cements showed them to be unacceptable because they contained silicon and had regions of porosity. Reactive glasses were then synthesized and evaluated. The results showed that the nature of surface reactions of the glasses with the cell components was more important than an exact match in coefficients of expansion. At the cell edge, the sealant can be relatively rigid because it has to tolerate the thermal expansion behavior of, at most, only two different cell ceramics. At the stack-to-manifold junction, however, a somewhat compliant seal is required to accommodate the thermal expansion of four ceramics plus the manifold. Fifteen novel sealants having varying degrees of rigidity were synthesized, bonded to ceramics typical of the fuel cell, and tested by heating to 1000-1350°C. Optical and scanning electron microscopy showed that two were superior—a rigid cementitious material and a glass-ceramic composite that was expected to be somewhat compliant. Further thermal and electrical testing showed that the composite material formed a rigid, elastic, gas-tight seal. Further work culminated in successful glass-ceramic sealants based on a five-component oxide system: SrO, La_2O_3, Al_2O_3, SiO_2, and B_2O_3. These seals were subjected to long-term tests for materials compatibility, physical integrity, and electrical properties. Samples of the high-temperature sealants were sent to several fuel-cell manufacturers for evaluation and received favorable reviews.

A model developed in 1990 that simulated the electrochemistry and thermal hydraulics of an MSOFC stack with a cross-flow design yielded the temperature distribution and the fuel and oxidant compositions for the entire system. This model could also simulate operation with different fuel gases, such as hydrogen, coal gas, and methanol. In 1991, much of the electrochemical modeling work was transferred to the University of Oklahoma.

Molten Carbonate Fuel Cells. The molten carbonate fuel cell (MCFC), as mentioned previously, consists of a porous nickel anode, a porous nickel oxide cathode, a liquid Li_2CO_3-K_2CO_3 electrolyte in a $LiAlO_2$ matrix, and appropriate metal separator sheets. Anode material studies continued on $LiFeO_2$ and MnO, which had been shown earlier to be stable in the anode environment and to have bulk resistivities at 650°C of 0.2 and 20 Ω-cm, respectively, which are in an acceptable range for use in the anode. Cell tests of doped and undoped $LiFeO_2$ were conducted at 140-650°C in a stream of humidified anode gas. Niobium doping did decrease the resistivity with a minimum at a Nb/Fe mole ratio of 0.06. The performance was also improved by a new method of synthesizing $LiFeO_2$ with starting material of a higher surface area. Further full- and half-cell tests were made with $LiFeO_2$ anodes (undoped and doped with Mn, Co, and Ni), and Nb-doped CeO_2. Polarization curves (voltage *vs.* current density) indicated that none of these electrodes performed as well as the state-of-the-art Ni-Cr anode, and none could sustain a current density higher than about 80 A/cm^2. In the case of CeO_2, the electrode performance decreased on exposure to the reducing atmosphere but could be restored temporarily by exposure to an oxidant.

A large proportion of the development work on the MCFCs was on the cathode. Tests were made both with 25 cm^2 full-size cells and with diagnostic half-cells. Effort was focused on $LiFeO_2$ cathodes as a thermodynamically stable replacement for NiO, which had been experiencing lifetime-limiting dissolution/precipitation problems. The performance of $LiFeO_2$ electrodes appeared to be restricted by slow reaction kinetics. Modification of the synthetic method for $LiFeO_2$ to increase the surface area resulted in some, but not sufficient, improvement in the performance. An increase in the total surface area of the cathode, with little attention to the pore structure, produced some improvement in the performance, but it was probably limited by access of gas to the bulk electrode, which depends on the pores. The best-performing electrode in full-cell tests was Co-doped $LiFeO_2$. Data on the kinetics of the reactions suggest that different mechanisms are involved in the operation of NiO and $LiFeO_2$ electrodes. As a result of ensuing detailed studies on the effects of temperature, oxygen partial pressure, carbon dioxide partial pressure, electrolyte thickness, and electrolyte composition, a new oxidant mixture (95% air-5% CO_2) was selected. With this oxidant, the performance of Co-doped $LiFeO_2$ cathodes in half-cell tests at 675°C was similar to that of NiO electrodes at 650°C. Tests of $LiFeO_2$ cathodes with other dopants (Cu-, Mn-, Ni- and Mg-doped Li_3MnO_3) resulted in less improvement than that from the cobalt-doped material. Seebeck coefficient measurements led to the conclusion that the alternative cathode must be a p-type conductor with little or no electronic component.

There was an incentive to increase the normally ambient operating pressure of MCFCs to about 10 atm to increase the power density. The increased chemical activities of the gases at the higher pressures shorten the life of the NiO cathode, which is a major reason for seeking alternative cathode materials. The resistivities of $LiFeO_2$, $LiCoO_2$, and NiO are about 300, 1, and 0.1 Ω-cm, respectively. Thus, to match the performance of NiO, the resistivities of the two alternative materials would have to be decreased significantly. Individual dopants did not accomplish this goal, but a new, proprietary double dopant was found, which reduced the resistivity from 300 to about 0.3 Ω-cm. This double dopant improved the performance of $LiFeO_2$ and $LiCoO_2$ cathodes over a wide range of cathode-gas compositions. At this stage, NiO and the two doubly doped cathode materials showed about the same performance, and lifetime became the issue. The relative lifetimes of NiO, doubly doped (D.D.) $LiFeO_2$, and $LiCoO_2$ were evaluated, using mixtures of CO_2 and O_2 as the cathode gas. The results, shown in Fig. 6-6, as well as other data, indicated that both $LiCoO_2$ and $LiFeO_2$ have a longer life than NiO.

Although $LiCoO_2$, with a low resistivity of about 1 Ω-cm, could serve as a longer-lived direct substitute for NiO, its high material cost could prevent its large-scale use.

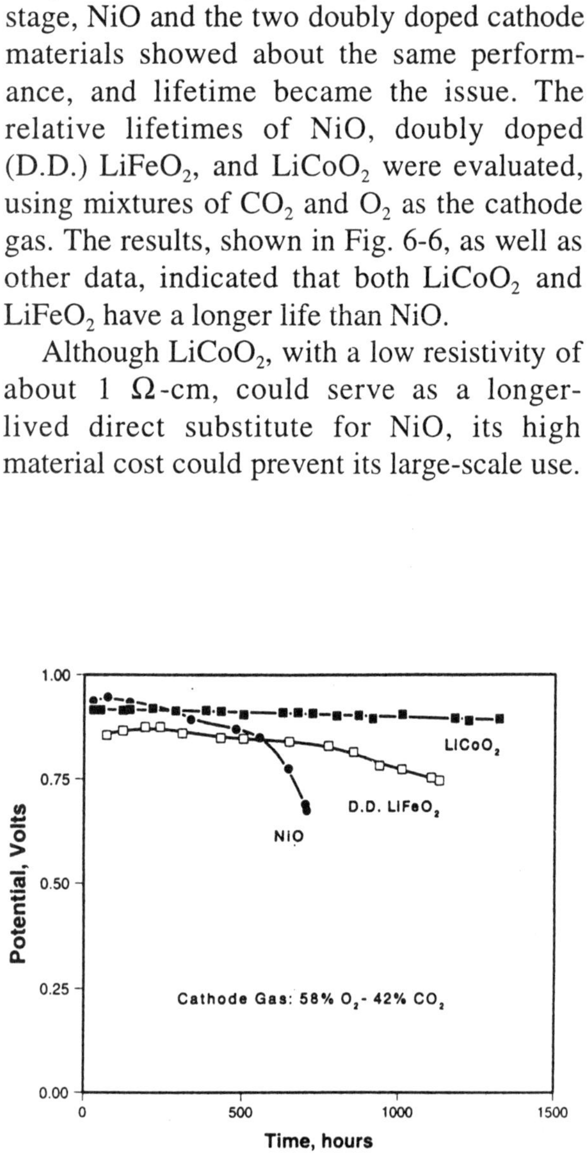

Fig. 6-6. Potential *vs.* Operating Time for Experimental Cathodes in MCFCs

Therefore, attention was directed toward lowering the resistivity of $LiFeO_2$ (300 Ω-cm). By adding a combination of cation substituents, the resistivity of a $LiFeO_2$-based material (designated "Material 1") was lowered to 1.3 Ω-cm at 650°C, close to the desired value for a MCFC. This material appeared promising for use in a practical fuel cell, but a remaining problem was identified when preliminary studies indicated that it had a higher creep rate than the lithiated NiO used in state-of-the-art cells.

Some work was done on ways to reduce the solubility of the NiO cathode by modifying the 70 mol% Li_2CO_3-30 mol% K_2CO_3 electrolyte. Additions of $CaCO_3$ lowered the solubility of NiO and decreased the corrosion rates of stainless steel cell components, but $CaCO_3$ segregated during cell operation due to concentration polarization, thereby removing it from further consideration. Cells having Li_2CO_3-K_2CO_3 electrolyte tended to form short circuits during extended operation at high current densities due to segregation of the electrolyte. Recent studies with carbonate-wetted $LiAlO_2$ strips and 100-cm^2 cells have indicated that it may be possible to minimize this problem by using a Li_2CO_3-Na_2CO_3 electrolyte.

As of this writing, the current molten carbonate fuel cell consists of a porous nickel anode, a semi-fluid $(Li,K)_2CO_3$ + $LiAlO_2$ electrolyte, and a metal bipolar plate. An ongoing effort is being made to develop alternative cathodes, non-segregating electrolytes, and more corrosion-resistant cell components.

Polymer Electrolyte Fuel Cell. As of 1992, CMT was becoming more involved with the technical management of DOE-supported research and development of the polymer electrolyte fuel cell (PEFC) and the phosphoric acid fuel cell (PAFC) for transportation applications. Some in-house work that was supported on the polymer-electrolyte program involved research on the proton-exchange membrane fuel cells (PEMFC), which consist of a carbon-supported platinum electrode for oxidation of hydrogen fuel, an identical cathode for reduction of oxygen, and a polymer electrolyte membrane that transports hydrogen ions from the anode to the cathode. Because of the interest in renewable fuel sources, the hydrogen fuel gas was obtained by the reforming of alcohols, and it contained substantial amounts of CO_2 and traces of CO. At the relatively low operating temperature of 80°C, these impurities decrease the efficiency of the platinum catalyst and, therefore, the power output. The high cost of the platinum catalyst was an incentive to decrease the amount that was needed.

The initial research was aimed at platinum-based alloys that would decrease the amount of platinum while retaining the high electrochemical activity for hydrogen oxidation. The candidate alloys were first screened for catalytic activity under simulated operating conditions of a cell. They were exposed to CO-containing electrolyte just long enough to reach a platinum surface poisoning about equivalent to that of a platinum electrode operating on reformer output. The promising alloys were then placed in an actual cell, and the potential resulting from a current pulse was measured as a function of time. This procedure permitted separation of the reaction kinetics from other effects contributing to the measured potential. The results showed that platinum alone was the best anode when operating with pure hydrogen, but with CO in the fuel stream the alloy was superior to platinum.

An evaluation was made of two methods for fabricating polymer electrolyte cells: in one method, the electrolyte was impregnated into the electrodes; in the other, the electrolyte was mixed directly with the platinum catalyst used in the anode ("ink technique"). Two different gas-flow configurations (serpentine and waffle) were also examined. The results showed that the ink technique and the serpentine gas-flow pattern gave better performance.

In 1995, the emphasis was shifted to the direct use of methanol fuel in a PEFC without a need for reforming. Such cells were having problems with high anodic polarization, methanol crossover across the electrolyte membrane, and a reduced cell voltage due to mixed cathode potentials. A study was undertaken on the effect of ruthenium on the electronic structure of platinum and its electrocatalytic behavior in oxidizing methanol. The methanol crossover problem results from its solvation of the protons that carry the cell current. Work was started on the possibility of using a basic system where hydroxide is the conducting ion, and it moves in the opposite direction, so there is no chance of methanol transfer.

The major obstacle to developing a polymer electrolyte fuel cell that operates directly on methanol has been the low electrocatalytic activity of the anode catalyst for the methanol oxidation reaction. In 1997, a systematic study of platinum-based alloy catalysts showed that Pu-Ru alloys gave the best results.

Fuel Reformer Technology. In 1990, a study was conducted on fuel reformers to assess the state of the art for converting alternative fuels (methanol, ethanol, and compressed and liquefied natural gas) into a hydrogen-rich gas mixture for powering fuel cells in automotive applications. The existing reformers did not have the start-up and dynamic response capability required for a stand-alone fuel cell unit in an automotive application. In a separate project, the State of Illinois and DOE agreed to sponsor a program at ANL to develop reformer technology to convert ethanol to hydrogen for use in fuel-cell-powered vehicles.

A search was begun for catalysts to enhance the ethanol-steam reaction

$$C_2H_5OH + 3\ H_2O \rightarrow 2\ CO_2 + 6\ H_2$$

This is the ideal reaction. In actuality, some other reactions are involved, which lead to the formation of methane (CH_4). A Ni-Al-Zn catalyst gave a higher gas production rate, but a Cu-Zn-Ni catalyst produced a higher concentration of hydrogen and less methane in the product gas. Further studies of this reaction with a nickel-based catalyst showed that a high hydrogen yield with a low CH_4 concentration would be possible through the use of a water-gas shift converter after the reformer.

A possible alternative to the ethanol-steam reaction is partial oxidation. Methanol, for example, undergoes the reaction

$$CH_3OH + 1/2\ O_2 \rightarrow 2\ H_2 + CO_2$$

With a Cu-Zn-O catalyst, the methanol began to reform at about 150°C, and 100% conversion was achieved at 350°C. In 1994, a bench-top [10 kW(e)] methanol reformer based on partial oxidation was built and tested. The catalyst was copper zinc oxide (in the oxidized form). Startup was rapid, and the reaction product composition of 50% H_2, 20% CO_2, and <5% CO_2 was typical. This reformer was much lighter and smaller than existing units and was a radical departure from current reformer technology at that time. The next step was to look at the partial oxidation reforming of hydrocarbon fuels.

An idealized general reaction can be written for partial-oxidation reforming of a generic hydrocarbon fuel:

$$C_nH_mO_p + x\,(O_2{+}3.76N_2) + (2n{-}2x{-}p)\ H_2O$$
$$\rightarrow nCO_2 + (2n{-}2x{-}p{+}m/2)\ H_2 + 3.76x\ N_2$$

where x is the oxygen/fuel molar ratio. Conventional catalysts (including Fischer-Tropsch types) and a new class of proprietary catalysts developed by CMT scientists were tested to determine their effectiveness in promoting this reaction. The best performance was obtained with the Argonne catalyst. This catalyst was then tested with a variety of hydrocarbons that are typically present in gasoline. The results were very good. For example, partial-oxidation reforming of iso-octane at 630°C resulted in nearly complete reaction of the isooctane and a hydrogen concentration over 60%. Similar experiments were done with pipeline natural gas, which typically contains over 90% methane. Partial-oxidation reforming of natural gas at 690°C with the ANL catalyst left only 0.5% of the methane unreacted, and the hydrogen concentration was 70%. This research on reforming catalysts has been highly productive and is continuing. Work is also in progress on methods of removing CO from the reformate by a regenerative sorption process that uses CuCl-impregnated porous materials such as bauxite.

Modeling Studies. Mathematical modeling was used extensively in all of the fuel cell programs, and it took many forms. Detailed modeling was used to characterize the electrochemical and chemical reactions and rate-controlling kinetic factors in the electrodes and electrolytes. Engineering-type modeling involving mass and heat transfer, current and voltage distribution, flow patterns and pressure drops, and other properties of that type were used to optimize cell and stack designs. Finally, systems analysis studies of practical applications such as integrated systems involving reformers, fuel cells, and batteries for bus propulsion were a form of modeling.

The following are specific examples typical of the modeling work in the fuel-cell programs. Modeling of the MSOFC thermochemistry and thermal hydraulics characterized the temperature distribution, fuel and oxidant compositions, energy efficiency, fuel utilization, and power density for cells fueled by hydrogen, coal gas, and methanol. A modeling study was used to determine the factors affecting the electrochemical kinetics of the reactions in the MSOFC. In another study, the efficiency of fuel cell/battery-powered buses was determined as a function of fuel and oxidant flow rates. A computer simulation was developed for polymer electrolyte fuel cells used in a bus propulsion system. In the molten carbonate fuel cell program, the effectiveness of distributed manifolds having different flow patterns was evaluated by modeling. Finally, a generic steam reformer for methanol was modeled to determine its behavior under various operating conditions.

Technical Management. The CMT Division, through the ANL Electrochemical Technology Program, provided support to the DOE Office of Transportation Technologies and to the DOE Office of Buildings Technologies in the form of technical management of R&D contracts with industrial developers of fuel cells and related components. This activity involved the preparation of work statements, proposal evaluations, and progress reviews.

Because of the large number of contracts and wide scope of the work, the following is only a brief summary of these activities. An Allied Signal Aerospace/AiResearch program on monolithic solid oxide fuel cells extended into the 1990s. A major activity was the Fuel Cell Bus Project, which was supported jointly by DOE and the California South Coast Air Quality Management District (SCAQMD) and managed jointly by CMT and Georgetown University. The objective of this program was to demonstrate an urban bus powered by a methanol-fueled phosphoric acid fuel cell combined with a battery. Two industrial contractors were selected: Booz-Allen & Hamilton, Inc., and Energy Research Corp. In the second phase of this project, H-Power Corp. was contracted to assemble and test fuel cell/battery systems prior to their installation in buses. Subcontractors on the H-Power team included Transportation Manufacturing Corp., Bus Manufacturing USA, Inc., Booz-Allen Hamilton, Inc., Fuji Electric Co., and Soleq Corp. In 1994, this program resulted in the first U.S.-built bus powered by a fuel cell, and H-Power Corp. built two more buses in 1995. These buses are still in operation.

Polymer electrolyte fuel cells (PEFC) were also of interest for transportation applications. The Allison Gas Turbine Division of General Motors Corp. built and tested a 10-kW PEFC in 1993 during the first phase of a three-year contract. Supporting subcontractors included Los Alamos National Laboratory, Dow Chemical Co., Ballard Power Systems, General Motors Research Laboratories, and General Motors Advanced Engineering Staff. Ford Motor Co. had an ongoing contract to develop a 50-kW PEFC fuel-cell stack, which included subcontracts with International Fuel Cells and Mechanical Technology, Inc. Chrysler also had a contract for similar work, using Pentastar Electronics, Inc., as a subcontractor.

Arthur D. Little had a major role in the development and engineering of reformer systems for converting ethanol and natural gas into hydrogen for fuel cells, and they engaged SETA Corp. in environmental assessments of fuel-cell-powered vehicles, and Engineering Systems Management in life-cycle and cost comparisons. Other miscellaneous contracts included Ovonics work on nickel/metal hydride batteries and investigations of thermoelectrochemical power sources by the Delco-Remy and Hughes Aircraft divisions of General Motors Corp. In 1996 six new contracts were issued to Energy Partners, Texas A.&M. University, International Fuel Cells, Allied Signal, Arthur D. Little, and Vairex Corp. to develop innovative designs and materials to improve the performance of PEFC systems.

Personnel. The fuel cell programs in the 1990s were under the general purview of Mike Myles and directed by Mike Krumpelt. Ira Bloom was the leader of the solid oxide fuel cell work, which also included David Carter, Dennis Dees, Rajiv Dosi, Mark Hash, Tim Kueper, Frank Mrazek, Deborah Myers, and Xiaoping Wang. The research on molten carbonate fuel cells was led by Gene Kucera and involved Ira Bloom, Alan Brown, Kevin Byrne, Gary Henriksen, Sheldon Lee, Mike Roche, Jim Smith, and Joe Stapay. Research on the polymer electrolyte fuel cell was conducted by Shabbir Ahmed, Khalil Amine, Ira Bloom, Deborah Myers, and Tony Tummillo under the leadership of Romesh Kumar. Technical assessments and modeling were performed by Shabbir Ahmed, Ira Bloom, Gary Henriksen, Ted Krause, Mike Krumpelt, Romesh Kumar, Jim Miller, Deborah Myers, Walt Podolski, and Bill Swift.

Fossil Fuel Research

The fossil fuel research program, often referred to by people in the Division as the "coal program," continued into the 1990s, but waned rather quickly in the early part of the decade, due to loss of funding, until it had disappeared by the mid-1990s.

FLUIDIZED-BED COMBUSTION STUDIES

As described earlier, fluidized-bed combustion is a process for burning coal under atmospheric or pressurized conditions in a fluidized bed of limestone or dolomite, which removes most of the SO_2 from the off-gases during combustion. As of 1990, studies were in progress on the following:

1. Loss of metal from heat exchanger surfaces.
2. Materials performance in a fluidized-bed cogeneration air heater.
3. Alkali vapor emissions from pressurized fluidized-bed combustion (PFBC).
4. Combustion of cleaned coals.
5. Use of Illinois limestones to reduce SO_2 and HCl emissions during combustion of Illinois coal containing high levels of sulfur and chlorine.

Metal Loss in Fluidized-Bed Combustors. Loss of metal from heat-transfer tubes in FBCs was an ongoing problem that was frustrating efforts to commercialize this technology. The following organizations joined into a cooperative effort to address this problem: DOE, ANL, Electric Power Research Institute (EPRI), Illinois Center for Research on Sulfur in Coal (CRSC), Tennessee Valley Authority (TVA), ASEA Babcock PFBC, Foster Wheeler Development Corp. (FWDC), and ABB Combustion Engineering Systems (ABB/CE). The objective was to develop models and computer codes to predict hydrodynamics and metal loss of tubes in fluidized beds, and to translate these models into designs and operating conditions that would minimize metal loss. The ANL divisions involved in this effort were CMT, Energy Systems (ES), and Materials and Components Technology (MCT).

Data from fluidized-bed experiments at the University of Illinois at Urbana-Champaign (UI-UC) and at FWDC were used in ANL-developed models to analyze the fluidized-bed hydrodynamics. Babcock and Wilcox (B&W) developed a three-dimensional model. Room-temperature (no combustion) tests by FWDC characterized the erosion and hydrodynamic behavior in which the conditions could be altered by changing the geometry of the equipment. The data were analyzed at ANL to evaluate bubble formation and its effect on erosion patterns. In the experiments at UI-UC, particle velocity was determined by use of a radioactive tracer particle. By putting together all this information, it was possible to calculate the directional distribution and velocity distribution of the particles impinging on the tube surfaces. Bubble velocities in the fluidized bed were computed from differential pressure measurements in the FWDC experiment and were found to agree with computed values. Babcock and Wilcox then completed a three-dimensional hydrodynamic model, which was validated by comparing its output with that from the ANL two-dimensional model.

Experimental work then continued, with testing of a "cold-model" fluidized bed at FWDC and studies of erosion and particle velocity at UI-UC. In the UI-UC work, particle velocities of soda-lime glass beads were determined by the tracking technique, and the unsurprising conclusion was that the high erosion regions of the immersed tubes coincided with those of high particle speed and high hitting frequency. The experimental data from FWDC and UI-UC were compared with predictions from hydrodynamic and erosion models (FLUFIX/MOD2 and EROSION/MOD) developed at ANL. The maximum erosion rates in the FWDC experiments were determined to be 0.5 mm/1,000 h for aluminum tubes and 2.9 mm/1,000 h for polyvinyl chloride (PVC) tubes. Similar, but not identical, results were obtained by UI-UC.

Atmospheric Fluidized-Bed Cogeneration Air Heater Experiment. Argonne managed the Atmospheric Fluidized-Bed Cogeneration Air Heater Experiment for DOE. The ANL divisions involved with this project were CMT and MCT. The purpose of the project was to assess materials and process performance of in-bed air heaters for cogeneration of electricity and hot air in an atmospheric fluidized-bed combustor (AFBC). Rockwell International was awarded a contract for experimental testing of heat-exchanger tube bundles in a DOE-owned AFBC unit in El Segundo, California. Test specimens of a wide variety of alloys were supplied by ANL and three organizations under subcontracts with Westinghouse Electric Corp. The objective of the tests was to determine metal-loss information that would permit lifetime projections for the alloys in the AFBC environment. The tests had a duration of about 2,000 h, and the temperature was approximately 880°C. Oxygen partial pressures, which were normally about 10^{-3} Pa, were measured by means of a cell with an yttria-stabilized zirconia solid electrolyte. In general, the results of these tests showed that deposits of the bed material on the alloy surfaces caused significant, sometimes catastrophic corrosion. The same alloys, however, had acceptable corrosion rates when their surfaces were free of such deposits.

The overall conclusions of this investigation were the following:

1. Some stainless steels and other alloys exhibited little corrosion.
2. Long-term (3,000-h) laboratory tests revealed that others underwent catastrophic corrosion at sustained low oxygen partial pressures and in the presence of bed-material deposits.
3. Some alloys were susceptible to unacceptable corrosion when in contact with the bed material.
4. Among the weldments, a variety of fillers had acceptable corrosion resistance.
5. Of the coated specimens, aluminized coatings were very poor. In general, the development of crack-free coatings was difficult.

6. Among the cladding alloys, Type 310 stainless steel on Type 304 stainless steel or Alloy 800H exhibited superior performance.

Regenerable Activated Bauxite Sorbent Alkali Monitor. Vapor-phase alkalis in the PFBC exhaust can corrode the turbine-blade materials. Earlier work had shown that a time-averaged value of the alkali-vapor concentration in the exhaust gas could be measured by using an analytical activated-bauxite sorber bed. The adsorbed alkalis were recovered by water leaching and analyzing the solution by atomic absorption spectrometry. The objective of this work was to develop a regenerable activated bauxite sorber alkali monitor (RABSAM) for *in situ* measurement of the alkali vapor in the PFBC exhaust gas. The commercial-grade activated bauxite that was used contained some clay impurities that could react with the alkali vapors. These impurities were deactivated by impregnating fresh activated bauxite with NaCl solution, followed by heat treating, water leaching, acid leaching, and Soxhlet extraction with deionized water.

Lithium chloride (LiCl) was then considered as a possible substitute for NaCl because it is more reactive than NaCl and would decrease the amount of non-chloride sodium produced by deactivation of the clay impurities. Test results showed that the clay materials in raw activated bauxite were completely deactivated by impregnating the material with 6 *M* LiCl solution, followed by heat treatment at 700°C for about 50 h.

Two test runs were made, using fresh activated bauxite treated by the LiCl-solution technique, preconditioned at 895-900°C and 10 atm for 3.5 h with a simulated PFBC off-gas containing no NaCl vapor. The preconditioned material was then tested for NaCl capture efficiency under essentially the same conditions as the preconditioning step, but with the addition of 5-9 ppm (by weight) NaCl. Each test run was for 10 h. The NaCl capture efficiencies were 94.5% in the first test and 96.7% in the second one. Surprisingly, under similar test conditions "fresh" (untreated) activated bauxite had a capture efficiency of >99%. The conclusions from these tests were that (1) the treated activated bauxite behaves similarly to untreated activated bauxite in terms of the capture mechanisms for NaCl vapor in simulated PFBC off-gas, and (2) exposure to NaCl vapor in the off-gas might almost completely deactivate the clay impurities in the untreated activated bauxite.

Measurement of Alkali Metal Vapor/Aerosol in FBC Exhaust Gases. Illinois has the largest bituminous coal deposits of any state in the union. Unfortunately, Illinois coal, although abundant, is "dirty," with a high sulfur content (much of it around 3%) and other undesirable impurities such as chlorine. The specific objective of this effort, which was sponsored by the Illinois Center for Research on Sulfur in Coal, was to evaluate the alkali-vapor corrosivity of PFBC off-gas generated by burning Illinois coals of high-chlorine, high-sulfur content. Two typical Illinois coals were burned in the ANL laboratory-scale fluidized-bed facility, and the alkali-vapor emission was measured by the activated bauxite sorber technique. The results showed that sodium was the major alkali metal species present, and that the alkali metal concentrations were 2.5 times greater than the specified limit of 24 ppb for an industrial gas turbine. It was clear that further work was needed to develop a method to control alkali vapors for turbine protection and that the effectiveness of washing procedures to remove sodium compounds from coal should be evaluated.

A joint project between ANL and Sargent & Lundy was undertaken to determine the effect of coal washing on emissions, costs, and amounts of solid-waste products for an

80-MW(e) circulating fluidized-bed combustor. Tests were conducted by burning these coals in the ANL laboratory-scale FBC facility at atmospheric pressure. A Sargent & Lundy evaluation of the results, when applied to the 80-MW facility, indicated that washing the coal resulted in a greater than 50% reduction in SO_2 emissions, and that 95% sulfur removal was possible for both coals with a limestone sorbent. Nitrogen oxide (NO_x) emissions from the washed coal were about 33% less than those for the untreated coal, and both values were below the acceptable standard. Washing the coal would also reduce by 69% the solid-waste handling burden and area of run-off pond needed. In fact, laboratory analyses showed that toxic metal concentrations in the storage area or leachate would not exceed acceptable levels. A cost analysis based on this information indicated that a plant using washed coal would have an economic advantage.

Because of the success of the ANL activated-bauxite technique for determining the alkali-vapor content of exhaust gases from fluidized-bed combustion units, it attracted interest from several organizations. At the request of a Swedish firm, ABB Carbon, ANL was contracted to measure the alkali vapor/aerosol content of off-gases from a 15-MW(t) Component Test Facility in Finspong, Sweden. The results of these tests were considered proprietary and were not reported. A similar request from Electric Power Development Co. and Ishikawajima-Harima Heavy Industries Co. in Japan involved measurements of the alkali metal/aerosol content of off-gases at the Wakamatsu 71-MW(e) PFBC Demonstration Plant in Kitakyushu City, Japan. The necessary instrumentation was designed and built and sent to Japan along with a documentation of the operating procedure. In the U.S., ANL was contracted to measure the alkali vapor/aerosol content of flue gas from an atmospheric circulating fluidized-bed combustor fabricated by Ahlstrom Pyropower. This was a subcontract to Ahlstrom Pyropower's contract with the Illinois Clean Coal Institute. Tests with three Illinois No. 6 coals indicated time-averaged concentrations of 49, 47, and 44 ppb, which are about twice the acceptable limit of 24 ppb for turbine use. Further tests showed that in-bed getter materials (emathlite and kaolin) reduced the levels to 19 and 23 ppb, respectively.

Evaluation of Illinois Limestones for Reducing SO_2 and HCl Emissions. In this CRSC study, Illinois limestones and dolomites were evaluated for reducing SO_2 and HCl emissions from the ANL fluidized-bed combustor (FBC) with a 15-cm dia. The test was under excess air conditions typical of a single-stage, bubbling-bed, pressurized FBC. The sorbents were a Fredonia limestone, a Racine/Sugar Run dolomite, and a Racine dolomite from Illinois quarries. The fuel was a Herron No. 6 coal, which contained 2.6 wt% sulfur, and the operating conditions were 8 atm absolute pressure, 3.5% O_2 in the fuel gas, 0.9-m bed height, Ca/S mole ratios of 0.75 to 4.6, and temperatures of approximately 800-900°C. The sulfur retention ranged from 67.4% with the Fredonia limestone at a Ca/S ratio of 1.3 to a high of 98.0% with the same limestone at a Ca/S ratio of 4.6. As expected, the sulfur retention increased markedly with increasing Ca/S ratio, but only slightly with increasing temperature. The two dolomites were more effective than the limestone.

Continuing studies showed that these results were consistent with the physical characteristics of the sorbents, as determined by mercury porosimetry and nitrogen BET measurements of surface area and pore-size distribution. Measurements of HCl in the flue gases during the combustion tests indicated that the sorbent was not effective in reducing HCl emissions, which was an expected result.

The Fredonia limestone sorbent was also tested under fuel-rich combustion conditions, *i.e.*, an air/fuel ratio <1.0. Sulfur retention, which varied from 80 to 90%, increased with increasing Ca/S ratio and with decreasing air/fuel ratio. An important conclusion from these findings is that a staged FBC may offer significant environmental and operational advantages over conventional FBC systems.

Selection of Non-Alkali-Adsorbing Components. During the on-line, time-averaged measurements of alkali-vapor concentrations in PFBC off-gases, the results were found to be distorted through capture of the alkali vapor by the Type 304 stainless steel sampling line. A joint study was undertaken with the MCT Division to investigate this phenomenon and to identify materials that do not interact with the alkali vapors and would therefore be suitable for sampling lines and other components such as valves, piping, filters, and separation membranes. Thermogravimetric analysis (TGA) was used to evaluate candidate materials, which included Fe-, Ni-, and Co-based alloys as well as noble-metal-coated and ceramic-lined alloys. After the thermogravimetric analysis, the specimens were examined in detail by scanning electron microscopy (SEM) coupled with energy-dispersive X-ray analysis (EDAX) and Auger electron spectroscopy. In nearly all cases, an oxide scale had formed on the surfaces of the specimens. Coatings of noble metals (Pt, Au, Ag) did not prevent the scale formation. Aluminizing of the alloys followed by heating in air produced a coating that remained intact after the 100-h exposure to the baseline PFBC environment. In none of these cases was any sulfur found in the surface layers.

Parametric Study of N_2O Emissions from FBC. The amount of nitrous oxide (N_2O) present in FBC off-gases and the effect of operating conditions on the amount were not well known. Nitrous oxide was thought to be one of the pollutants contributing to the greenhouse effect and depletion of ozone in the upper atmosphere. Some confusion existed as to whether it was subject to emissions regulations; moreover, the analytical method used to determine NO_x emissions compliance is not sensitive to N_2O. A parametric study was done with the laboratory-scale (15-cm-dia) FBC at ANL to shed more light on N_2O emissions.

Two Illinois coals from Kerr-McGee Coal Co. and Peabody Coal Co. were burned in the FBC, and the N_2O emissions were measured by means of an on-line infrared analyzer and gas chromatograph. The SO_2 and NO_x levels were also determined. Measurements were made over a range of temperatures, excess air levels, fluidizing velocities, and pressures. The results were as follows:

1. The N_2O concentration decreased substantially as the temperature was raised.
2. The concentration increased with increasing excess air.
3. Increased residence time (the inverse of fluidizing velocity) tended to decrease the N_2O emissions.
4. Increasing pressure appeared to increase the N_2O levels somewhat.
5. Increasing the SO_2 level (by decreasing the sorbent feed rate) tended to increase the N_2O level.
6. In general, N_2O levels observed in these experiments ranged from about 40 to 150 ppm.

Personnel. Full-time members of CMT who worked on this program were Erv Carls, Sheldon Lee, Walt Podolski, Bill Swift, and Gale Teats. Several members from other divisions and Shelby Miller, an STA, also contributed to the effort. Swift was in charge of the work during the 1990s.

MAGNETOHYDRODYNAMIC STUDIES

Chemical Technology continued as the lead division for the ANL program on magnetohydrodynamics (MHD); the work on materials was done jointly with the MCT Division, and the mathematical modeling was a joint effort with the ES Division.

Materials Evaluations. Work continued on the development of open-cycle MHD systems in an effort to improve the electrical efficiency of coal-fired power plants and to reduce their environmental impact. Although the primary objective of the MHD plant is to extract heat from the combustion gases to produce high-pressure steam, it must also separate the seed material (usually a potassium salt) from the ash and recover it for reuse, preheat the primary combustion air to about 700°C or higher, lower nitrous oxide (N_2O) concentrations to acceptable levels, and inject secondary air to complete combustion of the fuel. The fuel-rich mixture leaving the topping cycle at about 2000°C enters a steam-bottoming cycle similar to that in a conventional coal-fired power plant.

Corrosion of boiler tubes by sulfur and alkalis in the combustion gases is a concern in conventional coal plants, and the problem is magnified in MHD plants because of the higher alkali content of the gases. Laboratory autoclave tests were performed to study the corrosion of candidate materials for components of the bottoming cycle of an MHD plant. Examinations were conducted on specimens of tube materials used in long-duration (2000-h) proof-of-concept tests with Illinois No. 6 coal in the low mass flow train (LMFT) at the Coal Fired Flow Facility (CFFF). The CFFF, which was operated for DOE by the University of Tennessee Space Institute, simulated the MHD bottoming cycle. Analyses of deposits on the tube samples taken at various times and locations indicated that the chemistry of the deposit did not change significantly with variations in the K_2/S ratio in the potassium seed material. The deposits in all locations were predominantly soluble potassium sulfate with no carbonate or bicarbonate. Of the materials tested, Type 310 stainless steel showed the least corrosion, while Types 304 and 316 were attacked severely.

Materials-evaluation studies were also made on the topping cycle. A piece of a cooling panel made of SA 387, a low chromium-molybdenum steel, that was tested for 250 h showed no sulfur in the gas-side scale, even though the coal being burned contained 3.6 wt% sulfur. Among 11 candidate alloys that were corrosion-tested for cooling circuits in the MHD channel, tungsten-copper alloys showed the best resistance. Tests were also made on the corrosion resistance of boron nitride and silicon nitride insulating materials in a potassium environment at temperatures of 538 and 760°C. Also measured were the electrical conductivities of the specimens before and after exposure to the potassium. The boron nitride was attacked and its conductivity increased markedly, whereas the silicon nitride showed no ill effects. Therefore, silicon nitride appears to be the material of choice.

Power Train Modeling. The Component Development and Integration Facility (CDIF) is an engineering-scale test facility for MHD topping cycle components that is operated by MSE, Inc., for DOE. In an effort to interpret and understand the MHD generator operation at CDIF, the CMT and ES Divisions began a joint effort to develop and use computer-generated simulations to investigate flow patterns and nonuniformities (*e.g.*, in temperature and plasma conductivity) in the major components of the MHD power train.

The combustion code that was developed solved transport equations for multiple gaseous species and solid particles of various sizes. Conservation laws were used in conjunction with rate equations governing the mass, momentum, chemical species composition, turbulent kinetic energy, and turbulent dissipation for a two-phase reacting flow. A sample calculation was made by using both the combustion and channel codes to simulate the second-stage combustor and MHD channel at CDIF. Results were obtained for the predicted properties (power, conductivity, and velocity) of the plasma flow in the MHD channel.

Personnel. Bill Swift and Dai-Kai Sze of CMT were involved in the MHD work, but most of the experimental effort was provided by individuals from the Energy Technology (ET) and Technology Development (TD) Divisions.

Hazardous and Mixed Waste Research

This program was concerned with the management and disposal of a wide variety of troublesome waste materials, including municipal wastes, toxic materials, and transuranium radioactive wastes that had been produced during defense-related work at DOE sites. Jim Cunnane, Jim Helt, and George Vandegrift had management responsibilities for these programs.

MODELING STUDIES

A mathematical model was developed to simulate the pyrolysis and combustion of municipal solid waste (MSW) in a water-wall mass burner. This process involved the heatup, pyrolysis, and burnout of the waste material as it was moved by a traveling grate from the inlet of the incinerator to the exit. The model permitted calculation of the temperatures and mass-flow profiles over the length of the bed. The model showed that: (1) the grate length in the reference design was longer than necessary to heat the material to the pyrolysis temperature, (2) an increased solids feed rate caused a slowdown of the heating and conversion process, and (3) bed height had a negligible effect on the solids heating rate. The simulations also showed that the solids processing rate could be increased by reducing the length of the burnout zone and by distributing the air flow, so it was kept low in the heatup and reaction zones to permit rapid heating and conversion of the solids. A two-dimensional model was considered to be a good beginning for the study, and the researchers concluded that to proceed to the next level more data were needed on the pyrolysis and combustion behavior of MSW components such as plastics, hemicellulose, *etc.* Mechanisms of the combustion reactions of char also needed to be included.

During the combustion process, radiant heat from the flame zone heats up the bed as it is carried across the grate, then the moisture in the MSW is vaporized and carried up into the flame zone by the air flowing in from under the grate. The rate of drying can be expressed by an Arrhenius function and depends upon the temperature and the moisture content. With a further rise of temperature, the cellulose fraction of the MSW begins to pyrolyze, yielding several gases (hydrogen, water, carbon oxides, and methane), which are transported into the flame zone by the air. Pyrolysis of the cellulose also yields a char (carbon) and ash residue in the bed. After a review of the literature on char combustion, empirical equations were developed to estimate the rate of char combustion, which turned out to be higher that the rate of cellulose pyrolysis. Consequently, char combustion was a minor consideration in the combustor design. The main investigators in this program were Jack Demirgian, Christine Snyder, and Milton Haas.

MICROWAVE-ASSISTED DETOXIFICATION

Safe and effective disposal methods were needed for chlorinated hydrocarbons such as 1,1,1-trichloroethane (TCA) and trichloroethylene (TCE), which had become a major hazardous waste problem at DOE Defense Program sites. The objective of this project was to evaluate the technical and economic aspects of using microwaves to detoxify these materials by converting them to easily disposable compounds such as CO_2, CO, HCl, H_2O, and H_2. In previous studies, a 6-kW, 2.45-GHz microwave generator with a 2.1-cm-dia quartz reactor was used to produce an oxygen/argon plasma (20.13% oxygen) for investigation of the oxidative degradation of TCA and TCE. Destructive removal efficiencies of 95.9-98.3% were achieved by a single pass of the stream through the reactor. In the case of the TCA, however, a significant portion of the chlorine was not totally converted to the desired products.

Experiments were then conducted in which water vapor was added to the argon/oxygen with the idea that hydrogen in the water would tend to remove the chlorine as HCl, and the oxygen would remove carbon as CO or CO_2:

$$C_2H_3Cl_3 + 4\ H_2O \leftrightarrow 2\ CO_2 + 4\ H_2 + 3\ HCl$$

Single-pass destructive removal efficiencies were >97% for TCA.

This work continued with a study of the reaction of TCA and oxygen, using a 7-kW microwave generator. The operating conditions were varied (O_2/TCA molar feed ratios from 1.5 to 165 and microwave power inputs from 300 to 600 W). The TCA conversion ranged from 75% at 300 W and an O_2/TCA ratio of 1.5 to >99% at 600 W and a ratio of 150. The products of this reaction with excess oxygen and high microwave power input (600 W) were CO_2 and HCl. A reduction of the O_2 concentration or power input resulted in the formation of soot, CO, and hydrocarbons. In the case of TCE, power inputs of 600-800 W and O_2/TCE molar ratios of 13 to 190 gave conversions of >96%. Because the TCE contains insufficient hydrogen to convert all the chlorine to HCl, the excess chlorine is removed as Cl_2. This problem was solved by adding water vapor to the oxygen in the argon plasma. With the water addition, the only reaction products were CO_2 and HCl.

In 1992, work was started on a modification of the process whereby an air plasma with water as a co-reactant was substituted for the argon/oxygen mixture. (One potential application of the process was to detoxify air streams containing volatile organics that were generated by air stripping of contaminated soils.) A major concern with this approach, however, was the possibility that NO_x could be a problem because nitrogen fixation from air is more likely at the high plasma temperatures than at the more modest temperatures in a conventional thermal combustor. This project, however, was terminated at the end of 1992.

Jim Helt, Ravi Varma, and Ted Krause were responsible for much of this work.

AQUEOUS BIPHASE PROCESSES

When a suitable combination of soluble inorganic salts and water-soluble polymers such as polyethylene glycol (PEG) is added to water, the system separates into two liquid phases with the polymer phase on top and the salt phase on the bottom. In some respects, this kind of a system, which is called an "aqueous biphase," is similar to the combination of aqueous and organic solvents used in solvent-extraction processes. Separations in biphase systems, however, generally depend upon differences in the distribution of insoluble solids, rather than soluble species between the liquid phases. Very fine (colloid-size) solid particles, when suspended in an aqueous biphase system, will seek one phase or the other, depending on the particle interactions with the particular solvent. The water-soluble polymers used in biphase separations, unlike the organic compounds used in solvent-extraction processes, are nonflammable, inexpensive, nontoxic, and biodegradable.

The initial objective of this research was to develop a scheme for the recovery of plutonium from solid wastes stored at DOE sites to levels where the plutonium concentration would meet or exceed "economic discard limits," and the recovered plutonium would be compatible with existing plutonium recovery processes. As the work progressed, however, several other potential applications of aqueous biphase separations became apparent.

Preliminary Studies. The initial experiments showed that single-stage separation factors $>10^6$ could be obtained between silica and a wide variety of metal oxides. For example, with a biphase system consisting of 15% PEG/7.5% Na_2SO_4 at pH = 3, >99.99% of metal oxides such as Fe_2O_3 and Al_2O_3 partitioned to the Na_2SO_4 (bottom) phase, and >99.99% of the silica particles partitioned to the PEG-containing (top) phase. The partitioning behavior of the solid particles depended upon a number of variables, including the heat of immersion (ΔH_i), the pH of the system, the nature of the bonding of sorbed species, and the effects of surfactants.

Plutonium Recovery. The plutonium wastes usually consisted largely of silica or graphite resulting from foundry operations. To prepare the solid waste material for the biphasic extraction, it was first wet ground to a particle size of approximately 1 µm. The heterogeneous nature of the plutonium-bearing solid wastes and their physical properties were such that most of the PuO_2 was expected to be in the form of micron-size grains after the ultrafine grinding step. In the initial plutonium studies, polymeric Pu(IV) was used as a stand-in for particulate PuO_2. (Detailed characterization studies had shown that these two forms of plutonium shared the same chemical and physical properties that are of importance in this type of separation.) Preliminary studies of the partitioning behavior of monomeric Pu(IV), polymeric Pu(IV), and micron-size PuO_2 particles between 15% PEG/7.5% Na_2SO_4 and 15% PEG/15% $(NH_4)_2SO_4$ showed that efficient plutonium recovery should be possible in alkaline solutions (pH >3).

Experiments were then conducted with polymeric Pu(IV) and silica or graphite, which are the major constituents of foundry residues and incinerator ashes. In a test with SiO_2 ground to 0.2-1.0 µm, the silica distributed to the upper (PEG) phase and the Pu to the lower phase, with a separation factor of $>10^6$ in a single-stage separation. In a similar experiment with graphite (1-2 µm), the graphite distributed strongly to the PEG phase, but the plutonium polymer apparently tended to adsorb on the graphite particles and accompany the graphite into the PEG phase. Through the use of a suitable surfactant, however, the partition coefficient of plutonium was decreased from 4.0 to 0.04, thereby increasing the separation factor to a value of $>10^4$.

In addition to selective partitioning between the liquid phases in these systems, selective flocculation occurs at high ionic strengths. In the 15% PEG/7.5% Na_2SO_4 system, CeO_2 particles became flocculated in the salt phase, and the SiO_2 particles flocculated in the PEG phase. Solid-liquid separations were then achieved readily either by mild centrifugation or by microfiltration. When the two phases were centrifuged in the same tube, the solids appeared as sharply defined bands at the bottom of the tube.

The foregoing observations suggested that several options were available for designing a process flowsheet for plutonium recovery from process wastes. Samples of such wastes were obtained from the Rocky Flats Plant, Los Alamos National Laboratory, the Hanford site, and the Savannah River Laboratory. To aid in the flowsheet development, phase diagrams were determined for PEG/salt systems of interest. A flowsheet for the aqueous biphase separation (ABS) process would include wet grinding of the waste material to a micron-size level to liberate the refractory PuO_2, selective partitioning between the two immiscible aqueous phases, a solid-liquid separation, and recycle of the aqueous phases. The goal was to recover most of the plutonium in a concentrate that was only 1-5% of the initial waste volume. When the concentration of the extractant was increased, the plutonium-containing particles distributed to the PEG phase with partition coefficients up to about 300, and the particles without plutonium remained in the salt phase.

Uranium Recovery. Aqueous biphase separations were also considered for possible use in remediation of uranium-contaminated soil, a typical example being soil at the DOE Defense Production Facility at Fernald, Ohio. The objective of such a process was to remove sub-micron particles of UO_2 from the clay soil and recover a uranium concentrate having a volume less than 2 to 4% of that of the feed, and to leave a "clean" soil fraction containing less than 35 pCi of uranium per gram of soil.

Laboratory experiments were conducted by use of a model system consisting of kaolinite and submicron-size UO_2. When feed containing 200,000 ppm uranium was suspended in a 15% PEG/10% Na_2CO_3 system and scrubbed several times with a Na_2CO_3 solution, the recovered clay product contained 45 ppm uranium, which was below the 35 pCi/g limit. Batch extractions were then made with a <50-μm size fraction of Fernald soil, which represented about 70-80% of the total. "Clean" soil containing about 125 ppm uranium was recovered from the top (PEG) phase, while the uranium concentrated in the bottom in a volume equal to about 1% of the initial soil volume. These results suggested that the goal of 50 ppm uranium or less in the soil was achievable through the use of countercurrent extractions and pretreatment of the soil by attrition scrubbing.

Further test-tube-scale tests were conducted with the Fernald soils. The ABS process worked reasonably well with soils collected near the incinerator at Fernald, but those samples from storage areas did not respond as well.

In 1993, scale-up tests were made with uncontaminated clay feeds, in the pilot-plant facilities of the Otto H. York Co. in Houston, Texas. Clay was fed into a Karr column (2.54-cm ID and 3.6-m high) at a rate of 70 g/h. The column used countercurrent feed solutions of 30% PEG and 20% Na_2CO_3. The results indicated that a full-scale column could achieve solid throughputs of about 600 to 1200 kg/h, which should make this type of processing less expensive than landfill disposal. In a pilot-scale test with the Fernald soil in the Karr column in 1994, the uranium content of the soil was reduced from 471 mg/kg to about 78 mg/kg. Electron microscopy revealed that some of the uranium particles had remained in the soil. Because one of the major concerns with the process was the capital equipment and processing costs for separating the soil product from the effluent stream, the possibility of using flocculants was investigated with some success.

A preliminary process flowsheet was then developed for uranium removal from soils by an ABS system having a capacity of 20 ton/h of dry soil. The treatment cost was projected as $120/ton.

Waste Pretreatment. A collaborative effort was undertaken with Northern Illinois

University to investigate the use of biphasic systems for the extraction of long-lived radionuclides such as technetium-99, iodine-129, and selenium-75 from caustic high-level liquid wastes. Initial experiments were performed to examine the ability of aqueous PEG solutions to extract iodide and pertechnetate (TcO_4^{2-}) selectively from simulated supernatant solutions in the underground waste storage tanks at the DOE Hanford site. Because of the high ionic strengths of the caustic waste solutions, aqueous biphasic systems could be generated simply by adding aqueous PEG solutions directly to the waste solutions.

Tests were conducted with three Hanford waste-tank simulants. In all cases the separation was rapid, and the single-stage separation factors for pertechnetate ranged from 79 to 26,500, depending on the anionic species in the system. In general, separation factors were higher in the presence of polyvalent anions such as carbonate (CO_3^{2-}). These results indicated that it should be possible to achieve technetium decontamination factors well above 10^3 in a multistage, countercurrent extraction. Neither irradiation nor temperature appeared to affect the partition coefficients of the technetium.

Either PEG or PPG can be used to form the biphasic systems. Separation factors for technetium in the PPG system, listed in Table 6-6, are considerably higher than those in the PEG system. The high separation factor between pertechnetate and nitrate (NO_3^-) is significant because that has always been recognized as a particularly difficult separation.

Metal Separations. The biphasic systems used in most of the separations consisted of immiscible PEG and salt solutions. Some inorganic salts that promote biphase formation are the sodium and potassium salts of sulfate, carbonate, phosphate, and hydroxide. A major problem that arose in using biphasic separations for metal-ion extraction was an inability to recover solutes efficiently from the polymer-rich phase by back-extraction. This problem was solved by a discovery that the metal complex could be extracted efficiently from the PEG phase by a water-immiscible alcohol such as n-butanol at an elevated temperature (60-80°C). The partition coefficient for PEG in the butanol-water system was typically about 15, while that for the salts was 10^{-3} or less. Although the PEG partition coefficient is marginal, mass transfer of the PEG into the alcohol causes a large volume change, which results in >99.4% PEG recovery. The PEG can then be recovered by back-extraction into a dilute salt solution at 20°C. When this procedure is incorporated into a process flowsheet involving actinides, the actinide product is concentrated in a small volume of low-salt solution, which is suitable

Table 6-6. Separation Factors for Pertechnetate in PPG Systems

Ionic Species	Separation Factor
NO_3^-	1,049
NO_2^-	2,297
CO_3^{2-}	3,148
Na^+	2,576
$Al(OH)_x^{3-x}$	10,500

for vitrification and disposal. The original salt solution that was fed to the process, now free of the actinides, can be recycled.

Biocatalytic Destruction of Nitrate. The presence of high nitrate (NO_3^-) and nitrite (NO_2^-) concentrations, often of the order of 4 *M*, in DOE mixed waste streams creates problems in establishing suitable final waste forms and disposal modes. Disposal of low-level wastes by near-surface grouting requires grout formulations that will minimize the leachability of the nitrates, and the vitrification process for high-level wastes to form ceramic glasses produces volatile nitrogen oxides in the off-gases. The objective of this work was to convert the nitrates and nitrites to oxygen and water by use of nitrate, nitrite, and nitrous oxide reductase enzymes. Such enzymes result in very high specific catalytic activity without a need for additional chemical reagents or production of secondary waste streams.

In this process, an aqueous biphasic system consisting of a salt-rich aqueous phase and a PEG-rich aqueous phase is used to protect the enzymes and to facilitate removal of the nitrate/nitrite from the waste stream. The enzymes stay in the PEG phase, where they are not exposed to the high ionic strength of the waste stream. The nitrates and nitrites transfer into the PEG phase while the radionuclides remain in the salt phase. A schematic illustration of the process is shown in Fig. 6-7.

The enzymes are immobilized on the surface of an electrode, and an electron-transfer agent, which is also immobilized on the electrode surface, shuttles electrons from the electrode to the enzyme. The overall chemical reaction for the reduction of nitrate to nitrogen is

$$2\,NO_3^- + 12\,H^+ + 10\,e^- \rightarrow N_2 + 6\,H_2O$$

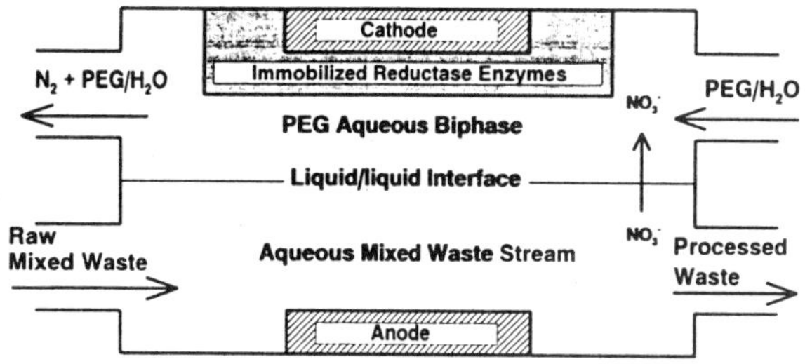

Fig. 6-7. Schematic for Biocatalytic Destruction of Nitrate

The reaction occurs through a sequence of steps, $NO_3^- \rightarrow NO_2^- \rightarrow N_2O \rightarrow N_2$, so any nitrite or N_2O in the waste stream will also be reduced. The reductase and electron-transfer agent were immobilized on the gold electrode surface by an electrodeposition procedure. The enzymes retained their catalytic activity over an extended period. Plans were made to demonstrate the process on a surrogate waste stream. The aqueous biphase studies were led by Dave Chaiko. Others involved in the work were Adam Ellison, Gary Henriksen, John Kopasz, Ted Krause, Robert Mensah-Biney, Carol Mertz, and Erv Van Deventer.

ELECTROKINETIC PROCESSES

A major problem at DOE sites is the presence of metals, particularly actinides and transuranics, and/or hazardous organic compounds such as chlorinated and aromatic hydrocarbons in the soil. Electrokinetics, which refers to the movement of charged particles through a porous medium under the influence of an applied electric field, offers the possibility of *in situ* remediation of contaminated soils. The two principal forms of electrokinetics are electroosmosis and electrophoresis. The DOE had asked ANL to investigate a soil decontamination process being developed by Isotron Coating Technology of New Orleans, Louisiana. This procedure, called the "Electro-sorb Soil Decontamination Process," involved the superimposition of oscillating waves, either electromagnetic (*e.g.*, microwave or electric power) or mechanical (*e.g.*, sonic or ultrasonic), on a normal dc electric field. Argonne was to determine whether the superimposed waves did, in fact, enhance the electrokinetic recovery process.

Argonne proposed a separate study to develop electrokinetic processes to enhance the molecular transport rates in systems of this type. The ANL effort was to investigate novel catalysts to destroy or detoxify hazardous organic chemicals in contaminated soils and groundwaters. A major engineering challenge was to bring the catalyst and the contaminant in contact in the soils, which are complex structures that inhibit transport rates and complicate flow patterns. These effects can be minimized by applying an electrical potential to the system to generate an electroosmotic flow.

The overall efficiency of the electrokinetic process depended on a multiplicity of variables, including the mobilities of ions and charged particles, viscosity of the groundwater, ion concentration, hydration of ions and charged particles, dielectric constant of the medium, and the temperature.

Experiments were initiated on the effect of temperature on the efficiency of electrokinetic remediation for the extraction of potassium dichromate ($K_2Cr_2O_7$) from kaolinite samples. This system was selected in anticipation of a future field demonstration on a chromate waste site at Sandia National Laboratories. In general, the rate of chromium migration was increased by raising the temperature (from 21 to 55°C). The overall results were complicated, however, by the many physical and chemical mechanisms involved in the system.

Some work was done on the use of silica gel barriers to inhibit the migration of contaminants, such as radionuclides, heavy metals, and hazardous organic compounds, through soil subsurface as a means of environmental remediation and waste management. The silica gel barriers are formed by destabilizing a colloidal silica sol by two steps—lowering the pH and increasing the salt concentration. Hydraulic pumping is used to inject the sol mixture into the soil, but it presents a number of difficulties, including the need for multiple cycles to fill all the pores and problems in controlling the flow of the sol solution. Electrokinetic methods of injecting the sol appeared to be an attractive alternative to pumping because the electro-osmotic effect fills all the pores uniformly, and the transport can be controlled by the positions of the electrodes in the soil. In bench-scale electrokinetic tests at Lehigh University, the procedure resulted in a great decrease in the permeability of the soil, but in kaolinite soil the permeability actually increased. The kaolinite fibers were not cemented, and silica gel fibers were extruded from the sample.

CHEMICALLY BONDED WASTE FORMS

A cooperative project between the Energy Technology (ET) Division and CMT was undertaken to evaluate the suitability of phosphate inorganic binders for solidification and stabilization of mixed wastes (heavy metals and actinides) that are not suitable for thermal processes. The rationale was that certain phosphate mineral phases such as monazites and apatite occur in natural environments and can incorporate thorium, uranium, and rare earth elements. Samples of magnesium and zirconium phosphate waste forms containing Cd, Cr, Ni, and Pb were subjected to standard leaching tests. Untreated ash greatly exceeded the EPA limits, while the treated material was well below the limits.

The fact that these phosphate wastes contain water of crystallization led to concerns that hydrogen might be produced. In addition to hydrogen production, oxygen depletion occurred in the vessels, which were filled with air. Much of the oxygen was in the form of CO and CO_2, and more was consumed by formation of water and other condensable molecules.

An alternative approach to converting the zeolite wastes into an acceptable waste form for storage was to incorporate the zeolite into a low-temperature phosphate glass. Zinc phosphate, which has a glass-transition temperature in the range of 325-375°C, was a candidate glass for this purpose. Small-scale composite monoliths containing cesium-loaded crystalline silicotitanate, zeolite, and sodalite were prepared by cold pressing and sintering at 460°C. Cesium release from these materials in standard 3-day leach tests was extremely low. Other test indicated that cesium has about the same fractional release rate as Zn, Na, K, and Li from the salt-free glass. These studies were led by Don Reed. Others involved in the effort were Allen Bakel, Edgar Buck, Jim Cunnane, Don Fischer, Terry Johnson, and Michele Lewis.

RADIOLYTIC GAS-GENERATION STUDIES

Radiolytic gas-generation studies are being conducted on a chemically bonded phosphate ceramic called "Ceramicrete," which is being developed for encapsulation of TRU-bearing wastes. Radiation from the radionuclides produces hydrogen and smaller amounts of other gases by radiolysis of water and other materials present in the waste form. Among the experiments that were performed was a comparison of the hydrogen produced in a plutonium-loaded Ceramicrete sample. The sample, which originally contained 18.9 wt% H_2O, had a G value of 0.37 molecules H_2/100 eV, but when the sample was dried to a water content of about 1 wt%, the G value was only 0.04 molecules of H_2/100 eV. Other types of samples gave somewhat mixed results, and further work is in progress.

ACTINIDE STABILITY/SOLUBILITY STUDIES

The Waste Isolation Pilot Plant (WIPP), located in Carlsbad, New Mexico, is a possible disposal site for TRU radioactive wastes. Studies were undertaken to provide experimental data for WIPP on actinide solubilities and oxidation-state distributions in WIPP brines having pH values between 5 and 10. Steady-state concentrations of Pu(VI) were maintained in the brine over the 18-month duration of the experiments. Additions of reducing agents relevant to the WIPP reduced the Pu(VI) to Pu(IV) and lowered the steady-state concentrations of plutonium in the brine solutions.

Organic chelating agents expected to be present in the brines also caused the reduction of Pu(VI) and Pu(IV). Work is continuing in collaboration with Sandia National Laboratories in which the long-term data that are generated are being used to challenge their high-ionic-strength models for actinide solubility.

ACTINIDE SPECIATION IN GROUNDWATER

These studies, which involved at various stages the collaboration of Pacific Northwest Laboratory, Brookhaven National Laboratory, Lawrence Berkeley Laboratory, and Northwestern University, were concerned with the speciation (*i.e.*, complexation, oxidation states, and aggregation) of radionuclides under conditions relevant to the subsurface groundwaters at DOE installations. The initial investigations were aimed at the hydrolysis of Pu(VI), which was known to be much more soluble than Pu(IV) and was therefore more likely to have high migration rates in the soil. Those studies were followed by investigations of the interactions of Pu(VI) with organic complexing agents that were likely to be present in the soil as a consequence of plutonium processing operations. That work led into studies of the effects of microbes on the plutonium reactions and the effects of plutonium on the microbes. These studies employed high-intensity laser spectroscopic methods supplemented by several more conventional spectroscopic and radiochemical methods.

Because of some uncertainty as to whether the hydrolysis product of Pu(VI) was PuO_2OH^+ or $(PuO_2OH)_2^{2+}$, spectroscopic measurements were made on the hydrolysis products of Pu(VI) in 0.1 *M* sodium perchlorate solution as a function of pH, Pu(VI) concentration, and temperature. Increasing the pH above 3 resulted in the formation of hydrolysis products; four distinct species were found in the pH range of 3 to 7. The extent of hydrolysis and the contribution of polynuclear hydrolysis products were much larger than previously thought. The temperature variable measurements (10-45°C) led to the first experimental determinations of the thermochemical properties of the hydrolysis products. The effect of temperature was much less than previous estimates in the literature.

The interactions of Pu(VI) with organic complexing agents (citric acid and acetohydroxamic acid) in aqueous systems were examined as a function of pH. At pH greater than 2, the Pu(VI) was reduced to Pu(III) by the acetohydroxamic acid and to Pu (IV) by the citric acid. In further work on the citrate system, the interaction of Pu(VI) with citric acid was investigated as a function of ligand-to-metal ratio (1:1 to 100:1) and pH (2 to 10). At ligand-to-metal ratios of less than 10:1, citrate reduced the Pu(VI) to form a stable Pu(IV) complex, and the reaction rate became higher with increasing pH. At ligand-to-metal ratios of about 100:1, no Pu(VI) reduction occurred, even at near-neutral pH, which suggests that multiligand complexation of the Pu(VI) stabilizes the oxidation state.

Some preliminary studies were conducted on the degradation of the plutonium citrate and nitrilotriacetic acid (NTA) complexes by a *Pseudomonas* strain microbe. A stable Pu(IV)-citrate complex that was not biotically oxidized or reduced was observed, and about 20% of the plutonium initially present was associated with the microbe. In the case of the NTA complex, the existence of the complex did not prevent conversion of the NTA to CO_2, but it decreased the reaction rate slightly. The NTA complex was of particular interest because NTA was a complexing agent associated with plutonium processing. At near-neutral pH (6 to 8), the Pu(IV)-NTA complex was stable over a period of months. In the presence of the *chelatobacter* microbe, however, the NTA was metabolized, destabilizing the complex and forming plutonium polymers. A set of conditions was established in such a way that the rate of the biodegradation process could be monitored by monitoring the CO_2 formation. Up to 80% of the plutonium was associated with the microbial fraction after the reaction.

The radiation tolerance of the microbes was a factor of importance to this work. In the NTA growth medium, the presence of plutonium proved to have a detrimental effect, with a survivability of less than 1% at the highest plutonium concentrations (10^{-5} *M*). The radiation tolerance of the *halobium* microbe, which was being considered for use in the bioremediation of DOE sites, was evaluated by measuring the survivability as a function of absorbed doses of gamma radiation. The radiation tolerance of this microbe was about 1 Mrad, which led to the conclusion that only radiation from the uptake of plutonium caused radiation damage when plutonium was the sole source of ionizing radiation. This study was followed by an investigation of the interactions of the plutonium-NTA complex with the *Chelatobacter heintzii,* which was known to be a degrader of the NTA complexant. Increasing concentrations of plutonium up to 0.01 m*M* resulted in a lower rate of NTA degradation. The NTA tied up in the plutonium complex was not readily accessible to the microbe for degradation, and the ionizing radiation reduced the survivability of the microbes.

In a continuation of this work, gamma irradiation experiments showed that over 99% of the *C. heintzii* bacteria was killed by an exposure of 18.5×10^3 rad. When Pu-242 was substituted for Pu-239, the cell death rate was proportional to the activity of the plutonium rather than to its concentration. The plutonium radiation dose to kill the bacteria was far less than the gamma dose. Apparently, association of plutonium with the bacteria was a critical step in the loss of cell viability in that alpha-particle deposition in or near the cell was needed to account for the toxicity. At low plutonium concentrations ($<10^{-4}$ m*M*), bioassociation of the plutonium was rapid and occurred even in the presence of the NTA complexing agent. At higher plutonium concentrations, association of the plutonium with the biomass was slow and was attributed to destabilization of the Pu-NTA complex and subsequent plutonium polymer formation. An effort was undertaken at Northwestern University to develop a biodegradation model that would assist in correlating data and developing a more comprehensive picture of the biodegradation processes.

In 1997, work continued on the interactions of multivalent actinide species with bacteria. Investigations were conducted on the interaction of the actinide-nitrilotriacetic (NTA) complex with the NTA-degrading microbe *Chelatobacter heintzii* in aqueous solution. This research involved extensive integration of computer modeling with laboratory experiments. Results to date show that actinide toxicity depends upon the oxidation state and isotope of the actinide. For example, Pu(IV) toxicity is radiolytic in nature, while Np(V) toxicity is chemical and similar to that of other heavy metals.

The actinide speciation work, under Don Reed, included Del Bowers, Jim Cunnane, Jim Helt, Shigeo Okajima, and Mike Richmann.

OFFICE OF WASTE MANAGEMENT

The Office of Waste Management in CMT assisted DOE in a national applied R&D program of environmental restoration and waste management. The objective was to improve the efficiency, safety, and timeliness of environmental cleanup activities so as to meet DOE's 30-year environmental compliance and cleanup goals. This activity consisted of technical and management support to the Minimum Additive Waste Stabilization (MAWS) Program and the Hazardous Substance Research Center (HSRC) Program, which was co-funded by DOE and the Environmental Protection Agency (EPA). The objective of the MAWS program was to separate the hazardous from the nonhazardous components of the waste and vitrify the hazardous portion in a small volume. The MAWS technology had potential application to a wide variety of DOE wastes such as soils, tank wastes, sludge pits, ashes, and asbestos at various sites, including Hanford, Savannah River, Rocky Flats, Oak Ridge, and Fernald. The HSRC program involved five HSRCs in different regions of the country, each one being located at the site of a main university in a consortium of universities. The research in this program was wide-ranging, but all aimed at developing a better understanding of the nature of the contaminated areas and different approaches to remedial actions. A third task was then added to this program in the form of technical assistance to DOE and coordination of information from the Integrated Program for Development of In-Situ Remediation Technologies. This office was disbanded in 1995. Jim Helt bore the primary responsibility for this effort.

PERSONNEL

Other participants in the above programs include the following: Scott Aase, Jim Banaszak, John Basco, John Bates, Nick Beskid, Dave Chamberlain, Cliff Conner, Jim Cunnane, Jas Devgun, Bill Ebert, Jeff Emery, Bob Finch, Pat Finn, Jeff Fortner, Margaret Goldberg, Lohman Hafenrichter, Mark Hash, Joe Hoh, Lester Morss, Michael Nole, Kevin Quigley, John Quinn, Lin Simpson, Shiu-Wing Tam, and Stephen Wolf. Kevin Byrne, Martin Clemens, Sheldon Lee, Jian Shu Luo, and Paul Nelson worked on the design of an unvented reactor for the incineration or vitrification of hazardous and mixed radioactive wastes. In addition to these people, a large number of part-time and temporary personnel were involved in this effort.

Nuclear Waste Management

In the 1990s, CMT investigators continued their studies of the corrosion behavior of nuclear waste slated for disposal in the proposed Yucca Mountain Repository in southwestern Nevada. In these studies, a variety of tests were performed on the corrosion of simulated high-level waste glasses and spent nuclear fuel under an unsaturated water environment typical of that in the proposed repository. Tests were also conducted to support development of vitrified forms for the disposal of low-level waste generated at the DOE Hanford site, as well as glasses for disposal of surplus plutonium and plutonium-bearing wastes. Contaminated soils at various sites, such as the Fernald plant in Ohio, were characterized to aid in the remediation of those sites.

John Bates was in charge of this work in the early 1990s, reporting first to Jim Battles, then to Jim Laidler, followed by Terry Johnson, and much of the success of the program can be attributed to John's influence. In 1994, Denis Strachan was appointed as Acting Manager of the Waste Management Department, which included Hazardous Waste Management under Jim Cunnane, Nuclear Waste Management under Denis Strachan, and Separation Science and Technology under George Vandegrift. In 1998, Robert Einziger became the head of this department.

TESTING OF HIGH-LEVEL WASTE GLASSES

In long-term test programs, some of which were begun in the 1980s, the corrosion behavior of high-level waste glasses upon exposure to water or water vapor has been followed continuously. As of 1996, some glasses had been under test for 12 years. The purpose of this work was to assist DOE in the qualification of high-level waste glasses for disposal in the Yucca Mountain Repository. The work was aimed primarily at glass compositions similar to those produced at the Savannah River Site in the Defense Waste Processing Facility (DWPF) and at the West Valley Demonstration Project. Other glasses under test are an Environmental Assessment glass, a benchmark glass produced in the DWPF, and a radioactive glass incorporating a Hanford tank sludge. The information obtained from these tests is to be used to predict the long-term (>10,000-y) behavior of high-level glasses under conditions typical of what may be expected in the Yucca Mountain Repository.

Static Dissolution Tests of High-Level Waste Glasses. Borosilicate glass has been chosen as the vitrification material for confining transuranic elements and fission products. Static dissolution tests with simulated high-level wastes are in progress at 90°C and at several sample surface area/solution volume (S/V) ratios. The corrosion mechanism is inferred from the release rate of the glass matrix elements (Li, Na, B, and Si) and the radionuclides, solution pH, and alteration phases that form on the glass surface. (Alteration phases are a variety of minerals, such as hydrates, silicates, and zeolites that form on the glass surface as a result of reaction of the base material with water and dissolved constituents in the water.) In tests up to five years at S/V ratios of 340 and 2,000 m^{-1}, alteration layers formed on both radioactive and nonradioactive Savannah River glasses (SRL 131), but two other glasses of different composition (SRL 165 and 200) had not reached the alteration formation stage. At a S/V ratio of 20,000 m^{-1}, alteration phases had formed on the nonradioactive SRL 200 glass, but not on its radioactive homologue, a difference attributed to a lower pH (by 0.5 to 1.0 pH unit) in the leachate for the radioactive glass. This lower pH is believed to have resulted from radiolysis in the air and solution. An important finding in these tests is that the dissolution rate of the glass increases markedly after alteration phases have formed on the glass.

The disposition of radionuclides is determined primarily by their solution chemistries. Both plutonium and americium are sparingly soluble in the alkaline solutions generated during these tests, while uranium and neptunium are moderately soluble. Significant amounts of these actinides become associated with colloidal material resulting from spallation of the alteration phases, and settle out over time. The solubility of technetium is high in short-term tests, but decreases as test times are extended to longer than one year.

In tests of an Environmental Assessment glass, alteration phases formed within one year at a S/V ratio of 2000 m^{-1} and within 14 days at a S/V ratio of 20,000 m^{-1}. Unlike the SRL glasses, the dissolution rate decreased with time after formation of the alteration phases for unknown reasons.

The results of various tests were interpreted in terms of a reaction progress variable that reflects the transformation of glass to more thermodynamically stable alteration phases. In the first stage of corrosion, the glass dissolves at a rate controlled by its composition, the solution pH, and temperature. In the second stage, the rate is controlled by the buildup of dissolved glass components. The formation of alteration phases may affect the dissolution rate in the third stage. Several glasses have shown corrosion increasing to a high rate after zeolites and other phases form. Preliminary results from a model that has been developed to relate glass composition to its long-term corrosion behavior are promising.

Drip Tests of Savannah River and West Valley Glasses. Drip tests of an actinide- and technetium-doped Savannah River glass (SRL 165) were begun and were still in progress in 1998. Designed to simulate potential conditions in the proposed Yucca Mountain Repository, groundwater representative of the site (EJ-13 well water equilibrated with volcanic tuff) is slowly dripped onto glass samples at 90°C in a closed stainless steel vessel. Release of the glass components, including the actinides, is monitored by periodic analyses of the leachate. The procedure and equipment were described in Chapter 5. Insoluble elements, including U, Pu, and Am, are incorporated into alteration phases as the glass reacts and are subsequently released with particulate or colloidal matter as the alteration phases spall from the glass. Recent trends have shown that the release rates of Pu and Am, while initially low, accelerate as the alteration phases spall from the surface. As shown in Fig. 6-8, the increase in their release rates began at about the eighth year of testing. Ultimately, the transport of these actinides from the waste form will depend on the transport of particles and colloids suspended in solutions.

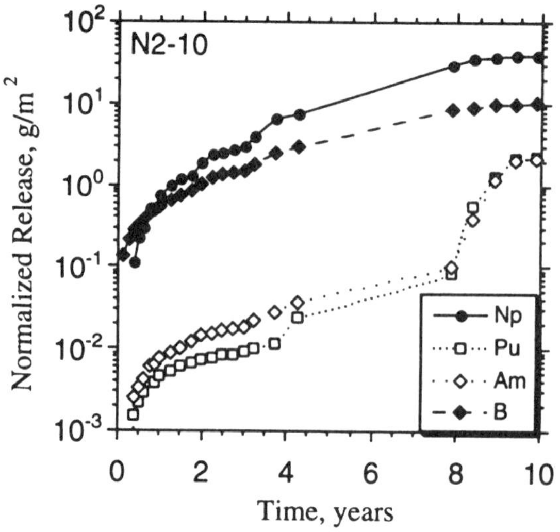

Fig. 6-8. Release of Boron and Actinide Elements from Waste Glass by Leaching

Similar drip tests were conducted with a West Valley glass (ATM 10), beginning in 1987. This glass contains a substantial amount of thorium, which was incorporated into the alteration phases as $CaThPO_4$ and as thorium-titanium-iron silicates. These alteration phases appear to play a substantial role in corrosion of the glass as the element thorium was released to the test solution at a rate nearly three orders of magnitude lower than the water-soluble elements, lithium and boron.

As of 1998, after more than ten years of drip testing, the data indicate that the insoluble elements (U, Pu, and Am) were incorporated into alteration phases as the glass reacted and were subsequently released with particulate or colloidal matter as the alteration products spalled from the glass. The more recent trends showed that releases of Pu and Am, although quite low initially compared to those of the soluble elements such as B and Np, become accelerated as the alteration phases spall off the glass surfaces and enter the test solution. Release of technetium, which is a fission product of concern because of its migration characteristics, was similar to those of B and Np. In glasses containing thorium, the thorium concentrated in the alteration phases.

Development of Low-Temperature Phosphate Glasses. In the electrometallurgical treatment of spent nuclear fuels, fission products and transuranic (TRU) elements accumulate in the molten LiCl-KCl electrolyte. These elements can be removed from the molten salt by ion exchange in zeolite. In the current process flowsheet, the spent zeolite is converted into a composite waste form consisting of the zeolite and borosilicate glass. This composite is fabricated by hot isostatic pressing. An alternative process that avoids the scaleup limitations of hot pressing incorporates the zeolite into a low-temperature phosphate glass to produce a ceramic waste form.

Zinc phosphate glasses, which have a low glass transition temperature in the range of 325 to 375°C, permit relatively low processing temperatures of about 450-700°C. The low process temperatures reduce losses of volatile metals (*e.g.*, ^{137}Cs, ^{99}Tc, Cd), and the zinc phosphate system can incorporate chloride salts at loading levels of 10-15%. Leaching tests of salt-loaded phosphate glass samples showed about the same losses of Cs, Na, K, and Li as salt-free glass. This material is a possible candidate for the disposal of fission-product and TRU elements in the electrometallurgical process for EBR-II fuel.

Critical Review and Testing of Natural Analogs. During the 1990s, CMT workers conducted a comprehensive critical review on the use of natural glasses as analogs for the long-term behavior of nuclear waste glasses. Parameters that affect the reactions between glass and repository groundwater include temperature, glass composition, radiation, and ratio of glass surface area to water volume. A review was also prepared on the status of glass-reaction modeling.

Obsidian, a rhyolitic glass, and tektite are naturally occurring glasses that have survived in nature for millions of years. Studies of their reactions with water vapor, started in the

1980s, were continued into the early 1990s to see if their behavior could offer insights into the long-term corrosion of nuclear waste glass in a geologic setting and the development of more corrosion-resistant glasses. Obsidian and tektite are enriched in silica and depleted in alkali metals and boron, relative to nuclear waste glasses.

The reactions of both obsidian and tektite with water vapor have a common mechanism—molecular water diffusion. The isothermal hydration rate of both depends on the intrinsic water content, even though tektite contains a very small amount of water, about 0.01%. Extrapolation of hydration rates of tektite to repository-relevant conditions resulted in much lower rates than those observed with nuclear waste glasses, indicating that water diffusion might not be the controlling reaction mechanism for nuclear waste glasses. However, at the lower repository temperatures that would occur over time as the nuclear waste cools, other possible reaction mechanisms, such as ion exchange or dissolution, may diminish, allowing water diffusion to emerge as the dominant long-term mechanism.

Natural obsidian and basalt were subjected to vapor hydration tests at 75°C and relative humidities of 95 and 100% for periods up to seven years. The results indicated that vapor hydration tests of a few years duration can replicate corrosion behavior over several thousand years in nature. This test method can be used to forecast the long-term behavior of nuclear waste glasses under similar conditions.

Immobilization of Plutonium in Glass. One of several alternatives under consideration by DOE for the long-term disposal of plutonium from dismantled nuclear weapons and cleanup of weapon production sites is fixation in a glass or ceramic. In a program begun in 1993, CMT investigators developed a low melting (1150°C) alkali-tin-silicate glass (ATS) for immobilizing plutonium. This glass was engineered to accommodate a high plutonium loading (up to 7.0 wt% plutonium as PuO_2) together with gadolinium (as Gd_2O_3) added as a neutron absorber to mitigate against criticality problems during preparation, storage, and ultimate disposal of the glass. In ANL vapor-hydration tests, the durability of this glass was better than that of an Environmental Assessment reference glass. No significant reaction had occurred on exposure of the glass to water-saturated air at 200°C for 56 days.

Another glass that will be tested in the future is a lanthanide borosilicate glass that is also chemically durable and can dissolve substantial amounts of plutonium oxide, as well as the neutron absorbers, gadolinium and hafnium (as their oxides).

Argonne participated in the Fissile Materials Disposition Program by (1) characterizing a ceramic waste form being developed at Lawrence Livermore National Laboratory (LLNL), (2) corrosion testing of the LLNL ceramic, and (3) corrosion testing of plutonium-loaded glasses prepared at Savannah River, Pacific Northwest, and Argonne. The characterization study of the LLNL ceramic showed a number of titanate-based phases, and other types of phases such as perovskites and undissolved PuO_{2-x}. Both Pu(IV) and Pu(III) appeared to be present in a zirconolite phase. Titanate ceramic was selected as the waste form for the disposition of surplus plutonium. Advantages cited for the titanate ceramics are their ability to accept various cations as substitutes in their structure and their high resistance to aqueous corrosion.

Analytical Support. A sophisticated analytical capability was developed for analyzing the products of reactions of nuclear waste glasses with water and water vapor in a simulated Yucca Mountain Repository environment and for following their behavior (transformation, loss) with time. Applied were various analytical techniques that fall under the general heading of analytical electron microscopy (AEM). These include transmission electron microscopy (TEM), X-ray energy dispersive spectroscopy (EDS), electron energy loss spectroscopy (EELS), and electron diffraction (ED). Point-to-point resolution of images approaching 3 Å could be achieved with TEM, 200 Å with EDS, and 20 Å with EELS and ED. Extremely thin samples, 500 Å or thinner, were required for these analyses. A specially designed diamond knife was effective in preparing ultrathin sections of a wide variety of reacted glasses.

Personnel. Many individuals, including temporary personnel, were involved in this program. Full-time CMT people included Teofilo Abrajano, Allen Bakel, John Bates, Dick Biwer, Charles Bradley, Edgar Buck, Jim Cunnane, Bill Ebert, Jeff Emery, Xiangdong Feng, Bob Finch, Tom Gerding, Joe Hoh, Jim Mazer, Denis Strachan, Shiu-Wing Tam, Stephen Wolf, Dave Wronkiewicz, and John Young.

TESTING OF SPENT FUEL

The reference design for limiting radionuclide release from spent UO_2 fuel stored in the proposed Yucca Mountain Repository is an engineered, multiple-barrier system surrounded by volcanic tuff. After long storage of the spent fuel in the repository (several hundred years), the barriers are breached, allowing exposure of the fuel material to the repository environment. Ongoing tests were started in the mid-1980s with both unirradiated and irradiated UO_2 (spent fuel) to compare their behavior and to determine the radionuclide release characteristics of the spent fuel in a water-unsaturated environment typical of that in the repository.

Unsaturated Tests with Unirradiated UO_2. In these tests, Zircaloy-clad UO_2 pellets in a stainless steel test vessel supported by a Teflon stand were exposed to water at 90°C dripping at a rate of 0.75 mL every 3.5 days. The water was a Nevada well water (EJ 13) that had been equilibrated with volcanic tuff of the repository. The release rate of the uranium was rapid for the first two years of exposure because of preferential corrosion along the grain boundaries and spallation of UO_2 grain particles from the sample surface (see Fig. 6-9). After two years, the UO_2 release rate decreased by an order of magnitude, remaining at a relatively constant value thereafter. This rate reduction was caused by the formation of a dense mat of uranium silicates that enveloped the loosened UO_2 grains at the sample surface. The uranium release rate ranged between 0.1 and 0.3 mg/m^2·day. Most of the released uranium (86 to 97%) was sorbed on the walls of the stainless steel test vessel and the Teflon support stand. From 1 to 12% of the uranium was present in the leachate as >5-nm particles composed mainly of alkali or alkaline earth uranium silicates. Only about 2% of the uranium was dissolved in the leachate.

Tests with Irradiated UO_2 Spent Fuel. In these tests, small chunks (0.3 to 1 g) of irradiated fuel having a burnup of 30-40 MW·day/kg U were exposed to dripping

Fig. 6-9. Intergranular Corrosion and Spallation in UO_2 Samples

Nevada well water at 90°C. A layer of uranium silicates analogous to those found on unirradiated UO_2 formed on the surface. Over the first 580 days, a small fraction of the actinides (10^{-6} to 10^{-7}) was released to the leachate, appearing as dissolved species, colloids, and precipitated materials on the test vessel walls.

Of particular interest was the release of technetium (^{99}Tc), which did not change over a four-year test period and was more than 100 times greater than the release rate of uranium. The fraction of technetium released provides a lower limit for the extent of fuel reaction. Moreover, the constant rate of technetium release suggests that a surface reaction may be the rate-controlling step and that through-grain dissolution may dominate over grain-boundary-enhanced dissolution.

Drip tests of UO_2 have been in progress for over 13 years. Uranium release was rapid during the first two years, but the rates were lower in the following ten years. The rapid release was attributed to preferential corrosion along the UO_2 grain boundaries and spallation of very small (1 µm or less) UO_2 particles. After two years of reaction, a dense mat of U(VI) alteration phases had enveloped the UO_2 particles, decreasing the release rate.

Another test method called "petri-dish tests" was developed to measure the compositions of thin films of solutions in contact with the fuel specimens. The resulting data are used in model development and confirmation. Some information was obtained on the amounts of released plutonium and the percentage in colloidal form.

Personnel. Pat Finn played a major role in some of these activities. Others were Allen Bakel, John Bates, Edgar Buck, Dave Chamberlain, Jim Cunnane, Jeff Emery, Bob Finch, Tom Gerding, Lohman Hafenrichter, Joe Hoh, Susan Slater, Mark Surchik, Ben Tani, Ewald Veleckis, Stephen Wolf, and Dave Wronkiewicz.

WASTE ISOLATION PILOT PLANT PROJECT

In a program begun in 1989, CMT researchers investigated the effect of ionizing radiation on aspects of importance to the long-term performance of the Waste Isolation Pilot Plant (WIPP) repository. The work was part of an overall effort, coordinated by Sandia National Laboratories, to assess the performance of WIPP for long-term disposal of TRU waste. Measurements were made of the gas evolved over a 160-day test period from WIPP brine spiked with Pu(VI). Hydrogen was the primary gaseous product of radiolytic origin. Gas generation rates were 0.14, 0.002, and <0.0003 mol/($m^3 \cdot y$) for Pu(VI) concentrations of 10^{-4}, 10^{-6}, and 10^{-8} *M*, respectively. Such data enable an assessment of the contribution of radiolytic gas generation to the overall gas generation rate in the repository. The ANL personnel in this project included Don Reed, Peter Lindahl, and Shigeo Okajima.

VITRIFICATION OF LOW-LEVEL WASTE

The safe disposal of radioactive and toxic waste elements requires durable waste forms that effectively isolate these components from the biosphere. In 1992, work was begun on vitrification for treatment and disposal of low-level and mixed wastes. Glass or glass/ceramic waste forms were sought for inexpensive remediation of contaminated materials, including soils, waste tank sludges, evaporator sludges, industrial catalysts, incinerator residues, High Efficiency Particulate Air (HEPA) filters, and equipment.

Development of Low-Level Glass Waste Forms. The objective of this work was to identify and produce glass formulations that could be used to vitrify low-level wastes such as those listed above. A composition envelope was developed for satisfactorily incorporating these wastes in a range of alkali boroalumino-silicate glasses. Glasses within this envelope could be produced with relatively small (7 to 28%) additions of glass-forming materials, *e.g.*, borax, diatomaceous earth, boric acid, and sodium carbonate. The overall waste volume was reduced by a factor of four to eight, and the glasses so produced met EPA requirements. The work provided compelling evidence that a wide variety of ANL waste streams could be vitrified in borosilicate glasses.

Glass-Crystal Compositions. Many DOE sites have large volumes of radioactive waste that are not amenable to disposal in glass waste forms. These wastes contain large amounts of metallic elements such as Cr, Ni, Ti, Fe, Ca, and Mg, which have low solubilities in glass. Glass-crystal composite waste forms consisting of crystals encapsulated in a glass matrix were evaluated for their potential in disposal of these mixed low-level and hazardous waste materials. Crystal formation was induced from the melts by proper blending of the simulated waste streams, use of chemical additives such as Zr, Ti, and P, and slow cooling of the melts from the 1400°C melting temperature. Hazardous elements such as U, Pu, Sr, Cd, Pb, As, and Ba can be incorporated into corrosion-resistant crystalline phases. The glass-crystal composites were very resistant to aqueous corrosion. In fact, the uranium release rate from two composite samples was 100- and 300-fold lower than that from a Savannah River borosilicate glass, a result attributed mainly to the durability of the crystalline phases and to the durable nature of the glass matrix.

It was concluded that glass-crystal composites combine the excellent durability of predominantly crystalline waste forms, such as Synroc, with the ease of processing offered by the use of commercially available vitrification facilities.

Testing of Glass Waste Forms for Low-Level Wastes. Several glasses developed at Pacific Northwest Laboratory as part of the Hanford low-level waste project were provided to ANL for testing. Three-year tests of two sodium-rich borosilicate glasses, designated LD6-5412 and FLLW-1, were begun in 1994. After the first two years, sodium release was higher than that of silicon or boron, and its release was by a diffusion-controlled, ion-exchange process. The formulation of alteration glasses on the surface of the LD6-5412 glass caused a dramatic increase in the dissolution rate of that glass. The alteration phases also affected the release of radionuclides present in the FLLW-1 glass. Final conclusions await the completion of the third year of tests.

Personnel. The vitrification work involved Allen Bakel, John Bates, Edgar Buck, Dave Chamberlain, Cliff Conner, Jim Cunnane, Bill Ebert, Adam Ellison, Jeff Emery, Xiangdong Feng, Bob Finch, Jeff Fortner, Jim Helt, Joe Hoh, Terry Johnson, Ted Krause, Jian Shu Luo, Jim Mazer, Carol Mertz, Luis Nuñez, Dale Smith, Denis Strachan, Ben Tani, Stephen Wolf, Dave Wronkiewicz, and Dave Wygmans.

CHARACTERIZATION OF CONTAMINATED SOILS AND OTHER MATERIALS

Analytical electron microscopy, described earlier, was brought to bear on the characterization of the following contaminated soils and materials to assist in the remediation of the sites.

Contaminated Soils at the Fernald Site. The Fernald site had become heavily contaminated with uranium as a result of uranium processing operations. The major uranium-bearing phase in the soil was identified as a calcium uranyl phosphate. Also found in the soil were phosphate phases that resulted from the use of tributyl phosphate in the uranium purification process.

Johnson Atoll Soil. In the early 1960s, uranium and plutonium contamination of Johnson Atoll in the Pacific resulted from an abort of a nuclear weapon. Analyses of fragments of bomb components showed that plutonium and uranium in these fragments had oxidized and formed carbonate complexes, facilitating their migration to surrounding soils.

Rocky Flats Incinerator Residues. Plutonium in Rocky Flats incinerator residues was found to exist mainly in the form of a reduced oxide, PuO_{2-x}. Some plutonium was also incorporated into a silicate glass in the ash.

Cement Mortar Formulations. Leach tests were conducted on cement mortar formulations from Applied Innovation Inc., to evaluate their ability to retain radioactive elements from radioactive waste, as well as elements in toxic and mixed waste. According to the Nuclear Regulatory Commission, the effective diffusivity of an element in an acceptable low-level waste form should be less than 10^{-6} (leachability index of six). Leachability indices measured for Tc, Cs, Ni, U, Se, As, and Cd were 15, 14, 21, 22, 15, 18, and 18, respectively. These excellent results showed that Portland-cement mortars are promising for immobilizing these wastes.

Personnel. Various aspects of this work were carried out by John Bates, Edgar Buck, Dave Chamberlain, Mike Chen, Cliff Conner, Jim Cunnane, Jack Demirgian, Jeff Fortner, Jim Helt, Ted Krause, Don Reed, John Schneider, Ian Tasker, George Vandegrift, and Stephen Wolf.

OTHER PERSONNEL

Besides the people listed specifically in the above programs, the following also played an important role in the Nuclear Waste Management effort: John Basco, Mark Clark, Tom DiSanto, John Falkenberg, Margaret Goldberg, Mike Goss, Mark Hash, Joe Hoh, Terry Johnson, Mike Kaminski, Lester Morss, Mike Nole, Kevin Quigley, Wally Seefeldt, Chuck Seils, Dan Shull, and Sy Vogler. Secretaries were Lauren Ambrose, Norma Barrett, Elaine Estand, Susan McKinney, Sue Rura, and Virginia Strezo. Roberta Riel was in charge of quality assurance (QA).

Separation Science and Technology

One of the major programs in this area through the mid 1990s was the Transuranic Extraction (TRUEX) process for recovering TRU elements from a wide variety of aqueous systems and for lowering disposal costs of nuclear wastes. Three other solvent-extraction processes introduced in the 1990s for cleanup of nuclear wastes were the SREX (strontium extraction) and CSEX (cesium extraction) processes for removing strontium and cesium, and the NEPEX (neptunium-plutonium extraction process) for selectively partitioning Np(IV), plutonium, and uranium. These processes could be used in different combinations to achieve various recovery and separation objectives. Improvements continued to be made in ANL-designed centrifugal contactors for specific process applications.

Other programs were:

1. Concentration of plutonium solids in pyrochemical residues by aqueous biphase extractions.
2. Membrane-assisted solvent extraction for treating natural and process wastes contaminated by volatile organic compounds.

3. Evaporation/concentration technology for concentrating radioactive waste and product streams.
4. A process for separating ^{99}Mo from low-enriched uranium targets for nuclear medicine uses.
5. Technical support to ANL management for treating transuranic and organic-containing wastes.

DEVELOPMENT AND APPLICATION OF TRUEX TECHNOLOGY

The TRUEX process was outlined in Chapter 5. The use of trichloroethylene or carbon tetrachloride as a diluent for the TRUEX extractant, CMPO [octyl(phenyl)-N,N-diisobutyl carbamoyl phosphine oxide], was discontinued because of growing environmental, safety, and health concerns. Instead, a normal paraffinic hydrocarbon (C_{12}-C_{14}) mixture or normal dodecane was used. Degradation of this solvent combination by hydrolysis and radiolysis was shown to be very low, the fraction of the CMPO decomposing in one year's operation (100 processing cycles) being only 8.6×10^{-4}.

Generic TRUEX Model. Improvements continued to be made in the Generic TRUEX Model, a computer program first issued in 1989 as a tool for (1) designing TRUEX flowsheets for specific waste stream compositions, process constraints, and goals; (2) estimating space and cost requirements for installing the TRUEX process; and (3) monitoring and controlling the process. These functions were performed by executing different modules within a computer code, illustrated in Fig. 6-10. The success of the GTM was dependent on underlying data that were sound, correct, and defensible. A data base for generating the code, which was begun in 1987, consists of information from the literature, laboratory results from CMT, and information on the extraction behavior of all important feed components over a wide range of possible waste-stream and processing compositions. An operating manual for the generic TRUEX process was prepared by Scott Aase, Dave Chaiko, Lorac Chow, Dave Chamberlain, Jaqueline Copple, Don Fredrickson, Ray Jaskot, Ralph Leonard, Luis Nuñez, Monica Regalbuto, Jake Sedlet, Ian Tasker, Verne Trevorrow, George Vandegrift, and Erv Van Deventer. Updated versions of the Generic TRUEX Model were published annually through 1994 to incorporate new information and capabilities that were introduced. Development of the TRUEX process was terminated at the end of September 1995.

Validation of the Generic TRUEX Model. When there was the opportunity, the GTM was tested experimentally. One opportunity was the processing of analytical wastes from the New Brunswick Laboratory, using a flowsheet and equipment of a GTM design. The TRU content of the raffinate was 1.8 nCi/mL, a reduction of 22,000 from the concentration in the feed. The calculated raffinate concentration using the GTM was 1.0 nCi/mL. This remarkable agreement provided confidence in using the GTM for designing flowsheets and equipment for processing the much larger volumes of TRU wastes generated at the Rocky Flats Plant, Los Alamos National Laboratory, Hanford, and the Idaho Chemical Processing Plant.

Treatment of New Brunswick Laboratory Analytical Wastes. Over 200 L of waste solution from the analysis of plutonium samples had accumulated at the New Brunswick Laboratory and at Argonne. The Chemical Technology Division was given the project of converting the bulk of this material into a non-TRU (low-level) waste. The design of a flowsheet and equipment was based on

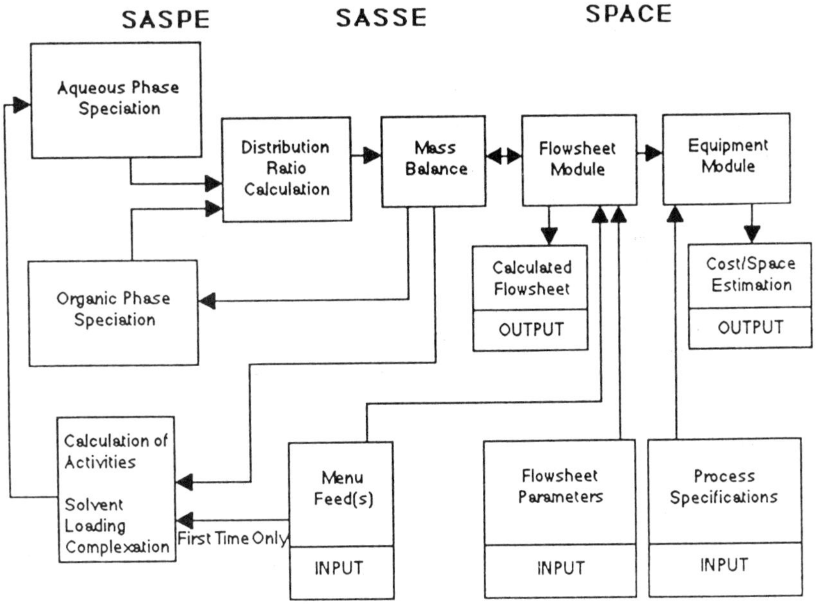

SASSE = Spreadsheet Algorithm for Stagewise Solvent Extraction
SASPE = Spreadsheet Algorithm for Speciation and Partitioning Equilibria
SPACE = Spreadsheet for Calculating Equipment Size, Plant Space, and Capital Costs

Fig. 6-10. Interactions among Computer Program Modules of the Generic TRUEX Model for Flowsheet Calculations

an initial goal of less than 10 nCi/mL of TRU in the low-level waste. The flowsheet for this operation is shown in Fig. 6-11.

The TRUEX process was conducted in a 20-stage centrifugal contactor bank (4-cm-dia rotors) installed in a glove box. Six stages were devoted to extraction, two to scrubbing, six to stripping americium, four to stripping plutonium, and two to washing the solvent to remove degradation products and permit recycle of the solvent. The goal of 10 nCi/mL or less of TRU activity was easily achieved, but after the first batch of New Brunswick waste solution had been processed, the ANL Waste Management Operation changed the goal from 10 nCi/mL to 0.1 nCi/mL or less, a reduction of 100 from the original goal. Achieving this very low TRU concentration in the waste would have necessitated radical changes in the flowsheet. It was decided to finish the campaign, using the original flowsheet. Raffinate TRU concentrations of 10 nCi/mL or less (generally between 2 and 5 nCi/mL) were achieved in processing the remaining batches. Note that these concentrations are well below the 100 nCi/mL of TRU below which solutions are classified as non-TRU, allowing them to be treated as low-level waste.

Demonstration of SREX Process. The SREX (strontium extraction) version of the TRUEX process was selected as the reference process for the removal of strontium from high-level sodium-bearing and calcined wastes produced at the Idaho Chemical Processing Plant. The process was invented by researchers in the ANL Chemistry Division, and the CMT Division provided engineering and modeling support, including design, fabrication, and testing of a 24-stage "minicontactor" to be used for a process demonstration at Idaho

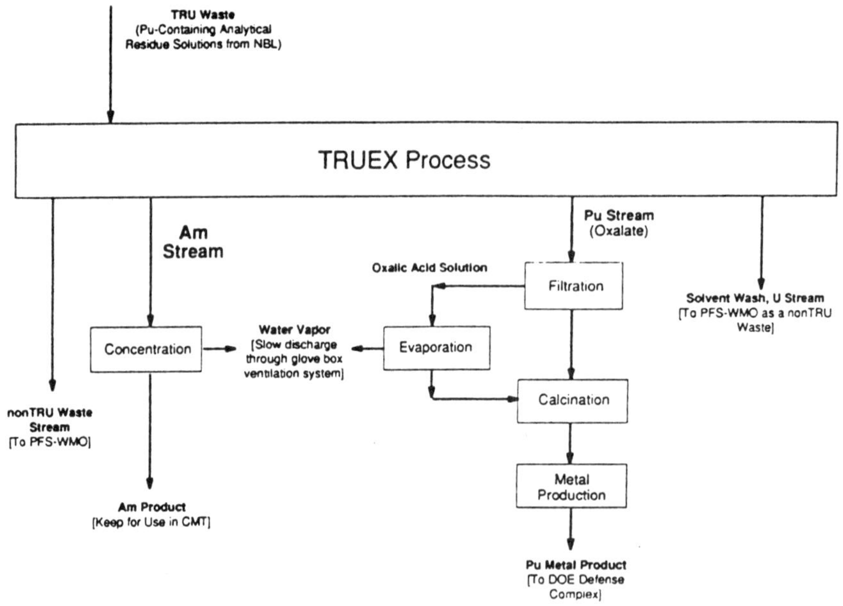

A bank of 20 centrifugal contactors with 4-cm-dia rotors was employed, six for extraction, two for scrubbing the solvent extract, six for extracting americium, four for extracting plutonium, and two for washing the solvent. Conversion of the plutonium to metal by precipitation of the oxalate, calcination of the oxalate to yield PuO_2, and reduction of the PuO_2 to metal was not performed.

Fig. 6-11. General Flowsheet for Processing Plutonium Waste Solutions of New Brunswick Laboratory

National Engineering and Environmental Laboratory (INEEL). The flowsheet is shown schematically in Fig. 6-12.

The hot demonstration of the SREX process at INEEL went very smoothly, with 99.995% of the ^{90}Sr being removed from the aqueous raffinate. With this removal efficiency, the ^{90}Sr activity was decreased to 0.0089 Ci/m^3, which is well below the limit for low-level waste.

Demonstration of SRTALK Process. A collaborative program was undertaken with Oak Ridge and Pacific Northwest National Laboratories on the SRTALK (Sr, Tc, and Cs alkaline-side extraction) process. This is a solvent-extraction process that can separate alkali-metal pertechnetate salts from the alkaline supernate that comes directly from nuclear waste-storage tanks. Two advantages of the SRTALK process over the ion-exchange columns currently used for this purpose are (1) the technetium is recovered free of other salts in the tank supernate, and (2) no added chemicals (*e.g.*, tin, ethylene diamine) are required to strip technetium from ion-exchange columns. These features would reduce greatly the number of glass canisters needed for waste disposal.

The SRTALK solvent consists of 0.04 *M* crown ether and 1.8 *M* tributyl phosphate in Isopar®L, which extracts technetium from the feed with a distribution ratio of 6 to 10, and can be stripped of technetium with water or dilute nitric acid. The process concept was demonstrated first in a test tube, then in a minicontactor. The final flowsheet is shown in Fig. 6-13.

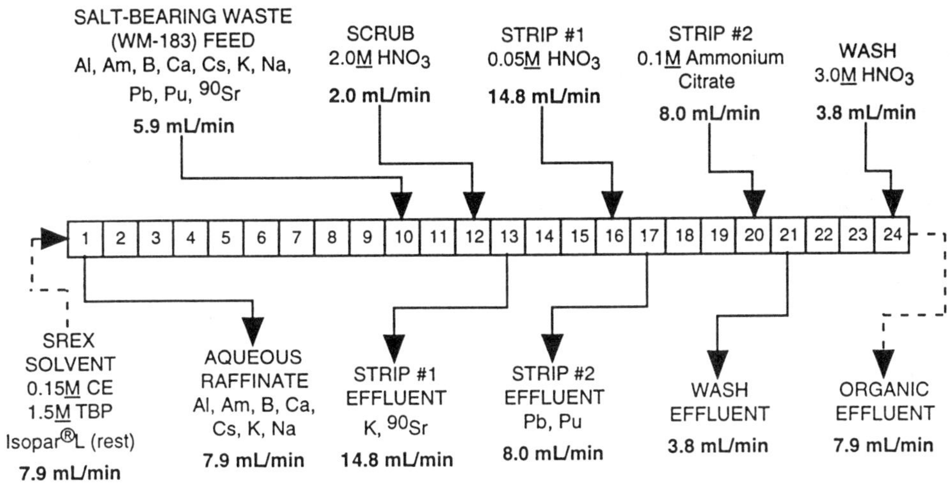

The tests were conducted in a shielded cell at ICPP in Idaho. The SREX solvent was a crown ether and tributyl phosphate in Isopar®L. The ^{90}Sr in the sodium-bearing waste feed was removed by the solvent in the extraction section (stages 1-10). Other components in the feed were washed back by the scrub feed in stages 11 and 12. In the first strip section (stages 13-16), dilute acid stripped ^{95}Sr and any potassium out of the solvent. In the second strip section (stages 17-20), a complexing agent (ammonium citrate) stripped lead and actinides from the solvent. The solvent was reacidified in stages 21-24 so it could be recycled to stage 1.

Fig. 6-12. Flowsheet for a Test of the SREX Process with Hot Tank Waste at INEEL

Tests of the SRTALK process were very successful, and plans were made to extend the process to Cs and possibly Sr extractability.

CENTRIFUGAL CONTACTOR DEVELOPMENT

Ralph Leonard continued to adapt the basic design of the centrifugal contactor to various applications by use of computer models for (1) the flow of organic and aqueous phases through the contactor and (2) the vibrational parameters of the spinning motor/rotor combinations. By use of these models, a new contactor design was developed that, without increasing rotor diameter, would double the throughput. The high throughput design was developed for use in the proposed PUREX facility at the Hanford Plutonium Finishing Plant.

This work was later extended to the design of a super high-throughput centrifugal contactor that would increase the throughput by a factor of four to eight, yet keep the unit criticality safe by use of rotor diameters of less than 10 cm. Thus, a single bank of contactors could handle a throughput that otherwise would have required four to eight banks of contactors. Figure 6-14 is a schematic drawing of a two-stage super high-throughput contactor.

Twelve-stage banks of contactors were built for Pacific Northwest Laboratories and the Lockheed Idaho Technology Co. so that they could carry out pilot-plant tests of solvent-extraction processes for cleaning up their nuclear wastes. An ANL-type contactor fabricated of Kynar®, which has high resistance to attack by HCl, was designed for removal of plutonium and uranium from Los Alamos chloride salt wastes.

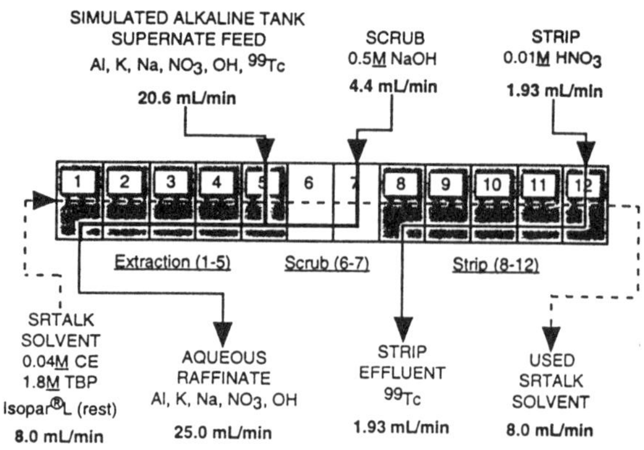

In stages 1-5, the technetium is extracted into the organic phase and a scrub section in stages 6 and 7 is used to free the tank solution of other salts. In stages 6-12, the technetium is back-extracted into a 0.01 *M* nitric acid solution. The strip effluent is essentially free of added chemicals. By suitable adjustments of the organic-to-aqueous flow ratios in the extraction and strip sections, the technetium concentration in the strip effluent can be increased by a factor of 10 compared to that of the aqueous feed. At the same time, the decontamination factor for technetium in the aqueous raffinate can be kept over 6.4, as required.

Fig. 6-13. Flowsheet for SRTALK Demonstration

Hydraulic and vibrational analyses were also used to assist in the design of a high-temperature (500-800°C) centrifugal contactor for contacting molten chloride salt wastes (produced in the electrometallurgical process) with liquid cadmium to extract plutonium (see Chapter 5).

An outgrowth of the centrifugal contactor development was a concept for a continuous countercurrent solid-liquid contactor for mixing and then separating liquid and solid phases. Such a contactor could be used for washing sludges from the Hanford and Savannah River waste tanks to, in the case of the Savannah River sludges, remove selected

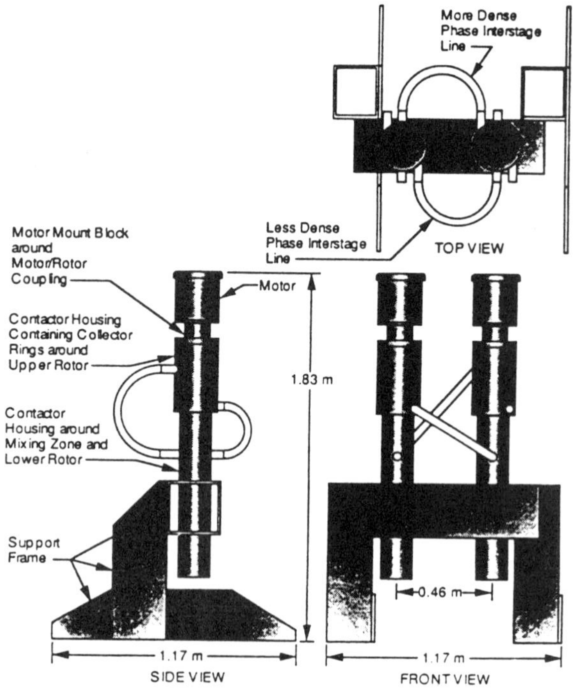

Fig. 6-14. Two-Stage Super High-Throughput Centrifugal Contactor

constituents such as cesium and strontium. Figure 6-15 shows two stages of a solid-liquid contactor. Solid particles, which have a higher energy density than the liquid, move by gravity, countercurrent to the liquid flow.

Ralph Leonard was the leader of the work on centrifugal contactors. Others who participated in this work were Dave Chamberlain, Cliff Conner, Monica Regalbuto, George Vandegrift, and Dave Wygmans. The high-temperature molten salt contactors were developed primarily by Terry Johnson with John Ackerman, John Basco, and Lorac Chow.

RECOVERY AND SEPARATION OF MOLYBDENUM-99

Work on the extraction of ^{99}Mo from low-enriched uranium (LEU), which was begun in 1985 by George Vandegrift, was resumed in 1993 after a dormant period of four years.

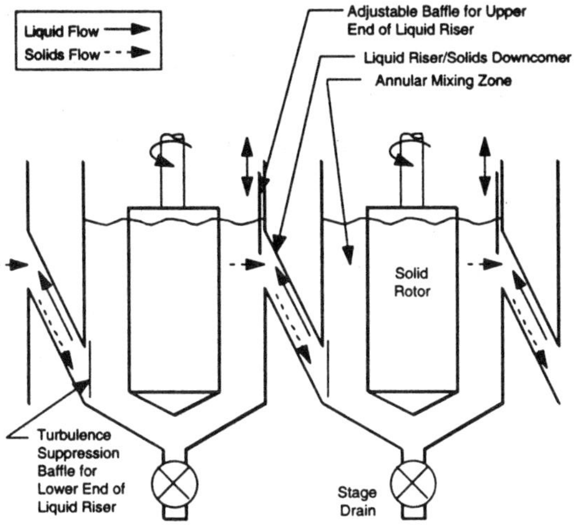

Fig. 6-15. Two Stages in Continuous Countercurrent Solid-Liquid Contactor

This four-year hiatus resulted from low budgets for the program on Reduced Enrichment for Research and Test Reactors (RERTR) from which the CMT program obtained its funding. Revival of the program was prompted by 1993 legislation that severely restricted the export of high-enriched uranium (HEU) and placed emphasis on the use of LEU.

To overcome the low-concentration of ^{235}U in LEU, while maintaining the same target geometries, targets having high concentrations of uranium were sought. Uranium silicide (U_3Si_2) and uranium metal foil were proposed as substitutes for the HEU target materials previously used (UAl_x and a thin layer of UO_2 on the inside of a cylindrical target). After they are irradiated, the target materials must be dissolved, after which molybdenum is precipitated quantitatively from the solution by alpha-benzoin oxime (α-BO). Purified molybdenum containing the ^{99}Mo isotope is deposited on a column of alumina, from which its decay daughter, metastable technetium-99 (^{99}Tc), may be eluted periodically with a saline solution. The ^{99}Tc, which has a half-life of 6.02 h, is used in medical diagnoses.

The CMT program was focused on target designs for the two LEU target materials and on methods of dissolving the irradiated targets. Because irradiated targets were processed in other countries (Indonesia, Argentina, and South Africa), CMT developments were designed for application in those facilities.

Use of uranium foil was found to require thin metal barriers to prevent bonding with the target container material. Barriers of iron, copper, and nickel about 10-μm thick performed well.

In the Cintichem process used in Indonesia, a 3 *M* H_2SO_4-2 *M* HNO_3 solution was used to dissolve irradiated UO_2 targets. While this acid mixture could also be used to dissolve uranium metal foil and metal barriers of copper, nickel, or iron, the presence of sulfate in the acidic waste solution complicated the uranium recovery, volume reduction, and waste disposal. The CMT researchers found that dissolution of uranium foil and any barrier materials could be accomplished at the same rate in 8 *M* HNO_3. Elimination of sulfuric acid did not significantly affect the recovery of molybdenum by precipitation with α-BO or the purity of the final product.

Dissolution of uranium foils with alkaline peroxide (*e.g.*, 5 *M* H_2O_2-1.5 *M* NaOH) was investigated as a replacement for processing HEU aluminate targets. Unfortunately, the uranium surface catalyzes the destruction of H_2O_2, resulting in the consumption of a very large amount of peroxide. By sequential small additions of peroxide, the consumption was reduced 20 fold (from 100 to 5 times the stoichiometric amount required for dissolution). The barrier metal candidates of copper, nickel, and iron are not dissolved by the peroxide solution. For this situation, investigations of other barrier materials, such as zinc, were started.

Although dissolution of uranium silicide, proposed as an alternative to uranium aluminate, was investigated, work on it was suspended due to difficulties in dissolving the silicide, problems caused by dissolved silicate in recovery of the molybdenum, and the relatively high degree of success achieved with uranium foil targets.

The current studies are aimed at development of a process for separating molybdenum from LEU (and its fission and absorption products) that is as similar as possible to the Cintichem process used in Indonesia with highly enriched uranium. Two LEU target designs, uranium metal foil and UO_2/Al dispersion fuel, are being investigated. The CMT work in this area consisted of three major areas: (1) the behavior of fission products (I, Rh, and Ag) in the Cintichem process, (2) the effects of zinc fission barriers for uranium-foil targets, and (3) a procedure that will measure alpha contamination in the ^{99}Mo product.

The investigators concluded that contamination of the molybdenum product by the three fission products would not be a problem. An effort was mounted on a procedure for separating and recovering ^{99}Mo that will allow low-dose, facile, and effective measurement of alpha contamination in the ^{99}Mo. In the development of metal-foil targets, the use of thin metal barriers (to fission recoil) between the uranium foil and the target walls was investigated, zinc being a major contender. Studies were also conducted on the use of alkaline peroxide solutions for dissolution of UO_2 and metal foil targets.

This work, led by George Vandegrift, was done by Scott Aase, Cliff Conner, Joe Hutter, Jane Kwok, Ralph Leonard, Samson Marshall, Jake Sedlet, "Chino" Srinivasan, Don Vissers, and Dave Wygmans.

MISCELLANEOUS PROGRAMS

In the area of Separation Science and Technology, a number of other small programs of varying duration were conducted in the 1990s. They are the following.

1. Aqueous biphase extraction in which colloid-size particles partition between two immiscible aqueous phases. A possible application is the separation of ultra-fine PuO_2 particles from other materials.
2. Separation of volatile organic compounds from groundwater by means of membrane-assisted solvent extraction in which the organics are transferred through a porous membrane to an extractant such as sunflower oil.
3. An advanced evaporator technology for concentrating radioactive waste and product streams and for achieving very high decontamination factors in the condensate (in the range of 10^6 to 10^7). To advance this technology, CMT procured a laboratory-scale evaporator from LICON Inc., the design of which showed promise of achieving the requisite high decontamination factors.
4. Magnetically assisted chemical separation, which combines the selective separation afforded by chemical sorption with the physical separation provided by magnetic recovery of ferromagnetic particles. This work was aimed at separating cesium, strontium, and TRU elements from the waste tank sludges and supernatants in underground storage tanks at Hanford.
5. A closed-loop off-gas system that will prevent the direct release of toxic gases during incineration or vitrification of hazardous and mixed radioactive wastes.
6. The *in situ* decontamination of equipment (*e.g.*, piping systems and storage tanks) in production plants.
7. Technical support to ANL Waste Management in the treatment of nuclear wastes for disposal, treatment of mixed wastes that contain hazardous metals (Pb, Cd, Cr, Ag, and Hg), and evaporator technology.

Integral Fast Reactor Pyrochemical Process

Work on development of a pyrochemical process for recovering and purifying fuel discharged from the proposed Integral Fast Reactor (IFR) continued until October 1995, at which time the IFR program was terminated.

PROCESS FLOWSHEET AND CHEMISTRY STUDIES

The evolution of the IFR pyrochemical process was described earlier in Chapter 5. In the process, shown schematically in Fig. 6-16, spent fuel in stainless steel cladding is chopped and dissolved anodically in molten LiCl-KCl at 500°C. During the dissolution, the uranium is electrotransported to solid cathodes. The plutonium and some uranium are then electrotransported to a liquid cadmium cathode. More than 99.9% of the actinides are removed from the cladding shells. The noble metal fission products either remain in the basket or fall as a particulate metal into the cadmium pool at the bottom of the electrorefiner. The alkali and alkaline earth metals in the spent fuel are converted to their chlorides and remain in the electrorefiner salt, as do most of the rare-earth fission products. Figure 6-17 is an artist's rendering of a commercial-scale facility.

Computer Code. A computer code, PYRO, which was developed by John Ackerman, accurately predicts the distribution of the elements in the electrorefiner during the individual operations of the process. This code was used extensively to calculate the compositions of product and waste streams under various conditions and to study fuel and waste processing strategies.

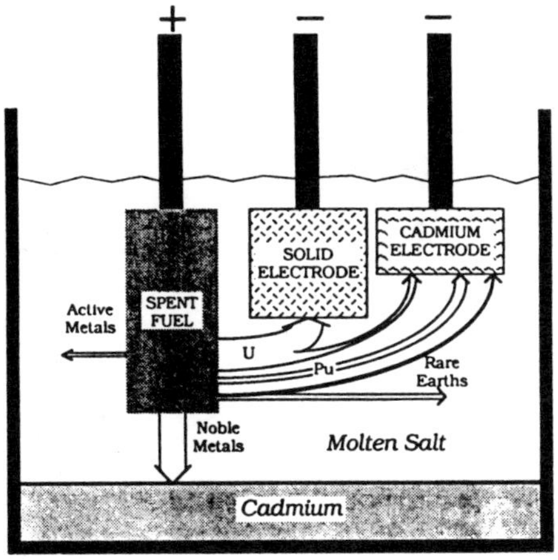

Quarter-inch-long segments of IFR core fuel are loaded into a porous stainless steel basket, which is suspended in the molten LiCl-KCl electrolyte at 500°C. Application of a dc voltage causes uranium to transport to a solid iron mandrel. Plutonium and the other transuranic elements and the alkaline earth, rare earth, and iodine fission products report to the electrolyte salt phase. Noble metal fission products remain either in the anode basket or are collected in the liquid cadmium pool at the bottom of the electrochemical cell. Gaseous fission products are released to the cell cover gas, from which they are removed for concentration and storage. Plutonium and other transuranic elements are subsequently transported to a liquid cadmium cathode along with some uranium and 10-20% of the rare earth elements. Contamination of the plutonium by rare earth elements provides a very strong deterrent to diversion of the plutonium for weapons use. The uranium and plutonium products are recovered by retorting to vaporize cadmium and drive off adhering electrolyte salt.

Fig. 6-16. Electrorefining Cell for IFR Pyroprocess

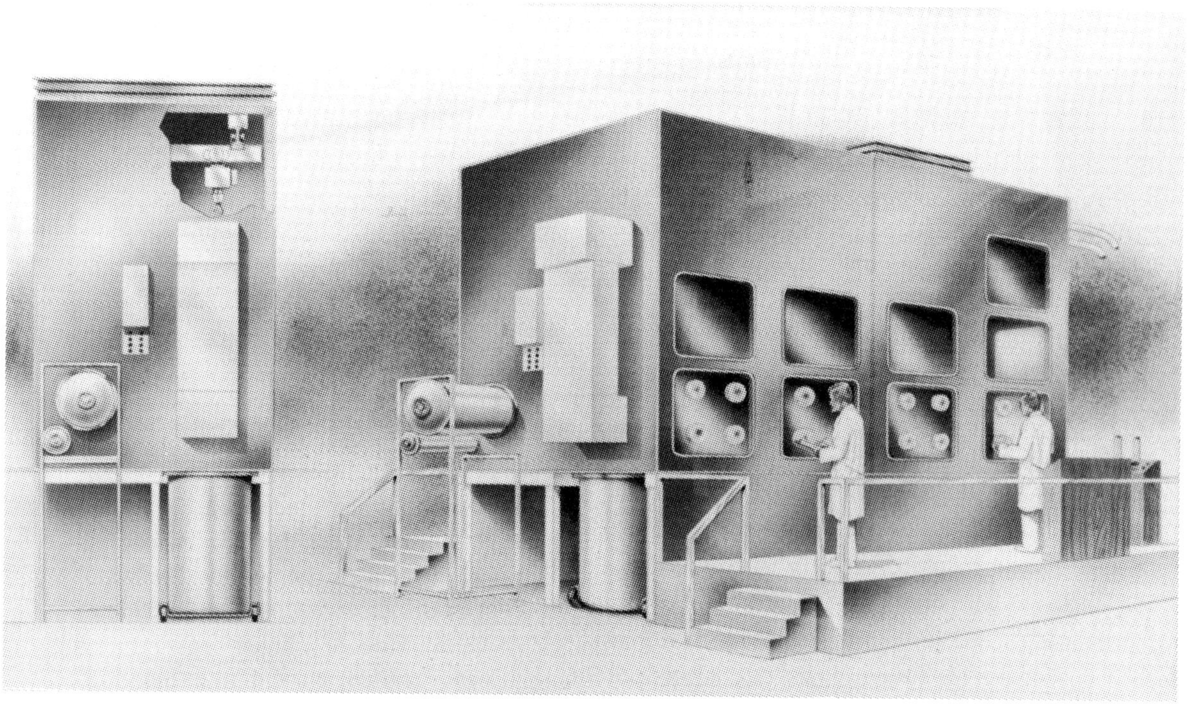

Fig. 6-17. Drawing of Commerical-Scale Facility for Processing of Spent Metal Fuel

Separation Factors. Part of the supporting chemistry studies for verifying the code were measurements of separation factors, which are the ratios of distribution coefficients of pairs of elements. The distribution coefficient of an element is its concentration in the salt phase divided by its concentration in the metal phase. Separation factors are useful in predicting salt and metal compositions, decontamination factors for impurity elements in the uranium and plutonium products, and the potential of countercurrent operations to amplify separations of various elements. Tables 6-7 and 6-8 give separation factors for several actinides and rare earth elements between cadmium and the LiCl-KCl molten salt electrolyte and an alkaline earth-rich salt used in the early process development. The latter salt was of the nominal composition (in wt%) 35 $BaCl_2$, 32 $CaCl_2$, 23 LiCl, and 10 NaCl. The rare earths, europium and samarium, behaved like the divalent alkaline earths, barium and strontium. These elements remained almost entirely in the salt phase. As shown by the separation for the Y-La and Y-Nd pairs, yttrium was also more easily oxidized than were the rare earth elements. Jack Settle did this work under the direction of John Ackerman.

PROCESS DEVELOPMENT STUDIES

Salt/Atmosphere Reactions. For safety purposes, it was important to determine the consequences of a hypothetical accident that would result from inleakage of argon cell gas containing oxygen and water at concentrations up to 100 ppm into an electrorefiner. Ziggy Tomczuk and Bill Miller conducted some experiments to investigate this matter. On admission of the impure argon gas into an electrorefiner crucible, uranium was preferentially oxidized to UO_2, its concentration decreasing much more rapidly than that of plutonium. The finding that plutonium was not preferentially concentrated in the precipitate eliminated concern that a situation favorable to nuclear criticality would occur.

Table 6-7. Mean Separation Factors at 500°C

Element	Separation Factor[a]	
	LiCl-KCl Salt	Alkaline Earth-Rich Salt
Neptunium	2.12 ± 0.42	-
Plutonium	1.88 ± 0.09	1.32 ± 0.14
Americium	3.08 ± 0.78	2.85 ± 0.98
Curium	3.52 ± 0.59	2.3 ± 1.2
Cerium	45 ± 6	48 ± 13
Neodymium	39 ± 6	33 ± 5

[a] Relative to uranium. Quoted uncertainty is standard deviation of the sample population.

Table 6-8. Measured Separation Factors for Pairs of Elements at 500°C

Element Pair	Separation Factor[a]	Element Pair	Separation Factor[a]
Nd-Pu	23.4 ± 1.2	Dy-La	3.79 ± 1.41
Am-Pu	1.54 ± 0.15	La-Ce	2.72 ± 0.37
Nd-Am	15.5 ± 1.1	La-Nd	3.01 ± 0.71
Nd-Pr	1.02 ± 0.04	Y-La	42.5 ± 9.3
Gd-La	1.10 ± 0.33	Y-Nd	140 ± 21

[a] Quoted uncertainty is twice the standard deviation of the mean.

Corrosion Behavior of Steel Electrorefiner Container. After four years of operation at 500°C and continuous exposure to the electrolyte and cadmium phases of the electrorefining process, the low-carbon steel electrorefining crucible was destructively examined. There was no detectable corrosion of the vessel in either the salt- or cadmium-wetted areas, although, in the cadmium-wetted area, there was a thin U-Fe layer about 1-3 μm thick. No grain-boundary attack was evident. Low-carbon steel was concluded to be an excellent material of construction for the electrorefining operation.

Zirconium Behavior. Zirconium constitutes about 30 vol% of IFR fuel. During electrorefining, most of it remained in the anode basket, but some of it escaped as particulate metal and was collected in the cadmium pool. At high voltages (0.5-0.7 V), some of it electrotransported with the uranium.

After 2.5 years of operation, a total of 11 kg of zirconium had been charged to the engineering-scale electrorefiner. Of this, about 3 kg had been removed in the uranium cathode deposits. (In the engineering-scale electrorefiner, uranium was electrotransported from the cadmium pool to an iron mandrel cathode.) About 0.5 kg of zirconium was in solution in the cadmium pool, leaving about 7.5 kg elsewhere in the electrorefiner. Samples of the cadmium pool showed no zirconium in suspension, nor was it vaporized as $ZrCl_4$ and deposited on the upper surfaces of the electrorefiner. Further searching revealed that it had deposited as $ZrCd_2$ on cadmium-wetted metal surfaces (principally the electrorefiner walls) near the salt-cadmium interface. This work was done by Erv Carls with the assistance of the electrorefiner staff—Eddie Gay, Nick Quattropani, and Jack Arntzen.

Removal of the zirconium by electrotransport requires that any uranium and plutonium dissolved in the cadmium be oxidized to their chlorides and transferred to the electrolyte. When this was done, zirconium was electrotransported to the iron mandrel cathode with the $ZrCd_2$ decomposing as the electrotransport proceeded.

The co-transport of uranium and zirconium to the solid cathode mandrel was also demonstrated. At low voltages (0.2 to 0.3 V), co-transport of zirconium with uranium was negligible, but at high voltages (0.5 to 0.7 V), the rate of zirconium transport with uranium was appreciable, and essentially complete transport of the zirconium could be achieved. This procedure has the advantage of direct recycle of zirconium for fabrication of new fuel.

Uranium-Plutonium Drawdown Procedure. A uranium-plutonium drawdown procedure was required to remove these and other actinide elements from the electrolyte salt before it was removed to waste for further processing. The drawdown procedure consisted of loading 5.8 wt% Li-94.2 wt% Cd (a solid) into an anode basket and, by ion displacement, transporting uranium to a solid cathode and uranium and plutonium to a liquid cadmium cathode. The reaction is

$$3\,Li + M^{3+} \rightarrow 3\,Li^{+} + M$$

where M is uranium or plutonium. The net effect is to displace M^{3+} ions by Li^{+} ions. A schematic representation of the drawdown procedure is shown in Fig. 6-18. The crucible below the cathode caught the deposits that slipped off the cathode.

In a series of eight drawdown runs, the uranium concentration in the electrolyte was reduced from 6.7 to 0.01 wt%. By the final run, 99.9% of the uranium had been removed along with only 13% of the rare earths. As the ion-displacement process continued, it was possible to reduce the rare earth concentration to less than 0.01 wt%.

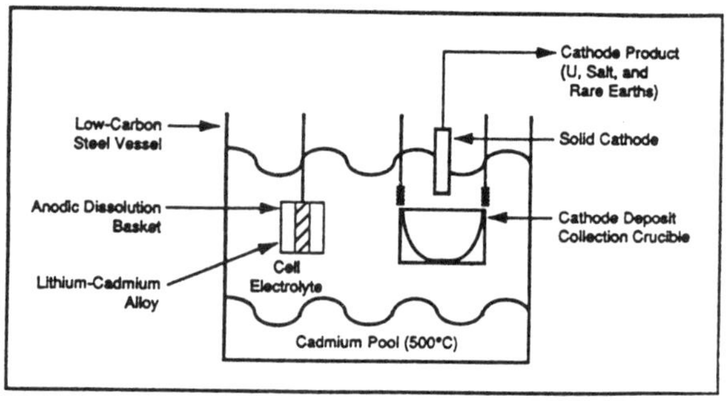

In the uranium-plutonium drawdown procedure, a solid lithium-cadmium alloy (5.8 wt% Li-94.2 wt% Cd) is used as an anode. Under an applied voltage, lithium displaces uranium and plutonium in the salt. The crucible below the cathode is used to catch uranium that slips off the cathode. Plutonium is transferred to a liquid cadmium cathode. Usually, uranium and plutonium are deposited at different times, but in theory uranium and plutonium depositions could be made at the same time.

Fig. 6-18. Schematic Representation of Electrorefiner Drawdown

The feasibility of drawing down the $PuCl_3$ concentration by electrotransport of plutonium to a liquid cadmium cathode was also demonstrated in five scouting runs.

Electrotransport of Plutonium to Liquid Cadmium Cathodes. Work was done in the laboratory-scale electrorefiner to demonstrate the transport of plutonium to a liquid cadmium cathode. Some uranium was also electrotransported along with the plutonium. A problem with this operation was the growth of uranium dendrites from the surface of the cadmium once it became saturated with uranium. In uranium-saturated cadmium, the uranium has a chemical activity of unity, the same as that of uranium metal. If the uranium dendrites come in contact with the structure supporting the cadmium cathode, the cell short-circuits, and the transport of uranium and plutonium ceases.

The solid phase in equilibrium with plutonium-saturated cadmium is $PuCd_6$, and, if the liquid cadmium cathode was to be of a reasonable size, the plutonium had to be "pumped" into the cathode until a large amount of $PuCd_6$ had been formed. The goal was a plutonium concentration of 10 wt% in the final product.

To eliminate the growth of uranium dendrites, a "pounder" was designed by Bill Miller (Fig. 6-19) to break up and push emerging dendrites back into the cadmium. The pounder was a cylindrical ceramic insulator (1/2-in. thick) that fit loosely within the crucible containing the cadmium. It was moved up and down and rotated around the cathode electrical lead to a new position for each up-and-down cycle. A wedge-shaped notch provided access to the salt electrolyte, which was pumped in and out of the volume between the cadmium surface and the upper position of the pounder.

Forty runs were made by Bill Miller, Ziggy Tomczuk, and John Heiberger to investigate

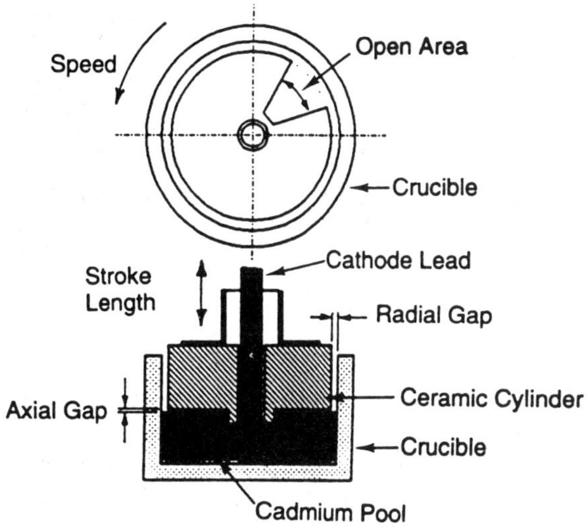

Fig. 6-19. Pounder Design for Liquid Cadmium Cathode

variables that might affect pounder performance, such as thickness of the radial gap, stroke length and frequency, spacing between the cadmium surface and the pounder at the low point of its stroke, rotation speed, transport current, and the ratio of plutonium to uranium in the electrolyte salt. Combinations of these variables were found that consistently controlled dendrite growth and prevented electrical short-circuiting.

In laboratory runs employing a 5-cm-dia ceramic crucible containing about 200 g of cadmium, collection efficiencies for uranium and plutonium were excellent, ranging between 95 and 100%. Figure 6-20 shows a typical product ingot. The goal of 10% heavy metal (U plus Pu) in the product was achieved routinely. The highest loading was 19 wt% (16 wt% Pu, 3 wt% U). Ceramic crucibles of either beryllia (BeO) or high purity AlN (no Al_2O_3) are not wetted by the cadmium and, when provided with a 4° taper, permit good release of the product ingots.

Retention of Noble Metals in Anode Baskets. Tests using simulated EBR-II fuel were run to determine whether metal screens (200 or 325 mesh stainless steel) were needed in

Fig. 6-20. Product Ingot of Deposit from Liquid Cadmium Cathode

anode baskets to retain the noble metals. These tests showed that 100% of the zirconium, 99% of the molybdenum, 60% of the palladium, 100% of the rhodium, and 92% of the ruthenium were retained in the anode baskets, whether or not screen retainers were employed. On average, 92% of the noble metal content was retained in the anode baskets. Yet to be determined is the retention of noble metals when actual spent fuel is processed. The minor fission products, selenium, antimony, and tellurium, were found to behave like the noble metals and, therefore, are also likely to be left in the anode baskets on electrotransport of the uranium.

Process Instrumentation. In the early 1990s, Jim Willit developed voltammetric techniques for measuring *in situ* the concentrations of uranium, plutonium, and rare earths in the electrolyte. In electrotransport of plutonium to the cadmium cathode, monitoring of the electrolyte composition is critical for controlling the composition of the cathode product. *In situ* concentration measurements were also very useful in following the concentrations of uranium and plutonium during drawdown operations.

Making such *in situ* measurements possible was the development of a reference electrode based on the use of a sparingly soluble salt, K_xZrCl_{x+2} where x equals 1 or 2. A single-probe design that incorporated two reference electrodes, as well as working and counter electrodes, was developed. In all voltammetric techniques, a potential waveform is applied to the working electrode and the resulting current is measured. Current peaks occur at different voltages for the various elements. The size of the peak for a particular element is proportional to its concentration in the electrolyte. An investigation of waveforms disclosed that square-wave voltammetry was the preferred technique.

ENGINEERING-SCALE PROCESS DEVELOPMENT

Electrotransport of Uranium. Improved performance was achieved in the electrotransport of uranium from the cadmium pool to the single iron mandrel cathode. With mixing of the electrolyte discontinued during the latter portion of the two electrotransports, 9.5 and 9.3 kg of uranium were transferred in 37 and 26.5 h, respectively. Figure 6-21 shows one of the uranium deposits.

Dissolution of Spent Fuel. Anodic dissolution of U-Zr-fissium fuel pins was demonstrated at the plant scale (10 kg of uranium). With a coulombic efficiency of 50%, 99.8% of the fuel was removed from the cladding. Uranium was simultaneously deposited on a solid iron cathode. Anodic dissolution was concluded to be a practical operation for plant-scale application.

Equipment Testing. Harvesting of the uranium cathode deposit was achieved by pulling the iron mandrel through a die and collecting the stripped uranium in a steel basket. Three devices were tested for providing the stripping force—a torque wrench, an impact wrench, and a modified pipe threader. Good performance was realized

Fig. 6-21. Uranium Deposit (9.3 kg) on Single Mandrel Cathode

with the torque wrench and the modified pipe threader.

Cover Gas Treatment System. During operation of the engineering-scale electro-refiner, cadmium aerosols were released from the electrorefiner and deposited on the interior of the glove-box windows and on the equipment. Such deposition of cadmium could not be tolerated in the EBR-II Fuel Cycle Facility. To reduce the release of cadmium from the electrorefiner, a treatment system was designed for removal of cadmium vapor from the cover gas of the electrorefiner. The cadmium vapor was removed by condensation in a mass-transfer unit (MTU) consisting of countercurrent heat-exchanger trays. This unit was designed under the supervision of Erv Carls of CMT by members of the Engineering Division. In this system, a portion of the cover gas in the electrorefiner was removed, then passed through the MTU. Using the MTU, the concentration of cadmium in the cover gas within the electrorefiner was reduced from 2,000-4,000 ppm by weight to less than 100 ppm. In the advanced IFR process flowsheet, which eliminates the cadmium pool, the concentration of cadmium vapor in the electrorefiner cover gas would be greatly reduced, possibly negating a need for the MTU.

WASTE TREATMENT PROCESSES

After an electrorefining campaign, which consists of processing about 100 kg of spent fuel, the metal and salt phases in the electrorefiner must be processed to remove accumulated fission products and to recover the TRU elements for recycle. Before salt is removed from the electrorefiner, about 90% of the actinides is removed from the salt by the uranium-plutonium drawdown procedure described previously. The reference flowsheet for treatment of the wastes is shown in Fig. 6-22.

Salt Extraction. For the salt extraction step, a high-temperature centrifugal contactor similar to that developed by ANL for aqueous/organic solutions was designed and fabricated from Type 304 stainless steel. Good performance of a single-stage contactor with a 4-cm-dia rotor was demonstrated in a glove-box facility in which the contactor was installed along with feed and receiver vessels. During countercurrent flow of the cadmium and salt, the contactor operated very smoothly—no vibration was observed and feed rates could be controlled. In tests in which cerium was used as a stand-in for uranium, lanthanum for plutonium, and yttrium for rare earths, because of the similarities in the relative separation factors, stage efficiencies of 90% were found at rotor speeds of 2,700 rpm. The results show that it should be possible to achieve good separation of the TRU elements

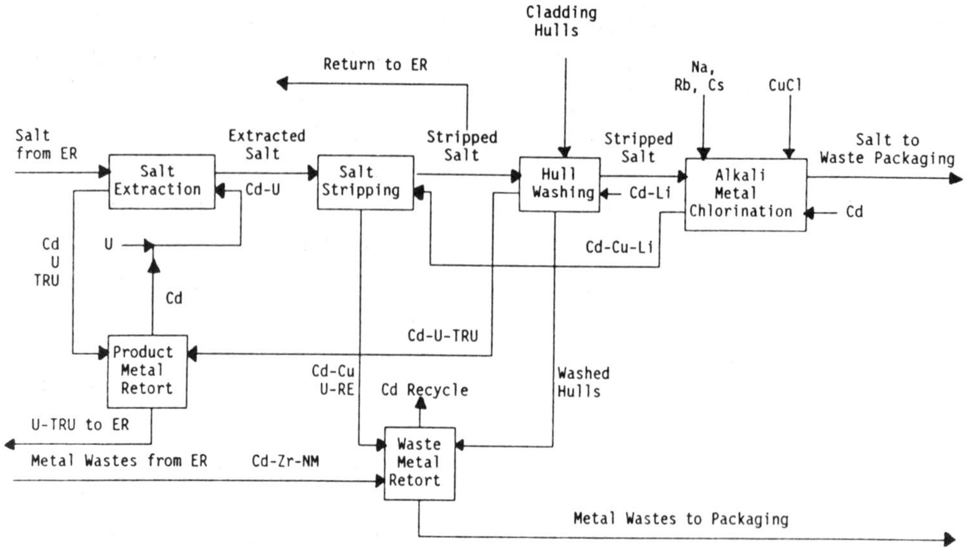

Plutonium and uranium are separated from the rare earth and alkaline earth elements by a countercurrent extraction process in which the uranium and TRU elements (Pu, Am, Cm) are extracted into a cadmium-2.2 wt% uranium solution. About five or six extraction stages are required to remove 99.9% of the plutonium. Essentially all the uranium, any residual TRU elements, and most of the rare earth elements are then removed from the salt raffinate by contacting it with a Cd-0.1 wt% Li alloy in a salt stripping step. The salt at this point contains all of the alkali metal, alkaline earth, and halide fission products, and only traces of actinides (<10 nCi/g). The cadmium from the salt stripping step is combined with spent metal from the electrorefiner and the cladding hulls, and the excess cadmium is removed by distillation and returned to the electrorefiner (ER). The salt and metal wastes are then converted to suitable waste forms for disposal in a high-level waste repository.

Fig. 6-22. Waste Treatment Flowsheet

from rare earths in a multistage unit. In 1995, a four-stage unit was built and installed in the glove-box facility. Flow tests showed satisfactory performance of the contactor, and the extraction tests, completed in 1997, were successful. This work was done under the direction of Terry Johnson and Lorac Chow.

Salt Stripping. Demonstrations of the salt-stripping operation were performed by use of the apparatus shown in Fig. 6-23. This apparatus was installed in an argon-filled glove box adjacent to the engineering-scale electrorefiner. Electrolyte salt from the electrorefiner was pumped into the apparatus through a heated transfer line by a submersible centrifugal pump. Because uranium had not been removed from the salt by a drawdown procedure, a cadmium-15 wt% lithium alloy was used for reduction of the actinides, the rare earth elements, and yttrium, rather than the cadmium-0.1 wt% lithium alloy called for in the flowsheet. The experimental results shown in Fig. 6-24 agree well with predictions based on laboratory-scale experimental data. Successive additions of the reductant alloy are represented by S4 through S9 on the abscissa.

Waste Salt Immobilization. The waste salt from the salt-stripping operation will contain the fission products, cesium and strontium, and is, therefore, a high-level waste which must be disposed of in a geologic repository. Michele Lewis and others, under the supervision of Terry Johnson, investigated the use of zeolites for immobilizing the salt waste. To immobilize the salt waste, it is

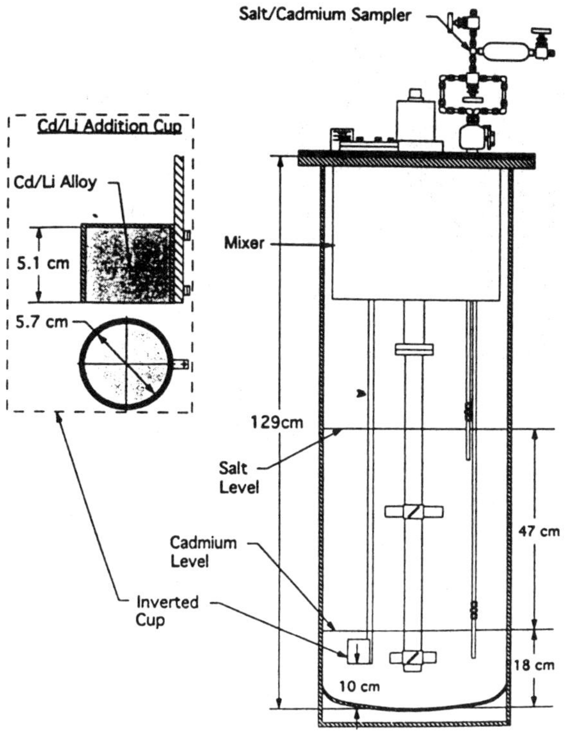

Fig. 6-23. Salt Stripping Apparatus

sorbed into a zeolite, which must then be converted to a solid waste form. The zeolite of choice is zeolite A, $M_{12}(AlO_2{\cdot}SiO_2)_{12}$, where M is a cation such as sodium. Treatment of this zeolite with a pure LiCl-KCl eutectic produces $M_{24}(AlO_2{\cdot}SiO_2)_{12}Cl_{12}$. When this latter material is exposed to the IFR waste salt, the cations exchange with the fission products, removing them from the salt. Blending this material with "anhydrous" zeolite, $Na_{12}(AlO_2{\cdot}SiO_2)_{12}{\cdot}mH_2O$, where m is 3 or 4, removes surface salt by absorption or occlusion in cages within the zeolite structure. This blended material is the starting material for preparation of the waste form. This material is stable to radiation at doses up to 10^9 rad, the maximum dose rate tested, and also to heat.

Two approaches were investigated for producing the final waste form. The first was to blend the salt-occluded zeolite powders with glass frit and to fabricate, under heat and pressure, a monolithic waste form. The second was conversion to sodalite, $Na_6(AlO_2)_6(SiO_2)_6{\cdot}2NaCl$, in which, for stoichiometric reasons, less waste salt is occluded in the zeolite. The addition of glass frit promotes conversion of the zeolite to sodalite. Under heat and pressure, a hard, monolithic waste form that is highly resistant to water leaching is produced. Accordingly, sodalite was concluded to be the superior waste form.

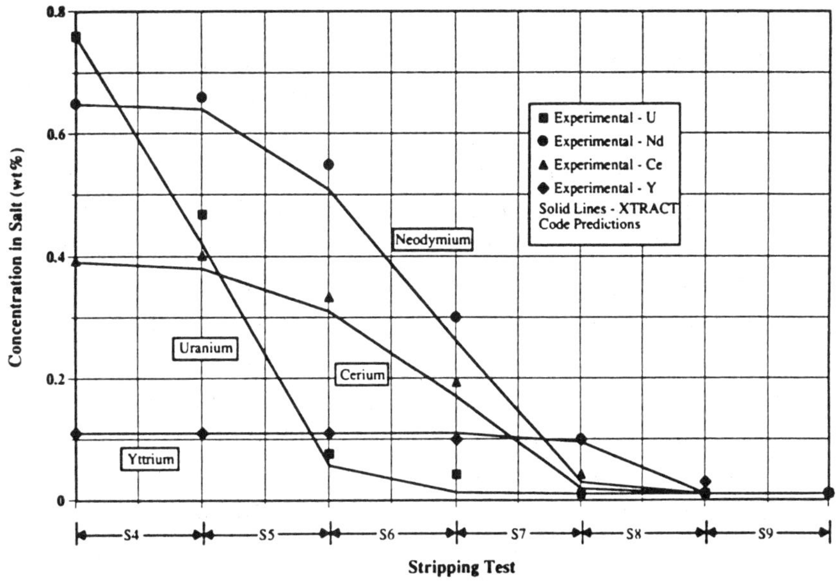

Fig. 6-24. Results of Salt Stripping Tests

Waste Metal Immobilization. The metal waste consists of cladding hulls, the cadmium-rare earth solution from the salt-stripping step, and cadmium from the electrorefiner. Removal of cadmium from this mixture for recycle by retorting was slow and difficult. The cadmium recycle load was diminished by elimination of the cadmium pool in the electrorefiner. It was found that rare earths could be extracted from the cadmium-rare earth solution into a copper-aluminum alloy, thereby allowing recycle of most of the cadmium and reducing greatly the amount going into the metal waste, but this increased the complexity of the process. Eventually, as outlined in the advanced IFR flowsheet below, the salt-stripping step was eliminated, and the rare earth elements were removed on zeolites along with the alkali and alkaline earth elements. The metal waste then consisted of cladding hulls and stainless steel filter cartridges used to remove particulate matter from the salt electrolyte. These stainless steel components containing all the noble metal fission-product elements were then simply melted together to produce a compact waste form that could easily be packaged for disposal in a geologic repository.

ADVANCED IFR FLOWSHEET

Several major flowsheet changes, listed below, were devised from work done in the early 1990s and were incorporated into an "advanced" IFR flowsheet (Fig. 6-25). They would have been demonstrated on a large scale had development of the IFR process continued.

A major feature of the advanced flowsheet is elimination of the cadmium pool to avoid problems of cadmium vaporization and condensation on surfaces of the electrorefiner. Zirconium and noble metal fission products are expected to remain largely within the anode basket. Any of them that do escape to

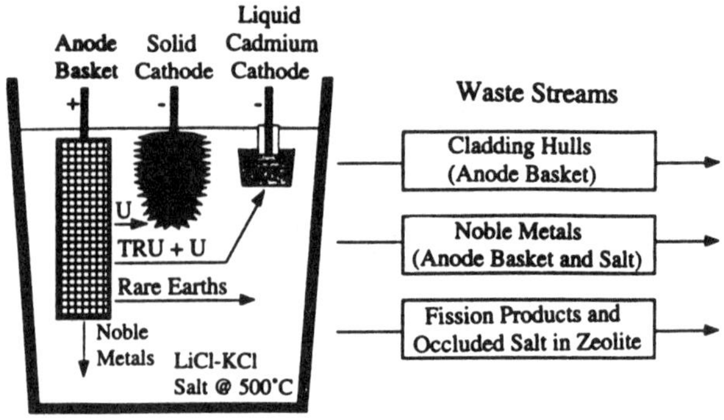

Fig. 6-25. Schematic Representation of Advanced IFR Process

the salt, where they would exist as particulates, would have to be removed by filtration.

The principal flowsheet changes are the following:

1. The volatile metals, sodium, cesium, and rubidium, would be retorted from chopped fuel pins before they are loaded into an anode basket for charging to the electrorefiner. This would permit the volatile metals to be handled separately and would prevent buildup of sodium to a point where disposal of the entire electrolyte would be required.
2. Uranium trichloride (UCl_3) rather than $CdCl_2$ would be used to oxidize chemically active fission products (alkali, alkaline earth, and rare earth metals) to their chlorides in order to maintain an actinide concentration of 2 mol% in the salt.
3. The lower cadmium pool would be eliminated from the electrorefiner. This would avoid condensation of cadmium vapor on the upper surfaces of the electrorefiner and would eliminate contamination of cladding hulls and noble metal fission product wastes by cadmium.
4. The small fraction (~10%) of the noble metal fission products that escape as particulates from the cladding hulls to

the electrolyte salt would be removed periodically from the salt by filtration.

5. Countercurrent extraction would be used for reduction and removal of TRU elements from the salt (drawdown), for separation of the rare earth elements from the TRU elements, and for oxidation of the TRU elements back into the salt for the next electrorefining campaign.
6. The salt-stripping step would be eliminated. The rare earth elements, as well as the alkali and alkaline earth elements, would be sorbed on zeolite for disposal. Not having to deal with a cadmium metal waste would greatly simplify the production of a metal waste form.

PERSONNEL

Jim Battles was in charge of the IFR program until he became the Division Director, whereupon Jim Laidler took over that responsibility, and he, in turn, was succeeded by Chuck McPheeters. Various portions of the program were headed up by John Ackerman, Erv Carls, Eddie Gay, Terry Johnson, and Bill Miller. Irv Johnson made major contributions to the effort by drawing on his extensive background on the chemistry and thermodynamics of liquid metal and molten salt systems. A large number of other CMT people were involved in the program: Dan Abraham, Jack Arntzen, John Basco, Bob Blaskovitz, Bob Blomquist, Lorac Chow, Jerry Dewey, Bob Everhart, Don Fischer, Greg Fletcher, Eddie Gay, John Heiberger, Len Leibowitz, Michele Lewis, Dusan Lexa, Dick Malecha, Sean McDeavitt, Chuck McPheeters, Bali Misra, Tom Mulcahey, Henry Myers, Aron Newman, Candido Pereira, Nick Quattropani, Mike Richmann, Jack Settle, Susan Slater, Mike Slawecki, Jim Smith, Ziggy Tomczuk, Verne Trevorrow, Dave Warren, and Jim Willit. A number of Japanese scientists and engineers who participated in this effort under an international agreement made significant contributions to this research and development program. These were: Masatoshi Iizuka, Katsumasa Kanasugi, T. Kobayashi, Tadafumi Koyama, Tomohiro Nishimura, Takanari Ogata, Moriyasu Tokiwai, Yasuo Tsuchie, Takeshi Yokoo, and Takuma Yoshida. Secretaries were Lilia Barbosa and Janet Carothers.

Electrometallurgical Treatment of Spent Reactor Fuels

When development of the IFR process was terminated, work was initiated on development of electrometallurgical processes for treatment of a variety of spent nuclear fuels for disposal in a repository. The objective was a reduction of risk of chemical reactions of groundwater with fuel emplaced directly in the repository. The treatment was aimed at separating actinides, principally plutonium, and fission products and immobilizing them in waste forms that are highly resistant to attack by groundwater. Much of the technology developed for the IFR process is applicable to the electrometallurgical treatment of the spent nuclear fuels, especially electrorefining for removal of a pure uranium product. To put it simply, the difference in objectives of the IFR process and the electrometallurgical processes is that the IFR process was designed to recycle fuel to a reactor while the objective of electrometallurgical processes is the safe disposal of spent reactor fuel and other nuclear materials.

The generic electrometallurgical process consists of dismantling fuel subassemblies, chopping the fuel into segments, and charging these into an anode basket in an electrorefiner. Uranium is then transported to a solid cathode. The fission products and actinides are then placed in two stable waste forms for disposal in a repository. One is a glass-bonded sodalite into which actinides and highly reactive fission products have been incorporated. The other is a metal waste form, principally the cladding hulls, which also contain nearly all of the noble metal fission products. There are no other high-level wastes, and only small amounts of low-level wastes are produced. Variations in this basic process had to be developed to accommodate the different types of spent fuels. In the case of reactor fuels, separation of a bulk constituent such as uranium greatly reduces the volume of waste produced for repository storage and, hence, the costs of waste disposal.

ADVANCED ELECTROREFINER CONCEPTS

To accommodate large quantities of spent fuel requiring treatment for disposal, an effort was mounted on the development of an advanced electrorefiner capable of high uranium throughput (>40 kg/h) and a batch-size capacity of 100 kg. Illustrated in Fig. 6-26 are anodic dissolution baskets in a high-throughput electrorefiner constructed in 1995 for testing. The four baskets were loaded with uranium from the lithium reduction step for converting oxide fuel from the Hanford "N" reactor to metal. The baskets are lined with 325-mesh stainless steel screen to retain uranium particles from the reduction step. The cathode is a steel crucible with a salt drainage hole at the bottom. During electrorefining, the uranium is electrodeposited on the walls of the cathode crucible, scraped off the crucible by steel or beryllia scrapers attached to the ends of the rotating anodic dissolution baskets, and collected in the bottom of the crucible. After transport of the uranium has been completed, the anodic dissolution baskets are replaced by a compactor to compact the uranium to a density greater than 10 g/cm^3. The uranium is removed and melted to produce a uranium product ingot.

Fig. 6-26. Anodic Dissolution Baskets in Advanced Electrorefiner

Short circuits plagued the initial runs using U-Zr-fissium pin segments (EBR-II-type fuel) as a result of which the number of ampere-hours required to transport a kilogram of uranium was very high. Uranium transport rates were very low—only about 60 g of uranium per hour. Design improvements were needed to achieve high uranium throughput.

Effort continued on a high-throughput electrorefiner (HTER) design with a uranium throughput capacity of >40 kg U/h and a batch size of 100 kg. The HTER consists of 20 stainless steel anode baskets and has a batch size of 150 kg. This electrorefiner, with a diameter of 25 in. (0.6 m), accommodated a uranium batch size of 150 kg in tests conducted with unirradiated "N" reactor fuel. The number of ampere-hours (31,219) was nearly twice the number of ampere-hours passed through the electrorefiner in all the earlier runs. About 75 kg of uranium was collected from the tests, and about 50 kg of that uranium was transferred to a cathode processor for consolidation. The limiting factor in the scaleup was the scraper arrangement for removing the uranium product from the cathode tubes. These initial scaleup tests were considered successful, and more favorable operating conditions along with improved scraper blade designs are expected to extend the operating time of the unit.

PROCESS DEVELOPMENT STUDIES

Process flowsheets are being developed for the following types of reactor fuels.

Spent Fuel from Experimental Breeder Reactor-II. When plutonium disposal, rather than recovery, is the objective, the treatment process becomes much simpler than the IFR process, which was designed for recovery and recycle of plutonium from the U-Pu-Zr fuel alloy. In particular, the salt-extraction and salt-stripping steps can be eliminated. Following a campaign in which uranium is removed from ten batches of fuel (about 100 kg) by electrotransport to a solid cathode mandrel, the actinides can be removed from the salt (drawdown) by the lithium-ion displacement procedure described previously under the IFR pyrochemical process. (An alternative to drawdown by electrotransport, which is gaining favor, is a chemical extraction of uranium and plutonium into cadmium by a countercurrent extractor.) The uranium and plutonium concentrations in the salt electrolyte would be restored by converting them to the chlorides with $CdCl_2$. The alkali, alkaline earth, and rare earth fission products can then be sorbed on a zeolite, which is converted to a waste form for disposal. Alternatively, uranium alone can be removed from the electrolyte, and plutonium, together with the highly reactive fission products, can be sorbed on the zeolite. Experimental results showed that the amount of plutonium sorbed on Zeolite A was substantial—up to 22 wt%. Thus, Zeolite A appears to be a good candidate for removing plutonium from waste electrolyte salt.

The dendritic uranium deposit resulting from the electrotransport of uranium is wetted by the electrolyte salt, which contains plutonium and fission products. To produce a pure uranium product that does not have to be sent to a geologic repository, it is necessary to remove the contaminated salt. It was found that washing with a salt that is in contact with lithium dissolved in a molten metal such as cadmium is effective. Two successive washings resulted in complete removal of the rare earth elements. The final rare earth concentrations in the washed salt containing the uranium were generally below the limits of analytical detection (<0.01 wt%). Plutonium was not present in these experiments, but its behavior should be the same as that of the rare earths.

The metal waste consists of (1) the cladding hulls along with most of the noble metal fission products (about 92% according to experimental results) and (2) filter cartridges, which remove residual noble metals and uranium oxide from the molten salt. The metal waste would be melted to form an ingot.

The "Mark V" high-throughput electrorefiner, with a diameter of 10 in. (0.2 m) and a capacity of 150 kg/month, is being readied for a demonstration of EBR-II blanket processing at ANL-West.

"N" Reactor Fuel. Located at Hanford, the "N" reactor uses a uranium oxide fuel clad with Zircaloy. After reduction with lithium to convert the uranium oxide to the metal, which is discussed later, the reduced metal is made the anode of an electrorefiner and electrotransported to a solid cathode. Treatment for disposal of the fission products can be effected with only slight modifications of the process described above for EBR-II fuel.

Single-Pass Reactor Fuel. Single-pass reactor (SPR) fuel consists of aluminum-clad uranium slugs bonded by an Al-Si braze. Opening up the cladding to permit removal of the uranium by electrotransport was accomplished by drawing a cutting blade longitudinally along the surface of the SPR fuel element and laying open the cladding. With a simulated SPR fuel element used as an anode in a test cell, the uranium was successfully transported to a solid cathode. The mass of the aluminum sheath remained constant, and there was no evidence of corrosion. This work demonstrated that uranium can be selectively oxidized in the presence of aluminum at a reasonable current density and a current efficiency of 75%.

Aluminum-Based Spent Fuels. Over the next forty years, 128 metric tons of spent aluminum-matrix reactor fuel will be shipped to the Savannah River Site from U.S. and foreign research reactors. When fabricated initially, this fuel contained over 55 metric tons of uranium at an average enrichment of approximately 20% U-235. This fuel is now corroding in wet storage and must be stabilized. The fuel would normally be processed in the Savannah River processing facilities, but these are slated for decommissioning in 2005.

Electrometallurgical treatment was regarded favorably by a special DOE task team formed to look into the treatment, packaging, and disposal of these fuels because the method has the potential to separate aluminum and uranium, thus greatly reducing the amount of high-level waste with an estimated cost saving of $200 million. The enriched uranium has a commercial value of about $400 million.

The proposed flowsheet for aluminum-based fuels is shown in Fig. 6-27.

Waste Salt from the Molten Salt Reactor Experiment (MSRE). The Molten Salt Reactor, which was operated at Oak Ridge in the 1960s, used a uranium fluoride fuel that was dissolved in a mixture of fluoride salts. Initially, ^{235}U was the fissionable material, but this was changed later to ^{233}U. The fuel salt has been in storage at ORNL since 1969 when the reactor was shut down. A small amount of ^{232}U, which has a half-life of 70 years, was produced during reactor irradiation. In its decay chain is thallium-208, which emits hard (2.6-MeV) gamma radiation. Irradiation of the fluoride salts by this gamma and that resulting from fission-product decay cause decomposition of the fluoride salts with the production of fluorine gas (F_2). Some of the fluorine reacts with UF_3 to form UF_6, whose high volatility leads to migration of uranium and the spread of activity throughout the storage system. Cleanup of the salt by removal of the fission products and recovery of the ^{233}U has become a matter of high priority.

One of the methods under consideration for salt cleanup is an electrometallurgical process proposed by the CMT Division. In this process, a lithium-bismuth alloy is used as the anode in an electrochemical cell, and constituents of the salt are successively reduced in four steps by lithium-ion displacement. In the first step, noble metal fission products and most of the zirconium are deposited onto a solid cathode. In the second step, uranium, TRU elements, and the rest of the zirconium are deposited on a solid cathode. The bismuth of this step is made the anode of a second electrorefiner to purify and isolate U-233. In the third and fourth steps, rare earths, alkaline earths, thorium, cesium, and barium are also removed to liquid bismuth cathodes by using increasingly higher voltages. This leaves the salt essentially devoid of activity and disposable as a low-level waste. Construction was begun in 1995 on an electrorefining cell suitable for work with fluorides at temperatures up to 700°C.

The remediation process proposed by ANL for the MSRE salt is to separate the salt into actinides for recovery, fission products for disposal as high-level waste, and bulk salt for disposal as low-level waste. The total salt volume of about 2000 L would be processed in 30-L batches in a lightly shielded facility. Seventeen tests were made. A solid cathode and simulated MSRE salt were employed to investigate the removal of zirconium and its separation from uranium. Lithium-bismuth alloy was used in the anode baskets. The

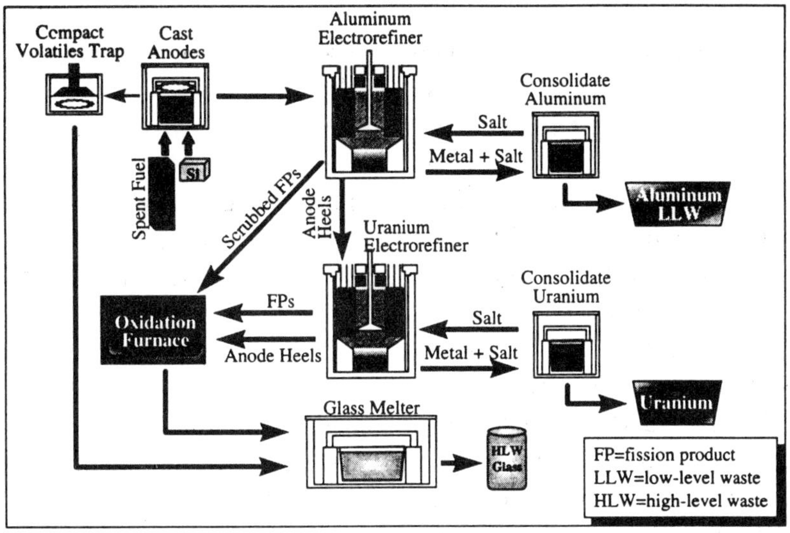

Following removal of the end hardware, the fuel is melted and cast into shapes for use as anodes in the succeeding electrorefining operations. Silicon is added to form stable intermetallic compounds with the uranium and enhance the separation of aluminum in the first electrorefining step. A cryolite electrolyte (Na_3AlF_5) is used for the electrotransport of aluminum. The anode basket is then transferred to a uranium electrorefiner in which the electrolyte is the LiCl-KCl eutectic, and uranium is transported to a solid cathode. High rates for the product metals are achieved by use of the high-throughput advanced electrorefiner described earlier. The product metal dendrites are consolidated by melting. The lower melting electrolytes that adhere to the dendrites are poured off and recycled after the product metals have solidified. Fission products from various sources are converted into oxides, combined with glass formers, and melted to produce a glass waste form, which is placed in a canister for emplacement in a geologic repository.

Fig. 6-27. Flowsheet for Electrometallurgical Treatment of Aluminum-Based Fuel

results indicated that zirconium (along with other noble metals in the salt) can be separated from the salt by deposition on an iron cathode. This material could be converted into a stable metal waste form. Any remaining zirconium and all the uranium can then be deposited on a second solid cathode. The process concept shows promise, but further work is needed on the details, one of which is removal of cesium from the salt.

ACTINIDE RECOVERY FROM SPENT OXIDE FUELS

Spent fuel from commercial, water-cooled power reactors (LWRs) contains significant amounts of unreacted fissile uranium as well as TRU elements (Np, Pu, Am, and Cm). Separation and recovery of these elements from the spent fuel would offer two major benefits: (1) these actinide elements could be a useful source of additional energy if burned in a fast reactor and (2) removal of these elements, which have long half-lives, from the spent fuel would considerably shorten the time required for assured confinement in a geologic repository. In the early 1990s, this was a natural companion program with the pyrochemical process for IFR fuel.

Because of a national policy decision to discontinue all work on fast reactors in 1993, actinide recycle became a dead issue, and these programs were phased out. It was realized, however, that the capabilities of such processes could be redirected toward much-needed methods for treating various types of oxide fuels for safe disposal. Adaptation of the lithium reduction process (to be described later) for this purpose is the subject of the next section on "Treatment of Spent Oxide Fuels for Safe Disposal."

The work on this process drew heavily from previous technology that had been developed during the development of processes for EBR-II fuel and subsequent work. Four processing schemes were investigated, all of which involved a reduction of the UO_2 fuel to uranium metal, followed by separation steps to recover the uranium and TRU elements. Three reactive metals—magnesium, calcium, and lithium—were evaluated as candidate reductants for the UO_2. All three are capable of reducing the UO_2 in molten salt solutions, but each one presents its own set of problems. The reduction reactions are

$$2\,Mg + UO_2 \rightarrow 2\,MgO + U$$

$$2\,Ca + UO_2 \rightarrow 2\,CaO + U$$

$$4\,Li + UO_2 \rightarrow 2\,Li_2O + U$$

To avoid the production of large amounts of radioactive MgO, CaO, or Li_2O by-product waste, an electrowinning step is considered necessary to regenerate the Mg, Ca, or Li for recycle. Electrowinning of these metals from the oxides is a difficult operation and has been avoided by industry, but initial studies indicated that it might be feasible for this purpose. Direct electrolytic reduction of UO_2 to the metal has also been tried with limited success. In all cases, the main problem appears to be the lack of a satisfactory anode for oxygen release. Although earlier work had shown that magnesium-zinc alloys could be used to reduce UO_2, only calcium and lithium were used as reductants in the actinide recycle program.

After the UO_2 reduction, several approaches are available for separation of the uranium, TRU elements, fission products, and other fuel constituents. Among these are salt transport, magnesium extraction, precipitation in zinc-magnesium, halide slagging, and electrorefining.

Salt Transport Process. In this process, the oxide fuel was first reduced by calcium in a Mg-Cu/$CaCl_2$-CaF_2 system to produce a Cu-Mg alloy containing the TRU elements and some fission products. The reduced uranium precipitated because of its low solubility in the alloy. Alkali, alkaline earth, and halide fission products remained in the salt, and the gaseous fission products were released. In the ensuing salt transport step, the Cu-Mg alloy and a Zn-Mg alloy in mutual contact with a molten salt were equilibrated, causing the actinide elements to transfer to the Zn-Mg alloy. The more noble metals and the uranium remained in the Cu-Mg alloy. The TRU elements and rare earths were recovered by evaporating the Zn and Mg and were transferred to the IFR electrorefiner, where they were introduced into the IFR fuel. The CaO from the UO_2 reduction was electrolyzed to produce metallic calcium for recycle.

In initial tests using UO_2, PuO_2, NpO_2, and oxides of representative fission-product elements, the reduction step was effective, and the salt transport separation was as expected. However, significant coprecipitation of neptunium with the uranium necessitated the addition of a halide slagging step to recover the neptunium from the uranium when it was melted to form a U-Fe product ingot.

Investigations of this process were discontinued in favor of the lithium reduction process in 1992. However, the significant body of information on the thermodynamic, chemical, and electrochemical processes involved in the salt-transport process and on materials compatibility that had been developed during the work by Irv Johnson and others on this process became a highly useful database for the ongoing development programs in this area.

Magnesium Extraction Process. The head-end steps of the magnesium extraction process were essentially the same as those for the salt-transport process; that is, the UO_2 fuel was reduced by calcium in a $CaCl_2$-CaF_2 molten salt at about 800°C. To obtain a liquid phase, the reduced uranium was alloyed with iron. The U-Fe alloy was then contacted with liquid magnesium, which extracted the TRU elements and some of the rare earth metals, leaving behind the U-Fe, noble metals, and the rest of the rare earths. The magnesium was removed from the TRU elements by retorting, and the TRU product was used as a feed to the IFR fuel cycle.

Two runs were made at 790°C with synthetic LWR fuel containing UO_2, PuO_2, and NpO_2, along with oxides of fission product elements, to evaluate the process. The reductions were satisfactory, but the extraction results were somewhat inconclusive, due mainly to poor material balances. Further work showed that multiple extractions would be required, and, as in the salt-transport process, a halide slagging step would be needed to recover neptunium from the uranium. Because of these problems and difficulties with materials compatibility, work on this process was discontinued in 1991.

Zinc-Magnesium Process. Again, the UO_2 fuel was reduced by calcium in $CaCl_2$-CaF_2, but in this case the TRU elements and uranium metal were collected in a magnesium-rich Mg-Zn alloy in which the uranium has a low solubility and precipitates as the metal. The Mg-Zn alloy containing the TRU elements was separated from the precipitated uranium and then retorted to vaporize the zinc and magnesium, leaving an ingot of the TRU elements to be added to the IFR electrorefiner. The uranium precipitate was cast into an ingot for storage. Because about 20% of the neptunium coprecipitated with the uranium in the process, a $CaCl_2$-UCl_3 salt was used in the uranium casting step to extract the neptunium into the salt phase, which was recycled to the reduction step to complete the neptunium recovery. Work on this process was terminated in 1992.

Lithium Reduction Process. In 1992, work began on another alternative process which employs lithium as the UO_2 reductant. A flowsheet for the lithium reduction process is shown in Fig. 6-28. In 1992, the lithium reduction process was selected for continuing development work on actinide recycle. As mentioned earlier, however, the program was redirected to the treatment of oxide fuels for disposal, rather than recovery of the actinides.

Laboratory-scale (0.5-kg) experiments showed that 99.99% reduction of the actinide oxides could be achieved. The reduction, uranium electrotransport, and uranium consolidation steps were performed by using simulated oxide fuel containing oxides of uranium, plutonium, neptunium, curium, and fission products. Reasonable material balances were obtained, and the basic feasibility of the process was established. Engineering-scale equipment capable of

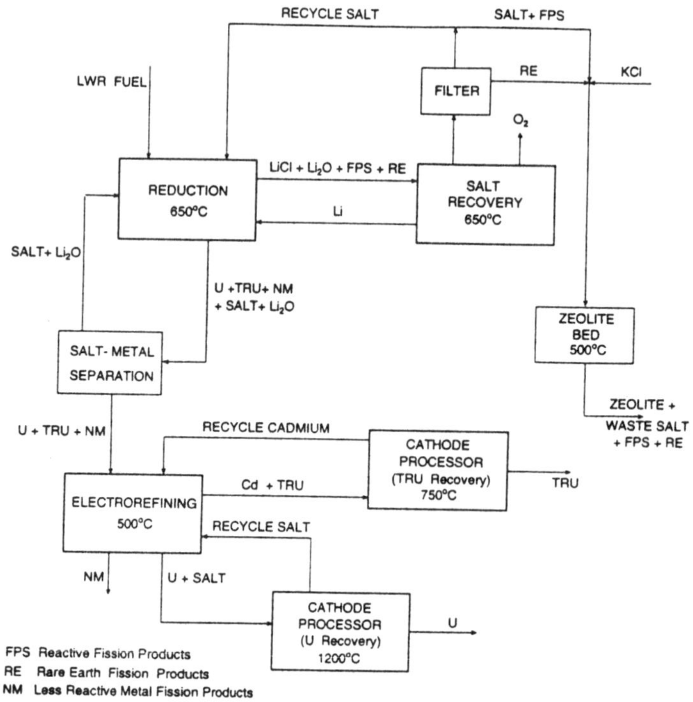

The oxide fuel is reduced with lithium in the presence of molten LiCl-KCl at 500°C (later changed to LiCl alone at 650°C). The spent reduction salt is then processed electrochemically to decompose the Li_2O reduction product and recover lithium and LiCl-KCl. A small portion of the recovered salt is removed to limit the accumulation of unreduced fission products (*e.g.*, Cs, Sr, I, and rare earths), and the bulk of the salt and the recovered lithium are recycled. The removed salt is processed for disposal by methods outlined under the IFR process. After separation from the reduction salt, reduced fuel and less reactive fission products (*e.g.*, Zr, Mo, and Pd), which precipitate during the reduction step, are processed in an electrorefiner to recover purified uranium on a solid cathode. The TRU elements are recovered in a liquid cadmium cathode. The uranium is stripped from the cathode, melted, and recovered as an ingot. The TRU product is recovered as an ingot by distilling off the cadmium for recycle. The uranium is placed in long-term storage for future use as a reactor blanket material, and the TRU ingot is fed into the IFR fuel cycle.

Fig. 6-28. Lithium Process Flowsheet

handling up to 20-kg batches of simulated fuel was designed, fabricated, and installed in controlled-atmosphere glove boxes.

TREATMENT OF SPENT OXIDE FUELS FOR SAFE DISPOSAL

In the electrometallurgical processing of spent reactor fuels, oxide fuels consisting primarily of uranium dioxide (UO_2) present a problem because the oxides must first be reduced to the metals before the fuel can be processed in an electrorefiner. Lithium metal, an exceptionally strong reducing agent, was selected for this purpose to assure as rapid and complete reduction as possible. Molten lithium chloride (LiCl) at a temperature of about 650°C was chosen for the process medium. To avoid the production of wastes containing large quantities of Li_2O produced by this reaction, a lithium electrowinning step is incorporated into the process flowsheet:

$$Li_2O \rightarrow 2\,Li + 1/2\,O_2$$

By recycle of the lithium, the net effect is an electrolytic reduction of UO_2 to produce uranium metal and oxygen. In theory, direct electrowinning of uranium from UO_2 is possible, but previous attempts to do so had shown that it was difficult, so the use of lithium as an intermediate reductant was adopted for the process development effort. It is possible, however, to carry out the two reactions simultaneously in a single process vessel. Subsequent electrorefining purifies the uranium metal product, which can then either be enriched for recycle or be converted to a form suitable for storage or disposal as a low-level waste. In the electrorefining step, the TRU elements (Np, Pu, Am, Cm) can be recovered and alloyed with cladding material and combined with fission products to produce a leach-resistant and diversion-resistant form for ultimate disposal. An alternative is to combine the TRU elements and fission products in a ceramic waste form.

Processing of the TMI-2 Core Debris. The emphasis in this program was largely on a process for the treatment of the debris resulting from the accident at the Three Mile Island-2 reactor in March 1979. This would be a stringent test of the process capabilities in view of the fact that the debris is a highly heterogeneous mixture of oxides, mainly of uranium and zirconium, and a variety of metals from the fuel cladding and other reactor hardware. The debris was divided into four parts, depending on the location in the core and the degree of damage at that location: (1) essentially undamaged fuel near the bottom of the reactor, (2) a portion in the center of the core, which had melted, producing a mixture mainly of $(U,Zr)O_2$ ceramic phases, (3) an upper crust region, surrounding the part of the core that had melted and consisting of the ceramic phases and metallic components that had melted, and (4) a similar lower crust region. If a process can be developed to handle this material, it should be able to cope with almost any UO_2-based system.

Four process flowsheets were developed for treating the TMI-2 debris. The first was similar to one developed earlier for processing normal spent fuel from a light water reactor. The second flowsheet was designed to accommodate the mixed ceramic melt $(U,Zr)O_2$, which contained uranium- and zirconium-rich phases. The third and fourth flowsheets were intended to handle agglomerates of metallic and ceramic phases of widely different ceramic/metal ratios. A suitable substitute for TMI-2 core debris that could be used for some laboratory- and engineering-scale experiments was available from other ANL reactor safety programs in the form of "corium," which is a synthetic molten core material consisting (in wt%) of

58 UO_2, 11 ZrO_2, 14 Zr, 14 Fe, and 3 Cr. This material was used to test the process flowsheets.

Anode Materials and Designs. In the lithium electrowinning operation, the oxygen produced at the anode in the molten LiCl is an aggressive oxidizing agent. This condition severely limits the choice of materials for the anode, which must be both resistant to oxidation and electronically conductive. Carbon, or graphite, which is sometimes used as an anode in corrosive environments as a sacrificial electrode (producing CO or CO_2, rather than oxygen), is unsatisfactory because the molten salt becomes contaminated by carbon particles. Platinum or a platinum-rhodium alloy was used in most of the experimental work, but it is attacked by metallic lithium. An effort is underway to develop an electronically conductive oxide ceramic (similar to those used in fuel cells) or other alternative materials. Possible oxide ceramics include Fe_3O_4, antimony-doped tin oxide, and lithium-doped nickel oxide.

During the electrowinning of lithium from Li_2O in the LiCl salt, gaseous oxygen is generated at the anode. Ideally, the oxygen would be removed from the system immediately by forming bubbles that escape from the molten salt. What actually happened was that the oxygen generated at the anode surface formed very small bubbles that were slow to coalesce and disengage. Although a shroud had been installed around the anode to confine the oxygen, the very small oxygen bubbles appeared to form a quasi-homogeneous lighter phase with the salt, which created a stagnant region within the shroud, thereby forcing new bubbles into the bulk salt as they were formed. This oxygen was then transported by the salt flow to the cathode where it recombined with the metallic lithium. Design studies of this system resulted in a modified anode design that was used in engineering-scale studies (see Fig. 6-29).

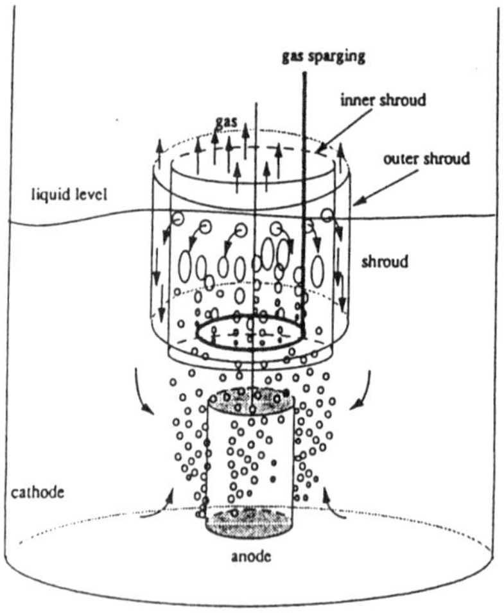

Fig. 6-29. Anode Design for Electrowinning Lithium from Lithium Oxide

Cathode Materials and Designs. Two types of cathodes, liquid bismuth and porous metal, were investigated with the objective of decreasing or eliminating the recombination of oxygen in the electrolyte with the lithium metal product. When lithium is electrodeposited into liquid bismuth, its chemical activity is decreased markedly (by a factor of 10^3 to 10^6) due to the formation of intermetallic compounds. The densities of the bismuth and the intermetallic compounds are much higher than that of the electrolyte salt. The limited surface area of the bismuth electrode and the low activity of the lithium tend to protect the lithium from the oxygen in the electrolyte. The confinement of lithium in the bismuth and its low activity also tend to protect platinum anodes from lithium attack. The bismuth cathode, however, requires an additional process step to recover the lithium and a suitable ceramic containment material.

Porous metals are available in a variety of materials and forms, including screens, felts, and foams. Other work had shown that porous stainless steel was readily wet by liquid lithium and that repeated electrolytic deposition and removal of lithium from porous stainless steel electrodes was a viable process operation. The lithium remains within the electrode in the molten salt medium, so there is no need for crucibles or other ceramic parts.

Uranium Dioxide Reduction. In 1995, two preliminary experiments were performed to determine whether the chemical uranium reduction and the lithium electrowinning steps could be combined into a single operation that would be, in effect, a direct electrolytic reduction of the UO_2. A sketch of the cell that was used is shown in Fig. 6-30. The anode was a Pt-Rh tube, housed in a MgO shroud that was open at the bottom. The cathode was a cylindrical basket made of stainless steel screen. The cathode was connected to a stainless steel tube that served both as the electrical lead and as a helium-gas purge tube for the cell to minimize the oxygen concentration in the gas phase during the electrolysis. The cell container was a high-density MgO crucible. The two reduction experiments were similar, except 100 g of UO_2 chips was used in the first one and only 55 g was used in the second. The cell was operated at a temperature of 650°C, a constant potential of 3.0 V, and a time period of about 30 h. In the first experiment, in which about

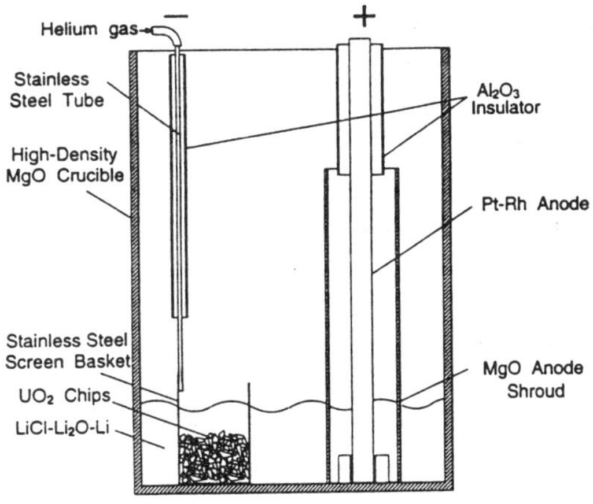

Fig. 6-30. Experimental Cell for Uranium Oxide Reduction

105% of the theoretical current was passed, some of the interiors of the UO_2 chips still contained unreacted UO_2. In the second experiment, in which 140% of the theoretical current was passed, the reduction to uranium metal was complete. Thus, the direct electrolytic reduction of UO_2 was shown to be possible.

The reduction studies, using steel-clad UO_2 pellets as the feed, were continued. The pellets, which were 1.27-cm long and 0.95-cm dia, were completely reduced in approximately 50 h; unclad pellets of the same size required 30 h. During the reduction, the pellets retained their original size and shape, and voids were created due to the fact that the uranium product has a higher density than the UO_2. The voids were filled with the LiCl electrolyte salt, which contained dissolved Li_2O. In a process, the Li_2O concentration would have to be kept below 3.3 wt% to achieve complete reduction of any PuO_2 that may be present. In the case of TMI-2 fuel debris, Li_2ZrO_3 can be formed unless the Li_2O concentration in the salt is held below 1.4 wt%. Although the direct electrolytic reduction of UO_2 had been demonstrated in the earlier experiments, it was clear that development work would be required to achieve higher reduction rates and to manage the Li_2O concentration in the electrolyte. Among the specific problems to be addressed are physical pretreatments of the feed material such as crushing or grinding, the design of the process cell, and suitable materials of construction. None of these problems appears to be "show-stoppers," but a significant effort may be required to solve them.

Fission-Product Behavior. At the anode of the electrolysis cell, tellurium, which is a fission product, tended to form an insulating layer of compounds on the Pt-Rh electrode. Removal of the tellurium by a sacrificial molybdenum electrode prior to the lithium electrowinning step proved to be effective.

The behavior of rare earth fission products was complicated by the fact that the rare earths did not all behave similarly. The sesquioxides (RE_2O_3) were converted to the mixed oxides ($LiREO_2$) under certain conditions, and the rare earths had a strong mutual influence on one another in the molten salt solution. Studies were conducted in which rare earth oxides were equilibrated with molten LiCl that was saturated with lithium and contained varying concentrations of Li_2O at 650°C. The concentrations of Nd and Ce increased, and those of Eu and Sm decreased with increasing Li_2O concentration. When the LiCl salt did not contain elemental lithium, the concentrations of Eu and Sm were decreased significantly. Mixed oxides, $LiREO_2$, formed in all cases when the Li_2O concentration reached a high enough level. Free energies of formation of $LiREO_2$ compounds were calculated from the transition points (Li_2O concentrations) for the change from sesquioxides to mixed oxides.

Engineering-Scale Demonstration. The engineering-scale facilities consist of a large (7.6 × 2.4 × 2.6 m) glove box with an argon atmosphere, and a smaller one (2.4 × 1.1 × 2.6 m) with a nitrogen atmosphere. The two boxes are connected by a transfer lock. The equipment is designed to demonstrate the reduction, electrowinning, casting, oxidation, drying, and retorting steps of the process. A cutaway view of the equipment is shown in Fig. 6-31.

An initial reduction run using the lithium process and LiCl-KCl salt at 500°C was made in 1993 with 4.4 kg of UO_2-based simulated fuel. The product was a very fine metal precipitate that could not be melted into ingots or recovered by electrorefining. New process equipment was installed in the argon glove box in 1994, and two experiments were run in 1995. The first runs, conducted at 650°C with LiCl salt and 10 kg of UO_2 simulated fuel, were successful in reducing the uranium, but not in regenerating the spent salt by the lithium electrowinning step. Engineering-scale reductions of synthetic corium (mainly UO_2 and ZrO_2) and simulated TMI-2 fuel were conducted in 1996. Neither reduction was complete, apparently due to an underestimation of the required time based on the laboratory data.

Development work was then continued both on the reduction and the electrowinning

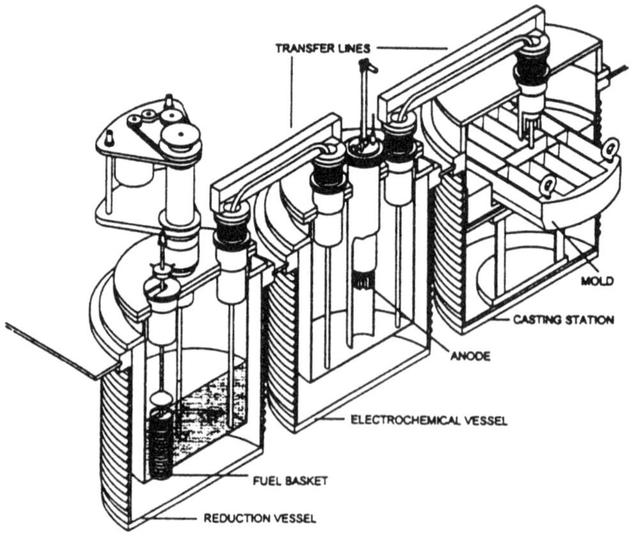

Fig. 6-31. Engineering-Scale Electrometallurgical Processing Facility

steps of the lithium reduction process, as well as scaleup and materials studies. In the electrowinning step, recombination of the oxygen and metallic lithium products was identified as a problem, which was solved by electrodepositing the lithium in a porous metal cathode. The lithium wets the porous metal (stainless steel) and is retained in the pores. Another major problem with the electrowinning step is the cost of the Pt-Rh anodes that were being used. A study was initiated on the possibility of replacing the Pt-Rh alloy with electronically conductive ceramics such as Fe_3O_4, SnO_2, or NiO. Other continuing experiments showed that complete reductions of about 3-4 kg of UO_2 could be achieved in the engineering-scale equipment, but that design improvements were needed to increase the reduction rates.

Treatment of INEEL Spent Fuel. The inventory of spent fuel at the Idaho National Engineering and Environmental Laboratory contains a variety of fuel types that are unsuitable for direct disposal. Examples of these are metallic fuels, ceramic-matrix oxides, hydride and carbide fuels, and components of metallic fuels that may have reacted with water while in storage. Studies are being made to determine whether the electrometallurgical treatment could convert these materials into a uniform set of three waste streams (metallic uranium, a metal waste form, and a ceramic waste form).

PERSONNEL

This program, under Chuck McPheeters's part of the Division, was directed by Dean Pierce, until his retirement, when Jerry Johnson assumed the responsibility. Jerry, unfortunately, died of cancer in 1997, and Eric Karell took his place. Other members of the group included Bob Everhart, Karthick Gourishankar, Tom Mulcahey, Dave Poa, Jim Smith, Mike Vest, Dave Warren, and Eugene Wesolowski.

Fusion-Related Research

Work continued on fusion-reactor research, with emphasis on the feasibility of using lithium-containing oxide ceramics as the blanket fuel. Safety and cost considerations make it desirable to minimize the inventory of tritium in a fusion reactor system, but sufficient amounts must be available to start up the reactor and operate it in a self-sufficient manner. These conflicting requirements are complex and require modeling, which, in turn, depends upon reliable experimental information. Most of the CMT work in the 1990s was oriented toward tritium generation, recovery, and handling in the breeder blanket and associated equipment.

DESORPTION MEASUREMENTS FOR $LiAlO_2$-H_2-H_2O(g) SYSTEM

Temperature-programmed desorption (TPD) was used to measure activation energies and other kinetic factors relating to desorption of H_2O and H_2 from $LiAlO_2$ in helium and helium-hydrogen streams. The hydrogen was expected to promote tritium release in a helium purge gas. In TPD tests, which were conducted at temperatures up to about 600°C, H_2O was evolved, both in pure helium and in helium containing up to 990 ppm H_2. Three peaks in plots of the H_2O evolution were attributed to different sites on the $LiAlO_2$. Activation energies were ~18, 23, and 28 kcal/mol, respectively, for the three sites. The first-order kinetics of the reactions were not consistent with the conventional wisdom that desorption was a bimolecular process involving the recombination of chemisorbed OH^- ions. An alternative possibility is that the process consists of bimolecular combination of OH^- and H_2O on the surface, followed by unimolecular desorption of H_2O. In a continuation of this study, additional experimental results indicated that the rates and amounts of H_2O desorption were substantially higher with He-H_2 mixtures than with helium alone.

TRITIUM RELEASE STUDIES

Because there was some conflict between the CMT results on the desorption mechanism and those from other work, the validity of the CMT model was tested by comparing its predictions with tritium release data from in-pile and laboratory experiments.

In-Pile Experiments. In a reactor experiment, the effects of neutron flux, and therefore the tritium generation rates, on tritium release from Li_4SiO_4 and Li_2ZrO_3 samples, were determined. The tritium residence times (inventory/generation rate) were independent of flux, and the rate-controlling mechanism was not second order in tritium. Other factors indicated that desorption was the rate-controlling step for these samples, and desorption was first order under these conditions. In other tests, the tritium inventory was found to increase with increasing grain size of the samples. An analysis of the data showed that the rate of tritium release into a He-1% H_2 purge gas was not controlled by diffusion processes. Other tests employing purge gases consisting of helium with 1.0% H_2, 0.1% H_2, or 50 ppm H_2 + 100 ppm H_2O indicated that tritium release was better for the gas containing 1.0% H_2 than the one with 0.1% H_2, but the activation energies were the same. The activation energy for the He-H_2-H_2O mixture, however, was lower. These studies suggest that the activation energy for tritium desorption depends on the hydroxide surface concentration.

The SIBELIUS experiment was a joint CMT/European effort to identify performance-limiting factors connected with the use of beryllium neutron multipliers in ceramic breeder blankets. These reflector materials generally remain in place for years at low temperatures (<100°C), and when they are annealed they tend to release the tritium as bursts at high temperatures. Experiments were conducted in the Siloe reactor at Grenoble, France, in which the behavior of beryllium disks located adjacent to stainless steel was compared with that of disks adjacent to ceramic blanket material. Because of the many variables involved, the results were not unequivocal, but it appeared that a lower density beryllium and operating temperatures above about 500°C would be desirable. In a continuation of these studies in 1994, tests were made on the compatibility of Li_2ZrO_3, Li_2O, Li_4SiO_4, and $LiAlO_2$ with beryllium components when they were in close contact in a neutron environment. No detrimental effects were observed on the performance of the ceramics, and chemical interactions of the two materials were negligible. It was concluded that the tritium inventories for all the ceramics should not cause any safety problems.

Laboratory Experiments. In a joint study with Commissariat á l'Energie Atomique, Centre de Saclay, France, tritium release experiments were conducted on sintered $LiAlO_2$ pellets similar to those used in the in-pile studies. The kinetics and rate constants of the release behavior were essentially identical to those observed in the in-pile experiments. One difference, however, was that tritium was retained in the solid at low temperatures, even after extended annealing times at 538°C, and was released only when heated to 850°C. The retained tritium suggests that it may exist in two different forms in the solid material.

The TPD method was used to investigate tritium release from single crystals of $LiAlO_2$, Mg-doped $LiAlO_2$, and Pt-coated $LiAlO_2$. The TPD plot for $LiAlO_2$ showed five different peaks, each with a different activation energy. The magnesium doping and the platinum coating both shifted the peaks to lower activation energies. This finding suggested that the rate-controlling step involved the surface and not bulk diffusion. These TPD results provided a basis for the development of a tritium release model that, unlike earlier models, required no adjustable input parameters. Calculated results from the model showed excellent agreement with measurements from an in-pile tritium release experiment with $LiAlO_2$.

Further studies were made to obtain diffusivity and desorption rate constants for Mg-doped and undoped samples of $LiAlO_2$ at temperatures of 528-785°C. The results showed that tritium transport was controlled by desorption when the grain radii in the sample were less than 100 μm. Lithium titanate (Li_2TiO_2) is an alternative low-activation ceramic that was of interest as a candidate blanket material for fusion reactors, but very little tritium-release work had been done on it. Several tritium-release tests were conducted to extend the database on Li_2TiO_2. General conclusions from the results were as follows:

1. Tritium was released at temperatures as low as 300°C, which compares favorably with the behavior of other ceramic blanket materials.
2. Several different types of sites in the crystals were involved in the release.
3. Addition of hydrogen to the helium purge gas was detrimental to tritium recovery; this was not the case for the other ceramic breeder materials.

More detailed studies of tritium desorption from Li_2O were made by a combination of TPD and diffuse reflectance infrared Fourier transform spectroscopy (DRIFTS). These experiments were performed with a purge gas of argon containing 0.1% hydrogen (or deuterium) at temperatures between 100 and 400°C. The results indicated that the desorption was a relatively complicated process that involved different reaction sites, adsorbed species, and activation energies. In addition to the experimental work, theoretical studies of the desorption mechanism were undertaken by using *ab initio* Hartree-Fock calculations and a slab model. The model produced a rational explanation for the enhancement of tritium release by hydrogen. The hydrogen present in "pure" helium is at a level of a few parts per million. At this level, the surface coverage by chemisorbed hydrogen is too low to be effective in tritium removal. On the other hand, the extent to which tritium removal can be increased by adding hydrogen to the helium is limited due to a saturation effect when the surface sites favorable for hydrogen chemisorption become filled.

In another study, experimental and calculational methods were used to evaluate the probability and consequences of a water vapor reaction with Li_2ZrO_3 blanket material. The conclusion was that the thermodynamic stability of the Li_2ZrO_3 makes it an excellent candidate for use as a tritium breeder blanket.

Conceptual Processes for Tritium Recovery. A detailed conceptual design of a tritium processing system for the Li_2O/Be blanket was prepared for the design team of the International Thermonuclear Experimental Reactor. The functions of the system were to (1) recover and purify tritium from the two purge streams, (2) remove waste products from the streams and recover the tritium in the waste products, and (3) recirculate the purge streams to the blanket zones.

LITHIUM VAPORIZATION AND CORROSION

Tritium may be present in the form of LiOT within a ceramic breeder blanket and, under certain conditions of concentration and temperature gradients, may be transported to cooler regions of the blanket. Preliminary experiments were performed to determine whether stainless steel components could be corroded by LiOT(g) and/or LiOH(g) from the blanket material. Lithium oxide (Li_2O) was placed under a stainless steel specimen, and helium containing a measured water vapor concentration was passed through the system at 750°C. Most of the LiOH(g) formed by reaction of the water vapor with the Li_2O reacted with the stainless steel sheet, and the remainder condensed in a cooler region downstream. The data showed that LiOH(g) does corrode structural stainless steel under fusion reactor conditions, and the reaction rate was inversely proportional to the distance between the stainless steel and the Li_2O breeder blanket material, and directly proportional to the LiOH(g) diffusivity in the carrier gas. The results of these experiments, together with other available information, were used to calculate possible corrosion of stainless steel by LiOT(g) from the Li_2O blanket planned for the International Thermonuclear Experimental Reactor.

Further studies on LiOH(g) vaporization showed that the potential problem was greatest for Li_2O breeder material, and less for other candidate materials ($LiAlO_2$, Li_4SiO_4, and Li_2ZrO_3). To assess the need for further experimental work in this area, existing literature data were expressed as a ratio of what would be predicted thermodynamically for equilibrium at the same temperatures and partial pressures of H_2O(g). The conclusions from this comparison were that diffusional contributions to mass loss would be negligible and that a condition of thermodynamic equilibrium would prevail. Due to the consistency and replicability of the existing studies, further experimental work was deemed unnecessary.

Some additional work was done in 1995 on the compatibility of vanadium alloys for the containment of ceramic breeder blanket materials (Li_2O, Li_2TiO_3, and Li_2ZrO_3). Vanadium and its alloys were of interest because it is a low-activation material. A thermodynamic analysis was made in which equilibrium calculations were performed for the Li_2O/V system. The results showed very little potential for the oxidation of vanadium. Similar results were found for the oxidation of vanadium by TiO_2 or ZrO_2. On the basis of these results, no significant oxidation of the vanadium should occur by reaction with the lithium-bearing ceramic materials. There was some concern, however, that tritium could be released from the ceramic as condensable "tritiated water," which might react with the lithium oxide to form LiOH. Calculations using the SOLGAS computer code were applied to the various reactions that could be postulated for the system, and the general conclusion was that the partial pressure of moisture that could be formed could be very close to that required thermodynamically for vanadium oxidation.

PERSONNEL

The fusion work was under the direction of Carl Johnson in the 1990s. Other CMT personnel included Paul Blackburn, Rob Clemmer, Larry Curtiss, Pat Finn, Al Fischer, Larry Greenwood, John Kopasz, Vic Maroni, Shiu-Wing Tam, Erv Van Deventer, and Dai-Kai Sze. Much of this work was a cooperative effort with foreign scientists, who were coauthors of many of the publications. The secretary was Jan Carothers.

Applied Physical Chemistry

Work continued in the 1990s on several projects under this general category. Carl Johnson had the overall responsibility for these programs.

PHYSICAL PROPERTIES OF CORE-CONCRETE MIXTURES

Work continued up through 1992 on the thermophysical properties of molten mixtures of reactor core materials and concrete that could be formed in a severe accident involving the meltdown of a water-cooled nuclear reactor. The molten core-concrete interaction (MCCI) would be preceded by several other events: loss of coolant, heating and degradation of the core caused by fission-product decay, melting of the core debris (uranium oxide, fission products, Zircaloy cladding, control rods, and structural materials) through the stainless steel reactor vessel, and deposition of the core debris and molten steel on the concrete floor beneath the vessel. During the ensuing MCCI phase, the concrete basemat would react with the hot (~2700°C) core debris and steel to form a molten mixture of phases, which is vigorously agitated by the evolution of gaseous products (CO_2, CO, H_2O, and H_2) from decomposition and reduction of the concrete. Understanding and modeling the consequences of this kind of accident require a knowledge of the thermophysical properties of the solids, liquids, and gases that are involved. Earlier studies of this subject were described in the previous chapter.

The results of this research were to be incorporated into thermal-hydraulic codes such as CORCON, which is an integral part of the Source-Term Code Package for the Nuclear Regulatory Commission. In brief, CMT contributed two important improvements to the computer models. The CORCON model had assumed that ideal liquid and solid solutions were formed in the reactions of the core material and concrete. The CMT experimental work showed that the system was much more complex than the ideal model, and that the temperature differences between the liquidus and solidus phases were much larger than had previously been thought. The other contribution had to do with the viscosities of the mixtures. Experimental results showed that the viscosities of the mixtures were typically two orders of magnitude higher than the estimates that were being used in the CORCON code.

Differential Thermal Analysis Studies. Differential thermal analysis (DTA) measurements were used to determine the solidus-liquidus temperatures for three calcined concretes (limestone, limestone-sand, and siliceous) and mixtures of each of these with 73 wt% UO_2-ZrO_2. Measured differences between the solidus and liquidus temperatures for the three core-concrete mixtures were more than 1200°C, whereas the estimated differences using the CORCON code (for 27 wt% concrete) ranged between 500 and 700°C.

Experimental solidus and liquidus temperatures from the DTA measurements and calculated values from the literature are listed in Table 6-9 for mixtures of 27 wt% concrete with UO_2-ZrO_2 (mole ratio 1.6:1). In this case, the calculations were not based on ideal behavior. The measured and calculated results for the solidus values were in reasonable agreement. The disagreement between experimental and calculated liquidus values appeared to be roughly proportional to the calcia (CaO) content of the calcined concrete. The phase diagram of the UO_2-CaO system was redetermined (see Fig. 6-32), and incorporation of these new results into the calculations was expected to resolve the differences between the experimental and calculated results.

Viscosity Experiments. It is important to have reliable data on the viscosities of the products of core-concrete reactions because viscosity affects a variety of factors such as heat transfer, fission-product release, melt spreading, and cooling. Viscosity measurements were made on eight mixtures of

Table 6-9. Liquidus and Solidus Temperatures for Concrete-UO_2-ZrO_2 Mixtures

	Solidus Temp., K		Liquidus Temp., K	
Concrete with UO_2-ZrO_2	DTA	Calculated	DTA	Calculated
Limestone	1520	1550	>2723	2320
Limestone-Sand	1360	1450	>2638	2490
Siliceous	1412	1434	2549	2395

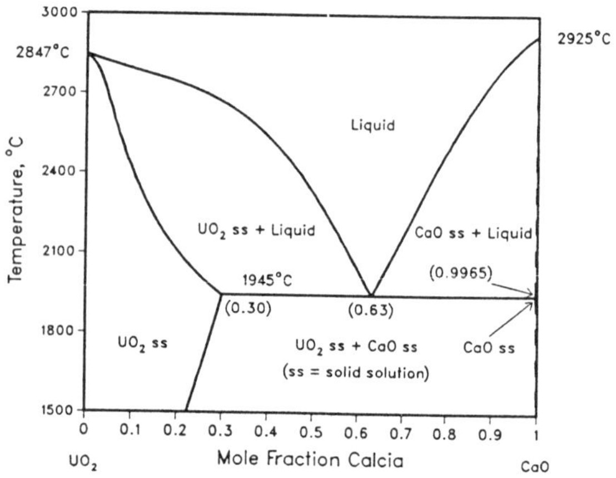

Fig. 6-32. Phase Diagram for the UO_2-CaO System

UO_2-ZrO_2 and calcined concrete—one with siliceous concrete, four with limestone concrete, and three with limestone-sand concrete. The measurements were made with a Brookfield programmable viscometer equipped with a 70Mo-30W alloy rotating spindle. The measurements were made at temperatures extending from above the liquidus down into the liquidus-solidus range. In the liquidus region or with only a small volume fraction of solids, the fluid exhibited Newtonian behavior, *i.e.*, the viscosity was independent of the spindle rotation rate. At lower temperatures, where there was a higher volume fraction of solids, the behavior became non-Newtonian, *i.e.*, the viscosity decreased significantly with increasing rotation rate (an effect known as "shear thinning").

The results of these studies showed that the viscosities of core-concrete mixtures were significantly higher than values calculated by the thermal-hydraulic code, CORCON, which was being used to compute the consequences of hypothetical reactor accidents.

Personnel. Individuals involved in this research were Joanne Fink, Len Leibowitz, Mike Roche, Jack Settle, and Dave Steidl.

SUPPORT STUDIES FOR NEW PRODUCTION REACTOR (NPR)

In the U.S. nuclear weapons program, tritium is a vital material that can be produced in sufficient quantity and quality only by the irradiation of lithium in a nuclear reactor. Because tritium undergoes radioactive decay, a continuing supply is necessary not only to produce new weapons, but also to maintain existing ones. The DOE initiated the New Production Reactor (NPR) Program to plan, design, and build safe and environmentally acceptable new reactor capacity to ensure an adequate supply of tritium. The CMT Division supported this program by performing studies on a heavy-water-reactor NPR (HWR-NPR). This work involved the development of seals to meet stringent requirements for tritium containment in process equipment and facilities, calculations of detritiation requirements in the event of accidental tritium releases into the containment structures, and calculational and experimental studies of fission-product release in a severe NPR accident.

Seal Design. Seal designs that are acceptable in normal commercial reactors fall far short of meeting the tritium-control requirements for an NPR. To achieve a desirably low level of tritium escape, the leak rates would have to be a factor of thirty or so lower than those for a commercial power reactor. The nuclear industry was already replacing asbestos valve packing and gaskets to meet EPA requirements, and graphite packing was coming into favor. A detailed study was made of state-of-the-art seal technology both in the U.S. and elsewhere. Detailed guidelines derived from this information were developed and published by Pat Finn (ANL-90/29). It appeared that additional capital costs of following these guidelines would be offset by lower operating costs and other benefits in long-term reactor operations.

Containment Habitability. The objective of this work was to develop an analytical approach to meeting the detritiation (decontamination from tritium activity) requirements in the event of an accidental tritium release within the containment structure. A critical factor is the time needed to reduce the activity in the containment structure to a level acceptable for human entry to perform maintenance or repair work without elaborate protective measures. The tritium activity in this situation was expected to be primarily in the chemical form of DOT. For the purpose of modeling, the containment structure was assumed to consist of concrete walls with an epoxy paint coating. The model could, however, accommodate other barrier materials. The mathematical model, which was called the Dynamic Tritium Release and Analysis Model (DTRAM), simulated detritiation for either burst or chronic releases of tritiated species. The interactions included exchange, desorption, adsorption, dissolution, and bulk diffusion. The calculations showed that the time necessary to reduce the tritium levels to an acceptable value was dominated by the retention capability of the tritiated species in the structural materials. In spite of the fact that the tritiated species could, in principle, diffuse through the barrier materials to the external environment, the study indicated that virtually no such escape would occur.

Fission-Product Release from Fuels and Target in Severe Accidents. The purpose of this study was to develop a thermodynamic model that would predict the release of fission products and tritium from the Al-U fuel and Al-Li targets if the HWR-NPR had a severe accident. A review of the literature and preliminary transpiration experiments were conducted on aluminum alloys and an ingot prepared by Savannah River Laboratory that contained Al, U, Ba, C, Ce, Mo, Sr, and Zr in a helium stream to which 3% water vapor was added. At 873°C, the sample began to increase in weight, and when the sample was held at about 975°C for four hours, it oxidized at a decreasing rate with time.

Design Study on Modification of ANL-W Facilities for Tritium Recovery. A design concept was developed for modifications of a hot cell at ANL-W to handle irradiated fuel and target tubes in a tritium-controlled environment. No further work was done because of funding termination of NPR in the early 1990s.

Personnel. This work was performed by Rob Clemmer, Pat Finn, and Dai-Kai Sze.

THERMOPHYSICAL PROPERTY STUDIES

The fact that the driver (core) and blanket materials of the Integral Fast Reactor (IFR) were both U-Zr-Pu alloys made it highly desirable to have detailed, accurate information on the phase relations of this ternary system. Moreover, it was recognized that appreciable quantities of other actinides, including neptunium, americium, and curium, would grow into the fuel under irradiation, so data were also needed on the effects of small concentrations of these elements on the phase relations and other physical properties of the alloys. Possible interactions of constituents in the fuel cladding material (stainless steel) and the fuel components were of interest. The demise of the IFR program in the early 1990s lessened the urgency of these studies, and they were discontinued by 1996.

The U-Pu-Zr System. Calculation of the ternary phase diagram for the U-Zr-Pu system requires reliable information on the existence of all the phases present in the three binary subsystems, *i.e.*, U-Pu, U-Zr, and Zr-Pu, and their thermodynamic properties. In 1990, an extensive review of the literature was conducted to make a critical evaluation of the information that was available at the time. The calculations made use of the Facility for the Analysis of Chemical Thermodynamics (FACT) computer system, which was located in Montreal, Canada. Recalculation of the binary U-Pu phase diagram did not result in a major revision of the existing diagram, which is quite complex, but it did require some detailed changes to achieve thermodynamic consistency.

Subsequent investigations of the binary Pu-Zr system were done in much the same way as the U-Pu studies—a comprehensive evaluation of phase-equilibrium and thermodynamic data that were available in the literature on the Pu-Zr system. All the literature data were well correlated by the thermodynamic analysis, using the FACT system, and reasonable, simple thermodynamic expressions were found for all the phases.

By 1995, a few new experimental results had become available on the U-Pu and Zr-Pu binaries. These were incorporated into a study that was aimed toward developing a set of model equations for the Gibbs energies of all phases as functions of composition and temperatures. All the thermodynamic properties and the phase diagrams could be calculated, and interpolations or extrapolations could be made in a thermodynamically correct manner. The Gibbs energy of the ternary solution phases can be represented by the Kohler equation, which is based upon an extension of regular solution theory:

$$\begin{aligned} G = {} & (X_{Pu}G^0{}_{Pu} + X_U G^0{}_U + X_{Zr}G^0{}_{Zr}) \\ & + RT\,(X_{Pu}\ln X_{Pu} + X_U \ln X_U \\ & + X_{Zr}\ln X_{Zr}) + (X_{Pu} + X_U)^2\, G^E{}_{PuU} \\ & + (X_U + X_{Zr})^2\, G^E{}_{UZr} \\ & + (X_{Zr} + X_{Pu})^2\, G^E{}_{ZrPu} + G^E{}_{PuUZr} \end{aligned}$$

where X is the atom fraction of species i, $G^0{}_i$ is the free energy of species i, $G^E{}_{AB}$ is the excess free energy from the A-B binary system, and $G^0{}_{ABC}$ is the excess free energy from the ternary A-B-C system. All the available phase-equilibrium data in the Pu-U-Zr system were analyzed to determine optimized equations for the Gibbs free energies of all phases as functions of composition and temperature. The resulting calculated phase diagram was believed to be the best available estimate of the U-Pu-Zr system at the time. This was not only a highly useful tool for practical application of the data; it was also a very satisfying intellectual achievement.

The U-Fe-Zr System. As indicated previously, this system was of interest from the standpoint of possible fuel-cladding interactions in the IFR, with particular concern about reactions of iron in the stainless steel cladding (HT9, D9, or Type 316) with zirconium in the fuel alloy. The Fe-U phase diagram had been determined in considerable detail, but there were still some areas of uncertainty, particularly in the location of the liquidus curve. After some DTA measurements, the new data were incorporated into a recalculation of the U-Fe diagram, which then appeared to be a reasonable representation of the system.

A literature study of the binary Fe-Zr system showed two somewhat different phase diagrams. This system is fairly complex because it involves a number of solid-phase transitions and intermetallic compounds. The phase diagram was optimized by calculations based on Gibbs free energy data and the necessity for thermodynamic consistency. In general, all the available experimental data were incorporated successfully into the model, but some modifications were necessary to achieve thermodynamic consistency.

Studies were then extended to the ternary system, U-Fe-Zr. Again, the diagram for the ternary system was determined by calculating the excess Gibbs energies in the ternary liquid and the solid phases from values for the three subsidiary binary systems. Used for this calculation was the Kohler equation or the somewhat similar Toop equation. The following phases were considered in the calculations:

Liquid	U(γ)/Zr(β)	Fe_2Zr
Fe(γ)	Fe_2U	$FeZr_2$
Fe(δ)	FeU_6	$FeZr_3$
Zn(β)	Fe_3Zr	

Two diagrams are possible, depending on whether Fe_2U and Fe_2Zr form a solid solution. The diagram based on the premise that such a solid solution does not exist is shown in Fig. 6-33. If such a solution does exist, the isotherms and phase boundaries in the lower left portion of the diagram are shifted slightly. It may be the case that reality lies somewhere between these two possibilities, that is, a partial solid solubility may be involved.

Minor Actinide Systems. During the operation of an IFR, the transuranic elements (Np, Pu, Am, and Cm) are formed in the fuel due to neutron capture. The lighter elements (Np, Pu), because their 5f electrons interact strongly with the 6d and 7s electrons, tend to behave chemically like uranium. In the heavier elements (Am and Cm), however, the 5f electrons are more localized, and they tend to act like the rare earths. This effect also influences the metallurgical as well as the chemical properties of Am and Cm, and this is reflected in the crystal habits of the metals. Although some lanthanide and actinide binary phase diagrams pertinent to the IFR were available, the absence of experimental data in this area forestalled application of the usual thermodynamic modeling and phase-diagram optimization methods, and the situation was

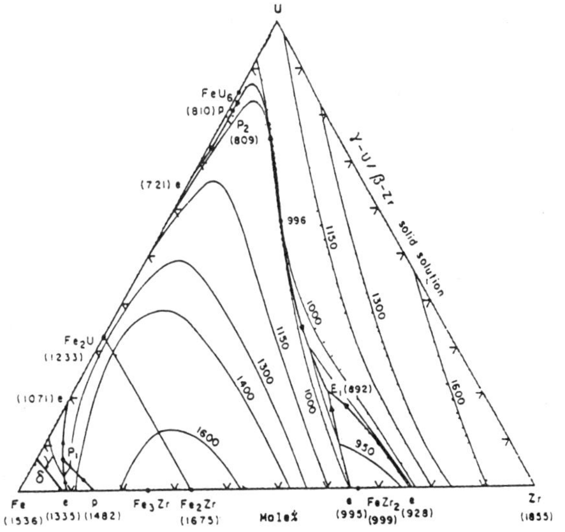

Fig. 6-33. Liquidus Surface of Calculated Fe-U-Zr Phase Diagram

made worse by the extraordinary complexity of the actinides. Application of several alternative modeling approaches met with mixed success, but the results did not engender a high confidence level.

Materials Data Base. A data base on the properties of IFR materials that had been generated in collaboration with the Fuels and Engineering (FE) Division in 1988 was updated in 1993. Data on vapor pressure, heat capacity, density, compressibility, surface tension, thermal conductivity, viscosity, and related properties of sodium were added to the data base. Updates were also made on plutonium phase transitions and thermal conductivity and thermal diffusivity of U, Pu, Zr, and alloys of these elements. A significant conclusion from this effort was that much more information was needed on thermal conductivities, diffusivities, and certain other properties of plutonium and its alloys.

Personnel. Involved in this work were Bob Blomquist, Joanne Fink, Len Leibowitz, and Ewald Veleckis.

DEVELOPMENT OF SYNCHROTRON ANALYTICAL CAPABILITIES

In 1984, CMT began collaborating with Louisiana State University (LSU) to build a synchrotron X-ray beamline at the Center for Advanced Microstructures and Devices (CAMD). This program grew out of earlier work in which a team of Argonne researchers developed methods to investigate small-scale structures, using the National Synchrotron Light Source (NSLS) at Brookhaven National Laboratory. By using a combination of X-ray fluorescence and diffraction, information on sample composition and structure can be determined simultaneously. Hardware and software for the system were developed and tested. This work was expected to lead to a unique facility for microscopic X-ray examinations in a wide variety of applications. Various people were involved in these studies, including Len Leibowitz, Carlos Melendres, and Zoltan Nagy.

Basic Chemistry Research

In the early 1990s, Vic Maroni and Milt Blander were in charge of the basic research programs. Vic was in charge of all the basic studies, which fell under the general purview of the CMT department headed by Mike Myles. In 1998, Zoltan Nagy and Larry Curtiss were transferred to other ANL divisions, at considerable loss to the CMT basic research programs.

PHYSICAL ORGANIC CHEMISTRY

The fluid catalysis work continued under the leadership of Jerry Rathke in the 1990s, and Hal Feder, although retired to Florida, also kept in touch with the program, largely through e-mail, essentially up to the time of his death in 1997. The fluid catalysis work included four types of investigations: (1) catalytic reaction chemistry of small molecules (CO, CO_2, CH_4, and H_2) that serve as precursors in the industrial production of useful products, (2) hydrocarbon activation to achieve the controlled catalytic reactions of methane and other hydrocarbons by activation of their C-H bonds, (3) organometallic C-H activation chemistry in the production of carbide-containing fibers, films, or larger objects, and (4) the mechanism(s) of ion transport in lithium-polymer batteries.

Catalytic Chemistry in Supercritical Media. Over 90% of industrial chemical processes entail the use of catalysts, and nearly half of these are homogeneous catalytic processes in which the catalyst is dissolved in a fluid (typically an organic solvent). Although these systems offer high reactivity, selectivity, and relative ease of modification for special purposes, catalyst recovery and product separation often present problems. Supercritical fluids such as CO_2 or H_2O appear to be environmentally benign alternative solvents that would eliminate the use, storage, handling, and disposal problems associated with organic solvents. Supercritical fluids, because of their gas-like transport properties, complete miscibilities with gases, and sharp changes in dissolving power with fluid density, have the potential both for accelerating diffusion-controlled reactions of gases with liquid and solid substrates and separating catalysts or catalytic products by energy-efficient pressure alterations.

The "oxo" process for the hydroformylation of olefins is the largest scale production process using homogeneous catalysis. In this process, CO and H_2 react with an olefin in the presence of a soluble catalyst to produce aldehydes and alcohols for detergents, plastics, and agricultural products. *In situ* nuclear magnetic resonance (NMR) studies were conducted to determine the effects that the use of supercritical media would have on the various equilibrium and kinetic factors in the process. The NMR probe is capable of operating at temperatures up to 300°C and pressures up to 34 MPa (5,000 psi). The solubility of $Co_2(CO)_8$ in supercritical CO_2 was measured, and the entropy and enthalpy of dissolution were derived: 20 ± 1.7 kJ/mol and 30 ± 4 kJ/mol·K, respectively. It is interesting that these experimental measurements, unlike those in most catalysis research, were made much easier by the fact that only a single phase was present. Additional detailed information was obtained on intermediate species, rate constants, *etc*., and some experimental information was obtained on the formation of butyraldehydes by hydroformylation of propylene. A significant result of this study was that the ratio of n-butyraldehyde to i-butyraldehyde in the product was different with supercritical CO_2 than it was for conventional solvents.

During the investigations of propylene hydroalkylation in supercritical CO_2 with the $Co_2(CO)_8$ catalyst, the NMR results suggested that a tetracarbonyl radical, $HCo(CO)_4\cdot$, may have had a role in the reactions. Spectra that were recorded at varied temperatures during the hydrogenation of $Co_2(CO)_8$ provided enthalpy and entropy data for the reaction

$$Co_2(CO)_8 + H_2 \leftrightarrow 2\ HCo(CO)_4$$

Broadening and coalescence of the resonance peaks for $HCo(CO)_4$ and $Co_2(CO)_8$ were attributed to a second step in the process:

$$(CO)_4Co\cdot + HCo(CO)_4 \leftrightarrow (CO)_4CoH + \cdot Co(CO)_4$$

Even more convincing evidence for the existence of this radical was obtained by isotopic exchange of ^{13}C between CO and $Co_2(CO)_8$, which probably occurs through the radical. Additional studies were made with $Mn_2(CO)_{10}$, which is similar to $Co_2(CO)_{10}$, although less effective in its catalytic behavior. It had been studied as a model for certain steps in the commercial cobalt carbonyl process. The results of these studies were consistent with those from the cobalt compound, and an interesting additional finding was that $Co_2(CO)_8$ and $Mn_2(CO)_{10}$ form the mixed dimer $(CO)_5Mn\text{-}Co(CO)_4$.

Results from NMR measurements of magnetic susceptibilities of solutions containing $Co_2(CO)_8$ at different temperatures were used to obtain thermochemical data on pertinent bond energies in the system. The enthalpy and entropy changes for the Co-Co bond were 80 ± 8 kJ/mol and 121 ± 17 J/mol·K, respectively. The H-CO bond energy was 248 ± 4 kJ/mol, which seemed more reasonable than an electrochemically determined value of 260 kJ/mol.

The total world capacity of the oxo process for the hydroformylation of olefin is about

5 million tons per year. The phosphine-modified cobalt hydroformylation process was invented by, and is used exclusively by, the Shell Chemical Company. Preliminary high-pressure NMR experiments showed that many of the organometallic species could be observed and quantified by NMR with phosphorus-31. One of the conclusions from these measurements was that the Co-Co bond in the phosphine-substituted dimer is stronger than that in $Co_2(CO)_8$. These *in situ* NMR measurements were made at temperatures of 75 to 225°C and initial H_2 and CO partial pressures of 6.9 MPa. Further studies were made on the effect of CO pressure on the compositions of the cobalt species present in the catalytic solutions of the tributylphosphine-modified catalyst, $[Bu_3PCo(CO)_3]_3$. High CO pressures result in the formation of a salt, $[(Bu_3P)_2Co(CO)_3]^+ + [Co(CO)_4]^-$. Because the salt would be expected to be inert, use of high CO pressures to encourage catalyst formation would be counterproductive.

In recent studies, the first thermodynamic measurements were completed on the key equilibria associated with phosphine-modified hydroformylation catalysis of the type used commercially. Equilibria were measured not only on the catalyst hydrogenation step, but also carbon monoxide-induced salt formation, which has so far prevented the reaction from being conducted in a supercritical fluid. The results suggest that phosphines having lower basicity might make such reactions in supercritical fluids possible for the first time. This discovery could have important commercial applications.

Hydrocarbon Activation Chemistry. The objective of this continuing research was to investigate the activation of hydrocarbons by using soluble metallophthalocyanines (MPc) as the catalysts. Although the MPc catalysts were well known for their catalytic activities on the autoxidation of hydrocarbons, the mechanisms for these reactions were not well understood. A Rh-Rh bonded dimer, $[(R_8Pc)Rh]_2$, where R_8Pc^{2-} is the dianion of 1,4,8,11,15,18,22,25-octapentylphthalocyanine, was synthesized and shown to react with H_2 and CH_4 to form the products $(R_8Pc)RhH$ and $(R_8Pc)RhCH_3$, respectively. The lower the Rh-Rh bond energy, the more active the dimer $[(R_8Pc)Rh]_2$ is for activating CH_4 and other hydrocarbons. Homolysis of the rhodium dimer to form $(R_8Pc)Rh\cdot$ radicals was proposed to explain the experimental observations.

At 180°C, the rhodium dimer catalyzed the H-D exchange reaction between D_2O and both the aromatic and the alkyl hydrogens of the R_8Pc^{2-} ligand to form the $(R_8Pc)RhD$ complex. The catalytic chain mechanism for the H-D exchange between D_2O and the alkyl hydrogens was worked out. Further studies of the $(R_8Pc)RhH$ complex showed the existence of two isomers, monomeric $(R_8Pc)RhH$ and dimeric $(R_8PcH)Rh\text{-}Rh(R_8PcH)$. Also prepared and isolated were $(R_8Pc)LRh\cdot$ radicals where L represents trimethyl or tributyl phosphine.

An effort was then focused on the use of catalytic complexes containing rhodium centers bound to solubilized phthalocyanine (Pc) ligands. The rhodium center has an ability to form reasonably strong, but still reactive, metal-carbon and metal-hydrogen bonds. Methane was activated by using a metal-metal bonded rhodium Pc dimer that interacted with methane by the equilibrium reaction

$$[(\text{n-pentyl})_8PcRh]_2 + CH_4 \leftrightarrow (\text{n-pentyl})_8PcRhCH_3 + (\text{n-pentyl})_8PcRhH$$

Several approaches to the synthesis of phthalocyanines, as well as the use of molecular mechanics, were combined with semiempirical molecular orbital calculations to evaluate various types of substituents for use as spacers. (Alkyl substitutes on the phthalocyanine nucleus serve both to solubilize the complexes and to provide spacers that weaken, and thereby activate, the Rh-Rh bond in the dimer by steric repulsions between the linked macrocycles.) Figure 6-34 compares models of phthalocyanine frameworks determined by X-ray measurement (upper figure) and by molecular modeling calculations (lower figure). The two models are in close agreement. The X-ray model, however, shows an unusual feature in that the presence of substituents results in distortions such that one of the benzpyrrole groups is twisted out of the molecular plane by 15°. Apparently, the stress introduced by long-chain alkyl substituents is relieved by out-of-plane bending of one pyrrole group.

Further studies showed that soluble rhodium complexes in which the n-pentyl groups were replaced by trifluoromethyl groups can be synthesized, for example, $(CF_3)_8PcRhOCH_3$. With these fluorocarbon constituents, the metallophthalocyanines can be tailored to very high electrophilicities, which, together with their high thermal and oxidative stabilities, open up the potential for a wide variety of homogeneous catalytic oxidations, oxidative couplings, substitution reactions, and other useful hydrocarbon conversion processes. A dimeric oxo-bridged iron complex of a highly fluorinated phthalocyanine ligand was synthesized, and it proved to be an effective catalyst for the epioxidation of olefins and the hydroxylation of alkanes.

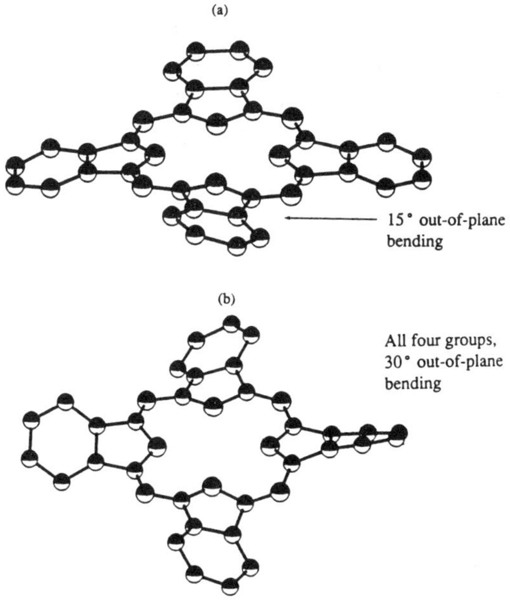

Fig. 6-34. Models of Phthalocyanine Frameworks

Ceramic Precursor Research. The objective of this research was to explore novel catalytic methods for the synthesis of organometallic polymers from small precursor molecules. The idea was to apply C-H bond activation chemistry to organometallic substrates and to explore the chemistry of organometallic polymers and their ceramic-forming mechanisms. This type of work could lead to improved processes for the preparation of coatings, whiskers, fibers, or even larger three-dimensional ceramic objects. Carbides of Si, B, W, Ti, and other metals have superior hardness, high tensile strength, and other desirable properties for many applications, but they are difficult to fabricate in useful shapes. An attractive approach to this problem would be to synthesize pre-ceramic polymers that could be formed readily into the desired shapes, and then fire the material to form the ceramic. This general approach is being used commercially for the production of high-strength SiC fibers.

A catalytic strategy was devised for the polymerization of small organometallic molecules such as tetramethylsilane. In principle, this approach should lead to polymers having an alternate arrangement of metal and carbon atoms. This strategy is based on two reactions:

$$M\text{-}CH_3 + (CH_3)_4Si \leftrightarrow MCH_2Si(CH_3)_3 + CH_4$$

$$MCH_2Si(CH_3)_3 + (CH_3)_4Si \leftrightarrow M\text{-}CH_3 + (CH_3)_3SiCH_2Si(CH_3)_3$$

where M is an electrophilic metal center in an early transition metal or lanthanide or actinide complex. Both of these reactions were well-known, and the product has the desired alternating arrangement of Si and C atoms. The catalyst, $M\text{-}CH_3$, is consumed in the first reaction and regenerated in the second one. Steps analogous to those in these reactions can bring about further reactions of dimer and higher oligomers. Such reactions lead to polymerization and cross-linking or branching that can be controlled by the steric requirements of the catalyst. Oligomerization of both tetramethylsilane and trimethylaluminum was achieved, and the products had the desired alternating arrangement of metal and carbon atoms.

A magnetic resonance imaging device that appears to offer the means for excellent characterization of advanced materials and materials chemistry processes in this technology was based on the toroid detectors used with high-pressure NMR probes. This toroid cavity imager proved to be extremely useful in obtaining (1) NMR spectra on different regions of a flame, (2) structure or composition on the different layers of a thin film or coating, (3) the penetration depth or diffusion rate of a fluid into a porous substrate at high pressures, and (4) concentration gradients of various species near an electrode of an electrochemical cell. In 1994, some new 180° pulse sequences were developed, which almost completely compensate for the radio-frequency field gradient within the torus. This development greatly extended the scope of applications for NMR imaging.

Determination of Ion Transport by NMR Imaging. Although nuclear magnetic resonance is a powerful tool for determining chemical structure and reaction dynamics, the distance resolution of standard magnetic resonance imaging is gained only at the expense of sacrificing the chemical structure and reaction dynamics information. For this reason, it was not well suited to examination of the processes occurring in electrochemical systems on a micro scale very close to the electrode surfaces, which is the region of most interest. A new analytical technique was developed, in which the distance resolution of the NMR imaging method was close to the micron scale, which is sufficient to resolve most of the electrochemical details of interest while retaining all the NMR chemical shift, coupling constant, and relaxation time information. This new technique was based on the toroid cavity detector that was used earlier in the *in situ* NMR work on fluid catalysis at high temperature and pressure.

A schematic diagram of the electrochemical NMR imaging device is shown in Fig. 6-35. The central conductor served both as the electrochemical working electrode and as the central wire in the NMR detector circuit. The first application of this device was to investigate the mechanism(s) of lithium ion transport in lithium-polymer electrolyte batteries, which were discussed earlier. Lithium-ion mobility depends upon its diffusion coefficient, the overall ionic conductivity, and the lithium-ion transference

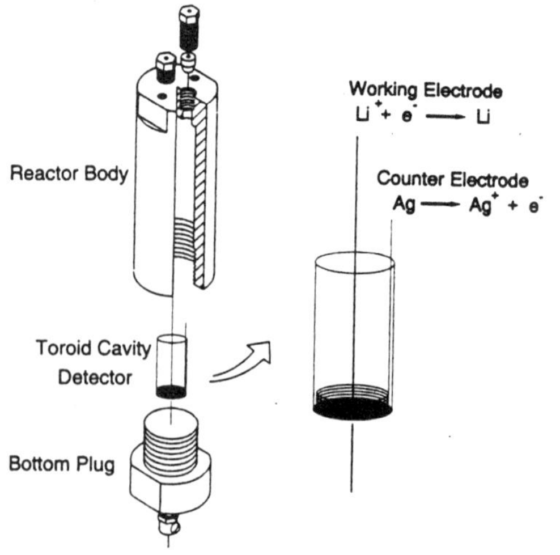

Fig. 6-35. Electrochemical-NMR Imaging Device

number (the fraction of the current carried by lithium ions). The transference number was the primary focus of this work because it was the item of greatest uncertainty.

The electrochemical NMR imaging technique was used to measure lithium-ion concentration profiles in the electrolyte depletion area at the electrode surface as a charge was passed through the cell containing a sample of CF_3SO_3Li in polyethylene oxide. From the amount of electrical charge passed through the cell and measurements of the electrolyte depletion zone (which was about 0.5 mm), the lithium-ion transference numbers could be calculated. In addition to transference numbers, lithium-ion diffusion coefficients could be measured by imaging the salt depletion zone after various time intervals. The results of ongoing studies of the polymer electrolyte batteries showed that the lithium-ion transference numbers and lithium-ion diffusion rates were both very low. This condition severely limited the thickness of the polymer film that was utilized. To develop a battery with adequate performance, it appeared that modifications would be necessary in the electrolyte systems.

Personnel. Jerry Rathke was in charge of this work, which also included Mike Chen, Bob Klingler, and others.

HIGH T_C SUPERCONDUCTORS

The excitement that had been generated by the discovery of solid oxide systems that are superconductive at considerably higher temperatures than the earlier metallic systems continued into the 1990s. Unlike the earlier materials, which were superconductive only up to a temperature of about 10 K, these oxides showed superconductivity at temperatures approaching 100 K. This means that they could find practical application in devices cooled by liquid nitrogen, which is widely used in industry for cryogenic purposes. The CMT work in this area was in cooperation with the American Semiconductor Corp., Radiation Monitoring Devices, Inc., and the ANL Materials and Components Technology Division.

Structural Transitions and Thermodynamic Behavior. This research was focused on the structural transformations and the nonstoichiometric and thermodynamic behavior of high-T_C superconducting oxide systems such as $YBa_2Cu_3O_x$. These properties were determined as a function of oxygen partial pressure, oxygen stoichiometry, and temperature by coulometric titration in which the oxygen content was varied by well-defined small amounts. Earlier electromotive force (emf) measurements on $YBa_2Cu_3O_x$ had shown a miscibility gap centered at x = 6.65 with a consolute temperature of 473 K. The objective of the present study was to investigate the effects of ionic size of Ln^{3+} (where Ln = Y, Gd, or Nd) on the structural transitions and thermodynamic and nonstoichiometric behavior in the $LnBa_2Cu_3O_x$ system.

Equilibrium oxygen pressures were calculated from the emf data for isotherms in the temperature range of 673-873 K as a function of x in $NdBa_2Cu_3O_x$. The partial pressures turned out to be higher than those reported earlier for the $YBa_2Cu_3O_x$ system. These results were consistent with the change in the observed composition dependence of T_c on ionic radii for Y and Nd in $YBa_2Cu_3O_x$ and $NdBa_2Cu_3O_x$. Two plateaus of T_C suggest the presence of phase separation in a miscibility gap, which would operate for the yttrium phase, but not for the neodymium phase. If a miscibility gap were present in the neodymium system, it would occur at a much lower temperature and higher value of x than in the yttrium system.

The thermodynamic and structural studies were then turned toward the high T_C system, $(Bi_{2-x}Pb_2)Sr_2Ca_2Cu_3O_y$ (Bi-2223), which was being used for the fabrication of silver-sheathed superconducting wire. Electromotive force measurements were used to measure oxygen fugacities as a function of oxygen stoichiometry, and a coulometric titration was made to investigate the thermodynamic behavior of the material as a function of oxygen partial pressure, oxygen stoichiometry, and temperature. The limiting stability condition at low oxygen partial pressure was the CuO-Cu_2O phase boundary.

Further investigations of Bi-2223 were performed with a sample of the material prepared at Ames Laboratory, and another one prepared at ANL. Oxygen partial pressures were determined as a function of oxygen stoichiometry, the value of x in $(Bi,Pb)_2Sr_2Ca_2Cu_3O_x$. The measurements gave reproducible results that defined the stability region for single-phase Bi-2223 as existing between oxygen partial pressures of about 0.02 and 0.20 atm. Differential thermal analysis and X-ray diffraction indicated that decomposition of the material at oxygen pressures below about 0.001 atm produces $(Bi,Pb)Sr_2CuO_x$, $(Ca,Sr)_2CuO_3$, and Cu_2O. The conclusion from these results was that Pb-doped Bi-2223 should remain as a single-phase material during processing or annealing at 750-800°C and oxygen partial pressures of about 0.02 to 0.20 atm.

A comparison was made between the oxygen partial pressures of lead-doped and lead-free Bi-2212. The results showed a narrower range of stability for lead-free Bi-2212 than that for lead-doped Bi-2212 and Bi-2223.

Development of High-T_C $(Bi,Pb)_2Sr_2Ca_2Cu_3O_x$ Wires. Collaborative research with the American Superconductor Corp. was performed on the growth and stability of high-T_C phases in the lead-doped Bi-Sr-Ca-Cu-O system. The goal of this work was to establish the synthesis and processing procedures that would have the greatest effect on the critical current properties of metal-sheathed high-T_C conductors. The growth and stability of the Bi-2223 phase fabricated into silver-sheathed wires were investigated by a combination of X-ray diffraction, scanning electron microscopy, energy dispersive X-ray analysis, and transmission electron microscopy. During the fabrication, a precursor powder, $(Bi,Pb)Sr_2CaCu_2O_x$ (Bi-2212), was converted to the Bi-2223. Modeling of the equilibration data indicated that the Bi-2212→Bi-2223 conversion was controlled by two sequential kinetic regimes—a liquid-phase diffusion process and a solid-state process. Data for fractional conversion to the Bi-2223 phase *vs.* time correlated best with a two-dimensional, diffusion-controlled mechanism that had an unusually large activation energy (>1 MJ/mol). The temperature and the oxygen pressure had a somewhat complex competing effect, and the XRD data showed grain alignment of the Bi-2223 crystals in the silver-clad wires, and the alignment was greatly enhanced during rolling or pressing of the wire. Tests of silver-sheathed Bi-2223 wires prepared by the ANL Materials and Components Technology Division showed that the conditions used for calcining the precursor Bi-2212 plus secondary phases had an effect on the alkaline earth cuprate secondary phases, which, in turn, had a pronounced effect on the rate and course of Bi-2223 formation.

Preparation of High-T_C Films by Alloy Oxidation. This effort was aimed toward the development of methods to prepare high-T_C superconducting films by oxidizing liquid alloy precursors. Earlier studies had shown that liquids having the composition MBa_2Cu_3, where M = Eu or Yb, and a melting point <1173 K could be oxidized to produce textured films of $MBa_2Cu_3O_7$ (123) if the liquid metal was supported on a suitable single-crystal substrate such as $SrTiO_3$ (100). The resulting high-T_C films showed a preferential orientation of the 123 phase a-b planes parallel to the substrate surface and achieved critical current densities in the range of 10^4 to 10^5 A/cm^2. A goal of the study was to find less expensive substrate materials and 123 materials more abundant than europium or ytterbium.

A collaborative research effort was established with American Superconductor Corp., and a specially designed high-temperature, high-vacuum apparatus was used for this work. When a highly oriented film of $YbBa_2Cu_3$ was applied to the single-crystal substrate by dipping, the high-T_C film was about 10 μm thick and had a strong c-axis orientation perpendicular to the substrate surface when the substrate was $SrTiO_3$ (100) or MgO (100). Microscopic and microprobe examinations revealed regions of near-epitaxial ordering of the high-T_C material within about 5 μm of the interface. These research studies were completed in 1991, when the processing parameters had been refined to the stage where high-quality, epitaxial $YbBa_2Cu_3O_x$ films, 10 to 20 μm in thickness, could be prepared. The critical temperature and current density were ≥80 K and about 1000 A/cm^2, respectively.

Personnel. This was an interdivisional project. The CMT portion of the work was led by Vic Maroni, and included the efforts of Milt Blander, Ira Bloom, Alan Brown, Larry Curtiss, Al Fischer, Dieter Gruen (CHM), Mark Hash, Stan Johnson, Jian Shu Luo, Shiu-Wing Tam, and Ben Tani, as well as some part-time people.

ORDERING AND ASSOCIATION IN LIQUIDS

Solubilities of NiO in Molten Carbonates. In molten carbonate fuel cells, dissolution of the NiO cathode and its transport to the cathode in the 62 mol% Li_2CO_3-38 mol% K_2CO_3 molten salt electrolyte were a frequent cause of cell failure. Measurements of the solubility of NiO in alkali carbonates were analyzed to develop a better understanding of the chemistry involved and thereby aid in improving cell performance. The following equilibrium exists for an alkali carbonate:

$$M_2CO_3 \text{ (soln.)} \leftrightarrow M_2O \text{ (soln.)} + CO_2$$

where the activity of the alkali oxide, M_2O (where M is Li, Na, K, or Rb), increases with a decrease in CO_2 pressure. At a given pressure, the oxide concentration is highest in Li_2CO_3 and lowest in Rb_2CO_3. The NiO solubility data were consistent with the presence of the three species Ni^{2+}, NiO, and NiO_2^{2-}. The major species in fuel cell electrolytes at 700°C and CO_2 pressures ≥1 atm is Ni^{2+}, *i.e.*, $NiCO_3$. With solid NiO present at the electrode and a constant CO_2 pressure, the $NiCO_3$ concentrations and NiO solubilities are inversely proportional to the activity coefficient of $NiCO_3$, which is highest in Li_2CO_3 and lowest in Rb_2CO_3.

Conformal Ionic Solution Theory for Aqueous Electrolytes. The Conformal Ionic Solution (CIS) Theory, a statistical mechanical perturbation theory that had been applied primarily to molten salt systems, was mentioned in previous chapters. When applied to aqueous systems, the simplest reciprocal salt system consists of two cations (A^+, B^+), two anions (X^-, Y^-), and the dielectric solvent, H_2O. Equations were derived for the Gibbs energies and enthalpies of mixing for the quaternary solutions. These equations can be used to calculate the thermodynamic properties of quaternary solutions from those of the lower-order binary and ternary subsystems.

Structures of Molten Salts. Experimental studies were continued on the melting mechanisms of trivalent metal halides. The two compounds, YCl_3 and $AlCl_3$, are interesting in that they have the same crystalline structure, but differ markedly in many of their physical properties, including melting points (YCl_3 melts at 721°C, while $AlCl_3$ sublimes at 193°C). Large differences also exist in the entropies of melting, the electrical conductivities, and the volume changes on melting. The Intense Pulsed Neutron Source (IPNS) at ANL was used to study the structure of YCl_3. The compound $AlCl_3$ was well known to form a liquid structure totally different from the crystal structure. The liquid consists of molecular Al_2Cl_6 units with two tetrahedra sharing a common Cl-Cl edge and an Al-Cl coordination number of four. The YCl_3, upon melting, retains its octahedral character in a loose arrangement with a coordination number of 5.8.

This work was a cooperative effort of CMT researchers, in particular, Milt Blander, plus David Price of the ANL Materials Science Division (MSD), and M. Tosi of the International Center of Theoretical Physics, Trieste, Italy.

Personnel. Milt Blander and Marie-Louise Saboungi have provided the impetus for these investigations. Other CMT members involved in the work were Ira Bloom, Larry Curtiss, and Jerry Johnson. As in most of the other basic programs, much of the effort was by part-time people and collaborators at other institutions.

QUANTUM CHEMICAL STUDIES

The use of quantum mechanics, in the form of *ab initio* molecular orbital calculations, to determine structural and thermochemical characteristics of compounds and other entities began in CMT in the calorimetry and thermochemical properties programs and soon found application in many of the Division's other research and development programs. This discussion concerns studies that were of more general scientific interest. In 1990, the Gaussian-2 (G2) theoretical procedure was developed for calculating molecular energies of compounds containing first row (Li→F) and second-row (Na→Cl) atoms. The G2 procedure is a refinement of the G1 procedure, which is based on Hartree-Fock theory and includes high levels of correlation energy, as well as quadratic configuration interactions and large basis sets. The G2 procedure involves corrections to various basis set extensions and gives improved calculated values in comparison with well-established experimental values for atomization energies, ionization potentials, electron affinities, and proton affinities for a large number of species. Larry Curtiss performed many of these calculations, which contributed to a much better understanding of structural aspects of the chemistry involved in a wide variety of CMT programs. Larry's work in this area received wide recognition, and he has been a collaborator and coauthor of many publications with Dr. John A. Pople of Northwestern University, who was awarded a Nobel prize in chemistry for this type of work in 1998.

The new procedure was tested on atomization energies of 125 molecules with an average deviation from experimental data of 5.06 kJ/mol, and it was used successfully to test uncertain data on 79 molecules. Methods were also developed for third-row non-transition metals (Ga→Kr). The increase in binding energy of silicon clusters (Si_n) as more silicon atoms were added to the cluster was calculated, and the results agreed with experimental data for n = 2-3. In work related to studies of methanol in the ANL Chemistry Division, calculations by CMT researchers showed that the literature value for the ionization potential of the methoxy radical was in error by about 3.5 eV, and that the established experimental C-H bond dissociation energy of methanol was in error. The G2 theory was also used to investigate silicon hydrides, Si_2H_n, where n = 0-6 and $Si_2H_n^+$, where n = 0-7. The calculated thermodynamic properties of the neutral species were in agreement with experimental data reporting the existence of stable silicon hydrides having n values of 2-5.

An interesting study was made in collaboration with the Central Institute of Inorganic Chemistry, Berlin, Germany, on $MAlF_4$ complexes where M = H, Li, or Na. These complexes are involved in high-temperature industrial processes, catalysis, chemical synthesis, and metal halide lamps. Investigations were conducted on the three structures shown in Fig. 6-36, in which the Al-M atoms are bridged at the corners, edges, or faces of the structure. Edge and corner bridging in $HAlF_4$ are of about equal energy (0.4 kJ/mol); $LiAlF_4$ has an edge-bridged structure; and the edge- and face-bridged structures of $NaAlF_4$ are of nearly equal stability. The $HAlF_4$ complex is much less stable than the alkali complexes, which may be a reason why difficulties have been encountered in observing this molecule.

In 1992, the CMT Annual Report stated that "The Gaussian-2 (G2) theoretical procedure, developed in a collaborative effort with AT&T Bell Laboratories and Carnegie Mellon University, is beginning to receive (1) wide acclaim for its accuracy in calculating molecular energies, and (2) extensive use in the field of computational chemistry." The G2 theory was being used widely by universities and industry for calculations of thermochemical data. The method is accurate to ±2 kcal/mol for calculating reaction energies. In 1992, the procedure was used in the modeling of thin-film diamond growth mechanisms, based on carbon dimer as the growth species. The model involved the formation of diamond-like carbon clusters by reactions of carbon with hydrocarbons.

By 1995, the G2 procedure had been extended to the elements Ga→Kr of the third row of the periodic table. Spin-orbit corrections for atoms and molecules having partially degenerate states were included explicitly in the G2-calculated energies. The average absolute deviation from experimental values for 40 test reactions was 1.37 kcal/mol.

An enlarged G2 neutral test set of 148 molecules was used to assess the performance of the G2 theory and two modified versions of it, G2(MP2) and G2(MP2,SVP). The mean absolute deviations for the three theories were 1.58, 2.04, and 1.93 kcal/mol, respectively. All three theories achieve the desired chemical accuracy of 2.0 kcal/mol. The modified versions, G2(MP2) and G2(MP2,SVP), are slightly less accurate than the original G2 theory, but they require significantly less computer time and disk space.

Personnel. Larry Curtiss was the Division's leader in this area. Some of the others who had a particular interest in this work were Milt Blander, Vic Maroni, Chris Marshall, Zoltan Nagy, Jerry Rathke, Marie-Louise Saboungi, and Shiu-Wing Tam.

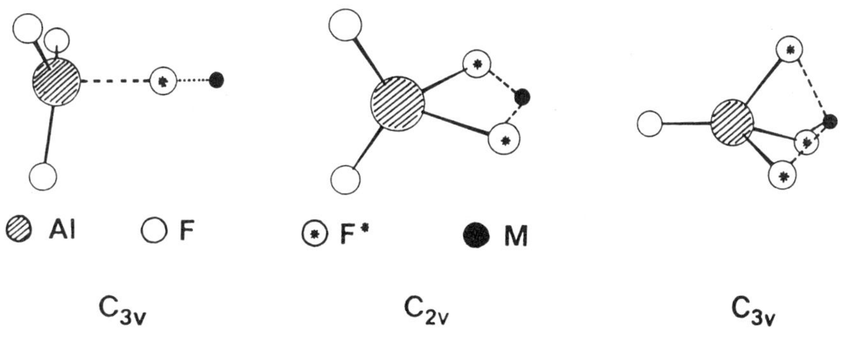

Fig. 6-36. Illustration of MA1F_4 Structures

MOLECULAR SIEVE RESEARCH

Advances in the synthesis of novel molecular sieve materials had paved the way for product-selective catalysis in the processing of fuels and chemical feedstocks. The CMT work in this area consisted mainly of theoretical studies of molecular sieve synthesis. Much of this work was based on the type of molecular orbital calculations discussed in the preceding section.

Ab initio molecular orbital calculations and inelastic neutron scattering techniques were used to determine the effects of the molecular sieve framework on the vibrational frequencies and rotational diffusion of occluded template molecules. The potential energy surface of tetraethyl-ammonium cation for rotation about the C-C and C-N bonds showed no distinct configurations that corresponded to local minima of approximately equal energy. Another study was made to investigate the Brønsted acidity of $Si_n(OH)_m$ clusters similar to those in molecular sieves. Proton affinities, which are an important factor in defining Brønsted acidity, were calculated with accuracies of 5-10%. This made possible a direct correlation of calculated O-H bond strength with observed catalytic activity.

The characteristics of clusters containing up to 160 atoms were calculated by the G2 theoretical procedure—some of the largest clusters ever used in electronic structure calculations on zeolites. Proton affinities, ammonium-ion affinities, and ammonia desorption energies were determined for these clusters. The calculated values for the properties of zeolites and molecular sieve materials were in good agreement with experimentally determined values that were available.

Continuing studies produced two significant findings. First, the energetics of the Brønsted acid site in TSi_nO_m clusters, where T = Si or Al, that have zeolite-type structures were calculated, and key properties of structures that are strongly correlated with the acid catalyst performance of zeolite materials were determined with unprecedented accuracy. Second, zeolite frameworks having sulfided metal clusters in void regions of the crystal structure were found to catalyze the conversion of methane to higher hydrocarbons. The reaction rates were low, but the selectivity to produce C_{2+} hydrocarbons was close to 100%.

In the hydrocarbon cracking process used by the petroleum industry, proton transfer between the adsorbed molecule and the zeolite catalyst was known to be a key mechanism. To understand the energetics of proton transfer and acid catalysis in zeolites, the equilibrium geometries of protonated alkanes (carbonium ions) were investigated, and their proton affinities were calculated using the G2 theory. For protonated ethane, the calculated proton affinity of the lowest energy structure was 141.0 kcal/mol, which agreed well with the experimental value of 139.6 kcal/mol. Proton transfer from the Brønsted acid site to an adsorbed molecule was believed to be a key mechanism not only for hydrocarbons, but also for other molecules such as water, ammonia, and methanol. In the case of water, the structure resulting from proton transfer to an adsorbed water molecule was found to exist only as a transition state, but the adsorption of a second water molecule stabilized the proton transfer structure, creating a local minimum in the potential energy surface (a more stable arrangement). The study was then extended to the interaction of the water dimer with the acid site in the zeolite ZSM-5 to a five-metal atom cluster ($AlSi_4O_4H_{12}$) model. Equilibrium geometries of the neutral $[ZH...(OH_2)_2]$ and ion-pair $[Z^-...H(OH_2)_2^+]$ adsorption complexes were calculated, as well as the transition geometries between these two species. Density function theory was used for these calculations. While not quite as accurate as G2 theory, it is more cost-effective. The relative energies of the two species and the transition state were small, indicating free movement of the proton.

Recent work in this area had shown some particularly noteworthy results on interactions of ethane with zeolite clusters. The $H_3SiO(H)Al(OSiH_3)_4$ cluster model, which includes five tetrahedral atoms, allows an adsorbed molecule to interact with three oxygen atoms adjacent to the substituted aluminum site. Calculations have identified several stationary points on the potential energy surface for the interaction of methane with the zeolite acid site (ZH). The calculations provide evidence for the existence of a stable ion-pair structure ($Z^-...C_2H_7^+$), whose existence had been contested. In addition, a new calculated value for the proton transfer barrier from the zeolite to the adsorbed hydrocarbon is in much closer agreement with typical experimental values than had been obtained previously.

Personnel. This work was done by Vic Maroni, Larry Curtiss, L. Hon, and Stan Zygmunt (an STA).

ELECTROCHEMICAL AND CORROSION STUDIES

Spectroelectrochemical and Synchrotron Studies. As mentioned earlier, a combination of spectroscopic and electrochemical techniques can be highly synergistic in understanding the detailed mechanisms of corrosion reactions and other chemical processes. The spectroelectrochemical investigations described in this section used a battery of such techniques: (1) laser Raman and UV-visible spectroscopy to determine *in situ* the structures and compositions of anodically formed corrosion films on metals, (2) photoelectrochemical and ac impedance measurements to characterize the electronic band structures of the films and to study their transport properties and conduction mechanisms, and (3) ac and dc polarization, cyclic potentiodynamic sweeps, and other transient techniques to examine the interfacial processes involved in metal corrosion. The synchrotron studies in this program were collaborative efforts with outside organizations, including Exxon Research and Engineering Co., the Naval Surface Warfare Center, the University of Poitiers in France, the University of Auckland in New Zealand, and members of the ANL Materials Science Division. Some of the research in this area was oriented toward possible future use of the Advanced Photon Source, which was under construction at ANL.

Surface enhancement of Raman scattering (SERS) by electrodeposition of silver showed that the passive films on iron consist of an inner layer of Fe_3O_4 and $Fe(OH)_2$ and outer layers of α-Fe_2O_3 and FeOOH, depending on the potential. Small amounts of thiocyanate (SCN^-) were found to break down the passivating oxidic layers on iron by complexing Fe^{2+} and Fe^{3+} ions, with subsequent precipitation of iron thiocyanate phases. At thiocyanate concentrations below about 0.001 *M*, however, the iron oxide films were self-healing, thereby preventing catastrophic breakdown. A similar study of the passive film formed on nickel showed that the film was composed of both $Ni(OH)_2$ and NiO, probably as a bilayer structure with the NiO closer to the metal. In subsequent studies at temperatures from ambient to 95°C, $Fe(OH)_2$ and Fe_3O_4 formed in the prepassive region, while Fe_3O_4 and FeOOH predominated at passive potentials.

In the case of chromium, a near-monolayer oxide film of Cr_2O_3 was formed. At more positive potentials (>0 V *vs.* the saturated calomel electrode), there was evidence of Cr(VI) in the form of a soluble chromate. A copper electrode in NaOH solution at -0.05 V *vs.* the standard calomel electrode formed a surface film of Cu_2O. Other work on copper involved studies of electrochemically induced pitting and morphological studies of its electrodeposition, and the behavior of copper in dilute solutions of cyanate (OCN^-) and thiocyanate (SCN^-) in perchlorate supporting electrolyte.

X-ray near-edge spectroscopy (XANES) and extended X-ray absorption fine structure (EXAFS) studies were made on the higher oxide forms of nickel (β-NiOOH, γ-NiOOH, and NiO_2). These materials were of interest both for corrosion protection and as battery electrode materials, and there was confusion in the literature as to the structure and valency of the nickel in these compounds. X-ray absorption measurements showed that all three of the higher oxide forms of nickel had the same disordered, layered structure, which can be represented as NiO_2H_x ($0 < x < 2.0$). Spectroscopic results indicated that the β and γ forms of NiOOH should be regarded as Ni^{3+} compounds. In subsequent studies, synchrotron X-ray absorption was used to characterize further the various phases that could be formed during corrosion of the metal or during the cyclic charging and discharging of nickel electrodes in batteries. The compounds that were characterized included NiO, α- and β-$Ni(OH)_2$, β- and γ-NiOOH, $Ni_3O_2(OH)_4$, NiO_2, Ni_3O_4, Ni_2O_3, and $KNiO_6$. Three distinct Ni-O bond distances were characteristic of Ni^{2+}, Ni^{3+}, and Ni^{4+}. In other work on nickel, an electrochemically deposited nickel oxide film on graphite was discovered to increase the capacitance of the graphite fourfold.

In situ laser Raman spectroscopy was used to investigate the compositions and structures of anodic corrosion films on titanium in various aqueous environments (H_2SO_4, Na_2SO_4, H_3BO_3-$Na_2B_4O_7$, and NaCl solutions). Bands characteristic of the anatase form of TiO_2 were observed at highly anodic potentials, while a disordered amorphous film was formed under cathodic conditions.

Synchrotron-generated far infrared radiation studies of water adsorbed on gold or platinum crystals permitted detection for the first time of the hindered translational mode of water and suggested the formation of water clusters at low coverage and an ice-like layer at higher coverages. Continuing studies showed that water adsorbed molecularly and non-dissociatively on the single-crystal platinum (111) under ultra-high vacuum and low temperatures (< -73°C). The far infrared spectrum of water on Pt (111) was very similar to that of gold.

Electrode Structure and Kinetics Studies. This research was concentrated on the electrode kinetics and atomistic mechanisms involved in metallic corrosion in aqueous solutions over a wide range of temperatures and pressures. This was a new area of investigation, since very little electrode kinetics research had been performed at temperatures over 60°C and practically none above 100°C in aqueous solutions. Much of the interest in aqueous corrosion at higher temperatures and pressures stemmed from corrosion effects in light water reactors (LWRs). The corrosion processes are electrochemical in nature and involve anodic dissolution of the metal together with the reduction of some component of the aqueous solution, such as dissolved oxygen. In reactions of this type, the essential elementary step is the charge transfer between the solid surface and the solution species at the interfacial solution layer (the "interphase"). The kinetics of the charge-transfer step are influenced by the molecular structure of the interphase and potential gradients within it. The electrode kinetics measurements were made by a combination of techniques, including galvanostatic, coulostatic, or potentiostatic pulse transients; ac capacitance methods; and rotating disk-electrode techniques.

Earlier work had shown that a simple charge-transfer reaction does not behave anomalously at high temperatures, so discrepancies reported in the literature must have been caused by something else. The ferrous/ferric reaction was known to be catalyzed by trace levels of anionic impurities, and precautions were taken in the experiments to avoid this effect. A catalytic effect on the Fe^{2+}/Fe^{3+} electron-transfer rates was observed after a single layer of copper had been plated on a gold working electrode. This effect was believed to result from the fact that the solvated ion could approach the plated surface more closely than the gold electrode.

Studies were extended to the Cu^{2+}/Cu electrode reaction, which entails two consecutive electron transfers followed by incorporation of a copper atom into the metal phase. There were two reasons for selecting this reaction for study: (1) copper deposition had been one of the most troublesome cathodic reactions in the corrosion of LWR cooling systems, and (2) the system presented a special challenge in the extension of the electron-transfer theory methods to an ion for which Jahn-Teller distortion of the hydration shell had to be accommodated. A computer modeling analysis indicated that it would be possible to measure both the fast and the slow steps of the Cu^{2+}/Cu reaction sequence. Work continued on theoretical models of Cu^{+} and Cu^{2+} ions in water. Preliminary experimental measurements on Cu^{2+}/Cu^{0} in perchloric acid solution reproduced some data in the literature. However, it became clear in the modeling studies that the calculated rate of the fast reaction in the two-step reaction sequence needed to be reevaluated.

Copper deposition is an important cathodic reaction during stress corrosion cracking in LWRs, and the mechanism was poorly understood. Experimental measurements confirmed earlier reports that the reaction rate is controlled by the Cu^{2+}/Cu^{+} charge-transfer step. It was also found that traces of chloride ion strongly catalyzed this reaction. The catalytic effect was attributed to the adsorption of chloride ions on the electrode surface and promotion of the charge transfer through an anion bridge. As the concentration of the chloride ion was increased, the exchange current density of the Cu^{2+}/Cu^{+} step doubled, while that of the Cu^{+}/Cu step remained essentially constant. When the importance of the chloride catalytic effect was recognized, rigorous purification procedures were applied to the perchloric acid solutions, and the rate constant of the reaction decreased by two orders of magnitude. In later studies, experimental measurements indicated that the rate of the Cu^{2+}/Cu^{+} charge-transfer reaction increased about a hundred fold when the temperature was raised from 25°C to 200°C. An Arrhenius plot for the temperature dependence of this reaction resulted in an activation energy of 30 kJ/mol.

In 1997, the kinetic studies of the Cu^{2+}/Cu electrode reaction were completed. The high-temperature/high pressure studies of the Cu^{2+}/Cu^{+} reaction, which is the rate-determining step in the overall copper deposition/dissolution process, were extended from a maximum temperature of 100°C to 200°C. The activation energy was found to be 32 ± 5 kJ/mol, and the transfer coefficient was independent of temperature.

In 1994, a collaborative effort with the ANL Materials Science Division was continued on the use of synchrotron X-ray methods to probe electrified interfaces that were buried under or within condensed phases. X-ray reflectivity techniques had been used to look at the metal side of the interface and the structure of the solution side of the electrical double layer. A similar approach was undertaken, using a complementary technique involving X-ray standing waves (XSW). This combination of techniques permitted studies of the double layer over a concentration range of five orders of magnitude at a distance from the electrode surface ranging from a few angstroms to tens of thousands of angstroms. A special thin-layer, X-ray/electrochemical cell capable of XSW measurements was developed. This apparatus was used to study incipient pore formation in luminescent silicon layers, which were of interest for future industrial applications due to their luminescence at visible frequencies.

Ultracapacitor Studies. A new program was started recently on the surface phenomena involved in ultracapacitor energy-storage devices. The characteristics of these devices are between those of classical "dielectric" capacitors and those of batteries. The capacity of ultracapacitors stems from three phenomena: (1) the capacitance of the electrical double layer, (2) the adsorption/desorption pseudocapacitance, and (3) the capacitance caused by oxidation/reduction processes of the electrode material occurring at or near the interface.

All of these phenomena can be investigated by means of synchrotron X-ray techniques, and initial studies are in progress on RuO_2 surfaces because this oxide, alone or with other oxides, is one of the most promising candidates for this application.

Theoretical Studies of Electrode Reactions. The investigations performed under this task were concerned with the more theoretical aspects of the electrochemical studies, but were so closely intertwined with the work in the preceding section that a separation may be rather artificial. Much of this work was done in cooperation with John Halley, a member of the Corrosion Institute at the University of Minnesota.

Theoretical models were developed for Cu^+ and Cu^{2+} ions in water. These were necessary for studying the Cu^+/Cu^{2+} electron-transfer process. A pair potential was derived for Cu^+/H_2O from *ab initio* molecular orbital calculations. This potential, with one derived previously for Cu^{2+}/H_2O, was tested successfully in molecular dynamics simulations of these ions in water. Two- and three-body interaction energies were calculated for tetrahedral, octahedral, and cubic arrangements of H_2O molecules around Cu^+ and Cu^{2+} ions in $[Cu(H_2O)_n]q^+$ clusters. Previous studies of copper ions in water had not taken into account the Jahn-Teller distortion of water in the first solvation shell about Cu^{2+}. X-ray diffraction data indicated that this distortion is significant and must be included in accurate calculations. Some disagreement had arisen as to the stability of metallic copper (Cu^0) in water. High-level calculations indicated that the interaction energy was very small (<0.2 eV).

In further studies of $Cu^{2+}(H_2O)_n$ clusters, the molecular dynamics results did not jibe with those from neutron-diffraction experiments. A possible explanation was a difference in the aqueous solutions used in the work. Detailed calculations were performed to elucidate the mechanism of the $Cu^{2+} + 2\,e^- \rightarrow Cu^0$ reaction in water. Molecular modeling studies of the electron-transfer reaction involved the multibody interactions of $Cu^{2+}(H_2O)_n$ clusters.

The electron-transfer rates are dependent upon electron coupling and a weighted density of states. The electronic coupling was calculated for three possible electron-transfer reactions in the copper deposition: (1) direct interaction, (2) outer-sphere electron transfer (water-water bridge), and (3) inner-sphere electron transfer (chloride bridge). The greatest coupling was found with the chloride bridge as a result of the shorter copper-copper distance rather than intrinsic chemical effects. Research continued in this area on the effects of other anions on the electronic coupling in Cu^{2+}/Cu^0 electron-transfer reactions. Electronic couplings for bromine- and fluorine-bridged reactions were similar to those for water-water bridged reactions. In 1996, investigations were conducted on the homogeneous inner-shell electron-transfer reaction $Fe^{3+} + e^- \rightarrow Fe^{2+}$ with the bridging halide anions, F^-, Br^-, and I^-. These anions gave increased electronic coupling due to a closer approach distance than is possible for an outer-sphere water bridge.

Personnel. These programs were under the general direction of Vic Maroni. Other CMT staff personnel contributing to this work included Wally Calaway, Larry Curtiss, Stan Johnson, Xiandong Feng, Carlos Melendres, Zoltan Nagy, George Papatheadorou, Gerry Reedy, and Ben Tani. This, like most of the other basic programs, made use of postdocs and other temporary personnel.

POLYMER ELECTROLYTES

This project is a fundamental study of lithium polymer electrolytes used in lithium batteries. The studies focus on the effects of the polymer host on ion pairing, which strongly affects the ionic transport in these systems. In recent work, *ab initio* molecular orbital theory has been applied to the energetic, structural, and dynamical properties of ion-ion and ion-polymer interactions at the molecular level, in combination with molecular dynamics simulations performed at the University of Minnesota. It was found that the binding energies of lithium cation complexes with alkyl oxide chains increase with coordination of the cation by oxygen, while the binding energy of the Li-O bond decreases. This theoretical study was conducted by Larry Curtiss.

GEOCHEMISTRY

The geochemistry research work was focused initially on two areas: (1) geochemistry and evolution of hydrothermal systems associated with volcanic areas, and (2) isotopic and organic geochemistry of carbon in sedimentary basins. In 1992, a more basic effort was added on mineral-fluid interactions.

Hydrothermal Systems. The extent of disequilibrium between ^{238}U and ^{230}Th was used to investigate the rates and mechanisms of element redistribution and the time scale of hydrothermal activity. One application was to determine the ages of travertine deposits in the northern part of Yellowstone Park. Travertines are calcium carbonate deposits that are formed when saturated or supersaturated groundwater emerges at the Earth's surface. The alpha decay rate of ^{234}U (half-life = 245,000 y) to ^{230}Th (half-life = 75,400 y) provided a "clock" for determining the age of the travertine, which contained a negligible amount of ^{230}Th when it was formed, so the $^{230}Th/^{234}U$ ratio is essentially zero. The time for the ^{234}U and ^{230}Th to reach equilibrium is about 500,000 y. The age of the travertine can thus be determined from the $^{230}Th/^{234}U$ activity ratio, with an upper limit of about 500,000 y. By using this approach, travertine formation was shown to have occurred about 130,000 to 170,000 years ago. This type of information should lead to a better understanding of the timing of the last glaciation in the region and the underlying thermal systems. These studies were done in collaboration with Kenneth Pierce of the U.S. Geological Survey and M. T. Murrell, a geochemist at Los Alamos National Laboratory (LANL).

Further studies of the Yellowstone Park area included the geochemistry and isotopy of radium in the hydrothermal systems. The radium concentration proved to be controlled by barite saturation and zeolite-water ion exchange.

Other field-based studies showed that the ages of silica samples from volcanic areas in the northern Kenya Rift Valley correlated with high paleolake levels, which were associated with humid climatic periods during the past 150,000 years. The main significance of this work is that it showed clearly, for the first time, the influence of climatic variations on geothermal activity in a continental rift zone.

Thermal springs along the shores of the Gulf of Suez and along the Nile River just south of Cairo were sampled for chemical and isotopic analysis in a collaborative study with Egyptian colleagues from Washington University and the Egyptian Geological Survey and Mining Authority. The objective was to determine the geothermal potential of these areas and their hydrogeochemical characteristics. The results showed that the Egyptian thermal waters contain water from either the Nubian sandstone aquifer or the Nile River, derived their solutes from Tertiary marine sedimentary rocks, and were heated conductively at depths of 3 to 4 km under a normal regional thermal gradient.

Mineral-Fluid Interactions. In 1992, a new project was initiated on the application of synchrotron radiation techniques to *in situ* studies of mineral-fluid interactions. A special X-ray-transparent cell that was developed for these studies is illustrated in Fig. 6-37. The cell performed well in tests at the National Synchroton Light Source at Brookhaven National Laboratory, and initial studies were made on changes in the surface structure of calcite during dissolution and growth. These experiments on a calcite/fluid interface indicated that the structure of the calcite cleavage surface during dissolution was determined by the near-ideal, atomic-scale termination of this surface and the long-range atomic order of the underlying bulk crystal. A multistep mathematical model was developed to characterize the surface.

X-ray reflectivity and diffraction measurements were used to characterize an ovatite over-growth as it was precipitated from aqueous solution onto a calcite cleavage surface. Scans were made during 4,585 min (76.4 h) of growth. Reflectivity was used to determine the properties of thin films, including the thickness, interfacial roughness, and real-space electron density profile across interfaces. A growth rate of 15 Å/h during the first 540 min decreased considerably thereafter. An analysis of the results from these measurements indicated that the ovatite had grown heteroepitaxially on the calcite cleavage surface.

X-ray standing wave measurements were made at mineral-water interfaces to provide information on the structures of metal adsorbates. Such information is needed to understand contaminant transport in groundwater aquifers. When lead was adsorbed from an aqueous solution onto the (104) cleavage surface of calcite, and the solution was then removed, approximately a 5% monolayer of lead was found adsorbed on the calcite. This lead occupied the same position as calcium relative to the (104) lattice plane. Additional experiments were conducted with solutions containing known concentrations of uncomplexed Pb^{2+} or $PbOH^{+}$ ions that were available for adsorption on the calcite surface. The conclusions from these experiments were:

1. About 5-17% of a monolayer equivalent of lead was adsorbed from the solution onto the calcite surface.
2. The amount adsorbed was independent of time from 8 to 93 h of reaction.
3. The lead occupied calcium sites.
4. The coverage and ordering of lead were about the same for *in situ* and *ex situ* measurements.
5. The adsorbed lead was stable in both aqueous solution and flowing helium for the duration of the measurements (about 96 h).

Isotope and Organic Geochemistry. A new gas chromatography/isotope ratio mass

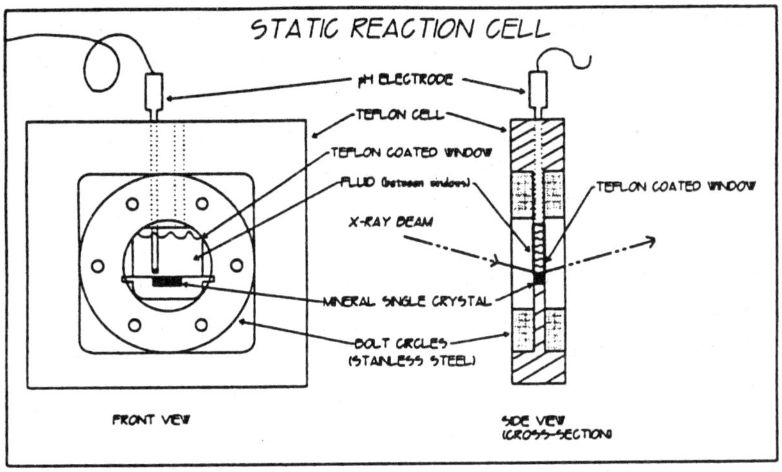

Fig. 6-37. Reaction Cell for Studies of Mineral-Fluid Reactions

spectrometer (GC/IRMS) was acquired, and a research program was undertaken in which this instrument was used for compound-specific carbon isotopic analysis in studying petroleum geochemistry. Suites of well-characterized petroleum samples were obtained from scientists at major oil companies, the U.S. Geological Survey, the Illinois Geological Survey, and several universities. Each suite was selected to address one or more of the following questions: (1) the relationship between source rock quality and the depositional environment, (2) the relationship between oil composition and source rock type, (3) the nature of chemical changes that occur within organic matter during diagenesis and catagenesis, (4) the effects of migration, mixing, and biodegradation on oil composition, and (5) the validity of oil-source rock and oil-oil correlations. Before analyzing the samples, CMT researchers spent considerable effort to test the GC/IRMS procedure, using n-alkanes of known carbon isotopic composition.

The carbon isotopic compositions of n-alkanes in oil samples from different reservoirs in an oil field near Phillipstown, Illinois, were similar. This finding is consistent with biomarker data indicating that these oils were generated from a similar source rock, but they appeared to have undergone varying degrees of biodegradation.

Although the chemical and isotopic compositions of hydrocarbon compounds in petroleum had been well studied, the geochemistry of nitrogen compounds in petroleum was another story. Therefore, some studies were conducted on nitrogen-rich crude oils in California. Nitrogen isotope determinations on 25 well-characterized nitrogen-rich crude oils showed that the sulfur-rich oils derived from marine-dominated organic matter had relatively low ^{15}N contents, while the sulfur-poor oils from organic matter containing some terrigenous material was high in ^{15}N. Variations in the nitrogen isotopic compositions were attributed to differences in the depositional environment, rather than to biodegradation or maturation processes.

Personnel. This effort, as in the past, was led by Neil Sturchio, and it included Teofilo Abrajano, Greg Archart, Allen Bakel, Ron Chiarello, Ben Holt (STA), Wally Calaway, Pete Lindahl, and Francis Markun, and several postdocs were also involved in certain aspects of the work. This was a highly collaborative program with many other institutions, several of which are in foreign countries.

ENHANCED METALLURGICAL PROCESSES

A small effort was continued on the use of molten sulfide mixtures to extract copper from scrap steel. For the production of sheet steel from scrap, the copper content of the material must be reduced to a level below 0.1 wt%. Measurements were performed at temperatures of 1365-1400°C on high carbon steel in equilibrium with mattes consisting of 80-90 wt% Al_2S_3, 10-20 wt% FeS, and 0-10 wt% MgS or CaS. For a steel/matte weight ratio of 4, the copper contents of the steel ranged from 0.04 to 0.09 wt%, where the initial values were 0.4 wt%. This procedure met the requirement of <0.1 wt% in the product, but required a large amount of matte. Investigations were then started on possible silicate-based slags that would remove copper from the steel, since silicate slags are already used widely in the industry.

Molten salts were also developed for the removal of lead from solid brass and copper surfaces. Treatment with the molten salts removed >90% of the surface lead from polished lead-bearing brass. Tests of treated brass samples by the EPA showed that this method of lead removal could be used for the pretreatment of brass parts in water systems.

This work, led by Milt Blander, was done by temporary personnel.

Analytical Chemistry Laboratory (ACL)

The Analytical Chemistry Laboratory continued to operate much the same as before, serving the Division, other projects at the Laboratory, various Federal agencies, and some private organizations. Dave Green continued to manage the organization. Fred Martino provided technical and quality assurance support for ACL and was the ANL Quality Assurance Representative for the Division. He was also responsible for a variety of administrative functions for ACL. Dave has become recognized as an authority on management of an analytical laboratory, and he has addressed various aspects of the subject in seminars and publications.

The organization consists of four groups—Chemical Analysis, Instrumental Analysis, Organic Analysis, and Environmental Analysis. The Group Leaders, respectively, are Don Graczyk, Del Bowers, Amrit Boparai, and Peter Lindahl. The ACL, like the rest of the Division, felt the effects of a shrinking budget. From 1990 to 1998, the technical staff level of the ACL decreased from about 50 to 35 or so. During that period, there was also a shifting emphasis in the nature of the analytical work that was needed. In the early 1990s, processing and recycling of IFR fuel was a major development effort that required extensive analytical work to evaluate the performance of the process operations. At the time that the IFR program was terminated, there was a rapidly growing interest in environmental and waste processing programs. The Division's programs up to that time had been concerned largely with inorganic and metallurgical materials. Many of these new program areas involved organic chemistry, and the ACL responded with a wide variety of new analytical capabilities for organic materials.

Upgrading and expansion of the ACL facilities and equipment continued in the 1990s. A computer-assisted bar code system was adopted for sample tracking and handling. A new laboratory was made available for analytical work on radioactive organic samples. In 1993, quality assurance and control procedures were developed to satisfy DOE and EPA reporting requirements. Among the new equipment items were a microwave digestion system, an automated C/H/N analyzer, a new mass spectrometer for organic analysis, an interfacing liquid chromatography/mass spectrometer for determining organic compounds, an inductively coupled plasma/mass spectrometer, a laboratory robotics system, a mercury analyzer, an oxygen combustion vessel for determining trace metals in water and environmental samples, a liquid scintillation counter for α and β activity, a germanium detector for γ-emitting nuclides, an upgraded scanning electron microscope user facility, a new X-ray diffractometer system, and Fourier transform infrared and Raman spectrometers.

In 1996, the ACL, as mentioned earlier, placed a home page on the Internet, which can be accessed on the World Wide Web by the address:

www.cmt.anl.gov/acl/acl.htm

This page is well done and contains a lot of information about the organization, as well as a color picture of the staff members.

The following is a selected list of the activities of the ACL in the 1990s; many of these relate to other CMT programs that were described earlier. This list gives a good sense of the wide range of ACL capabilities.

Engineering Studies of Pyrochemical Processes for Integral Fast Reactor Fuels
Rare-Earth Characterization Studies
Zeolite Immobilization of IFR Waste Salt
Studies of Alkalis in Hot Off-Gas from Pressurized Fluid-Bed Combustors
Analysis of Environmental Samples
Isolation of Strontium from Geological Samples for Isotopic Analysis
Plutonium Residue Recovery Program
Decomposition Products from Thermally Unstable Complexants
Molten Corium-Concrete Interaction Studies
Preparation of Samples for the National Energy Agency Committee on Reactor Physics Study
JANUS Water Analysis
Plutonium (VI) Speciation Study
Analysis of Cotton and Waters
ANL-West Soils, Vegetations, and Waters
Radiation Damage Studies for Superconductive Super Collider Detector Research
Synchrotron Thrust Group Activities
Radiological Environmental Analyses for Rocky Flats Plant
Analysis of Cores from Pressure Vessel of Experimental Boiling Water Reactor
Analytical Support Using X-Ray Diffraction and Scanning Electron Microscopy
Fourier Transform Infrared Microscopic Analysis of Fluid Inclusions
Studies of Organic Fluid Fouling in Heat Exchange Systems
Fourier Transform Infrared (FTIR) Analysis of Semivolatiles in Soils
Remote Detection of Chemical Agents with FTIR
Incinerator Monitoring for Organic Analytes
Determination of Volatile Organic Compounds for Lake Michigan Study
Walter Reed Army Institute of Research Site Characterization
Method Development and Testing for EPA
High Performance Liquid Chromatography Separation of Explosive and Other High Molecular Weight Compounds
Supercritical Fluid Chromatography/Matrix Isolation-Infrared Spectroscopy
Competitive Exchange Experiments in IFR Chemical and Engineering Support Studies
X-Ray Diffraction Studies of Zeolite Materials
Phase Composition/Solubility Studies on Equilibrated Mixtures of LiCl and KCl in Water
Procedures for Determining Uranium and Uranium Isotopes in High Silica Soils
Battery Program Support

Analysis of a Novel Neutron Absorber
Study of N_2O Emissions from Fluidized Bed Combustors
Environmental Monitoring Program
National Acid Precipitation Assessment Program
ANL Geoscience Programs
Radium Measurements (Geochemical)
Rocky Flats Environmental Restoration Program
Radiochemical Analysis of Experimental Boiling Water Reactor Vessel
Detection of Illegal Drug Laboratories
Computer Generation of Reporting Forms for Volatile Organic and PCB/Pesticide Environmental Analysis
Analysis of Waste Isolation Pilot Plant (WIPP) Samples by the Solid Adsorbent Method
Analytical Method Development for WIPP
Canister Cleaning and Certification for the WIPP
Analysis of Headspace Gases for the WIPP
Supercritical Fluid Extraction/Gas Chromatography for Analysis of Organic Compounds
Analysis of Environmental Samples for the U.S. Department of Agriculture
Environmental Assessment Study for Contaminants at Hohenfel Training Area Site, Germany
Development of Infrared Aerosol Analyzer
Support for CMT Nuclear Waste Programs
Monitoring of Antimony Levels in Scrap Aluminum
Chemical Transformation of Isotopically Enriched Tin Oxide
Advanced Photon Source Component Materials
Grafenwöhr Soil Samples
Stability of Calibration Standards for ICP/AES
Radiochemical Method Evaluation and New Method Development
Applications of FTIR Microscope to Identification of Wastes, Spill Residues, and Contaminants
Characterization of Products from Automobile Shredder "Fluff" Recycling
Determination of Rare Earth Elements and Calcium in Autopsy Tissues
Determination of Platinum in Tissues
Determination of ^{226}Ra in Human Tissues
Isolation of Uranium and Thorium from Soils for Isotopic Analysis
Characterization of Nitrogen in Lime-Scrubber Waste
Properties of High-Temperature Superconductors
Identification of Hazardous Wastes
Development of Integrated Performance Evaluation Program
Determination of Americium in Environmental Samples
Rietveld Analysis of X-Ray Powder Diffraction Data
Stand-Off (Remote) Detection
Analytical Certification of IFR Special Reference Materials
Incinerator Monitoring
X-Ray Diffraction Support (ANL Programs)
Studies of TRU·Spec and RE·Spec Chromatography
Improved Methods to Determine Actinides in Environmental Samples
Animal Orphan Waste
Methodology for Characterizing Chlorofluorocarbons in Polyurethane Foam
Analysis of Nuclear Waste Glasses and Slags
Calcium Isotopic Determination in Canine Bone and Blood Serum
Automated Real-Time Analysis of Chemical Sensor Data

Analysis of Soils for Explosives
Treatment of Cesium-Contaminated Milk
Engineering Studies of Electrometallurgical Treatment of Nuclear Fuel
Thermal Treatment Systems for Organic Effluents
Characterization of Used HEPA Filters for Disposal
Development of "Smart" Chemical Sensors
Analytical Procedures for Waste Minimization and Pollution Prevention
Gross α/β Analysis of Environmental and Mixed-Waste Samples
Microwave-Assisted Extraction of Semivolatile Organic Compounds from Soil
Analysis of Highly Acidic Mixed Waste
Determination of Arsenic Species in Soil Samples from the Rocky Mountain Arsenal
Transition Metal Speciation in Textile Mill Wastewater
Determination of Uranium in Dissolver Solutions for Production of ^{99}Mo
Development of an Analytical Scheme to Determine Actinides in Soil
Methods for Determining Metals in Waste Oil
Support for High-Temperature Superconductor Development
Wastewater Treatability Studies
Measurement of Calcium Isotopic Ratios in Canine Blood Serum
Development of a Continuous Emission Monitor for Thermal Treatment Systems
Determination of Platinum in Inner-Ear Fluid of Guinea Pigs
Radon Remediation
Separation of Mercury from Americium for Radioactive Waste Analysis
Determination of Reaction Products from Molten Sodium and Chlorinated Hydrocarbons
Dissolution Procedure for Low-Enriched Uranium Metal Targets
Automated Target Recognition Using Multispectral Images
Project in Support of Counternarcotics Efforts
Development of Solid-Phase Extraction Disks for Radiochemical Analysis
Analysis of Samples from Process to Recycle Aluminum Salt Cake
Continuous Monitoring of Plasma Arc Furnace at ANL-West
Preparation of Simulated Solidified Waste Samples for WIPP Performance Demonstration Program
Automated Data Cataloging Procedure for Objects in AMPS Multispectral Imagery
Development of Method for Radium Determination in Aqueous Samples
Domestic Nuclear Smuggling Exercise
Development of a No-Moving-Parts FTIR Sensor
JANUS Reactor Characterization
Study of Dry-Storage Casks for Spent Nuclear Fuel

THE DIVISION'S FUTURE

One certainty about the future is that the Division will continue to face fierce competition in seeking and keeping funding for programs. This competition comes from industry, the universities, other national laboratories, independent research organizations, and even other divisions at ANL. It is also affected by the changing perceptions of national technical needs and budget pressures in Washington. The hope is that CMT will be able to meet this competition because of its excellent staff people, unique facilities, and extensive experience in a variety of research areas and technology development programs. In recent years, DOE funding of the Division's programs has declined, but much of the slack has been taken up by cooperative work with industries in the form of CRADAs (Cooperative Research and Development Agreements) and "work for others" agreements, which may involve other Government agencies or private organizations.

With the demise of the Integral Fast Reactor program and the ban on commercial fuel processing in the United States, program support on processes for the recovery of uranium and plutonium from spent nuclear fuel has vanished. In its stead are programs to treat spent fuels to separate transuranic (TRU) elements (neptunium, plutonium, americium, and curium) for disposal in a waste depository. Electrometallurgical processes show promise for this application if current throughput rates can be increased. Preliminary results suggest that it may be possible to fix TRU elements on a zeolite bed in a waste form suitable for repository disposal.

For over ten years the Division has been conducting a variety of tests on simulated waste glasses and spent reactor fuels under conditions similar to those in the Yucca Mountain Repository. An exciting recent discovery is that even though neptunium, on release from a waste form, will migrate rapidly through the volcanic tuff of a repository, it will be retained by alteration products, principally zeolites, that form on the surface of the glass by reaction with water vapor. This finding must be explored further. In addition, continuation of long-term (5- to 10-year) corrosion tests with waste samples would be desirable.

Throughout its history, the Division has been at the forefront of development work on advanced solvent-extraction processes and equipment. An exciting new development is aqueous biphase extraction, which shows promise for a wide range of applications. Typical examples are cleanup of industrial wastes and purification of low-grade ores such as bauxite for aluminum production. The Division intends to pursue this emerging technology vigorously because of its potentially great versatility.

The deregulation of electricity may bring about some major changes. Utilities are likely to reduce their involvement in the generation of electricity and concentrate on its distribution. Fuel cells operating on natural gas are expected to be finding increased use for local generation of electricity by cities, industrial organizations, office buildings, large housing complexes, and even homes. There is also much interest in the use of fuel cells in transportation applications, where they would most likely be used in hybrid vehicle systems. Competition will be heavy for funding to support research for these applications. The Division is well positioned to contribute to this effort due to its past work on molten carbonate and solid oxide fuel cells, and especially its leading-edge work in polymer-electrolyte fuel cells. The best opportunities for work on fuel cells may be where highly focused, sophisticated research efforts can be directed toward niche areas that need attention, rather than complete systems.

The situation with batteries is somewhat similar to that for fuel cells. The Division has had large programs on high-temperature lithium-alloy/metal sulfide batteries for electric vehicle propulsion and load leveling on electric utility systems. This technology was developed to the stage where it was ready for commercialization. The battery work evolved into a number of other areas, including sodium/metal chloride, lithium/polymer electrolyte, lithium-ion, and nickel/metal hydride systems. The lithium systems, in particular, have a potentially huge market in consumer electronics. Research in this field is highly competitive. The Division is enjoying considerable success in using sophisticated basic research tools to investigate the detailed chemistry and electrochemistry of these systems, and might have a similar role in other specialized areas of industrial research.

The Division has, for years, operated a large facility for conducting performance and cycle-life tests on a variety of commercial and experimental batteries. Diagnostic procedures for failure analysis or performance degradation are also conducted when needed. Indications are that similar testing of fuel cells may be a significant program

When the Division came into being in 1948, the staff consisted of many young scientists and engineers. Most if not all of these people have retired by now. The Division has been able to attract as replacements outstanding young graduates and has also hired distinguished leaders in particular areas of research. In addition, the Division makes extensive use of postdoctoral people who are seeking R&D experience. Some of these are retained when their appointments expire. The Division takes very seriously its responsibility to train young scientists to do imaginative, high-quality research.

The consensus is that the future of the Chemical Technology Division can be viewed with cautious optimism.

APPENDIXES

General References

American Nuclear Society, *Controlled Nuclear Chain Reaction: The First 50 Years*, American Nuclear Society, LaGrange Park, IL (1992).

Clarfield, Gerald H., and William M. Wiecek, *Nuclear America, Military and Civilian Nuclear Power in the United States 1940-1980*, Harper & Row, New York (1984).

Evans, H., S. Vogler, and H. H. Hyman, *Studies in Counter-Current Solvent Extraction for the Decontamination of Enriched Uranium from the Proposed High Flux Research Pile*, ANL-4110, Argonne National Laboratory (Jan. 28, 1948).

Fermi, Laura, *Atoms in the Family; My Life with Enrico Fermi*, University of Chicago Press, Chicago, IL (1954).

Gillette, Robert, "Nuclear Safety (I): The Roots of Dissent," *Science* **177,** 771 (1972).

Gillette, Robert, "Nuclear Safety (II): The Years of Delay," *Science* **177**, 867 (1972).

Goldschmidt, Bertrand, *The Atomic Complex: A Worldwide Political History of Nuclear Energy*, American Nuclear Society, LaGrange Park, IL (1982).

Groueff, Stephane, *Manhattan Project: The Untold Story of the Making of the Atomic Bomb*, Little, Brown & Co., Boston (1967).

Hewlett, Richard G., and Oscar E. Anderson, Jr., *The New World, 1939-1946*, Pennsylvania State University Press, University Park, PA (1962).

Hewlett, Richard G., and Francis Duncan, *Atomic Shield, 1947/1952, Vol. II, History of the United States Atomic Energy Commission*, Pennsylvania State University Press, University Park, PA (1969).

Holl, Jack M., *Argonne National Laboratory, 1946-96*, University of Illinois Press, Urbana & Chicago, IL (1997).

Johnson, Otto (ed.), *1997 Information Please Almanac*, Houghton Mifflin Company, Boston (1997).

Massachusetts Institute of Technology Faculty, Jack Dennis (ed.), *The Nuclear Almanac—Confronting the Atom in War and Peace*, Addison-Wesley, Reading, MA (1984).

Orlans, Harold, *Contracting for Atoms*, Brookings Institution, Washington, DC (1967).

Rhodes, Richard , *The Making of the Atomic Bomb*, Simon & Schuster, New York (1986).

Seaborg, Glenn T., *The Plutonium Story: The Journals of Professor Glenn T. Seaborg, 1939-1946*, Ronald L. Kathren, Jerry B. Gough, and Gary T. Benedict, eds., Battelle Press, Columbus, OH (1994).

Smith, Alice K., *A Peril and a Hope: The Scientists' Movement in America 1945-1947*, University of Chicago Press, Chicago, IL (1965).

Smith, Alice K., and Charles Weiner (eds.), *Robert Oppenheimer: Letters and Recollections*, Harvard University Press, Cambridge, MA (1980).

Smyth, Henry D., *Atomic Energy for Military Purposes*, Princeton University Press, Princeton, NJ (1945).

United Nations, *Proceedings of the Second United Nations International Conference on the Peaceful Uses of Atomic Energy*, September 1-15, 1958, United Nations, Geneva, Switzerland (1958).

Wright, John W., *The New York Times 1998 Almanac*, Penguin Putnam Inc., New York (1998).

Acronyms and Definitions

A

ABS	Aqueous Biphase Separation (process)
ACL	Analytical Chemistry Laboratory
ACS	American Chemical Society; American Ceramics Society
ACU	Advisory Committee on Uranium
ADF	Aqueous Dissolution Fluorination (process)
AEC	Atomic Energy Commission
AEM	Analytical Electron Microscopy
AES	Atomic Emission Spectrometry
AFBC	Atmospheric Pressure Fluidized Bed Combustion
AIChE	American Institute of Chemical Engineers
ALPR	Argonne Low Power Reactor
AMD	ANL Applied Mathematics Division
AMF	American Machine and Foundry
AMPEL	Argonne MHD Process Engineering Laboratory
AMPS	Apparatus for Monitoring and Purifying Sodium
ANL	Argonne National Laboratory
ANL-E	ANL-East (Illinois Site)
ANL-W	ANL-West (Idaho Site)
ANS	American Nuclear Society
APS	Advanced Photon Source
APST	Alkali Metal Particulate Sampling Train
ASR	Area Specific Resistance
ATR	Attenuated Total Internal Reflection (spectroscopy)
ATS	Alkali-Tin-Silicate Glass (for Pu)
AVIDAC	Argonne's Version of the Institute's Digital Automatic Computer

B

B&W	Babcock and Wilcox
BAPL	Bettis Atomic Power Laboratory
BASIC	A computer programming language
BBI	Breeder Blanket Interface
BCSC	Blanket Comparison and Selection Study
BET	Brunauer-Emmet-Teller (method of surface area analysis)
BNL	Brookhaven National Laboratory
BORAX	A series of ANL reactors used for safety research
BWR	Boiling Water Reactor

C

CAMD	Center for Advanced Microstructures and Designs
CANDU	Heavy water Canadian reactor
CCT	Coordinated Cluster Theory (for molecular structure)
CD-ROM	Compact Disk-Read Only Memory
CDC	Control Data Corporation
CDIF	Component Development and Integration Facility
CEN	ANL Chemical Engineering Division
CENHAM	A standardized glove-box design used in CMT
CFFF	Coal Fired Flow Facility
CHM	ANL Chemistry Division
Cintichem	Process used in Indonesia for ^{99}Tc Recovery
CIS	Conformal Ionic Solution Theory
CMPO	Extractant used in TRUEX process
CMT	ANL Chemical Technology Division
COBOL	A business-oriented computer language
COMMIX	Computer code for liquid metal feed system
CORCON	Thermal hydraulic code for source-term packages
CP	Chicago Pile (several reactors)
CPC	Compound Parabolic Collector
CPM	Critical Path Method
CRADA	Cooperative Research and Development Agreement
CRBR	Clinch River Breeder Reactor

CREX Solvent Extraction Process for Sr, Pb, Tc
CRSC Center for Research on Sulfur in Coal (Illinois)
CRT Cathode Ray Tube
CSDP Continental Scientific Drilling Program
CSEX Cesium Extraction Process
CSP Chloride Silent Power (British battery manufacturer)
CT ANL Components Technology Division
CTF Components Test Facility
CTIF Components Test and Integration Facility
CTIU Component and Test Integration Unit
CTR Controlled Thermal Reactor (program)

D

DADAC Dosimetry And Damage Analysis Center
DARPA Defense Advanced Research Projects Agency
DEC Digital Equipment Corporation
DIGICALC A computer program language
DOE Department of Energy
DRIFTS Diffusion Reflectance Infrared Fourier Transfer Spectroscopy
DTA Differential Thermal Analysis
DTRAM Dynamic Tritium Release and Analysis Model
DWPP Defense Waste Processing Plant

E

EADL Electrochemical Analysis and Diagnostics Laboratory
EBR Experimental Breeder Reactor
EBR-I Experimental Breeder Reactor No. I
EBR-II Experimental Breeder Reactor No. II
EBWR Experimental Boiling Water Reactor
ECN Netherlands Research Foundation
ED Electron Diffraction
EDAX Energy Dispersive X-ray Analysis
EDS Energy-Dispersive Spectroscopy (X-ray)
EDTA Ethylene Diamine Tetraacetic Acid
EELS Electron Energy Loss Spectroscopy
EES ANL Energy and Environmental Sciences Division
EIA Electrochemical Industrial Associates
EIS ANL Environmental Impact Studies Division
EMF Electromotive Force
ENG ANL Engineering Division
EPA Environmental Protection Agency
EPRI Electric Power Research Institute
ERC Energy Research Corp.
ERDA Energy Research and Development Administration
EROSION/MOD Computer Code for Erosion in FBC
ES ANL Energy Systems Division
ET ANL Energy Technology Division
ESCA Electron Spectroscopy for Chemical Analysis
ETF Engineering Test Facility (MHD)
EXAFS Extended X-ray Absorption Fine Structure

F

FACT Facility for the Analysis of Chemical Thermodynamics
FARET Fast Reactor Test Facility
FBC Fluidized Bed Combustion
FCF Fuel Cycle Facility (at ANL-W)
FE ANL Fuels and Energy Division
FEF Fuel Examination Facility (at ANL-W)
FEUL Fossil Energy Users' Laboratory
FFTF Fast Flux Test Facility
FLUFIX/MOD2 Computer Code for Fluidized Bed Hydrodynamics
FORTRAN A computer language
FTIR Fourier Transform Infrared (spectroscopy)

FWDC Foster Wheeler Development Corp.
FY Government Fiscal Year

G

GAC General Advisory Committee
GC Gas Chromatography
GCFR Gas-Cooled Fast Reactor
GE General Electric Co.
GPU General Public Utilities Corp.
GTM Generic TRUEX Model

H

HDPE High Density Polyethylene
HEDL Hanford Engineering Development Laboratory
HEPA High Efficiency Particulate Air (filter)
HETP Height Equivalent to a Theoretical Plate (or stage)
HEU High Enriched Uranium
HFEF Hot Fuel Examination Facility
HFIR High Flux Isotope Reactor
HPLC High Performance Liquid Chromatography
HSRC Hazardous Substance Research Center
HTER High Throughput Electrorefiner
HTGR High Temperature Gas-Cooled Reactor
HTM Hydrogen Titration Method
HWR-NPR Heavy Water Reactor-NPR (see NPR)

I

IAEA International Atomic Energy Authority
IBM International Business Machines
ICCI Illinois Clean Coal Institute
ICBM Intercontinental Ballistic Missile
ICPP Idaho Chemical Processing Plant
IEA International Energy Agency
IFC International Fuel Cells
IFR Integral Fast Reactor
ILRR Interlaboratory Reaction Rate (program)
INEEL Idaho National Engineering and Environmental Laboratory
INEL Idaho National Engineering Laboratory (now INEEL)
INOR International Tokamak Reactor
IPMA Ion Probe Microanalyzer
IPNS Intense Pulsed Neutron Source
IR Infrared
IR-100 Award by *Industrial Research Magazine*
IRMS Isotopic Ratio Mass Spectrometry
ISOA Intermediate State of the Art
ITER International Thermonuclear Experimental Reactor

J

JAERI Japan Atomic Energy Institute
JANAF Joint Army-Navy-Air Force (thermodynamic reference tables)
JANUS Reactor used at ANL for biological research
JCAE Joint Committee on Atomic Energy
JPDR Japan Power Demonstration Reactor

K

KAPL Knolls Atomic Power Laboratory
KB Kilobyte
KGB Komitet Gosudarstrenney Bezopasnoski (Committee for State Security)

L

LAMPF Los Alamos Meson Physics Facility
LAMPRE Los Alamos Molten Plutonium Reactor Experiment
LANL Los Alamos National Laboratory
LASL Los Alamos Scientific Laboratory (now LANL)
LBL Lawrence Berkeley National Laboratory
LEU Low Enriched Uranium

LIPE	Lithium-Polymer Electrolyte (battery)
LITCO	Lockheed Idaho Technology Co.
LLNL	Lawrence Livermore National Laboratory
LM	Lanthanum Manganite (fuel cell cathode)
LMFBR	Liquid Metal Fast Breeder Reactor
LMFR	Liquid Metal Fuel Reactor
LMFT	Low-Mass Flow Train (at Coal Fired Test Facility)
LOFT	Loss Of Fluid Test
LPTL	Lithium Processing Test Loop
LSU	Louisiana State University
LWR	Light Water Reactor

M

MAP3S	Multistate Atmospheric Power Production Pollution Study
MASCOT	Computer code used for sodium trap design
MASS-11	A word-processing program
MAWS	Minimum Additive Waste Stabilization Program
MBO	Management By Objectives
MCCI	Molten Core-Concrete Interaction
MCFC	Molten Carbonate Fuel Cell
MCPU	Multiple Cycle Plutonium Utilization
MCT	ANL Materials and Components Technology Division
MD	Molecular Dynamics
MEDEC	Melt/Drain Evaporation Calcination Process
MERDC	Mobility Equipment Research and Development Command
MET	ANL Metallurgy Division (now SSS, MST and others)
METC	Morgantown Energy Technology Center
MeV	Million Electron Volts
MFTF	Mirror Fusion Test Reactor
MHD	Magnetohydrodynamics
MI	Matrix Isolation (spectrometry)
MIR	Russian Space Station
MISTT	Midwest Interstate Sulfur Transport and Transformation
MIT	Massachusetts Institute of Technology
MMS	Masked Multichannel Scaler
MNDO	Modified Neglect of Differential Overlap (molecular structure theory)
MNM	Metal-to-Nonmetal (transitions)
MOX	Mixed Uranium-Plutonium Oxide (reactor fuels)
MS	Mass Spectrometry
MS-DOS	Microsoft Disk Operating System
MSD	ANL Materials Science Division
MSID	Mass Spectrometric Isotopic Dilution
MSOFC	Monolithic Solid Oxide Fuel Cell
MSRE	Molten Salt Reactor Experiment (at ORNL)
MST	ANL Materials Science and Technology Division
MSW	Municipal Solid Waste
MTR	Materials Test Reactor
MTU	Mass Transfer Unit (in electrorefiner)
MW(e)	Megawatts Electrical (electricity produced by a reactor)
MW(t)	Megawatts Thermal (heat produced by a reactor)

N

N.S.	Nuclear Powered Ship (civilian)
NAA	Neutron Activation Analysis
NAPAP	National Acid Precipitation Assessment Program
NAS	National Academy of Sciences
NASA	National Aeronautics and Space Administration
NASICON	An experimental ionically conductive glass composition
NBL	New Brunswick Laboratory
NBS	National Bureau of Standards
NBTL	National Battery Test Laboratory
NDRC	National Defense Research Council
NEACRP	National Energy Agency Committee on Reactor Physics
NEPA	Nuclear Energy for Propulsion of Aircraft

NEPEX	Neptunium-Plutonium Extraction Process
NKVD	Predecessor of KGB (see)
NMFECC	National Magnetic Fusion Energy Computation Center
NMR	Nuclear Magnetic Resonance
NO_X	Generic term for nitrogen oxides, usually applied to air quality
NPR	New Production Reactor (for tritium production)
NPS	National Park Service
NRC	Nuclear Regulatory Commission
NRTS	National Reactor Testing Station (now INEEL)
NRX	Reactor at Chalk River, Canada
NSLS	National Synchrotron Light Source (at BNL)
NTA	Nitrilotriacetic Acid

O

O/M Ratio	Oxygen-to-Metal Ratio
OEPM	Office of Electrochemical Project Management
OPEC	Organization of Petroleum Exporting Countries
ORACLE	Oak Ridge Automatic Computer Logical Engine
ORNL	Oak Ridge National Laboratory
ORR	Oak Ridge Research Reactor
OSRD	Office of Scientific Research and Development

P

PAD	ANL Particle Accelerator Division
PAFC	Phosphoric Acid Fuel Cell
PAH	Polycyclic Aromatic Hyrdrocarbons
PAHR	Post-Accident Heat Removal
PASCAL	A computer language
PC	Personal Computer
PCB	Polychlorinated Biphenyl
PDP	Model of Digital Equipment Corp. computer
PEFC	Polymer Electrolyte Fuel Cell
PEG	Polyethylene Glycol
PEMFC	Polymer Exchange Membrane Fuel Cell
PEO	Polyethylene Oxide
PERT	Program Evaluation and Review Technique
PFBC	Pressurized Fluidized Bed Combustion
PFP	Plutonium Finishing Plant
PIFAG	Production Irradiated Fuel Assay Gauge
PL/1	A computer language
PNL	Pacific Northwest Laboratory
POB	Proof of Breeding
PPG	Polypropylene Glycol
PRDC	Power Reactor Development Corporation
PuRR	Plutonium Residue Recovery
PVC	Polyvinyl Chloride
PWR	Pressurized Water Reactor
PYRO	Computer Model for IFR Process

Q

QA	Quality Assurance; ANL Quality Assurance Division
Quad	One quadrillion (10^{15}) Btu

R

RABSAM	Regenerable Activated Bauxite Sorber Alkali Monitor
RAM	Random Access Memory
RAS	ANL Reactor Analysis and Safety Division
RBMK	Pressurized water reactor of Russian design
RDF	Refuse Derived Fuel
RDT	Reactor Development and Technology (an AEC division)
RE	ANL Reactor Engineering Division; Rare Earths
RERTR	Reduced Enrichment for Research and Test Reactors
RFP	Rocky Flats Plant
RMC	Reverse Monte Carlo
ROM	Read Only Memory

RTNS	Rotating Target Neutron Source

S

SACRD	Safety Analysis Computerized Reactor Data
SAFT	Societe des Accumulateurs Fixe et de Traction
SASPE	Algorithm for Speciation and Partitioning Equilibria
SASSE	Algorithm for Stagewise Solvent Extraction
SCAQMD	California South Coast Air Quality Management
SCTI	Sodium Components Test Installation
SEM	Scanning Electron Microscopy
SEM/EDS	Scanning Electron Microscopy/ Energy Dispersive Spectrometry
SERI	Solar Energy Research Institute
SFC	Supercritical Fluid Chromatography
SIBELIUS	Joint CMT-European Study of Ceramic Fusion Breeder Blankets
SIR	Sodium Intermediate Reactor
Site A	Site of CP-2 in Argonne Woods of Palos Park Forest Preserve
Site D	Current Location of ANL in DuPage County
Site W	Hanford, Washington
Site X	Oak Ridge, Tennessee
Site Y	Hanford, Washington
SNAP	Systems for Nuclear Auxiliary Power (in space satellites)
SNL	Sandia National Laboratories
SNM	Special Nuclear Materials
SOFC	Solid Oxide Fuel Cell
SOLGASMIX	Computer code for calculating equilibrium gas compositions
SPACE	Worksheet calculation for plant equipment size, plant space, costs
SPECTER	Computer code for radiation damage
SPM	ANL Special Materials Division
SPR	Single-Pass Reactor
SREX	Strontium Extraction Process
SRL	Savannah River Laboratory (now SRNL)
SRNL	Savannah River National Laboratory
SRP	Savannah River Plant
SRTALK	ORNL-PNL Solvent Extraction Process for Sr, Tc, Cs extraction
SS	Stainless Steel
SSC	Superconducting Super Collider
SSS	ANL Solid State Science Division
SSTR	Solid State Track Recorder
STA	Special Term Appointment
STARFIRE	A commercial fusion reactor concept (ANL)
STEP	Source Term Experiments Program
STR	Submarine Thermal Reactor

T

TBP	Tributyl Phosphate
TCA	1,1,1-Trichloroethylene
TCE	Trichloroethylene
TD	ANL Technology Development Division
TEM	Transmission Electron Microscopy
TEPR	Tokamak Experimental Power Reactor
T_EX	A word processing computer program
TGA	Thermogravimetric Analysis
THORP	Thermal Oxide Reprocessing Plant
TMI	Three Mile Island
TNS	The Next Step (a nuclear fusion reactor concept)
TPD	Temperature Programmed Desorption
TREAT	Transient Reactor Test Facility
TRIO	Tritium Recovery In Pile
TRU	Transuranium or transuranic
TRUEX	Transuranium Extraction (process)
TSTA	Tritium Systems Test Assembly
TVA	Tennessee Valley Authority

U

UCLA — University of California at Los Angeles
UI-UC — University of Illinois at Urbana-Champaign
USABC — United States Advanced Battery Consortium
U.S.S. — U. S. Navy Ship
U.S.S.R. — Union of Soviet Socialist Republics
UKAEA — United Kingdom Atomic Energy Authority
UTC — United Technologies Corp.

V

VAX — Model of Digital Equipment Corp. computer
VRLA — Valve-Regulated Lead-Acid (battery)

W

WIPP — Waste Isolation Pilot Plant
WRAIR — Walter Reed Army Institute of Research
WSRC — Westinghouse Savannah River Co. (type of glass)
WVDP — West Valley Demonstration Plant (type of glass)
WWII — World War II

X

X-10 — Clinton Laboratory at Oak Ridge
XANES — X-ray Absorption Near-Edge Spectroscopy
XRD — X-Ray Diffraction
XSW — X-ray Standing Waves

Y

Y-12 — Isotope Separations Plant at Oak Ridge
YSZ — Yttria-Stabilized Zirconia (fuel-cell electrolyte)

Z

ZGS — Zero Gradient Synchrotron
ZPPR — Zero Power Plutonium Reactor
ZPR — Zero Power Reactor
ZSM-5 — Type of zeolite

Subject Guide

1940-1950: THE BEGINNING.......... 1

The Manhattan Project.......... 1
The Atomic Energy Commission and the National Laboratories.......... 7
The Met Lab.......... 8
The Chemical Engineering Division.......... 16

1950-1960: THE NUCLEAR PROMISE.......... 23

The National and International Scene.......... 23
Reactors.......... 26
The Move to Site D.......... 29
New Employees.......... 31
Services.......... 34
Tools of the Trade.......... 37
The New Buildings.......... 39
Building D-205.......... 39
Building D-310.......... 46
Division Organization and Management.......... 47
Technical Programs.......... 50
Aqueous Processes.......... 50
Determination of Breeding Gain in EBR-I.......... 50
The Halex Process.......... 51
Aqueous Processing of Alloy Fuels.......... 53
Contactor Development.......... 53
Pyrometallurgical Processes.......... 54
Pyrometallurgical Research.......... 54
Pyrometallurgical Process Development.......... 56
The EBR-II Fuel Cycle Facility (FCF).......... 62
Fluoride Volatility Processes.......... 65
Feed Materials Processing.......... 71
Conversion of Uranyl Nitrate Solution to Uranium Oxide.......... 71
Production of Refined Uranium Hexafluoride.......... 72
Waste Treatment.......... 75
Incineration of Radioactive Solid Wastes.......... 75
Fluidized Bed Calcination of Aqueous Wastes.......... 76

Ion-Exchange Studies 77
Gamma Irradiation Facility 78
The "Hot Rock" 79
Fluidized Bed Technology 80
Reactor Safety 81
Metal-Air Reactions 81
Metal-Water Reactions 81
Calorimetry 82
The Analytical Chemistry Laboratory 83
Analytical Research and Nuclear Constant Measurements 85
Especially Significant Publications 86

1960-1970: TURBULENT TIMES 89

The Domestic and International Climate 89
Reactors 92
The Division 92
Management 96
Building Addition 98
Tools of the Trade 99
CENHAM Glove Boxes 101
Technical Programs 102
EBR-II Fuel Cycle Facility 102
Pyrochemical Process Development 106
Chemistry of Liquid Metals 108
Fluoride Volatility Processes 112
Laboratory Investigations 112
Process Development 114
Physical and Thermodynamic Properties of High Temperature Materials 121
Reactor Safety 123
Metal Oxidation and Ignition Kinetics 123
Metal-Water Reactions 124
Fast Reactor Safety Studies 124
Chemistry of Irradiated Fuels 125
Preparation of Fast Reactor Fuels 125
Preparation of High-Purity Plutonium-238 Metal 127
Determination of Nuclear Constants 127
Calorimetry 128

Energy Conversion and Storage 130
Thermoelectric Research 130
Thermally Regenerative Galvanic Cells 131
Lithium-Chalcogen Batteries 133
Fluidized Bed Combustion of Coal 134
Analytical Chemistry 135
Waste Processing and Gamma-Irradiation Facility 137
Studies and Evaluations 137

1970-1980: TRANSITIONS 139

The National Scene 140
Nuclear Power 140
The Laboratory 145
The Division 146
General Information 146
Division Administration 151
Personnel 152
Procurement 153
Budget Management 154
Building Facilities and Services 154
Computers 155
Miscellany 158
Technical Programs 160
Lithium/Metal Sulfide Batteries 160
Introduction 160
Industrial Participants, Consultants, and Contractors 161
Exploratory Studies 162
Materials Research and Development 165
Cell Development 167
"Status Cells" 168
Batteries 169
Personnel 171
Development of Aqueous Batteries for Electric Vehicles 172
Industrial Contractors' Programs 173
ANL Supporting Research 174
Fuel Cells 176
Electrolytic Generation of Oxygen for Space Applications 176

- Aqueous Electrolytes for Hydrocarbon Fuel Cells 177
- Molten Carbonate Fuel Cells 177

Coal Combustion 178

- Introduction 178
- Classes of Coal 180
- Fluidized Bed Combustion of Coal 180
- Fossil Fuel Utilization and Conversion 187
- Magnetohydrodynamics 188
- Personnel 189

Sodium Technology 191

- Analytical Standards Program 192
- National Meter Program 192
- Steam Generator Leak Detection 195
- Post-1974 Sodium Technology Program 196
- Design and Regeneration of Cold Traps 197
- Bulk Sodium Processing 198
- Personnel 199

Fuel Element Failure Detection 199

Solar Energy 199

Fuel-Cycle Studies 202

- Development of Advanced Fuel Recovery Operations 203
- Post-1973 Fuel-Cycle Programs 206

Nuclear Fusion Energy Research 209

- Lithium Hydride Systems 210
- Lithium Blanket Processing 213
- Neutron Dosimetry 214
- Tritium Handling and Containment Studies 216

Materials Chemistry and Thermodynamics 217

- Calorimetry 217
- High-Temperature Thermodynamic Studies 218
- Reactor Fuel Studies 219
- Irradiated Mixed Oxide Fuels 222

Molten Salt Chemistry 225

- Phase Equilibria 226
- Molten Salt Spectroscopy 227
- Theoretical Studies 228

Liquid Metal Chemistry 229

Other Basic Research 230
Nucleation Studies 230
Meteoritic Studies 230
Thermal Conductivity of Associating Gases 231
Fluid Catalysis 232
Environmental Chemistry 232
Analytical Chemistry Laboratory 234
Nuclear Safeguards 234

1980-1990: CMT—NEW NAME, NEW MISSIONS 237

National Affairs 238
The Nuclear Industry 241
The Division 241
Technical Programs 249
Lithium-Alloy/Iron Sulfide Batteries 249
Electrode Chemistry 250
Materials Engineering 252
Cell Development 253
Cell Testing 256
Modules, Batteries, and Auxiliary Components 257
Personnel 259
Other High-Temperature Battery Research 260
Sodium/Sulfur Batteries 260
Glass Electrolyte Systems 261
Composite Solid Electrolyte Systems 262
Alternative Cell Systems 262
Aqueous Battery Research and Development 264
Development of Electric-Vehicle Batteries by Contractors 264
National Battery Test Laboratory (NBTL) 265
Supporting Battery Research 266
Other Battery Applications 267
Industrial Electrochemical Technology 268
Personnel 268
Fuel Cells 269
Molten Carbonate Fuel Cells 269
Solid Oxide Fuel Cells 271
Personnel 274

Coal Utilization 274
Argonne Premium Coal Sample Program 274
Separation of Coal Macerals 275
Fluidized-Bed Combustion of Coal 275
Cleanup of Hot Gases from Pressurized Fluidized-Bed Combustors 277
Management Activities 278
Personnel 279
Magnetohydrodynamic Heat and Seed Recovery Technology 279
Control of NO_x and SO_x Emissions 280
Fouling of Steam Generator Tubes in the Radiant Boiler 280
Follow-Up Applications of Feul 281
Recovery of Potassium Carbonate Seed Material 281
Supporting Studies 282
Personnel 282
Energy and Municipal Waste Technology 282
Energy from Municipal Waste 282
Hazardous Organic Wastes 283
Treatment/Disposal of Reactive Wastes 284
Solar Energy 284
Solar Collector Development 284
Ice Storage 285
Thermal Storage 286
Personnel 286
Desiccant Studies 286
Nuclear Fusion Energy Research 287
Systems Analysis Studies 287
Materials Research and Development 289
Solid Breeder Blankets 289
Aqueous Breeder Blanket 292
Dosimetry and Damage Analysis 292
"Cold Fusion" 294
Steel Research Initiative 294
Electromagnetic Casting of Steel 295
Steel Scrap Beneficiation 296
Sodium Technology 297
Optimization of Cold Trap Design 297
In Situ Regeneration of Cold Traps 299
Sodium Purification by Distillation 299
Sodium Waste Technology 300

Sodium-Concrete Reactions 301
Personnel 301
Nuclear Separations Technology 301
Analytical Support for LWBR Proof-of-Breeding 301
TRUEX Process Development 302
Recovery and Separation of Molybdenum-99 306
Integral Fast Reactor Electrometallurgical Process 307
The Integral Fast Reactor Concept 307
Process Flowsheet Development 308
Process Freasibility and Demonstration 308
Large-Scale Demonstration of Electrorefining 310
IFR Waste Management 311
Personnel 312
Plutonium Residue Recovery (PuRR) Process 312
Nuclear Waste Management 314
Yucca Mountain Project 314
Analog Experiments for Proposed Hanford Repository 317
Brittle Fracture Studies of Solid Wastes 318
Nuclear Methods of Oil Well Logging 318
Personnel 318
Applied Physical Research 319
Oxide Reactor Fuels 319
Metallic Reactor Fuels 321
Fission-Product Behavior 323
Personnel 329
Basic Research 329
Homogeneous Catalysis 329
Thermodynamics 334
Thermochemistry 338
Electrochemistry 343
Other Spectroscopic and Structural Studies 346
Interfacial Chemistry 346
Special Basic Research Investigations 348
Solid Oxide Superconductors 351
Environmental Chemistry 352
Geochemistry of Thermal Systems 355
Analytical Chemistry Laboratory 357

1990-1998: INTO THE FUTURE 361

National Affairs 361
Energy and the Environment 363
The Laboratory 365
The Division 366
  Organization 366
  Personnel 368
  In Retrospect 368
Technical Programs 371
  Electrochemical Technology 371
    Lithium Alloy/Metal Sulfide Batteries 371
    Sodium/Metal Chloride Batteries 373
    Other Advanced Batteries 375
    Electrochemical Analysis and Diagnostics Laboratory 378
    Fuel Cell Research and Development 381
  Fossil Fuel Research 388
    Fluidized-Bed Combustion Studies 388
    Magnetohydrodynamic Studies 393
  Hazardous and Mixed Waste Research 394
    Modeling Studies 394
    Microwave-Assisted Detoxification 395
    Aqueous Biphase Processes 395
    Electrokinetic Processes 399
    Chemically Bonded Waste Forms 400
    Radiolytic Gas-Generation Studies 401
    Actinide Stability/Solubility Studies 401
    Actinide Speciation in Groundwater 401
    Office of Waste Management 403
    Personnel 404
  Nuclear Waste Management 404
    Testing of High-Level Waste Glasses 404
    Testing of Spent Fuel 408
    Waste Isolation Pilot Plant Project 409
    Vitrification of Low-Level Waste 409
    Characterization of Contaminated Soils and Other Materials 410
    Other Personnel 411
  Separation Science and Technology 411
    Development and Application of TRUEX Technology 412

Centrifugal Contactor Development 415
Recovery and Separation of Molybdenum-99 416
Miscellaneous Programs 418
Integral Fast Reactor Pyrochemical Process 419
Process Flowsheet and Chemistry Studies 419
Process Development Studies 420
Engineering-Scale Process Development 424
Waste Treatment Processes 425
Advanced IFR Flowsheet 428
Personnel 429
Electrometallurgical Treatment of Spent Reactor Fuels 429
Advanced Electrorefiner Concepts 430
Process Development Studies 431
Actinide Recovery from Spent Oxide Fuels 433
Treatment of Spent Oxide Fuels for Safe Disposal 437
Personnel 441
Fusion-Related Research 441
Desorption Measurements for $LiAlO_2$-H_2-H_2O(g) System 441
Tritium Release Studies 442
Lithium Vaporization and Corrosion 443
Personnel 444
Applied Physical Chemistry 444
Physical Properties of Core-Concrete Mixtures 445
Support Studies for New Production Reactor (NPR) 446
Thermophysical Property Studies 448
Development of Synchrotron Analytical Capabilities 450
Basic Chemistry Research 450
Physical Organic Chemistry 450
High T_c Superconductors 455
Ordering and Association in Liquids 457
Quantum Chemical Studies 458
Molecular Sieve Research 460
Electrochemical and Corrosion Studies 461
Polymer Electrolytes 465
Geochemistry 465
Enhanced Metallurgical Processes 468
Analytical Chemistry Laboratory (ACL) 468
The Division's Future 471

APPENDIXES

General References 475
Acronyms and Definitions 477
Subject Guide 485
Name Index 495

Name Index

Aase, Scott 404, 412, 418
Abraham, Bernard 20
Abraham, Daniel 429
Abrajano, Teofilo 319, 357, 408, 467
Ackerman, John 82, 122, 159, 171, 177, 178, 191, 246, 259, 268, 274, 312, 416, 419, 420, 429
Ackerman, Ray 122
Adams, Max 43, 71, 125, 192, 199
Ader, Milton 19, 22, 30, 37, 43, 55, 130
Agnew, Spiro[1] 140
Agrawal, Rajiv 279
Ahmed, Shabbir 284, 388
Aikens, Alex 10, 20, 22
Alamillo, Anita (Franchini) 369
Alford, Kurt 368
Alfredson, Peter 114
Alkire, Richard 161
Allen, John 133, 149, 171, 193, 199, 202, 286
Allen, Theodore 35
Althoetmar, James 111
Ambrose, Lauren 246, 368, 369, 411
Amine, Khalil 376, 377, 388
Ammerpohl, Laurie (see Carbaugh)
Anastasia, Louis 191
Anderson, Doris (see Danielson)
Anderson, Karl 55, 110, 111, 125, 171, 259
Anderson, Margaret 246
Andropov, Yuri 239
Anker-Johnson, Betsy 151
Anthers, David 84, 133
Archart, Gregory 467
Aristide, Jean Bertrand 362
Armstrong, Don 124, 127
Armstrong, Neil 91
Arntzen, John 37, 108, 171, 259, 312, 329, 422, 429
Arzate, Vanessa (see Mendez) 246, 359
Asanovich, George 22
Askew, Barry 161, 170
Aznavoorian, Peter 359
Bacher, Robert 7
Baerns, Manfred 121
Bailey, Jody 369
Bakel, Allen 401, 408-410, 467
Baker, Louis 38, 82, 95, 100, 124
Baker, Thomas 114
Banacek, Steve 84, 133
Banaszak, James 123, 404
Bane, Ralph 148, 359
Banks, Eunice 22, 52
Barbosa, Lilia (Mojica) 246, 359, 369, 429
Barghusen, John 124, 173, 268
Barnes, Charles 124
Barney, Duane 146, 161, 172, 243, 244, 246, 254, 259
Barr, Susan 37, 245
Barrett, Norma 369, 411
Barsky, Murray 137, 149, 192, 199
Bartholme, Louis 171, 193, 195, 199, 259
Basco, John 312, 404, 411, 416, 429
Bates, John 228, 318, 404, 408-411
Battaglia, Vincent 373, 376, 377
Battles, James 94, 95, 97, 122, 159, 161, 163, 165, 171, 244, 246, 257, 259, 294, 312, 361, 366, 367, 404, 429
Baudino, Gerald 359
Baxman, Horace 22
Bean, Charles 234
Beatty, Dick 172
Beck, Raymond 61
Beck, Theodore 162
Becker, Joseph 268
Bednarick, Helen 22
Beederman, Morris 20, 22
Begin, Menachem 140
Bell, Alexander 3
Benedek, Roy 376
Bennett, George 55, 61, 108, 159
Bennion, Douglas 162, 172
Benson, Roy 279, 302
Bentson, Lloyd 238
Beres, Karen 150
Bergland, Paul 55
Beria, Lavrenti 24
Berkman, Michael 22

[1] Names of non-ANL personnel in italics.

Bernstein, George 15, 19, 20, 22, 54, 63, 105, 125, 171, 205, 306, 312
Berrill, Edward 160, 171, 268
Bertoletti, Jacqueline 246
Beskid, Nick 404
Beyerlein, Robert 348
Binelli, Maureen 136, 149
Bingle, James 70, 81, 84, 124, 133, 159
Binz, Carl 355
Birk, James 161
Birmingham, Alice 150, 246, 274, 359, 369
Biwer, Richard 268, 408
Blackburn, Paul 122, 147, 191, 217, 218, 274, 282, 329, 444
Blaedel, Walter 22
Blake, Natalie 306
Blander, Marie-Louise (see Saboungi)
Blander, Milton 158, 171, 227-230, 259, 279, 287, 296, 338, 351, 352, 450, 457-459, 468
Blaskovitz, Robert 429
Blomquist, Robert 199, 219, 220, 229, 292, 329, 338, 429, 450
Bloom, Ira 259, 268, 334, 338, 346, 352, 373, 388, 457, 458
Bloom, William 136, 149
Bogusch, Eugene 149
Bohlke, John 357
Bohr, Niels 1, 3
Boland, James 124
Boparai, Amrit 279, 284, 359, 468
Böthe, Walter 5
Bouquist, James 171
Bourne, Sidney 286
Bowden, Marion 127
Bowers, Delbert 291, 306, 312, 359, 403, 468
Bowyer, Dale 191
Box, Steven 171
Boxberger, L. Michael 367, 368
Boyd, Aaron 22
Bozisch, Eugene 136
Bradbury, Norris 7
Bradley, Charles 408
Brady, James 238
Breeden, John 20, 22
Bresee, James 22
Brewer, Kenneth 78
Breyne, Ronald 70, 151, 152, 159, 243, 245, 246, 367
Brezhnev, Leonid 239
Briggs, Lyman 2
Bringle, Joyce 369
Brock, Ronald 191, 302
Brown, Alan 259, 268, 274, 279, 284, 346, 352, 388, 457
Brown, Harrison 21
Brown, Herbert 20, 22, 47, 51, 76, 99, 101, 149, 151, 152, 154, 158, 160, 243, 245, 246
Brown, Norma (Pinches) 22, 158
Brown, Philip 155
Brun, Torben 348
Buck, Edgar 401, 408-411
Buckner, Leila (see Settle)
Bukowski, Marlene 246, 269, 274
Bullinger, C. 76
Bump, Thomas 191
Burke, Mary 246, 369
Burkel, Louis 268
Burnett, L. 191
Burris, Leslie 17-20, 22, 31, 51, 52, 57, 61, 87, 93, 106, 139, 146, 147, 149, 159, 192, 199, 237, 243, 244, 246, 307, 312
Bush, Barbara 362
Bush, George 237-239, 361, 362, 365
Bush, Vannevar 2
Bussey, Susan 359
Butler, Jerry 369
Butschelder, Arthur 22
Butz, Susan 150
Buzzelli, Ed 162
Byers, Charles 319
Bykowski, Susan (see Rura)
Byrne, Kevin 388, 404

Cafasso, Fred 93, 111, 147, 191-193, 199, 227, 243, 244, 246, 287, 301, 329
Cahan, Boris 162
Cairns, Elton 93, 95, 112, 133, 134, 147, 161, 164, 172, 369
Calaway, Wallis 216, 292, 352, 465, 467
Canaday, Patricia (Wood) 150
Cannon, Thomas 61, 108, 127, 191, 302
Carbaugh, Laurie (Malak, Ammerpohl) 246, 369

Carls, Erwin 135, 153, 180, 183, 279, 312, 392, 422, 425, 429
Carlson, Chester 100
Carlson, Neill 102
Carothers, Earl 268
Carothers, Janet (Stratton) 247, 369, 429, 444
Carter, David 388
Carter, Jimmy 139, 140, 144, 237-239
Cassulo, John 124, 134
Caster, Jack 22
Cescato, Loretta 246, 369
Chaiko, David 306, 342, 399, 412
Chamberlain, David 306, 404, 409-412, 416
Chambers, Whittaker 24
Chandrasekharaiah, Malleseety 121
Chang, Yoon 365
Chaplin, Ann 150
Chapman, Glen 156, 171, 248, 368
Chapman, Stefanie 246
Chasanov, Martin 95, 110, 124, 127, 219, 220, 329
Che, Stanley 191, 292
Cheek, Diane 369
Chellew, Norman 22, 43, 55, 96, 125, 199, 216, 292
Chen, Horng-Yih 123
Chen, Michael 334, 411, 455
Chen, Tsewei 191
Chen, Yat-Ching 338
Chernenko, Konstantin 239
Chiarello, Ronald 467
Chilenskas, Albert 63, 105, 115, 159, 161, 168, 169, 171, 259, 268, 373
Cho, Dae 124
Chow, Lorac 306, 412, 416, 426, 429
Christianson, Clinton 173, 259, 268
Cissel, Donald 194
Claar, Terry 274
Clark, Mark 411
Clark, Sharon 37
Clark, Teresa 246
Clemens, Martin 279, 404
Clemmer, Robert 216, 289, 291, 292, 294, 444, 447
Clifton, Carol 122
Clinton, William 361, 362, 365
Cohen, Stanley 100
Cole, Brandon 381
Cole, Terry 162
Coleman, Lester 15, 17, 19, 20, 22, 30, 51, 63, 96, 105, 125, 147, 152, 199
Colunga, Maria 246
Comey, David 92
Compton, Arthur 2, 3, 9
Conant, James 2, 3, 7
Conner, Clifford 241, 306, 404, 410, 411, 416, 418
Consiglio, Joseph 264
Contos, Maria (Scaropoulos) 37, 149, 245, 246, 368, 369
Cook, Glenn 141, 142, 173. 268, 274
Cooper, Thomas 141, 171
Copple, Jacqueline (Leddin) 248, 306, 368, 412
Corp, David 268
Couture, Rex 317, 319
Cramer, Thomas 37
Creamer, Edward 268
Crewe, Albert 93, 94, 110, 111
Crouthamel, Carl 38, 51, 83, 85-87, 93, 95, 112, 125, 128, 131, 133, 136, 147, 156, 369
Crudup, Valerie 246
Cunnane, James 394, 401, 403, 404, 408-411
Cunningham, Paul 82, 148, 149, 171, 226, 228, 232, 355, 357
Curtiss, Larry 292, 334, 338, 342, 343, 348, 352, 370, 444, 450, 457-459, 461, 465
Cushing, Beatrice (Hjelte) 84, 158
Cushing, Carl 158

Dangerfield, Rodney 240
Danielson, Doris (Anderson) 158
Danielson, Philip 121, 122, 131, 158, 218
Davis, Kathleen 112, 133
Deeken, Philip 302
Deerwester, Marshall 82, 124
Dees, Dennis 274, 294, 346, 373, 376, 388
DeGrande, Virginia 22, 40
DeKany, John 61
Deloria, Nancy 246
DeLuca, William 171, 173, 260, 268, 381
Demirgian, Jack 284, 286, 359, 394, 411
Dempster, Arthur 3
Denst, Thomas 36, 155
DeNuccio, Dennis 171
Deutsch, Lee 22, 84
Devgun, Jas 404

Devine, Barry 22
Dewell, Edgar 128
Dewey, Gerald 429
Deziehl, Chester 22
Dhar, Sanat 111, 199
Dillon, Ira 20, 22, 51, 57, 79, 87, 111, 159
DiSanto, Tom 411
Dokmanovic, Branko 94
Dole, Robert 361
Donnely, Wendy (see Lamb)
Doolin, Patrick 155, 160
Dorsey, Lester 20, 22, 51, 61, 151, 154
Dosi, Rajiv 388
Drabek, Vera 136, 149
Draley, Joseph 147, 192, 199
Drapcho, David 355
Driskell, Marie 22, 37, 149
DuBridge, Lee 7
Dudey, Norman 128, 147, 148
Duffield, Robert 42, 93, 145
Dukakis, Michael 237, 238
Dunning, John 162
Duoba, Michael 260
Dusek, John 171
DuTemple, Octave 49, 70
Dyrkacz, Gary 279

Eastman, Dean 361, 366
Eberhart, James 111, 159, 171, 199, 230
Ebersole, Earl 103
Ebert, William 319, 404, 408, 410
Eckels, Thomas 63, 105
Edison, Thomas 264
Edwards, Harry 125, 192, 199
Edwards, Russell 95, 121-123, 130, 131
Eident, Paul 152
Einstein, Albert 2
Einziger, Robert 404
Eisenhower, Dwight 6, 24, 49
Elison, Marilyn 369
Elliott, Guy 55
Elliott, Ronald 171
Ellis, Linda 246
Ellison, Adam 399, 410
Emery, Jeffrey 404, 408- 410
Engelkemeier, Antoinette 51, 84, 279, 355, 359
Eppley, Richard 133
Essling, Alice 84, 359
Estand, Elaine 246, 269, 369, 411
Evans, Harold 15, 19, 20, 22, 53
Everhart, Robert 429, 441

Fagan, James 302
Fager, Di Ann 150
Fairgreve, Douglas 84
Falkenberg, John 191, 411
Fan, Liang-Shih 276
Farahat, Magdy 171, 199, 260, 279, 329
Feay, Bruce 171
Feder, Harold 17, 19, 22, 30, 35, 43, 48, 54, 55, 66, 67, 82, 95, 108, 111, 121, 130, 147, 199, 232, 244, 334, 369, 450
Fee, Darrell 191, 224, 225, 274, 276, 279, 329
Feng, Xiangdong 408, 410, 465
Fenske, Merrill 54
Fermi, Enrico 1, 3, 7, 8
Fermi, Laura 9
Ferraro, Geraldine 238, 241
Ferry, Florence 84, 136, 141, 142, 149
Filewicz, Erwin 199
Finch, Robert 404, 408, 409, 410
Fineman, Olga (Giacchetti) 15, 22, 158
Fineman, Philip 15, 19, 22, 46, 63, 78, 103, 158
Fink, Joanne 199, 220, 301, 320, 329, 446, 450
Finn, Patricia 178, 216, 289, 291, 292, 404, 409, 444, 447
Fischer, Albert 112, 133, 171, 191, 199, 260, 291, 292, 444, 457
Fischer, Donald 81, 124, 159, 199, 219, 312, 319, 401, 429
Fischer, Jack 65, 70, 71, 96, 112, 127, 159, 220, 229, 279
Fleischmann, Martin 294
Fletcher, Gregory 429
Flikkema, Stanley 84, 86
Florin, Al 70
Flotow, Howard 343, 359
Folke, Eva 268
Ford, Gerald 139, 140, 144
Fortner, Jeffrey 404, 410, 411
Foster, Frank 173
Foster, Melvin 100, 112, 133, 155, 229

Fox, Irene 84, 149, 359
Fradin, Frank 361, 365, 366
Fraioli, Anthony 274, 286
Franchini, Anita (see Alamillo)
Franchini, Guadalupe 246, 369
Frank, Robert 151, 155, 173, 260, 269, 306
Fredrickson, Donald 43, 55, 83, 130, 191, 219, 260, 268, 306, 412
Freeman, Artie 154
Freeman, David 51, 179
Freund, Erwin 8
Frigo, Arthur 368
Frisch, Otto 1
Frost, William 171
Frurip, David 338
Fuchs, Klaus 24
Fuchs, Louis 230, 355

Gabelnick, Stephen 156, 220, 248, 368
Gabor, John 95, 120, 121, 312
Gabrovic, Dorothea 246
Gaines, Valerie 245, 246
Gandhi, Indira 239
Gardner, Marian (see Harkins)
Garsky, Carol 84
Gatrousis, Christopher 128
Gaumer, Lee 15, 20, 22, 70
Gay, Eddie 153, 161, 171, 254, 260, 268, 306, 310, 312, 422, 429
Gehrke, Carol 150
Geller, Jay 172, 173
Genge, Michael 173, 268
Gentry, Kenneth 161
Gerald, Rex 376
Gerding, Thomas 114, 206, 319, 408, 409
Giacchetti, Olga (see Fineman)
Gilbreath, James 22
Gillie, Kenneth 260, 381
Glassner, Alvin 35, 55, 87
Gofman, John 92
Goldberg, Margaret 404, 411
Goldwater, Barry 90
Goodspeed, Harvey 359
Gorbachev, Mikhail 237, 239
Gore, Albert 361, 362
Goring, Geoffrey 70, 71
Gorski, Anthony 286
Gorth, G. 127
Goss, Michael 411
Gourishankar, Karthick 441
Graae, Johan 36, 40, 63, 64, 105, 155, 160, 171
Graczyk, Alice 29, 40, 149, 151
Graczyk, Donald 209, 294, 302, 355, 359, 468
Gray, Hanna 241
Green, David 149, 153, 159, 220, 245, 246, 329, 357, 359, 367, 468
Green, Raymond 191
Greenberg, Elliott 83, 130
Greenberg, Sherman 15, 22, 199
Greenglass, David 24
Greenwood, Lawrence 216, 291, 292, 294, 444
Grimmett, E. 76
Grisko, Sherry 150, 246, 260, 269, 369
Grosvenor, David 127, 204, 205
Groves, Leslie 3-5, 8
Gruen, Dieter 287, 352, 457
Gunderson, Gordon 122, 123
Gunther, William 114, 124, 329
Guzman, Gualalupe 246

Haas, Milton 135, 191, 394
Hafenrichter, Lohman 404, 409
Haglund, Robert 192, 199
Hahn, Otto 1
Halerz, Patricia 368, 369
Haley, James 127, 220
Hall, Cynthia 20, 22
Halley, John Woods 345, 464
Hamilton, Robert 161
Hampson, Donald 20, 22, 43, 51, 55, 63, 76, 103
Hamrin, David 245
Hanna, Len 158
Hanna, Ruth (Peterson) 84, 136, 149, 158
Harast, Joseph 78
Hariharan, Alleppey 114
Harkins, Marian (Gardner) 246, 260, 369
Harmon, Geraldine 22
Harmon, Joseph 37, 245, 368
Harrison, Lawrence 124
Harvey, Paul 34, 359
Hash, Mark 268, 352, 373, 388, 404, 411, 457
Hataburda, Paula 150

Hathaway, Ellen 37, 112, 133, 134, 171, 226-228
Hausman, Eugene 15, 22
Hayes, Edward 171, 260, 268
Hebditch, David 199
Heiberger, John 199, 205, 229, 274, 312, 334, 423, 429
Heinrich, Robert 112, 128, 133, 136, 149, 216, 279, 312, 359
Hejny, Mary Ann 369
Helt, James 279, 284, 394, 395, 403, 410, 411
Hendricks, Carol (Richie) 150
Henriksen, Gary 373, 376, 388, 399
Henry, Richard 279, 284
Hepperly, John 191
Herceg, Joseph 312
Heredy, Laszlo 163, 172
Hess, David 51
Hesson, James 61, 63, 124, 133
Hickenlooper, Bourke 7
Hilberry, Norman 49, 93
Hildebrand, Joel 49
Hildebrandt, Robert 20, 22, 70
Hill, Helen 246, 369
Hill, Orbille 21
Hinckley, John 238
Hirsch, Gideon 110
Hiss, Alger 24
Hitler, Adolph 5
Hjelte, Beatrice (see Cushing)
Hlavnicka, Marilyn 84
Hoekstra, Jodie 15, 22, 84
Hoff, Marcian 155
Hoffer, Eric 110
Hogrefe, Roger 260, 268, 381
Hoh, Joseph 302, 404, 408-411
Holifield, Timothy 260
Holl, Jack 365
Holloway, Joy (see Swoboda)
Holmes, John 121, 127, 147, 193, 195, 199
Holt, Ben 149, 232, 355, 357, 467
Homa, Myron 20, 22, 51, 83, 84, 136, 149, 171, 291
Hon, L. 461
Hon, Ping-Kay 125
Honesty, Charles 125, 199
Hoover, Alberta 43, 83
Hornstra, Fred 170, 171, 173, 260, 268
Horwitz, Philip 302
Hsu, Chen-Chao 191, 268
Huang, Hann-Sheng 191, 274, 279
Hubbard, Ward 38, 82, 95, 130, 147, 159, 217, 343
Hubble, William 191, 355
Hudson, Ray 161
Huff, Doris 359
Huff, Edmund 306, 312, 359
Huggins, Robert 162
Hulet, Donald 76
Humphrey, Hubert 90
Humphrey, Jimmy 199
Hunt, Paul 171
Hunt, Peter 110, 111, 123
Hussein, Sadam 362
Hutter, Joseph 418
Hykan, Edwin 20, 22, 76
Hyman, Herbert 7, 15, 17, 21, 22, 65, 70

Iizuka, Masatoshi 429
Isaacson, H. Ronald 355
Ivins, Richard 82, 95, 124, 161, 172

Jacobson, David 15, 22
Jacobson, Joseph 22
Jaffey, Arthur 51
Jansen, Andrew 373, 375-377
Janz, George 162
Jardine, Leslie 199, 206, 319
Jarry, Roger 65, 114, 135, 191
Jaskot, Raymond 306, 412
Javdani, Kambiz 191
Jensen, Kenneth 149, 312, 359
Jensen, Lucille 246
Jivery, Barbara 150
Jobs, Steven 155
Johnson, Andrew 377
Johnson, Carl 86, 114, 125, 133, 147, 171, 191, 210, 225-227, 245, 246, 282, 289, 292, 329, 444
Johnson, Carter 97
Johnson, Christine 369
Johnson, Christopher 376
Johnson, Gerald 83, 130, 312, 313, 338, 441, 458
Johnson, Irving 38, 55, 95, 100, 107-111, 156, 180, 191, 229, 276, 279, 312, 313, 329, 429
Johnson, Lyndon 90

Johnson, Stanley 112, 133, 171, 226, 232, 282, 352, 355, 457, 465
Johnson, Terry 30, 61, 108, 127, 191, 204, 279, 282, 312, 401, 404, 410, 411, 416, 426, 429
Johnston, Ernest 61, 151
Joliot-Curie, Frederick 1
Jones, Loretta 111
Jones, M. 119
Jonke, Albert 15, 22, 53, 72, 74, 76, 93, 112, 120, 135, 147, 180, 243, 244, 246, 279
Juco, Eleanor (see Kafalas)
Jurlow, Miriam 84
Juvinall, Ruth 78, 84, 136, 149, 206

Kacinskas, Henry 171
Kafalas, Eleanor (Juco) 158
Kafalas, Peter 86, 158
Kahaian, Arthur 376
Kaminski, Michael 411
Kanasugi, Katsumasa 429
Kann, William 312
Kaplan, Bettye 22
Kaplan, Louis 22
Karell, Eric 441
Kartunnen, John 149
Kassner, Thomas 191, 199
Katz, Joseph 21, 65, 70
Kaun, Thomas 167, 171, 256, 260, 268, 274, 372, 373
Kazadi, Msiri 248
Keday, Alec 22
Keener, Richard 245
Keigher, Mary 150
Keil, Klaus 230
Keledus, Tony 155
Keller, Rudolf 206
Kelley, John 260
Kelsheimer, Pleasant 151, 154
Kemp, Jack 361
Kennedy, Carolyn 15, 22
Kennedy, John 89, 90
Kennedy, Robert F. 90
Kepler, Keith 376
Kesser, Gwendolyn 37, 84, 85, 149, 245
Kessie, Robert 70, 71, 74, 114, 118, 156, 171, 199, 206, 248, 312
Khanna, V. 111
Khrushchev, Nikita 25, 89
Kilbourne, Byron 31
Kilian, Kerri 246, 269
Kilpatrick, Martin 82
Kilsdonk, Dennis 171, 260
Kincinas, John 120, 171, 302
King, James 15, 22
King, Judith 150
King, Martin Luther 90
Kinoshita, Kim 178, 274
Kirby, Jack 99
Kleb, Robert 130, 348
Klein, Milton 15, 22
Klemmer, Robert 291
Klinger, Jiri 173, 268, 282, 286
Klingler, Robert 334, 376, 455
Kloska, Corinne (see Thompson)
Kmiec, Richard 268
Knapp, William 125
Knighton, James 55, 84, 108-110
Knudsen, Irving 119, 123
Kobayashi, Masaru 121
Kobayashi, T. 429
Kolba, Verne 171, 193, 195, 199, 260, 274, 301
Kopasz, John 399, 444
Koppel, Lowell 121, 191
Kosner, Carol 136
Kotora, James 154
Koura, Nobuyuki ("Nick") 171
Koyama, Tadafumi 429
Kraft, Theodore 274
Krause, Douglas 20, 22, 31, 78, 83, 84
Krause, Paul 127
Krause, Theodore 334, 388, 395, 399, 410, 411
Kremesec, Victor 301
Kremsner, Walter 111, 171, 199, 229, 230, 260, 286, 355
Kronenberg, Marvin 268
Krumpelt, Michael 205, 229, 274, 294, 346, 388
Kucera, Eugene 84, 112, 128, 133, 136, 149, 158
Kucera, Gene (McCloud) 84, 158, 171, 178, 227, 229, 230, 260, 274, 388
Kueper, Timothy 388
Kulaga, Joseph 248, 368
Kullen, Bruce 78, 206

Kumar, Romesh 199, 206, 232, 284, 355, 388
Kummer, Joseph 260
Kurchatov, Igor 24
Kwok, Jane 306, 418
Kyger, Jack 146
Kyle, Martin 30, 108, 153, 165, 171

Laidler, James 361, 366-368, 404, 429
Lamb, Wendy (Donnelly) 246, 269
Lamoureux, Laura (Chromizky) 359
Land, Robert 156, 216, 248, 292
Landis, Dawn 246, 269
Lane, Linda 369
Laney, Robert 40, 42, 146
Lang, Teresa 246, 355
Langmuir, Irving 6
Lapp, Ralph 25
Larsen, Karen 367
Larsen, Robert 30, 35, 37, 43, 53, 76, 84-86, 136, 147-149, 160
Latimer, Trent 171
Lavendel, Henry 61
Lawrence, Ernest 2, 8, 10
Lawroski, Harry 50
Lawroski, Stephen 13, 15-19, 22, 30, 31, 39, 40, 43, 47-50, 71, 72, 74, 76, 79, 87, 89, 92, 96, 149, 234, 364
Leddin, Jacqueline (see Copple)
Lee, Henry 22
Lee, Johnsee 268
Lee, Sheldon 191, 279, 388, 392, 404
Lee, Tien 173, 268
Legerski, Lee 246
LeGrand, Charlotte 246, 269
Lehmann, Clarence 127
Lehmann, Jacqueline 150
Leibowitz, Leonard 81, 95, 124, 199, 219, 220, 292, 301, 312, 320, 329, 429, 446, 450
Lemke, Vernon 136, 149
Lenc, John 127, 191, 205, 276, 279
Leonard, Ralph 198, 199, 205, 289, 291, 302, 306, 412, 415, 416, 418
Leonard, Sally 150
Lescarret, Alain 191
Levenson, Milton 17, 19-22, 45, 51, 58, 59, 61, 63, 70, 95, 98, 102, 105, 137, 369
Levison, O. 76
Levitz, Norman 20, 22, 51, 74, 119, 147, 202, 204, 206, 207, 209, 302
Lewandowski, Edward 155
Lewin, Janice (see Muller)
Lewis, Michele 269, 302, 312, 401, 426, 429
Lexa, Dusan 429
Liimatainen, Robert 70, 71, 74, 82, 124
Lilienthal, David 7
Lincoln, Abraham 65
Lindahl, Peter 279, 359, 409, 467, 468
Link, Leonard 234
Linzer, Frederick 43
Litland, Howard 15
Littmen, Henry 22
Litty, Albin 74
Liu, Yung 329
Loeding, John 20, 22, 36, 51, 53, 72, 76, 79, 120, 159, 234
Long, Ray 35, 70
Loutfy, Ralph 206, 268, 346
Ludewig, Walt 70
Ludlow, John 105
Luner, Charles 111, 193, 199
Luo, Jian Shu 404, 410, 457

MacArthur, Douglas 23
Mack, Paul 108, 193, 195, 199
Madson, Allen 128
Madson, Sheila 150
Malak, Laurie (see Carbaugh)
Malak, Stephanie (Malm) 246, 369
Malcolm X 90
Malecha, Richard 36, 40, 63, 101, 155, 161, 172, 260, 291, 312, 368, 429
Malewicki, Russell 136, 149
Malm, John 70
Malm, Stephanie (see Malak)
Mandelstein, Les 22
Manevy, George 114
Manimelto, Michael 376
Marchetti, Frank 35
Marek, Lawrence 20, 22, 193, 199
Margolis, Asher 22
Markun, Francis 359, 467

Maroni, Victor 95, 112, 134, 159, 171, 210, 216, 226-228, 287, 292, 329, 348, 350, 352, 444, 450, 457, 459, 461, 465
Marr, Jane 171, 227, 229, 260, 268, 338, 359
Marshall, Christopher 459
Marshall, George 6, 24
Marshall, Samson 307, 418
Martin, Allan 55, 110, 122, 123, 147, 171, 227
Martin, Peter 82
Martino, Fredric 136, 149, 171, 260, 302, 359, 468
Maser, James 319
Mason, Donald 82
Mason, George 10, 22
Massey, Walter 139, 145, 237, 241
Masten, Frank 97, 151
Mathers, James 171
Matlak, Sharon 150
Mau, Cindy (see Meurer)
Mazer, James 408, 410
McAuliffe, Christa 239
McBeth, Robert 159
McBride, Peggy (see Parks)
McCann, Debra 150
McCarthy, Joseph 24
McCloud, Gene (see Kucera)
McCown, John 83, 84, 103
McDeavitt, Sean 429
McGarity, Arthur 286
McGovern, George 139, 140
McIntire, William 202
McKee, John 159, 193, 195, 199
McKinney, Susan 369, 411
McLaughlin, Malika 369
McMahon, John 344
McPheeters, Charles 159, 171, 195-199, 274, 297, 301, 367, 429, 441
Mead, Lee 151
Mecham, William 21, 22, 30, 70, 71, 74, 95, 137, 199, 206, 318, 319
Meisenhelder, Janet 205, 268
Meitner, Lise 1
Melendres, Carlos 171, 346, 352, 450, 465
Melton, Anne 149, 151, 246
Mendez, Vanessa (Arzate) 368, 369
Mensah-Biney, Robert 399
Mertz, Carol 399, 410
Metz, Clyde 131, 399, 410
Meurer, Cindy (Pospishil, Mau) 246
Meyer, Robert 85, 125, 136, 147, 149, 192, 199, 248, 302
Michalek, Doris (see Worthington)
Miller, James 173, 260, 388
Miller, John 269
Miller, Shelby 99, 279, 392
Miller, William 105, 147, 159, 161, 171, 199, 204, 260, 307, 308, 312, 420, 423, 429
Mingh, Nguyen 269
Mishler, Larry 81, 124
Misra, Balabhadra 199, 216, 260, 282, 289, 291, 292, 294, 301, 312, 429
Mojica, Lilia (see Barbosa)
Moncton, David 365
Mondale, Walter 238
Montagna, John 191
Moore, Mary 150
Morgan, Deborah (see O'Rourke) 246
Morrissey, Leo 152, 159, 368
Morss, Lester 404, 411
Moulton, David 191
Mowry, Russell 191, 279
Mrazek, Franklin 122, 123, 131, 171, 260, 269, 274, 306, 373, 388
Muehlenbachs, K. 356
Mueller, Catherine 150
Mulcahey, Thomas 191, 269, 312, 313, 429, 441
Muller, Janice (Lewin) 42, 154, 246, 368, 369
Mundy, John 171
Munnecke, Victor 19, 22, 47-49, 53, 92, 97
Murphy, William 159, 191
Murray, Richard 111, 149
Murrell, M.T. 465
Myers, Deborah (Zurawski) 388
Myers, Henry 429
Myles, Kevin Michael ("Mike") 163, 166, 171, 180, 227, 229, 260, 268, 269, 274, 276, 279, 282, 284, 294, 367, 388, 450

Nagy, Zoltan 338, 346, 450, 459, 465
Nandi, Satyendra 191
Napora, Sofia (Gawenda) 150, 153, 246, 368, 369
Natale, John 15, 19, 20, 22, 51, 101, 151
Nathans, Marcel 55

Neir, Alfred 3
Nelson, L. S. 230
Nelson, Paul 30, 61, 93, 96, 127, 146, 147, 153, 159, 161, 172, 192, 193, 195, 199, 243, 244, 246, 259, 268, 367, 373, 404
Nelson, Robert 206, 279, 302
Nevitt, Michael 146
Newman, Aron 429
Newman, John 162
Newnam, Steven 359
Newsam, John 348
Nicholson, E. 68
Nielson, Eugene 355, 376
Nishimura, Tomohiro 429
Nishio, Kazuo 108, 124
Nixon, Richard 89, 90, 139, 140, 179
Nolan, Bonnie 150
Nole, Michael 404, 411
Noriega, Manuel 239, 362
North, Oliver 238
Novey, Theodore 51
Nuñes, Franklin 279
Nuñez, Luis 191, 410, 412
Nuttall, Ralph 55, 56, 83, 130

O'Conner, Sandra Day 241
O'Hare, Patrick 95, 130, 217
O'Leary, Hazel 25
O'Neil, Florence 40
O'Rourke, Deborah (Morgan) 246
Ogata, Takanari 429
Okajima, Shigeo 403, 409
Oldham, Robert 136, 149
Olsen, William 39
Olszanski, Ruth (Voiland) 158
Olszanski, Theodore 158, 171
Oppenheimer, Robert 4, 7, 8
Osborn, Marilyn 246, 369
Osudar, John 156, 171, 248, 368
Oswald, Lee Harvey 90
Otto, Neil 171, 260

Pacholok, Maria 369
Panek, Allen 136, 149, 192, 199
Panek, Marge (Swanson) 150
Papatheodorou, George 346, 465
Parks, John 206, 286, 302
Parks, Peggy (McBride) 246
Paul, Nola Joy (Tasharski) 158
Paul, Ron 158, 269
Pauling, Linus 110
Pavlik, John 61, 124, 127, 191, 199
Payne, Charles 127
Peck, James 112, 124, 133, 134
Pehl, Wilfred 61, 76
Peierls, Rudolph 1
Pelto, Ralph 216, 306
Pelton, Arthur 279, 329, 338, 351
Pereira, Candido 429
Perot, Ross 362
Peterman, Ray 216
Peterson, Edwin 19, 22, 40, 158
Peterson, Elizabeth (Reilly) 20, 22, 31, 84, 158
Peterson, Laureen (Seils) 22, 158
Peterson, Ruth (see Hanna)
Petkus, Edward 72, 74, 79, 127
Petrick, Michael 191
Pewitt, Gale 146
Pfann, William 57
Pfeiffer, Carl 159
Pierce, Kenneth 465
Pierce, R. Dean 38, 61, 99, 100, 108, 127, 147, 159, 178, 204, 205, 274, 313, 441
Pike, Sumner 7
Pinches, Norma (see Brown)
Pizzolato, Philip 81
Planner, H. N. 230
Poa, David 173, 269, 312, 313, 441
Podolski, Walter 191, 279, 388, 392
Poeppel, Roger 274
Pollack, Israel 155, 172
Pollard, Richard 162
Pons, Stanley 294
Popek, Ray 83, 84, 136, 149, 294
Popik, Judith 246, 274, 368, 369
Pople, John 342, 346, 458
Porte, Howard 114, 130
Pospishil, Cindy (see Meurer)
Post, Roy 22, 250
Powell, Mary 369
Powers, Gary 89, 369
Prakash, Jai 375, 376

Preto, Sandra 159, 170, 171, 260
Preuss, Daniel 368
Price, David 458
Price, Melvin 25
Proud, Everett 93, 97, 99, 110, 146, 147, 149, 151, 154, 246
Prucha, Doreen 100, 149, 151, 154
Prucha, Lad 100
Pyle, Julitta 150

Quattropani, Nick 127, 204, 205, 312, 422, 429
Quayle, Dan 238, 361, 362
Quigley, Kevin 404, 411
Quinn, John 404

Rabi, Isador 7
Rafacz, Evelyn 40, 149
Rajan, John 173, 269, 274, 284, 306, 368
Ramaswami, Devabhaktuni 121, 127, 191
Raphaelian, Leo 279
Raptis, Penelope 245
Rashinskas, Anthony 108
Rathke, Jerome 232, 334, 450, 455, 459
Raue, Donald 119, 193, 195, 197-199, 301, 302, 312
Rauh, Everett 359
Rausch, Debra 369
Ray, James Earl 90, 108, 136, 301
Reagan, Ronald 237-239, 249, 365
Redding, George 67, 111, 114, 171, 229, 260
Redey, Laszlo 164, 171, 251, 260, 269, 294, 307, 346, 373, 375, 376, 377
Reed, Donald 401, 403, 409, 411
Reed, Kent 202
Reedy, Gerald 220, 291, 319, 359, 465
Regalbuto, Monica 306, 412, 416
Reichley-Yinger, Lucinda 306
Reilly, Elizabeth (see Peterson)
Reishus, John 121, 122
Reiter, Sandra (see Tummillo)
Renner, Thomas 197, 198, 199
Reuther, Walter 25
Rhode, Kenneth 43, 55
Richards, Darcell 150
Richardson, H. L. 230
Richie, Carol (see Hendricks)
Richmann, Michael 403, 429
Rickover, Hyman 7, 25, 28, 97, 140
Ride, Sally 241
Riel, Roberta 246, 368, 369, 411
Riha, James 114, 171, 191
Ritzman, Robert 329
Rizzi, Lysbeth 369
Roberts, John 243, 244
Robinson, Mary 84, 141
Roche, Michael 161, 171, 192, 193, 199, 260, 329, 388, 446
Rodger, Walton 17, 19, 20, 22, 35, 36, 48, 49, 70, 71, 74, 76, 79, 82, 91, 92
Rodighiero, Susan 150
Roessler, Susan (Cathimer) 150
Rogers, Chester 124, 134
Roosevelt, Franklin 2, 6, 24
Rosen, Charles 55
Rosenberg, Ethel 24
Rosenberg, Julius 24
Rosenthal, Seymour 22
Ross, Laurids 15, 22, 84, 85, 125, 136, 149, 171, 192, 199, 260, 359
Ross, Norman 101
Royal, Joseph 37, 96, 147
Rubischko, Richard 161
Ruby, Jack 90
Rudolph, Heidi 246
Rudzitis, Edgars 83, 89, 130
Ruppert, Lewis 269, 302
Rura, Susan (Bykowski) 246, 260, 369, 411
Rush, Robert 202
Ruther, Westly 193, 199
Ryberg, Carl 78

Sabau, Carmen 279, 306, 359
Saboungi, Marie-Louise (Blander) 158, 171, 227-229, 338, 459
Sachs, Robert 139, 145
Sadat, Anwar 140
Sandus, Oscar 74
Sano, Y. 356
Santelli, Dino 136, 149
Sato, George 84
Savage, James 114, 120
Saxena, Satish 191, 279

Scandora, Anthony 291
Scaropoulos, Maria (see Contos)
Schablaske, Robert 20, 22, 51, 83, 84, 110, 133, 136, 149, 171, 227
Schertz, William 156, 171, 202, 284, 286
Schierbeek, Gina 369
Schilb, Arden 78, 84
Schilb, John 76, 84, 108, 127
Schlessinger, James 145
Schlueter, Richard 193
Schneider, Alfred 61, 78
Schneider, John 279, 284, 359, 411
Schnizlein, Glenn 65, 70, 81, 206
Schnyders, Harold 229
Schoeneman, Karl 22, 43
Schoffstall, Charles 120, 121, 191, 279, 282
Schoij, Norman 155
Schraidt, John 15, 19, 20, 22, 33, 36, 39, 40, 51, 61, 63, 101, 102, 105, 151, 152, 154
Schriesheim, Alan 237, 241, 361, 365, 366
Schriesheim, Beatrice 247
Schulze, William 36
Seaborg, Glenn 3, 6-8, 11-13, 21, 110
Secord, Richard 238
Sedlet, Jacob 306, 412, 418
Seefeldt, Waldemar 22, 51, 70, 71, 74, 137, 206, 234, 306, 318, 411
Seils, Charles 20, 22, 43, 51, 84, 125, 136, 149, 158, 294, 329, 411
Seitz, Martin 206, 317, 319, 357
Selman, J. Robert ("Rob") 162, 171, 176
Settle, Jack 83, 130, 158, 163, 171, 312, 329, 338, 346, 420, 429, 446
Settle, Leila (Buckner) 158
Seufzer, Paul 70
Shablaske, Robert 112, 114
Shaffner, Irving 22
Sharma, Ram 226, 229
Shaw, Milton 97, 98, 145
Shearer, John 191, 276, 279, 292
Sheft, Irving 65, 70
Shepard, Alan 90
Sheth, Atul 191, 199, 282
Shields, Kathleen 246, 369
Shiffer, Vita 150
Shimotake, Hiroshi ("Paul") 133, 134, 161, 171, 260, 269
Shinn, William 70, 114, 122, 329
Shockley, William 37
Shor, Arthur 15, 22, 158
Shor, Roberta (Wagner) 22, 43, 53, 86, 114, 158
Shull, Daniel 411
Siczek, Aldona 205
Siegel, Stanley 136, 149, 171, 191, 355, 359
Siegfried, Yolantha 368
Silkwood, Karen 145
Silverman, Martin 124
Sim, James 171, 178, 274
Simmons, John 37, 245
Simon, Daniel 377
Simpson, Lin 404
Singleterry, Merril 269
Singleton, Ernest 155
Sirhan, Sirhan 90
Skladzien, Stanley 192, 193, 195, 198, 199, 301
Slater, Susan 409, 429
Slawecki, Michael 105, 195, 199, 291, 312, 368, 429
Smaga, John 165, 171, 257, 260, 269, 381
Smiley, Seymour 74
Smith, Charles 301
Smith, Dale 192, 410
Smith, Florence (Williams) 136, 149, 312, 359
Smith, Gregory 136, 191, 269, 276, 279
Smith, Harry 36, 40, 101, 155
Smith, James 274, 388, 429, 441
Smith, Joseph 339
Smith, Sherman 191, 279
Smith, Walter 35
Smither, Robert 294
Smyk, Eugene 191, 282
Smyth, H. 6
Snyder, Christine 359, 394
Snyder, Robert 191
Sobczak, Maureen 149
Sockman, Charles 57
Sovereign, William 36, 43, 83, 84, 103
Spedding, Frank 7
Spencer, Bruce 329
Spicer, William 51

Sridhar, Ramamritham 134, 171, 226
Srinivasan, Balasubrahmanyam ("Chino") 418
St. Germain, Renee 150
Staahl, Gustav 125
Stagg, Amos Alonzo 2
Stakem, Douglas 161
Stakkowski, Jackie 150
Stalica, Norbert 125, 136, 149
Stalin, Joseph 6, 25
Stapay, Joseph 274, 388
Starr, Kenneth 361, 362
Stavropoulos, George 193, 199
Steidl, David 43, 70, 71, 78, 114, 329, 446
Stein, Lawrence 65, 68-70, 83, 129
Steindler, Martin 30, 31, 43, 67, 71, 78, 86, 100, 112, 114, 147, 160, 205, 206, 237, 243, 244, 246, 247, 301, 318, 361, 366, 367
Steinquist, Janet 150, 246, 359
Sternglass, Ernest 92
Stethers, Howard 105
Steunenberg, Robert 49, 65, 66, 67, 70, 71, 74, 93, 96, 106, 110, 147, 161, 164, 171, 260, 274, 312
Stevens, Charles 76, 359
Stevens, J. 76
Stevenson, Adlai 24
Stevenson, Charles 18, 19, 22, 30, 39, 40, 49, 62, 76, 88, 102, 147, 209
Stimak, Raymond 155
Stimson, Henry 6
Stockbar, John 114, 191, 279
Stockman, Charles 240
Stockman, David 240
Stogsdill, Carol 246, 369
Story, Vincent 6, 20
Strachan, Denis 404, 408, 410
Strachan, Marietta 369
Stratton, Janet (see Carothers)
Strassman, Fritz 1
Strauss, Lewis 7
Streets, Elane 359
Stretch, Carl 124
Strezo, Virginia 246, 359, 369, 411
Strle, Wendy (see Zanelli)
Stupegia, Donald 128
Sturchio, Neil 317, 319, 357, 467
Sudar, Seymour 161
Sullivan, Kathryn 241
Surchik, Mark 409
Susman, Sherman 342
Swanson, Marge (see Panek)
Swanson, Stephen 359
Swaroop, Robert 171, 260, 279
Sweezer, Robert 84
Swift, David 124
Swift, William 180, 191, 279, 282, 388, 392, 394
Swoboda, Carl 158, 269
Swoboda, Joy (Holloway) 149, 158
Swope, Gladys 22, 77-79
Sy, Condeocita (Cajigas) 161, 171, 291
Sze, Dai-Kai 394, 444, 447
Szilard, Leo 1, 2, 3

Taecker, Rollin 49, 72
Tam, Shiu-Wing 291, 292, 329, 352, 404, 408, 444, 457, 459
Tamplin, Arthur 92
Tamura, Takeo 119
Tani, Benjamin 95, 110, 136, 149, 171, 352, 355, 359, 409, 410, 457, 465
Tasharski, Nola Joy (see Paul)
Tasker, Ian 306, 343, 411, 412
Taube, Henry 48
Taylor, Audrey 246
Teats, Gale 108, 191, 204, 279, 368, 392
Teitlus, Martha 368, 369
Telford, Ralph 149
Teller, Edward 2, 8
TenKate, Lynn 359
TenKate, Tony 359
Terdic, Heidi 368, 369
Terkel, Studs 110
Terry, Rosemary 108, 127, 130
Testa, Frank 82, 124
Tetenbaum, Marvin 19, 81, 95, 122, 123, 131, 199, 292, 301, 329, 338, 352
Tevebaugh, Arthur 70, 93, 95, 97, 112, 124, 133, 145, 146, 161
Thackeray, Michael 376, 377
Thigpen, Johnathan 70
Thompson, Augustella 37
Thompson, Corinne (Kloska) 20, 22
Thorn, Robert 122

Tiedemann, William 162
Till, Charles 243, 307, 365, 367
Tipton, Donna 246, 369
Tobias, Charles 95, 162
Tobias, Kenneth 108, 110
Todd, Mary 65
Tokiwai, M. 429
Tollner, Ronald 42, 368
Tomczuk, Zygmunt 136, 149, 159, 164, 171, 260, 308, 312, 420, 423, 429
Tomlinson, Glen 302
Tosi, M. 458
Towle, Warren 162
Trevorrow, LaVerne 70, 74, 114, 171, 206, 302, 306, 412, 429
Trice, Virgil 20, 22, 51, 53, 95, 137
Trischan, John 127
Truman, Harry 6, 24, 25
Tsai, Yifen 359
Tse, Pui-Kwan 306
Tsuchie, Yasuo 429
Tummillo, Anthony 260, 269, 381, 388
Tummillo, Sandra (Reiter) 246, 369
Tuohig, Wayne 171
Turk, Elton 19, 22, 53, 78
Turner, Clarence 191
Turner, Robert 376
Tyler, Homer 20, 22, 70

Uhle, Ronald 55
Urey, Harold 2, 21

Van Dahm, June 246
Van Deventer, Erven 110, 130, 292, 306, 399, 412, 444
Vandegrift, George 279, 302, 306, 307, 317, 319, 394, 404, 411, 412, 416, 418
Vargo, Noel 171, 191, 205
Varma, Ravi 260, 269, 279, 284, 395
Varteressian, Kegham 20, 22, 54, 124
Vasilopulos, George 269
Vaughey, John 376, 377
Veleckis, Ewald 110, 111, 159, 199, 216, 227, 229, 292, 319, 329, 338, 409, 450
Vest, Michael 441
Vilinskas, Peter 199
Vissers, Donald 111, 161, 164, 170, 171, 193, 195, 196, 199, 251, 260, 269, 279, 284, 294, 306, 307, 319, 373, 418
Vogel, John 74, 97, 117, 180, 247, 279
Vogel, Richard 19, 20, 22, 38, 49, 51, 55, 70, 71, 81, 86, 89, 92, 96, 139, 146, 147, 149, 150, 156
Vogler, Seymour 15, 22, 37, 43, 51, 53, 55, 71, 86, 127, 204, 291, 411
Voiland, Eugene 148
Voiland, Ruth (see Olszanski)
Volin, Kenneth 342
Von Winbush, Sam 176
Voss, Denise 246
Voss, William 20, 22, 36, 51, 155

Wach, Charles 55, 206, 302
Wade, Warren 70
Wagner, Bruce 162
Wagner, Roberta (see Shor)
Walker, David 171
Walker, Prince 302
Wall, Katherine 150, 237
Walling, Matthew 22
Walsh, William 97, 108, 159, 161, 171
Walters, William 15, 22
Wang, Xiaoping 388
Wantroba, Alvin 202, 286
Warchal, Raymond 82
Warinner, Douglas 282
Warren, David 312, 429, 441
Waters, Jodean 173
Watson, Thomas 3
Waughan, D. E. W. 348
Waymach, William 7
Weber, Carl 269
Weber, John 260, 269, 279
Weber, Neil 260
Webster, Carl 260, 381
Webster, Donald 19, 93, 146, 147, 161, 202, 205, 243, 244, 246
Weeks, Colleen 246
Weiler, Candace 150
Weinberg, Alvin 91
Weinstock, Bernard 70
Welch, Joseph 25
Wenz, Donald 108

Wesolowski, Cynthia 160, 246, 369
Wesolowski, Eugene 441
West, Elmo 2, 3, 10, 11, 13, 18, 20, 22, 191
Weston, James 216
Wetter, Lesa 359
Wheeler, John 1
White, George 126
White, Mary Lou 369
Wigeland, Ronald 306
Wigner, Eugene 2, 7, 28
Wilkinson, Cynthia 246
Williams, Florence (see Smith) 279, 312
Williams, Jacqueline 20, 22, 51, 84, 128, 149
Williamson, Kenneth 120
Williams, Samuel 356
Willis, Kathy 246
Willit, James 424, 429
Wilson, Melania 150
Wilson, Robert 124
Wilson, Tim 171
Wilson, W. Ira 191, 276, 279
Windsor, Paulette 37
Wingender, Ronald 158, 284, 359
Winsch, Irving 22, 61, 108, 127, 204, 206, 302
Winston, Roland 200
Wise, Stephen 83, 130
Wise, William 339
Wist, Thaddeus 155
Wolf, Stephen 404, 408, 409, 410, 411
Wolkoff, Jasper 61
Wolson, Raymond 42, 108, 198, 199, 301, 308, 368
Wong, Samuel Hei-Yu 276
Wood, Patricia (see Canaday)
Wood, Scott 161
Woods, Kenneth 35
Worrell, Wayne 162
Worthington, Doris (Michalek) 158, 246
Wozniak, Stephen 155
Wronkiewicz, David 408, 409, 410
Wygmans, David 410, 416, 418

Yaeger, Ernest 162
Yannopoulos, Lymperios 122
Yao, Neng-Ping ("Mike") 153, 163, 171-173, 258, 268, 269
Yasui, George 22, 79
Yeltsin, Boris 361
Yokoo, Takeshi 429
Yonco, Robert 55, 110, 111, 171, 216, 227, 229, 292, 338, 346, 350
Yoshida, Takuma 429
Yoshioka, Marion 22
Young, Hoylande 33
Young, John 180, 274, 279, 284, 408
Youngdahl, Arthur 198, 199

Zanelli, Wendy (Strle) 246, 269, 369
Zeno, Esmer 43, 86
Zepeda, Irma 368, 369
Ziegler, Anton 205, 279, 302, 306
Zielen Albin 172
Zinn, Walter 3, 7, 8, 17, 43, 49, 94
Zivi, Samuel 172
Zwick, Stanley 274, 282
Zygmunt, Stanislaus 461